Basic College Mathematics

Eighth Edition

ANNOTATED
INSTRUCTOR'S
EDITION

Margaret L. Lial
American River College

Stanley A. Salzman
American River College

Diana L. Hestwood
Minneapolis Community and Technical College

Addison-Wesley

Boston • New York • San Francisco
London • Toronto • Sydney • Tokyo • Singapore • Madrid
Mexico City • Munich • Paris • Cape Town • Hong Kong • Montreal

Editorial Director	Christine Hoag
Editor in Chief	Maureen O'Connor
Executive Project Manager	Kari Heen
Project Editor	Courtney Slade
Editorial Assistant	Mary Gallagher
Senior Managing Editor	Karen Wernholm
Senior Production Supervisor	Kathleen A. Manley
Senior Designer	Barbara T. Atkinson
Photo Researcher	Beth Anderson
Supplements Production	Marianne Groth and Kayla Smith-Tarbox
Media Producers	Ceci Fleming and Lin Mahoney
Software Development	Rebecca Williams, MathXL; Mary Durnwald, TestGen
Senior Marketing Manager	Michelle Renda
Marketing Assistant	Nathaniel Koven
Senior Author Support/Technology Specialist	Joseph K. Vetere
Senior Prepress Supervisor	Caroline Fell
Senior Media Buyer	Ginny Michaud
Rights and Permissions Advisor	Dana Weightman
Manufacturing Manager	Evelyn Beaton
Senior Manufacturing Buyer	Carol Melville
Composition/Production Coordination	Nesbitt Graphics, Inc.
Cover Image	Indian Summer Copyright © Lorraine Cota Manley

Library of Congress Cataloging-in-Publication Data

Lial, Margaret L.

Basic college mathematics.—8th ed. / Margaret L. Lial, Stanley A. Salzman, Diana L. Hestwood.

p. cm.

Includes bibliographical references and index.

ISBN-13: 978-0-321-55712-4 (student edition)   ISBN-13: 978-0-321-57234-9 (annotated instructor's edition)

ISBN-10: 0-321-55712-3 (student edition)   ISBN-10: 0-321-57234-3 (annotated instructor's edition)

1. Mathematics—Textbooks. I. Salzman, Stanley A. II. Hestwood, Diana. III. Title.

QA37.3.L48 2010

513'.1—dc22   2008024476

For permission to use copyrighted material, grateful acknowledgment is made to the copyright holders on page P-1, which is hereby made part of this copyright page.

Addison-Wesley
is an imprint of

www.pearsonhighered.com

ISBN 10: 0-321-55712-3
ISBN 13: 978-0-321-55712-4

In appreciation of your lasting support and never-ending enthusiasm: family, colleagues, and more than a generation of motivated students.

Stan Salzman

This book is dedicated to my dad, who always told me when I was young that girls could learn math, and to my students at Minneapolis Community and Technical College, who keep me in touch with the real world.

Diana Hestwood

▶▶▶ CONTENTS

▶▶▶ PREFACE

The eighth edition of *Basic College Mathematics* continues our ongoing commitment to provide the best possible text and supplements package that will help instructors teach and students succeed. To that end, we have addressed the diverse needs of today's students by integrating activities to help students improve their study skills, an attractive design, updated applications and graphs, helpful features, careful explanation of concepts, and an expanded package of supplements and study aids. We have also responded to the suggestions of users and reviewers and have added many new examples and exercises based on their feedback.

The text is designed to help students achieve success in a developmental mathematics program. It provides the necessary review and coverage of whole numbers, fractions, decimals, ratio and proportion, percent, and measurement, as well as an introduction to algebra and geometry, and a preview of statistics. This text is part of a series that also includes the following books:

- *Essential Mathematics,* Third Edition, by Lial and Salzman

- *Prealgebra,* Fourth Edition, by Lial and Hestwood

- *Introductory Algebra,* Ninth Edition, by Lial, Hornsby, and McGinnis

- *Intermediate Algebra,* Ninth Edition, by Lial, Hornsby, and McGinnis

- *Introductory and Intermediate Algebra,* Fourth Edition, by Lial, Hornsby, and McGinnis

- *Prealgebra and Introductory Algebra*, Third Edition, by Lial, Hestwood, Hornsby, and McGinnis

- *Developmental Mathematics: Basic Mathematics and Algebra*, Second Edition, by Lial, Hornsby, McGinnis, Salzman, and Hestwood

Hallmark Features

We believe students and instructors will welcome the following helpful features.

▶ *Chapter Openers* New and updated chapter openers feature real-world applications of mathematics that are relevant to students and tied to specific material within the chapters. Examples of topics include work/career applications, finding the best buy on cell phone service, home improvements, recipes, weather, fishing, and astronomy. (See Chapter 2, p. 113.)

▶ *Real-Life Applications* We are always on the lookout for interesting data to use in real-life applications. As a result, we have included many new or updated examples and exercises throughout the text that focus on real-life applications of mathematics. Students are often asked to find data in a table, chart, graph, or advertisement. (See pp. 158 and 160.) These applied problems provide an up-to-date flavor that will appeal to and motivate students.

▶ *Figures and Photos* Today's students are more visually oriented than ever. Thus, we have made a concerted effort to include mathematical figures, diagrams, tables, and graphs whenever possible. (See p. 158.) Many of the graphs use a style similar to that seen by students in today's print and electronic media. Photos have been incorporated to enhance applications in examples and exercises. (See p. 159.)

▶ *Emphasis on Problem Solving* Introduced at the end of Chapter 1, our six-step process for solving application problems is integrated throughout the text. The six steps, *Read, Plan,*

Estimate, Solve, State the Answer, and *Check,* are emphasized in boldface type and repeated in specific problem-solving examples in Chapters 1, 2, 3, 5, 6, 7, and 9. (See p. 164.)

▶ *Learning Objectives* Each section begins with clearly stated, numbered objectives, and the material within sections is keyed to these objectives so that students know exactly what concepts are covered. (See p. 114.)

▶ *Cautions and Notes* These color-coded and boxed comments, one of the most popular features of previous editions, warn students about common errors and emphasize important ideas throughout the exposition. (See pp. 145–146.) Cautions are highlighted in yellow and Notes are highlighted in purple.

▶ *Calculator Tips* These optional tips, marked with a red calculator icon, offer helpful information and instruction for students using calculators in the course. (See p. 270.)

▶ *Margin Problems* Margin problems, with answers immediately available on the bottom of the page, are found in every section of the text. (See pp. 121–122.) This key feature allows students to immediately practice the material covered in the examples in preparation for the exercise sets.

▶ *Ample and Varied Exercise Sets* The text contains a wealth of exercises to provide students with opportunities to practice, apply, connect, and extend the skills they are learning. Numerous illustrations, tables, graphs, and photos help students visualize the problems they are solving. Problem types include skill building, writing, estimation, and calculator exercises, as well as applications and correct-the-error problems. In the Annotated Instructor's Edition of the text, the writing exercises are marked with an icon for writing ✍ so that instructors may assign these problems at their discretion. Exercises suitable for calculator work are marked in both the student and instructor editions with a calculator icon ▦. (See pp. 177–180.) Students can watch an instructor work through the complete solution for all exercises marked with a DVD icon ◐ on the Videos on DVD.

▶ *Relating Concepts Exercises* These sets of exercises help students tie concepts together and develop higher level problem-solving skills as they compare and contrast ideas, identify and describe patterns, and extend concepts to new situations. (See p. 152.) These exercises make great collaborative activities for pairs or small groups of students.

▶ *Summary Exercises* With the addition of two more sets of summary exercises, all chapters now include this helpful mid-chapter review. These exercises provide students with the all-important *mixed* practice they need at these critical points in their skill development. (See pp. 141–142.)

▶ *Ample Opportunity for Review* Each chapter ends with a Chapter Summary featuring: Key Terms with definitions and helpful graphics, New Formulas, New Symbols, Test Your Word Power, and a Quick Review of each section's content with additional examples. Also included is a comprehensive set of Chapter Review Exercises keyed to individual sections, a set of Mixed Review Exercises, and a Chapter Test. Students can watch an instructor work out the full solutions to the Chapter Test problems on the new Chapter Test Prep Video CD. Beginning with Chapter 2, each chapter concludes with a set of Cumulative Review Exercises. (See pp. 183–194.)

▶ *Test Your Word Power* This feature, incorporated into each Chapter Summary, helps students understand and master mathematical vocabulary. Key terms from the chapter are presented along with four possible definitions in a multiple-choice format. Answers and examples illustrating each term are provided. (See p. 184.)

What's New in This Edition

The scope and sequence of topics in *Basic College Mathematics* has stood the test of time and rates highly with our reviewers. Therefore, you will find the table of contents intact, making the transition to the new edition easier.

- *Examples and Exercises* Throughout the text, examples and exercises have been adjusted or replaced to reflect current data and practices. Applications have been updated and cover a wider variety of topics such as the fields of technology, ecology, and health sciences.

- *Summary Exercises* Two new sets of Summary Exercises have been added; all chapters now include these helpful sets of review exercises. The Summary Exercises are positioned within the chapters to provide critically important mixed review so that students are better prepared to progress to new material.

- *Pointers* Pointers have been added to examples to provide students with important on-the-spot reminders and warnings about common pitfalls. (See pp. 133 and 164.)

- *Math in the Media* Each one-page activity presents a relevant look at how mathematics is used in the media. Designed to help instructors answer the often-asked question, "When will I ever use this stuff?," these activities ask students to read and interpret data from newspaper articles, the Internet, and other familiar, real-world sources. (See pp. 156 and 166.) The activities are well suited to collaborative work or they can be completed by individuals or used for open-ended class discussions.

- *Solutions* Solutions to selected section exercises have been added to the back of the book (following the Answers section). This provides students with easily accessible step-by-step help in solving the exercises that are most commonly missed. Solutions are provided for the exercises marked with a square of blue color around the exercise number, for example, **15.**

- *Study Skills* There is an increased emphasis on study skills. Poor study skills are a major reason why students do not succeed in mathematics. A few generic tips sprinkled here and there are not enough to help students change their behavior. This text includes 13 carefully designed activities, integrated into the text material, that cover note taking, homework, study cards, math anxiety, test preparation, test taking, preparing for a final exam, and more. (See pp. 139–140, 181–182, and 191–192.) Most are located within the first few chapters so that students can use the skills throughout the course. (See the Table of Contents for titles and locations.) The first activity, "Your Brain *Can* Learn Mathematics," explains *how* the brain actually learns and remembers so that students understand *why* the study skills will help them succeed in the course.

- *Chapter Test Prep Video CD* The Chapter Test Prep Video CD provides students with the opportunity to watch instructors work through step-by-step solutions to all the Chapter Test exercises from the textbook. The Chapter Test Video CD is included with each new student text.

What Supplements Are Available?

For a comprehensive list of the supplements and study aids that accompany *Basic College Mathematics,* Eighth Edition, see pages x and xi.

STUDENT SUPPLEMENTS

Student's Solutions Manual
- By Jeffrey A. Cole, *Anoka-Ramsey Community College*
- Provides detailed solutions to the odd-numbered section-level exercises and to all margin, Relating Concepts, Summary, Chapter Review, Chapter Test, and Cumulative Review Exercises
  ISBNs: 0-321-57464-8, 978-0-321-57464-0

Worksheets for Classroom or Lab Practice
- Extra practice exercises for every section of the text with ample space for students to show their work
- These lab- and classroom-friendly workbooks also list the learning objectives and key vocabulary terms for every text section, along with vocabulary practice problems
  ISBNs: 0-321-57470-2, 978-0-321-57470-1

Videos on DVD
- Features an engaging team of lecturers
- Complete set of lectures for each section of the text on DVD for student use at home or on campus
- Ideal for distance learning or supplemental instruction
- Include optional English and Spanish subtitles
- Students can watch an instructor work through the complete solutions for all exercises marked with a DVD icon ◉
  ISBNs: 0-321-60782-1, 978-0-321-60782-9

InterAct Math Tutorial Website *www.interactmath.com*
- Online practice and tutorial help
- Retry an exercise with new values each time for unlimited practice and mastery
- Every exercise is accompanied by an interactive guided solution that gives helpful feedback when an incorrect answer is entered
- View the steps of worked-out sample problems similar to those in the text

Chapter Test Prep Video CD
- Watch instructors work through step-by-step solutions to all the Chapter Test exercises from the textbook.
- Included with each new student text
- English subtitles available to accompany the videos

INSTRUCTOR SUPPLEMENTS

Annotated Instructor's Edition
- Provides answers to all text exercises in color next to the corresponding problems
- Icons identify writing ✍ and calculator ▦ exercises
  ISBNs: 0-321-57234-3, 978-0-321-57234-9

Instructor's Solutions Manual
- By Jeffrey A. Cole, *Anoka-Ramsey Community College*
- Provides complete solutions to all even-numbered section-level exercises
  ISBNs: 0-321-57460-5, 978-0-321-57460-2

Additional Teaching Resources
- Includes resources to help both new and adjunct faculty with course preparation and classroom management by offering helpful teaching tips correlated to the sections of the text.
  Available for download at *www.pearsonhighered.com*

Instructor's Resource Manual with Tests
- By James Ball, *Indiana State University*
- The resource manual contains a test bank with two diagnostic pretests, six free-response and two multiple-choice test forms per chapter, and two final exams.
- The manual also contains a mini-lecture for each section of the text with objectives, key examples, and teaching tips.
- A correlation guide from the seventh to the eighth edition and phonetic spellings for all key terms in the text are also included.
  ISBNs: 0-321-57461-3, 978-0-321-57461-9

PowerPoint Lecture Slides
- Present key concepts and definitions from the text
- Available for download at *www.pearsonhighered.com*

TestGen (*www.pearsonhighered.com/testgen*)
- Enables instructors to build, edit, print, and administer tests using a computerized bank of questions developed to cover all text objectives
- Algorithmically based, TestGen allows instructors to create multiple but equivalent versions of the same question or test with the click of a button
- Instructors can also modify test bank questions or add new questions
- Tests can be printed or administered online

Pearson Math Adjunct Support Center (*www. pearsontutor services.com/math-adjunct.html*) is staffed by qualified instructors with more than 50 years of combined experience at both the community college and university levels. Assistance is provided for faculty in the following areas:
- Suggested syllabus consultation
- Tips on using materials packed with your book
- Book-specific content assistance
- Teaching suggestions, including advice on classroom strategies

Available for Students and Instructors

MyMathLab MyMathLab is a series of text-specific, easily customizable online courses for Pearson Education's textbooks in mathematics and statistics. Powered by CourseCompass™ (our online teaching and learning environment) and MathXL® (our online homework, tutorial, and assessment system), MyMathLab gives you the tools you need to deliver all or a portion of your course online, whether your students are in a lab setting or working from home. MyMathLab provides a rich and flexible set of course materials, featuring free-response exercises that are algorithmically generated for unlimited practice and mastery. Students can also use online tools, such as video lectures, animations, and a multimedia textbook, to independently improve their understanding and performance. Instructors can use MyMathLab's homework and test managers to select and assign online exercises correlated directly to the textbook, and they can also create and assign their own online exercises and import TestGen tests for added flexibility. MyMathLab's online gradebook—designed specifically for mathematics and statistics—automatically tracks students' homework and test results and gives the instructor control over how to calculate final grades. Instructors can also add offline (paper-and-pencil) grades to the gradebook. MyMathLab also includes access to the **Pearson Tutor Center** (*www.pearsontutorservices.com*). The Tutor Center is staffed by qualified mathematics instructors who provide textbook-specific tutoring for students via toll-free phone, fax, e-mail, and interactive Web sessions. MyMathLab is available to qualified adopters. For more information, visit our Web site at *www.mymathlab.com* or contact your sales representative.

MathXL® MathXL is a powerful online homework, tutorial, and assessment system that accompanies Pearson Education's textbooks in mathematics or statistics. With MathXL, instructors can create, edit, and assign online homework and tests using algorithmically generated exercises correlated at the objective level to the textbook. They can also create and assign their own online exercises and import TestGen tests for added flexibility. All student work is tracked in MathXL's online gradebook. Students can take chapter tests in MathXL and receive personalized study plans based on their test results. The study plan diagnoses weaknesses and links students directly to tutorial exercises for the objectives they need to study and retest. Students can also access supplemental animations and video clips directly from selected exercises. MathXL is available to qualified adopters. For more information, visit our Web site at *www.mathxl.com*, or contact your sales representative.

 MathXL® Tutorials on CD This interactive tutorial CD-ROM provides algorithmically generated practice exercises that are correlated at the objective level to the exercises in the textbook. Every practice exercise is accompanied by an example and a guided solution designed to involve students in the solution process. Selected exercises may also include a video clip to help students visualize concepts. The software provides helpful feedback for incorrect answers and can generate printed summaries of students' progress.
ISBNs:
0-321-57463-X
978-0-321-57463-3

Acknowledgments

The comments, criticisms, and suggestions of users, nonusers, instructors, and students have positively shaped this textbook over the years, and we are most grateful for the many responses we have received. The feedback gathered for this revision of the text was particularly helpful, and we especially wish to thank the following individuals who provided invaluable suggestions for this and the previous edition:

George Alexander, *University of Wisconsin*
Sonya Armstrong, *West Virginia State College*
Vernon Bridges, *Durham Technical Community College*
Solveig R. Bender, *William Rainey Harper College*
Barbara Brown, *Anoka-Ramsey Community College*
Ernie Chavez, *Gateway Community College*
Terry Joe Collins, *Hinds Community College*
Martha Daniels, *Central Oregon Community College*
Matthew Flacche, *Camden Community College*
Donna Foster, *Piedmont Technical College*
Mark Gollwitzer, *Greenville Technical College*
Lourdes Gonzalez, *Miami-Dade Community College*
Lance Hemlow, *Raritan Valley Community College*
Joe Howe, *St. Charles County Community College*
Matthew Hudock, *St Philip's College*
Rose Kaniper, *Burlington County College*
Douglas Lewis, *Yakima Valley Community College*
Valerie Maley, *Cape Fear Community College*
Judy Mee, *Oklahoma City Community College*
Wayne Miller, *Lee College*
Kathy Peay, *Sampson Community College*
Thea Philliou, *College of Santa Fe*
Jane Roads, *Moberly Area Community College*
Richard D. Rupp, *Del Mar College*
Ellen Sawyer, *College of DuPage*
Lois Schuppig, *College of Mount St. Joseph*
Mary Lee Seitz, *Erie Community College—City Campus*
Kathryn Taylor, *Santa Ana College*
Mike Tieleman-Ward, *Anoka Technical College*
Sven Trenholm, *North Country Community College*
Bettie A. Truitt, *Black Hawk College*
Jackie Wing, *Angelina College*

Our sincere thanks go to the dedicated individuals at Pearson who have worked hard to make this revision a success: Maureen O'Connor, Kathy Manley, Barbara Atkinson, Michelle Renda, Beth Anderson, Kari Heen, Courtney Slade, Lin Mahoney, Ceci Fleming, Nathaniel Koven, and Mary Gallagher. We are also grateful to Janette Krauss and Bonnie Boehme of Nesbitt Graphics for their excellent production work; Lucie Haskins for producing a useful Index; Jeff Cole for writing the Solutions manuals; and Janis Cimperman, Paul Lorczak, De Cook, and Shannon d'Hemecourt for accuracy checking the manuscript.

Special thanks go to Linda Russell, who wrote the Study Skills activities that appear throughout the text. In her roles as both an instructor and a specialist in reading and study skills at Minneapolis Community and Technical College, she created and revised these activities based on her work with hundreds of developmental-level students.

The ultimate measure of this textbook's success is whether it helps students master basic skills, develop problem-solving techniques, and increase their confidence in learning and using mathematics. In order for us, as authors, to know what to keep and what to improve for the next edition, we need to hear from you, the instructor, and you, the student. Please tell us what you like and where you need additional help by sending an e-mail to math@pearson.com. We appreciate your feedback!

Margaret L. Lial
Stanley A. Salzman
Diana L. Hestwood

Study Skills

Your brain knows how to learn, just as your lungs know how to breathe; however, there are important things you can do to maximize your brain's ability to do its work. This short introduction will help you choose effective strategies for learning mathematics. This is a simplified explanation of a complex process.

Your brain's outer layer is called the **neocortex,** which is where higher level thinking, language, reasoning, and purposeful behavior occur. The neocortex has about 100 billion (100,000,000,000) brain cells called **neurons.**

▶ As you learn something new, threadlike branches grow out of each neuron. These branches are called **dendrites.**

▶ When the dendrite from one neuron grows close enough to the dendrite from another neuron, a connection is made. There is a small gap at the connection point called a **synapse.** One dendrite sends an electrical signal across the gap to another dendrite.

▶ *Learning = growth and connecting of dendrites.*

▶ When you practice a skill just once or twice, the connections between neurons are very weak. If you do not practice the skill again, the dendrites at the connection points wither and die back. You have forgotten the new skill!

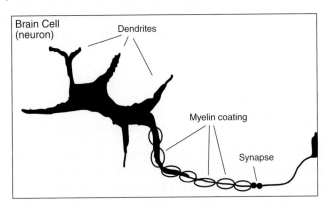

A neuron with several dendrites: one dendrite has developed a myelin coating through repeated practice.

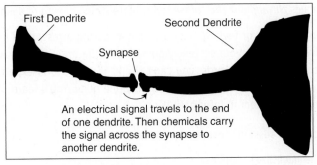

A close up view of the connection (synapse) between two dendrites.

OBJECTIVES

1 **Describe how practice fosters dendrite growth.**

2 **Explain the effect of anxiety on the brain.**

Learning Something New

Remembering New Skills

Become An Effective Student

▶ If you practice a new skill many times, the dendrites for that skill become coated with a fatty protein called **myelin.** Each time one dendrite sends a signal to another dendrite, the myelin coating becomes thicker and smoother, allowing the signals to move faster and with less interference. Thinking can now occur more quickly and easily, and *you will remember the skill for a long time* because the dendrite connections are very strong.

▶ You grow dendrites specifically for the thing you are studying. If you practice dividing fractions, you will grow specialized dendrites just for dividing fractions. If you *watch other people* solve fraction problems, *you will grow dendrites for watching, not for solving.* So, be sure you are actively learning and practicing.

▶ If you practice something the *wrong* way, you will develop strong dendrite connections for doing it the wrong way! So, as you study, check frequently that you are getting correct answers.

▶ As you study a new topic that is related to things you already know, you will grow new dendrites, but your brain will also send signals throughout the network of dendrites for the related topics. In this way, you build a complex **neural network** that allows you to apply concepts, see differences and similarities between ideas, and understand relationships between concepts.

In the first few chapters of this textbook you will find "brain friendly" activities that are designed to help you grow and develop your own reliable neural networks for mathematics. Since you must grow your own dendrites (no one can grow them for you), these activities show you how to

▶ develop new dendrites,

▶ strengthen existing ones, and

▶ encourage the myelin coating to become thicker so signals are sent with less effort.

When you incorporate the activities into your regular study routine, you will discover that you understand better, remember longer, and forget less.

Also remember that *it does take time for dendrites to grow.* Trying to cram in several new concepts and skills at the last minute is not possible. Your dendrites simply can't grow that quickly. You can't expect to develop huge muscles by lifting weights for just one evening before a body building competition! In the same way, practice the study techniques *throughout the course* to facilitate strong growth of dendrites.

When Anxiety Strikes

If you are under stress or feeling anxious, such as during a test, your body secretes **adrenaline** into your system. Adrenaline in the brain blocks connections between neurons. In other words, you can't think! If you've ever experienced "blanking out" on a test, you know what adrenaline does. You'll learn several solutions to that problem in later activities.

Start Your Course Right!

▶ Attend all class sessions (especially the first one).

▶ Gather the necessary supplies.

▶ Carefully read the syllabus for the course, and ask questions if you don't understand.

1

Whole Numbers

In 1965, seventeen-year-old Fred DeLuca had just graduated from high school. DeLuca and a family friend, Dr. Peter Buck, opened the first SUBWAY sandwich shop with a total investment of $1000. From the beginning, the freshest ingredients were used, with DeLuca driving many miles in his Volkswagen bug to purchase vegetables directly from the produce mart. Today, with over 27,000 stores in 83 countries, SUBWAY is the largest submarine sandwich chain in the world. The company operates more stores in the United States, Canada, and Australia than McDonald's. In this chapter, we discuss whole numbers, which are used daily in our lives. (See Exercises 97–100 in **Section 1.3**, Exercise 1 and 31 in **Section 1.10**, and Exercise 123 in the **Chapter 1 Review Exercises.**)

1

1.1 ▶▶▶ Reading and Writing Whole Numbers

OBJECTIVES

1. **Identify whole numbers.**
2. **Give the place value of a digit.**
3. **Write a number in words or digits.**
4. **Read a table.**

Knowing how to read and write numbers is an important step in learning mathematics.

OBJECTIVE 1 Identify whole numbers. The **decimal system** of writing numbers uses the ten digits

$$0, 1, 2, 3, 4, 5, 6, 7, 8, 9$$

to write any number. For example, these digits can be used to write **the whole numbers:**

$$0, 1, 2, 3, 4, 5, 6, 7, 8, 9, 10, 11, 12, 13 \ldots$$

The three dots indicate that the list goes on forever.

OBJECTIVE 2 Give the place value of a digit. Each digit in a whole number has a **place value**, depending on its position in the whole number. The following place value chart shows the names of the different places used most often and has the whole number 153,524 entered.

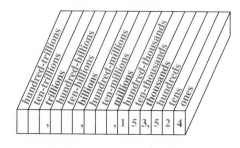

Starbucks sells 153,524 pounds of coffee each day. Each of the 5s in 153,524 represents a different amount because of its position, or **place value** within the number. The **place value** of the 5 on the left is 5 ten-thousands (50,000). The **place value** of the 5 on the right is 5 hundreds (500).

1 Identify the place value of the 4 in each whole number.

(a) 341

(b) 714

(c) 479

EXAMPLE 1 Identifying Place Values

Identify the place value of 8 in each whole number.

[Each "8" has a different value.]

(a) 28
　　└─ 8 ones

(b) 85
　　└─ 8 tens

(c) 869
　　└─ 8 hundreds

Notice that the value of 8 in each number is different, depending on its location (place) in the number.

◀ *Work Problem* **1** *at the Side.*

EXAMPLE 2 Identifying Place Values

Identify the place value of each digit in the number 725,283.

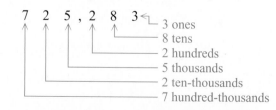

7 2 5 , 2 8 3 ← 3 ones
　　　　　　　── 8 tens
　　　　　　── 2 hundreds
　　　　── 5 thousands
　　── 2 ten-thousands
── 7 hundred-thousands

ANSWERS

1. **(a)** tens **(b)** ones **(c)** hundreds

Notice the comma between the hundreds and thousands position in the number 725,283 in Example 2.

Work Problem **2** *at the Side.* ▶

Using Commas

Commas are used to separate each group of three digits, starting from the right. This makes numbers easier to read. (An exception: Commas are frequently omitted in four-digit numbers such as 9748 or 1329.) Each three-digit group is called a **period.** Some instructors prefer to just call them **groups.**

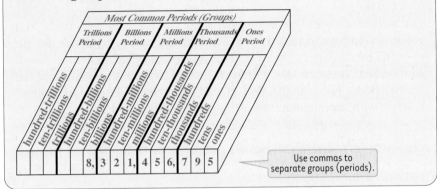

Use commas to separate groups (periods).

EXAMPLE 3 **Knowing the Period or Group Names**

Write the digits in each period of 8,321,456,795.

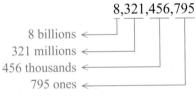

8,321,456,795

8 billions ←
321 millions ←
456 thousands ←
795 ones ←

Work Problem **3** *at the Side.* ▶

Use the following rule to read a number with more than three digits.

Writing Numbers in Words

Start at the left when writing a number in words or saying it aloud. Write or say the digit names in each period (group), followed by the name of the period, except for the period name "ones," which is *not* used.

OBJECTIVE **3** **Write a number in words or digits.**   The following examples show how to write names for whole numbers.

EXAMPLE 4 **Writing Numbers in Words**

Write each number in words.

(a) 57

This number means 5 tens and 7 ones, or 50 ones and 7 ones. Write the number as

fifty-seven.

Continued on Next Page

---

2 Identify the place value of each digit.

(a) 14,218

(b) 460,329

3 In the number 3,251,609,328 identify the digits in each period (group).

(a) billions period

(b) millions period

(c) thousands period

(d) ones period

ANSWERS

2. (a) 1 : ten-thousands
     4 : thousands
     2 : hundreds
     1 : tens
     8 : ones
  (b) 4 : hundred-thousands
     6 : ten-thousands
     0 : thousands
     3 : hundreds
     2 : tens
     9 : ones
3. (a) 3   **(b)** 251   **(c)** 609   **(d)** 328

4 Write each number in words.

(a) 18

(b) 36

(c) 418

(d) 902

5 Write each number in words.

(a) 3104

(b) 95,372

(c) 100,075,002

(d) 11,022,040,000

6 Rewrite each number using digits.

(a) one thousand, four hundred thirty-seven

(b) nine hundred seventy-one thousand, six

(c) eighty-two million, three hundred twenty-five

ANSWERS

4. (a) eighteen
   (b) thirty-six
   (c) four hundred eighteen
   (d) nine hundred two
5. (a) three thousand, one hundred four
   (b) ninety-five thousand, three hundred seventy-two
   (c) one hundred million, seventy-five thousand, two
   (d) eleven billion, twenty-two million, forty thousand
6. (a) 1437 (b) 971,006 (c) 82,000,325

(b) 94

ninety-four

(c) 874

eight hundred seventy-four

> Remember: Start at the *left* to read a number.

(d) 601

six hundred one

◄ *Work Problem* **4** *at the Side.*

CAUTION

The word *and* should never be used when writing whole numbers. You will often hear someone say "five hundred *and* twenty-two," but the use of "and" is not correct since "522" is a whole number. When you work with decimal numbers, the word *and* is used to show the position of the decimal point. For example, 98.6 is read as "ninety-eight *and* six tenths." Practice with decimal numbers is the topic of **Section 4.1.**

EXAMPLE 5 **Writing Numbers in Words by Using Period Names**

Write each number in words.

(a) 725,283

seven hundred twenty-five **thousand,** two hundred eighty-three

Number in period | Name of period | Number in period (not necessary to write "ones")

(b) 7252

> Careful: *Do not* use "and" when reading a whole number.

seven **thousand,** two hundred fifty-two

Name of period | No period name needed

(c) 111,356,075

one hundred eleven **million,** three hundred fifty-six **thousand,** seventy-five

(d) 17,000,017,000

seventeen **billion,** seventeen **thousand**

◄ *Work Problem* **5** *at the Side.*

EXAMPLE 6 **Writing Numbers in Digits**

Rewrite each number using digits.

(a) six **thousand,** twenty-two

6022

> With 4 digits or fewer, *no comma* is needed.

(b) two hundred fifty-six **thousand,** six hundred twelve

256,612

(c) nine **million,** five hundred fifty-nine

9,000,559

Zeros indicate there are no thousands.

◄ *Work Problem* **6** *at the Side.*

Calculator Tip Does your calculator show a comma between each group of three digits? Probably not, but try entering a long number such as 34,629,075. Notice that there is no key with a comma on it, so you do not enter commas. A few calculators may show the position of the commas *above* the digits, like this

34'629'075

Most of the time you will have to write in the commas where needed.

OBJECTIVE 4 Read a table. A common way of showing number values is by using a **table**. Tables organize and display facts so that they are more easily understood. The following table shows some past facts and future predictions for the United States.

These estimated numbers give us a glimpse of what we can expect in the 21st century.

NUMBERS FOR THE 21ST CENTURY

Year	1990	2010	2020*
U.S. population	261 million	307 million	338 million
Births	16 million	14 million	13 million
Household income	$42,936	$49,564	$53,375
Average salary	$21,129	$25,743	$28,050

*Estimated figures
Source: Family Circle magazine; U.S. Census Bureau.

If you read from left to right along the row labeled "U.S. population," you find that the population in 1990 was 261 million, then the population in 2010 was 307 million, and the estimated population for 2020 is 338 million.

EXAMPLE 7 **Reading a Table**

Use the table to find each number, and write the number in words.

(a) The estimated household income in the year 2020
Read from left to right along the row labeled "Household income" until you reach the 2020 column and find $53,375.

Fifty-three thousand, three hundred seventy-five dollars

(b) The average salary in 1990.
Read from left to right along the row labeled "Average salary." In the 1990 column you find $21,129.

Twenty-one thousand, one hundred twenty-nine dollars

Remember: use hyphens when necessary.

Work Problem **7** *at the Side.* ▶

Note
Notice in Example 7 that hyphens are used when writing numbers in words. A hyphen is used when writing the numbers 21 through 99 (twenty-one through ninety-nine), except for numbers ending in zero (20, 30, 40, ... 90).

7 Use the table to find each number, and write the number in digits when given in words, or write the number in words when given in digits.

(a) The number of births in 2010

(b) The estimated number of births in 2020

(c) Household income in 1990

(d) The estimated average salary in 2020

Math in the Media

THE TOLL OF WEDDING BELLS

The Wedding Report recently released statistics showing the average wedding costs in the United States in 2008. In 2004, the cost of the average wedding was $24,168. This cost has increased as the average number of wedding guests has grown to over 200. The graph gives most of the costs involved in a wedding. Use this information to answer the questions that follow.

Numbers in the News

'Til Debt Do You Part

With over 200 guests, the cost of an average wedding has continued to grow. Most of the money is spent on the following:

Transportation	$426
Wedding Stationery	$881
Music	$991
Favors and Gifts	$1166
Flowers	$2048
Wedding Jewelry	$2148
Ceremony	$2627
Wedding Attire	$2710
Photography and Video	$3846
Wedding Reception	$14,037

Source: The Wedding Report, *Wedding Statistics and Research for the Wedding Industry.*

1. What is the total of the costs shown in the graph? **$30,880**

2. How much more expensive was a wedding in 2008 compared with 2004? **$6712**

3. The groom pays for the photography and video, the flowers, the wedding jewelry, the clergy ($500), and the groom's formal wear ($95). What is the total amount spent by the groom? **$8637**

4. If you budgeted $65 per person for the wedding reception and you invited 225 guests to a wedding in 2004, how much would you have spent compared to the 2008 wedding reception costs? **$588 more**

5. If you budget $6000 for the wedding reception and the cost per person is $37, how many guests can you invite and how much of your budgeted amount will be left over? **162 guests; $6**

6. If you budget $1000 for the wedding reception and the cost per person is $15, how many guests can you invite and how much of your budgeted amount will be left over? **66 guests; $10**

7. What kind of an arithmetic problem did you work to get the answers to Problems 5 and 6? What is the mathematical term for the "left over" budget? **division; remainder**

1.1 ▶▶▶ Exercises

*Write the digit for the given **place value** in each whole number. See Examples 1 and 2.*

1. 3065
thousands **3**
tens **6**

2. 4681
thousands **4**
ones **1**

3. 18,015
ten-thousands **1**
hundreds **0**

4. 86,332
ten-thousands **8**
ones **2**

5. 7,628,592,183
millions **8**
thousands **2**

6. 1,700,225,016
billions **1**
millions **0**

*Write the digits for the given **period** (group) in each whole number. See Example 3.*

7. 3,561,435
millions **3**
thousands **561**
ones **435**

8. 28,785,203
millions **28**
thousands **785**
ones **203**

9. 60,000,502,109
billions **60**
millions **000**
thousands **502**
ones **109**

10. 100,258,100,006
billions **100**
millions **258**
thousands **100**
ones **6**

11. Do you think the fact that humans have four fingers and a thumb on each hand explains why we use a number system based on ten digits? Explain.
Evidence suggests that this is true. It is common to count using fingers.

12. The decimal system uses ten digits. Fingers and toes are often referred to as digits. In your opinion, is there a relationship here? Explain.
No doubt there is a relationship here. One answer might be that people could count using their fingers and toes and, therefore, thought of them as numbers or digits.

Write each number in words. See Examples 4 and 5.

13. 23,115
twenty-three thousand, one hundred fifteen

14. 37,886
thirty-seven thousand, eight hundred eighty-six

15. 346,009
three hundred forty-six thousand, nine

16. 218,033
two hundred eighteen thousand, thirty-three

17. 25,756,665
twenty-five million, seven hundred fifty-six thousand, six hundred sixty-five

18. 999,993,000
nine hundred ninety-nine million, nine hundred ninety-three thousand

Write each number using digits. See Example 6.

19. sixty-three thousand, one hundred sixty-three
63,163

20. ninety-five thousand, one hundred eleven
95,111

21. ten million, two hundred twenty-three
10,000,223

22. one hundred million, two hundred
100,000,200

Write the numbers from each sentence using digits. See Example 6.

23. There are three million, two hundred thousand parachute jumps in the United States each year. (*Source:* History Channel.)

3,200,000

24. The United States Postal Service set a record of two hundred eighty million, four hundred eighty-nine thousand postmarked pieces of mail on a single day. (*Source:* U.S. Postal Service.)

280,489,000

25. The number of cans of Pepsi Cola sold each day is fifty million, fifty-one thousand, five hundred seven. (*Source:* Andy Rooney, *60 Minutes.*)

50,051,507

26. The number of motorcycle owners in the world is twenty-three million, five hundred thirty-five thousand. (*Source:* Motorcycle Industry Council.)

23,535,000

27. There are fifty-four million, seven hundred fifty thousand Hot Wheels sold each year. (*Source:* Andy Rooney, *60 Minutes.*)

54,750,000

28. Burger King sells two million, four hundred thousand hamburgers each day, (*Source:* Andy Rooney, *60 Minutes.*)

2,400,000

29. Rewrite eight hundred trillion, six hundred twenty-one million, twenty thousand, two hundred fifteen by using digits.

800,000,621,020,215

30. Rewrite 70,306,735,002,102 in words

seventy trillion, three hundred six billion, seven hundred thirty-five million, two thousand, one hundred two

The table at the right shows various ways people get to work. Use the table to answer Exercises 31–34. See Example 7.

31. 🌐 Which method of transportation is least used? Write the number in words.

public transportation; six million, sixty-nine thousand, five hundred eighty-nine

32. Which method of transportation is most used? Write the number in words.

drive alone; eighty-four million, two hundred fifteen thousand, two hundred ninety-eight

33. Find the number of people who walk to work or work at home, and write it in words.

seven million, eight hundred ninety-four thousand, nine hundred eleven

34. Find the number of people who carpool, and write it in words.

fifteen million, three hundred seventy-seven thousand, six hundred thirty-four

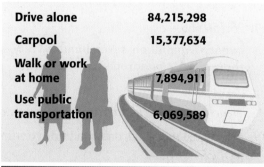

GETTING TO WORK	
How workers 16 and over get to work:	
Drive alone	84,215,298
Carpool	15,377,634
Walk or work at home	7,894,911
Use public transportation	6,069,589

Source: U.S. Census Bureau.

Study Skills

■■■ **USING YOUR TEXTBOOK**

Be sure to read Your Brain Can Learn Mathematics before this activity. You'll find out how your brain learns and remembers.

Your textbook can be very helpful. Find out what it has to offer. First, let's look at some general features that will help in all chapters.

Look at page v in the very front for the Table of Contents. Before you start Chapter 1, you should look at the Preface, which begins on page vii. On page x is a list of Supplementary Resources for students. If you are interested in any of these, ask your instructor if they are available.

Each chapter is divided into sections, and each section has a number, such as 1.3 or 3.5. Your instructor will use these numbers to assign readings and homework.

Chapter 3 → **3.5** ← Section 5 within Chapter 3

There are four features to pay special attention to as you work in your book.

▶ **Objectives.** Each section lists the objectives in the upper corner of the first page. The objectives are listed again as each one is introduced. An objective tells you *what you will be able to do after you complete the section*. An excellent way to check your learning is to go back to the list of objectives when you are finished with a section and ask yourself if you can do them all.

▶ **Margin Exercises.** The exercises in the shaded margins of the pages in your textbook give you immediate practice. **This is a perfect way to get your dendrites growing right away!** The answers are given at the bottom, so you can check yourself easily.

▶ **Cautions, Pointers, Notes, and Calculator Tips.**
  • **Caution!** A bright yellow box is a comment about a common error that students make or a common trouble spot you will want to avoid.
  • **Pointers** are little "clouds" next to worked examples. They point to specific places where common mistakes are made and give on-the-spot reminders.
  • Look for the specially marked purple **Note** boxes. They contain hints, explanations, or interesting side comments about a topic.
  • A small picture of a red calculator 🖩 appears several places. In the main part of the chapter, the icon means that there is a **Calculator Tip,** which helps you learn more about using your calculator. A calculator in an Exercise section is a recommendation to use your calculator to work that exercise.

Go back to the Table of Contents again. What is listed at the end of each chapter?

▶ **Chapter Summary.** Turn to page 99 to find the Summary for Chapter 1. It lists the chapter's **Key Terms** (arranged in the order that they appear in the chapter) and **New Symbols** and/or **New Formulas.** Then, **Test Your Word Power** checks your understanding of the math vocabulary. Next is a **Quick Review section.** It lists each topic in the chapter and shows a worked example, with tips.

OBJECTIVES

1 **Explain the meaning of text features such as section numbering, objectives, margin exercises.**

2 **Locate the Index, Answers, and Solutions sections.**

Table of Contents

Section Numbering

Chapter Features

End of Chapter Features

9

End of Chapter Features
(continued)

> How will you make good use of the features at the end of each chapter?
> _____
> _____

Answers

> *Flag the Answers section with a sticky note or other device, so that you can turn to it quickly.*

Solutions

Index

Why Are These Features Brain Friendly?

The textbook authors included text features that make it easier for you to understand the mathematics. **Your brain naturally seeks organization and predictability.** When you pay attention to the regular features of your textbook, you are allowing your brain to get familiar with all of the helpful tips, suggestions, and explanations that your book has to offer. You will make the best possible use of your textbook.

▶ **Review Exercises** Use these exercises as a way to check your understanding of all the concepts in the chapter. You can practice every type of problem. If you get stuck, the red numbers in brackets tell you which section of the chapter to go back to for more explanations. Make sure you do the **Mixed Review Exercises** to practice for tests.

▶ **Chapter Test.** Plan to take the test as a practice exam. That way you can be sure you really know how to work all types of problems without looking back at the chapter.

▶ **Cumulative Review (starting with chapter 2)** These exercises help you maintain the skills you've learned in all previous chapters. Working on previous skills throughout the course will be a big help on the final exam.

How do you find out if you've worked the exercises correctly? Your textbook provides many of the answers. Throughout each chapter you should work the sample problems in the **margins.** The answers for those are at the **bottom of each page** in the margin area.

For homework, you can find the answers to all of the **odd-numbered section exercises** in the **Answers to Selected Exercises** section near the end of your textbook. Also, *all* of the answers are given for the Chapter Review Exercises and Chapter Tests. Check your textbook now, and find the page on which the Answers section begins.

The Solutions section near the end of the book shows how to solve some of the harder odd-numbered exercises step by step. Look for the problem numbers with a square of blue shading around them. These are the ones that have a solution.

The Index is the last thing in your book. It lists all the topics, vocabulary, and concepts in alphabetical order. For example, look up the words below. Go to the page or pages listed and find each word. Write down the page that introduces or defines each one. There may be several subheadings listed under the main word, or several page numbers listed. Usually, the *first* place that a word appears in the textbook is where it is introduced and defined. So, the earliest page number is a good place to start.

> *Commutative property of multiplication* is defined on page _____.
> *Factors* of numbers are defined on page _____.
> *Rounding of mixed numbers* is explained on page _____.

> List a page number from Chapter 1 for each of these features:
>
> A *Caution* appears on page_____.
>
> A *Pointer* appears on page_____.
>
> A *Note* appears on page_____.
>
> A *Calculator Tip* appears on page_____.

1.2 ▶▶▶ Adding Whole Numbers

There are four soccer balls at the left and two at the right. In all, there are six soccer balls.

The process of finding the total is called **addition.** Here 4 and 2 were added to get 6. Addition is written with a + sign, so that

$$4 + 2 = 6.$$

OBJECTIVE 1 Add two single-digit numbers. In addition, the numbers being added are called **addends,** and the resulting answer is called the **sum** or **total.**

$$
\begin{array}{rl}
4 & \leftarrow \text{Addend} \\
+\,2 & \leftarrow \text{Addend} \\
\hline
6 & \leftarrow \text{Sum (total)}
\end{array}
$$

Addition problems can also be written horizontally, as follows.

$$
\underset{\underset{\text{Addend}}{\uparrow}}{4} \; + \; \underset{\underset{\text{Addend}}{\uparrow}}{2} \; = \; \underset{\underset{\text{Sum}}{\uparrow}}{6}
$$

> **Commutative Property of Addition**
>
> By the **commutative property of addition,** changing the order of the addends in an addition problem does not change the sum.

For example, the sum of $4 + 2$ is the same as the sum of $2 + 4$. This allows the addition of the same numbers in a different order.

EXAMPLE 1 Adding Two Single-Digit Numbers

Add, and then change the order of numbers to write another addition problem.

(a) $5 + 3 = 8$ and $3 + 5 = 8$

(b) $7 + 8 = 15$ and $8 + 7 = 15$ ◁ *Changing the order in addition does not change the sum.*

(c) $8 + 3 = 11$ and $3 + 8 = 11$

(d) $8 + 8 = 16$

——————————— *Work Problem* **1** *at the Side.* ▶

> **Associative Property of Addition**
>
> By the **associative property of addition,** changing the grouping of the addends in an addition problem does not change the sum.

For example, the sum of $3 + 5 + 6$ may be found as follows.

$$(3 + 5) + 6 = 8 + 6 = 14 \qquad \text{Parentheses tell us to add } 3 + 5 \text{ first.}$$

Another way to add the same numbers is   ◁ *Changing the grouping of addends does not change the sum.*

$$3 + (5 + 6) = 3 + 11 = 14. \qquad \text{Parentheses tell us to add } 5 + 6 \text{ first.}$$

Either grouping gives a sum of 14 because of the associative property of addition.

OBJECTIVES

1 Add two single-digit numbers.

2 Add more than two numbers.

3 Add when regrouping (carrying) is not required.

4 Add with regrouping (carrying).

5 Use addition to solve application problems.

6 Check the answer in addition.

---

1 Add, and then change the order of numbers to write another addition problem.

(a) $2 + 6$

(b) $9 + 5$

(c) $4 + 7$

(d) $6 + 9$

ANSWERS

1. (a) $8; 6 + 2 = 8$ **(b)** $14; 5 + 9 = 14$
   (c) $11; 7 + 4 = 11$ **(d)** $15; 9 + 6 = 15$

2 Add each column of numbers.

(a)   3
        8
        5
        4
      + 6

(b)   5
        6
        3
        2
      + 4

(c)   9
        6
        8
        7
      + 3

(d)   3
        8
        6
        4
      + 8

OBJECTIVE 2 Add more than two numbers.   To add several numbers, first write them in a column. Add the first number to the second. Add this sum to the third number. Continue until all the numbers are used.

EXAMPLE 2 Adding More Than Two Numbers

Add 2, 5, 6, 1, and 4.

$$
\begin{array}{r}
2 \\
5 \\
6 \\
1 \\
+\ 4 \\
\hline
18
\end{array}
$$

$2 + 5 = 7$
$7 + 6 = 13$
$13 + 1 = 14$
$14 + 4 = 18$

◀ *Work Problem* **2** *at the Side.*

> **Note**
> By the commutative and associative properties of addition, numbers may also be added starting at the bottom of a column. Adding from the top or adding from the bottom will give the same answer.

OBJECTIVE 3 Add when regrouping (carrying) is not required.
If numbers have two or more digits, you must arrange the numbers in columns so that the ones digits are in the same column, tens are in the same column, hundreds are in the same column, and so on. Next, you add column by column starting at the right.

EXAMPLE 3 Adding without Regrouping

Add 511 + 23 + 154 + 10.

First line up the numbers in columns, with the ones column at the right.

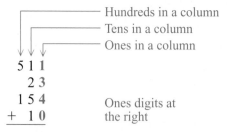

Hundreds in a column
Tens in a column
Ones in a column

Ones digits at the right

Now start at the right and add the ones digits. Add the tens digits next, and finally, the hundreds digits.

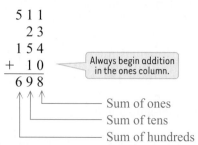

Always begin addition in the ones column.

Sum of ones
Sum of tens
Sum of hundreds

The sum of the four numbers is 698.

Work Problem **3** *at the Side.* ▶

OBJECTIVE **4** **Add with regrouping (carrying).**   If the sum of the digits in any column is more than 9, use **regrouping** (sometimes called **carrying**).

EXAMPLE 4 **Adding with Regrouping**

Add 47 and 29.

Add ones.

$$\begin{array}{r} 47 \\ + 29 \\ \hline \end{array}$$

7 ones and 9 ones = 16 ones

Regroup 16 ones as 1 ten and 6 ones. Write 6 ones in the ones column and write 1 ten in the tens column.

$$\begin{array}{r} 1 \\ 47 \\ + 29 \\ \hline 6 \end{array}$$

Write 1 ten in the tens column.
7 ones and 9 ones = 16 ones
Write 6 ones in the ones column.

Add the digits in the tens column, including the regrouped 1.

$$\begin{array}{r} 1 \\ 47 \\ + 29 \\ \hline 76 \end{array}$$

1 ten + 4 tens + 2 tens = 7 tens

Work Problem **4** *at the Side.* ▶

EXAMPLE 5 **Adding with Regrouping**

Add 324 + 7855 + 23 + 7 + 86.

Step 1   Add the digits in the ones column.

$$\begin{array}{r} 2 \\ 324 \\ 7855 \\ 23 \\ 7 \\ + 86 \\ \hline 5 \end{array}$$

Write 2 tens in the tens column.
Sum of the ones column is 25 ones.
Write 5 ones in the ones column.

Notice that 25 ones are regrouped as 2 tens and 5 ones.

Step 2   Add the digits in the tens column, including the regrouped 2.

$$\begin{array}{r} 12 \\ 324 \\ 7855 \\ 23 \\ 7 \\ + 86 \\ \hline 95 \end{array}$$

Write 1 hundred in the hundreds column.
Sum of the tens column is 19 tens.
Notice that 19 tens are regrouped as 1 hundred and 9 tens.
Write 9 in the tens column.

Continued on Next Page

3 Add.

(a) $\begin{array}{r} 26 \\ + 73 \\ \hline \end{array}$

(b) $\begin{array}{r} 534 \\ + 265 \\ \hline \end{array}$

(c) $\begin{array}{r} 42{,}305 \\ + 11{,}563 \\ \hline \end{array}$

4 Add with regrouping

(a) $\begin{array}{r} 66 \\ + 27 \\ \hline \end{array}$

(b) $\begin{array}{r} 58 \\ + 33 \\ \hline \end{array}$

(c) $\begin{array}{r} 56 \\ + 37 \\ \hline \end{array}$

(d) $\begin{array}{r} 34 \\ + 49 \\ \hline \end{array}$

ANSWERS
3. (a) 99 (b) 799 (c) 53,868
4. (a) 93 (b) 91 (c) 93 (d) 83

5 Add by regrouping as necessary.

(a) 42
     651
     396
   + 87

(b) 162
    4271
     372
  + 8976

(c) 57
      4
    392
    804
     51
   + 27

(d) 7821
     435
      72
     305
  + 1693

6 Add by regrouping mentally.

(a) 816
     363
      17
       2
       5
  + 7654

(b) 3305
     650
     708
      29
      40
       6
     + 3

(c) 15,829
       765
        78
        15
         9
         7
  + 13,179

Step 3 Add the hundreds column, including the regrouped 1.

```
  1 1 2
    324
   7855
     23
      7
 +   86
    295
```

Write 1 thousand in the thousands column.

Notice that 12 hundreds are regrouped as 1 thousand and 2 hundreds.

Sum of the hundreds column is 12 hundreds.

Write 2 hundreds in the hundreds column.

Step 4 Add the thousands column, including the regrouped 1.

```
  1 1 2
    324
   7855
     23
      7
 +   86
   8295
```

Sum of the thousands column is 8.

Finally, 324 + 7855 + 23 + 7 + 86 = 8295.

◀ *Work Problem* **5** *at the Side.*

Note

For additional speed, try to regroup mentally. Do not write the regrouped number, but just remember it as you move to the top of the next column. Try this method. If it works for you, use it.

◀ *Work Problem* **6** *at the Side.*

OBJECTIVE 5 Use addition to solve application problems. In Section 1.10 we will describe how to solve application problems in more detail. The next two examples are application problems that require adding.

EXAMPLE 6 Applying Addition Skills

On this map of the Walt Disney World area in Florida, the distance in miles from one location to another is written alongside of the road. Find the shortest route from Altamonte Springs to Clear Lake.

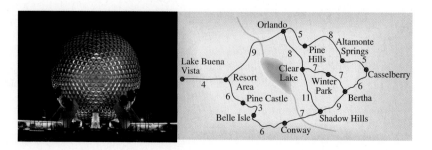

Continued on Next Page

Approach   Add the mileage along various routes to determine the distances from Altamonte Springs to Clear Lake. Then select the shortest route.

Solution   One way from Altamonte Springs to Clear Lake is through Orlando. Add the mileage numbers along this route.

$$
\begin{array}{rl}
8 & \text{Altamonte Springs to Pine Hills} \\
5 & \text{Pine Hills to Orlando} \\
+\,8 & \text{Orlando to Clear Lake} \\
\hline
21 \rightarrow & \text{miles from Altamonte Springs to} \\
& \text{Clear Lake, going through Orlando}
\end{array}
$$

> Remember: Shortest distance is the fewest total miles.

Another way is through Casselberry, Bertha, and Winter Park. Add the mileage numbers along this route.

$$
\begin{array}{rl}
5 & \text{Altamonte Springs to Casselberry} \\
6 & \text{Casselberry to Bertha} \\
7 & \text{Bertha to Winter Park} \\
+\,7 & \text{Winter Park to Clear Lake} \\
\hline
25 \rightarrow & \text{miles from Altamonte Springs to Clear Lake through} \\
& \text{Bertha and Winter Park}
\end{array}
$$

The shortest route from Altamonte Springs to Clear Lake is 21 miles through Orlando.

_____ Work Problem **7** at the Side. ▶

7 Use the map in Example 6 to find the shortest route from Lake Buena Vista to Conway.

(EXAMPLE 7)   **Finding a Total Distance**

Use the map in Example 6 to find the total distance from Shadow Hills to Casselberry to Orlando and back to Shadow Hills.

Approach   Add the mileage from Shadow Hills to Casselberry to Orlando and back to Shadow Hills to find the total distance.

Solution   Use the numbers from the map.

$$
\begin{array}{rl}
9 & \text{Shadow Hills to Bertha} \\
6 & \text{Bertha to Casselberry} \\
5 & \text{Casselberry to Altamonte Springs} \\
8 & \text{Altamonte Springs to Pine Hills} \\
5 & \text{Pine Hills to Orlando} \\
8 & \text{Orlando to Clear Lake} \\
+\,11 & \text{Clear Lake to Shadow Hills} \\
\hline
52 \rightarrow & \text{miles from Shadow Hills to Casselberry} \\
& \text{to Orlando and back to Shadow Hills}
\end{array}
$$

_____ Work Problem **8** at the Side. ▶

8 The road is closed between Orlando and Clear Lake, so this route cannot be used. Use the map in Example 6 to find the next shortest route from Orlando to Clear Lake.

(EXAMPLE 8)   **Finding a Perimeter**

Find the number of feet of floating pipe needed to contain the farm-raised salmon in the habitat shown.

The short way to write *feet* is *ft.*

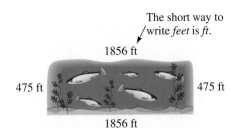

1856 ft

475 ft   475 ft

1856 ft

Approach   Find the **perimeter,** or total distance around the habitat, by adding the lengths of all the sides.

_____ **Continued on Next Page**

9 Solve the problem. Find the number of feet of fencing needed to enclose the solar electricity generating project shown.

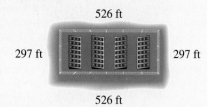

526 ft

297 ft   297 ft

526 ft

10 Check the following additions. If an answer is incorrect, find the correct answer.

(a)   63
          4
          9
      + 28
      ‾‾‾‾
        104

(b)   927
       395
        64
      + 251
      ‾‾‾‾‾
       1637

(c)   79
       218
         7
      + 639
      ‾‾‾‾‾
        953

(d)   21,892
      11,746
    + 43,925
    ‾‾‾‾‾‾‾‾
      79,563

ANSWERS

9. 1646 ft
10. (a) correct   **(b)** correct
      (c) incorrect; should be 943
      (d) incorrect; should be 77,563

Solution   Use the lengths shown.

          1856
           475
          1856
        + 475
        ‾‾‾‾‾‾
          4662 ft

The amount of floating pipe needed is 4662 ft, which is the perimeter of (distance around) the habitat.

◀ *Work Problem* **9** *at the Side.*

OBJECTIVE **6** **Check the answer in addition.**   Checking the answer is an important part of problem solving. A common method for checking addition is to re-add from bottom to top. This is an application of the commutative and associative properties of addition.

EXAMPLE 9 **Checking Addition**

Check the following addition.

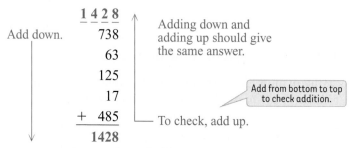

Here the answers agree, so the sum is probably correct.

EXAMPLE 10 **Checking Addition**

Check the following additions. Are they correct?

(a)   785   1 0 3 3   Correct, because both
         63   785   answers are the same.
      + 185   63
      ‾‾‾‾‾   + 185 ‾‾‾ To check, add up.
       1033   ‾‾‾‾‾
                      1033

(b)   635   2 4 5 4
         73   635   Error, because the
        831   73   answers are different.
      + 915   831
      ‾‾‾‾‾   + 915 ‾‾‾ To check, add up.   Avoid wrong answers by
       2444   ‾‾‾‾‾   checking your work.
                      2444

Re-add to find that the correct sum is 2454.

◀ *Work Problem* **10** *at the Side.*

1.2 ▶▶▶ Exercises

Add. See Examples 1–3.

1.	2.	⊙ 3.	4.	5.
43 + 54 **97**	18 + 11 **29**	56 + 33 **89**	83 + 15 **98**	317 + 572 **889**

6.	7.	8.	9.	10.
574 + 325 **899**	318 151 + 420 **889**	135 253 + 410 **798**	6310 252 + 1223 **7785**	121 5705 + 3163 **8989**

⊙ **11.** 932 + 44 + 613

1589

12. 517 + 131 + 250

898

13. 1251 + 4311 + 2114

7676

14. 3241 + 1513 + 2014

6768

15. 12,142 + 43,201 + 23,103

78,446

16. 41,124 + 12,302 + 23,500

76,926

17. 3213 + 5715

8928

18. 6344 + 1655

7999

19. 38,204 + 21,020

59,224

20. 63,251 + 36,305

99,556

Add, regrouping as necessary. See Examples 4 and 5.

21.	22.	23.	24.	25.
87 + 63 **150**	19 + 92 **111**	86 + 69 **155**	37 + 85 **122**	47 + 74 **121**

26.	27.	28.	29.	30.
97 + 79 **176**	67 + 78 **145**	96 + 47 **143**	73 + 29 **102**	68 + 37 **105**

31.	32.	33.	34.	35.
746 + 905 **1651**	621 + 359 **980**	306 + 848 **1154**	798 + 206 **1004**	278 + 135 **413**

36.	37.	38.	39.	40.
172 + 156 **328**	928 + 843 **1771**	686 + 726 **1412**	526 + 884 **1410**	116 + 897 **1013**

41.
$$\begin{array}{r} 3574 \\ + 2817 \\ \hline 6391 \end{array}$$

42.
$$\begin{array}{r} 6871 \\ + 7528 \\ \hline 14,399 \end{array}$$

43.
$$\begin{array}{r} 7896 \\ + 3728 \\ \hline 11,624 \end{array}$$

44.
$$\begin{array}{r} 9382 \\ + 7586 \\ \hline 16,968 \end{array}$$

45.
$$\begin{array}{r} 9625 \\ + 7986 \\ \hline 17,611 \end{array}$$

46.
$$\begin{array}{r} 5718 \\ 5623 \\ + 7436 \\ \hline 18,777 \end{array}$$

47.
$$\begin{array}{r} 9056 \\ 78 \\ 6089 \\ + 731 \\ \hline 15,954 \end{array}$$

48.
$$\begin{array}{r} 4022 \\ 709 \\ 8621 \\ + 37 \\ \hline 13,389 \end{array}$$

49.
$$\begin{array}{r} 18 \\ 708 \\ 9286 \\ + 636 \\ \hline 10,648 \end{array}$$

50.
$$\begin{array}{r} 1708 \\ 321 \\ 61 \\ + 8926 \\ \hline 11,016 \end{array}$$

51.
$$\begin{array}{r} 422 \\ 6074 \\ 435 \\ + 8663 \\ \hline 15,594 \end{array}$$

52.
$$\begin{array}{r} 6505 \\ 173 \\ 7044 \\ + 168 \\ \hline 13,890 \end{array}$$

53.
$$\begin{array}{r} 321 \\ 9603 \\ 8 \\ 21 \\ + 1604 \\ \hline 11,557 \end{array}$$

54.
$$\begin{array}{r} 7631 \\ 5983 \\ 7 \\ 36 \\ + 505 \\ \hline 14,162 \end{array}$$

55.
$$\begin{array}{r} 2109 \\ 63 \\ 16 \\ 3 \\ + 9887 \\ \hline 12,078 \end{array}$$

56.
$$\begin{array}{r} 322 \\ 6508 \\ 93 \\ 745 \\ 18 \\ + 2005 \\ \hline 9691 \end{array}$$

57.
$$\begin{array}{r} 553 \\ 97 \\ 2772 \\ 437 \\ 63 \\ + 328 \\ \hline 4250 \end{array}$$

58.
$$\begin{array}{r} 3187 \\ 810 \\ 527 \\ 76 \\ 2665 \\ + 317 \\ \hline 7582 \end{array}$$

⊙ 59.
$$\begin{array}{r} 413 \\ 85 \\ 9919 \\ 602 \\ 31 \\ + 1218 \\ \hline 12,268 \end{array}$$

60.
$$\begin{array}{r} 576 \\ 7934 \\ 60 \\ 781 \\ 5968 \\ + 371 \\ \hline 15,690 \end{array}$$

Check each addition. If an answer is incorrect, find the correct answer. See Examples 9 and 10.

61.
$$\begin{array}{r} 832 \\ 468 \\ + 791 \\ \hline 2091 \end{array}$$
correct

62.
$$\begin{array}{r} 326 \\ 852 \\ + 679 \\ \hline 1857 \end{array}$$
correct

63.
$$\begin{array}{r} 179 \\ 214 \\ + 376 \\ \hline 759 \end{array}$$
incorrect;
should be 769

64.
$$\begin{array}{r} 17 \\ 296 \\ 713 \\ + 94 \\ \hline 1220 \end{array}$$
incorrect;
should be 1120

65.
$$\begin{array}{r} 4713 \\ 28 \\ 615 \\ + 64 \\ \hline 5420 \end{array}$$
correct

66.
$$\begin{array}{r} 3\ 628 \\ 72 \\ 564 \\ + 7\ 319 \\ \hline 11,583 \end{array}$$
correct

67.
$$\begin{array}{r} 678 \\ 7\ 952 \\ 56 \\ 718 \\ + 2\ 173 \\ \hline 11,377 \end{array}$$
incorrect;
should be
11,577

68.
$$\begin{array}{r} 516 \\ 8\ 760 \\ 24 \\ 189 \\ + 1\ 723 \\ \hline 11,212 \end{array}$$
correct

⊙ 69.
$$\begin{array}{r} 4\ 714 \\ 27 \\ 77 \\ 8\ 878 \\ + 636 \\ \hline 14,332 \end{array}$$
correct

70.
$$\begin{array}{r} 6\ 715 \\ 283 \\ 9\ 617 \\ 13 \\ + 81 \\ \hline 16,719 \end{array}$$
incorrect;
should be
16,709

71. Explain the commutative property of addition in your own words. How is this used when checking an addition problem?

Changing the order in which numbers are added does not change the sum. You can add from bottom to top when checking addition.

72. Explain the associative property of addition. How can this be used when adding columns of numbers?

Grouping the addition of numbers in any order does not change the sum. You can add numbers in any order. For example, you can add pairs of numbers that add to 10.

For Exercises 73–76, use the map to find the shortest route between each pair of cities. See Examples 6 and 7.

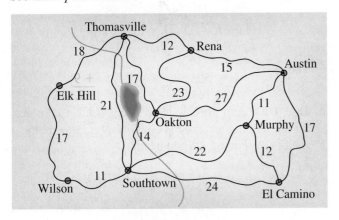

73. Southtown and Rena   33 miles

74. Elk Hill and Oakton   35 miles

75. Thomasville and Murphy   38 miles

76. Murphy and Thomasville   38 miles

Solve each application problem.

77. The Twin Lakes Food Bank raised $3482 at a flea market and $12,860 at their annual auction. Find the total amount raised at these two events.

$16,342

78. A clothing store ordered 75 tops and 52 pairs of shorts. How many items were ordered?

127 items

79. There are 413 women and 286 men on the sales staff. How many people are on the sales staff?

699 people

80. One department in an office building has 283 employees while another department has 218 employees. How many employees are in the two departments?

501 employees

81. This semester there are 13,786 students enrolled in on-campus day classes, 3497 students enrolled in night classes, and 2874 student's enrolled in on-line classes. Find the total number of students enrolled.

20,157 students

82. Robert and Crystal Hernandez have a car loan balance of $10,329 and a balance owed on their credit card of $2685. After receiving a home loan of $169,760, find the total amount of their loans.

$182,774

Solve each problem involving perimeter. See Example 8.

83. Find the total distance around a lot that has been
developed as a go-cart track.

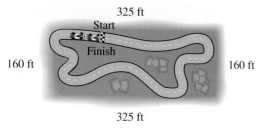

325 ft
Start
Finish
160 ft · 160 ft
325 ft

970 ft

84. Because of heavy snowfall this winter, Maria needs
to put new rain gutters around her entire roof. How
many feet of gutters will she need?

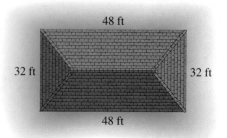

48 ft
32 ft · 32 ft
48 ft

160 ft

85. Martin plans to frame his back patio with redwood
lumber. How many feet of lumber will he need?

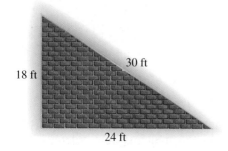

30 ft
18 ft
24 ft

72 ft

86. Due to a recent tornado, Carl Jones needs to replace
all the fencing around his cow pasture. How many
meters of fencing will he need?

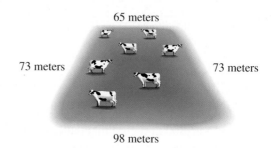

65 meters
73 meters · 73 meters
98 meters

309 meters

Relating Concepts (Exercises 87–94) For Individual or Group Work

Recall the place values of digits discussed in Section 1.1 and **work Exercises 87–94 in order.**

87. Write the largest four-digit number possible using
the digits 4, 1, 9, and 2. Use each digit once.

9421

88. Using the digits 4, 1, 9, and 2, write the smallest
four-digit number possible. Use each digit once.

1249

89. Write the largest five-digit number possible using
the digits 6, 2, and 7. Use each digit at least once.

77,762

90. Using the digits 6, 2, and 7, write the smallest five-
digit number possible. Use each digit at least once.

22,267

91. Write the largest seven-digit number possible using
the digits 4, 3, and 9. Use each digit at least twice.

9,994,433

92. Using the digits 4, 3, and 9, write the smallest
seven-digit number possible. Use each digit at least
twice.

3,334,499

93. Explain your rule or procedure for writing the
largest number in Exercise 91.

Write the largest digits on the left, using the
smaller digits as you move right.

94. Explain your rule or procedure for writing the
smallest number in Exercise 92.

Write the smallest digits on the left, using the
larger digits as you move right.

1.3 ▶▶▶ Subtracting Whole Numbers

Suppose you have $9, and you spend $2 for parking. You then have $7 left. There are two different ways of looking at these numbers.

As an addition problem:

$$\$2 \ + \ \$7 \ = \ \$9$$

↑   ↑   ↑

Amount   Amount   Original
spent   left   amount

As a subtraction problem:

$$\$9 \ - \ \$2 \ = \ \$7$$

↑   ↑   ↑   ↑

Original   Subtraction   Amount   Amount
amount   symbol   spent   left

OBJECTIVE **1** **Change addition problems to subtraction and subtraction problems to addition.** As shown in the box above, an addition problem can be changed to a subtraction problem and a subtraction problem can be changed to an addition problem.

EXAMPLE 1 **Changing Addition Problems to Subtraction**

Change each addition problem to a subtraction problem.

(a) $4 + 1 = 5$

Two subtraction problems are possible:

$$5 - 1 = 4 \quad \text{or} \quad 5 - 4 = 1$$

These figures show each subtraction problem.

$$5 - 1 = 4 \qquad\qquad 5 - 4 = 1$$

(b) $8 + 7 = 15$

$$15 - 7 = 8 \quad \text{or} \quad 15 - 8 = 7$$

Work Problem 1 *at the Side.* ▶

EXAMPLE 2 **Changing Subtraction Problems to Addition**

Change each subtraction problem to an addition problem.

(a) $8 - 3 = 5$

↓   ↓   ↓

$$8 = 3 + 5$$

It is also correct to write $8 = 5 + 3$.

Continued on Next Page

---

1 Write two subtraction problems for each addition problem.

(a) $7 + 2 = 9$

(b) $7 + 4 = 11$

(c) $15 + 22 = 37$

(d) $23 + 55 = 78$

ANSWERS

1. **(a)** $9 - 2 = 7$ or $9 - 7 = 2$
   (b) $11 - 4 = 7$ or $11 - 7 = 4$
   (c) $37 - 22 = 15$ or $37 - 15 = 22$
   (d) $78 - 55 = 23$ or $78 - 23 = 55$

2 Write an addition problem for each subtraction problem.

(a) $7 - 5 = 2$

(b) $9 - 4 = 5$

(c) $21 - 15 = 6$

(d) $58 - 42 = 16$

3 Subtract.

(a)
$$\begin{array}{r} 74 \\ -43 \\ \hline \end{array}$$

(b)
$$\begin{array}{r} 68 \\ -24 \\ \hline \end{array}$$

(c)
$$\begin{array}{r} 429 \\ -318 \\ \hline \end{array}$$

(d)
$$\begin{array}{r} 3927 \\ -2614 \\ \hline \end{array}$$

(e)
$$\begin{array}{r} 5464 \\ -324 \\ \hline \end{array}$$

ANSWERS

2. (a) $7 = 5 + 2$ or $7 = 2 + 5$
  (b) $9 = 4 + 5$ or $9 = 5 + 4$
  (c) $21 = 15 + 6$ or $21 = 6 + 15$
  (d) $58 = 42 + 16$ or $58 = 16 + 42$
3. (a) 31 **(b)** 44 **(c)** 111
  (d) 1313 **(e)** 5140

(b) $18 - 13 = 5$

$18 = 13 + 5$ or $18 = 5 + 13$

(c) $29 - 13 = 16$

$29 = 13 + 16$ or $29 = 16 + 13$

◀ *Work Problem* **2** *at the Side.*

OBJECTIVE 2 Identify the minuend, subtrahend, and difference. In subtraction, as in addition, the numbers in a problem have names. For example, in the problem $8 - 5 = 3$, the number 8 is the **minuend,** 5 is the **subtrahend,** and 3 is the **difference** or answer.

$$8 \quad - \quad 5 \quad = \quad 3 \leftarrow \text{Difference}$$
Minuend Subtrahend

The answer in subtraction is the difference.

$$\begin{array}{r} 8 \leftarrow \text{Minuend} \\ -5 \leftarrow \text{Subtrahend} \\ \hline 3 \leftarrow \text{Difference} \end{array}$$

OBJECTIVE 3 Subtract when no regrouping (borrowing) is needed. Subtract two numbers by lining up the numbers in columns so the digits in the ones place are in the same column. Next, subtract by columns, starting at the right with the ones column.

EXAMPLE 3 Subtracting Two Numbers

Subtract.

Ones digits are lined up in the same column.

(a)
$$\begin{array}{r} 53 \\ -21 \\ \hline 32 \end{array}$$
3 ones − 1 one = 2 ones
5 tens − 2 tens = 3 tens

Ones digits are lined up.

(b)
$$\begin{array}{r} 385 \\ -165 \\ \hline 220 \end{array}$$
Subtract from right to left.
5 ones − 5 ones = 0 ones
8 tens − 6 tens = 2 tens
3 hundreds − 1 hundred = 2 hundreds

(c)
$$\begin{array}{r} 9437 \\ -210 \\ \hline 9227 \end{array}$$
7 ones − 0 ones = 7 ones
3 tens − 1 ten = 2 tens
4 hundreds − 2 hundreds = 2 hundreds
9 thousands − 0 thousands = 9 thousands

◀ *Work Problem* **3** *at the Side.*

OBJECTIVE 4 Check subtraction answers by adding. Use addition to check your answer to a subtraction problem. For example, check $8 - 3 = 5$ by *adding* 3 and 5.

$$3 + 5 = 8, \quad \text{so} \quad 8 - 3 = 5 \quad \text{is correct.}$$

EXAMPLE 4   **Checking Subtraction by Using Addition**

Use addition to check each answer. If the answer is incorrect, find the correct answer.

(a)   89
        − 47
        ──────
          42

Rewrite as an addition problem, as shown in Example 2.

Subtraction problem
$$\left. \begin{array}{r} 89 \\ -\ 47 \\ \hline 42 \\ 89 \end{array} \right\}$$
Addition problem
$$\begin{array}{r} 47 \\ +\ 42 \\ \hline 89 \end{array}$$

Because $47 + 42 = 89$, the subtraction was done correctly.

> Avoid errors by checking answers.

(b) $72 - 41 = 21$

Rewrite as an addition problem.

$$72 = 41 + 21$$

But, $41 + 21 = 62$, **not** 72, so the subtraction was done **incorrectly.** Rework the original subtraction to get the correct answer, 31. Then, $41 + 31 = 72$.

(c)   374 ⟵── Match ──┐
        − 141   │
        ──────   ↓
          233    $141 + 233 = 374$

The answer checks.

─────────────── *Work Problem* **4** *at the Side.* ▶

OBJECTIVE **5** **Subtract with regrouping (borrowing).** When a digit in the minuend is less than the one directly below it, **regrouping** is necessary (also called **borrowing**).

EXAMPLE 5   **Subtracting with Regrouping**

Subtract 19 from 57.

Write the problem vertically.

$$\begin{array}{r} 57 \\ -\ 19 \end{array}$$

In the ones column, 7 is **less** than 9, so in order to subtract, we must regroup 1 ten as 10 ones.

5 tens − 1 ten = 4 tens ⟶ 4 17 ⟵ 1 ten = 10 ones, and
$$\begin{array}{r} \overset{4\ \ 17}{\cancel{5}\ \cancel{7}} \\ -\ 1\ 9 \end{array}$$ 10 ones + 7 ones = 17 ones

Now subtract 9 ones from 17 ones in the ones column. Then subtract 1 ten from 4 tens in the tens column.

$$\begin{array}{r} \overset{4\ \ 17}{\cancel{5}\ \cancel{7}} \\ -\ 1\ 9 \\ \hline 3\ 8 \end{array}$$ Difference

Finally, $57 - 19 = 38$. Check by adding 19 and 38; you should get 57.

─────────────── *Work Problem* **5** *at the Side.* ▶

---

4 Use addition to determine whether each answer is correct. If incorrect, what should it be?

(a)   76
        − 45
        ──────
          31

(b)   53
        − 22
        ──────
          21

(c)   374
        − 251
        ──────
          113

(d)   7531
        − 4301
        ──────
          3230

5 Subtract.

(a)   58
        − 19

(b)   86
        − 38

(c)   41
        − 27

(d)   863
        − 47

(e)   762
        − 157

ANSWERS

4. **(a)** correct   **(b)** incorrect; should be 31
   (c) incorrect; should be 123   **(d)** correct
5. **(a)** 39   **(b)** 48   **(c)** 14   **(d)** 816
   (e) 605

6 Subtract.

(a) 927
    − 43

(b) 675
    − 86

(c) 477
   − 389

(d) 1417
    − 988

(e) 8739
   − 3892

EXAMPLE 6   **Subtracting with Regrouping**

Subtract by regrouping when necessary.

(a)   7856
    − 137

Regroup 1 ten as 10 ones. ——— 10 ones + 6 ones = 16 ones

$$
\begin{array}{r}
{\scriptstyle 4\ 16}\\
7\ 8\ \cancel{5}\ \cancel{6}\\
-\ \ \ 1\ 3\ 7\\
\hline
7\ 7\ 1\ 9
\end{array}
$$
Difference

(b)   635
    − 546

Regroup 1 ten as 10 ones. ——— 10 ones + 5 ones = 15 ones

$$
\begin{array}{r}
{\scriptstyle 2\ 15}\\
6\ \cancel{3}\ \cancel{5}\\
-\ 5\ 4\ 6\\
\hline
9
\end{array}
$$
Need to regroup further because 2 is less than 4 in the tens column.

Regroup 1 hundred as 10 tens. ——— 10 tens + 2 tens = 12 tens

$$
\begin{array}{r}
{\scriptstyle 5\ 12\ 15}\\
\cancel{6}\ \cancel{3}\ \cancel{5}\\
-\ 5\ 4\ 6\\
\hline
8\ 9
\end{array}
$$
Difference

(c)   647
    − 489

$$
\begin{array}{r}
{\scriptstyle 3\ 17}\\
6\ \cancel{4}\ \cancel{7}\\
-\ 4\ 8\ 9\\
\hline
8
\end{array}
$$
Need to regroup further because 3 is less than 8 in the tens column.

$$
\begin{array}{r}
{\scriptstyle 5\ 13\ 17}\\
\cancel{6}\ \cancel{4}\ \cancel{7}\\
-\ 4\ 8\ 9\\
\hline
1\ 5\ 8
\end{array}
$$
Difference

◀ *Work Problem* **6** *at the Side.*

Sometimes a minuend has zeros in some of the positions. In such cases, regrouping may be a little more complicated than what we have shown so far.

EXAMPLE 7   **Regrouping with Zeros**

Subtract.

      4607
    − 3168

There are no tens that can be regrouped into ones. So you must first regroup 1 hundred as 10 tens.

Regroup 1 hundred as 10 tens. ——— Write 10 tens.

$$
\begin{array}{r}
{\scriptstyle 5\ 10}\\
4\ \cancel{6}\ \cancel{0}\ 7\\
-\ 3\ 1\ 6\ 8
\end{array}
$$

Now we may regroup from the tens position.

Regroup 1 ten as 10 ones; 10 tens − 1 ten = 9 tens.

$$
\begin{array}{r}
{\scriptstyle 5\ \cancel{10}\ 17}\\
4\ \cancel{6}\ \cancel{0}\ \cancel{7}\\
-\ 3\ 1\ 6\ 8\\
\hline
9
\end{array}
$$

10 ones + 7 ones = 17 ones

Continued on Next Page

Complete the problem.

$$\begin{array}{r} \overset{9}{}\\ 5\ \overset{9}{\cancel{10}}\ 17\\ 4\ \cancel{6}\ \cancel{0}\ \cancel{7}\\ -\ 3\ 1\ 6\ 8\\ \hline 1\ 4\ 3\ 9 \end{array}$$ Difference

Check by adding 1439 and 3168; you should get 4607.

Work Problem **7** *at the Side.* ▶

EXAMPLE 8 **Regrouping with Zeros**

Subtract.

(a) 708
− 149

Write 10 tens. ——— Regroup 1 ten as 10 ones.

Regroup 1 hundred as → 6 $\overset{9}{\cancel{10}}$ 18 ← 10 ones + 8 ones = 18 ones
10 tens.

$$\begin{array}{r} 7\ \cancel{0}\ \cancel{8}\\ -\ 1\ 4\ 9\\ \hline 5\ 5\ 9 \end{array}$$

(b) 380
− 276

Regroup 1 ten as 10 ones. ——— Write 10 ones.

$$\begin{array}{r} 7\ 10\\ 3\ \cancel{8}\ \cancel{0}\\ -\ 2\ 7\ 6\\ \hline 1\ 0\ 4 \end{array}$$

(c) 9000
− 6999

$$\begin{array}{r} 8\ \overset{9}{\cancel{10}}\ \overset{9}{\cancel{10}}\ 10\\ \cancel{9}\ \cancel{0}\ \cancel{0}\ \cancel{0}\\ -\ 6\ 9\ 9\ 9\\ \hline 2\ 0\ 0\ 1 \end{array}$$

Be extra careful when zeros are involved.

Work Problem **8** *at the Side.* ▶

As we have seen, an answer to a subtraction problem can be checked by adding.

EXAMPLE 9 **Checking Subtraction by Using Addition**

Use addition to check each answer.

Check

(a) $\begin{array}{r}613\\-275\\\hline 338\end{array}$ Match $\begin{array}{r}275\\+338\\\hline 613\end{array}$ Correct

Continued on Next Page

---

7 Subtract.

(a) 206
− 177

(b) 703
− 415

(c) 7024
− 2632

8 Subtract.

(a) 308
− 159

(b) 570
− 368

(c) 1570
− 983

(d) 7001
− 5193

(e) 4000
− 1782

ANSWERS

7. (a) 29 **(b)** 288 **(c)** 4392
8. (a) 149 **(b)** 202 **(c)** 587
   (d) 1808 **(e)** 2218

9 Use addition to check each answer. If the answer is incorrect, find the correct answer.

(a) $\begin{array}{r} 357 \\ -168 \\ \hline 189 \end{array}$

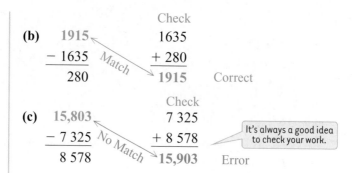

(b)
$$\begin{array}{r} 1915 \\ -1635 \\ \hline 280 \end{array}$$
Match
Check
$$\begin{array}{r} 1635 \\ +280 \\ \hline 1915 \end{array}$$ Correct

(c)
$$\begin{array}{r} 15{,}803 \\ -7\,325 \\ \hline 8\,578 \end{array}$$
No Match
Check
$$\begin{array}{r} 7\,325 \\ +8\,578 \\ \hline 15{,}903 \end{array}$$ Error

It's always a good idea to check your work.

Rework the original problem to get the correct answer, 8478. Then, 7325 + 8478 **does** give 15,803.

◀ *Work Problem* **9** *at the Side.*

(b) $\begin{array}{r} 570 \\ -328 \\ \hline 252 \end{array}$

OBJECTIVE 6 Solve application problems with subtraction. As shown in the next example, subtraction can be used to solve an application problem.

EXAMPLE 10 **Applying Subtraction Skills**

Use the table to find how much more, on average, a person with an Associate of Arts degree earns each year than a high school graduate.

(c) $\begin{array}{r} 14{,}726 \\ -8\,839 \\ \hline 5\,887 \end{array}$

EDUCATION PAYS

The more education adults get, the higher their annual earnings.

Education Level	Average Earnings
Not a high school graduate	$27,401
High school graduate	$32,860
Some college, no degree	$39,744
Associate of Arts degree	$43,094
Bachelor's degree	$56,655
Master's degree	$67,543
Doctoral degree	$92,035
Professional degree	$118,785

Note: Average annual earnings for workers between ages 25 and 64.

Source: U.S. Census Bureau.

10 Use the table from Example 10 to find, on average,

(a) how much more a person with a Bachelor's degree earns each year than a person with an Associate of Arts degree.

Approach The average earnings for a person with an Associate of Arts degree is $43,094 each year and the average for a high school graduate is $32,860. Find how much more a college graduate earns by subtracting $32,860 from $43,094.

Solution
$$\begin{array}{r} \$43{,}094 \\ -\$32{,}860 \\ \hline \$10{,}234 \end{array}$$
← Associate of Arts degree
← High school graduate
← More earnings

Education pays.

(b) how much more a person with an Associate of Arts degree earns each year than a person who is not a high school graduate.

On average, a person with an Associate of Arts degree earns $10,234 more each year than a high school graduate.

◀ *Work Problem* **10** *at the Side.*

1.3 ▶▶▶ **Exercises**

FOR EXTRA HELP

 **MyMathLab**   Math **XL** PRACTICE   WATCH   DOWNLOAD   READ   REVIEW

Work each subtraction problem. Use addition to check each answer. See Examples 3 and 4.

1. 48 − 32 16	**2.** 17 − 13 4	**3.** 86 − 53 33	**4.** 78 − 35 43	**5.** 77 − 60 17
6. 87 − 63 24	**7.** 335 − 122 213	**8.** 602 − 301 301	**9.** 552 − 451 101	**10.** 888 − 215 673
☉ **11.** 7352 − 241 7111	**12.** 4420 − 310 4110	**13.** 5546 − 2134 3412	**14.** 1875 − 1362 513	☉ **15.** 6259 − 4148 2111
16. 9654 − 4323 5331	**17.** 24,392 − 11,232 13,160	**18.** 57,921 − 34,801 23,120	**19.** 46,253 − 5 143 41,110	**20.** 75,904 − 3 702 72,202

Use addition to check each subtraction problem. If an answer is not correct, find the correct answer. See Example 4.

21. 54 − 42 12 correct	**22.** 87 − 43 44 correct	☉ **23.** 89 − 27 63 incorrect; should be 62	**24.** 47 − 35 13 incorrect; should be 12	**25.** 382 − 261 131 incorrect; should be 121
26. 754 − 342 412 correct	**27.** 4683 − 3542 1141 correct	**28.** 5217 − 4105 1132 incorrect; should be 1112	**29.** 8643 − 1421 7212 incorrect; should be 7222	**30.** 9428 − 3124 6324 incorrect; should be 6304

Subtract, regrouping when necessary. See Examples 5–8.

31. 75 − 37 38	**32.** 86 − 28 58	**33.** 94 − 49 45	**34.** 68 − 39 29	**35.** 57 − 38 19
36. 47 − 29 18	**37.** 828 − 547 281	**38.** 916 − 618 298	**39.** 771 − 252 519	**40.** 973 − 788 185

41. 7538
− 479
7059

42. 5863
− 1295
4568

43. 9988
− 2399
7589

44. 3576
− 1658
1918

45. 38,335
− 29,476
8859

46. 82,731
− 14,826
67,905

47. 40
− 37
3

48. 80
− 73
7

49. 60
− 37
23

50. 70
− 27
43

51. 308
− 289
19

52. 600
− 599
1

53. 4041
− 1208
2833

54. 4602
− 2063
2539

55. 9305
− 1530
7775

56. 7120
− 6033
1087

57. 1580
− 1077
503

58. 3068
− 2105
963

59. 2006
− 1850
156

60. 8203
− 5365
2838

61. 8240
− 6056
2184

62. 7050
− 6045
1005

63. 8503
− 2816
5687

64. 16,004
− 5 087
10,917

65. 80,705
− 61,667
19,038

66. 81,000
− 55,456
25,544

67. 66,000
− 34,444
31,556

68. 77,000
− 65,308
11,692

69. 20,080
− 13,496
6584

70. 80,056
− 23,869
56,187

Use addition to check each subtraction problem. If an answer is incorrect, find the correct answer. See Example 9.

71. 9428
− 4509
4919

correct

72. 1671
− 1325
1346

**incorrect;
should be 346**

73. 2548
− 2278
270

correct

74. 5274
− 1130
4144

correct

75. 93,758
− 52,869
40,889

correct

76. 82,357
− 14,396
68,961

**incorrect;
should be 67,961**

77. 36,778
− 17,405
19,373

correct

78. 34,821
− 17,735
17,735

**incorrect;
should be 17,086**

79. An addition problem can be changed to a subtraction problem and a subtraction problem can be changed to an addition problem. Give two examples of each to demonstrate this.

Possible answers are:

1. **3 + 2 = 5 could be changed to 5 − 2 = 3 or 5 − 3 = 2.**
2. **6 − 4 = 2 could be changed to 2 + 4 = 6 or 4 + 2 = 6.**

80. Can you use the commutative and the associative properties in subtraction? Explain.

No, you cannot. Numbers must be subtracted in the order given. The difference found in subtraction is the result of subtracting the subtrahend from the minuend. Changing the order of the minuend and subtrahend *does* change the answer.

Solve each application problem. See Example 10.

81. A man burns 187 calories during 60 minutes of sitting at a computer while a woman burns 140 calories at the same activity. How many fewer calories does a woman burn than a man? (*Source:* www.cookinglight.com)

47 calories

82. A woman burns 302 calories during 60 minutes of walking, while a man burns 403 calories doing the same activity. How many more calories does a man burn than a woman? (*Source:* www.cookinglight.com)

101 calories

83. Toronto's skyline is dominated by the CN Tower, which rises 1821 ft. The Sears Tower in Chicago is 1454 ft high. Find the difference in height between the two structures. (*Source:* Trizec Properties; *World Almanac.*)

1821 ft

d

1454 ft

CN Tower   Sears Tower

367 ft

84. The fastest animal in the world, the peregrine falcon, dives at 217 miles per hour (mph). A Boeing 747 cruises at 580 mph. How much faster is the plane?

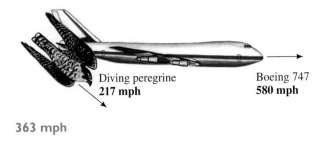

Diving peregrine
217 mph

Boeing 747
580 mph

363 mph

85. A cruise ship has 1815 passengers. When in port at Grand Cayman, 1348 passengers go ashore for the day while the others remain on the ship. How many passengers remain on the ship?

467 passengers

86. In a recent three-month period there were 81,465 Ford Explorers and 70,449 Jeep Grand Cherokees sold. Which vehicle had greater sales? By how much? (*Source:* J. D. Power and Associates.)

11,016 more Ford Explorers than Jeep Grand Cherokees

87. Six years ago there were 6970 bridge and lock-tender jobs across the United States. Today there are 3700 that remain. How many of these jobs have been eliminated? (*Source:* Bureau of Labor Statistics.)

3270 jobs eliminated

88. In 1964, its first year on the market, the Ford Mustang sold for $2500. In 2009, the Ford Mustang sold for $33,065. Find the increase in price. (*Source:* eBay.)

$30,565

89. Patriot Flag Company manufactured 14,608 U.S. flags and sold 5069. How many flags remain unsold?

9539 flags

90. Eye exams have been given to 14,679 children in the school district. If there are 23,156 students in the school district, how many have not received eye exams?

8477 students

91. The Jordanos now pay rent of $650 per month. If they buy a house, their housing expense will be $913 per month. How much more will they pay per month if they buy a house?

$263

92. A retired couple who used to receive a Social Security payment of $1479 per month now receives $1568 per month. Find the amount of the monthly increase.

$89

93. On Monday, 11,594 people visited Arcade Amusement Park, and 12,352 people visited the park on Tuesday. Which day had more people visit the park? How many more?

758 more people visited on Tuesday.

94. In the year 2020 it is predicted that we will need 2,820,000 nurses in the United States, while only 1,810,000 nurses will be available. Find the shortage in the number of nurses. (*Source:* American Hospital Association.)

1,010,000 nurses short

Solve each application problem. Add or subtract as necessary.

95. A survey of large hotels found that the average salary for a general manager of a deluxe spa and tennis resort is one hundred one thousand, five hundred dollars per year, while spa and tennis directors earn $44,000. How much more does a general manager earn than a spa and tennis director?

$57,500

96. This year there were 555,800 knee surgeries performed in the United States. The number of knee surgeries performed six years ago was 328,900. How many more of these surgeries were performed this year than six years ago? (*Source:* Agency for Healthcare Research and Quality.)

226,900 more surgeries

SUBWAY promotes healthy food choices by offering eight sandwiches that are low in fat. The nutritional information, printed on every SUBWAY napkin, appears below and includes information to answer Exercises 97–100. (Source: SUBWAY.)

OUR 6" SANDWICHES:	CALORIES	FAT(g)
VEGGIE DELITE®	230	3
HAM	290	5
TURKEY BREAST	280	4.5
ROAST BEEF	290	5
SUBWAY CLUB	320	6
TURKEY BREAST & HAM	290	5
ROASTED CHICKEN BREAST	330	5
SWEET ONION CHICKEN TERIYAKI	370	5

SUBWAY® regular 6" subs include italian or wheat bread, veggies and meat. Addition of condiments or cheese alters nutrition content.

MUSTARD (2 tsp.)	5	0
CHEESE TRIANGLES (2)	40	3.5
OLIVE OIL (1 tsp.)	45	5
VERSUS:		
BIG MAC®	560	30
WHOPPER®	670	39

97. How many fewer calories and grams of fat are in a 6-inch Veggie Delite sandwich than a Big Mac?

330 fewer calories; 27 fewer fat grams

98. How many fewer calories and grams of fat are in a 6-inch Turkey Breast and Ham sandwich than a Whopper?

380 fewer calories; 34 fewer fat grams

99. Find the total number of calories and grams of fat in a Roasted Chicken Breast sandwich with mustard and olive oil.

380 calories; 10 fat grams

100. A customer ate two sandwiches, one with the least calories and one with the most calories. Find the total number of calories and grams of fat in the two sandwiches.

600 calories; 8 fat grams

1.4 ▶▶▶ Multiplying Whole Numbers

Suppose we want to know the total number of exercise bicycles available at the gym. The bicycles are arranged in four columns with three stations in each column. Adding the number 3 a total of 4 times gives 12.

$$3 + 3 + 3 + 3 = 12$$

This result can also be shown with a figure.

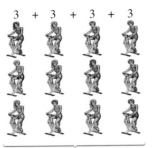

3 + 3 + 3 + 3

3 bicycles in each column

4 columns

OBJECTIVE 1 Identify the parts of a multiplication problem.
Multiplication is a shortcut for repeated addition. In the exercise bicycle example, instead of *adding* $3 + 3 + 3 + 3$ to get 12, we can *multiply* 3 by 4 to get 12. The numbers being multiplied are called **factors.** The answer is called the **product.** For example, the product of 3 and 4 can be written with the symbol $\times$, a raised dot, or parentheses, as follows.

$$
\begin{array}{r}
3 \\
\times\ 4 \\
\hline
12
\end{array}
$$

3 ← Factor (also called *multiplicand*)
× 4 ← Factor (also called *multiplier*)
12 ← Product (answer)

$3 \times 4 = 12$ *or* $3 \cdot 4 = 12$ *or* $(3)(4) = 12$ *or* $3(4) = 12$

Work Problem ① *at the Side.* ▶

Commutative Property of Multiplication

> Multiply numbers in any order.

By the **commutative property of multiplication,** the product (answer) remains the same when the order of the factors is changed. For example,
$$3 \times 5 = 15 \quad \text{and} \quad 5 \times 3 = 15.$$

CAUTION
Recall that addition also has a commutative property. For example, $4 + 2$ gives the same sum as $2 + 4$. Subtraction, however, is *not* commutative.

EXAMPLE 1 Multiplying Two Numbers

Multiply.

(a) $3 \times 4 = 12$

(b) $6 \cdot 0 = 0$ (The product of any number and 0 is 0; if you give no money to each of 6 relatives, you give no money.)

(c) $4(8) = 32$

─── **Continued on Next Page**

OBJECTIVES

1 Identify the parts of a multiplication problem.

2 Do chain multiplication.

3 Multiply by single-digit numbers.

4 Use multiplication shortcuts for numbers ending in zeros.

5 Multiply by numbers having more than one digit.

6 Solve application problems with multiplication.

1 Identify the factors and the product in each multiplication problem.

(a) $8 \times 5 = 40$

(b) $6(4) = 24$

(c) $7 \cdot 6 = 42$

(d) $(3)(9) = 27$

ANSWERS

1. (a) factors: 8, 5; product: 40
(b) factors: 6, 4; product: 24
(c) factors: 7, 6; product: 42
(d) factors: 3, 9; product: 27

2 Multiply.

(a) 7×4

(b) 0×9

(c) $8(5)$

(d) $6 \cdot 5$

(e) $(3)(8)$

3 Multiply.

(a) $3 \times 2 \times 5$

(b) $4 \cdot 7 \cdot 1$

(c) $(8)(3)(0)$

You may want to review this multiplication table.

Multiplication Table

×	1	2	3	4	5	6	7	8	9
1	1	2	3	4	5	6	7	8	9
2	2	4	6	8	10	12	14	16	18
3	3	6	9	12	15	18	21	24	27
4	4	8	12	16	20	24	28	32	36
5	5	10	15	20	25	30	35	40	45
6	6	12	18	24	30	36	42	48	54
7	7	14	21	28	35	42	49	56	63
8	8	16	24	32	40	48	56	64	72
9	9	18	27	36	45	54	63	72	81

Know the "Times Table."

◀ *Work Problem* **2** *at the Side.*

OBJECTIVE 2 Do chain multiplication. Some multiplications involve more than two factors.

> **Associative Property of Multiplication**
> By the **associative property of multiplication,** grouping the factors differently does not change the product.

EXAMPLE 2 **Multiplying Three Numbers**

Multiply $2 \times 3 \times 5$.

$$(2 \times 3) \times 5 \qquad \text{Parentheses show what to do first.}$$
$$6 \quad \times 5 = 30$$

Also,

$$2 \times (3 \times 5)$$
$$2 \times \quad 15 = 30$$

Either grouping results in the same product.

◀ *Work Problem* **3** *at the Side.*

⊞ Calculator Tip   The calculator approach to Example 2 uses chain calculations.

$$2 \ \text{ⓧ} \ 3 \ \text{ⓧ} \ 5 \ \text{⊜} \ 30$$

A problem with more than two factors, such as the one in Example 2, is called a **chain multiplication** problem.

OBJECTIVE 3 Multiply by single-digit numbers. Regrouping may be needed in multiplication problems with larger factors.

EXAMPLE 3 **Multiplying with Regrouping**

Multiply.

(a)   53
       $\times$ 4

Start by multiplying in the ones column.

$$\begin{array}{r}\overset{1}{5}3 \\ \times\ 4 \\ \hline 2 \end{array}$$

Write 1 ten in the tens column.

$4 \times 3 = \mathbf{12}$ ones

Write 2 ones in the ones column.

Next, multiply 4 times 5 tens.

$$\begin{array}{r}\overset{1}{5}\,3 \\ \times\ 4 \\ \hline 2 \end{array}$$

4×5 tens $= \mathbf{20}$ tens

Add the 1 ten that was written at the top of the tens column.

$$\begin{array}{r}\overset{1}{5}\,3 \\ \times\ 4 \\ \hline 212 \end{array}$$

20 tens + 1 ten = 21 tens

(b)   724
       $\times$ 5

Work as shown.

$$\begin{array}{r}\overset{1\ 2}{7}24 \\ \times\ \ 5 \\ \hline 3620 \end{array}$$

Use regrouping here.

$5 \times 4 = \mathbf{20}$ ones; write 0 ones; write 2 tens in the tens column.

$5 \times 2 = \mathbf{10}$ tens; add the 2 regrouped tens to get 12 tens; write 2 tens; write 1 hundred in the hundreds column.

5×7 hundreds $= \mathbf{35}$ hundreds; add the 1 regrouped hundred to get 36 hundreds.

Work Problem 4 *at the Side.* ▶

OBJECTIVE 4 Use multiplication shortcuts for numbers ending in zeros. The product of two whole number factors is also called a **multiple** of either factor. For example, since $4 \cdot 2 = 8$, the whole number 8 is a multiple of both 4 and 2. *Multiples of 10 are very useful when multiplying.* A **multiple of 10** is a whole number that ends in 0, such as 10, 20, or 30; 100, 200, or 300; 1000, 2000, or 3000; and so on. There is a short way to multiply by these multiples of 10. Look at the following examples.

$$26 \times 1 = 26$$
$$26 \times 10 = 260$$
$$26 \times 100 = 2600$$
$$26 \times 1000 = 26{,}000$$

Do you see a pattern? These examples suggest the rule that follows.

4 Multiply.

(a)   53
       $\times$ 5

(b)   79
       $\times$ 0

(c)   758
       $\times$ 8

(d)   2831
       $\times$ 7

(e)   4714
       $\times$ 8

5 Multiply.

(a) 63×10

(b) 305×100

(c) 714×1000

Multiplying by Multiples of 10

To multiply a whole number by 10, 100, or 1000, attach one, two, or three zeros, respectively, to the right of the whole number.

EXAMPLE 4 **Using Multiples of 10 to Multiply**

Multiply.

(a) $59 \times 10 = 590$
— Attach 0.

(b) $74 \times 100 = 7400$
— Attach 00.

(c) $803 \times 1000 = 803{,}000$ ← Attach 000.

◀ Work Problem **5** at the Side.

You can also find the product of other multiples of 10 by attaching zeros.

EXAMPLE 5 **Using Multiples of 10 to Multiply**

Multiply.

(a) 75×3000
Multiply 75 by 3, and then attach three zeros.

$$
\begin{array}{r}
75 \\
\times\ 3 \\
\hline
225
\end{array}
$$

$75 \times 3000 = 225{,}000$

Use useful shortcuts.

— Attach 000.

(b) 150×70
Multiply 15 by 7, and then attach two zeros.

$$
\begin{array}{r}
15 \\
\times\ 7 \\
\hline
105
\end{array}
$$

$150 \times 70 = 10{,}500$ ← Attach 00.

◀ Work Problem **6** at the Side.

---

6 Multiply.

(a) 16×50

(b) 73×400

(c) $\begin{array}{r} 180 \\ \times\ 30 \\ \end{array}$

(d) $\begin{array}{r} 4200 \\ \times\ 80 \\ \end{array}$

(e) $\begin{array}{r} 800 \\ \times\ 600 \\ \end{array}$

OBJECTIVE 5 Multiply by numbers having more than one digit.
The next example shows multiplication when both factors have more than one digit.

EXAMPLE 6 **Multiplying with More Than One Digit**

Multiply 46 and 23.

First multiply 46 by 3.

$$
\begin{array}{r}
\overset{1}{46} \\
\times\ 3 \\
\hline
138
\end{array}
$$

Regrouping is needed here.

← $46 \times 3 = 138$

Continued on Next Page

ANSWERS

5. (a) 630 (b) 30,500 (c) 714,000
6. (a) 800 (b) 29,200 (c) 5400
   (d) 336,000 (e) 480,000

Now multiply 46 by 20.

$$
\begin{array}{r}
\overset{1}{46} \\
\times\ 20 \\
\hline
920
\end{array} \longleftarrow 46 \times 20 = 920
$$

Add the results.

$$
\begin{array}{r}
46 \\
\times\ 23 \\
\hline
138 \\
+\ 920 \\
\hline
1058
\end{array}
\begin{array}{l}
\\
\longleftarrow 46 \times 3 \\
\longleftarrow 46 \times 20 \\
\\
\end{array}
$$

— Add.

Both 138 and 920 are called **partial products.** To save time, the 0 in 920 is usually not written.

$$
\begin{array}{r}
46 \\
\times\ 23 \\
\hline
138 \\
92 \\
\hline
1058
\end{array}
$$

← { 0 not written. Be very careful to place the 2 in the tens column.

────────────── Work Problem **7** at the Side. ▶

EXAMPLE 7 **Using Partial Products**

Multiply.

(a)
$$
\begin{array}{r}
2\,3\,3 \\
\times\ 1\,3\,2 \\
\hline
4\,6\,6 \\
6\,9\,9 \\
2\,3\,3 \\
\hline
3\,0{,}7\,5\,6
\end{array}
$$

(Tens lined up)
(Hundreds lined up)
← Product

[Be certain to align numbers in columns.]

(b)
$$
\begin{array}{r}
538 \\
\times\ 46 \\
\hline
\end{array}
$$

First multiply by 6.

$$
\begin{array}{r}
\overset{2\,4}{538} \\
\times\ \ 46 \\
\hline
3228
\end{array}
$$

← Regrouping is needed here.

Now multiply by 4, being careful to line up the tens.

$$
\begin{array}{r}
\overset{1\,3}{\overset{2\,4}{5\,3\,8}} \\
\times\ \ \ 4\,6 \\
\hline
3\,2\,2\,8 \\
2\,1\,5\,2 \\
\hline
2\,4{,}7\,4\,8
\end{array}
$$

— Finally, add the partial products.

────────────── Work Problem **8** at the Side. ▶

7 Complete each multiplication.

(a)
$$
\begin{array}{r}
35 \\
\times\ 54 \\
\hline
140 \\
175 \\
\hline
\end{array}
$$

(b)
$$
\begin{array}{r}
76 \\
\times\ 49 \\
\hline
684 \\
304 \\
\hline
\end{array}
$$

8 Multiply.

(a)
$$
\begin{array}{r}
52 \\
\times\ 16 \\
\hline
\end{array}
$$

(b)
$$
\begin{array}{r}
81 \\
\times\ 49 \\
\hline
\end{array}
$$

(c)
$$
\begin{array}{r}
75 \\
\times\ 63 \\
\hline
\end{array}
$$

(d)
$$
\begin{array}{r}
234 \\
\times\ 73 \\
\hline
\end{array}
$$

(e)
$$
\begin{array}{r}
835 \\
\times\ 189 \\
\hline
\end{array}
$$

ANSWERS

7. (a) 1890   **(b)** 3724
8. (a) 832   **(b)** 3969   **(c)** 4725
    (d) 17,082   **(e)** 157,815

9 Multiply.

(a)   28
    × 60

(b)   728
    × 50

(c)   562
    × 109

(d)   3526
    × 6002

10 Find the total cost of the following items.

(a) 289 redwood planters at $12 per planter

(b) 58 compound miter saws priced at $129 each

(c) 12 delivery vans at $28,300 per van

When 0 appears in the multiplier, be sure to move the partial products to the left to account for the position held by the 0.

EXAMPLE 8 **Multiplying with Zeros**

Multiply.

(a)   1 3 7
    × 3 0 6
       8 2 2
     0 0 0   (Tens lined up)
    4 1 1   (Hundreds lined up)
    4 1,9 2 2

(b)   1 4 0 6
    × 2 0 0 1
       1 4 0 6
     0 0 0 0 ← (Zeros to line up tens)
    0 0 0 0 ← (Zeros to line up hundreds)
    2 8 1 2
    2,8 1 3,4 0 6

> Use extra caution when working with 0s.

       1 4 0 6
    × 2 0 0 1
       1 4 0 6
    2 8 1 2 0 0 ← Zeros are written so this partial product starts in the thousands column.
    2,8 1 3,4 0 6

Note

In Example 8(b) in the alternative method on the right, zeros were inserted so that thousands were placed in the thousands column. This is a commonly used shortcut.

◀ Work Problem **9** *at the Side.*

OBJECTIVE **6** **Solve application problems with multiplication.** The next example shows how multiplication can be used to solve an application problem.

EXAMPLE 9 **Applying Multiplication Skills**

Find the total cost of 53 portable DVD players priced at $78 each.

Approach   To find the cost of all the DVD players, multiply the number of players (53) by the cost of one player ($78).

Solution   Multiply 53 by 78.

           53
    ×   78
         424
         371
       $4134

The total cost of the portable DVD players is $4134.

📇 **Calculator Tip**   If you are using a calculator for Example 9, you will do this calculation.

53 ⊗ 78 ⊜ 4134

◀ Work Problem **10** *at the Side.*

1.4 ▶▶▶ **Exercises**

FOR EXTRA HELP

 **MyMathLab**

Math XL
PRACTICE

WATCH

DOWNLOAD

READ

REVIEW

Work each chain multiplication. See Example 2.

1. $2 \times 6 \times 2$
24

2. $3 \times 5 \times 3$
45

3. $8 \times 6 \times 1$
48

4. $2 \times 4 \times 5$
40

5. $7 \cdot 8 \cdot 0$
0

6. $9 \cdot 0 \cdot 5$
0

7. $4 \cdot 1 \cdot 6$
24

8. $1 \cdot 5 \cdot 7$
35

9. $(4)(5)(2)$
40

10. $(4)(1)(9)$
36

11. $(3)(0)(7)$
0

12. $(0)(9)(4)$
0

13. Explain in your own words the commutative property of multiplication. How do the commutative properties of addition and multiplication compare to each other?

Factors may be multiplied in any order to get the same answer. They are the same; you may add or multiply numbers in any order.

14. Explain in your own words the associative property of multiplication. How do the associative properties of addition and multiplication compare to each other?

You may shift the parentheses in a multiplication problem. Just as in addition, the different grouping results in the same answer.

Multiply. See Example 3.

15. 35
× 6
——
210

16. 53
× 7
——
371

17. 34
× 7
——
238

18. 76
× 5
——
380

19. 642
× 5
——
3210

20. 472
× 4
——
1888

21. 624
× 3
——
1872

22. 852
× 7
——
5964

23. 2153
× 4
——
8612

24. 1137
× 3
——
3411

25. 2521
× 4
——
10,084

26. 2544
× 3
——
7632

27. 2561
× 8
——
20,488

28. 7326
× 5
——
36,630

29. 36,921
× 7
——
258,447

30. 28,116
× 4
——
112,464

Multiply. See Examples 4 and 5.

31. 40
× 7
——
280

32. 20
× 7
——
140

33. 80
× 6
——
480

34. 70
× 5
——
350

35. 740
× 3
——
2220

36. 400
× 8
——
3200

37. 600
× 6
——
3600

38. 860
× 7
——
6020

39. 125
× 30
——
3750

40. 246
× 50
——
12,300

41. 1635
$\times$ 40
65,400

42. 7311
$\times$ 50
365,550

◑ **43.** 900
$\times$ 300
270,000

44. 400
$\times$ 700
280,000

45. 43,000
$\times$ 2 000
86,000,000

46. 11,000
$\times$ 9 000
99,000,000

47. 970 · 50
48,500

48. 730 · 40
29,200

49. 800 · 900
720,000

50. 850 · 700
595,000

51. 9700 · 200
1,940,000

52. 10,050 · 300
3,015,000

Multiply. See Examples 6–8.

53. 28
$\times$ 17
476

54. 16
$\times$ 34
544

◑ **55.** 75
$\times$ 32
2400

56. 82
$\times$ 32
2624

57. 83
$\times$ 45
3735

58. (75)(21)
1575

59. (58)(41)
2378

60. (82)(67)
5494

61. (67)(92)
6164

62. (26)(33)
858

63. (28)(564)
15,792

64. (58)(312)
18,096

65. (619)(35)
21,665

66. (681)(47)
32,007

67. (55)(286)
15,730

68. 286
$\times$ 574
164,164

69. 735
$\times$ 112
82,320

70. 621
$\times$ 415
257,715

71. 538
$\times$ 342
183,996

72. 3228
$\times$ 751
2,424,228

73. 9352
$\times$ 264
2,468,928

74. 528
$\times$ 106
55,968

◑ **75.** 215
$\times$ 307
66,005

76. 218
$\times$ 106
23,108

77. 428
$\times$ 201
86,028

78. 3706
$\times$ 208
770,848

79. 6310
$\times$ 3078
19,422,180

80. 3533
$\times$ 5001
17,668,533

81. 2195
$\times$ 1038
2,278,410

82. 1502
$\times$ 2009
3,017,518

83. A classmate of yours is not clear on how to use a shortcut to multiply a whole number by 10, by 100, or by 1000. Write a short note explaining how this can be done.

To multiply by 10, 100, or 1000, just add one, two, or three zeros, respectively, to the number you are multiplying and that's your answer.

84. Show two ways to multiply when a 0 is in the multiplier. Use the problem 291×307 to show this.

```
    2 9 1          2 9 1
  × 3 0 7        × 3 0 7
  ─────────      ─────────
    2 0 3 7        2 0 3 7
    0 0 0        8 7 3 0
  8 7 3          ─────────
  ─────────      8 9,3 3 7
  8 9,3 3 7
```

Solve each application problem. See Example 9.

85. Carepanian Company, a health care supplier, purchased 300 cartons of Thera Bond Gym Balls. If there are 10 balls in each carton, find the total number of balls purchased.

3000 balls

86. A medical supply house has 30 bottles of vitamin C tablets, with each bottle containing 500 tablets. Find the total number of vitamin C tablets in the supply house.

15,000 tablets

87. Annie's Restaurant buys 15 cartons of eggs. If each carton contains 36 eggs, find the number of eggs purchased.

540 eggs

88. A hummingbird's wings beat about 65 times per second. How many times do the hummingbird's wings beat in 30 seconds?

1950 times

89. The average amount of water used per person each day in the United States is 66 gallons. How much water does the average person use in one year? (1 year = 365 days). (*Source:* Okfam.)

24,090 gallons

90. Squid are being hauled out of the Santa Barbara Channel by the ton. They are then processed, renamed calamari, and exported. Last night 27 fishing boats each hauled out 40 tons of squid. What was the total catch for the night? (*Source: Santa Barbara News Press.*)

1080 tons

Find the total cost of the following items. See Examples 7–9.

91. 75 first-aid kits at $8 per kit

$600

92. 38 gardeners at $64 per day

$2432

93. 65 rebuilt alternators at $24 per alternator

$1560

94. 62 wheelchair cushions at $44 per cushion

$2728

95. 206 desktop computers at $548 per computer

$112,888

96. 520 printers at $219 per printer

$113,880

Multiply.

97. $21 \cdot 43 \cdot 56$ **50,568**

98. $(600)(8)(75)(40)$ **14,400,000**

Use addition, subtraction, or multiplication to solve each application problem.

99. In a forest-planting project, 450 trees are planted on each acre. Find the number of trees needed to plant 85 acres.

38,250 trees

100. The largest living land mammal is the African elephant, and the largest mammal of all time is the blue whale. An African elephant weighs 15,225 pounds and a blue whale weighs 28 times that amount. Find the weight of the blue whale.

426,300 pounds

101. New York City has a population of 8,214,426, the largest in the country. Boston, in twenty-second place, has a population of 590,763. How many more people live in New York City than in Boston? (*Source:* Analysis of Census Bureau Estimates.)

7,623,663 people

102. Los Angeles, the second largest city in the country, has a population of 3,849,378. Dallas, at ninth largest, has a population of 1,232,940. Find the difference in the population of these two cities. (*Source:* Analysis of Census Bureau Estimates.)

2,616,438 people

103. A medical center purchased six laptop computers at $880 each, six printers at $235 each, and five fax machines at $140 each. Find the total cost of this equipment.

$7390

104. A motorcycle club traveled 640 miles on Saturday, 438 miles on Sunday, and 535 miles on Monday. Find the total number of miles traveled.

1613 miles

Relating Concepts (Exercises 105–114) For Individual or Group Work

Work Exercises 105–114 in order.

105. Add.

(a) $189 + 263$ **452**

(b) $263 + 189$ **452**

106. Your answers to Exercise 105(a) and (b) should be the same. This shows that the order of numbers in an addition problem does not change the sum. This is known as the __commutative__ property of addition.

107. Add. Recall that parentheses show you what to do first.

(a) $(65 + 81) + 135$ **281**

(b) $65 + (81 + 135)$ **281**

108. Since the answers to Exercise 107(a) and (b) are the same, we see that grouping the numbers differently when adding does not change the sum. This is known as the __associative__ property of addition.

109. Multiply.

(a) 220×72 **15,840**

(b) 72×220 **15,840**

110. Since the answers to Exercise 109(a) and (b) are the same, we see that the product remains the same when the order of the factors is changed. This is known as the __commutative__ property of multiplication.

111. Multiply. Recall that parentheses tell you what to do first.

(a) $(26 \times 18) \times 14$ **6552**

(b) $26(18 \times 14)$ **6552**

112. Since the answers to Exercise 111(a) and (b) are the same, we see that grouping the numbers differently when multiplying does not change the product. This is known as the __associative__ property of multiplication.

113. Do the commutative and associative properties apply to subtraction? Explain your answer using several examples.

No. Some examples are

1. $7 - 5 = 2$, but $5 - 7$ does not equal 2.

2. $12 - 6 = 6$, but $6 - 12$ does not equal 6.

3. $(8 - 2) - 5 = 1$, but $8 - (2 - 5)$ does not equal 1.

114. Do you think that the commutative and associative properties will apply to division? Explain your answer using several examples.

No. Some examples are

1. $10 \div 2 = 5$, but $2 \div 10$ does not equal 5.

2. $(16 \div 8) \div 2 = 1$, but $16 \div (8 \div 2)$ does not equal 1.

1.5 ▷▷▷ Dividing Whole Numbers

Suppose the cost of lunch at a SUBWAY is $18 and is to be divided equally by three friends. Each person would pay $6, as shown here.

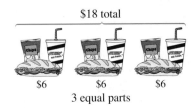

$18 total

$6 $6 $6
3 equal parts

OBJECTIVE 1 Write division problems in three ways.   Just as
$3 \cdot 6$, 3×6, and $(3)(6)$ are different ways of indicating the multiplication
of 3 and 6, there are several ways to write 18 divided by 3.

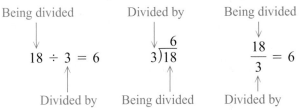

Being divided   Divided by   Being divided

$$18 \div 3 = 6 \qquad 3\overline{)18}^{\,6} \qquad \frac{18}{3} = 6$$

Divided by   Being divided   Divided by

We will use all three division symbols, $\div$, $\overline{)}\;$, and —. In courses such as
algebra, a slash symbol, /, or a fraction bar, —, is most often used.

EXAMPLE 1   **Using Division Symbols**

Write each division problem using two other symbols.

(a) $18 \div 6 = 3$
This division can also be written as shown below.

$$6\overline{)18}^{\,3} \quad \text{or} \quad \frac{18}{6} = 3 \qquad \triangleleft \text{Remember the three division symbols.}$$

(b) $\dfrac{15}{5} = 3$

$$15 \div 5 = 3 \quad \text{or} \quad 5\overline{)15}^{\,3}$$

(c) $5\overline{)20}^{\,4}$

$$20 \div 5 = 4 \quad \text{or} \quad \frac{20}{5} = 4$$

Work Problem **1** at the Side. ▶

OBJECTIVE 2 Identify the parts of a division problem.   In
division, the number being divided is the **dividend,** the number divided by
is the **divisor,** and the answer is the **quotient.**

$$\textbf{dividend} \div \textbf{divisor} = \textbf{quotient}$$

$$\text{divisor}\overline{)\text{dividend}}^{\,\text{quotient}} \qquad \frac{\textbf{dividend}}{\textbf{divisor}} = \textbf{quotient}$$

OBJECTIVES

1 Write division problems in three ways.

2 Identify the parts of a division problem.

3 Divide 0 by a number.

4 Recognize that a number cannot be divided by 0.

5 Divide a number by itself.

6 Divide a number by 1.

7 Use short division.

8 Use multiplication to check the answer to a division problem.

9 Use tests for divisibility.

---

1 Write each division problem using two other symbols.

(a) $24 \div 6 = 4$

(b) $9\overline{)36}^{\,4}$

(c) $48 \div 6 = 8$

(d) $\dfrac{42}{6} = 7$

ANSWERS

1. (a) $6\overline{)24}^{\,4}$ and $\dfrac{24}{6} = 4$

(b) $36 \div 9 = 4$ and $\dfrac{36}{9} = 4$

(c) $6\overline{)48}^{\,8}$ and $\dfrac{48}{6} = 8$

(d) $6\overline{)42}^{\,7}$ and $42 \div 6 = 7$

2 Identify the dividend, divisor, and quotient.

(a) $15 \div 3 = 5$

(b) $18 \div 6 = 3$

(c) $\dfrac{28}{7} = 4$

(d) $9\overline{)27}$ with quotient 3

3 Divide.

(a) $0 \div 5$

(b) $\dfrac{0}{9}$

(c) $\dfrac{0}{24}$

(d) $37\overline{)0}$

EXAMPLE 2   Identifying the Parts of a Division Problem

Identify the dividend, divisor, and quotient.

(a) $35 \div 7 = 5$

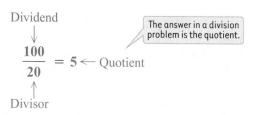

$$35 \div 7 = 5 \leftarrow \text{Quotient}$$
Dividend   Divisor

(b) $\dfrac{100}{20} = 5$

Dividend
$$\dfrac{100}{20} = 5 \leftarrow \text{Quotient}$$
Divisor

> The answer in a division problem is the quotient.

(c) $8\overline{)72}$ with quotient 9

$$9 \leftarrow \text{Quotient}$$
$$8\overline{)72} \leftarrow \text{Dividend}$$
Divisor

◄ *Work Problem* **2** *at the Side.*

OBJECTIVE 3   Divide 0 by a number.   If no money, or $0, is divided equally among five people, each person gets $0. The general rule for dividing 0 follows.

> **Dividing 0 by a Number**
> The number **0** divided by any nonzero number is **0**.

EXAMPLE 3   Dividing 0 by a Number

Divide.

(a) $0 \div 12 = 0$

(b) $0 \div 1728 = 0$

(c) $\dfrac{0}{375} = 0$ — Zero divided by any nonzero number is zero.

(d) $129\overline{)0}$

◄ *Work Problem* **3** *at the Side.*

Just as a subtraction such as $8 - 3 = 5$ can be written as the addition $8 = 3 + 5$, any division can be written as a multiplication. For example, $12 \div 3 = 4$ can be written as

$$3 \times 4 = 12 \quad \text{or} \quad 4 \times 3 = 12.$$

EXAMPLE 4 **Changing Division Problems to Multiplication**

Change each division problem to a multiplication problem.

(a) $\dfrac{20}{4} = 5$ becomes $4 \cdot 5 = 20$ or $5 \cdot 4 = 20$

(b) $8\overline{)48}$ (quotient 6)   becomes $8 \cdot 6 = 48$ or $6 \cdot 8 = 48$

(c) $72 \div 9 = 8$ becomes $9 \cdot 8 = 72$ or $8 \cdot 9 = 72$

Work Problem **4** *at the Side.* ▶

OBJECTIVE 4 **Recognize that a number cannot be divided by 0.**
Division of any number by 0 cannot be done. To see why, try to find

$$9 \div 0 = ?$$

As we have just seen, any division problem can be converted to a multiplication problem so that

divisor • quotient = dividend.

If you convert the preceding problem to its multiplication counterpart, it reads as follows.

$$0 \cdot ? = 9$$

You already know that 0 times any number must always be 0. Try any number you like to replace the "**?**" and you'll always get 0 instead of 9. Therefore, the division problem $9 \div 0$ cannot be done. Mathematicians say it is ***undefined*** and have agreed never to divide by 0. However, $0 \div 9$ *can* be done. Check by rewriting it as a multiplication problem.

$$0 \div 9 = 0 \quad \text{because} \quad 9 \cdot 0 = 0 \text{ is true}.$$

> **Dividing a Number by 0**
> Since dividing any number by 0 cannot be done, we say that division by **0** is ***undefined.*** It is impossible to compute an answer.

EXAMPLE 5 **Dividing Numbers by 0**

All the following are undefined.

(a) $\dfrac{6}{0}$ is undefined.

(b) $0\overline{)8}$ is undefined.

(c) $18 \div 0$ is undefined.   ◁ You **cannot** divide a number by zero.

(d) $\dfrac{3}{0}$ is undefined.

---

4 Write each division problem as a multiplication problem.

(a) $5\overline{)15}$ (quotient 3)

(b) $\dfrac{32}{4} = 8$

(c) $45 \div 9 = 5$

---

5 Divide. If the division is not possible, write "undefined."

(a) $\dfrac{4}{0}$

(b) $\dfrac{0}{4}$

(c) $0\overline{)36}$

(d) $36\overline{)0}$

(e) $100 \div 0$

(f) $0 \div 100$

6 Divide.

(a) $8 \div 8$

(b) $15\overline{)15}$

(c) $\dfrac{37}{37}$

Division Involving 0

$$0 \div \text{nonzero number} = 0 \quad \text{and} \quad \dfrac{0}{\text{nonzero number}} = 0$$

but

$$\text{nonzero number} \div 0 \quad \text{and} \quad \dfrac{\text{nonzero number}}{0} \quad \text{are } \textbf{undefined.}$$

CAUTION
When 0 is the divisor in a problem, you write "undefined" as the answer. Never divide by 0.

◀ Work Problem **5** at the Side.

⌨ **Calculator Tip**   Try these two problems on your calculator. Jot down your answers.

$$9 \; ÷ \; 0 \; = \; \underline{\qquad} \qquad\qquad 0 \; ÷ \; 9 \; = \; \underline{\qquad}$$

When you try to divide by 0, the calculator cannot do it, so it shows the word "Error" or the letter "E" (for error) in the display. But, when you divide 0 by 9 the calculator displays 0, which is the correct answer.

OBJECTIVE 5 Divide a number by itself.   What happens when a number is divided by itself? For example, what is $4 \div 4$ or $97 \div 97$?

Dividing a Number by Itself
Any *nonzero* number divided by itself is **1.**

EXAMPLE 6 | **Dividing a Nonzero Number by Itself**

Divide.

(a) $16 \div 16 = 1$

(b) $32\overline{)32}^{\,1}$

(c) $\dfrac{57}{57} = 1$ ← A nonzero number divided by itself is 1.

◀ Work Problem **6** at the Side.

OBJECTIVE 6 Divide a number by 1.   What happens when a number is divided by 1? For example, what is $5 \div 1$ or $86 \div 1$?

Dividing a Number by 1
Any number divided by 1 is itself.

┌───┐
EXAMPLE 7 **Dividing Numbers by 1**

Divide.

(a) $5 \div 1 = 5$

(b) $1\overline{)26}$ with quotient 26

(c) $\dfrac{41}{1} = 41$ ◁── A number divided by one is itself.

───────────── Work Problem **7** at the Side. ▶
└───┘

OBJECTIVE **7** **Use short division.** **Short division** is a method of dividing a number by a one-digit divisor.

┌───┐
EXAMPLE 8 **Using Short Division**

Divide: $3\overline{)96}$.

First, divide 9 by 3.

$$3\overline{)96}^{\,3} \quad \leftarrow \frac{9}{3} = 3$$

Next, divide 6 by 3.

$$3\overline{)96}^{\,32} \quad \leftarrow \frac{6}{3} = 2$$

───────────── Work Problem **8** at the Side. ▶
└───┘

When two numbers do not divide exactly, the leftover portion is called the **remainder.** The remainder must always be less than the divisor.

┌───┐
EXAMPLE 9 **Using Short Division with a Remainder**

Divide 147 by 4.
Write the problem.

$$4\overline{)147}$$

Because 1 cannot be divided by 4, divide 14 by 4. Notice that the 3 is placed over the 4 in 14.

$$4\overline{)14^27}^{\,3} \qquad \frac{14}{4} = 3 \text{ with 2 left over}$$

Next, divide 27 by 4. The final number left over is the remainder. Use **R** to indicate the remainder, and write the remainder to the side.

$$4\overline{)14^27}^{\,3\,6\ \mathbf{R}3} \qquad \frac{27}{4} = 6 \text{ with 3 left over}$$

───────────── Work Problem **9** at the Side. ▶
└───┘

7 Divide.

(a) $9 \div 1$

(b) $1\overline{)18}$

(c) $\dfrac{43}{1}$

8 Divide.

(a) $2\overline{)24}$

(b) $3\overline{)93}$

(c) $4\overline{)88}$

(d) $2\overline{)624}$

9 Divide.

(a) $2\overline{)125}$

(b) $3\overline{)215}$

(c) $4\overline{)538}$

(d) $\dfrac{819}{5}$

10 Divide.

(a) $4\overline{)523}$

(b) $\dfrac{515}{7}$

(c) $3\overline{)1885}$

(d) $6\overline{)1415}$

EXAMPLE 10   **Dividing with a Remainder**

Divide 1809 by 7.

Divide 18 by 7.

$$7\overline{)18^409}^{\ 2} \qquad \dfrac{18}{7} = 2 \text{ with 4 left over}$$

Divide 40 by 7.

$$7\overline{)18^40^59}^{\ 2\ 5} \qquad \dfrac{40}{7} = 5 \text{ with 5 left over}$$

Divide 59 by 7.

$$7\overline{)18^40^59}^{\ 2\ 5\ 8\ \textbf{R3}} \qquad \dfrac{59}{7} = 8 \text{ with 3 left over}$$

> The remainder must be less than the divisor.

◀ *Work Problem* 10 *at the Side.*

Note
Short division takes practice but is useful in many situations.

OBJECTIVE 8 **Use multiplication to check the answer to a division problem.** **Check** the answer to a division problem as follows.

Checking Division

(divisor × quotient) + remainder = dividend

Parentheses tell you what to do first: Multiply the divisor by the quotient, then add the remainder.

EXAMPLE 11   **Checking Division by Using Multiplication**

Check each answer.

(a) $5\overline{)458}^{\ 91\ \textbf{R3}}$

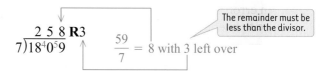

(divisor × quotient) + remainder = dividend

$(5 \times 91) + 3$

> Be careful! Always add the remainder when checking division.

$455 + 3 = \textbf{458}$

Matches original dividend, so the division was done correctly.

Continued on Next Page

(b) $6\overline{)1437}$ $\overset{239\ \textbf{R4}}{}$

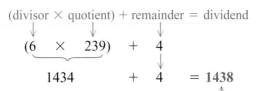

(divisor $\times$ quotient) + remainder = dividend

$(6 \quad \times \quad 239) \quad + \quad 4$

$1434 \qquad + \quad 4 \quad = 1438$

Does not match original dividend.

The answer does **not** check. Rework the original problem to get the correct answer, 239 **R**3. Then, $(6 \times 239) + 3$ **does** give 1437.

> **CAUTION**
> A common error when checking division is to forget to add the remainder. Be sure to add any remainder when checking a division problem.

―――――――――――――――― *Work Problem* ⑪ *at the Side.* ▶

OBJECTIVE ⑨ **Use tests for divisibility.**   It is often important to know whether a number is *divisible* by another number. You will find this useful in Chapter 2 when writing fractions in lowest terms.

> **Divisibility**
> One whole number is **divisible** by another if the remainder is 0.

Use the following tests to decide whether one number is divisible by another number.

> **Tests for Divisibility**
> **A number is divisible by**
>
>   2   if it ends in 0, 2, 4, 6, or 8. These are the even numbers.
>
>   3   if the sum of its digits is divisible by 3.
>
>   4   if the last two digits make a number that is divisible by 4.
>
>   5   if it ends in 0 or 5.
>
>   6   if it is divisible by both 2 and 3.
>
>   7   has no simple test.
>
>   8   if the last three digits make a number that is divisible by 8.
>
>   9   if the sum of its digits is divisible by 9.
>
>   10   if it ends in 0.

> The most often used rules are for 2, 5, 10, and occasionally 3.

The most commonly used tests are those for 2, 3, 5, and 10.

⑪ Use multiplication to check each division. If an answer is incorrect, give the correct answer.

(a) $2\overline{)65}$ $\overset{32\ \textbf{R}1}{}$

(b) $7\overline{)586}$ $\overset{83\ \textbf{R}4}{}$

(c) $3\overline{)1223}$ $\overset{407\ \textbf{R}2}{}$

(d) $5\overline{)2383}$ $\overset{476\ \textbf{R}3}{}$

12 Which numbers are divisible by 2?

(a) 258

(b) 307

(c) 4216

(d) 73,000

Divisibility by 2

A number is divisible by **2** if the number ends in 0, 2, 4, 6, or 8. All even numbers are divisible by 2.

EXAMPLE 12   **Testing for Divisibility by 2**

Are the following numbers divisible by 2?

(a) 986

⤴ Ends in 6

> All even numbers are divisible by 2.

Because the number ends in 6, which is an even number, the number 986 is divisible by 2.

(b) 3255 is not divisible by 2.

⤴ Ends in 5, and not in 0, 2, 4, 6, or 8

◀ *Work Problem* **12** *at the Side.*

Divisibility by 3

A number is divisible by **3** if the sum of its digits is divisible by **3**.

---

13 Which numbers are divisible by 3?

(a) 743

(b) 5325

(c) 374,214

(d) 205,633

EXAMPLE 13   **Testing for Divisibility by 3**

Are the following numbers divisible by 3?

(a) 4251
Add the digits.

> If the sum of the digits is divisible by 3, the number is divisible by 3.

Because 12 is divisible by 3, the number 4251 is also divisible by 3.

(b) 29,806
Add the digits.

Because 25 is *not* divisible by 3, the number 29,806 is *not* divisible by 3.

CAUTION

Be careful when testing for divisibility by adding the digits. This method works only for the numbers 3 and 9.

◀ *Work Problem* **13** *at the Side.*

ANSWERS

12. all but b
13. b and c

> **Divisibility by 5 and by 10**
> A number is divisible by **5** if it ends in 0 or 5.
> A number is divisible by **10** if it ends in 0.

14 Which numbers are divisible by 5?

(a) 180

EXAMPLE 14 **Testing for Divisibility by 5**

Are the following numbers divisible by 5?

(a) 12,900 ends in 0 and is divisible by 5.

> If the number ends in 0 or 5, it's divisible by 5.

(b) 635

(b) 4325 ends in 5 and is divisible by 5.

(c) 392 ends in 2 and is not divisible by 5.

(c) 8364

—— *Work Problem* **14** *at the Side.* ▶

EXAMPLE 15 **Testing for Divisibility by 10**

Are the following numbers divisible by 10?

> If the number ends in 0, it's divisible by 10.

(a) 700 and 9140 both end in 0 and are divisible by 10.

(d) 206,105

(b) 355 and 18,743 do not end in 0 and are not divisible by 10.

—— *Work Problem* **15** *at the Side.* ▶

15 Which numbers are divisible by 10?

(a) 270

(b) 495

(c) 5030

(d) 14,380

Math in the Media

STUDY TIME

As a college student, you need to plan a schedule to accommodate many responsibilities. These will include: preparing for and attending class, studying for exams, traveling to and from college, part-time work, family responsibilities, and personal time. Learning how to better manage your time is a skill presented by Stephen R. Covey at his Web site, stephencovey.com, and in his book *First Things First*. Dr. Covey says, "We're constantly making choices about the way we spend our time, from the major seasons to the individual moments of our lives. We're also living with the consequences of those choices." One of the first things you must do is calculate the amount of time dedicated to each of your obligations.

As an example, suppose that you are a full-time student enrolled in 12 credit hours of class: 4 credits of biology, 4 credits of computer science, 3 credits of mathematics, and 1 credit of physical education. Biology and computer science each have 3 hours of lecture and a 3-hour lab each week. Your biology and mathematics instructors recommend that you spend an additional 2 hours per week of study time for each hour of lecture time. Your physical education class is only 1 credit, but you are in class 3 hours each week.

Activity	Hours per Week
Class time	12
Lab time	6
Study time for mathematics and biology	12
Travel time to and from college	5
Part-time work (including travel time)	25
Sleep (8 hours per day)	56
Meals (3 hours per day)	21
Hygiene (showers, dressing, etc.)	7
Other (housecleaning, laundry, etc.)	14

1. How many total hours are in one week?   **168 hours**

2. Fill in the table entries for the number of hours in a week spent on class time, lab time, study time for biology and mathematics, sleep, and meals.   **See table.**

3. How many hours per week are spent on college-related activities?
   35 hours

4. How much more time is required for personal time (sleeping, eating, hygiene) than for college-related activities?
   49 hours

5. Based on the table data, how many hours per week are spent on the activities listed?   **158 hours**

6. How many hours per week are available for other activities, such as dating, shopping, family responsibilities, and so on?
   10 hours

1.5 ▶▶▶ **Exercises**

Write each division problem using two other symbols. See Example 1.

1. $24 \div 4 = 6$

$4\overline{)24}\,^{6}$ $\dfrac{24}{4} = 6$

2. $36 \div 3 = 12$

$3\overline{)36}\,^{12}$ $\dfrac{36}{3} = 12$

3. $\dfrac{45}{9} = 5$

$9\overline{)45}\,^{5}$ $45 \div 9 = 5$

4. $\dfrac{56}{8} = 7$

$56 \div 8 = 7$

$8\overline{)56}\,^{7}$

5. $2\overline{)16}\,^{8}$

$16 \div 2 = 8$

$\dfrac{16}{2} = 8$

6. $8\overline{)48}\,^{6}$

$48 \div 8 = 6$

$\dfrac{48}{8} = 6$

Divide. If the division is not possible, write "undefined." See Examples 3–7.

7. $9 \div 9$

1

8. $36 \div 9$

4

9. $\dfrac{14}{2}$

7

10. $\dfrac{10}{0}$

undefined

11. $22 \div 0$

undefined

12. $6 \div 6$

1

13. $\dfrac{24}{1}$

24

14. $\dfrac{12}{1}$

12

15. $15\overline{)0}$

0

16. $\dfrac{0}{12}$

0

17. $0\overline{)43}$

undefined

18. $\dfrac{8}{0}$

undefined

19. $\dfrac{15}{1}$

15

20. $\dfrac{6}{0}$

undefined

21. $\dfrac{8}{1}$

8

22. $\dfrac{0}{5}$

0

Divide by using short division. Use multiplication to check each answer. See Examples 8–10.

23. $3\overline{)75}\,^{25}$

24. $5\overline{)85}\,^{17}$

25. $7\overline{)126}\,^{18}$

26. $6\overline{)168}\,^{28}$

27. $4\overline{)1216}\,^{304}$

28. $5\overline{)2305}\,^{461}$

29. $4\overline{)2509}\,^{627\,R1}$

30. $8\overline{)1335}\,^{166\,R7}$

31. $6\overline{)9137}\,^{1522\,R5}$

32. $9\overline{)8371}\,^{930\,R1}$

33. $6\overline{)1854}\,^{309}$

34. $8\overline{)856}\,^{107}$

35. 12,020 ÷ 4
3005

36. 8012 ÷ 4
2003

37. 30,036 ÷ 6
5006

38. 32,008 ÷ 8
4001

39. 2434 ÷ 3
811 R1

40. 5993 ÷ 7
856 R1

41. 12,947 ÷ 5
2589 R2

42. 33,285 ÷ 9
3698 R3

43. 29,298 ÷ 4
7324 R2

44. 17,937 ÷ 6
2989 R3

45. 12,630 ÷ 4
3157 R2

46. 46,560 ÷ 7
6651 R3

47. 21,040 ÷ 8
2630

48. $\dfrac{8199}{9}$
911

49. $\dfrac{74,751}{6}$
12,458 R3

50. $\dfrac{72,543}{5}$
14,508 R3

51. $\dfrac{71,776}{7}$
10,253 R5

52. $\dfrac{77,621}{3}$
25,873 R2

53. $\dfrac{128,645}{7}$
18,377 R6

54. $\dfrac{172,255}{4}$
43,063 R3

Use multiplication to check each answer. If an answer is incorrect, find the correct answer.
See Example 11.

55. 5)1877   375 R2
correct

56. 3)1282   427 R1
correct

57. 3)5725   1908 R2
incorrect;
should be 1908 R1

58. 5)2158   432 R3
incorrect;
should be 431 R3

59. 7)4692   650 R2
incorrect;
should be 670 R2

60. 9)5974   663 R5
incorrect;
should be 663 R7

61. 6)21,409   3 568 R2
incorrect;
should be 3568 R1

62. 6)3192   532
correct

63. 8)16,019   2 002 R3
correct

64. 8)33,664   4 208
correct

65. 6)69,140   11,523 R2
correct

66. 3)82,598   27,532 R1
incorrect;
should be 27,532 R2

67. 9)86,655   9 628 R7
incorrect;
should be 9628 R3

68. 7)50,809   7 258 R4
incorrect;
should be 7258 R3

69. 8)222,576   27,822
correct

70. 4)311,216   77,804
correct

✐ **71.** Explain in your own words how to check a division problem using multiplication. Be sure to include what must be done if the quotient includes a remainder.

Multiply the quotient by the divisor and add any remainder. The result should be the dividend.

✐ **72.** Describe the three divisibility rules that you feel might be most useful to you and tell why.

Three choices might be

1. **A number is divisible by 2 if it ends in a 0, 2, 4, 6, or 8.**

2. **A number is divisible by 5 if it ends in 0 or 5.**

3. **A number is divisible by 10 if it ends in 0.**

Solve each application problem.

73. The Carnival Cruise Line has 2624 linen napkins. If it takes eight napkins to set each table, find the number of tables that can be set. (*Source: USA Today.*)

328 tables

74. A school district will distribute 1620 new science books equally among 12 schools. How many books will each school receive?

135 books

75. In one 8-hour day Dreyer's Edy's can produce 76,800 ice cream drumsticks. How many are produced each hour? (*Source:* History Channel, *Modern Marvels: Snack Food Tech.*)

9600 each hour

76. Tootsie Roll Industries produces 415,000,000 Tootsie Rolls in a 5-day week. Find the number of Tootsie Rolls produced each day. (*Source:* History Channel, *Modern Marvels: Snack Food Tech.*)

83,000,000 or 83 million each day

77. Lottery winnings of $436,500 are divided equally among nine Starbucks employees. Find the amount received by each employee.

$48,500

78. How many 5-pound bags of organic whole wheat flour can be filled from a 17,175-pound bin of flour?

3435 bags

79. If 8 gallons of fertilizer are needed for each acre of land, find the number of acres that can be fertilized with 1080 gallons of fertilizer.

135 acres

80. A roofing contractor has purchased 1134 squares of roofing material. If each cabin needs 9 squares of material, find the number of cabins that can be roofed. (1 square measures 10 ft by 10 ft.)

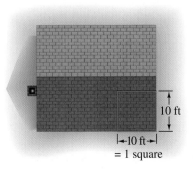

10 ft

|←10 ft→|
= 1 square

126 cabins

81. A class-action lawsuit settlement of $6,825,000 will be divided evenly among six injured people. Find the amount received by each person.

$1,137,500

82. A 12,000-square foot condominium at the edge of Central Park in Manhattan sold for a record $45,000,000. If the buyer pays for the condominium in eight equal payments, find the amount of each payment. (*Source: USA Today.*)

$5,625,000

83. The state of Maryland has the highest annual median household income of $65,148. How much income is this each month? (*Source:* U.S. Census Bureau.)

$5429

84. A professional basketball player has signed a 4-year contract for $21,937,500. How much is this each year?

$5,484,375

Put a ✓ mark in the blank if the number at the left is divisible by the number at the top.
Put an X in the blank if the number is not divisible by the number at the top.
See Examples 12–15.

	2	3	5	10			2	3	5	10
85. 60	✓	✓	✓	✓	**86.** 35	X	X	✓	X
87. 92	✓	X	X	X	**88.** 96	✓	✓	X	X
89. 445	X	X	✓	X	**90.** 897	X	✓	X	X
91. 903	X	✓	X	X	**92.** 500	✓	X	✓	✓
93. 5166	✓	✓	X	X	**94.** 8302	✓	X	X	X
95. 21,763	X	X	X	X	**96.** 32,472	✓	✓	X	X

1.6 ▶▶▶ Long Division

If the total cost of 42 Olympus digital cameras is $3066, we can find the cost of each camera using **long division.** Long division is used to divide by a number with more than one digit.

OBJECTIVE 1 Do long division.   In long division, estimate the various numbers by using a **trial divisor,** which is used to get a **trial quotient.**

OBJECTIVES

1 Do long division.

2 Divide numbers ending in 0 by numbers ending in 0.

3 Use multiplication to check division answers.

EXAMPLE 1 **Using a Trial Divisor and a Trial Quotient**

Divide. $42\overline{)3066}$

Because 42 is closer to 40 than to 50, use the first digit of the divisor as a trial divisor.

$$42$$

Trial divisor ⟶   ↑

> Using a trial divisor is a helpful tool.

Try to divide the first digit of the dividend by 4. Since 3 cannot be divided by 4, use the first *two* digits, 30.

$$\frac{30}{4} = 7 \text{ with remainder } 2$$

$$7 \leftarrow \text{Trial quotient}$$

$$42\overline{)3066}$$

⟶ 7 goes over the 6, because $\frac{306}{42}$ is about 7.

Multiply 7 and 42 to get 294; next, subtract 294 from 306.

$$
\begin{array}{r}
7 \\
42\overline{)3066} \\
294 \leftarrow 7 \times 42 \\
12 \leftarrow 306 - 294
\end{array}
$$

Bring down the 6 at the right.

$$
\begin{array}{r}
7 \\
42\overline{)3066} \\
294\downarrow \\
126 \leftarrow 6 \text{ brought down}
\end{array}
$$

Use the trial divisor, 4.

First two digits of 126 ⟶ $\frac{12}{4} = 3$

$$
\begin{array}{r}
73 \\
42\overline{)3066} \\
294 \\
\hline
126 \\
126 \leftarrow 3 \times 42 = 126 \\
\hline
0
\end{array}
$$

The cost of each camera is $73.
Check the answer by multiplying 42 and 73. The product should be 3066.

1 Divide.

(a) $28\overline{)2296}$

(b) $16\overline{)1024}$

(c) $61\overline{)8784}$

(d) $\dfrac{2697}{93}$

2 Divide.

(a) $24\overline{)1344}$

(b) $72\overline{)4472}$

(c) $65\overline{)5416}$

(d) $89\overline{)6649}$

CAUTION
The *first digit* of the quotient in long division must be placed in the proper position over the dividend.

◀ *Work Problem* **1** *at the Side.*

EXAMPLE 2 **Dividing to Find a Trial Quotient**

Divide. $58\overline{)2730}$

Use 6 as a trial divisor, since 58 is closer to 60 than to 50.

First two digits of dividend ⟶ $\dfrac{27}{6} = 4$ with 3 left over

$$\begin{array}{r} 4 \leftarrow \text{Trial quotient} \\ 58\overline{)2730} \\ \underline{232} \leftarrow 4 \times 58 = 232 \\ 41 \leftarrow 273 - 232 = 41 \text{ (smaller than 58, the divisor)} \end{array}$$

Bring down the 0.

$$\begin{array}{r} 4 \\ 58\overline{)2730} \\ \underline{232}\downarrow \\ 410 \leftarrow 0 \text{ brought down} \end{array}$$

First two digits of 410 ⟶ $\dfrac{41}{6} = 6$ with 5 left over

$$\begin{array}{r} 46 \leftarrow \text{Trial quotient} \\ 58\overline{)2730} \\ \underline{232} \\ 410 \\ \underline{348} \leftarrow 6 \times 58 = 348 \\ 62 \leftarrow \text{Greater than 58} \end{array}$$

Do not leave a remainder that is **greater** than the divisor.

The remainder, 62, is greater than the divisor, 58, so 7 should be used instead of 6.

$$\begin{array}{r} 47\ \textbf{R4} \leftarrow \\ 58\overline{)2730} \\ \underline{232} \\ 410 \\ \underline{406} \leftarrow 7 \times 58 = 406 \\ 4 \leftarrow 410 - 406 \end{array}$$

Now the remainder, 4, is *less* than the divisor, 58.

◀ *Work Problem* **2** *at the Side.*

Sometimes it is necessary to write a 0 in the quotient.

> **EXAMPLE 3** **Writing Zeros in the Quotient**

Divide: $34\overline{)7068}$

Start as in Examples 1 and 2.

$$
\begin{array}{r}
2 \\
34\overline{)7068} \\
\underline{68} \leftarrow 2 \times 34 = 68 \\
2 \leftarrow 70 - 68 = 2
\end{array}
$$

Bring down the 6.

$$
\begin{array}{r}
2 \\
34\overline{)7068} \\
\underline{68}\!\downarrow \\
26 \leftarrow 6\ \text{brought down}
\end{array}
$$

Since 26 cannot be divided by 34, write a 0 in the quotient as a placeholder.

$$
\begin{array}{r}
\mathbf{20} \leftarrow 0\ \text{in quotient} \\
34\overline{)7068} \\
\underline{68} \\
26
\end{array}
$$

Use a zero to hold a place in the quotient.

Bring down the final digit, the 8.

$$
\begin{array}{r}
20 \\
34\overline{)7068} \\
\underline{68}\downarrow \\
268 \leftarrow 8\ \text{brought down}
\end{array}
$$

Complete the problem.

$$
\begin{array}{r}
207\ \mathbf{R}30 \\
34\overline{)7068} \\
\underline{68} \\
268 \\
\underline{238} \\
30
\end{array}
$$

The quotient is 207 **R**30.

CAUTION
There *must be a digit* in the quotient (answer) above every digit in the dividend once the answer has begun. Notice in Example 3 that a **0** was used to assure a digit in the quotient above every digit in the dividend.

Work Problem **3** *at the Side.* ▶

OBJECTIVE 2 Divide numbers ending in 0 by numbers ending in 0.
When the divisor and dividend both contain zeros at the far right, recall that these numbers are multiples of 10. As with multiplication, there is a short way to divide these multiples of 10. Look at the following examples.

$$26{,}000 \div 1 = 26{,}000$$
$$26{,}000 \div 10 = 2600$$
$$26{,}000 \div 100 = 260$$
$$26{,}000 \div 1000 = 26$$

Do you see a pattern? These examples suggest the following rule.

---

3 Divide.

(a) $17\overline{)1823}$

(b) $23\overline{)4791}$

(c) $39\overline{)15{,}933}$

(d) $78\overline{)23{,}462}$

4 Divide.

(a) $70 \div 10$

(b) $2600 \div 100$

(c) $505,000 \div 1000$

Dividing a Whole Number by 10, 100, or 1000

Divide a whole number by 10, 100, or 1000 by dropping the appropriate number of zeros from the whole number.

EXAMPLE 4 Dividing by Multiples of 10

Divide.

(a) $60 \div 10 = 6$

One 0 in divisor

0 dropped

(b) $3500 \div 100 = 35$

Two zeros in divisor

00 dropped

(c) $915,000 \div 1000 = 915$

Three zeros in divisor

000 dropped

◀ *Work Problem* **4** *at the Side.*

Now we'll find the quotients for other multiples of 10 by dropping zeros.

EXAMPLE 5 Dividing by Multiples of 10

Divide:

(a) $40\overline{)11,000}$ Drop one zero from the divisor and the dividend.

$$
\begin{array}{r}
275 \\
4\overline{)1100} \\
8 \\
\hline
30 \\
28 \\
\hline
20 \\
20 \\
\hline
0
\end{array}
$$

Since $1100 \div 4$ is 275, then $11,000 \div 40$ is also 275.

(b) $3500\overline{)31,500}$ Drop two zeros from the divisor and the dividend.

$$
\begin{array}{r}
9 \\
35\overline{)315} \\
315 \\
\hline
0
\end{array}
$$

Since $315 \div 35$ is 9, then $31,500 \div 3500$ is also 9.

Note

Dropping zeros when dividing by multiples of 10 **does not** change the quotient (answer).

---

5 Divide.

(a) $50\overline{)6250}$

(b) $130\overline{)131,040}$

(c) $3400\overline{)190,400}$

ANSWERS

4. (a) 7 (b) 26 (c) 505
5. (a) 125 (b) 1008 (c) 56

◀ *Work Problem* **5** *at the Side.*

OBJECTIVE **3** **Use multiplication to check division answers.**
Answers in long division can be checked just as answers in short division were checked.

EXAMPLE 6   **Checking Division by Using Multiplication**

Check each answer.

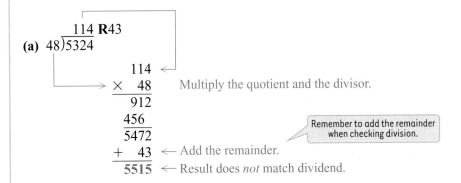

(a)

$$\begin{array}{r} 114 \text{ R43} \\ 48\overline{)5324} \end{array}$$

$$\begin{array}{r} 114 \\ \times\quad 48 \\ \hline 912 \\ 456 \\ \hline 5472 \\ +\quad 43 \\ \hline 5515 \end{array}$$

Multiply the quotient and the divisor.

Remember to add the remainder when checking division.

← Add the remainder.

← Result does *not* match dividend.

The answer does *not* check. Rework the original problem to get 110 **R44**.
Then $(110 \times 48) + 44$ **does** give 5324.

(b)

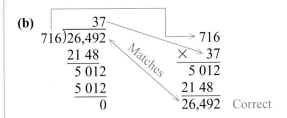

$$\begin{array}{r} 37 \\ 716\overline{)26{,}492} \\ 21\ 48 \\ \hline 5\ 012 \\ 5\ 012 \\ \hline 0 \end{array}$$

Matches

$$\begin{array}{r} 716 \\ \times\quad 37 \\ \hline 5\ 012 \\ 21\ 48 \\ \hline 26{,}492 \end{array}$$ Correct

---

▦ **Calculator Tip**   To check the answer to Example 6(a), don't forget
to add the remainder.

$$48 \ \text{⊗}\ 110 \ \text{⊕}\ 44 \ \text{⊜}\ \mathbf{5324}$$

Add the remainder.

CAUTION
When checking a division problem, first multiply the quotient and the
divisor. Then be sure to *add any remainder* before checking it against
the original dividend.

Work Problem **6** *at the Side.* ▶

---

6 Decide whether each answer
is correct. If the answer is
incorrect, find the correct
answer.

(a)

$$\begin{array}{r} 38 \\ 16\overline{)608} \\ 48 \\ \hline 128 \\ 128 \\ \hline 0 \end{array}$$

(b)

$$\begin{array}{r} 42 \text{ R}178 \\ 426\overline{)19{,}170} \\ 17\ 040 \\ \hline 1\ 130 \\ 952 \\ \hline 178 \end{array}$$

(c)

$$\begin{array}{r} 57 \text{ R}18 \\ 514\overline{)29{,}316} \\ 25\ 700 \\ \hline 3\ 616 \\ 3\ 598 \\ \hline 18 \end{array}$$

Math in the Media

LIGHTS ARE STILL ON AT THE DRIVE-IN MOVIES

It's just beginning to get dark, you're with family or friends in your car, you have plenty of pizza, popcorn, and other snacks, and the movie is about to start. You are at a drive-in movie theater.

A recent article in *USA Today* included some interesting facts about drive-ins. The first drive-in movie theater was opened in 1933 by the owner of an auto parts company and gas station to attract customers at night. It consisted of a sheet hanging between two trees. The popularity of drive-ins peaked in 1960, declined in the years that followed, but has remained fairly stable in recent years. The biggest reason for the decline of drive-ins has been the rising value of land and not a decline in popularity.

Use the drive-in movie theater facts below to answer the questions that follow.

Drive-in Movie Facts	States with the Most Open Drive-ins in 2008
1933: First drive-in movie	Pennsylvania: 35
1941: 12 open drive-ins	Ohio: 33
1958: 4063 open drive-ins	New York: 32
1960: 5000 open drive-ins	Indiana: 21
2008: 407 open drive-ins	California: 19
Typical screen size: 35 feet by 70 feet	Texas: 18
Typical popcorn machine cost: $12,500	Tennessee: 17
	Michigan: 14
	Illinois: 13
	Missouri: 13

(*Source: USA Today*)

1. What was the increase in the number of drive-in movies from 1941 to 1958?
   4051 drive-in movies

2. Find the increase in the number of drive-in movies from 1958 to 1960.
   937 drive-in movies

3. How many less drive-ins were there in 2008 than in the peak year?   **4593 drive-ins**

4. Find **(a)** the total number of drive-ins in the ten states with the most open drive-ins and **(b)** the number of open drive-ins in the remaining 40 states.   **(a) 215 drive-ins   (b) 192 drive-ins**

5. Find the cost of ten typical popcorn machines.   **$125,000**

6. What is the area of the typical drive-in movie screen. (*Hint:* Area = width × height. The answer is to be expressed in square feet.)   **2450 square feet**

7. Using your answer to Question 4(a), find the average number of open drive-ins in the ten states with the most open drive-ins. (The answer includes a remainder.)
   21 R5 open drive-ins

1.6 ▷▷▷ Exercises

Decide where the first digit in the quotient would be located. Then without finishing the division, you can tell which of the three choices is the correct answer. Circle your choice. See Examples 1 and 2.

1. $50\overline{)2650}$
  5 (53) 530

2. $14\overline{)476}$
  3 (34) 304

3. $18\overline{)4500}$
  2 25 (250)

4. $35\overline{)5600}$
  16 (160) 1600

5. $86\overline{)10,327}$
  12 (120 R7) 1200

6. $46\overline{)24,026}$
  5 52 (522 R14)

7. $26\overline{)28,735}$
  11 110 (1105 R5)

8. $12\overline{)116,953}$
  974 R2 (9746 R1) 97,460

9. $21\overline{)149,826}$
  71 713 (7134 R12)

10. $64\overline{)208,138}$
  325 R2 (3252 R10) 32,521

11. $523\overline{)470,800}$
  9 R100 90 R100 (900 R100)

12. $230\overline{)253,230}$
  11 110 (1101)

Divide by using long division. Use multiplication to check each answer. See Examples 1–3, 5, and 6.

13. $18\overline{)1319}$ 73 R5

14. $58\overline{)3654}$ 63

15. $23\overline{)10,963}$ 476 R15

16. $83\overline{)39,692}$ 478 R18

17. $26\overline{)62,583}$ 2407 R1

18. $28\overline{)84,249}$ 3008 R25

19. $74\overline{)84,819}$ 1146 R15

20. $238\overline{)186,948}$ 785 R118

21. $153\overline{)509,725}$ 3331 R82

22. $308\overline{)26,796}$ 87

23. $420\overline{)357,000}$ 850

24. $900\overline{)153,000}$ 170

Use multiplication to check each answer. If an answer is incorrect, find the correct answer.
See Example 6.

25. $35\overline{)3549}$ 101 **R**4

incorrect; should be 101 R14

26. $64\overline{)2712}$ 42 **R**26

incorrect; should be 42 R24

27. $28\overline{)18,424}$ 658 **R**9

incorrect; should be 658

28. $145\overline{)34,776}$ 239 **R**121

correct

29. $614\overline{)38,068}$ 62 **R**3

incorrect; should be 62

30. $557\overline{)97,286}$ 174 **R**368

correct

31. Describe in your own words a shortcut you can use to divide multiples of 10 by 10, by 100, or by 1000. Write an example problem and solve it.

When dividing by 10, 100, or 1000, drop the same number of zeros from the dividend as there are in the divisor to get the quotient. One example is 2500 ÷ 100 = 25.

32. Suppose you have a division problem with a remainder in the answer. Explain how to check your answer by writing an example problem that has a remainder.

Multiply the quotient and the divisor and add any remainder. One example is

17 ÷ 5 = 3 R2
Check: (3 × 5) + 2 = 15 + 2 = 17

Solve each application problem by using addition, subtraction, multiplication, or division as needed. See Examples 3–5.

33. Scientists using high-tech instruments have traced the travels of a tiger shark from Australia to South Africa—a total of 4950 miles in 99 days. On average, how far did the shark travel each day? (*Source:* Discovery Channel, *Shark Week.*)

50 miles

34. A new bridge from Owensboro, Kentucky, to Rockport, Indiana, is 2200 feet long and cost $55,998,800. Find the construction cost per foot. (*Source:* Federal Highway Administration.)

$25,454

35. Don Gracey, the Mountain Timesmith, has serviced and repaired 636 clocks this year. He has worked on 272 wall clocks and 308 table clocks. The rest were standing floor clocks. Find the number of floor clocks he worked on this year.

56 floor clocks

36. There are 24,000,000 business enterprises in the United States. If 7000 of these are larger businesses (over 500 employees), find the number of businesses that are small to mid-size. (*Source:* U.S. Census Bureau.)

23,993,000 small to mid-size businesses

37. To complete her college education, Judy Martinez received education loans of $34,080 including interest. Find her monthly payment if the loan is to be paid off in 96 months (8 years).

$355

38. A consultant charged $13,050 for evaluating a school's compliance with the Americans with Disabilities Act. If the consultant worked 225 hours, find the rate charged per hour.

$58

39. Each minute there is one diamond ring sold on eBay's U.S. site. Find the number of diamond rings sold in 30 days. (*Source: Time Style and Design.*)

43,200 rings

40. A retired milkman in Indianapolis has eaten a Twinkie every day for the last 60 years. How many Twinkies has he eaten over this time period? *Hint:* Use a 365-day year, ignoring leap years. (*Source:* History Channel, *Modern Marvels: Snack Food Tech.*)

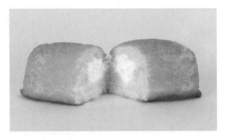

21,900 Twinkies

41. The average U.S. household of 2.5 people spent $2028 eating away from home last year. Find the average weekly household cost of eating away from home. *Hint:* 1 year equals 52 weeks. (*Source:* U.S. Bureau of Labor Statistics consumer expenditure surveys.)

$39 per week

42. Former professional basketball player Junior Bridgeman now owns 120 Wendy's restaurants with 4080 employees. Find the average number of employees at each restaurant. (*Source:* National Basketball Retired Players Association.)

34 employees

Relating Concepts (Exercises 43–50) For Individual or Group Work

Knowing and using the rules of divisibility is necessary in problem solving.
Work Exercises 43–50 in order.

43. If you have $0 and you divide this amount among three people, how much will each receive?

$0

44. When 0 is divided by any nonzero number, the result is __0__.

45. Divide.

$8 \div 0$

undefined

46. We say that division by 0 is *undefined* because it is ___impossible___ to compute the answer. Give an (possible/impossible) example involving cookies that will support your answer.

If you have 6 cookies, it is not possible to divide them among 0 people.

47. Divide.

(a) $14 \div 1$ **14**

(b) $1\overline{)17}$ **17**

(c) $\dfrac{38}{1}$ **38**

48. Any number divided by 1 is the number itself. Is this also true when multiplying by 1? Give three examples that support your answer.

Yes. Some examples are 18 · 1 = 18; 26 · 1 = 26; and 43 · 1 = 43.

49. Divide.

(a) $32,000 \div 10$ **3200**

(b) $32,000 \div 100$ **320**

(c) $32,000 \div 1000$ **32**

50. Write a rule that explains the shortcut for doing divisions like the ones in Exercise 49.

Drop the same number of zeros that appear in the divisor. The result is the quotient. With a divisor of 10, drop one zero; with 100, drop two zeros; with 1000, drop three zeros.

Math in the Media

POST OFFICE FACTS

The United States Postal Service posts a Web page on the Internet that gives a list of facts about their service. Some of those facts are listed in the table below.

Resource/Service	Number
1. Mail collection boxes	326,000
2. Post offices	36,895
3. Delivery points	146 million
4. Pieces of First Class mail delivered each year	213 billion
5. Processing plants sorting and shipping the mail	331
6. Pounds of mail carried on commercial airline flights annually	2.7 billion
7. Number of new deliveries each day	3500
8. Miles driven to move the mail annually	1.2 billion
9. Vehicles to pick up, transport, and deliver the mail	216,450
10. Customers each day who transact business at a post office	9 million

Source: United States Postal Service, www.usps.com

1. Write the number of delivery points entirely in digits. (Do not use the word "million" in your answer.) **146,000,000**

2. Write the number of miles driven annually to move the mail entirely in digits. **1,200,000,000**

3. Round the number of post offices to the nearest thousand. **37,000 post offices**

4. Use your rounded number from Question 3 to compute the average number of customers who transact business each day per post office. Round the answer to the nearest person. (*Hint:* To find the average, divide the total number of customers per day by the number of post offices). **243 people per day per office (rounded)**

5. Find the total number of new deliveries each year. Write the result entirely in words. (*Hint:* Assume that new deliveries are added 365 days per year.) **one million, two hundred seventy-seven thousand, five hundred**

6. Mail is sorted and shipped from processing plants to the post offices. Which of the following is a rough estimate of the number of post offices served by each processing plant: 10, 100, or 1000? Explain your choice. **100; There are approximately 300 processing plants and 37,000 post offices. The product of 300 times 100 is 30,000 which is closest to 37,000.**

UNITED STATES POSTAL SERVICE®

Summary Exercises on Whole Numbers

Write the digit for the given **place value** *in each whole number.*

1. 631,548

ten-thousands   **3**

tens   **4**

2. 76,047,309

millions   **6**

hundred-thousands   **0**

3. 9,181,576,423

hundred-millions   **1**

thousands   **6**

Write each number in words.

4. 86,002

eighty-six thousand, two

5. 425,208,733

four hundred twenty-five million, two hundred eight thousand, seven hundred thirty-three

Add or subtract as indicated.

6.   46
     + 51
     ———
      97

7. 166 + 739

    905

8.   82
     − 61
     ———
      21

9.   798
     − 389
     ———
      409

10. 6382 + 4062 + 7129

   17,573

11. 75 + 81,579 + 506 + 4

   82,164

12.   1704
      − 1027
      ———
       677

13.   55,000
      − 17,326
      ———
       37,674

14.   70,552
      − 34,663
      ———
       35,889

Multiply using the shortcut for multiples of 10.

15. 56 × 10

   560

16. 140 × 40

   5600

17.   500
      × 700
      ———
      350,000

18.   3600
      ×   70
      ———
      252,000

Write the numbers in each sentence using digits.

19. The world population is six billion, six hundred seventeen million, four hundred eighteen thousand, three hundred fifty-one people. (*Source:* U.S. Census Bureau.)

6,617,418,351

20. Each day there are twenty-four thousand, six hundred fifty-seven bags of Whiskas Cat Food sold. (*Source: Time* magazine.)

24,657

Multiply or divide as indicated. If the division cannot be done, write "undefined."

21. 8 ÷ 8
1

22. 0 ÷ 9
0

23. $\dfrac{12}{0}$ undefined

24. $\dfrac{15}{1}$ 15

25. 7 × 8
56

26. (6)(0)(5)
0

27. 8 · 4 · 3
96

28. $\dfrac{608}{2}$ 304

29. 3)8252
2750 R2

30. 4569 ÷ 6
761 R3

31.   65
× 52
3380

32.   507
× 435
220,545

33. (28)(72)
2016

34. (41)(36)
1476

35. 78
25)1950

36. 210
18)3780

37.   3602
× 5008
18,038,816

38. 506 R28
62)31,400

39. 52
630)32,760

40. 1208 R3
351)424,011

41. 4587 ÷ 8
573 R3

42. 41
72)2952

43.   662
× 315
208,530

44.   2186
× 504
1,101,744

1.7 ▷▷▷ Rounding Whole Numbers

One way to get a quick check on an answer is to *round* the numbers in the problem. **Rounding** a number means finding a number that is close to the original number, but easier to work with.

For example, the county planning commissioner might be discussing the need for more affordable housing. To demonstrate this, she probably would not need to say that the county is in need of 8235 more affordable housing units—she probably could say that the county needs 8200 or even 8000 housing units.

OBJECTIVE 1 Locate the place to which a number is to be rounded. The first step in rounding a number is to locate the *place to which the number is to be rounded.*

> **EXAMPLE 1** **Finding the Place to Which a Number Is to Be Rounded**

Locate and draw a line under the place to which each number is to be rounded.

(a) Round 83 to the nearest ten. Is 83 closer to 8$\underline{0}$ or to 9$\underline{0}$?

83 is closer to 80. ◁── | 83 is closer to 80 than to 90. |

Tens place ────┘

(b) Round 54,702 to the nearest thousand. Is it closer to 5$\underline{4}$,000 or to 5$\underline{5}$,000?

54,702 is closer to 55,000.

Thousands place ────┘

(c) Round 2,806,124 to the nearest hundred-thousand. Is it closer to 2,$\underline{8}$00,000 or to 2,$\underline{9}$00,000?

2,806,124 is closer to 2,800,000.

└──── Hundred-thousands place

─────────── Work Problem **1** at the Side. ▶

OBJECTIVE 2 Round numbers. Use the following rules for rounding whole numbers.

> **Rounding Whole Numbers**
>
> **Step 1** Locate the *place* to which the number is to be rounded. Draw a line under that place.
>
> **Step 2(a)** Look only at the next digit to the right of the one you underlined. If it is *5 or more, increase* the underlined digit by 1.
>
> **Step 2(b)** If the next digit to the right is *4 or less, do not change* the digit in the underlined place.
>
> **Step 3** *Change* all digits to the right of the underlined place to zeros.

> **EXAMPLE 2** **Using Rounding Rules for 4 or Less**

Round 349 to the nearest hundred.

Step 1 Locate the place to which the number is being rounded. Draw a line under that place.

3$\underline{4}$9

└── Hundreds place

Continued on Next Page

OBJECTIVES

1. Locate the place to which a number is to be rounded.

2. Round numbers.

3. Round numbers to estimate an answer.

4. Use front end rounding to estimate an answer.

---

1 Locate and draw a line under the place to which each number is to be rounded. Then answer the question.

(a) 373 (nearest ten)

Is it closer to 370 or to 380?

_____

(b) 1482 (nearest thousand)

Is it closer to 1000 or to 2000?

_____

(c) 89,512 (nearest hundred)

Is it closer to 89,500 or to 89,600?

_____

(d) 546,325 (nearest ten-thousand)

Is it closer to 540,000 or to 550,000?

_____

ANSWERS

1. **(a)** 3$\underline{7}$3 is closer to 3$\underline{7}$0.
   (b) $\underline{1}$482 is closer to $\underline{1}$000.
   (c) 89,$\underline{5}$12 is closer to 89,$\underline{5}$00.
   (d) 5$\underline{4}$6,325 is closer to 5$\underline{5}$0,000.

2 Round to the nearest ten.

  (a) 62

  (b) 94

  (c) 134

  (d) 7543

3 Round to the nearest thousand.

  (a) 3683

  (b) 6502

  (c) 84,621

  (d) 55,960

Step 2 Because the next digit to the right of the underlined place is 4, which is 4 or less, do *not* change the digit in the underlined place.

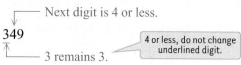

     Next digit is 4 or less.

     349

     3 remains 3.

     4 or less, do not change underlined digit.

Step 3 Change all digits to the right of the underlined place to zeros.

     3̲49 rounded to the nearest hundred is 300.

In other words, 349 is closer to 300 than to 400.

◀ *Work Problem* **2** *at the Side.*

EXAMPLE 3 **Using Rounding Rules for 5 or More**

Round 36,833 to the nearest thousand.

Step 1 Find the place to which the number is to be rounded and draw a line under that place.

     36,833

     Thousands

Step 2 Because the next digit to the right of the underlined place is 8, which is 5 or more, add 1 to the underlined place.

     Next digit is 5 or more.

     36,833

     5 or more, add 1 to underlined digit.

     Change 6 to 7.

Step 3 Change all digits to the right of the underlined place to zeros.

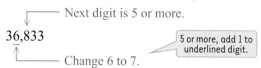

     Change to 0.

     36,833 rounded to the nearest thousand is 37,000.

     Change 6 to 7.

In other words, 36,833 is closer to 37,000 than to 36,000.

◀ *Work Problem* **3** *at the Side.*

EXAMPLE 4 **Using Rounding Rules**

(a) Round 2382 to the nearest ten.

Step 1 2382

     Tens place

Step 2 The next digit to the right is 2, which is 4 or less.

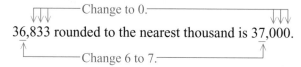

     Next digit is 4 or less.

     2382

     Leave 8 as 8.

Step 3 2382   Change to 0.

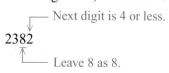

2382 rounded to the nearest ten is 2380.
In other words, 2382 is closer to 2380 than to 2390.

Continued on Next Page

(b) Round 13,961 to the nearest hundred.

Step 1 13,961
└── Hundreds place

Step 2 The next digit to the right is 6.

┌── Next digit is 5 or more.
13,961
├── Change 9 to 10; write 0 and regroup 1 into thousands place.
└── 3 + regrouped 1 = 4
┌── Change to 0

Step 3 14,061

13,961 rounded to the nearest hundred is 14,000.
In other words, 13,961 is closer to 14,000 than to 13,900.

> **Note**
> In Step 2 of Example 4(b), notice that the first three digits increased from 139 to 140 when we added 1 to the hundreds place.
> (13,9)61 rounded to (14,0)00

Work Problem **4** *at the Side.* ▶

EXAMPLE 5 **Rounding Large Numbers**

(a) Round 37,892 to the nearest ten-thousand.

Step 1 37,892
└── Ten-thousands place.
┌ Remember to *underline* the place to which you are rounding.

Step 2 The next digit to the right is 7.

┌── Next digit is 5 or more.
37892
├── Change 3 to 4.
└── Change 0.

Step 3 47,892

37,892 rounded to the nearest ten-thousand is 40,000.

(b) Round 528,498,675 to the nearest million.

Step 1 528,498,675
└── Millions place

┌── Next digit is 4 or less.
Step 2 528,498,675
└── Leave 8 as 8
┌── Change to 0. ┌ Remember to change *everything* to the right of the place you have rounded to 0.

Step 3 528,498,675

528,498,675 rounded to the nearest million is 528,000,000.

Work Problem **5** *at the Side.* ▶

4 Round each number as indicated.

(a) 3458 to the nearest ten

(b) 6448 to the nearest hundred

(c) 73,077 to the nearest hundred

(d) 85,972 to the nearest hundred

5 Round each number as indicated.

(a) 14,598 to the nearest ten-thousand

(b) 724,518,715 to the nearest million

ANSWERS
4. **(a)** 3460 **(b)** 6400
   (c) 73,100 **(d)** 86,000
5. **(a)** 10,000 **(b)** 725,000,000

Sometimes a number must be rounded to different places.

6 Round each number to the nearest ten and to the nearest hundred.

(a) 458

EXAMPLE 6 **Rounding to Different Places**

Round 648 **(a)** to the nearest ten and **(b)** to the nearest hundred.

(a) to the nearest ten

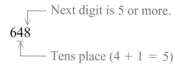

648 rounded to the nearest ten is 650.

(b) to the nearest hundred

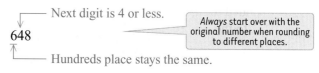

(b) 549

648 rounded to the nearest hundred is 600.

Notice that if 648 is rounded to the nearest ten (650), and then 650 is rounded to the nearest hundred, the result is 700. If, however, 648 is rounded directly to the nearest hundred, the result is 600 (not 700).

◀ Work Problem **6** at the Side.

CAUTION
Before rounding to a different place, always go back to the *original, unrounded number.*

EXAMPLE 7 **Applying Rounding Rules**

Round each number to the nearest ten, nearest hundred, and nearest thousand.

(a) 4358
First round 4358 to the nearest ten.

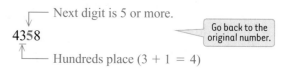

(c) 9308

4358 rounded to the nearest ten in 4360.

Now go back to 4358, the *original* number, before rounding to the nearest hundred.

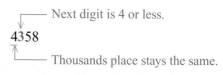

4358 rounded to the nearest hundred is 4400.

Again, go back to the *original* number before rounding to the nearest thousand.

Next digit is 4 or less.

4358

Thousands place stays the same.

4358 rounded to the nearest thousand is 4000.

Continued on Next Page

(b) 680,914

First, round to the nearest ten.

Next digit is 4 or less.

680,91~~4~~

Tens place stays the same.

680,914 rounded to the nearest ten is 680,910.
   Go back to 680,914, the *original* number, to round to the nearest hundred.

Next digit is 4 or less.

680,9~~14~~

Hundreds place stays the same.

Remember to go back to the original number.

680,914 rounded to the nearest hundred is 680,900.
   Go back to the *original* number to round to the nearest thousand.

Next digit is 5 or more.

680,~~914~~

Thousands place (0 + 1 = 1)

680,914 rounded to the nearest thousand is 681,000.

Work Problem **7** *at the Side.* ▶

OBJECTIVE 3 Round numbers to estimate an answer.   Numbers may be rounded to **estimate** an answer. An estimated answer is one that is close to the exact answer and may be used as a check when the exact answer is found. The "≈" sign is often used to show that an answer has been rounded or estimated and is almost equal to the exact answer; ≈ means "approximately equal to."

(EXAMPLE 8) Using Rounding to Estimate an Answer

Estimate each answer by rounding to the nearest ten.

(a)
$$
\begin{array}{rcl}
76 & \longrightarrow & 80 \\
53 & \longrightarrow & 50 \\
38 & \longrightarrow & 40 \\
+\,91 & \longrightarrow & +\,90 \\
\hline
 & & 260
\end{array}
$$
Rounded to the nearest ten

260   Estimated answer

(b)
$$
\begin{array}{rr}
27 & 30 \\
-\,14 & -\,10 \\
\hline
 & 20
\end{array}
$$
Rounded to the nearest ten

20   Estimated answer

(c)
$$
\begin{array}{rr}
16 & 20 \\
\times\,21 & \times\,20 \\
\hline
 & 400
\end{array}
$$
Rounded to the nearest ten

400   Estimated answer

Work Problem **8** *at the Side.* ▶

---

7 Round each number to the nearest ten, nearest hundred, and nearest thousand.

(a) 4078

(b) 46,364

(c) 268,328

8 Estimate the answers by rounding each number to the nearest ten.

(a)
$$
\begin{array}{r}
16 \\
74 \\
58 \\
+\,31 \\
\end{array}
$$

(b)
$$
\begin{array}{r}
53 \\
-\,19 \\
\end{array}
$$

(c)
$$
\begin{array}{r}
46 \\
\times\,74 \\
\end{array}
$$

ANSWERS

7. (a) 4080; 4100; 4000
   (b) 46,360; 46,400; 46,000
   (c) 268,330; 268,300; 268,000
8. (a) 20 + 70 + 60 + 30 = 180
   (b) 50 − 20 = 30
   (c) 70 × 50 = 3500

9 Estimate the answers by rounding each number to the nearest hundred.

(a)   358
      743
      822
    + 978

(b)   842
    − 475

(c)   723
    × 478

10 Use front end rounding to estimate each answer.

(a)   36
     3852
      749
    + 5474

(b)   2583
    − 765

(c)   648
    × 67

EXAMPLE 9   **Using Rounding to Estimate an Answer**

Estimate each answer by rounding to the nearest hundred.

(a)   252 ⟶ 300
       749 ⟶ 700
       576 ⟶ 600   } Rounded to the nearest hundred
     + 819 ⟶ + 800
              2400   Estimated answer ← The hundreds position is 3 places to the left.

(b)   780   800
    − 536   − 500   } Rounded to the nearest hundred
               300   Estimated answer

(c)   664   700
    × 834   × 800   } Rounded to the nearest hundred
             560,000   Estimated answer

◀ *Work Problem* **9** *at the Side.*

OBJECTIVE **4** **Use front end rounding to estimate an answer.**
A convenient way to estimate an answer is to use *front end rounding*. With **front end rounding,** we round to the highest possible place so that all the digits become 0 except the first one. For example, suppose you want to buy a big flat-screen television for $2449, a home theater system for $1759, and a reclining chair for $525. Using front end rounding, you can estimate the total cost of these purchases.

      Television $2449 →   2000
      Home theater system $1759 →   2000
      Reclining chair $525 → + 500
                                      $4500 ← Estimated total cost

EXAMPLE 10   **Using Front End Rounding to Estimate an Answer**

Estimate each answer using front end rounding.

(a)   3825   4000
        72   70   } All digits changed to 0 except first digit, which is rounded
       565   600
    + 2389   + 2000
               6670   Estimated answer

(b)   6712   7000   } First digit rounded and all others changed to 0   ← Notice: Front end rounding leaves *only* one nonzero digit.
    − 825   − 800
               6200   Estimated answer

(c)   725   700
    × 86   ×   90
             63,000   Estimated answer

Note

When using front end rounding, all the digits become 0 except the highest-place digit (the first digit).

◀ *Work Problem* **10** *at the Side.*

1.7 ▶▶▶ Exercises

Round each number as indicated. See Examples 1–5.

1. 624 to the nearest ten
620

2. 509 to the nearest ten
510

3. 855 to the nearest ten
860

4. 946 to the nearest ten
950

5. 6771 to the nearest hundred
6800

6. 5847 to the nearest hundred
5800

7. 86,813 to the nearest hundred
86,800

8. 17,211 to the nearest hundred
17,200

9. 28,472 to the nearest hundred
28,500

10. 18,273 to the nearest hundred
18,300

11. 5996 to the nearest hundred
6000

12. 4452 to the nearest hundred
4500

13. 15,758 to the nearest thousand
16,000

14. 28,465 to the nearest thousand
28,000

15. 78,499 to the nearest thousand
78,000

16. 14,314 to the nearest thousand
14,000

17. 7,760,058,721 to the nearest billion
8,000,000,000

18. 44,706,892 to the nearest ten-million
40,000,000

19. 12,987 to the nearest ten-thousand
10,000

20. 6599 to the nearest ten-thousand
10,000

21. 595,008 to the nearest ten-thousand
600,000

22. 725,182 to the nearest ten-thousand
730,000

23. 4,860,220 to the nearest million
5,000,000

24. 13,713,409 to the nearest million
14,000,000

Round each number to the nearest ten, nearest hundred, and nearest thousand. See Examples 6 and 7.

		Ten	Hundred	Thousand			Ten	Hundred	Thousand
25.	4476	4480	4500	4000	**26.**	6483	6480	6500	6000
27.	3374	3370	3400	3000	**28.**	7632	7630	7600	8000
29.	6048	6050	6000	6000	**30.**	7065	7070	7100	7000

	Ten	Hundred	Thousand			Ten	Hundred	Thousand
31. 5343	5340	5300	5000	**32.** 7456	7460	7500	7000
33. 19,539	19,540	19,500	20,000	**34.** 59,806	59,810	59,800	60,000
35. 26,292	26,290	26,300	26,000	**36.** 78,519	78,520	78,500	79,000
37. 93,706	93,710	93,700	94,000	**38.** 84,639	84,640	84,600	85,000

39. Write in your own words the three steps that you would use to round a number when the digit to the right of the place to which you are rounding is 5 or more.

1. Locate the place to be rounded and underline it.

2. Look only at the next digit to the right. If this digit is 5 or more, increase the underlined digit by 1.

3. Change all digits to the right of the underlined place to zeros.

40. Write in your own words the three steps that you would use to round a number when the digit to the right of the place to which you are rounding is 4 or less.

1. Locate the place to be rounded and underline it.

2. Look only at the next digit to the right. If this digit is 4 or less, do not change the underlined digit.

3. Change all digits to the right of the underlined place to zeros.

Estimate the answer by rounding each number to the nearest ten. Then find the exact answer. See Example 8.

41. *Estimate:*   *Exact:*

```
        Rounds to
 30 ←——————————  25
 60 ←——————————  63
 50 ←——————————  47
+ 80 ←—————————  + 84
 220             219
```

42. *Estimate:*   *Exact:*

```
 60      56
 20      24
 90      85
+ 70   + 71
 240    236
```

43. *Estimate:*   *Exact:*

```
 80      78
− 40   − 43
 40      35
```

44. *Estimate:*   *Exact:*

```
 60      57
− 20   − 24
 40      33
```

45. *Estimate:*   *Exact:*

```
 70      67
× 30   × 34
2100    2278
```

46. *Estimate:*   *Exact:*

```
 50      53
× 80   × 75
4000    3975
```

Estimate the answer by rounding each number to the nearest hundred. Then find the exact answer. See Example 9.

47. *Estimate:*   *Exact:*

```
        Rounds to
 900 ←—————————  863
 700 ←—————————  735
 400 ←—————————  438
+ 800 ←————————  + 792
2800            2828
```

48. *Estimate:*   *Exact:*

```
 600     623
 400     362
 200     189
+ 700   + 736
1900    1910
```

49. *Estimate:*   *Exact:*

$$\begin{array}{r} 900 \\ -\ 400 \\ \hline 500 \end{array} \qquad \begin{array}{r} 883 \\ -\ 448 \\ \hline 435 \end{array}$$

50. *Estimate:*   *Exact:*

$$\begin{array}{r} 600 \\ -\ 300 \\ \hline 300 \end{array} \qquad \begin{array}{r} 614 \\ -\ 276 \\ \hline 338 \end{array}$$

51. *Estimate:*   *Exact:*

$$\begin{array}{r} 800 \\ \times\ 400 \\ \hline 320{,}000 \end{array} \qquad \begin{array}{r} 752 \\ \times\ 375 \\ \hline 282{,}000 \end{array}$$

52. *Estimate:*   *Exact:*

$$\begin{array}{r} 800 \\ \times\ 400 \\ \hline 320{,}000 \end{array} \qquad \begin{array}{r} 845 \\ \times\ 396 \\ \hline 334{,}620 \end{array}$$

Estimate each answer using front end rounding. Then find the exact answer. See Example 10.

53. *Estimate:*   *Exact:*

$$\begin{array}{r} 8000 \\ 60 \\ 700 \\ +\ 4000 \\ \hline 12{,}760 \end{array} \xleftarrow{\text{Rounds to}} \begin{array}{r} 8215 \\ 56 \\ 729 \\ +\ 3605 \\ \hline 12{,}605 \end{array}$$

54. *Estimate:*   *Exact:*

$$\begin{array}{r} 3000 \\ 70 \\ 600 \\ +\ 7000 \\ \hline 10{,}670 \end{array} \qquad \begin{array}{r} 2685 \\ 73 \\ 592 \\ +\ 7183 \\ \hline 10{,}533 \end{array}$$

55. *Estimate:*   *Exact:*

$$\begin{array}{r} 700 \\ -\ 500 \\ \hline 200 \end{array} \qquad \begin{array}{r} 687 \\ -\ 529 \\ \hline 158 \end{array}$$

56. *Estimate:*   *Exact:*

$$\begin{array}{r} 500 \\ -\ 200 \\ \hline 300 \end{array} \qquad \begin{array}{r} 543 \\ -\ 174 \\ \hline 369 \end{array}$$

57. *Estimate:*   *Exact:*

$$\begin{array}{r} 900 \\ \times\ 30 \\ \hline 27{,}000 \end{array} \qquad \begin{array}{r} 939 \\ \times\ 29 \\ \hline 27{,}231 \end{array}$$

58. *Estimate:*   *Exact:*

$$\begin{array}{r} 900 \\ \times\ 70 \\ \hline 63{,}000 \end{array} \qquad \begin{array}{r} 864 \\ \times\ 74 \\ \hline 63{,}936 \end{array}$$

59. The number 3492 rounded to the nearest hundred is 3500, and 3500 rounded to the nearest thousand is 4000. But when 3492 is rounded directly to the nearest thousand it becomes 3000. Why is this true? Explain.

Perhaps the best explanation is that 3492 is closer to 3500 than to 3400, but 3492 is closer to 3000 than to 4000.

60. The use of rounding is helpful when estimating the answer to a problem. Why is this true? Give an example using either addition, subtraction, multiplication, or division to show how this works.

Rounding numbers usually allows for faster calculation and results in an estimated answer prior to getting an exact answer. One example is

$$\begin{array}{r} 400 \\ -\ 200 \\ \hline \textbf{200} \end{array} \qquad \begin{array}{r} 432 \\ -\ 209 \\ \hline \textbf{223} \end{array}$$

Estimate: ***Exact:***

61. In 1900, the population of the United States was 76 million. Today it's 303 million. Round each of these numbers to the nearest ten-million. (*Source:* Reiman Publications and U.S. Census Bureau.)

80 million people; 300 million people

62. In 1900, the average workweek in the United States was 59 hours. Today it's 38 hours. Round each of these numbers to the nearest ten. (*Source:* Reiman Publications.)

60 hours; 40 hours

63. There are 348,900 streets named Elm Street in the United States. Round this number to the nearest thousand and nearest ten-thousand. (*Source:* Expo Design Center.)

349,000; 350,000

64. The most expensive item ever sold at a Costco Wholesale warehouse was a $235,000 diamond ring. Round this number to the nearest ten-thousand and the nearest hundred-thousand. (*Source:* Costco Stores magazine.)

$240,000; $200,000

65. In Chicago, the Sears Tower tenants recycled 1,667,300 pounds of paper last year. Round this number to the nearest ten-thousand, nearest hundred-thousand, and nearest million. (*Source:* Trizec Properties.)

1,670,000 pounds; 1,700,000 pounds; 2,000,000 pounds

66. The population of India is 1,129,866,154 people. Round the number to the nearest ten-thousand, nearest hundred-thousand, and nearest million. (*Source:* U.S. Census Bureau.)

1,129,870,000; 1,129,900,000; 1,130,000,000

67. American pharmaceutical companies spent $25,765,475,000 last year to develop new products. Round this amount to the nearest hundred-thousand, nearest hundred-million, and nearest billion. (*Source:* American Demographics.)

$25,765,500,000; $25,800,000,000; $26,000,000,000

68. In one year the U.S. Federal Food Assistance Program paid out $18,915,762,568 in food stamps. Round this amount to the nearest hundred-thousand, nearest hundred-million, and nearest ten-billion. (*Source:* U.S. Department of Agriculture.)

$18,915,800,000; $18,900,000,000; $20,000,000,000

Relating Concepts (Exercises 69–75) For Individual or Group Work

To see how both rounding and front end rounding are used in solving problems, **work Exercises 69–75 in order.**

69. A number rounded to the nearest thousand is 72,000. What is the *smallest* whole number this could have been before rounding?

71,500

70. A number rounded to the nearest thousand is 72,000. What is the *largest* whole number this could have been before rounding?

72,499

71. When front end rounding is used, a whole number rounds to 8000. What is the *smallest* possible original number?

7500

72. When front end rounding is used, a whole number rounds to 8000. What is the *largest* possible original number?

8499

The graph below shows the number of personal injuries in the United States each year for people participating in common activities.

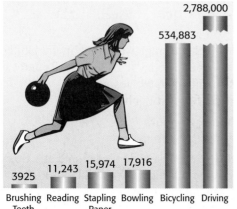

WATCH YOUR STEP

Accidents while participating in common activities

2,788,000
534,883
3925 11,243 15,974 17,916

Brushing Teeth | Reading | Stapling Paper | Bowling | Bicycling | Driving

Source: AARP Magazine.

73. Round the number of accidents occurring in each activity to the nearest ten.

3930; 11,240; 15,970; 17,920; 534,880; 2,788,000

74. Use front end rounding to round the number of accidents in each activity.

4000; 10,000; 20,000; 20,000; 500,000; 3,000,000

75. (a) What is one advantage of using front end rounding instead of rounding to the nearest ten?

When using front end rounding, all digits are 0 except the first digit. These numbers are easier to work with when estimating answers.

(b) What is one disadvantage?

Sometimes when using front end rounding, the estimated answer can vary greatly from the exact answer.

1.8 ▶▶▶ Exponents, Roots, and Order of Operations

OBJECTIVE 1 Identify an exponent and a base. The product $3 \cdot 3$ can be written as 3^2 (read as "3 squared"). The small raised number 2, called an **exponent**, says to use 2 factors of 3. The number 3 is called the **base**. Writing 3^2 as 9 is called *simplifying the expression.*

EXAMPLE 1 Simplifying Expressions

Identify the exponent and the base, and then simplify each expression.

(a) 4^3

The small raised number is the exponent.

Base $\rightarrow 4^3 \leftarrow$ Exponent $4^3 = 4 \times 4 \times 4 = 64$

(b) $2^5 = 2 \times 2 \times 2 \times 2 \times 2 = 32$
The base is 2 and the exponent is 5.

Work Problem **1** *at the Side.* ▶

OBJECTIVE 2 Find the square root of a number. Because $3^2 = 9$, the number 3 is called the **square root** of 9. The square root of a number is one of two identical factors of that number. Square roots of numbers are written with the symbol $\sqrt{}$.

$$\sqrt{9} = 3$$

Square Root

$$\sqrt{\text{number} \cdot \text{number}} = \sqrt{\text{number}^2} = \text{number}$$

For example: $\sqrt{36} = \sqrt{6 \cdot 6} = \sqrt{6^2} = 6$ is the square root of 36.

To find the square root of 64 ask, "What number can be multiplied by itself (that is, *squared*) to give 64?" The answer is 8, so

$$\sqrt{64} = \sqrt{8 \cdot 8} = \sqrt{8^2} = 8.$$

A **perfect square** is a number that is the square of a *whole number.* The first few perfect squares are listed here.

Perfect Squares Table

$0 = 0^2$	$16 = 4^2$	$64 = 8^2$	$144 = 12^2$
$1 = 1^2$	$25 = 5^2$	$81 = 9^2$	$169 = 13^2$
$4 = 2^2$	$36 = 6^2$	$100 = 10^2$	$196 = 14^2$
$9 = 3^2$	$49 = 7^2$	$121 = 11^2$	$225 = 15^2$

EXAMPLE 2 Using Perfect Squares

Find each square root.

(a) $\sqrt{16}$ Because $4^2 = 16$, $\sqrt{16} = 4$.   **(b)** $\sqrt{49} = 7$
(c) $\sqrt{0} = 0$ **(d)** $\sqrt{169} = 13$

Work Problem **2** *at the Side.* ▶

OBJECTIVE 3 Use the order of operations. Frequently problems may have parentheses, exponents, and square roots, and may involve more than one operation. Work these problems by following the **order of operations.**

1 Identify the exponent and the base, and then simplify each expression.

(a) 4^2

(b) 5^3

(c) 3^4

(d) 2^6

2 Find each square root.

(a) $\sqrt{4}$

(b) $\sqrt{25}$

(c) $\sqrt{36}$

(d) $\sqrt{225}$

(e) $\sqrt{1}$

ANSWERS
1. (a) 2; 4; 16   **(b)** 3; 5; 125
   (c) 4; 3; 81   **(d)** 6; 2; 64
2. (a) 2   **(b)** 5   **(c)** 6   **(d)** 15   **(e)** 1

3 Simplify each expression.

(a) $4 + 5 + 2^2$

(b) $3^2 + 2^3$

(c) $4 \cdot 6 \div 12 - 2$

(d) $60 \div \sqrt{36} \div 2$

(e) $8 + 6(14 \div 2)$

4 Simplify each expression.

(a) $12 - 6 + 4^2$

(b) $2^3 + 3^2 - (5 \cdot 3)$

(c) $2 \cdot \sqrt{64} - 5 \cdot 3$

(d) $20 \div 2 + (7 - 5)$

(e) $15 \cdot \sqrt{9} - 8 \cdot \sqrt{4}$

Order of Operations

1. Do all operations inside *parentheses* or *other grouping symbols.*
2. Simplify any expressions with *exponents* and find any *square roots.*
3. *Multiply* or *divide,* proceeding from left to right.
4. *Add* or *subtract,* proceeding from left to right.

EXAMPLE 3 Understanding the Order of Operations

Use the order of operations to simplify each expression.

(a) $8^2 + 5 + 2$

$$8^2 + 5 + 2$$
$$8 \cdot 8 + 5 + 2 \qquad \text{Evaluate exponent first; } 8^2 \text{ is } 8 \cdot 8.$$
$$64 + 5 + 2 \qquad \text{Add from left to right.}$$
$$69 + 2 = 71$$

(b) $35 \div 5 \cdot 6 \qquad$ Divide first (start at left).
$7 \cdot 6 = 42 \qquad$ Multiply.

(c) $9 + (20 - 4) \cdot 3 \qquad$ Work inside parentheses first.
$9 + 16 \cdot 3 \qquad$ Multiply.
$9 + 48 = 57 \qquad$ Add last.

(d) $12 \cdot \sqrt{16} - 8(4) \qquad$ Find the square root first.
$12 \cdot 4 - 8(4) \qquad$ Multiply from left to right.
$48 - 32 = 16 \qquad$ Subtract last.

*◄ Work Problem **3** at the Side.*

EXAMPLE 4 Using the Order of Operations

Use the order of operations to simplify each expression.

(a) $15 - 4 + 2 \qquad$ Subtract first (start at left).
$11 + 2 = 13 \qquad$ Add.

(b) $8 + (7 - 3) \div 2 \qquad$ Work inside parentheses first.
$8 + 4 \div 2 \qquad$ Divide.   *Add or subtract last.*
$8 + 2 = 10 \qquad$ Add last.

(c) $4^2 \cdot 2^2 + (7 + 3) \cdot 2 \qquad$ Work inside parentheses first.
$4^2 \cdot 2^2 + 10 \cdot 2 \qquad$ Evaluate exponents.
$16 \cdot 4 + 10 \cdot 2 \qquad$ Multiply from left to right.
$64 + 20 = 84 \qquad$ Add last.

(d) $4 \cdot \sqrt{25} - 7 \cdot 2 + \dfrac{0}{5} \qquad$ Find the square root first.   *Remember: Zero divided by any nonzero number is zero.*

$4 \cdot 5 - 7 \cdot 2 + \dfrac{0}{5} \qquad$ Multiply or divide from left to right.

$20 - 14 + 0 = 6 \qquad$ Add or subtract last.

Note

Getting a correct answer depends on following the order of operations.

*◄ Work Problem **4** at the Side.*

1.8 ▶▶▶ Exercises

Identify the exponent and the base, and then simplify each expression. See Example 1.

1. 3^2 2; 3; 9

2. 2^3 3; 2; 8

3. 5^2 2; 5; 25

4. 4^2 2; 4; 16

5. 8^2 2; 8; 64

6. 10^3 3; 10; 1000

7. 15^2 2; 15; 225

8. 11^3 3; 11; 1331

Use the Perfect Squares Table on page 77 to find each square root. See Example 2.

9. $\sqrt{16}$ 4

10. $\sqrt{25}$ 5

11. $\sqrt{64}$ 8

12. $\sqrt{36}$ 6

13. $\sqrt{100}$ 10

14. $\sqrt{49}$ 7

15. $\sqrt{144}$ 12

16. $\sqrt{225}$ 15

Fill in each blank. See Example 2.

17. $6^2 = $ __36__ so $\sqrt{36} = 6$

18. $9^2 = $ __81__ so $\sqrt{81} = 9$

19. $20^2 = $ __400__ so $\sqrt{400} = 20$

20. $30^2 = $ __900__ so $\sqrt{900} = 30$

21. $35^2 = $ __1225__ so $\sqrt{1225} = 35$

22. $38^2 = $ __1444__ so $\sqrt{1444} = 38$

23. $25^2 = $ __625__ so $\sqrt{625} = 25$

24. $50^2 = $ __2500__ so $\sqrt{2500} = 50$

25. $100^2 = $ __10,000__ so $\sqrt{10,000} = 100$

26. $60^2 = $ __3600__ so $\sqrt{3600} = 60$

27. Describe in your own words a perfect square. Of the two numbers 25 and 50, identify which is a perfect square and explain why.

A perfect square is the square of a whole number. The number 25 is the square of 5 because 5 · 5 = 25. The number 50 is not a perfect square. There is no whole number that can be squared to get 50.

28. Use the following list of words and phrases to write the four steps in the order of operations.

add	square root
exponents	subtract
multiply	divide

parentheses or other grouping symbols

1. **Do all operations inside parentheses or other grouping symbols.**

2. **Simplify any expressions with exponents and find any square roots.**

3. **Multiply or divide proceeding from left to right.**

4. **Add or subtract proceeding from left to right.**

Simplify each expression by using the order of operations. See Examples 3 and 4.

29. $3^2 + 8 - 5$ 12

30. $5^2 + 5 - 6$ 24

31. $3 \cdot 7 - 6$ 15

32. $5 \cdot 7 - 7$ 28

33. $8 \cdot 5 \div 10$ 4

34. $6 \cdot 8 \div 8$ 6

35. $25 \div 5(8 - 4)$ 20

36. $36 \div 18(7 - 3)$ 8

37. $5 \cdot 3^2 + \dfrac{0}{8}$ 45

38. $8 \cdot 3^2 - \dfrac{10}{2}$ 67

39. $4 \cdot 1 + 8(9 - 2) + 3$ 63

40. $3 \cdot 2 + 7(3 + 1) + 5$ 39

41. $2^2 \cdot 3^3 + (20 - 15) \cdot 2$ 118

42. $4^2 \cdot 5^2 + (20 - 9) \cdot 3$ 433

43. $5\sqrt{36} - 2(4)$ 22

44. $2 \cdot \sqrt{100} - 3(4)$ 8

45. $8(2) + 3 \cdot 7 - 7 =$ 30

46. $10(3) + 6 \cdot 5 - 20$ 40

47. $2^3 \cdot 3^2 + 3(14 - 4)$ **102**

48. $3^2 \cdot 4^2 + 2(15 - 6)$ **162**

49. $7 + 8 \div 4 + \dfrac{0}{7}$ **9**

50. $6 + 8 \div 2 + \dfrac{0}{8}$ **10**

51. $3^2 + 6^2 + (30 - 21) \cdot 2$ **63**

52. $4^2 + 5^2 + (25 - 9) \cdot 3$ **89**

53. $7 \cdot \sqrt{81} - 5 \cdot 6$ **33**

54. $6 \cdot \sqrt{64} - 6 \cdot 5$ **18**

55. $8 \cdot 2 + 5(3 \cdot 4) - 6$ **70**

56. $5 \cdot 2 + 3(5 + 3) - 6$ **28**

57. $4 \cdot \sqrt{49} - 7(5 - 2)$ **7**

58. $3 \cdot \sqrt{25} - 6(3 - 1)$ **3**

59. $7(4 - 2) + \sqrt{9}$ **17**

60. $5(4 - 3) + \sqrt{9}$ **8**

61. $7^2 + 3^2 - 8 + 5$ **55**

62. $3^2 - 2^2 + 3 - 2$ **6**

63. $5^2 \cdot 2^2 + (8 - 4) \cdot 2$ **108**

64. $5^2 \cdot 3^2 + (30 - 20) \cdot 2$ **245**

65. $5 + 9 \div 3 + 6 \cdot 3$ **26**

66. $8 + 3 \div 3 + 6 \cdot 3$ **27**

67. $8 \cdot \sqrt{49} - 6(9 - 4)$ **26**

68. $8 \cdot \sqrt{49} - 6(5 + 3)$ **8**

69. $5^2 - 4^2 + 3 \cdot 6$ **27**

70. $3^2 + 6^2 - 5 \cdot 8$ **5**

71. $8 + 8 \div 8 + 6 + \dfrac{5}{5}$ **16**

72. $3 + 14 \div 2 + 7 + \dfrac{8}{8}$ **18**

73. $6 \cdot \sqrt{25} - 7(2)$ **16**

74. $8 \cdot \sqrt{36} - 4(6)$ **24**

75. $9 \cdot \sqrt{16} - 3 \cdot \sqrt{25}$ **21**

76. $6 \cdot \sqrt{81} - 3 \cdot \sqrt{49}$ **33**

77. $7 \div 1 \cdot 8 \cdot 2 \div (21 - 5)$ **7**

78. $12 \div 4 \cdot 5 \cdot 4 \div (15 - 13)$ **30**

79. $15 \div 3 \cdot 2 \cdot 6 \div (14 - 11)$ **20**

80. $9 \div 1 \cdot 4 \cdot 2 \div (11 - 5)$ **12**

81. $6 \cdot \sqrt{25} - 4 \cdot \sqrt{16}$ **14**

82. $10 \cdot \sqrt{49} - 4 \cdot \sqrt{64}$ **38**

83. $5 \div 1 \cdot 10 \cdot 4 \div (17 - 9)$ **25**

84. $15 \div 3 \cdot 8 \cdot 9 \div (12 - 8)$ **90**

85. $8 \cdot 9 \div \sqrt{36} - 4 \div 2 + (14 - 8)$ **16**

86. $3 - 2 + 5 \cdot 4 \cdot \sqrt{144} \div \sqrt{36}$ **41**

87. $2 + 1 - 2 \cdot \sqrt{1} + 4 \cdot \sqrt{81} - 7 \cdot 2$ **23**

88. $6 - 4 + 2 \cdot 9 - 3 \cdot \sqrt{225} \div \sqrt{25}$ **11**

89. $5 \cdot \sqrt{36} \cdot \sqrt{100} \div 4 \cdot \sqrt{9} + 8$ **233**

90. $9 \cdot \sqrt{36} \cdot \sqrt{81} \div 2 + 6 - 3 - 5$ **241**

Study Skills

▶▶▶ **TAKING LECTURE NOTES**

Study the set of sample math notes in this section, and read the comments about them. Then try to incorporate the techniques into your own math note taking in class.

OBJECTIVES

1 Apply note taking strategies, such as writing problems as well as explanations.

2 Use appropriate abbreviations in notes.

▶ The **date and title** of the day's lecture topic are always at the top of every page. **Always begin a new day with a new page.**

▶ Note the **definitions** of base and exponent are written in parentheses—don't trust your memory!

▶ **Skipping lines** makes the notes easier to read.

▶ See how the **direction word** (*simplify*) is emphasized and explained.

▶ A **star marks an important concept.** This is a warning to avoid future mistakes. **Note the underlining,** too, which highlights the importance.

▶ Notice the two columns, which allow for the example and its explanation to be close together. **Whenever you know you'll be given a series of steps to follow, try the two-column method.**

▶ Note the **brackets and arrows,** which clearly show how the problem is set up to be simplified.

January 2   Exponents

Exponents used to show repeated multiplication.

$3 \cdot 3 \cdot 3 \cdot 3$ can be written 3^4 exponent (how many times it's multiplied)
base (the number being multiplied)

Read 3^2 as 3 to the 2nd power or 3 squared
3^3 as 3 to the 3rd power or 3 cubed
3^4 as 3 to the 4th power
etc.

Simplifying an expression with exponents
→ actually do the repeated multiplication

2^3 means $2 \cdot 2 \cdot 2$ and $2 \cdot 2 \cdot 2 = 8$

☆ Careful! 5^2 means $5 \cdot 5$ NOT $5 \cdot 2$
so $5^2 = 5 \cdot 5 = 25$ BUT $5^2 \neq 10$

Example	Explanation
Simplify ②⁴ · ③²	Exponents mean multiplication.
$2 \cdot 2 \cdot 2 \cdot 2$ · $3 \cdot 3$	Use 2 as a factor 4 times. Use 3 as a factor 2 times.
16 · 9	$2 \cdot 2 \cdot 2 \cdot 2$ is 16 16·9 is 144
	$3 \cdot 3$ is 9
144	Simplified result is 144 (no exponents left)

83

Now Try This ▶▶▶

Why Are These Notes Brain Friendly?

The notes are **easy to look at,** and you know that the brain responds to things that are visually pleasing. Other techniques that are visually memorable are the use of spacing (the two columns), stars, underlining, and circling. All of these methods **allow your brain to take note of important concepts and steps.**

The notes are also **systematic,** which means that they use certain techniques regularly. This way, your brain easily recognizes the topic of the day, the signals that show an important point, and the steps to follow for procedures. When you develop a system that you always use in your notes, your notes are easy to understand later when you are reviewing for a test.

Find one or two people in your math class to work with. Compare each other's lecture notes over a period of a week or so. Ask yourself the following questions as you examine the notes.

1. What are you doing in your notes to show the **main points** or larger concepts? (Such as underlining, boxing, using stars, capital letters, etc.)

2. In what ways do you **set off the explanations** for worked problems, examples, or smaller ideas (subpoints)? (Such as indenting, using arrows, circling or boxing)

3. What does **your instructor do** to show that he or she is moving from one idea to the next? (Such as saying "Next" or "Any questions," "Now," or erasing the board, etc.)

4. **How do you mark** that in your notes? (Such as skipping lines, using dashes or numbers, etc.)

5. What **explanations (in words) do you give yourself** in your notes, so when those new dendrites you grew in lecture are fading, you can read your notes and still remember the new concepts later when you try to do your homework?

6. What **did you learn** by examining your classmates' notes?

   * _____
   * _____
   * _____

7. What **will you try** in your own note taking? List **four** techniques that you will use next time you take notes in math class.

   * _____
   * _____
   * _____
   * _____

1.9 ▶▶▶ Reading Pictographs, Bar Graphs, and Line Graphs

We have all heard the saying "A picture is worth a thousand words," and there may be some truth in this. Today, so much information and data are being presented in the form of pictographs, circle graphs, bar graphs, and line graphs that it is important to be able to read and understand these tools.

OBJECTIVE 1 Read and understand a pictograph. A **pictograph** is a graph that uses pictures or symbols. It displays information that can be compared easily. However, since a symbol is used to represent a certain quantity, it can be difficult to determine the amount represented by a fraction of a symbol.

The Elvis Presley U.S. postage stamp retains the title as the most popular stamp of all time. The pictograph below compares the number of U.S. postage stamps sold in the five top releases. In this pictograph, it is difficult to determine what fractional amount is represented by the partial stamps.

ELVIS IS STILL KING

Five of the most popular U.S. postage stamps of all time.

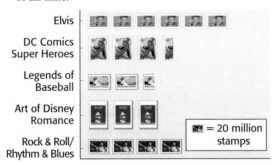

■ = 20 million stamps

Source: United States Postal Service.

EXAMPLE 1 Using a Pictograph

Use the pictograph to answer each question.

> Each stamp symbol represents 20 million stamps.

(a) Which of the U.S. postage stamps shown has the lowest number sold?

The row representing Legends of Baseball has the fewest symbols. This means that the lowest number of U.S. postage stamps sold was Legends of Baseball.

(b) Approximately how many more Elvis stamps were sold than the Art of Disney Romance stamps?

The row representing the Elvis stamps has three more symbols than that for the Art of Disney stamps. This means that 3 • 20 million or 60,000,000 more Elvis stamps were sold than the Art of Disney Romance stamps.

Work Problem ⬚1⬚ *at the Side.* ▶

OBJECTIVE 2 Read and understand a bar graph. Bar graphs are useful for showing comparisons. For example, the following bar graph shows how many people out of every 100 fans surveyed chose each sport as their favorite.

OBJECTIVES

1 Read and understand a pictograph.

2 Read and understand a bar graph.

3 Read and understand a line graph.

⬚1⬚ Use the pictograph to answer each question.

(a) Which of the U.S. stamps had the second greatest number of sales?

(b) Approximately how many more Rock and Roll/Rhythm and Blues stamps were sold than Art of Disney Romance stamps?

ANSWERS

2. (a) Rock and Roll/Rhythm & Blues
   (b) about 20 million (or 20,000,000) more stamps

2 Use the bar graph to find the approximate number of fans who picked each sport as their favorite.

(a) College football

(b) Pro baseball

(c) Pro basketball

(d) College basketball

(e) Golf

3 Use the line graph to find the predicted population of the United States for each year.

(a) 2050

(b) 2075

(c) 2100

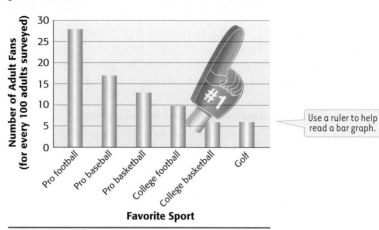

FAN APPEAL

Source: The Harris Poll.

EXAMPLE 2   Using a Bar Graph

Use the bar graph to find the number of fans who picked pro football as their favorite sport.

Use a ruler or straightedge to line up the top of the bar labeled "Pro football," with the numbers on the left edge of the graph, labeled "Number of Adult Fans." We see that 28 out of 100 adult fans picked pro football as their favorite sport.

◀ *Work Problem* **2** *at the Side.*

OBJECTIVE 3 Read and understand a line graph. A **line graph** is often used for showing a trend. The following line graph shows the U.S. Census Bureau predictions for U.S. population growth to the year 2100.

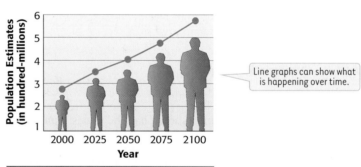

ANOTHER CENTURY OF GROWTH

Source: U.S. Census Bureau.

EXAMPLE 3   Using a Line Graph

Use the line graph to answer each question.

(a) What trend or pattern is shown in the graph?
The population will continue to increase.

(b) What is the estimated population for 2025?
Use a ruler or straightedge to line up the dot above the year labeled 2025 on the horizontal line with the numbers along the left edge of the graph. Notice that the label on the left side says "in hundred-millions." Since the 2025 dot is halfway between 3 and 4, the population in 2025 is halfway between 3 • 100,000,000 and 4 • 100,000,000 or 300,000,000 and 400,000,000. That means that the predicted population in 2025 is about 350,000,000 people.

◀ *Work Problem* **3** *at the Side.*

ANSWERS

2. (a) 10 out of 100   **(b)** 17 out of 100
    (c) 13 out of 100   **(d)** 6 out of 100
    (e) 6 out of 100
3. (a) 400,000,000   **(b)** 475,000,000
    (c) 575,000,000

FOR EXTRA HELP

The following pictograph shows the number of retail stores for the seven companies with the greatest number of outlets. Use the pictograph to answer Exercises 1–6. See Example 1.

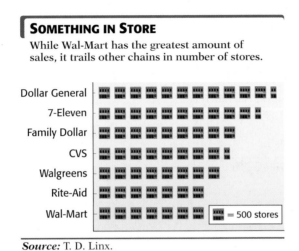

SOMETHING IN STORE
While Wal-Mart has the greatest amount of sales, it trails other chains in number of stores.

Dollar General
7-Eleven
Family Dollar
CVS
Walgreens
Rite-Aid
Wal-Mart

▦ = 500 stores

Source: T. D. Linx.

1. Find the number of Family Dollar retail stores.

   4500 stores

2. Approximately how many retail stores does 7-Eleven have?

   about 5250 stores

3. Which company has the greatest number of retail stores? How many is that?

   Dollar General; about 5750 stores

4. Which companies have the least number of retail stores? How many does each one have?

   Rite-Aid, Wal-Mart; 3500 stores

5. How many fewer stores does Family Dollar have than Dollar General?

   1250 fewer stores

6. How many more retail stores does Walgreens have than Wal-Mart?

   500 more stores

The following bar graph shows the results of a survey that was taken of 100 working adults to determine how they chose their careers. Use the bar graph to answer Exercises 7–12. See Example 2.

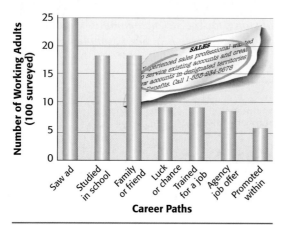

HOW DID YOU CHOOSE YOUR CAREER?

Source: Market Facts/TeleNation for Career Education Corporation.

7. How many people found their careers as a result of training for a job?

9 people

8. How many people found their careers because they studied for the career in school?

18 people

9. (a) Which career path was taken by the greatest number of people?

saw ad

(b) How many people used this path?

25 people

10. (a) Which career path was taken by the least number of people?

promoted within

(b) How many people used this path?

6 people

11. How many more people found their careers as a result of "Studied in school" than "Luck or chance"?

9 people

12. Find the total number of people who found their careers as a result of either "Studied in school" or "Trained for a job."

27 people

Foxworthy Reforestation Company collected tree planting data and prepared the following line graph. Remembering that the tree planting data is shown in thousands of trees, use the line graph to answer Exercises 13–18. See Example 3.

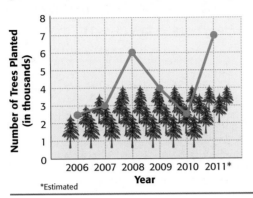

FOXWORTHY REFORESTATION COMPANY

*Estimated

13. Which year had the greatest number of trees planted? How many trees were planted?

2011; 7000 trees

14. Which two years had the least number of trees planted? How many trees were planted in each of those years?

2006 and 2010; 2500 trees planted

15. Find the increase in the number of trees planted from 2010 to 2011.

4500 trees

16. Find the decrease in the number of trees planted from 2009 to 2010.

1500 trees

17. Give three possible explanations for the decrease in trees planted in 2009.

Possible answers are

1. **shortage of trees to plant**

2. **lack of qualified workers**

3. **poor economy**

4. **less demand for planting**

18. Give three possible explanations for the increase in trees planted in 2011.

Possible answers are

1. **greater demand**

2. **more trees available**

3. **many qualified workers**

4. **availability of money for reforestation projects.**

Relating Concepts (Exercises 19–23) For Individual or Group Work

Getting a correct answer in mathematics always depends on following the order of operations. Insert grouping symbols (parentheses) so that each given expression will result in the given number when simplified. **Work Exercises 19–23 in order.**

19. $7 - 2 \cdot 3 - 6$; simplifies to 9

$(7 - 2) \cdot 3 - 6$

20. $4 + 2 \cdot 5 + 1$; simplifies to 36

$(4 + 2) \cdot (5 + 1)$

21. $36 \div 3 \cdot 3 \cdot 4$; simplifies to 16

$36 \div (3 \cdot 3) \cdot 4$

22. $56 \div 2 \cdot 2 \cdot 2 + \dfrac{0}{6}$; simplifies to 7

$56 \div (2 \cdot 2 \cdot 2) + \dfrac{0}{6}$

23. The Good Shepherd Ranch owns one section of land (one mile by one mile) shown below as Parcel 1. Parcels 2 and 3 are leased from neighbors.

(a) Use the order of operations to write an expression for the distance around the combined parcels.

$7920 + 1320 + 2640 + (5280 - 1320 - 1320) + 2640 + 1320 + 7920 + 5280$

(b) How many feet of barbed wire are needed for a three-strand barbed wire fence around the parcels?

$31{,}680 \times 3 = 95{,}040$ feet

(c) How many miles of barbed wire is this? (*Hint:* 1 mile $= 5280$ feet.)

18 miles

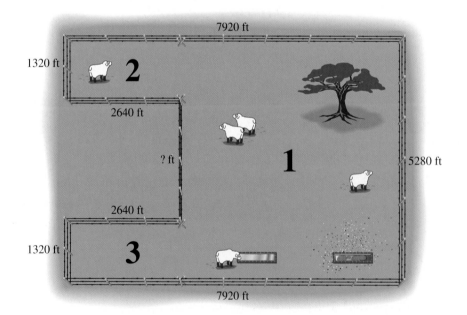

1.10 ▶▶▶   Solving Application Problems

Most problems involving applications of mathematics are written out in sentence form. You need to read the problem carefully to decide how to solve it.

OBJECTIVE 1   Find indicator words in application problems.
As you read an application problem, look for **indicator words** that help you determine whether to use addition, subtraction, multiplication, or division. Some of these indicator words are shown here.

Addition	Subtraction	Multiplication	Division	Equals
plus	less	product	divided by	is
more	subtract	double	divided into	the same as
more than	subtracted from	triple	quotient	equals
added to	difference	times	goes into	equal to
increased by	less than	of	divide	yields
sum	fewer	twice	divided equally	results in
total	decreased by	twice as much	per	are
sum of	loss of			
increase of	minus			
gain of	take away			

CAUTION
The word *and* does not always indicate addition, so it does not appear as an indicator word in the preceding table. Notice how the "and" shows the location of several different operation signs below.

The sum of 6 *and* 2 is 6 + 2.
The difference of 6 *and* 2 is 6 − 2.
The product of 6 *and* 2 is 6 · 2.
The quotient of 6 *and* 2 is 6 ÷ 2.

OBJECTIVE 2   Solve application problems.   Solve application problems by using the following six steps.

Solving an Application Problem

Step 1   **Read** the problem carefully and be certain you *understand* what the problem is asking. It may be necessary to read the problem several times.

Step 2   Before doing any calculations, **work out a plan** and try to visualize the problem. Draw a sketch if possible. Know which facts are given and which must be found. Use *indicator words* to help decide on the *plan* (whether you will need to add, subtract, multiply, or divide).

Step 3   **Estimate** a *reasonable answer* by using rounding.

Step 4   **Solve** the problem by using the facts given and your plan.

Step 5   **State the answer.**

Step 6   **Check** your work. If the answer does not seem reasonable, begin again by reading the problem.

Read and **understand** a problem before you begin.

Always **estimate** the final answer.

Check your answer to see if it is *reasonable*.

1 Pick the most reasonable answer for each problem.

(a) A grocery clerk's hourly wage: $1.40; $14; $140

(b) The total length of five sport-utility vehicles: 8 ft; 18 ft; 80 ft; 800 ft

(c) The cost of heart bypass surgery: $1000; $100,000; $10,000,000

2 Solve each problem.

(a) On a recent geology field trip, 84 fossils were collected. If the fossils are divided equally among John, Sean, Jenn, and Kara, how many fossils will each receive?

(b) This week there are 408 children attending a winter sports camp. If 12 children are assigned to each camp counselor, how many counselors are needed?

> **CAUTION**
> Be certain that you know what the problem is asking before you try to solve it.

OBJECTIVE 3 Estimate an answer. The six problem-solving steps give a systematic approach for solving word problems. Each of the steps is important, but special emphasis should be placed on Step 3, estimating a *reasonable answer.* Many times an "answer" just does not fit the problem.

What is a reasonable answer? Read the problem and try to determine the approximate size of the answer. Should the answer be part of a dollar, a few dollars, hundreds, thousands, or even millions of dollars? For example, if a problem asks for the cost of a man's shirt, would an answer of $20 be reasonable? $2000? $2? $200?

> **CAUTION**
> Always estimate the answer, then look at your final result to be sure it fits your estimate and is reasonable. This step will give greater success in problem solving.

◀ *Work Problem* **1** *at the Side.*

EXAMPLE 1 **Applying Division**

A community group has raised $8260 for charity. Equal amounts are given to the Food Bank, Children's Center, Boy Scouts of America, and the Women's Shelter. How much did each group receive?

Step 1 **Read.** A reading of the problem shows that the four charities divided $8260 equally.

Step 2 **Work out a plan.** The indicator words, *divided equally,* show that the amount each received can be found by dividing $8260 by 4.

Step 3 **Estimate.** Round $8260 to $8000. Then $8000 \div 4 = 2000, so a reasonable answer would be a little greater than $2000 each.

Step 4 **Solve.** Find the actual answer by dividing $8260 by 4.

$$\frac{2065}{4)\overline{8260}}$$

Step 5 **State the answer.** Each charity received $2065.

Step 6 **Check.** The exact answer of $2065 is reasonable, as $2065 is close to the estimated answer of $2000. Is the answer $2065 correct? Check by multiplying.

$$\$2065 \leftarrow \text{Amount received by each charity}$$
$$\times \quad 4 \leftarrow \text{Number of charities}$$
$$\$8260 \leftarrow \text{Total raised; matches number given in problem}$$

> Remember: Check your work.

◀ *Work Problem* **2** *at the Side.*

ANSWERS
1. (a) $14 (b) 80 ft (c) $100,000
2. (a) 21 fossils (b) 34 counselors

EXAMPLE 2 **Applying Addition**

One week, Andrea Abriani, operations manager, decided to total the stroller production at Safe T First Strollers. The daily production figures were 7642 strollers on Monday, 8150 strollers on Tuesday, 7916 strollers on Wednesday, 8419 strollers on Thursday, and 7704 strollers on Friday. Find the total production for the week.

Step 1 **Read.** In this problem, the production for each day is given and the total production for the week must be found.

Step 2 **Work out a plan.** Add the daily production figures to arrive at the weekly total.

Step 3 **Estimate.** Because the production was about 8000 strollers per day for a week of five days, a reasonable estimate would be $5 \cdot 8000 = 40,000$ strollers.

Step 4 **Solve.** Find the exact answer by adding the production numbers for the 5 days.

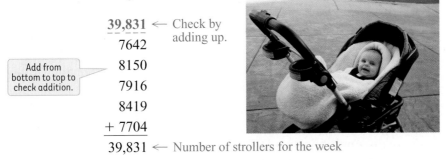

Add from bottom to top to check addition.

$$
\begin{array}{r}
\underline{39,831} \leftarrow \text{Check by adding up.} \\
7642 \\
8150 \\
7916 \\
8419 \\
+\ 7704 \\
\hline
39,831 \leftarrow \text{Number of strollers for the week}
\end{array}
$$

Step 5 **State the answer.** Abriani's total production figure for the week was 39,831 strollers.

Step 6 **Check.** The exact answer of 39,831 strollers is close to the estimate of 40,000 strollers, so it is reasonable. Add up the columns to check the exact answer.

📱 **Calculator Tip** The calculator solution to Example 2 uses chain calculations.

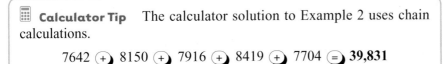

7642 ⊕ 8150 ⊕ 7916 ⊕ 8419 ⊕ 7704 ⊜ **39,831**

Work Problem ③ *at the Side.* ▶

EXAMPLE 3 **Determining Whether Subtraction Is Necessary**

The number of miles driven this year is 3028 fewer than the number driven last year. The miles driven last year was 16,735. Find the number of miles driven this year.

Step 1 **Read.** In this problem, the miles driven decreased from last year to this year. The miles driven last year and the decrease in miles driven are given. This year's miles driven must be found.

Step 2 **Work out a plan.** The indicator word, *fewer,* shows that subtraction must be used to find the number of miles driven this year.

Step 3 **Estimate.** Because the driving last year was about 17,000 miles, and the decrease in driving is about 3000 miles, a reasonable estimate would be $17,000 - 3000 = 14,000$ miles.

Continued on Next Page

③ Solve each problem.

(a) During the semester, Cindy received the following points on examinations and quizzes: 92, 81, 83, 98, 15, 14, 15, and 12. Find her total points for the semester.

(b) Stephanie Dixon works at the telephone order desk of a catalog sales company. One week she had the following number of customer contacts: Monday, 78; Tuesday, 64; Wednesday, 118; Thursday, 102; and Friday, 196. How many customer contacts did she have that week?

4 Solve each problem.

(a) A home has a living area of 1450 square feet, while an apartment has 980 square feet. Find the difference between the number of square feet in the two living areas.

(b) The Antique Military Vehicle Collectors (AMVC) had $14,863 in their club treasury bank account. After writing a check for $1180 to rent a display hall, find the amount remaining in the club account.

5 Solve each problem.

(a) Brenda is paid $685 for each kitchen remodeling job that she sells. If she sold 6 remodeling jobs and had $320 in sales expense deducted, how much did she make?

(b) An Internet book company had sales of 12,628 books with a profit of $6 for each book sold. If 863 books were returned, how much profit remains?

Step 4 **Solve.** Find the exact answer by subtracting 3078 from 16,735.

$$\begin{array}{r} 16{,}735 \\ -3078 \\ \hline 13{,}707 \end{array}$$

Step 5 **State the answer.** The driving this year is 13,707 miles.

Step 6 **Check.** The exact answer of 13,707 is reasonable, as it is close to the estimate of 14,000. Check by adding.

$$\begin{array}{r} 13{,}707 \\ +3028 \\ \hline 16{,}735 \end{array}$$ ← miles driven this year
← decrease in miles driven
← miles driven last year; matches number given in problem

◀ *Work Problem* **4** *at the Side.*

⎧ **EXAMPLE 4** **Solving a Two-Step Problem**

In May, a landlord received $720 from each of eight tenants. After paying $2180 in expenses, how much rent money did the landlord have left?

Step 1 **Read.** The problem asks for the amount of rent remaining after expenses have been paid.

Step 2 **Work out a plan.** The wording *from each of eight tenants* indicates that the eight rents must be totaled. Since the rents are all the same, use multiplication to find the total rent received. Then, subtract expenses.

Step 3 **Estimate.** The amount of rent is about $700, making the total rent received about $700 • 8 = $5600. The expenses are about $2000. A reasonable estimate of the amount remaining is $5600 − $2000 = $3600.

Step 4 **Solve.** Find the exact amount by first multiplying $720 by 8 (the number of tenants).

$$\begin{array}{r} \$720 \\ \times8 \\ \hline \$5760 \end{array}$$

Then subtract the $2180 in expenses from $5760.

$$\begin{array}{r} \$5760 \\ -\$2180 \\ \hline \$3580 \end{array}$$ The exact answer is close to the estimate and reasonable.

Step 5 **State the answer.** The amount remaining is $3580.

Step 6 **Check.** The exact answer of $3580 is reasonable, since it is close to the estimated answer of $3600. Check by adding the expenses to the amount remaining and then dividing by 8.

$3580 + $2180 = $5760 | Always check your work.

$$\begin{array}{r} \$720 \\ 8\overline{)5760} \end{array}$$ Matches the rent amount given in the problem

◀ *Work Problem* **5** *at the Side.*

ANSWERS
4. **(a)** 470 square feet **(b)** $13,683
5. **(a)** $3790 **(b)** $70,590

1.10 ▶▶▶ Exercises

FOR EXTRA HELP

MyMathLab   Math XL PRACTICE   WATCH   DOWNLOAD   READ   REVIEW

Solve each application problem. First use front end rounding to estimate the answer.
Then find the exact answer. See Examples 1–4.

1. Last week, SUBWAY sold 602 Veggie Delite sandwiches, 935 ham sandwiches, 1328 turkey breast sandwiches, 757 roast beef sandwiches, and 1586 SUBWAY club sandwiches. Find the total number of sandwiches sold.

Estimate: 600 + 900 + 1000 + 800 + 2000 = 5300 sandwiches

Exact: **5208 sandwiches**

2. During a recent week, Radio Flyer, Inc. manufactured 32,815 Model #18 wagons, 4875 steel mini-wagons, 1975 wood 40-inch wagons, 15,308 scooters, and 9815 new-design plastic wagons. Find the total number of units manufactured.

Estimate: 30,000 + 5000 + 2000 + 20,000 + 10,000 = 67,000 units

Exact: **64,788 units**

The graph shows the number of contaminated meat recalls by the U.S. Department of Agriculture over a five-year period. Use the graph for Exercises 3–4.

3. How many more meat recalls were there in 2003 than in 2007?

Estimate: 70 − 60 = 10 more recalls

Exact: **13 more recalls**

4. How many fewer meat recalls were there in 2006 than in 2008?

Estimate: 60 − 40 = 20 fewer recalls

Exact: **20 fewer recalls**

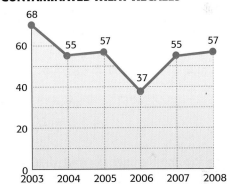

CONTAMINATED MEAT RECALLS

Source: U.S. Department of Agriculture.

5. A packing machine can package 236 first-aid kits each hour. At this rate, find the number of first-aid kits packaged in 24 hours.

Estimate: 200 × 20 = 4000 kits

Exact: **5664 kits**

6. If 450 admission tickets to a classic car show are sold each day, how many tickets are sold in a 12-day period?

Estimate: 500 × 10 = 5000 tickets

Exact: **5400 tickets**

7. Clarence Hanks, coordinator of Toys for Tots, has collected 2628 toys. If his group can give the same number of toys to each of 657 children, how many toys will each child receive?

Estimate: 3000 ÷ 700 ≈ 4 toys

Exact: **4 toys**

8. If profits of $680,000 are divided evenly among a firm's 1000 employees, how much money will each employee receive?

Estimate: $700,000 ÷ 1000 = $700

Exact: **$680**

9. The number of boaters and campers at the lake was 8392 on Friday. If this was 4218 more than the number of people at the lake on Wednesday, how many were there on Wednesday?

Estimate: 8000 − 4000 = 4000 people

Exact: **4174 people**

10. The community has raised $52,882 for the homeless shelter. If the total amount needed for the shelter is $75,650, find the additional amount needed.

Estimate: $80,000 − $50,000 = $30,000

Exact: **$22,768**

11. Turn down the thermostat in the winter and you can save money and energy. In the upper Midwest, setting back the thermostat from 68° to 55° at night can save $34 per month on fuel. Find the amount of money saved in five months.

Estimate: $30 × 5 = $150

Exact: $170

12. The cost of tuition and fees at a community college is $785 per quarter. If Gale Klein has five quarters remaining, find the total amount that she will need for tuition and fees.

Estimate: $800 × 5 = $4000

Exact: $3925

The table shows the average annual earnings of those in careers that do not require a four year degree. Refer to the table to answer Exercises 13–16. First use front end rounding to estimate the answer. Then find the exact answer.

13. How much more does a dental hygienist earn than a flight attendant?

Estimate: $50,000 − $40,000 = $10,000

Exact: $14,100

14. How much more does an air traffic controller earn than a court reporter?

Estimate: $90,000 − $40,000 = $50,000.

Exact: $47,520

No Degree? Apply Here?

Four years of college may be your best ticket to a high-paying career, but these solid jobs don't require an undergraduate degree:

Profession	Median Annual Earnings		
Air traffic controller	$87,930	Locomotive engineer	46,540
Nuclear power reactor operator	60,180	Telecom equipment installer/repairer*	46,390
Dental hygienist*	54,700	Funeral director*	42,010
Elevator installer/repairer	51,630	Aircraft mechanic*	41,990
Real estate broker	51,380	Brick mason	41,590
Commercial pilot (non-airline)	47,410	Police officer	40,970
Electrical power line installer/repairer	47,210	Electrician	40,770
		Flight attendant	40,600
		Court reporter*	40,410
		Real estate appraiser*	38,950

*Requires associate's degree or vocational diploma

Source: U.S. Department of Labor.

15. Mr. White is a locomotive engineer and Mrs. White is a court reporter. Mr. Easterly is an aircraft mechanic and Mrs. Easterly is a real estate broker.

(a) Which couple has higher earnings?

Estimate: $50,000 + $40,000 = $90,000, White; $40,000 + $50,000 = $90,000, Easterly

Exact: $86,950, White; $93,370, Easterly; Mr. and Mrs. Easterly have higher earnings.

(b) Find the difference in the earnings.

Estimate: $90,000 − $90,000 = $0

Exact: $6420

16. Mr. Means is a funeral director and Mrs. Means is an electrician. Mr. Strong is a police officer and Mrs. Strong is a commercial pilot.

(a) Which couple has higher earnings?

Estimate: $40,000 + $40,000 = $80,000, Means; $40,000 + $50,000 = $90,000, Strong

Exact: $82,780, Means; $88,380 Strong; Mr. and Mrs. Strong have higher earnings.

(b) Find the difference in the earnings.

Estimate: $90,000 − $80,000 = $10,000

Exact: $5600

17. Ronda Biondi decides to establish a monthly budget. She will spend $695 for rent, $340 for food, $435 for child care, $240 for transportation, $180 for other expenses, and she will put the remainder in savings. If her monthly take-home pay is $2240, find her monthly savings.

Estimate: $2000 − $700 − $300 − $400 − $200 − $200 = $200

Exact: $350

18. Robert Heisner had $2874 in his checking account. He wrote checks for $308 for auto repairs, $580 for child support, and $778 for an insurance payment. Find the amount remaining in his account.

Estimate: $3000 − $300 − $600 − $800 = $1300

Exact: $1208

19. There are 43,560 square feet in one acre. How many square feet are there in 138 acres?

Estimate: 40,000 × 100 = 4,000,000 square feet

Exact: 6,011,280 square feet

20. The number of gallons of water polluted each day in an industrial area is 209,670. How many gallons of water are polluted each year? (Use a 365-day year.)

Estimate: 200,000 × 400 = 80,000,000 gallons

Exact: 76,529,550 gallons

The Internet was used to find the following minivan optional features and the price of each feature. Use this information to answer Exercises 21–24.

Safety and Security Options		Convenience and Comfort Options	
Option	Cost	Option	Cost
Integrated child bench seats	$400	Power sliding door	$400
Supplemental side air bags	$1395	8-way power seat	$395
Security alarm	$170	Roof rack	$250
Hands-free communication	$360	Power door locks	$315
Power adjustable pedal	$195	Keyless entry	$150
Full-size spare tire	$160	AM/FM stereo and CD	$350

Source: www.edmunds.com

21. Find the total cost of all Safety and Security Options listed.

Estimate: $400 + $1000 + $200 + $400 + $200 + $200 = $2400

Exact: $2680

22. Find the total cost of all Convenience and Comfort Options listed.

Estimate: $400 + $400 + $300 + $300 + $200 + $400 = $2000

Exact: $1860

23. A new-car dealer offers an option value package that includes integrated child bench seats, security alarm, keyless entry, and a power sliding door at a cost of $980. If Jill buys the value package instead of paying for each option separately, how much will she save?

Estimate: $400 + $200 + $200 + $400 = $1200
$1200 − $1000 = $200

Exact: $140

24. A new-car dealer offers an option package that includes integrated child bench seats, security alarm, full-size spare tire, power door locks, 8-way power seat, and a roof rack for a total of $1550. How much will Samuel save if he buys the option package instead of paying for each option separately?

Estimate: $400 + $200 + $200 + $300 + $400 + $300 = $1800
$1800 − $1600 = $200

Exact: $140

25. The Enabling Supply House purchased 6 wheelchairs at $1256 each and 15 speech compression recorder-players at $895 each. Find the total cost.

Estimate: ($1000 × 6) + ($900 × 20) = $24,000

Exact: $20,961

26. A college bookstore buys 17 desktop computers at $506 each and 13 printers at $482 each. Find the total cost.

Estimate: ($500 × 20) + ($500 × 10) = $15,000

Exact: $14,868

27. Being able to identify indicator words is helpful in determining how to solve an application problem. Write three indicator words for each of these operations: add, subtract, multiply, and divide. Write two indicator words that mean equals.

Possible answers are
Addition: more; total; gain of
Subtraction: less; loss of; decreased by
Multiplication: twice; of; product
Division: divided by; goes into; per
Equals: is; are

28. Identify and explain the six steps used to solve an application problem. You may refer to the text if you need help, but use your own words.

1. Read the problem carefully.
2. Work out a plan.
3. Estimate a reasonable answer.
4. Solve the problem.
5. State the answer.
6. Check your work.

🖋 **29.** Write in your own words why it is important to estimate a reasonable answer. Give three examples of what might be a reasonable answer to a math problem from your daily activities.

Estimating the answer can help you avoid careless mistakes like decimal or calculation errors. Examples of reasonable answers in daily life might be a $35 bag of groceries, $50 to fill the gas tank, or $45 for a phone bill.

🖋 **30.** First estimate by rounding to thousands, then find the exact answer to the following problem.

$$7438 + 6493 + 2380$$

Do the two answers vary by more than 1000? Why? Will estimated answers always vary from exact answers?

Estimate: 7000 + 6000 + 2000 = 15,000

Exact: 16,311

Yes, the answers vary by more than 1000 as a result of the rounding. Yes, they usually do; however, in this example the estimated answer and the exact answer are close enough to give some assurance that the answer is reasonable.

Solve each application problem. See Examples 1–4.

31. Steve Edwards, manager, decided to total his sales at SUBWAY. The daily sales figures were $2358 on Monday, $3056 on Tuesday, $2515 on Wednesday, $1875 on Thurdsay, $3978 on Friday, $3219 on Saturday, and $3008 on Sunday. Find his total sales for the week. **$20,009**

32. The numbers of visitors at a war veteran's memorial during one week are 5318; 2865; 4786; 1998; 3899; 2343; and 7221. Find the total attendance for the week. **28,430 visitors**

33. A car weighs 2425 pounds. If its 582-pound engine is removed and replaced with a 634-pound engine, what will the car weigh? **2477 pounds**

34. Barbara has $2324 in her preschool operating account. She spends $734 from this account, and then the class parents raise $568 in a rummage sale. Find the balance in the account after she deposits the money from the rummage sale. **$2158**

35. In a recent survey of Reno/Lake Tahoe hotels, the cost per night at Harrah's in Reno was $45, while the cost at Harrah's in Lake Tahoe was $99 per night. Find the amount saved on a 7-night stay at Harrah's in Reno instead of staying at Harrah's in Lake Tahoe. (*Source: Harrah's Casinos and Hotels.*) **$378**

36. The most expensive hotel room in a recent study was the Ritz-Carlton at $645 per night, while the least expensive was Motel 6 at $74 per night. Find the amount saved in a 4-night stay at Motel 6 instead of staying at the Ritz-Carlton. (*Source: Ritz-Carlton/Motel 6.*) **$2284**

37. A youth soccer association raised $7588 through fund-raising projects. After expenses of $838 were paid, the balance of the money was divided evenly among the 18 teams. How much did each team receive?

$375

38. Feather Farms Egg Ranch collected 3545 eggs in the morning and 2575 eggs in the afternoon. If the eggs are packed in flats containing 30 eggs each, find the number of flats needed for packing.

204 flats

39. A theater owner wants to provide enough seating for 1250 people. The main floor has 30 rows of 25 seats in each row. If the balcony has 25 rows, how many seats must be in each balcony row to satisfy the owner's seating requirements? **20 seats**

40. Jennie makes 24 grapevine wreaths per week to sell to gift shops. She works 40 weeks a year and packages six wreaths per box. If she ships equal quantities to each of five shops, find the number of boxes each store will receive. **32 boxes**

Chapter 1 ▶▶▶ Summary

▶ Key Terms

1.1	**whole numbers**	The whole numbers are 0, 1, 2, 3, 4, 5, 6, 7, 8, and so on.
	place value	The place value of each digit in a whole number is determined by its position in the whole number.
	table	A table is a display of facts in rows and columns.
1.2	**addition**	The process of finding the total is addition.
	addends	The numbers being added in an addition problem are addends.
	sum (total)	The answer in an addition problem is called the sum.
	commutative property of addition	The commutative property of addition states that the order of numbers in an addition problem can be changed without changing the sum.
	associative property of addition	The associative property of addition states that grouping the addition of numbers differently does not change the sum.
	regrouping	The process of regrouping is used in an addition problem when the sum of the digits in a column is greater than 9.
	perimeter	The perimeter is the distance around the outside edges of a figure.
1.3	**minuend**	The number from which another number (the subtrahend) is being subtracted is the minuend.
	subtrahend	The subtrahend is the number being subtracted in a subtraction problem.
	difference	The answer in a subtraction problem is called the difference.
	regrouping	The process of regrouping is used in subtraction if a digit is less than the one directly below it.
1.4	**factors**	The numbers being multiplied are called factors. For example, in $3 \times 4 = 12$, both 3 and 4 are factors.
	product	The answer in a multiplication problem is called the product.
	commutative property of multiplication	The commutative property of multiplication states that changing the order of the factors in a multiplication problem does not change the product.
	associative property of multiplication	The associative property of multiplication states that grouping the numbers differently does not change the product.
	chain multiplication problem	A multiplication problem having more than two factors is a chain multiplication problem.
	multiple	The product of two whole number factors is a multiple of those numbers.
1.5	**dividend**	The number being divided by another number in a division problem is the dividend.
	divisor	The divisor is the number doing the dividing in a division problem.
	quotient	The answer in a division problem is called the quotient.
	short division	A method of dividing a number by a one-digit divisor is short division.
	remainder	The remainder is the number left over when two numbers do not divide exactly.
1.6	**long division**	The process of long division is used to divide by a number with more than one digit.
1.7	**rounding**	Rounding is used to find a number that is close to the original number, but easier to work with. Use the $\approx$ sign, which means "approximately equal to."
	estimate	An estimated answer is one that is close to the exact answer.
	front end rounding	Rounding to the highest possible place so that all the digits become zeros except the first one is front end rounding.
1.8	**square root**	The square root of a whole number is the number that can be multiplied by itself to produce the given number.
	perfect square	A number that is the square of a whole number is a perfect square.
	order of operations	For problems or expressions with more than one operation, the order of operations tells what to do first, second, and so on to get the correct answer.
1.9	**pictograph**	A graph that uses pictures or symbols to show data is a pictograph.
	bar graph	A graph that uses bars of various heights to show quantity is a bar graph.
	line graph	A graph that uses dots connected by lines to show trends is a line graph.
1.10	**indicator words**	Words in a problem that indicate the necessary operations—addition, subtraction, multiplication, or division—are indicator words.

▶ New Symbols

≈   This sign is used to show that an answer has been estimated. It means "is approximately equal to."

√   The symbol for square root.

5^2 The small raised 2 is an exponent; it tells how many times to use 5 (the base) as a factor in multiplication.

▶ Test Your Word Power

See how well you have learned the vocabulary in this chapter. Answers follow the Quick Review.

1. When using **addends** you are performing
   A. division
   B. subtraction
   C. addition
   D. multiplication.

2. The subtrahend is the
   A. number being multiplied
   B. number being subtracted
   C. number being added
   D. answer in division.

3. A **factor** is
   A. the answer in an addition problem
   B. one of two or more numbers being added
   C. one of two or more numbers being multiplied
   D. one of two or more numbers being divided.

4. The **divisor** is
   A. the number being rounded
   B. the number being multiplied
   C. always the largest number
   D. the number doing the dividing.

5. We use **rounding** to
   A. avoid solving a problem
   B. purely guess at the answer
   C. help estimate a reasonable answer
   D. find the remainder.

6. A **perfect square** is
   A. the square of a whole number
   B. the same as square root
   C. similar to a perfect triangle

▶ Quick Review

Concepts	Examples

[1.1] Reading and Writing Whole Numbers

Do not use the word *and* when writing a whole number. Commas help divide the periods or groups for ones, thousands, millions, and billions. A comma is not needed when a number has four digits or fewer.

795 is written *seven hundred ninety-five.*

9,768,002 is written *nine million, seven hundred sixty-eight thousand, two*

[1.2] Adding Whole Numbers

Add from top to bottom, starting with the ones column and working left. To check, add from bottom to top.

(Add up to check.)

$$
\begin{array}{r}
1\ 1\ 4\ 0 \\
6\ 8\ 7 \\
2\ 6 \\
9 \\
+\ 4\ 1\ 8 \\
\hline
1\ 1\ 4\ 0
\end{array}
$$

Addends

Sum

[1.2] Commutative Property of Addition

Changing the order of the addends in an addition problem does not change the sum.

$2 + 4 = 6$

$4 + 2 = 6$

By the commutative property, the sum is the same.

[1.2] Associative Property of Addition

Grouping the addends differently when adding does not change the sum.

$(2 + 3) + 4 = 5 + 4 = 9$

$2 + (3 + 4) = 2 + 7 = 9$

By the associative property, the sum is the same.

[1.3] Subtracting Whole Numbers

Subtract the subtrahend from the minuend to get the difference, using regrouping when necessary. To check, add the difference to the subtrahend to get the minuend.

Problem

$$
\begin{array}{r}
6\ 12\ 18 \\
4\ 7\ 3\ 8 \quad \leftarrow \text{Minuend} \\
-\ \ \ 6\ 4\ 9 \quad \text{Subtrahend} \\
\hline
4\ 0\ 8\ 9 \quad \text{Difference}
\end{array}
$$

Check

$$
\begin{array}{r}
4\ 0\ 8\ 9 \\
+\ \ \ 6\ 4\ 9 \\
\hline
4\ 7\ 3\ 8
\end{array}
$$

Concepts	Examples

1.4 Multiplying Whole Numbers

Use $\times$, $\bullet$ (a raised dot), or parentheses to indicate multiplication.

The numbers being multiplied are called *factors*. The multiplicand is being multiplied by the multiplier, giving the product. When the multiplier has more than one digit, partial products must be used and added to find the product.

3×4 or $3 \bullet 4$ or $(3)(4)$ or $3(4)$

$$
\begin{array}{r}
78 \quad \text{Multiplicand} \\
\times \quad 24 \quad \text{Multiplier} \\
\hline
312 \quad \text{Partial product} \\
156 \quad \text{Partial product (move one position left)} \\
\hline
1872 \quad \text{Product}
\end{array}
$$

Multiplicand, Multiplier $\}$ Factors

1.4 Commutative Property of Multiplication

The product in a multiplication problem remains the same when the order of the factors is changed.

$$3 \times 4 = 12$$
$$4 \times 3 = 12$$

By the commutative property, the product is the same.

1.4 Associative Property of Multiplication

Grouping the factors differently when multiplying does not change the product.

$$(2 \times 3) \times 4 = 6 \times 4 = 24$$
$$2 \times (3 \times 4) = 2 \times 12 = 24$$

By the associative property, the product is the same.

1.5 Dividing Whole Numbers

$\div$ and $\overline{)}$ mean divide.

Also a —, as in $\frac{25}{5}$, means to divide the top number (dividend) by the bottom number (divisor).

Divisor $\longrightarrow$ $4\overline{)88}$ $\leftarrow$ Dividend, $\quad$ $22 \leftarrow$ Quotient

$$
\begin{array}{r}
22 \quad \leftarrow \text{Quotient} \\
4\overline{)88} \quad \leftarrow \text{Dividend} \\
88 \\
\hline
0
\end{array}
$$

$88 \div 4 = 22$

Dividend, Quotient, Divisor

Dividend

$\dfrac{88}{4} = 22 \leftarrow$ Quotient

1.7 Rounding Whole Numbers

Rules for Rounding:

Step 1   Locate the place to be rounded, and draw a line under it.

Step 2   If the next digit to the right is 5 or more, increase the underlined digit by 1. If the next digit is 4 or less, do not change the underlined digit.

Step 3   Change all digits to the right of the underlined place to zeros.

Round 726 to the nearest ten.

Next digit is 5 or more.

726

Tens place increases by 1 ($2 + 1 = 3$).

726 rounds to 730.

Round 1,498,586 to the nearest million.

Next digit is 4 or less.

1,498,586

Millions place does not change.

1,498,586 rounds to 1,000,000.

1.7 Front End Rounding

Front end rounding is rounding to the highest possible place so that all the digits become 0 except the first digit.

Round each number using front end rounding.

76 rounds to 80.

348 rounds to 300.

6512 rounds to 7000.

23,751 rounds to 20,000.

652,179 rounds to 700,000.

Concepts	Examples

1.8 Order of Operations

Problems may have several operations. Work these problems using the order of operations.

1. Do all operations inside parentheses or other grouping symbols.
2. Simplify any expressions with exponents and find any square roots $\left(\sqrt{}\right)$.
3. Multiply or divide proceeding from left to right.
4. Add or subtract proceeding from left to right.

Simplify, using the order of operations.

$7 \cdot \sqrt{9} - 4 \cdot 5$ Find the square root.

$7 \cdot \underbrace{3}_{} - \underbrace{4 \cdot 5}_{}$ Multiply from left to right.

$\underbrace{21}_{} - \underbrace{20}_{} = 1$ Subtract.

1.9 Reading Pictographs, Bar Graphs, and Line Graphs

A *pictograph* uses pictures or symbols to show data.

A *bar graph* uses bars of various heights to show quantity.

A *line graph* uses dots connected by lines to show trends.

When reading a pictograph, be certain that you determine the quantity represented by each picture or symbol.

When reading a bar graph, use a straightedge to line up the top of the bar with the numbers along the left edge of the graph.

When reading a line graph, use a straightedge to line up the dot with the numbers along the left edge of the graph.

1.10 Application Problems

Steps for Solving an Application Problem

Step 1 **Read** the problem carefully, perhaps several times.

Step 2 **Work out a plan** before starting. Draw a sketch if possible.

Step 3 **Estimate** a reasonable answer.

Step 4 **Solve** the problem.

Step 5 **State the answer.**

Step 6 **Check** your work. If the answer is not reasonable, start over.

Manuel earns $118 on Sunday, $87 on Monday, and $63 on Tuesday. Find his total earnings for the 3 days.

Step 1 The earnings for each day are given, and the total for the 3 days must be found.

Step 2 Add the daily earnings to find the total.

Step 3 Since the earnings were about $100 + $90 + $60 = $250, a reasonable estimate would be approximately $250.

Step 4
$$\begin{array}{r} \mathbf{\$268} \\ \$118 \\ 87 \\ +63 \\ \hline \$268 \end{array}$$
Check by adding up

Total earnings

Step 5 Manuel's total earnings are $268.

Step 6 The exact answer is reasonable, because it is close to the estimate of $250.

ANSWERS TO TEST YOUR WORD POWER

1. C; *Example:* In $2 + 3 = 5$, the 2 and the 3 are addends.

2. B; *Example:* In $5 - 4 = 1$, the 4 is the subtrahend.

3. C; *Example:* In $3 \times 5 = 15$, the numbers 3 and 5 are factors.

4. D; *Example:* In $8 \div 4 = 2$, $\dfrac{8}{4} = 2$, and $4\overline{)8}$, the 4 is the divisor.

5. C; *Example:* We can use rounding to estimate our answer and then determine whether the exact answer is reasonable.

6. A; *Example:* 25 is a perfect square because $5^2 = 25$ and 5 is a whole number.

Chapter 1 ▶▶▶ Review Exercises

If you need help with any of these Review Exercises, look in the section indicated in brackets.

[1.1] *Write the digits for the given period or group in each number.*

1. 6573

thousands **6**

ones **573**

2. 36,215

thousands **36**

ones **215**

3. 105,724

thousands **105**

ones **724**

4. 1,768,710,618

billions **1**

millions **768**

thousands **710**

ones **618**

Rewrite each number in words.

5. 728

seven hundred twenty-eight

6. 15,310

fifteen thousand, three hundred ten

7. 319,215

three hundred nineteen thousand, two hundred fifteen

8. 62,500,005

sixty-two million, five hundred thousand, five

Rewrite each number in digits.

9. ten thousand, eight

10,008

10. two hundred million, four hundred fifty-five

200,000,455

[1.2] *Add.*

11.   72
      + 38
    ‾‾‾‾‾
      110

12.   54
      + 67
    ‾‾‾‾‾
      121

13.   807
        4606
      + 51
    ‾‾‾‾‾‾‾
      5464

14.   8215
            9
      + 7433
    ‾‾‾‾‾‾‾‾
      15,657

15.   2130
          453
         8107
      + 296
    ‾‾‾‾‾‾‾‾
      10,986

16.   5684
          218
         2960
      + 983
    ‾‾‾‾‾‾‾
      9845

17.   5 732
         11,069
             37
          1 595
      + 22,169
    ‾‾‾‾‾‾‾‾‾
      40,602

18.   3 451
         12,286
             43
          1 291
      + 32,784
    ‾‾‾‾‾‾‾‾‾
      49,855

[1.3] *Subtract.*

19. 64
     -28
     36

20. 46
     -19
     27

21. 375
     -186
     189

22. 573
     -389
     184

23. 7416
     -567
     6849

24. 5210
     -883
     4327

25. 2210
     -1986
     224

26. 99,704
     $-73,838$
     25,866

[1.4] *Multiply.*

27.   7
     $\times 7$
     49

28.   8
     $\times 0$
     0

29. 8(4)
     32

30. 8(8)
     64

31. (5)(9)
     45

32. (6)(7)
     42

33. 7 • 8
     56

34. 9 • 9
     81

Work each chain multiplication.

35. $5 \times 4 \times 2$
     40

36. $9 \times 1 \times 5$
     45

37. $4 \times 4 \times 3$
     48

38. $2 \times 2 \times 2$
     8

39. (6)(0)(8)
     0

40. (7)(1)(6)
     42

41. 6 • 1 • 8
     48

42. 7 • 7 • 0
     0

Multiply.

43.   28
     $\times 3$
     84

44.   46
     $\times 8$
     368

45.   58
     $\times 9$
     522

46.   98
     $\times 1$
     98

47.   625
     $\times 8$
     5000

48.   374
     $\times 8$
     2992

49.   1349
     $\times 4$
     5396

50.   9163
     $\times 5$
     45,815

51.   7456
     $\times 2$
     14,912

52.   2880
     $\times 7$
     20,160

53.   93,105
     $\times 5$
     465,525

54.   21,873
     $\times 8$
     174,984

55.   35
      × 25
     ———
      875

56.   74
      × 32
     ———
      2368

57.   98
      × 12
     ———
      1176

58.   68
      × 75
     ———
      5100

59.   472
      × 33
     ———
      15,576

60.   392
      × 77
     ———
      30,184

61.   4051
      × 219
     ———
      887,169

62.   1527
      × 328
     ———
      500,856

Find each total cost.

63. 30 scientific calculators at $12 per calculator
$360

64. 76 subscribers at $14 per subscription
$1064

65. 318 drill bit sets at $64 per set
$20,352

66. 114 earplugs at $6 per earplug
$684

Multiply by using the shortcut for multiples of 10.

67.   280
      × 50
     ———
      14,000

68.   340
      × 70
     ———
      23,800

69.   517
      × 400
     ———
      206,800

70.   637
      × 500
     ———
      318,500

71.   16,000
      ×   8 000
     ———
      128,000,000

72.   43,000
      ×   2 100
     ———
      90,300,000

[1.5] *Divide. If the division is not possible, write "undefined."*

73. $20 \div 4$ 5

74. $35 \div 5$ 7

75. $42 \div 7$ 6

76. $18 \div 9$ 2

77. $\dfrac{54}{9}$ 6

78. $\dfrac{36}{9}$ 4

79. $\dfrac{49}{7}$ 7

80. $\dfrac{0}{6}$ 0

81. $\dfrac{148}{0}$ undefined

82. $\dfrac{0}{23}$ 0

83. $\dfrac{64}{8}$ 8

84. $\dfrac{81}{9}$ 9

[1.5–1.6] *Divide.*

85. $4\overline{)328}$ 82

86. $3\overline{)294}$ 98

87. $6\overline{)26,532}$ 4422

88. $76\overline{)26,752}$ 352

89. $2704 \div 18$ 150 R4

90. $15,525 \div 125$ 124 R25

[1.7] *Round as indicated.*

91. 817 to the nearest ten
820

92. 15,208 to the nearest hundred
15,200

93. 20,643 to the nearest thousand
21,000

94. 67,485 to the nearest ten-thousand
70,000

Round each number to the nearest ten, nearest hundred, and nearest thousand.
Remember to round from the original number.

	Ten	**Hundred**	**Thousand**
95. 3487	3490	3500	3000
96. 20,065	20,070	20,100	20,000
97. 98,201	98,200	98,200	98,000
98. 352,118	352,120	352,100	352,000

[1.8] *Find each square root by using the Perfect Squares Table on page 77.*

99. $\sqrt{16}$ 4

100. $\sqrt{49}$ 7

101. $\sqrt{144}$ 12

102. $\sqrt{196}$ 14

Identify the exponent and the base, and then simplify each expression.

103. 7^3 3; 7; 343

104. 3^6 6; 3; 729

105. 5^3 3; 5; 125

106. 4^5 5; 4; 1024

Simplify each expression by using the order of operations.

107. $7^2 - 15$ 34

108. $6^2 - 10$ 26

109. $2 \cdot 3^2 \div 2$ 9

110. $9 \div 1 \cdot 2 \cdot 2 \div (11 - 2)$ 4

111. $\sqrt{9} + 2(3)$ 9

112. $6 \cdot \sqrt{16} - 6 \cdot \sqrt{9}$ 6

[1.9] *The bar graph shows the number of parents out of 100 surveyed who nag their children about performing certain household chores.*

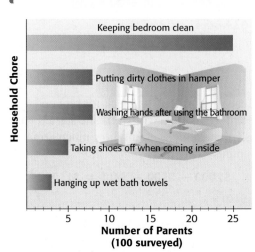

CLEAN UP YOUR ROOM

Keeping bedroom clean

Putting dirty clothes in hamper

Washing hands after using the bathroom

Taking shoes off when coming inside

Hanging up wet bath towels

Household Chore

5   10   15   20   25

**Number of Parents
(100 surveyed)**

Source: Opinion Research Corporation for the Soap and Detergent Association.

113. How many parents nagged their children about washing hands after using the bathroom?

8 parents

114. Find the number of parents who nagged their children about taking shoes off when coming inside.

5 parents

115. Which household chore was nagged about by the greatest number of parents? How many were there?

keeping bedroom clean; 25 parents

116. Which household chore was nagged about by the least number of parents? How many did this?

hanging up wet bath towels; 3 parents

[1.10] *Solve each application problem. First use front end rounding to estimate the answer. Then find the exact answer.*

117. Bank of America processes 40 million checks each day. Find the number of checks processed by the bank in a year. Use a 365-day year. (*Source:* Bank of America.)

Estimate: 40 million × 400 = 16,000 million
or 16,000,000,000 checks

Exact: 14,600 million or 14,600,000,000 checks

118. A pulley on an evaporative cooler turns 1400 revolutions per minute. How many revolutions will the pulley turn in 60 minutes?

Estimate: 1000 × 60 = 60,000 revolutions

Exact: 84,000 revolutions

119. There are 144 plastic forks in a box. Find the number of plastic forks in 15 boxes.

Estimate: 100 × 20 = 2000 forks

Exact: 2160 forks

120. A drum contains 6000 brackets. How many brackets are in 30 drums?

Estimate: 6000 × 30 = 180,000 brackets

Exact: 180,000 brackets

121. It takes 2000 hours of work to build one home. How many hours of work are needed to build 12 homes?

Estimate: 2000 × 10 = 20,000 hours

Exact: 24,000 hours

122. A Japanese bullet train travels 80 miles in 1 hour. Find the number of miles traveled in 5 hours.

Estimate: 80 × 5 = 400 miles

Exact: 400 miles

123. Find the total cost if SUBWAY buys 32 baking ovens at $1538 each and 28 warming ovens at $887 each.

Estimate: (30 × $2000) + (30 × $900) = $87,000

Exact: $74,052

124. A newspaper carrier has 62 customers who take the paper daily and 21 customers who take the paper on weekends only. A daily customer pays $16 per month and a weekend-only customer pays $7 per month. Find the total monthly collections.

Estimate: (60 × $20) + (20 × $7) = $1340

Exact: $1139

125. This holiday season, the average amount consumers plan to spend on holiday shopping for others is $620. If they plan to spend $107 on holiday shopping for themselves, how much more do they plan to spend on others than themselves? (*Source:* National Retail Federation.)

Estimate: **$600 − $100 = $500**

Exact: **$513**

126. A stamping machine produces 986 license plates each hour. How long will it take to produce 32,538 license plates?

Estimate: **30,000 ÷ 1000 = 30 hours**

Exact: **33 hours**

127. A food canner uses 1 pound of pork for every 175 cans of pork and beans. How many pounds of pork are needed for 8750 cans?

Estimate: **9000 ÷ 200 = 45 pounds**

Exact: **50 pounds**

128. Rachel Leach writes a $520 check for rent and a $385 check for her car payment. If she started with $1924 in her checking account, how much remains in her account.

Estimate: **$2000 − $500 − $400 = $1100**

Exact: **$1019**

129. Nitrogen sulfate is used in farming to enrich nitrogen-poor soil. If 625 pounds of nitrogen sulfate are spread per acre, how many acres can be spread with 32,500 pounds of nitrogen sulfate?

Estimate: **30,000 ÷ 600 = 50 acres**

Exact: **52 acres**

130. Each home in a subdivision requires 180 feet of fencing. Find the number of homes that can be fenced with 5760 feet of fencing material.

Estimate: **6000 ÷ 200 = 30 homes**

Exact: **32 homes**

▶▶▶ Mixed Review Exercises*

Perform the indicated operations.

131. 4(83)

332

132. 7(64)

448

133.
$$\begin{array}{r} 309 \\ -\ 56 \\ \hline 253 \end{array}$$

134.
$$\begin{array}{r} 835 \\ -\ 247 \\ \hline 588 \end{array}$$

135.
$$\begin{array}{r} 662 \\ +\ 379 \\ \hline 1041 \end{array}$$

136.
$$\begin{array}{r} 789 \\ +\ 872 \\ \hline 1661 \end{array}$$

137.
$$\begin{array}{r} 38,140 \\ -\ 6\ 078 \\ \hline 32,062 \end{array}$$

138.
$$\begin{array}{r} 29,156 \\ -\ 4\ 209 \\ \hline 24,947 \end{array}$$

139. 21 ÷ 7

3

140. $\frac{42}{6}$

7

141.
$$\begin{array}{r} 7\ 218 \\ 3 \\ 18 \\ 1\ 791 \\ 82,623 \\ +\ 1\ 982 \\ \hline 93,635 \end{array}$$

142.
$$\begin{array}{r} 3\ 812 \\ 5 \\ 22 \\ 1\ 836 \\ 75,134 \\ +\ 2\ 369 \\ \hline 83,178 \end{array}$$

143. $\frac{9}{0}$

undefined

144. $\frac{7}{1}$

7

145. 27,600 ÷ 4

6900

* The order of exercises in this final group does not correspond to the order in which topics occur in the chapter. This random ordering should help you prepare for the chapter test in yet another way.

146. 18,480 ÷ 8
2310

147. 8430
× 128
‾‾‾‾‾‾‾‾
1,079,040

148. 21,702
× 6
‾‾‾‾‾‾‾‾
130,212

🔲 **149.** 34)3672 ‾ → 108

🔲 **150.** 68)14,076 ‾ → 207

151. Rewrite 376,853 in words.

three hundred seventy-six thousand, eight hundred fifty-three

152. Rewrite 408,610 in words.

four hundred eight thousand, six hundred ten

153. Round 8749 to the nearest hundred.

8700

154. Round 400,503 to the nearest thousand.

401,000

Find each square root.

155. $\sqrt{64}$

8

156. $\sqrt{81}$

9

Find each total cost.

157. 308 pairs of knee guards at $18 per pair

$5544

158. 84 dishwashers at $370 per dishwasher

$31,080

159. 208 baseball hats at $11 per hat

$2288

160. 607 boxes of avocados at $26 per box

$15,782

Solve each application problem.

161. There are 52 playing cards in a deck. How many cards are there in nine decks?

468 cards

162. Your college bookstore receives textbooks packed 20 books per carton. How many textbooks are received in a delivery of 180 cartons?

3600 textbooks

163. Push-type gasoline-powered lawn mowers cost $100 less than self-propelled mowers that you walk behind. If a self-propelled mower costs $380, find the cost of a push-type mower.

$280

164. The Country Day School wants to raise $218,450 to construct and equip a computer lab. If $103,815 has already been raised, how much more is needed?

$114,635

American River Raft Rentals lists the following daily raft rental fees. Notice that there is an additional $2 launch fee payable to the park system for each raft rented. Use this information to solve Exercises 165 and 166.

AMERICAN RIVER RAFT RENTALS

Size	Rental Fee	Launch Fee
4-person	$28	$2
6-person	$38	$2
10-person	$70	$2
12-person	$75	$2
16-person	$85	$2

Source: American River Raft Rentals.

165. On a recent Tuesday the following rafts were rented: 6 4-person; 15 6-person; 10 10-person; 3 12-person; and 2 16-person. Find the total receipts, including the $2 per-raft launch fee.

$1905

166. On the 4th of July the following rafts were rented: 38 4-person; 73 6-person; 58 10-person; 34 12-person; and 18 16-person. Find the total receipts, including the $2 per-raft launch fee.

$12,420

Chicago plans to build the world's tallest structure, the Chicago Spire. The pictogram below shows the height of the tallest buildings in the United States. Use this information to solve Exercises 167–170.

167. How much taller is the Chicago Spire (planned) than the Trump Tower in Chicago?

869 ft

168. How much taller is the Freedom Tower in New York than the San Francisco Towers (planned)?

162 ft

169. (a) What is the combined height of all the buildings shown? Include the planned buildings.

(b) Is the combined height of these six buildings greater or less than a mile? How much greater or less than a mile? (*Hint:* 1 mile = 5280 ft)

(a) 8393 ft

(b) greater than a mile by 3113 ft

170. The height of the San Francisco Towers (planned) is equivalent to the length of how many football fields? (*Hint:* A football field is 100 yards long and 1 yard = 3 feet.)

4 football fields

Building Higher and Higher

Developers in San Francisco have submitted a proposal that includes two 1,200-foot towers, which would be the tallest buildings west of Chicago. How some planned buildings compare with the nation's tallest (in feet):

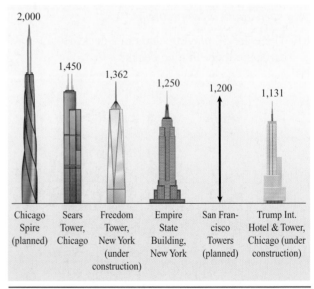

Source: USA Today research.

Chapter 1 ▶▶▶ Test

Test Prep VIDEO CD Use the Chapter Test Prep Video CD to see fully worked-out solutions to any of the exercises you want to review.

Write each number in words.

1. 9205

2. 25,065

3. Use digits to write four hundred twenty-six thousand, five.

Add.

4.
$$\begin{array}{r} 853 \\ 66 \\ 4022 \\ + 3589 \end{array}$$

5.
$$\begin{array}{r} 17,063 \\ 7 \\ 12 \\ 1\,505 \\ 93,710 \\ + \quad 333 \end{array}$$

Subtract.

6.
$$\begin{array}{r} 9009 \\ - 7964 \end{array}$$

7.
$$\begin{array}{r} 9075 \\ - 2869 \end{array}$$

Multiply

8. $7 \times 6 \times 4$

9. $57 \cdot 3000$

10. $85(19)$

11.
$$\begin{array}{r} 7381 \\ \times 603 \end{array}$$

Divide. If the division is not possible, write "undefined."

12. $16\overline{)112,752}$

13. $\dfrac{835}{0}$

14. $19,241 \div 42$

15. $280\overline{)44,800}$

Round as indicated.

16. 6347 to the nearest ten

17. 76,502 to the nearest thousand

1. nine thousand, two hundred five

2. twenty-five thousand, sixty-five

3. 426,005

4. 8530

5. 112,630

6. 1045

7. 6206

8. 168

9. 171,000

10. 1615

11. 4,450,743

12. 7047

13. undefined

14. 458 R5

15. 160

16. 6350

17. 77,000

Simplify each expression.

18. $5^2 + 8\,(2)$

18. <u>41</u>

19. $7 \cdot \sqrt{64} - 14 \cdot 2$

19. <u>28</u>

Solve each application problem. First use front end rounding to estimate the answer. Then find the exact answer.

20. Amy collects the following monthly rents from the tenants in her fourplex: $485, $500, $515, and $425. After she pays expenses of $785, how much does she have left?

20. *Estimate:* $500 + $500 + $500 + 400 − $800 = $1100

   Exact: $1140

21. The major producer of ethanol made from corn is the United States. If 374 gallons of ethanol can be produced from the corn grown on one acre of land, how many acres are needed to produce 86,394 gallons. (*Source:* Earth Policy Institute.)

21. *Estimate:* 90,000 ÷ 400 = 225 acres

   Exact: 231 acres

22. Barb Weaks paid $528 for tires, $195 for brakes, and $235 for a timing belt. If this money was withdrawn from her checking account, which had a balance of $1906, find her new balance.

22. *Estimate:* $2000 − $500 − $200 − $200 = $1100

   Exact: $948

23. The technicians at a chicken ranch identify the sex of 48 baby chicks each minute for 4 hours in the morning and 36 baby chicks each minute for 3 hours in the afternoon. Find the total number of baby chicks identified. (*Source:* Discovery Channel, *Dirty Jobs.*)

23. *Estimate:* (50 × 60 × 4) + (40 × 60 × 3) = 19,200 chicks

   Exact: 18,000 chicks

24. See sample answer at the right.

✎ **24.** Explain in your own words the rules for rounding numbers. Give an example of rounding a number to the nearest ten-thousand.

   1. Locate the place to which you are rounding and underline it.
   2. Look only at the next digit to the right. If this digit is a 4 or less, do not change the underlined digit. If the digit is a 5 or more, increase the underlined digit by 1.
   3. Change all digits to the right of the underlined place to zeros.
   Each person's rounding example will vary.

25. _____

✎ **25.** List the six steps for solving application problems.

   1. Read the problem carefully.
   2. Work out a plan.
   3. Estimate a reasonable answer.
   4. Solve the problem.
   5. State the answer.
   6. Check your work.

Multiplying and Dividing Fractions

M ost recipes include ingredients that use common fractions and mixed numbers in their measurements. The recipe shown below uses Jelly Belly jelly beans and will make $2\frac{1}{2}$ dozen (30) cookies. But suppose you wanted to make 3 dozen, 10 dozen, or even $3\frac{1}{2}$ dozen cookies? You would have to multiply or divide each of the ingredients to arrive at the proper amounts needed. In this chapter we discuss multiplication and division of fractions, which you need to know when cooking or baking. (See Exercise 39 in **Section 2.7** and Exercises 25–28 in **Section 2.8**.)

NEST COOKIES

$1\frac{1}{2}$ cups all purpose flour	$\frac{1}{2}$ cup sugar	2 cups shredded coconut
1 tsp. baking powder	1 egg white	5 oz. Jelly Belly
$\frac{1}{2}$ tsp. salt	1 tsp. vanilla	jelly beans
$\frac{1}{2}$ cup shortening	2 Tbs. milk	assorted flavors

H eat oven to 375° F. Sift together flour, baking powder, and salt and set aside. In large bowl beat shortening, sugar, egg white, and vanilla until well blended. Add flour mixture and milk until blended. Stir in coconut.

Roll dough into a ball, divide in half. Roll 15 one-inch balls from each half and place on ungreased baking sheet. Make thumb print depression in center of each ball to form nest. Bake 6 minutes. Remove from oven and place 4 Jelly Belly beans in center of each cookie. Return to oven and bake 5 more minutes. Transfer cookies to wire rack to cool. Makes 30 cookies.

2.1 ▶▶▶ Basics of Fractions

OBJECTIVES

1 Use a fraction to show which part of a whole is shaded.

2 Identify the numerator and denominator.

3 Identify proper and improper fractions.

In **Chapter 1** we discussed whole numbers. Many times, however, we find that parts of whole numbers are considered. One way to write parts of a whole is with **fractions.** Another way is with decimals, which is discussed in **Chapter 4.**

OBJECTIVE 1 Use a fraction to show which part of a whole is shaded. The number $\frac{1}{8}$ is a fraction that represents 1 of 8 equal parts. Read $\frac{1}{8}$ as "one eighth."

EXAMPLE 1 Identifying Fractions

> A fraction represents part of a whole.

Use fractions to represent the shaded portions and the unshaded portions of each figure.

(a) The figure on the left has 6 equal parts. The 1 shaded part is represented by the fraction $\frac{1}{6}$. The *un*shaded part is $\frac{5}{6}$.

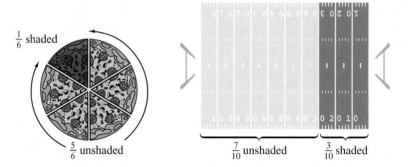

$\frac{1}{6}$ shaded

$\frac{5}{6}$ unshaded

$\frac{7}{10}$ unshaded $\frac{3}{10}$ shaded

(b) The 3 shaded parts of the 10-part figure on the right are represented by the fraction $\frac{3}{10}$. The *un*shaded part is $\frac{7}{10}$.

◀ *Work Problem* **1** *at the Side.*

 1 Write fractions for the shaded portions and the unshaded portions of each figure.

(a)

(b)

(c)

Fractions can be used to represent more than one whole object.

EXAMPLE 2 Representing Fractions Greater Than 1

Use a fraction to represent the shaded part of each figure.

(a) $\frac{1}{4}$

(b)

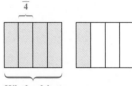

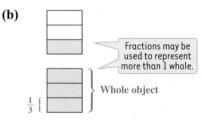

> Fractions may be used to represent more than 1 whole.

Whole object

Whole object

$\frac{1}{3}$

An area equal to 5 of the $\frac{1}{4}$ parts is shaded. Write this as $\frac{5}{4}$.

An area equal to 4 of the $\frac{1}{3}$ parts is shaded, so $\frac{4}{3}$ is shaded.

◀ *Work Problem* **2** *at the Side.*

 2 Write fractions for the shaded portions of each figure.

(a) $\frac{1}{7}$
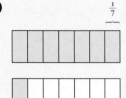

(b)
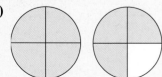

OBJECTIVE 2 Identify the numerator and denominator. In the fraction $\frac{2}{3}$, the number 2 is the **numerator** and 3 is the **denominator.** The bar between the numerator and the denominator is the *fraction bar*.

Fraction bar ⟶ $\dfrac{2 \; \leftarrow \text{Numerator}}{3 \; \leftarrow \text{Denominator}}$

ANSWERS

1. (a) $\frac{3}{4}, \frac{1}{4}$ (b) $\frac{1}{6}, \frac{5}{6}$ (c) $\frac{7}{8}, \frac{1}{8}$

2. (a) $\frac{8}{7}$ (b) $\frac{7}{4}$

Numerators and Denominators

The **denominator** of a fraction shows the number of equivalent parts in the whole, and the **numerator** shows how many parts are being considered.

Note

Remember that a bar, —, is one of the division symbols, and that division by 0 is undefined. A fraction with a denominator of 0 is also undefined.

EXAMPLE 3 **Identifying Numerators and Denominators**

Identify the numerator and denominator in each fraction.

(a) $\dfrac{3}{4}$ 　　　　　　　　　　**(b)** $\dfrac{8}{5}$

$\dfrac{3}{4}$ ← Numerator　　　　　$\dfrac{8}{5}$ ← Numerator
　　← Denominator　　　　　　　← Denominator

Work Problem **3** *at the Side.* ▶

OBJECTIVE **3** **Identify proper and improper fractions.** Fractions are sometimes called *proper* or *improper* fractions.

Proper and Improper Fractions

If the numerator of a fraction is *less* than the denominator, the fraction is a **proper fraction.** A proper fraction is less than 1 whole.

If the numerator is *greater than or equal to* the denominator, the fraction is an **improper fraction.** An improper fraction is greater than or equal to 1 whole.

Proper Fractions	**Improper Fractions**
$\dfrac{5}{8}\quad\dfrac{3}{5}\quad\dfrac{23}{24}$	$\dfrac{6}{5}\quad\dfrac{10}{10}\quad\dfrac{115}{112}$

EXAMPLE 4 **Classifying Types of Fractions**

(a) Identify all proper fractions in this list.

$$\dfrac{3}{4}\quad\dfrac{5}{9}\quad\dfrac{17}{5}\quad\dfrac{9}{7}\quad\dfrac{3}{3}\quad\dfrac{12}{25}\quad\dfrac{1}{9}\quad\dfrac{5}{3}$$

Proper fractions have a numerator that is smaller than the denominator. The proper fractions are shown below.

$\dfrac{3}{4}$ ← 3 is smaller than 4.　　$\dfrac{5}{9}\quad\dfrac{12}{25}\quad\dfrac{1}{9}$ ⎯ A proper fraction is less than 1.

(b) Identify all improper fractions in the list in part (a).

Improper fractions have a numerator that is equal to or greater than the denominator. The improper fractions are shown below.

$\dfrac{17}{5}$ ← 17 is greater than 5.　$\dfrac{9}{7}\quad\dfrac{3}{3}\quad\dfrac{5}{3}$ ⎯ An improper fraction is equal to or greater than 1.

Work Problem **4** *at the Side.* ▶

---

3 Identify the numerator and the denominator. Draw a picture with shaded parts to show each fraction. Your drawings may vary, but they should have the correct number of shaded parts.

(a) $\dfrac{2}{3}$

(b) $\dfrac{1}{4}$

(c) $\dfrac{8}{5}$

(d) $\dfrac{5}{2}$

4 From the following group of fractions:

$$\dfrac{2}{3}\quad\dfrac{4}{3}\quad\dfrac{3}{4}\quad\dfrac{8}{8}\quad\dfrac{3}{1}\quad\dfrac{1}{3}$$

(a) list all proper fractions;

(b) list all improper fractions.

ANSWERS

3. (a) N: 2; D: 3

(b) N: 1; D: 4

(c) N: 8; D: 5

(d) N: 5; D: 2

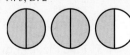

4. (a) $\dfrac{2}{3},\dfrac{3}{4},\dfrac{1}{3}$ **(b)** $\dfrac{4}{3},\dfrac{8}{8},\dfrac{3}{1}$

Math in the Media

QUILT PATTERNS

People who make quilts often base their designs on a block that is cut into a grid of 4, 9, 16, 25, or 49 squares. The quilter chooses various colors for the pieces. Each quilt design shown was selected from Antique Quilt Designs at their Web site http://earlywomanmasters.net/quilts/index.html.

1. Identify the makeup of the block as 4, 9, 16, etc. Each color is what fractional part of the block?

Aunt Eliza's Star

Block size: __9__
Tan: $\frac{2}{3}$ __ Purple: $\frac{2}{9}$ __
Green: $\frac{1}{9}$ __

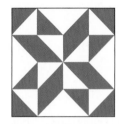

Barbara Fritchie Star

Block size: __16__
Blue: $\frac{1}{2}$ ____ White: $\frac{1}{2}$ _____

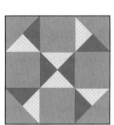

Cobwebs

Block size: __9__
Mauve: $\frac{2}{3}$ __ Purple: $\frac{1}{6}$ __
Yellow: $\frac{1}{6}$ ____

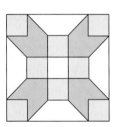

Farmer's Daughter

Block size: __25__
Lavender: $\frac{8}{25}$ __ White: $\frac{9}{25}$ __
Yellow: $\frac{8}{25}$ __

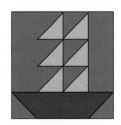

Tall Ships

Block size: __16__
Blue: $\frac{3}{16}$ ____ Green: $\frac{5}{8}$ _____
Light gray: $\frac{3}{16}$ __

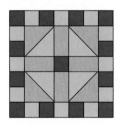

Lincoln's Platform

Block size: __49__
Blue: $\frac{13}{49}$ ____ Aqua: $\frac{20}{49}$ __
Gold: $\frac{16}{49}$ __

2. Use the blocks to design and color your own quilt patterns. Tell the fractional part of the block that is represented by each color.

3. Find the next two numbers in this pattern: 4, 9, 16, 25, __36__, __49__.

4. Explain how the pattern works: **sequence of perfect squares: 2^2, 3^2, 4^2, 5^2 6^2, 7^2**

116

Write fractions to represent the shaded and unshaded portions of each figure. See Examples 1 and 2.

1.

$$\frac{3}{4}; \frac{1}{4}$$

2.

$$\frac{5}{8}; \frac{3}{8}$$

3.

$$\frac{1}{3}; \frac{2}{3}$$

4.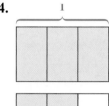

$$\frac{5}{3}; \frac{1}{3}$$

5.

$$\frac{7}{5}; \frac{3}{5}$$

6.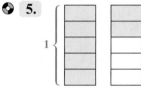

$$\frac{11}{6}; \frac{1}{6}$$

7. What fraction of these 6 bills has a life span of 2 years or greater? What fraction has a life of 4 years or less? What fraction has a life of 9 years?

$$\frac{5}{6}; \frac{4}{6}; \frac{2}{6}$$

A BILL'S LIFE

A $1 bill lasts about 18 months as compared with the average lifespan of other denominations:

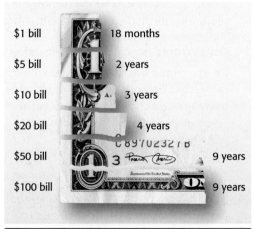

$1 bill	18 months
$5 bill	2 years
$10 bill	3 years
$20 bill	4 years
$50 bill	9 years
$100 bill	9 years

Source: Federal Reserve System; Bureau of Engraving and Printing.

8. What fraction of the 5 pets shown are dogs? What fraction are cats? What fraction of the pets are tan?

$$\frac{2}{5}; \frac{3}{5}; \frac{1}{5}$$

9. In an American Sign Language (ASL) class of 25 students, 8 are hearing impaired. What fraction of the students are hearing impaired?

$\dfrac{8}{25}$

10. Of 98 bicycles in a bike rack, 67 are mountain bikes. What fraction of the bicycles are not mountain bikes?

$\dfrac{31}{98}$

11. There are 520 rooms in a hotel. If 217 of the rooms are reserved for nonsmokers, what fraction of the rooms are for smokers?

$\dfrac{303}{520}$

12. Of the 46 employees at the college bookstore, 15 are full-time while the rest are part-time student help. What fraction of the bookstore employees work part-time?

$\dfrac{31}{46}$

Identify the numerator and denominator. See Example 3.

	Numerator	Denominator		Numerator	Denominator
13. $\dfrac{4}{5}$	4	5	**14.** $\dfrac{5}{6}$	5	6
15. $\dfrac{9}{8}$	9	8	**16.** $\dfrac{7}{5}$	7	5

List the proper and improper fractions in each group. See Example 4.

		Proper	Improper
17.	$\dfrac{8}{5}\ \dfrac{1}{3}\ \dfrac{5}{8}\ \dfrac{6}{6}\ \dfrac{12}{2}\ \dfrac{7}{16}$	$\dfrac{1}{3},\dfrac{5}{8},\dfrac{7}{16}$	$\dfrac{8}{5},\dfrac{6}{6},\dfrac{12}{2}$
18.	$\dfrac{1}{3}\ \dfrac{3}{8}\ \dfrac{16}{12}\ \dfrac{10}{8}\ \dfrac{6}{6}\ \dfrac{3}{4}$	$\dfrac{1}{3},\dfrac{3}{8},\dfrac{3}{4}$	$\dfrac{16}{12},\dfrac{10}{8},\dfrac{6}{6}$
19.	$\dfrac{3}{4}\ \dfrac{3}{2}\ \dfrac{5}{5}\ \dfrac{9}{11}\ \dfrac{7}{15}\ \dfrac{19}{18}$	$\dfrac{3}{4},\dfrac{9}{11},\dfrac{7}{15}$	$\dfrac{3}{2},\dfrac{5}{5},\dfrac{19}{18}$
20.	$\dfrac{12}{12}\ \dfrac{15}{11}\ \dfrac{13}{12}\ \dfrac{11}{8}\ \dfrac{17}{17}\ \dfrac{19}{12}$	none	$\dfrac{12}{12},\dfrac{15}{11},\dfrac{13}{12},\dfrac{11}{8},\dfrac{17}{17},\dfrac{19}{12}$

✏ 21. Write a fraction of your own choice. Label the parts of the fraction and write a sentence describing what each part represents. Draw a picture with shaded parts showing your fraction.

One possibility is

$\dfrac{3}{4}$ ← Numerator
← Denominator

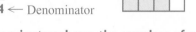

The denominator shows the number of equal parts in the whole and the numerator shows how many of the parts are being considered.

✏ 22. Give one example of a proper fraction and one example of an improper fraction. What determines whether a fraction is proper or improper? Draw pictures with shaded parts showing these fractions.

An example is $\dfrac{1}{2}$ as a proper fraction and $\dfrac{3}{2}$ as an improper fraction. A proper fraction has a numerator smaller than the denominator. An improper fraction has a numerator that is equal to or greater than the denominator. Drawings will vary.

Fill in the blanks to complete each sentence.

23. The fraction $\frac{3}{8}$ represents __3__ of the __8__ equal parts into which a whole is divided.

24. The fraction $\frac{7}{16}$ represents __7__ of the __16__ equal parts into which a whole is divided.

25. The fraction $\frac{5}{24}$ represents __5__ of the __24__ equal parts into which a whole is divided.

26. The fraction $\frac{24}{32}$ represents __24__ of the __32__ equal parts into which a whole is divided.

Study Skills

▶▶▶ HOMEWORK: HOW, WHY, AND WHEN

It is best for your brain if you keep up with the reading and homework in your math class. Remember that the more times you work with the information, the more dendrites you grow! So, give yourself every opportunity to read, work problems, and review your mathematics.

You have two choices for reading your math textbook. Read the short descriptions below and decide which will be best for you.

Maddy learns best by listening to her teacher explain things. She "gets it" when she sees the instructor work problems on the board. She likes to ask questions in class and put the information in her notes. She has learned that it helps if she has *previewed* the section before the lecture, so she knows generally what to expect in class. *But after the class instruction*, when Maddy gets home, she finds that she can understand the math textbook easily. She remembers what her teacher said, and she can double-check her notes if she gets confused. So, Maddy does her **careful** reading of the section in her text **after** hearing the classroom lecture on the topic.

De'Lore, on the other hand, feels he learns well by reading on his own. He prefers to read the section and try working the example problems before coming to class. That way, he already knows what the teacher is going to talk about. Then, he can follow the teacher's examples more easily. It is also easier for him to take notes in class. De'Lore likes to have his questions answered right away, which he can do if he has already read the chapter section. So, De'Lore **carefully** reads the section in his text **before** he hears the classroom lecture on the topic.

Notice that there is **no one right way** to work with your textbook. You always must figure out what works best for you. Note also that both Maddy and De'Lore work with one section at a time. **The key is that you read the textbook regularly!** The rest of this activity will give you some ideas of how to make the most of your reading.

Try the following steps as you **read** your math textbook.

▶ Read slowly. Read only one section—or even part of a section—at a time.

▶ Do the sample problems in the margins **as you go.** Check them right away. The answers are at the bottom of the page.

▶ If your mind wanders, work problems on separate paper and write explanations in your own words.

▶ Make study cards as you read each section. Pay special attention to the yellow and blue boxes in the book. Make cards for new vocabulary, rules, procedures, formulas, and sample problems.

▶ **NOW**, you are ready to do your homework assignment!

OBJECTIVES

1 Select an appropriate strategy for homework.

2 Use textbook features effectively.

Preview before Class; Read Carefully after Class

Read Carefully before Class

Why Are These Reading Techniques Brain Friendly?

The steps at the left encourage you to be **actively working with the material** in your text. Your brain grows dendrites when it is doing something.

These methods require you to **try several different techniques,** not just the same thing over and over. Your brain loves variety!

Also, the techniques allow you to **take small breaks** in your learning. Those rest periods are crucial for good dendrite growth.

Now Try This ▶▶▶

Which steps for reading this book will be most helpful for you?

1. _____

2. _____

3. _____

Homework

Why Are These Homework Suggestions Brain Friendly?

Your brain will grow dendrites as you study the worked examples in the text and **try doing them yourself** on separate paper. So, when you see similar problems in the homework, you will already have dendrites to work from.

　Giving yourself a practice test by trying to remember the steps (without looking at your card) is an excellent way to reinforce what you are learning.

　Correcting errors right away is how you learn and reinforce the correct procedures. It is hard to unlearn a mistake, so always check to see that you are on the right track.

Teachers assign homework so you can grow your own dendrites (learn the material) and then coat the dendrites with myelin through practice (remember the material). Really! In learning, you get good at what you practice. So, completing homework every day will strengthen your neural network and prepare you for exams.

If you have read each section in your textbook according to the steps above, you will probably encounter few difficulties with the exercises in the homework. Here are some additional suggestions that will help you succeed with the homework.

▶ If you **have trouble with a problem,** find a similar worked example in the section. Pay attention to *every line* of the worked example to see how to get from step to step. Work it yourself too, on separate paper; don't just look at it.

▶ If it is **hard to remember the steps** to follow for certain procedures, write the steps on a separate card. Then write a short explanation of each step. Keep the card nearby while you do the exercises, but try *not* to look at it.

▶ If you **aren't sure you are working the assigned exercises correctly,** choose two or three odd-numbered problems that are a similar type and work them. Then check the answers in the Answers section of your book and see if you are doing them correctly. If you aren't, go back to the section in the text and review the examples and find out how to correct your errors. Finally, when you are sure you understand, try the assigned problems again.

▶ **Make sure you do some homework every day,** even if the math class does not meet each day!

Now Try This ▶▶▶

What are your biggest homework concerns?

List your two main concerns and a **brain friendly solution** for each one.

1. Concern: _____

　Solution: _____

2. Concern: _____

　Solution: _____

2.2 ▶▶▶ Mixed Numbers

Suppose you had three whole trays of muffins and half of another tray. You would state this as a whole number and a fraction.

OBJECTIVE 1 Identify mixed numbers. When a whole number and a fraction are written together, the result is a **mixed number.** For example, the mixed number

$$3\frac{1}{2} \quad \text{represents} \quad 3 + \frac{1}{2}$$

or 3 wholes and $\frac{1}{2}$ of a whole. Read $3\frac{1}{2}$ as "three and one-half." As this figure shows, the mixed number $3\frac{1}{2}$ is equal to the improper fraction $\frac{7}{2}$.

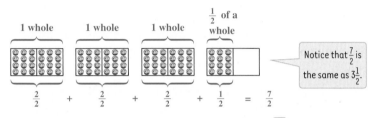

1 whole	1 whole	1 whole	$\frac{1}{2}$ of a whole

Notice that $\frac{7}{2}$ is the same as $3\frac{1}{2}$.

$$\frac{2}{2} \quad + \quad \frac{2}{2} \quad + \quad \frac{2}{2} \quad + \quad \frac{1}{2} \quad = \quad \frac{7}{2}$$

Work Problem ▶ 1 *at the Side.* ▶

OBJECTIVE 2 Write mixed numbers as improper fractions. Use the following steps to write $3\frac{1}{2}$ as an improper fraction without drawing a figure.

Step 1 Multiply 3 and 2.

$$3\frac{1}{2} \qquad 3 \cdot 2 = 6$$

Step 2 Add 1 to the product.

$$3\frac{1}{2} \qquad 6 + 1 = 7$$

Step 3 Use 7, from Step 2, as the numerator and 2 as the denominator.

$$3\frac{1}{2} = \frac{7}{2}$$

Same denominator

In summary, use the following steps to *write a mixed number as an improper fraction.*

Writing a Mixed Number as an Improper Fraction
Step 1 *Multiply* the denominator of the fraction and the whole number.

Step 2 *Add* to this product the numerator of the fraction.

Step 3 Write the result of Step 2 as the *numerator* and the original denominator as the *denominator.*

OBJECTIVES

1 Identify mixed numbers.

2 Write mixed numbers as improper fractions.

3 Write improper fractions as mixed numbers.

1 (a) Use these diagrams to write $1\frac{2}{3}$ as an improper fraction.

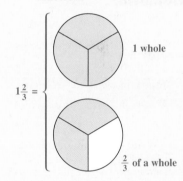

$1\frac{2}{3} =$ {

1 whole

$\frac{2}{3}$ of a whole

(b) Use these diagrams to write $2\frac{1}{4}$ as an improper fraction.

$2\frac{1}{4} =$ {

2 wholes

$\frac{1}{4}$ of a whole

ANSWERS

1. (a) $\frac{5}{3}$ (b) $\frac{9}{4}$

2 Write as improper fractions.

(a) $6\dfrac{1}{2}$

(b) $7\dfrac{3}{4}$

(c) $4\dfrac{7}{8}$

(d) $8\dfrac{5}{6}$

3 Write as whole or mixed numbers.

(a) $\dfrac{6}{5}$

(b) $\dfrac{9}{4}$

(c) $\dfrac{35}{5}$

(d) $\dfrac{78}{7}$

EXAMPLE 1 **Writing a Mixed Number as an Improper Fraction**

Write $7\frac{2}{3}$ as an improper fraction (numerator greater than denominator).

Step 1 $7\dfrac{2}{3}$ $\qquad 7 \cdot 3 = 21$ $\qquad$ Multiply 7 and 3.

Step 2 $7\dfrac{2}{3}$ $\qquad 21 + 2 = 23$ $\qquad$ Add 2. The numerator is 23.

Step 3 $7\dfrac{2}{3} = \dfrac{23}{3}$ $\qquad$ Use the same denominator.

> Always use the same denominator.

◀ *Work Problem* **2** *at the Side.*

OBJECTIVE **3** **Write improper fractions as mixed numbers.**
Write an improper fraction as a mixed number as follows.

> **Writing an Improper Fraction as a Mixed Number**
> Write an **improper fraction** as a mixed number by dividing the numerator by the denominator. The quotient is the whole number (of the mixed number), the remainder is the numerator of the fraction part, and the denominator remains unchanged.

EXAMPLE 2 **Writing Improper Fractions as Mixed Numbers**

Write each improper fraction as a mixed number.

> Divide numerator by denominator.

(a) $\dfrac{17}{5}$ $\qquad$ Divide 17 by 5. $\qquad \longrightarrow$

$$
\begin{array}{r} 3 \\ 5\overline{)17} \\ 15 \\ \hline 2 \end{array}
$$
$\leftarrow$ **Whole number part**

$\leftarrow$ Remainder

The quotient **3** is the whole number part of the mixed number. The remainder **2** is the numerator of the fraction, and the denominator stays as **5**.

$$\dfrac{17}{5} = 3\dfrac{2}{5} \quad \leftarrow \text{Remainder}$$
$$\text{Same denominator}$$

We can check this by using a diagram in which $\frac{17}{5}$ is shaded.

$\frac{1}{5}$

$\frac{5}{5} = 1$ (whole) $\qquad$ $\frac{5}{5} = 1$ (whole) $\qquad$ $\frac{5}{5} = 1$ (whole) $\qquad$ $\frac{2}{5}$

3 wholes

(b) $\dfrac{24}{4}$ $\qquad$ Divide 24 by 4. $\qquad \longrightarrow$

$$
\begin{array}{r} 6 \\ 4\overline{)24} \\ 24 \\ \hline 0 \end{array}
$$
so $\dfrac{24}{4} = 6$

$\leftarrow$ No remainder

◀ *Work Problem* **3** *at the Side.*

Write each mixed number as an improper fraction. See Example 1.

1. $1\frac{1}{4}$ $\frac{5}{4}$

2. $2\frac{1}{2}$ $\frac{5}{2}$

3. $4\frac{3}{5}$ $\frac{23}{5}$

4. $5\frac{2}{3}$ $\frac{17}{3}$

5. $6\frac{1}{2}$ $\frac{13}{2}$

6. $5\frac{3}{5}$ $\frac{28}{5}$

7. $8\frac{1}{4}$ $\frac{33}{4}$

8. $8\frac{1}{2}$ $\frac{17}{2}$

9. $1\frac{7}{11}$ $\frac{18}{11}$

10. $4\frac{3}{7}$ $\frac{31}{7}$

11. $6\frac{1}{3}$ $\frac{19}{3}$

12. $8\frac{2}{3}$ $\frac{26}{3}$

13. $10\frac{1}{8}$ $\frac{81}{8}$

14. $12\frac{2}{3}$ $\frac{38}{3}$

15. $10\frac{3}{4}$ $\frac{43}{4}$

16. $5\frac{4}{5}$ $\frac{29}{5}$

17. $3\frac{3}{8}$ $\frac{27}{8}$

18. $2\frac{8}{9}$ $\frac{26}{9}$

19. $8\frac{3}{5}$ $\frac{43}{5}$

20. $3\frac{4}{7}$ $\frac{25}{7}$

21. $4\frac{10}{11}$ $\frac{54}{11}$

22. $11\frac{5}{8}$ $\frac{93}{8}$

23. $32\frac{3}{4}$ $\frac{131}{4}$

24. $15\frac{3}{10}$ $\frac{153}{10}$

25. $18\frac{5}{12}$ $\frac{221}{12}$

26. $19\frac{8}{11}$ $\frac{217}{11}$

27. $17\frac{14}{15}$ $\frac{269}{15}$

28. $9\frac{5}{16}$ $\frac{149}{16}$

29. $7\frac{19}{24}$ $\frac{187}{24}$

30. $9\frac{7}{12}$ $\frac{115}{12}$

Write each improper fraction as a whole or mixed number. See Example 2.

31. $\frac{4}{3}$ $1\frac{1}{3}$

32. $\frac{11}{9}$ $1\frac{2}{9}$

◗ **33.** $\frac{9}{4}$ $2\frac{1}{4}$

34. $\frac{7}{2}$ $3\frac{1}{2}$

35. $\frac{48}{6}$ 8

36. $\frac{64}{8}$ 8

37. $\frac{38}{5}$ $7\frac{3}{5}$

38. $\frac{33}{7}$ $4\frac{5}{7}$

39. $\frac{39}{8}$ $4\frac{7}{8}$

40. $\frac{40}{9}$ $4\frac{4}{9}$

41. $\frac{27}{3}$ 9

42. $\frac{78}{78}$ 1

43. $\frac{63}{4}$ $15\frac{3}{4}$

44. $\frac{19}{5}$ $3\frac{4}{5}$

45. $\frac{47}{9}$ $5\frac{2}{9}$

46. $\frac{65}{9}$ $7\frac{2}{9}$

47. $\frac{65}{8}$ $8\frac{1}{8}$

48. $\frac{37}{6}$ $6\frac{1}{6}$

49. $\frac{84}{5}$ $16\frac{4}{5}$

50. $\frac{92}{3}$ $30\frac{2}{3}$

51. $\frac{112}{4}$ 28

52. $\frac{117}{9}$ 13

53. $\frac{183}{7}$ $26\frac{1}{7}$

54. $\frac{212}{11}$ $19\frac{3}{11}$

✐ **55.** Your classmate asks you how to change a mixed number to an improper fraction. Write a couple of sentences and give an example to show her how this is done.

Multiply the denominator by the whole number and add the numerator. The result becomes the new numerator, which is placed over the original denominator.

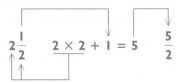

$$2\frac{1}{2} \qquad 2 \times 2 + 1 = 5 \qquad \frac{5}{2}$$

✐ **56.** Explain in a sentence or two how to change an improper fraction to a mixed number. Give an example to show how this is done.

Divide the numerator by the denominator. The quotient is the whole number of the mixed number and the remainder is the numerator of the fraction part. The denominator is unchanged.

$$\frac{5}{2} \qquad 2\overline{)5}^{\,2\,R1} \qquad 2\frac{1}{2}$$

Write each mixed number as an improper fraction.

57. $250\frac{1}{2}$ $\dfrac{\mathbf{501}}{\mathbf{2}}$

58. $185\frac{3}{4}$ $\dfrac{\mathbf{743}}{\mathbf{4}}$

⊙ **59.** $333\frac{1}{3}$ $\dfrac{\mathbf{1000}}{\mathbf{3}}$

60. $138\frac{4}{5}$ $\dfrac{\mathbf{694}}{\mathbf{5}}$

61. $522\frac{3}{8}$ $\dfrac{\mathbf{4179}}{\mathbf{8}}$

62. $622\frac{1}{4}$ $\dfrac{\mathbf{2489}}{\mathbf{4}}$

Write each improper fraction as a whole or mixed number.

63. $\dfrac{617}{4}$ $154\frac{1}{4}$

64. $\dfrac{760}{8}$ 95

▦ **65.** $\dfrac{2565}{15}$ 171

▦ **66.** $\dfrac{2915}{16}$ $182\frac{3}{16}$

▦ **67.** $\dfrac{3917}{32}$ $122\frac{13}{32}$

▦ **68.** $\dfrac{5632}{64}$ 88

Relating Concepts (Exercises 69–74) For Individual or Group Work

Knowing the basics of fractions is necessary in problem solving. **Work Exercises 69–74** *in order.*

69. Which of these fractions are proper fractions?

$$\frac{2}{3} \quad \frac{4}{5} \quad \frac{8}{5} \quad \frac{3}{4} \quad \frac{6}{6} \quad \frac{7}{10}$$

$$\frac{2}{3}, \frac{4}{5}, \frac{3}{4}, \frac{7}{10}$$

70. (a) The proper fractions in Exercise 69 are the ones where the __numerator__ is less than the __denominator__.

(b) Draw a picture with shaded parts to show each proper fraction in Exercise 69.

(c) The proper fractions in Exercise 69 are all ___less___ than 1.
(less/greater)

71. Which of these fractions are improper fractions?

$$\frac{5}{5} \quad \frac{3}{4} \quad \frac{10}{3} \quad \frac{2}{3} \quad \frac{5}{6} \quad \frac{6}{5}$$

$$\frac{5}{5}, \frac{10}{3}, \frac{6}{5}$$

72. (a) The improper fractions in Exercise 71 are the ones where the __numerator__ is equal to or greater than the __denominator__.

(b) Draw a picture with shaded parts to show each improper fraction in Exercise 71.

(c) The improper fractions in Exercise 71 are all equal to or ___greater___ than 1.
(less/greater)

73. Identify which of these fractions can be written as whole or mixed numbers, and then write them as whole or mixed numbers.

$$\frac{5}{3} \quad \frac{7}{8} \quad \frac{7}{7} \quad \frac{11}{6} \quad \frac{4}{5} \quad \frac{15}{16}$$

$$\frac{5}{3} = 1\frac{2}{3}; \frac{7}{7} = 1; \frac{11}{6} = 1\frac{5}{6}$$

74. (a) The fractions that can be written as whole or mixed numbers in Exercise 73 are __improper__ fractions, and their value is
(proper/improper)
always __greater than or equal to__ 1.
(less than/greater than or equal to)

(b) Draw a picture with shaded parts to show each whole or mixed number in Exercise 73.

(c) Explain how to write an improper fraction as a whole or mixed number.

Divide the numerator by the denominator. Use the quotient as the whole number and place the remainder over the denominator.

2.3 ▶▶▶ Factors

OBJECTIVE 1 Find factors of a number. You will recall that numbers multiplied to give a product are called **factors.** Because 2 • 5 = 10, both 2 and 5 are factors of 10. The numbers 1 and 10 are also factors of 10, because

$$1 • 10 = 10$$

The various tests for divisibility show that 1, 2, 5, and 10 are the only whole number factors of 10. The products 2 • 5 and 1 • 10 are called **factorizations** of 10.

> **Note**
>
> The tests to decide whether one number is divisible by another number were shown in **Section 1.5.** You might want to review these. The tests that you will want to remember are those for 2, 3, 5, and 10.

EXAMPLE 1 | **Using Factors**

Find all possible two-number factorizations of each number.

(a) 12

$$1 • 12 = 12 \qquad 2 • 6 = 12 \qquad 3 • 4 = 12$$

The factors of 12 are 1, 2, 3, 4, 6, and 12.

(b) 60

$$
\begin{aligned}
1 • 60 &= 60 & 2 • 30 &= 60 \\
3 • 20 &= 60 & 4 • 15 &= 60 \\
5 • 12 &= 60 & 6 • 10 &= 60
\end{aligned}
$$

> The factors of a number all divide evenly into that number.

The factors of 60 are 1, 2, 3, 4, 5, 6, 10, 12, 15, 20, 30, and 60.

Work Problem **1** *at the Side.* ▶

> **Composite Numbers**
>
> A number with a factor other than itself or 1 is called a **composite number.**

EXAMPLE 2 | **Identifying Composite Numbers**

Which of the following numbers are composite?

(a) 6
  Because 6 has factors of **2** and **3**, numbers other than 6 or 1, the number 6 is composite.

(b) 11
  The number 11 has only two factors, 11 and 1. It is *not* composite.

(c) 25
  A factor of 25 is **5**, so 25 is composite.

Work Problem **2** *at the Side.* ▶

OBJECTIVE 2 Identify prime numbers. Whole numbers that are not composite are called **prime numbers,** except 0 and 1, which are neither prime nor composite.

> **Prime Numbers**
>
> A **prime number** is a whole number that has exactly *two different* factors, *itself* and *1.*

OBJECTIVES

1. **Find factors of a number.**
2. **Identify prime numbers.**
3. **Find prime factorizations.**

1 Find all the whole number factors of each number.

(a) 18

(b) 16

(c) 36

(d) 80

2 Which of these numbers are composite?
2, 4, 5, 6, 8, 10, 11, 13, 19, 21, 27, 28, 33, 36, 42

ANSWERS

1. **(a)** 1, 2, 3, 6, 9, 18   **(b)** 1, 2, 4, 8, 16
   (c) 1, 2, 3, 4, 6, 9, 12, 18, 36
   (d) 1, 2, 4, 5, 8, 10, 16, 20, 40, 80
2. 4, 6, 8, 10, 21, 27, 28, 33, 36, 42

3 Which of the following are prime?

4, 7, 9, 13, 17, 19, 29, 33

4 Find the prime factorization of each number.

(a) 8

(b) 28

(c) 18

(d) 40

The number 3 is a prime number, since it can be divided evenly only by itself and 1. The number 8 is not a prime number (it is composite), since 8 can be divided evenly by 2 and 4, as well as by itself and 1.

> **CAUTION**
> A prime number has **only two** different factors, itself and 1. The number 1 is not a prime number because it does not have *two different* factors; the only factor of 1 is 1.

EXAMPLE 3 Finding Prime Numbers

Which of the following numbers are prime?

Only the number itself and 1 will divide evenly into a prime number.

2   5   11   15   27

The number 15 can be divided by 3 and 5, so it is not prime. Also, because 27 can be divided by 3 and 9, then 27 is not prime. The other numbers in the list, 2, 5, and 11, are divisible only by themselves and 1, so they are prime.

◀ Work Problem **3** at the Side.

OBJECTIVE 3 Find prime factorizations. For reference, here are the prime numbers less than 50.

2	3	5	7	11
13	17	19	23	29
31	37	41	43	47

These are the prime numbers less than 50.

> **CAUTION**
> All prime numbers are odd numbers except the number 2. Be careful, though, because *all odd numbers are not prime numbers*. For example, 9, 15, and 21 are odd numbers but are *not* prime numbers.

The **prime factorization** of a number can be especially useful when we are adding or subtracting fractions and need to find a common denominator or write a fraction in lowest terms.

> **Prime Factorization**
> A **prime factorization** of a number is a factorization in which every factor is a *prime number*.

EXAMPLE 4 Determining the Prime Factorization

Find the prime factorization of 12.
Try to divide 12 by the first prime, 2.

$$12 \div 2 = 6,$$

↑ First prime

so

$$12 = 2 \cdot 6$$

Try to divide 6 by the prime, 2.

$$6 \div 2 = 3,$$

so

$$12 = 2 \cdot \underline{2 \cdot 3}$$

↑ Factorization of 6

Because all factors are prime, the prime factorization of 12 is

$$2 \cdot 2 \cdot 3$$

All these factors are prime.

◀ Work Problem **4** at the Side.

EXAMPLE 5 **Factoring by Using the Division Method**

Find the prime factorization of 48.

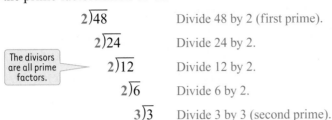

$2\overline{)48}$ Divide 48 by 2 (first prime).

$2\overline{)24}$ Divide 24 by 2.

The divisors are all prime factors. $2\overline{)12}$ Divide 12 by 2.

$2\overline{)6}$ Divide 6 by 2.

$3\overline{)3}$ Divide 3 by 3 (second prime).

1   Continue to divide until the quotient is 1.

Because all factors (divisors) are prime, the prime factorization of 48 is

$$2 \cdot 2 \cdot 2 \cdot 2 \cdot 3$$

In Chapter 1, we wrote $2 \cdot 2 \cdot 2 \cdot 2$ as 2^4, so the prime factorization of 48 can be written, using exponents, as

$$48 = 2 \cdot 2 \cdot 2 \cdot 2 \cdot 3 = 2^4 \cdot 3$$

——————— *Work Problem* **5** *at the Side.* ▶

Note

When using the division method of factoring, the last quotient found is 1. The "1" is never used as a prime factor because 1 is neither prime nor composite. Besides, 1 times any number is the number itself.

EXAMPLE 6 **Using Exponents with Prime Factorization**

Find the prime factorization of 225.

$3\overline{)225}$ 225 is not divisible by 2; use 3.

$3\overline{)75}$ Divide 75 by 3.

All the divisors are prime factors. $5\overline{)25}$ 25 is not divisible by 3; use 5.

$5\overline{)5}$ Divide by 5.

1 is **not** a prime factor.

1   Continue to divide until the quotient is 1.

Write the prime factorization.

$$255 = 3 \cdot 3 \cdot 5 \cdot 5$$

Or, using exponents,

$$255 = 3^2 \cdot 5^2$$

——————— *Work Problem* **6** *at the Side.* ▶

5 Find the prime factorization of each number. Write the factorization with exponents.

(a) 36

(b) 54

(c) 60

(d) 81

6 Write the prime factorization of each number using exponents.

(a) 48

(b) 44

(c) 90

(d) 120

(e) 180

ANSWERS

5. (a) $2^2 \cdot 3^2$ (b) $2 \cdot 3^3$ (c) $2^2 \cdot 3 \cdot 5$
   (d) 3^4
6. (a) $2^4 \cdot 3$ (b) $2^2 \cdot 11$ (c) $2 \cdot 3^2 \cdot 5$
   (d) $2^3 \cdot 3 \cdot 5$ (e) $2^2 \cdot 3^2 \cdot 5$

Another method of factoring is a *factor tree*.

7 Complete each factor tree and give the prime factorization.

(a) 28

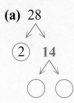

(b) 35

(c) 78

(**EXAMPLE 7**) **Factoring by Using a Factor Tree**

Find the prime factorization of each number using a factor tree.

(a) 30

Try to divide by the first prime, 2. Write the factors under the 30. Circle the 2, since it is a prime.

Since 15 cannot be divided evenly by 2, try the next prime, 3.

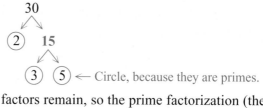

No uncircled factors remain, so the prime factorization (the circled factors) has been found.

$$30 = 2 \cdot 3 \cdot 5$$

(b) 24

Divide by 2.

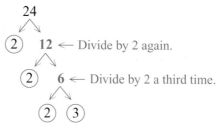

$24 = 2 \cdot 2 \cdot 2 \cdot 3$ or, using exponents, $24 = 2^3 \cdot 3$

(c) 45

Because 45 cannot be divided by 2, try 3.

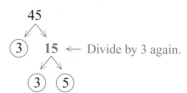

$45 = 3 \cdot 3 \cdot 5$ or, using exponents, $45 = 3^2 \cdot 5$

Note

The diagrams used in Example 7 look like tree branches, and that is why this method is referred to as using a factor tree.

◀ *Work Problem* **7** *at the Side.*

2.3 ▶▶▶ Exercises

FOR EXTRA HELP  PRACTICE   WATCH   DOWNLOAD   READ REVIEW

Find all the factors of each number. See Example 1.

1. 8 1, 2, 4, 8

2. 12 1, 2, 3, 4, 6, 12

3. 15 1, 3, 5, 15

4. 28 1, 2, 4, 7, 14, 28

5. 48 1, 2, 3, 4, 6, 8, 12, 16, 24, 48

6. 30 1, 2, 3, 5, 6, 10, 15, 30

7. 36 1, 2, 3, 4, 6, 9, 12, 18, 36

8. 20 1, 2, 4, 5, 10, 20

9. 40 1, 2, 4, 5, 8, 10, 20, 40

10. 60 1, 2, 3, 4, 5, 6, 10, 12, 15, 20, 30, 60

11. 64 1, 2, 4, 8, 16, 32, 64

12. 84 1, 2, 3, 4, 6, 7, 12, 14, 21, 28, 42, 84

Decide whether each number is prime *or* composite. *See Examples 2 and 3.*

13. 6 composite

14. 9 composite

15. 5 prime

16. 7 prime

17. 16 composite

18. 10 composite

19. 13 prime

20. 65 composite

21. 19 prime

22. 17 prime

23. 25 composite

24. 26 composite

25. 48 composite

26. 47 prime

27. 45 composite

28. 53 prime

Find the prime factorization of each number. Write answers with exponents when repeated factors appear. See Examples 4–7.

29. 8 2^3

30. 6 $2 \cdot 3$

31. 20 $2^2 \cdot 5$

32. 40 $2^3 \cdot 5$

33. 36 $2^2 \cdot 3^2$

34. 18 $2 \cdot 3^2$

35. 25 5^2

36. 56 $2^3 \cdot 7$

37. 68 $2^2 \cdot 17$

38. 70 $2 \cdot 5 \cdot 7$

39. 72 $2^3 \cdot 3^2$

40. 64 2^6

41. 44 $2^2 \cdot 11$

42. 104 $2^3 \cdot 13$

43. 100 $2^2 \cdot 5^2$

44. 112 $2^4 \cdot 7$

45. 125 5^3

46. 135 $3^3 \cdot 5$

47. 180 $2^2 \cdot 3^2 \cdot 5$

48. 300 $2^2 \cdot 3 \cdot 5^2$

49. 320 $2^6 \cdot 5$

50. 480 $2^5 \cdot 3 \cdot 5$

51. 360 $2^3 \cdot 3^2 \cdot 5$

52. 400 $2^4 \cdot 5^2$

53. Give a definition in your own words of both a composite number and a prime number. Give three examples of each. Which whole numbers are neither prime nor composite?

A composite number has a factor(s) other than itself or 1. Examples include 4, 6, 8, 9, 10. A prime number is a whole number that has exactly two _different_ factors, itself and 1. Examples include 2, 3, 5, 7, 11. The numbers 0 and 1 are neither prime nor composite.

54. With the exception of the number 2, all prime numbers are odd numbers. Nevertheless, all odd numbers are not prime numbers. Explain why these statements are true.

No even number other than 2 is prime because all even numbers have 2 as a factor. Many odd numbers are multiples of prime numbers and are not prime. For example, 9, 21, 33, and 45 are all multiples of 3.

55. Explain the difference between finding all possible factors of 24 and finding the prime factorization of 24.

All the possible factors of 24 are 1, 2, 3, 4, 6, 8, 12, and 24. This list includes both prime numbers and composite numbers. The prime factors of 24 include only prime numbers. The prime factorization of 24 is

$$24 = 2 \cdot 2 \cdot 2 \cdot 3 = 2^3 \cdot 3$$

56. Use the division method to find the prime factorization of 36. Can you divide by 3s before you divide by 2s? Does the order of division change the answers?

No. The order of division does not matter. As long as you use only prime numbers, your answers will be correct. However, it does seem easier to always start with 2 and then use progressively greater prime numbers. The prime factorization of 36 is

$$36 = 2 \cdot 2 \cdot 3 \cdot 3 = 2^2 \cdot 3^2$$

Find the prime factorization of each number. Write answers using exponents.

57. 350

$2 \cdot 5^2 \cdot 7$

58. 640

$2^7 \cdot 5$

59. 960

$2^6 \cdot 3 \cdot 5$

60. 1000

$2^3 \cdot 5^3$

61. 1560

$2^3 \cdot 3 \cdot 5 \cdot 13$

62. 2000

$2^4 \cdot 5^3$

63. 1260

$2^2 \cdot 3^2 \cdot 5 \cdot 7$

64. 2200

$2^3 \cdot 5^2 \cdot 11$

Relating Concepts (Exercises 65–70) For Individual or Group Work

An understanding of factors and factorization will be needed to solve fraction problems. **Work Exercises 65–70 in order.**

65. A prime number is a whole number that has exactly two different factors, itself and 1. List all prime numbers less than 50.

2, 3, 5, 7, 11, 13, 17, 19, 23, 29, 31, 37, 41, 43, 47

66. Explain what it is about the numbers in Exercise 65 that makes them prime.

A prime number is a whole number that is evenly divisible by itself and 1 only.

67. The number 2 is an even number and a prime number. Can any other even numbers be prime numbers? Explain.

No. Every other even number is divisible by 2 in addition to being divisible by itself and 1.

68. Can a multiple of a prime number be prime (for example, 6, 9, 12, and 15 are multiples of 3)? Explain.

No. A multiple of a prime number can never be prime because it will always be divisible by the prime number.

69. Find the prime factorization of 2100. Do not use exponents in your answer.

2 · 2 · 3 · 5 · 5 · 7

70. Write the answer to Exercise 69 using exponents for repeated factors.

$2^2 \cdot 3 \cdot 5^2 \cdot 7$

2.4 ▶▶▶ Writing a Fraction in Lowest Terms

When working problems involving fractions, we must often compare two fractions to determine whether they represent the same portion of a whole. Look at the two cases of soda.

$\frac{3}{4}$ full 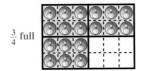 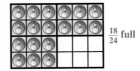 $\frac{18}{24}$ full

The cases of soda show areas that are $\frac{3}{4}$ full and $\frac{18}{24}$ full. Because the full areas are equivalent (the same portion), the fractions $\frac{3}{4}$ and $\frac{18}{24}$ are **equivalent fractions.** They each represent the same portion of the whole.

$$\frac{3}{4} = \frac{18}{24}$$

Because the numbers 18 and 24 both have 6 as a factor, 6 is called a **common factor** of the numbers. Other common factors of 18 and 24 are 1, 2, and 3.

Work Problem **1** *at the Side.* ▶

OBJECTIVE **1** **Tell whether a fraction is written in lowest terms.**
The fraction $\frac{3}{4}$ is written in *lowest terms* because the numerator and denominator have no common factor other than 1. However, the fraction $\frac{18}{24}$ is **not** in lowest terms because its numerator and denominator have common factors of 6, 3, 2, and 1.

> **Writing a Fraction in Lowest Terms**
> A fraction is written in **lowest terms** when the numerator and denominator have no common factor other than 1.

EXAMPLE 1 **Understanding Lowest Terms**

Are the following fractions in lowest terms?

(a) $\frac{3}{8}$ ◀ No common factors other than 1.
The numerator and denominator have no common factor other than 1, so the fraction is in lowest terms.

(b) $\frac{21}{36}$ ◀ 3 is a common factor of 21 and 36.
The numerator and denominator have a common factor of 3, so the fraction is **not in lowest terms.**

Work Problem **2** *at the Side.* ▶

OBJECTIVE **2** **Write a fraction in lowest terms using common factors.** There are two common methods for writing a fraction in lowest terms. These methods are shown in the next examples. The first method works best when the numerator and denominator are small numbers.

OBJECTIVES

1 Tell whether a fraction is written in lowest terms.

2 Write a fraction in lowest terms using common factors.

3 Write a fraction in lowest terms using prime factors.

4 Determine whether two fractions are equivalent.

1 Decide whether the number in blue is a common factor of the other two numbers.

(a) 6, 12; **2**

(b) 32, 64; **8**

(c) 32, 56; **16**

(d) 75, 81; **1**

2 Are the following fractions in lowest terms?

(a) $\frac{4}{5}$

(b) $\frac{6}{18}$

(c) $\frac{9}{15}$

(d) $\frac{17}{46}$

ANSWERS
1. **(a)** yes **(b)** yes **(c)** no **(d)** yes
2. **(a)** yes **(b)** no **(c)** no **(d)** yes

3 Write in lowest terms.

(a) $\dfrac{8}{16}$

(b) $\dfrac{9}{12}$

(c) $\dfrac{28}{42}$

(d) $\dfrac{30}{80}$

(e) $\dfrac{16}{40}$

EXAMPLE 2 **Writing Fractions in Lowest Terms**

Write each fraction in lowest terms.

(a) $\dfrac{18}{24}$

The greatest common factor of 18 and 24 is **6**. Divide both numerator and denominator by **6**.

$$\frac{18}{24} = \frac{18 \div 6}{24 \div 6} = \frac{3}{4}$$

Divide by the greatest common factor, 6.

(b) $\dfrac{30}{50} = \dfrac{30 \div 10}{50 \div 10} = \dfrac{3}{5}$ Divide both numerator and denominator by 10.

(c) $\dfrac{24}{42} = \dfrac{24 \div 6}{42 \div 6} = \dfrac{4}{7}$ Divide both numerator and denominator by 6.

(d) $\dfrac{60}{72}$

Suppose we thought that 4 was the greatest common factor of 60 and 72. Dividing by 4 would give

$$\frac{60}{72} = \frac{60 \div 4}{72 \div 4} = \frac{15}{18} \leftarrow \text{Not in lowest terms.}$$

But $\frac{15}{18}$ is not in lowest terms, because 15 and 18 have a common factor of 3. So we divide by 3.

$$\frac{15}{18} = \frac{15 \div 3}{18 \div 3} = \frac{5}{6}$$

Continue dividing until there is no common factor other than 1.

The fraction $\frac{60}{72}$ could have been written in lowest terms in one step by dividing by 12, the greatest common factor of 60 and 72.

$$\frac{60}{72} = \frac{60 \div 12}{72 \div 12} = \frac{5}{6} \leftarrow \text{Same answer as above.}$$

Continue dividing until the fraction is in lowest terms.

> **Note**
> Dividing the numerator and denominator by the same number results in an equivalent fraction.

In Example 2, we wrote fractions in lowest terms by dividing by a common factor. This method is summarized in the following steps.

> **The Method of Dividing by a Common Factor**
>
> *Step 1* Find the greatest number that will divide evenly into both the numerator and denominator. This number is a *common factor.*
>
> *Step 2* *Divide* both numerator and denominator by the common factor.
>
> *Step 3* *Check* to see whether the new fraction has any common factors (besides 1). If it does, repeat Steps 2 and 3. If the only common factor is 1, the fraction is in lowest terms.

ANSWERS

3. (a) $\dfrac{1}{2}$ (b) $\dfrac{3}{4}$ (c) $\dfrac{2}{3}$ (d) $\dfrac{3}{8}$ (e) $\dfrac{2}{5}$

OBJECTIVE 3 Write a fraction in lowest terms using prime factors. The method of writing a fraction in lowest terms by division works well for fractions with small numerators and denominators. For larger numbers, when common factors are not obvious, use the method of *prime factors,* which is shown in the next example.

EXAMPLE 3   **Using Prime Factors**

Write each fraction in lowest terms.

(a) $\dfrac{24}{42}$

Write the prime factorization of both numerator and denominator. See **Section 2.3** for help.

$$\frac{24}{42} = \frac{2 \cdot 2 \cdot 2 \cdot 3}{2 \cdot 3 \cdot 7}$$

Just as with the method used in Example 2, divide both numerator and denominator by any common factors. Write a **1** by each factor that has been divided.

$$2 \div 2 \text{ is } 1 \qquad \frac{24}{42} = \frac{\overset{1}{\cancel{2}} \cdot 2 \cdot 2 \cdot \overset{1}{\cancel{3}}}{\underset{1}{\cancel{2}} \cdot \underset{1}{\cancel{3}} \cdot 7} \qquad 3 \div 3 \text{ is } 1$$

Multiply the remaining factors in both numerator and denominator.

$$\frac{24}{42} = \frac{1 \cdot 2 \cdot 2 \cdot 1}{1 \cdot 1 \cdot 7} = \frac{4}{7} \leftarrow \text{Lowest terms}$$

Finally, $\frac{24}{42}$ written in lowest terms is $\frac{4}{7}$.

(b) $\dfrac{162}{54}$

Write the prime factorization of both numerator and denominator.

$$\frac{162}{54} = \frac{2 \cdot 3 \cdot 3 \cdot 3 \cdot 3}{2 \cdot 3 \cdot 3 \cdot 3}$$

Now divide by the common factors. ***Do not forget to write the 1s.***

$$\frac{162}{54} = \frac{\overset{1}{\cancel{2}} \cdot \overset{1}{\cancel{3}} \cdot \overset{1}{\cancel{3}} \cdot \overset{1}{\cancel{3}} \cdot 3}{\underset{1}{\cancel{2}} \cdot \underset{1}{\cancel{3}} \cdot \underset{1}{\cancel{3}} \cdot \underset{1}{\cancel{3}}}$$

Remember to write in the 1s when dividing by a common factor.

$$= \frac{1 \cdot 1 \cdot 1 \cdot 1 \cdot 3}{1 \cdot 1 \cdot 1 \cdot 1} = \frac{3}{1} = 3$$

(c) $\dfrac{18}{90}$

$$\frac{18}{90} = \frac{\overset{1}{\cancel{2}} \cdot \overset{1}{\cancel{3}} \cdot \overset{1}{\cancel{3}}}{\underset{1}{\cancel{2}} \cdot \underset{1}{\cancel{3}} \cdot \underset{1}{\cancel{3}} \cdot 5} = \frac{1 \cdot 1 \cdot 1}{1 \cdot 1 \cdot 1 \cdot 5} = \frac{1}{5}$$

CAUTION
In Example 3(c) above, all factors of the numerator were divided. But $1 \cdot 1 \cdot 1$ is still 1, so the final answer is $\frac{1}{5}$ (***not*** 5).

In Example 3, we wrote fractions in lowest terms using prime factors. This method is summarized as follows.

4 Use the method of prime factors to write each fraction in lowest terms.

(a) $\dfrac{12}{36}$

Step 1 Write the **prime factorization** of both numerator and denominator.

Step 2 Use slashes to show you are **dividing** both numerator and denominator by common factors.

Step 3 **Multiply** the remaining factors in the numerator and denominator.

(b) $\dfrac{32}{56}$

◀ *Work Problem* **4** *at the Side.*

OBJECTIVE **4** **Determine whether two fractions are equivalent.**
The next example shows how to decide whether two fractions are equivalent.

EXAMPLE 4 **Determining Whether Two Fractions Are Equivalent**

(c) $\dfrac{74}{111}$

Determine whether each pair of fractions is equivalent. In other words, do both fractions represent the same part of a whole?

(a) $\dfrac{16}{48}$ and $\dfrac{24}{72}$

Use the method of prime factors to write each fraction in lowest terms.

(d) $\dfrac{124}{340}$

$$\frac{16}{48} = \frac{\overset{1}{\cancel{2}} \cdot \overset{1}{\cancel{2}} \cdot \overset{1}{\cancel{2}} \cdot \overset{1}{\cancel{2}}}{\underset{1}{\cancel{2}} \cdot \underset{1}{\cancel{2}} \cdot \underset{1}{\cancel{2}} \cdot \underset{1}{\cancel{2}} \cdot 3} = \frac{1 \cdot 1 \cdot 1 \cdot 1}{1 \cdot 1 \cdot 1 \cdot 1 \cdot 3} = \frac{1}{3} \longleftarrow$$

Equivalent $\left(\dfrac{1}{3} = \dfrac{1}{3}\right)$

5 Is each pair of fractions equivalent?

(a) $\dfrac{24}{48}$ and $\dfrac{36}{72}$

$$\frac{24}{72} = \frac{\overset{1}{\cancel{2}} \cdot \overset{1}{\cancel{2}} \cdot \overset{1}{\cancel{2}} \cdot \overset{1}{\cancel{3}}}{\underset{1}{\cancel{2}} \cdot \underset{1}{\cancel{2}} \cdot \underset{1}{\cancel{2}} \cdot \underset{1}{\cancel{3}} \cdot 3} = \frac{1 \cdot 1 \cdot 1 \cdot 1}{1 \cdot 1 \cdot 1 \cdot 1 \cdot 3} = \frac{1}{3} \longleftarrow$$

(b) $\dfrac{32}{52}$ and $\dfrac{64}{112}$

(b) $\dfrac{45}{60}$ and $\dfrac{50}{75}$

$$\frac{32}{52} = \frac{\overset{1}{\cancel{2}} \cdot \overset{1}{\cancel{2}} \cdot 2 \cdot 2 \cdot 2}{\underset{1}{\cancel{2}} \cdot \underset{1}{\cancel{2}} \cdot 13} = \frac{2 \cdot 2 \cdot 2}{1 \cdot 1 \cdot 13} = \frac{8}{13} \longleftarrow$$

Not equivalent $\left(\dfrac{8}{13} \neq \dfrac{4}{7}\right)$

$$\frac{64}{112} = \frac{\overset{1}{\cancel{2}} \cdot \overset{1}{\cancel{2}} \cdot \overset{1}{\cancel{2}} \cdot \overset{1}{\cancel{2}} \cdot 2 \cdot 2}{\underset{1}{\cancel{2}} \cdot \underset{1}{\cancel{2}} \cdot \underset{1}{\cancel{2}} \cdot \underset{1}{\cancel{2}} \cdot 7} = \frac{1 \cdot 1 \cdot 1 \cdot 1 \cdot 2 \cdot 2}{1 \cdot 1 \cdot 1 \cdot 1 \cdot 7} = \frac{4}{7} \longleftarrow$$

(c) $\dfrac{20}{4}$ and $\dfrac{110}{22}$

(c) $\dfrac{75}{15}$ and $\dfrac{60}{12}$

$$\frac{75}{15} = \frac{\overset{1}{\cancel{3}} \cdot \overset{1}{\cancel{5}} \cdot 5}{\underset{1}{\cancel{3}} \cdot \underset{1}{\cancel{5}}} = \frac{1 \cdot 1 \cdot 5}{1 \cdot 1} = 5 \longleftarrow$$

Equivalent $(5 = 5)$

(d) $\dfrac{120}{220}$ and $\dfrac{180}{320}$

$$\frac{60}{12} = \frac{\overset{1}{\cancel{2}} \cdot \overset{1}{\cancel{2}} \cdot \overset{1}{\cancel{3}} \cdot 5}{\underset{1}{\cancel{2}} \cdot \underset{1}{\cancel{2}} \cdot \underset{1}{\cancel{3}}} = \frac{1 \cdot 1 \cdot 1 \cdot 5}{1 \cdot 1 \cdot 1} = 5 \longleftarrow$$

ANSWERS

4. (a) $\dfrac{1}{3}$ (b) $\dfrac{4}{7}$ (c) $\dfrac{2}{3}$ (d) $\dfrac{31}{85}$

5. (a) equivalent (b) not equivalent
   (c) equivalent (d) not equivalent

◀ *Work Problem* **5** *at the Side.*

2.4 ▶▶▶ Exercises

Put a ✓ mark in the blank if the number at the left is divisible by the number at the top. Put an ✗ in the blank if the number is not divisible by the number at the top. (For help, see **Section 1.5**.*)*

		2	3	5	10			2	3	5	10
1.	60	✓	✓	✓	✓	2.	90	✓	✓	✓	✓
3.	48	✓	✓	✗	✗	4.	36	✓	✓	✗	✗
5.	160	✓	✗	✓	✓	6.	175	✗	✗	✓	✗
7.	138	✓	✓	✗	✗	8.	150	✓	✓	✓	✓

Write each fraction in lowest terms. See Example 2.

9. $\dfrac{6}{8}$ $\dfrac{3}{4}$ 10. $\dfrac{6}{12}$ $\dfrac{1}{2}$ 11. $\dfrac{3}{12}$ $\dfrac{1}{4}$ 12. $\dfrac{4}{12}$ $\dfrac{1}{3}$

13. $\dfrac{15}{25}$ $\dfrac{3}{5}$ 14. $\dfrac{32}{48}$ $\dfrac{2}{3}$ 15. $\dfrac{36}{42}$ $\dfrac{6}{7}$ 16. $\dfrac{22}{33}$ $\dfrac{2}{3}$

17. $\dfrac{56}{64}$ $\dfrac{7}{8}$ 18. $\dfrac{21}{35}$ $\dfrac{3}{5}$ 19. $\dfrac{180}{210}$ $\dfrac{6}{7}$ 20. $\dfrac{72}{80}$ $\dfrac{9}{10}$

21. $\dfrac{72}{126}$ $\dfrac{4}{7}$ 22. $\dfrac{73}{146}$ $\dfrac{1}{2}$ 23. $\dfrac{12}{600}$ $\dfrac{1}{50}$ 24. $\dfrac{8}{400}$ $\dfrac{1}{50}$

25. $\dfrac{96}{132}$ $\dfrac{8}{11}$ 26. $\dfrac{165}{180}$ $\dfrac{11}{12}$ 27. $\dfrac{60}{108}$ $\dfrac{5}{9}$ 28. $\dfrac{112}{128}$ $\dfrac{7}{8}$

Write the numerator and denominator of each fraction as a product of prime factors and divide by the common factors. Then write the fraction in lowest terms. See Example 3.

29. $\dfrac{18}{24}$ $\dfrac{\cancel{2}\cdot\cancel{3}\cdot 3}{\cancel{2}\cdot 2\cdot 2\cdot\cancel{3}}=\dfrac{3}{4}$

30. $\dfrac{16}{64}$ $\dfrac{\cancel{2}\cdot\cancel{2}\cdot\cancel{2}\cdot\cancel{2}}{\cancel{2}\cdot\cancel{2}\cdot\cancel{2}\cdot\cancel{2}\cdot 2\cdot 2}=\dfrac{1}{4}$

31. $\dfrac{35}{40}$ $\dfrac{\cancel{5}\cdot 7}{2\cdot 2\cdot 2\cdot\cancel{5}}=\dfrac{7}{8}$

32. $\dfrac{20}{32}$ $\dfrac{\cancel{2}\cdot\cancel{2}\cdot 5}{\cancel{2}\cdot\cancel{2}\cdot 2\cdot 2\cdot 2}=\dfrac{5}{8}$

33. $\dfrac{90}{180}$ $\dfrac{\cancel{2}\cdot 3\cdot\cancel{3}\cdot\cancel{5}}{\cancel{2}\cdot 2\cdot 3\cdot\cancel{3}\cdot\cancel{5}}=\dfrac{1}{2}$

34. $\dfrac{36}{48}$ $\dfrac{\cancel{2}\cdot\cancel{2}\cdot 3\cdot 3}{\cancel{2}\cdot\cancel{2}\cdot 2\cdot 2\cdot\cancel{3}}=\dfrac{3}{4}$

35. $\dfrac{36}{12}$ $\dfrac{\overset{1}{\cancel{2}} \cdot \overset{1}{\cancel{2}} \cdot \overset{1}{\cancel{3}} \cdot 3}{\underset{1}{\cancel{2}} \cdot \underset{1}{\cancel{2}} \cdot \underset{1}{\cancel{3}}} = 3$

36. $\dfrac{192}{48}$ $\dfrac{\overset{1}{\cancel{2}} \cdot \overset{1}{\cancel{2}} \cdot \overset{1}{\cancel{2}} \cdot \overset{1}{\cancel{2}} \cdot 2 \cdot 2 \cdot \overset{1}{\cancel{3}}}{\underset{1}{\cancel{2}} \cdot \underset{1}{\cancel{2}} \cdot \underset{1}{\cancel{2}} \cdot \underset{1}{\cancel{2}} \cdot \underset{1}{\cancel{3}}} = 4$

37. $\dfrac{72}{225}$ $\dfrac{2 \cdot 2 \cdot 2 \cdot \overset{1}{\cancel{3}} \cdot \overset{1}{\cancel{3}}}{\underset{1}{\cancel{3}} \cdot \underset{1}{\cancel{3}} \cdot 5 \cdot 5} = \dfrac{8}{25}$

38. $\dfrac{65}{234}$ $\dfrac{5 \cdot \overset{1}{\cancel{13}}}{2 \cdot 3 \cdot 3 \cdot \underset{1}{\cancel{13}}} = \dfrac{5}{18}$

Write each fraction in lowest terms. Then state whether the fractions are equivalent or not equivalent. See Example 4.

39. $\dfrac{3}{6}$ and $\dfrac{18}{36}$

$\dfrac{1}{2} = \dfrac{1}{2}$; equivalent

40. $\dfrac{3}{8}$ and $\dfrac{27}{72}$

$\dfrac{3}{8} = \dfrac{3}{8}$; equivalent

41. $\dfrac{10}{24}$ and $\dfrac{12}{30}$

$\dfrac{5}{12} \neq \dfrac{2}{5}$; not equivalent

42. $\dfrac{15}{35}$ and $\dfrac{18}{40}$

$\dfrac{3}{7} \neq \dfrac{9}{20}$; not equivalent

43. $\dfrac{15}{24}$ and $\dfrac{35}{52}$

$\dfrac{5}{8} \neq \dfrac{35}{52}$; not equivalent

44. $\dfrac{21}{33}$ and $\dfrac{9}{12}$

$\dfrac{7}{11} \neq \dfrac{3}{4}$; not equivalent

45. $\dfrac{14}{16}$ and $\dfrac{35}{40}$

$\dfrac{7}{8} = \dfrac{7}{8}$; equivalent

46. $\dfrac{27}{90}$ and $\dfrac{24}{80}$

$\dfrac{3}{10} = \dfrac{3}{10}$; equivalent

47. $\dfrac{48}{6}$ and $\dfrac{72}{8}$

$8 \neq 9$; not equivalent

48. $\dfrac{33}{11}$ and $\dfrac{72}{24}$

$3 = 3$; equivalent

49. $\dfrac{25}{30}$ and $\dfrac{65}{78}$

$\dfrac{5}{6} = \dfrac{5}{6}$; equivalent

50. $\dfrac{24}{72}$ and $\dfrac{30}{90}$

$\dfrac{1}{3} = \dfrac{1}{3}$; equivalent

51. What does it mean when a fraction is expressed in lowest terms? Give three examples.

A fraction is in lowest terms when the numerator and the denominator have no common factors other than 1. Some examples are $\dfrac{1}{2}, \dfrac{3}{8},$ and $\dfrac{2}{3}$.

52. Explain what equivalent fractions are, and give an example of a pair of equivalent fractions. Show that they are equivalent.

Two fractions are equivalent when they represent the same portion of a whole. For example, the fractions $\dfrac{10}{15}$ and $\dfrac{8}{12}$ are equivalent.

$$\dfrac{10}{15} = \dfrac{2 \cdot \overset{1}{\cancel{5}}}{3 \cdot \underset{1}{\cancel{5}}} = \dfrac{2 \cdot 1}{3 \cdot 1} = \dfrac{2}{3} \longleftarrow$$

Equivalent $\left(\dfrac{2}{3} = \dfrac{2}{3}\right)$

$$\dfrac{8}{12} = \dfrac{\overset{1}{\cancel{2}} \cdot \overset{1}{\cancel{2}} \cdot 2}{\underset{1}{\cancel{2}} \cdot \underset{1}{\cancel{2}} \cdot 3} = \dfrac{1 \cdot 1 \cdot 2}{1 \cdot 1 \cdot 3} = \dfrac{2}{3} \longleftarrow$$

Write each fraction in lowest terms.

53. $\dfrac{160}{256}$ $\dfrac{5}{8}$

54. $\dfrac{363}{528}$ $\dfrac{11}{16}$

55. $\dfrac{238}{119}$ $\dfrac{2}{1} = 2$

56. $\dfrac{570}{95}$ $\dfrac{6}{1} = 6$

Study Skills

USING STUDY CARDS

You may have used "flash cards" in other classes before. In math, study cards can be helpful, too. However, they are different because the main things to remember in math are *not* necessarily terms and definitions; they are *sets of steps to follow* to solve problems (and how to know which set of steps to follow) and *concepts about how math works* (principles). So, the cards will look different but will be just as useful.

In this two-part activity, you will find four types of study cards to use in math. Look carefully at what kinds of information to put on them and where to put it. Then use them the way you would any flash card:

▶ to quickly review when you have a few minutes,

▶ to do daily reviews,

▶ to review before a test.

Remember, the most helpful thing about study cards is making them. While you are making them, you have to do the kind of thinking that is most brain friendly, which improves your neural network of dendrites. After each card description you will find an assignment to try. It is marked **NOW TRY THIS.**

For **new vocabulary cards,** put the word (spelled correctly) and the page number where it is found on the front of the card. On the back, write:

▶ the definition (in your own words if possible),

▶ an example, an exception (if there are any),

▶ any related words, and

▶ a sample problem (if appropriate).

OBJECTIVES

1 **Create study cards for all new terms.**

2 **Create study cards for new procedures.**

New Vocabulary Cards

Prime Numbers   *p. 122*

Front of Card

Definition: Whole numbers that can only be divided by themselves and 1.

★Must divide evenly, NO remainders!

Ex: 2, 3, 5, 7, 11, 13, 17 are the first few primes.

→ NOT 0 or 1
→ Used in factoring
→ Related word: composite number

Back of Card

Why Are Study Cards Brain Friendly?

• Making cards is **active.**

• Cards are **visually** appealing.

• **Repetition** is good for your brain.

For details see Using Study Cards Revisited following **Section 2.8.**

Now Try This ▶▶▶

> *List four new vocabulary words/concepts you need to learn right now.*
> *Make a card for each one.*
>
> _____ _____ _____ _____

Procedure ("Steps") Cards

For **procedure cards,** write the name of the procedure at the top on the front of the card. Then write each step *in words*. If you need to know abbreviations for some words, include them along with the whole words written out. On the back, put an example of the procedure, showing each step you need to take. You can review by looking at the front and practicing a new worked example, or by looking at the back and remembering what the procedure is called and what the steps are.

Writing a fraction in lowest terms using prime factors

Use this method with larger denominators.

Step 1: Write prime factorization of numerator and denominator.

Step 2: Divide out all common factors.

Step 3: Multiply remaining factors.

Front of Card

Example: Write this fraction in lowest terms.

$$\frac{64}{112} = \frac{2 \cdot 2 \cdot 2 \cdot 2 \cdot 2 \cdot 2}{2 \cdot 2 \cdot 2 \cdot 2 \cdot 7}$$

Prime factors of 64.
Prime factors of 112.

Divide out common factors of 2

$$= \frac{\overset{1}{2} \cdot \overset{1}{2} \cdot \overset{1}{2} \cdot \overset{1}{2} \cdot 2 \cdot 2}{\underset{1}{2} \cdot \underset{1}{2} \cdot \underset{1}{2} \cdot \underset{1}{2} \cdot 7} = \frac{1 \cdot 1 \cdot 1 \cdot 1 \cdot 2 \cdot 2}{1 \cdot 1 \cdot 1 \cdot 1 \cdot 7} = \frac{4}{7}$$

multiply remaining factors

lowest terms

Back of Card

Now Try This ▶▶▶

> *What procedure are you learning right now? Make a "steps" card for it.*
>
> *Procedure:* _____

Summary Exercises on Fraction Basics

Write fractions to represent the shaded and unshaded portions of each figure.

1. 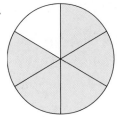 $\dfrac{5}{6}; \dfrac{1}{6}$

2. $\dfrac{1}{3}; \dfrac{2}{3}$

3. $\dfrac{5}{8}; \dfrac{3}{8}$

Identify the numerator and denominator.

	Numerator	**Denominator**

4. $\dfrac{3}{4}$ 3   4

5. $\dfrac{8}{5}$ 8   5

List the proper and improper fractions.

                       Proper   **Improper**

6. $\dfrac{8}{2} \quad \dfrac{3}{5} \quad \dfrac{16}{7} \quad \dfrac{8}{8} \quad \dfrac{4}{25} \quad \dfrac{1}{32}$ $\dfrac{3}{5}, \dfrac{4}{25}, \dfrac{1}{32}$ $\dfrac{8}{2}, \dfrac{16}{7}, \dfrac{8}{8}$

The bar graph below shows the countries that have produced the most men's winners in the New York City Marathon. Use this graph to answer Exercises 7–10.

Kenyans Run to Marathon Glory

Former champions and Kenyans Martin Lel, Rodgers Rop, and Stephen Kiogora are among the favorites in Sunday's ING New York City Marathon. Countries that have produced the most men's winners:

Kenya — 7
USA — 7
Italy — 3
Mexico — 3
South Africa — 2

Source: New York Road Runners.

7. What fraction of the winners were from Kenya?

$\dfrac{7}{22}$

8. What fraction of the winners were not from either Kenya or the United States?

$\dfrac{8}{22}$

9. What fraction of the winners were from either Mexico or South Africa?

$\dfrac{5}{22}$

10. What fraction of the winners were not from Kenya?

$\dfrac{15}{22}$

Write each improper fraction as a mixed number or whole number.

11. $\dfrac{5}{2}$ $2\dfrac{1}{2}$ **12.** $\dfrac{11}{8}$ $1\dfrac{3}{8}$ **13.** $\dfrac{9}{7}$ $1\dfrac{2}{7}$ **14.** $\dfrac{8}{3}$ $2\dfrac{2}{3}$

15. $\dfrac{64}{8}$ 8   **16.** $\dfrac{45}{9}$ 5   **17.** $\dfrac{36}{5}$ $7\dfrac{1}{5}$ **18.** $\dfrac{47}{10}$ $4\dfrac{7}{10}$

Write each mixed number as an improper fraction.

19. $3\frac{1}{3}$ $\frac{10}{3}$

20. $5\frac{3}{8}$ $\frac{43}{8}$

21. $6\frac{4}{5}$ $\frac{34}{5}$

22. $10\frac{3}{5}$ $\frac{53}{5}$

23. $12\frac{3}{4}$ $\frac{51}{4}$

24. $4\frac{10}{13}$ $\frac{62}{13}$

25. $11\frac{5}{6}$ $\frac{71}{6}$

26. $23\frac{5}{8}$ $\frac{189}{8}$

Write the prime factorization of each number using exponents.

27. 10 $2 \cdot 5$

28. 55 $5 \cdot 11$

29. 36 $2^2 \cdot 3^2$

30. 81 3^4

31. 280 $2^3 \cdot 5 \cdot 7$

32. 360 $2^3 \cdot 3^2 \cdot 5$

Write each fraction in lowest terms.

33. $\frac{4}{12}$ $\frac{1}{3}$

34. $\frac{7}{14}$ $\frac{1}{2}$

35. $\frac{4}{16}$ $\frac{1}{4}$

36. $\frac{15}{25}$ $\frac{3}{5}$

37. $\frac{18}{24}$ $\frac{3}{4}$

38. $\frac{30}{36}$ $\frac{5}{6}$

39. $\frac{56}{64}$ $\frac{7}{8}$

40. $\frac{6}{300}$ $\frac{1}{50}$

41. $\frac{25}{200}$ $\frac{1}{8}$

42. $\frac{125}{225}$ $\frac{5}{9}$

43. $\frac{88}{154}$ $\frac{4}{7}$

44. $\frac{70}{126}$ $\frac{5}{9}$

Write the numerator and denominator of each fraction as a product of prime factors and divide by the common factors. Then write the fraction in lowest terms.

45. $\frac{24}{36}$ $\dfrac{\cancel{2} \cdot \cancel{2} \cdot 2 \cdot \cancel{3}}{\cancel{2} \cdot \cancel{2} \cdot 3 \cdot \cancel{3}} = \dfrac{2}{3}$

46. $\frac{80}{160}$ $\dfrac{\cancel{2} \cdot \cancel{2} \cdot \cancel{2} \cdot \cancel{2} \cdot \cancel{5}}{\cancel{2} \cdot \cancel{2} \cdot \cancel{2} \cdot \cancel{2} \cdot 2 \cdot \cancel{5}} = \dfrac{1}{2}$

47. $\frac{126}{42}$ $\dfrac{\cancel{2} \cdot \cancel{3} \cdot 3 \cdot \cancel{7}}{\cancel{2} \cdot \cancel{3} \cdot \cancel{7}} = 3$

48. $\frac{96}{112}$ $\dfrac{\cancel{2} \cdot \cancel{2} \cdot \cancel{2} \cdot \cancel{2} \cdot 2 \cdot 3}{\cancel{2} \cdot \cancel{2} \cdot \cancel{2} \cdot \cancel{2} \cdot 7} = \dfrac{6}{7}$

2.5 ▶▶▶ Multiplying Fractions

OBJECTIVE 1 Multiply fractions. Suppose that you give $\frac{1}{2}$ of your Energy Bar to your kickboxing partner Jennifer. Then Jennifer gives $\frac{1}{2}$ of her share to Tony. How much of the Energy Bar does Tony get to eat?

Start with a sketch showing the Energy Bar cut in half (2 equal pieces).

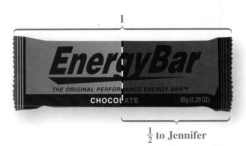

$\frac{1}{2}$ to Jennifer

Next, take $\frac{1}{2}$ of the shaded area. (Here we are dividing $\frac{1}{2}$ into 2 equal parts and shading one darker than the other.)

The sketch shows that Tony gets $\frac{1}{4}$ of the Energy Bar.

$\frac{1}{2}$ of $\frac{1}{2}$ to Tony

Tony gets $\frac{1}{2}$ of $\frac{1}{2}$ of the Energy Bar. When used between two fractions, the word **of** tells us to multiply.

$$\frac{1}{2} \text{ of } \frac{1}{2} \quad \text{means} \quad \frac{1}{2} \cdot \frac{1}{2}$$

Tony's share of the Energy Bar is

$$\frac{1}{2} \cdot \frac{1}{2} = \frac{1}{4}$$

Work Problem **1** *at the Side.* ▶

The rule for multiplying fractions follows.

> **Multiplying Fractions**
> Multiply two fractions by multiplying the numerators and multiplying the denominators.

OBJECTIVES

1. **Multiply fractions.**
2. **Use a multiplication shortcut.**
3. **Multiply a fraction and a whole number.**
4. **Find the area of a rectangle.**

1 Use these figures to find $\frac{1}{4}$ of $\frac{1}{2}$.

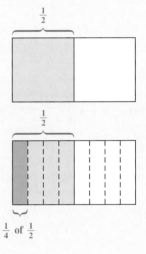

$\frac{1}{4}$ of $\frac{1}{2}$

ANSWER

1. $\frac{1}{8}$

2 Multiply. Write answers in lowest terms.

(a) $\dfrac{1}{2} \cdot \dfrac{3}{4}$

(b) $\dfrac{3}{5} \cdot \dfrac{1}{3}$

(c) $\dfrac{5}{6} \cdot \dfrac{1}{2} \cdot \dfrac{1}{8}$

(d) $\dfrac{1}{2} \cdot \dfrac{3}{4} \cdot \dfrac{3}{8}$

Use this rule to find the product of $\frac{2}{3}$ and $\frac{1}{3}$ (multiply $\frac{2}{3}$ by $\frac{1}{3}$).

$$\frac{2}{3} \cdot \frac{1}{3} = \frac{2 \cdot 1}{3 \cdot 3} \quad \begin{matrix} \leftarrow \text{Multiply numerators.} \\ \leftarrow \text{Multiply denominators.} \end{matrix}$$

Finish multiplying.

$$\frac{2}{3} \cdot \frac{1}{3} = \frac{2 \cdot 1}{3 \cdot 3} = \frac{2}{9} \quad \begin{matrix} \leftarrow 2 \cdot 1 = 2 \\ \leftarrow 3 \cdot 3 = 9 \end{matrix}$$

Check that the final result is in lowest terms. $\frac{2}{9}$ is in lowest terms because 2 and 9 have no common factor other than 1.

EXAMPLE 1 **Multiplying Fractions**

Multiply. Write answers in lowest terms.

(a) $\dfrac{5}{8} \cdot \dfrac{3}{4}$

Multiply the numerators and multiply the denominators.

$$\frac{5}{8} \cdot \frac{3}{4} = \frac{5 \cdot 3}{8 \cdot 4} = \frac{15}{32} \qquad \boxed{\text{Already in lowest terms}}$$

Notice that 15 and 32 have no common factors other than 1, so the answer is in lowest terms.

(b) $\dfrac{4}{7} \cdot \dfrac{2}{5}$

$$\frac{4}{7} \cdot \frac{2}{5} = \frac{4 \cdot 2}{7 \cdot 5} = \frac{8}{35} \quad \leftarrow \text{Lowest terms}$$

(c) $\dfrac{5}{8} \cdot \dfrac{3}{4} \cdot \dfrac{1}{2}$

$$\frac{5}{8} \cdot \frac{3}{4} \cdot \frac{1}{2} = \frac{5 \cdot 3 \cdot 1}{8 \cdot 4 \cdot 2} = \frac{15}{64} \quad \leftarrow \text{Lowest terms}$$

◀ *Work Problem* **2** *at the Side.*

OBJECTIVE **2** **Use a multiplication shortcut.** A multiplication shortcut that can be used with fractions is shown in Example 2.

EXAMPLE 2 **Using the Multiplication Shortcut**

Multiply $\frac{5}{6}$ and $\frac{9}{10}$. Write the answer in lowest terms.

$$\frac{5}{6} \cdot \frac{9}{10} = \frac{5 \cdot 9}{6 \cdot 10} = \frac{45}{60} \quad \leftarrow \text{Not in lowest terms}$$

The numerator and denominator have a common factor other than 1, so write the prime factorization of each number.

$$\frac{5}{6} \cdot \frac{9}{10} = \frac{5 \cdot 9}{6 \cdot 10} = \frac{5 \cdot 3 \cdot 3}{2 \cdot 3 \cdot 2 \cdot 5} \qquad \boxed{\text{Write the prime factorization of each number.}}$$

Continued on Next Page

ANSWERS

2. (a) $\dfrac{3}{8}$ (b) $\dfrac{1}{5}$ (c) $\dfrac{5}{96}$ (d) $\dfrac{9}{64}$

Next, divide by the common factors of 5 and 3.

$$\frac{5}{6} \cdot \frac{9}{10} = \frac{5 \cdot 9}{6 \cdot 10} = \frac{\overset{1}{\cancel{5}} \cdot \overset{1}{\cancel{3}} \cdot 3}{2 \cdot \underset{1}{\cancel{3}} \cdot 2 \cdot \underset{1}{\cancel{5}}}$$

Finally, multiply the remaining factors in the numerator and in the denominator.

$$\frac{5}{6} \cdot \frac{9}{10} = \frac{1 \cdot 1 \cdot 3}{2 \cdot 1 \cdot 2 \cdot 1} = \frac{3}{4} \quad \leftarrow \text{Lowest terms}$$

As a shortcut, instead of writing the prime factorization of each number, find the product of $\frac{5}{6}$ and $\frac{9}{10}$ as follows.

First, divide by 5, a common factor of both 5 and 10. $\dfrac{\overset{1}{\cancel{5}}}{6} \cdot \dfrac{9}{\underset{2}{\cancel{10}}}$

Next, divide by 3, a common factor of both 6 and 9. $\dfrac{\overset{1}{\cancel{5}}}{\underset{2}{\cancel{6}}} \cdot \dfrac{\overset{3}{\cancel{9}}}{\underset{2}{\cancel{10}}}$

Finally, multiply numerators and multiply denominators. $\dfrac{1 \cdot 3}{2 \cdot 2} = \dfrac{3}{4}$

CAUTION
When using the multiplication shortcut, you are dividing a numerator and a denominator by a common factor. Be certain that you divide a numerator and a denominator **by the same number.** If you do all possible divisions, your answer will be in lowest terms.

EXAMPLE 3   **Using the Multiplication Shortcut**

Use the multiplication shortcut to find each product. Write the answers in lowest terms and as mixed numbers where possible.

(a) $\dfrac{6}{11} \cdot \dfrac{7}{8}$

Divide both 6 and 8 by their common factor of 2. Notice that 7 and 11 have no common factor. Then multiply.

$$\dfrac{\overset{3}{\cancel{6}}}{11} \cdot \dfrac{7}{\underset{4}{\cancel{8}}} = \frac{3 \cdot 7}{11 \cdot 4} = \frac{21}{44} \quad \leftarrow \text{Lowest terms}$$

(b) $\dfrac{7}{10} \cdot \dfrac{20}{21}$

> Divide a numerator and a denominator by the same number.

Divide 7 and 21 by 7, and divide 10 and 20 by 10.

$$\dfrac{\overset{1}{\cancel{7}}}{\underset{1}{\cancel{10}}} \cdot \dfrac{\overset{2}{\cancel{20}}}{\underset{3}{\cancel{21}}} = \frac{1 \cdot 2}{1 \cdot 3} = \frac{2}{3} \quad \leftarrow \text{Lowest terms}$$

Continued on Next Page

3 Use the multiplication shortcut to find each product.

(a) $\dfrac{3}{4} \cdot \dfrac{2}{3}$

(b) $\dfrac{6}{11} \cdot \dfrac{33}{21}$

(c) $\dfrac{20}{4} \cdot \dfrac{3}{40} \cdot \dfrac{1}{3}$

(d) $\dfrac{18}{17} \cdot \dfrac{1}{36} \cdot \dfrac{2}{3}$

(c) $\dfrac{35}{12} \cdot \dfrac{32}{25}$

$$\dfrac{\overset{7}{\cancel{35}}}{\underset{3}{\cancel{12}}} \cdot \dfrac{\overset{8}{\cancel{32}}}{\underset{5}{\cancel{25}}} = \dfrac{7 \cdot 8}{3 \cdot 5} = \dfrac{56}{15} \quad \text{or} \quad 3\dfrac{11}{15} \;\leftarrow \text{Mixed number}$$

(d) $\dfrac{2}{3} \cdot \dfrac{8}{15} \cdot \dfrac{3}{4}$

$$\dfrac{\overset{1}{\cancel{2}}}{\underset{1}{\cancel{3}}} \cdot \dfrac{\overset{4}{\cancel{8}}}{15} \cdot \dfrac{\overset{1}{\cancel{3}}}{\underset{\underset{1}{2}}{\cancel{4}}} = \dfrac{1 \cdot 4 \cdot 1}{1 \cdot 15 \cdot 1} = \dfrac{4}{15} \;\leftarrow \text{Lowest terms}$$

This shortcut is especially helpful when the fractions involve large numbers.

> **Note**
>
> There is no specific order that must be used when dividing numerators and denominators, as long as both the numerator and the denominator are divided by the *same* number.

◀ *Work Problem* **3** *at the Side.*

OBJECTIVE **3** **Multiply a fraction and a whole number.** The rule for multiplying a fraction and a whole number follows.

> **Multiplying a Whole Number and a Fraction**
>
> Multiply a whole number and a fraction by writing the whole number as a fraction with a denominator of 1.

For example, write the whole numbers 8, 10, and 25 as follows.

$$8 = \dfrac{8}{1} \qquad 10 = \dfrac{10}{1} \qquad 25 = \dfrac{25}{1} \;\;\overset{\text{Write the whole}}{\underset{\text{number over 1.}}{\longleftarrow}}$$

EXAMPLE 4   **Multiplying by Whole Numbers**

Multiply. Write answers in lowest terms and as whole numbers where possible.

(a) $8 \cdot \dfrac{3}{4}$

Write 8 as $\frac{8}{1}$ and multiply.

$$8 \cdot \dfrac{3}{4} = \dfrac{\overset{2}{\cancel{8}}}{1} \cdot \dfrac{3}{\underset{1}{\cancel{4}}} = \dfrac{2 \cdot 3}{1 \cdot 1} = \dfrac{6}{1} = 6 \;\;\longleftarrow \boxed{\begin{array}{l}\frac{6}{1} \text{ is the same as}\\ 6 \div 1, \text{ which}\\ \text{equals } 6.\end{array}}$$

Continued on Next Page

ANSWERS

3. **(a)** $\dfrac{\overset{1}{\cancel{3}}}{\underset{2}{\cancel{4}}} \cdot \dfrac{\overset{1}{\cancel{2}}}{\underset{1}{\cancel{3}}} = \dfrac{1}{2}$ **(b)** $\dfrac{\overset{2}{\cancel{6}}}{\underset{1}{\cancel{11}}} \cdot \dfrac{\overset{3}{\cancel{33}}}{\underset{7}{\cancel{21}}} = \dfrac{6}{7}$

(c) $\dfrac{\overset{1}{\cancel{20}}}{4} \cdot \dfrac{\overset{1}{\cancel{3}}}{\underset{2}{\cancel{40}}} \cdot \dfrac{1}{\underset{1}{\cancel{3}}} = \dfrac{1}{8}$

(d) $\dfrac{\overset{1}{\cancel{18}}}{17} \cdot \dfrac{1}{36} \cdot \dfrac{\overset{1}{\cancel{2}}}{\underset{\underset{1}{2}}{\cancel{3}}} = \dfrac{1}{51}$

(b) $15 \cdot \dfrac{5}{6}$

$$15 \cdot \frac{5}{6} = \frac{\overset{5}{\cancel{15}}}{1} \cdot \frac{5}{\underset{2}{\cancel{6}}} = \frac{5 \cdot 5}{1 \cdot 2} = \frac{25}{2} = 12\frac{1}{2}$$

Work Problem **4** *at the Side.* ▶

OBJECTIVE 4 Find the area of a rectangle. To find the area of a rectangle (the amount of surface inside the rectangle), use the following formula.

Area of a Rectangle
The area of a rectangle is equal to the length multiplied by the width.
Area = length • width

For example, the rectangle shown here has an area of 12 square feet (ft²).

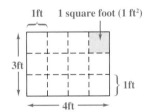

Area = length • width ◁ Area is the amount of surface.

Area = 4 ft • 3 ft

Area = 12 ft²

Other units for measuring area are square inches (in.²), square yards (yd²), and square miles (mi²). (See **Section 8.3** for more information on area.)

EXAMPLE 5 **Applying Fraction Skills**

Find the area of each rectangle.

(a) Find the area of each shower tile.

$\frac{3}{4}$ ft

$|\!\leftarrow \frac{11}{12} \text{ ft} \rightarrow\!|$

Area = length • width

Area = $\dfrac{11}{12} \cdot \dfrac{3}{4}$

$= \dfrac{11}{\underset{4}{\cancel{12}}} \cdot \dfrac{\overset{1}{\cancel{3}}}{4}$ Divide numerator and denominator by 3.

$= \dfrac{11}{16}$ ft² ◁ This is less than 1 ft².

Continued on Next Page

---

4 Multiply. Write answers in lowest terms and as whole numbers or mixed numbers where possible.

(a) $8 \cdot \dfrac{1}{8}$

(b) $\dfrac{3}{4} \cdot 5 \cdot \dfrac{5}{3}$

(c) $\dfrac{3}{5} \cdot 40$

(d) $\dfrac{3}{25} \cdot \dfrac{5}{11} \cdot 99$

ANSWERS

4. **(a)** 1 **(b)** $6\dfrac{1}{4}$ **(c)** 24 **(d)** $5\dfrac{2}{5}$

5 Find the area of each rectangle.

(a)

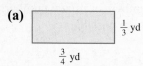

$\frac{1}{3}$ yd

$\frac{3}{4}$ yd

(b) a community college campus that is $\frac{3}{8}$ mile by $\frac{1}{3}$ mile

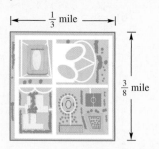

$\frac{1}{3}$ mile

$\frac{3}{8}$ mile

(c) a parcel of subdivided land that is $\frac{9}{7}$ mile by $\frac{7}{12}$ mile

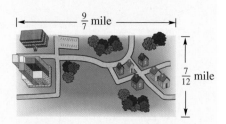

$\frac{9}{7}$ mile

$\frac{7}{12}$ mile

(b) Find the area of this polished diamond plate SUV running board.

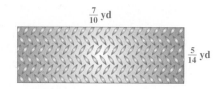

$\frac{7}{10}$ yd

$\frac{5}{14}$ yd

Multiply the length and width.

$$\text{Area} = \frac{7}{10} \cdot \frac{5}{14}$$

$$= \frac{\overset{1}{\cancel{7}}}{\underset{2}{\cancel{10}}} \cdot \frac{\overset{1}{\cancel{5}}}{\underset{2}{\cancel{14}}}$$

Divide 7 and 14 by 7.

Divide 10 and 5 by 5.

$$= \frac{1}{4} \text{ square yard (yd}^2)$$

◀ *Work Problem* **5** *at the Side.*

2.5 ▶▶▶ **Exercises**

FOR EXTRA HELP **MyMathLab** | Math XL PRACTICE | WATCH | DOWNLOAD | READ | REVIEW

Multiply. Write answers in lowest terms. See Examples 1–3.

1. $\frac{1}{3} \cdot \frac{3}{4}$ $\frac{1}{4}$

2. $\frac{2}{5} \cdot \frac{3}{4}$ $\frac{3}{10}$

3. $\frac{2}{7} \cdot \frac{1}{5}$ $\frac{2}{35}$

4. $\frac{2}{3} \cdot \frac{1}{2}$ $\frac{1}{3}$

5. $\frac{8}{5} \cdot \frac{15}{32}$ $\frac{3}{4}$

6. $\frac{5}{9} \cdot \frac{4}{3}$ $\frac{20}{27}$

7. $\frac{2}{3} \cdot \frac{7}{12} \cdot \frac{9}{14}$ $\frac{1}{4}$

8. $\frac{7}{8} \cdot \frac{16}{21} \cdot \frac{1}{2}$ $\frac{1}{3}$

9. $\frac{3}{4} \cdot \frac{5}{6} \cdot \frac{2}{3}$ $\frac{5}{12}$

10. $\frac{2}{5} \cdot \frac{3}{8} \cdot \frac{2}{3}$ $\frac{1}{10}$

11. $\frac{9}{22} \cdot \frac{11}{16}$ $\frac{9}{32}$

12. $\frac{5}{12} \cdot \frac{7}{10}$ $\frac{7}{24}$

13. $\frac{5}{8} \cdot \frac{16}{25}$ $\frac{2}{5}$

14. $\frac{6}{11} \cdot \frac{22}{15}$ $\frac{4}{5}$

15. $\frac{14}{25} \cdot \frac{65}{48} \cdot \frac{15}{28}$ $\frac{13}{32}$

16. $\frac{35}{64} \cdot \frac{32}{15} \cdot \frac{27}{72}$ $\frac{7}{16}$

17. $\frac{16}{25} \cdot \frac{35}{32} \cdot \frac{15}{64}$ $\frac{21}{128}$

18. $\frac{39}{42} \cdot \frac{7}{13} \cdot \frac{7}{24}$ $\frac{7}{48}$

Multiply. Write answers in lowest terms and as whole or mixed numbers where possible.
See Example 4.

19. $5 \cdot \frac{4}{5}$ 4

20. $20 \cdot \frac{3}{4}$ 15

21. $\frac{5}{8} \cdot 64$ 40

22. $\frac{5}{6} \cdot 24$ 20

23. $36 \cdot \frac{2}{3}$ 24

24. $30 \cdot \frac{3}{10}$ 9

25. $36 \cdot \frac{5}{8} \cdot \frac{9}{15}$ $13\frac{1}{2}$

26. $35 \cdot \frac{3}{5} \cdot \frac{1}{2}$ $10\frac{1}{2}$

27. $100 \cdot \frac{21}{50} \cdot \frac{3}{4}$ $31\frac{1}{2}$

28. $400 \cdot \dfrac{7}{8}$ 350

29. $\dfrac{2}{5} \cdot 200$ 80

30. $\dfrac{6}{7} \cdot 245$ 210

31. $142 \cdot \dfrac{2}{3}$ $94\dfrac{2}{3}$

32. $\dfrac{12}{25} \cdot 430$ $206\dfrac{2}{5}$

33. $\dfrac{28}{21} \cdot 640 \cdot \dfrac{15}{32}$ 400

34. $\dfrac{21}{13} \cdot 520 \cdot \dfrac{7}{20}$ 294

35. $\dfrac{54}{38} \cdot 684 \cdot \dfrac{5}{6}$ 810

36. $\dfrac{76}{43} \cdot 473 \cdot \dfrac{5}{19}$ 220

Find the area of each rectangle. See Example 5.

37. 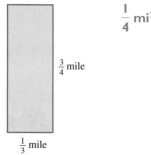 $\dfrac{1}{4}$ mi^2

$\frac{3}{4}$ mile

$\frac{1}{3}$ mile

38. 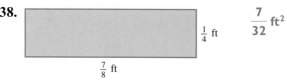 $\dfrac{7}{32}$ ft^2

$\frac{1}{4}$ ft

$\frac{7}{8}$ ft

39.  $\frac{3}{4}$ meter   **9 meters^2**

12 meters

40. 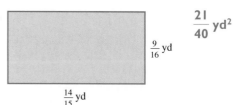 $\frac{3}{8}$ in.   **3 in.^{2}**

8 in.

41. $\frac{3}{14}$ in. $\dfrac{3}{10}$ in.^{2}

$\frac{7}{5}$ in.

42. $\dfrac{21}{40}$ yd^2

$\frac{9}{16}$ yd

$\frac{14}{15}$ yd

43. Write in your own words the rule for multiplying fractions. Make up an example problem to show how this works.

Multiply the numerators and multiply the denominators. An example is

$$\frac{3}{4} \cdot \frac{1}{2} = \frac{3 \cdot 1}{4 \cdot 2} = \frac{3}{8}$$

44. A useful shortcut when multiplying fractions is to divide a numerator and a denominator by the same number. Describe how this works and give an example.

You must divide a numerator and a denominator by the same number. If you do all possible divisions, your answer will be in lowest terms. One example is

$$\frac{3}{4} \cdot \frac{2}{3} = \frac{3 \cdot 2}{4 \cdot 3} = \frac{\overset{1}{\cancel{3}} \cdot \overset{1}{\cancel{2}}}{\underset{2}{\cancel{4}} \cdot \underset{1}{\cancel{3}}} = \frac{1}{2}$$

Find the area of each rectangle in these application problems. Write answers in lowest terms and as whole or mixed numbers where possible. See Example 5.

45. Find the area of a heating-duct grill having a length of 2 yd and a width of $\frac{3}{4}$ yd.

$1\frac{1}{2}$ yd²

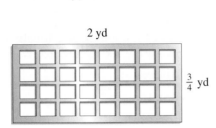

2 yd

$\frac{3}{4}$ yd

46. Find the area of the top of an office desk having a length of 2 yd and a width of $\frac{7}{8}$ yd.

$1\frac{3}{4}$ yd²

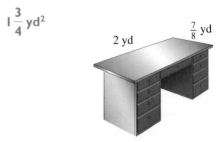

2 yd $\frac{7}{8}$ yd

47. A wildfire is contained in an area measuring $\frac{7}{8}$ mile by 4 miles. Find the total area of the containment.

$3\frac{1}{2}$ mi²

48. A motorcycle racecourse is $\frac{2}{3}$ mile wide by 4 miles long. Find the area of the racecourse.

$2\frac{2}{3}$ mi²

49. The Sunny Side Soccer Park is $\frac{1}{4}$ mile long and $\frac{3}{16}$ mile wide, while the Creek Side Soccer Park is $\frac{3}{8}$ mile long and $\frac{1}{8}$ mile wide. Which park has the larger area?

They are both the same size: $\frac{3}{64}$ mi².

50. The Rocking Horse Ranch is $\frac{3}{4}$ mile long and $\frac{2}{3}$ mile wide. The Silver Spur Ranch is $\frac{5}{8}$ mile long and $\frac{4}{5}$ mile wide. Which ranch has the larger area?

They are both the same size: $\frac{1}{2}$ mi².

Relating Concepts (Exercises 51–56) For Individual or Group Work

Front end rounding can be used to estimate an answer when multiplying fractions.
Work Exercises 51–56 in order. *Round exact answers to the nearest whole number.*

The bar graph shows the six largest fast-food chains by the number of stores.

Who's Hungry?
Largest Fast-Food Chains
(based on the number of stores)

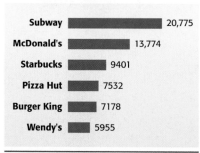

Subway	20,775
McDonald's	13,774
Starbucks	9401
Pizza Hut	7532
Burger King	7178
Wendy's	5955

Source: Technomic.

51. Use front end rounding to estimate the total number of stores for the six largest chains.

60,000 stores

52. Find the exact total number of stores for the six largest chains.

64,615 stores

53. If $\frac{2}{3}$ of the Starbucks stores are in medium to large population areas, use front end rounding to estimate, and then find the exact number of stores in these population areas.

Estimate: $\frac{2}{3} \cdot 9000 = 6000$

Exact: **6267 stores (rounded)**

54. If $\frac{2}{5}$ of the Pizza Hut stores are in shopping centers and food courts, use front end rounding to estimate, and then find the exact number of stores in these locations.

Estimate: $\frac{2}{5} \cdot 8000 = 3200$

Exact: **3013 stores (rounded)**

Compare the estimated answers and the exact answers in Exercises 53 and 54. How can you get an estimated answer that is closer to the exact answer? Try rounding the number of stores to some multiple of the denominator in the fraction that has two *nonzero digits.*

55. Refer to Exercise 53. Round the number of Starbucks locations to some multiple of the denominator in $\frac{2}{3}$ that has *two* nonzero digits. Now estimate the answer, showing your work.

$\frac{2}{3} \cdot 9300 = 6200$ **stores**

56. Refer to Exercise 54. Round the number of Pizza Hut stores to some multiple of the denominator in $\frac{2}{5}$ that has *two* nonzero digits. Now estimate the answer, showing your work.

$\frac{2}{5} \cdot 7500 = 3000$ **stores**

2.6 ▶▶▶ Applications of Multiplication

OBJECTIVE 1 Solve fraction application problems using multiplication. Many application problems are solved by multiplying fractions. Use the following indicator words for multiplication.

product
double
triple
times
of (when "of" follows a fraction)
twice
twice as much

Always look for indicator words.

Look for these indicator words in the following examples.

OBJECTIVE

1 Solve fraction application problems using multiplication.

EXAMPLE 1 Applying Indicator Words

Lois Stevens gives $\frac{1}{10}$ of her income to her church. One month she earned $2980. How much did she give to the church that month?

Step 1 **Read** the problem. The problem asks us to find the amount of money given to the church.

Step 2 **Work out a plan.** The indicator word is *of*: Stevens gave $\frac{1}{10}$ *of* her income. When it follows a fraction, the word *of* indicates multiplication, so find the amount given to the church by multiplying $\frac{1}{10}$ and $2980.

Step 3 **Estimate** a reasonable answer. Round the income of $2980 to $3000. Then divide $3000 by 10 to find $\frac{1}{10}$ of the income (one of 10 equal parts). Our estimate is $3000 ÷ 10 = $300. (Recall the shortcut for dividing by 10; drop one 0 from the dividend.)

Step 4 **Solve** the problem.

$$\text{amount} = \frac{1}{\cancel{10}_{1}} \cdot \frac{\overset{298}{\cancel{2980}}}{1} = \frac{298}{1} = 298$$

Step 5 **State the answer.** Stevens gave $298 to her church that month.

Step 6 **Check.** The exact answer, $298, is close to our estimate of $300.

Work Problem **1** *at the Side.* ▶

EXAMPLE 2 Solving a Fraction Application Problem

Of the 39 students in Carol Dixon's high school biology class, $\frac{2}{3}$ plan to go to college. How many plan to go to college?

Step 1 **Read** the problem. The problem asks us to find the number of students who plan to go to college.

Step 2 **Work out a plan.** Reword the problem to read

$\frac{2}{3}$ **of** the students plan to go to college.
   ↑
Indicator word for multiplication
when it follows a fraction

Continued on Next Page

1 Solve each problem.

(a) Eric and Sabrina Means are saving $\frac{3}{8}$ of their income for the down payment on their first home. If they have a combined annual income of $81,576, how much can they save in a year?

(b) A retiring firefighter will receive $\frac{5}{8}$ of her highest annual salary as retirement income. If her highest annual salary is $62,504, how much will she receive as retirement income?

ANSWERS

1. (a) $30,591 **(b)** $39,065

2 At Sid's Pharmacy, $\frac{5}{16}$ of the prescriptions are paid by a third party (insurance company). If 3696 prescriptions are filled, find the number paid by a third party.

Use the six problem-solving steps.

Step 3   **Estimate** a reasonable answer. Round the number of students in the class from 39 to 40. Then, $\frac{1}{2}$ of 40 is 20. Since $\frac{2}{3}$ is more than $\frac{1}{2}$, our estimate is that "more than 20 students" plan to go to college.

Step 4   **Solve** the problem. Find the number who plan to go to college by multiplying $\frac{2}{3}$ and 39.

$$\text{number who plan to go} = \frac{2}{3} \cdot 39$$

$$= \frac{2}{\overset{}{\underset{1}{3}}} \cdot \frac{\overset{13}{\cancel{39}}}{1} = \frac{26}{1} = 26$$

Step 5   **State the answer.** 26 students plan to go to college.

Step 6   **Check.** The exact answer, 26, fits our estimate of "more than 20."

◀ *Work Problem* **2** *at the Side.*

EXAMPLE 3   **Finding a Fraction of a Fraction**

In her will, a woman divides her estate into 6 equal parts. Five of the 6 parts are given to relatives. Of the sixth part, $\frac{1}{3}$ goes to the Salvation Army. What fraction of her total estate goes to the Salvation Army?

Step 1   **Read** the problem. The problem asks for the fraction of an estate that goes to the Salvation Army.

Step 2   **Work out a plan.** Reword the problem to read

the Salvation Army gets $\frac{1}{3}$ **of** $\frac{1}{6}$.

↑
Indicator word for multiplication when it follows a fraction

3 At our college, $\frac{1}{3}$ of the students speak a foreign language. Of those speaking a foreign language, $\frac{3}{4}$ speak Spanish. What fraction of the students speak Spanish?

Use the six problem-solving steps.

Step 3   **Estimate** a reasonable answer. If the estate is divided into 6 equal parts and each of these parts was divided into 3 equal parts, we would have 6 · 3 = 18 equal parts. Our estimate is $\frac{1}{18}$.

Step 4   **Solve** the problem. The Salvation Army gets $\frac{1}{3}$ **of** $\frac{1}{6}$.

↑
Indicator word

To find the fraction that the Salvation Army is to receive, multiply $\frac{1}{3}$ and $\frac{1}{6}$.

$$\text{fraction to Salvation Army} = \frac{1}{3} \cdot \frac{1}{6}$$

$$= \frac{1}{18}$$

Step 5   **State the answer.** The Salvation Army gets $\frac{1}{18}$ of the total estate.

Step 6   **Check.** The exact answer, $\frac{1}{18}$, matches our estimate.

> Remember to check your work.

◀ *Work Problem* **3** *at the Side.*

ANSWERS

2. 1155 prescriptions

3. $\frac{1}{4}$ speak Spanish

EXAMPLE 4 **Using Fractions with a Circle Graph**

The circle graph, or pie chart, shows where children 8 to 17 years of age make food purchases when away from home. If 2500 children were in the survey, find the number of children who buy food in the school cafeteria.

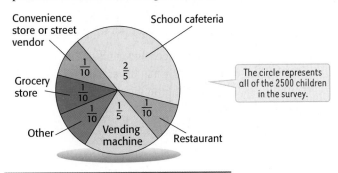

FINDING FOOD

Children ages 8 to 17 made food purchases at the following locations:

The circle represents all of the 2500 children in the survey.

Source: Pursuant Inc. for American Dietetic Association Foundation.

Step 1 **Read** the problem. The problem asks for the number of children who buy food in the school cafeteria.

Step 2 **Work out a plan.** Reword the problem to read

$\frac{2}{5}$ of 2500 children buy food in the school cafeteria.

Indicator word for multiplication when it follows a fraction

Step 3 **Estimate** a reasonable answer. $\frac{1}{2}$ of 2500 people is 1250 people. $\frac{2}{5}$ is less than $\frac{1}{2}$, so our estimate is "less than 1250 people."

Step 4 **Solve** the problem. Find the number who buy food in the school cafeteria by multiplying $\frac{2}{5}$ and 2500.

$$\text{number in school cafeteria} = \frac{2}{5} \cdot 2500$$

$$= \frac{2}{5} \cdot \frac{2500}{1}$$

$$= \frac{2}{\overset{1}{\cancel{5}}} \cdot \frac{\overset{500}{\cancel{2500}}}{1} \quad \text{Divide both numerator and denominator by 5.}$$

$$= \frac{1000}{1} = 1000$$

Step 5 **State the answer.** 1000 children buy food in the school cafeteria.

Step 6 **Check.** The exact answer, 1000 children, fits our estimate of "less than 1250 children."

Work Problem **4** *at the Side.* ▶

---

4 Solve each problem using the six problem-solving steps. Use the circle graph in Example 4.

(a) What fraction of the children buy food from vending machines?

(b) What number of children buy food from vending machines?

(c) What fraction of the children buy food from a convenience store or street vendor?

(d) What number of children buy food from a convenience store or street vendor?

ANSWERS

4. **(a)** $\frac{1}{5}$ **(b)** 500 children **(c)** $\frac{1}{10}$
   (d) 250 children

Math in the Media

HEART-RATE TRAINING ZONE

The Kaiser Permanente *Healthwise Handbook* says that "no one can prescribe the perfect fitness plan for you." However, your personal fitness plan should include aerobic conditioning that strengthens your heart and lungs. Activities that provide aerobic conditioning include brisk walking, running, stair climbing, biking, swimming, aerobic dance, or anything else that raises your heart rate to within your training zone for a minimum of 12 minutes.

If you train at the higher end of the training zone, you will burn glycogen and improve aerobic fitness. Training for longer periods at the lower end of the training zone results in your body using fat reserves for energy. Wickipedia, The Free Encyclopedia, at www.wickipedia.org, explains the calculations used to find maximum heart rates and training zones.

Example: The training zone (TZ) is based on your heart rate (HR) for one minute. To see if you are in the training zone, measure your heart rate for 15 seconds. Compare it to the 15-second training zone. Find the exact answer, and then round to the nearest whole number.

Instruction	Calculation	Example (age 22)
Calculate maximum heart rate (MHR).	$220 -$ your age	$220 - 22 = 198$
Calculate lower limit of training zone (TZ).	$\frac{3}{5} \times$ (MHR)	$\frac{3}{5} \times (198) = \frac{594}{5} = 118\frac{4}{5}$
Calculate upper limit of training zone (TZ).	$\frac{4}{5} \times$ (MHR)	$\frac{4}{5} \times (198) = \frac{792}{5} = 158\frac{2}{5}$
Calculate the exact 15-second training zone. Round the results to the nearest whole number.	$\left(\frac{1}{4} \times \text{lower TZ}, \frac{1}{4} \times \text{Upper TZ}\right)$	$\frac{1}{4} \times \frac{594}{5} = 29\frac{7}{10}; \frac{1}{4} \times \frac{792}{5} = 39\frac{3}{5}$ HR between $29\frac{7}{10}$ and $39\frac{3}{5}$ HR between 30 and 40

1. Suppose you work in a physical fitness center and decide to design a poster to remind the clients of the training zone for their age. Compute the exact 15-second training zone for people of each of the ages in the table. Write fractions in lowest terms. Then round the answers to the nearest whole number.

2. Explain why the lower and upper training zones (TZ) are multiplied by $\frac{1}{4}$.

$$\frac{15 \text{ seconds}}{60 \text{ seconds}} = \frac{1}{4}$$

Age	MHR	Lower Limit of TZ	Upper Limit of TZ	15-Second TZ (exact)	15-Second TZ (rounded)
18	202	$121\frac{1}{5}$	$161\frac{3}{5}$	HR between $30\frac{3}{10}$ and $40\frac{2}{5}$	HR between 30 and 40
25	195	117	156	HR between $29\frac{1}{4}$ and 39	HR between 29 and 39
30	190	114	152	HR between $28\frac{1}{2}$ and 38	HR between 29 and 38
40	180	108	144	HR between 27 and 36	HR between 27 and 36
50	170	102	136	HR between $25\frac{1}{2}$ and 34	HR between 26 and 34
60	160	96	128	HR between 24 and 32	HR between 24 and 32

2.6 ▶▶▶ Exercises

Solve each application problem. Look for indicator words. See Examples 1–4.

1. A file cabinet top is $\frac{3}{4}$ yd by $\frac{2}{3}$ yd. Find its area.

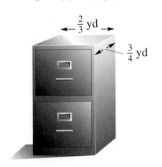

$\frac{1}{2}$ **yd²**

2. The rectangular floor of Darby's house measures $\frac{14}{15}$ yd by $\frac{3}{4}$ yd. Find its area.

$\frac{7}{10}$ **yd²**

3. A cookie sheet is $\frac{4}{3}$ ft by $\frac{2}{3}$ ft. Find its area.

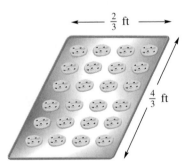

$\frac{8}{9}$ **ft.²**

4. Each day there are 16 million people who shop at flea markets. If $\frac{2}{5}$ of these people purchase produce at the flea market, how many purchase produce? (*Source:* National Flea Market Association.)

6,400,000 people

5. Pete is helping Colin make a rectangular mahogany lamp table for Carolyn's birthday. Find the area of the top of the table if it is $\frac{4}{5}$ yd long by $\frac{3}{8}$ yd wide.

$\frac{3}{10}$ **yd²**

6. A convenience store sells 1650 items, of which $\frac{12}{25}$ are classified as junk food. How many of the items are junk food?

792 items

7. Dan Crump had expenses of $6848 during one semester of college. His part-time job provided $\frac{3}{8}$ of the amount he needed. How much did he earn on his job?

$2568

8. Erin Hernandez produces $5680 in profits for her employer. If her personal earnings are $\frac{2}{5}$ of these profits, find the amount of her earnings.

$2272

9. The city with the most expensive daily parking fee is New York City (Midtown) at $45. The daily parking fee in Boston (third most expensive) is $\frac{4}{5}$ as much as New York City. Find the daily parking fee in Boston. (*Source:* Colliers International.)

$36

10. The daily parking fee in New York City (Downtown) is $36. In San Francisco (fifth most expensive), the daily parking fee is $\frac{3}{4}$ the cost of New York City (Downtown). How much is the daily parking fee in San Francisco? (*Source:* Colliers International.)

$27

11. At the Garlic Festival Fun Run, $\frac{7}{12}$ of the runners are women. If there are 1560 runners, how many are women?

910 women

12. A hotel has 408 rooms. Of these rooms, $\frac{9}{17}$ are for nonsmokers. How many rooms are for nonsmokers?

216 rooms

Almost $\frac{1}{3}$ of all television owners own a TiVo or some other DVR recording device. The circle graph below shows how the owners of this device have changed their television viewing time. Use this information to work Exercises 13–18.

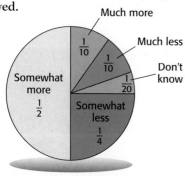

TV with TiVo

Do you watch more or less TV now that you have this device? 1000 people surveyed.

Source: Greenfield Online Omnibus

13. Which response was given by the least number of people? How many people gave this response?

Don't know; 50 people

14. Which response was given by the greatest number of people? How many gave this response?

Somewhat more; 500 people

15. Find the total number of people who said they watch less television.

100 much less + 250 somewhat less = 350 people

16. Find the total number of people who said they watch more television.

100 much more + 500 somewhat more = 600 people

17. Without actually adding the fractions given for all the groups, explain why their sum has to be 1.

Because everyone is included and fractions are given for *all* groups, the sum of the fractions must be 1, or *all* of the people.

18. Refer to Exercise 17. Suppose you added all the fractions for the groups and did not get 1 as an answer. List some possible explanations.

Answers will vary. Some possibilities are
1. You made an addition error.
2. The fractions on the pie graph are incorrect.
3. The fraction errors were caused by rounding.

The table shows the earnings for the Owens family last year and the circle graph shows how they spent their earnings. Use this information to answer Exercises 19–24.

Month	Earnings	Month	Earnings
January	$4575	July	$5540
February	$4312	August	$3732
March	$4988	September	$4170
April	$4530	October	$5512
May	$4320	November	$4965
June	$4898	December	$6458

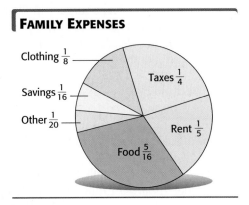

FAMILY EXPENSES

Clothing $\frac{1}{8}$

Savings $\frac{1}{16}$

Other $\frac{1}{20}$

Taxes $\frac{1}{4}$

Rent $\frac{1}{5}$

Food $\frac{5}{16}$

19. Find the Owens family's total income for the year.

$58,000

20. How much of their annual earnings went to taxes?

$14,500

21. Find the amount of their rent for the year.

$11,600

22. How much did they spend for food during the year?

$18,125

23. Find their annual savings.

$3625

24. How much of their annual income was spent on clothing?

$7250

25. Here is how one student solved a multiplication problem. Find the error and solve the problem correctly.

$$\frac{9}{10} \times \frac{20}{21} = \frac{\overset{3}{\cancel{9}}}{\cancel{10}_1} \times \frac{\overset{2}{\cancel{20}}}{\cancel{21}_3} = \frac{6}{3} = 2$$

The error was made when dividing 9 and 21 by the common factor of 3. The correct solution is

$$\frac{9}{10} \times \frac{20}{21} = \frac{\overset{3}{\cancel{9}}}{\cancel{10}_1} \times \frac{\overset{2}{\cancel{20}}}{\cancel{21}_7} = \frac{6}{7}$$

26. When two whole numbers are multiplied, the product is always larger than the numbers being multiplied. When two proper fractions are multiplied, the product is always smaller than the numbers being multiplied. Are these statements true? Why or why not?

Yes, the statements are true. Since whole numbers are 1 or greater, when you multiply, the product will always be greater than either of the numbers multiplied. But, when you multiply two proper fractions, you are finding a fraction of a fraction, and the product will be smaller than either of the two proper fractions.

Solve each application problem.

27. The cost of Lasik eye surgery in the United States is $2000 for each eye. The same surgery in Thailand is $\frac{3}{8}$ of this amount. What is the cost of this procedure in Thailand? (*Source: Reader's Digest.*)

$750

28. A knee replacement in the United States costs $36,300, while in Mexico the same procedure costs $\frac{3}{50}$ of this amount. Find the cost of a knee replacement in Mexico. (*Source: Reader's Digest.*)

$2178

29. A collector of scale model World War II ships wants to know the length of a $\frac{1}{128}$ scale model of a ship that was 256 feet in length. Find the length of the scale model. (*Source:* Lilliput Motor Company, LTD.)

2 ft

30. Howard Martin, manager of Bayside Fishing, is adding $\frac{3}{32}$ of a quart of 2-cycle oil to each gallon of gasoline. How many quarts of oil will he need to add to 760 gallons of gasoline?

$71\frac{1}{4}$ quarts

31. LaDonna Washington is running for city council. She needs to get $\frac{2}{3}$ of her votes from senior citizens and 27,000 votes in all to win. How many votes does she need from voters other than the senior citizens?

9000 votes

32. The start-up cost of a Subs and Sandwich Shop is $32,000. If the bank will loan you $\frac{9}{16}$ of the start up and you must pay the balance, how much more will you need to open a shop?

$14,000

33. A will states that $\frac{7}{8}$ of an estate is to be divided among relatives. Of the remaining estate, $\frac{1}{4}$ goes to the American Cancer Society. What fraction of the estate goes to the American Cancer Society?

$\frac{1}{32}$ of the estate

34. A couple has $\frac{2}{5}$ of their total investments in real estate. Of the remaining investments, $\frac{1}{3}$ is invested in bonds. What fraction of the total investments is in bonds?

$\frac{1}{5}$ of the total investment

2.7 ▷▷▷ Dividing Fractions

OBJECTIVE 1 **Find the reciprocal of a fraction.** To divide fractions, we need to know how to find the **reciprocal** of a fraction.

> **Reciprocal of a Fraction**
> Two numbers are reciprocals of each other if their product is 1. To find the reciprocal of a fraction, interchange the numerator and denominator.

For example, the reciprocal of $\frac{3}{4}$ is $\frac{4}{3}$.

$$\text{Fraction} \quad \frac{3}{4} \diagup\!\!\!\!\diagdown \frac{4}{3} \quad \text{Reciprocal}$$

> **Note**
> Notice that you invert, or "flip," a fraction to find its reciprocal.

OBJECTIVES

1 Find the reciprocal of a fraction.

2 Divide fractions.

3 Solve application problems in which fractions are divided.

EXAMPLE 1 **Finding Reciprocals**

Find the reciprocal of each fraction.

> Flip a fraction to find the reciprocal.

(a) The reciprocal of $\frac{1}{4}$ is $\frac{4}{1}$ because $\frac{1}{4} \cdot \frac{4}{1} = \frac{4}{4} = 1$

(b) The reciprocal of $\frac{2}{3}$ is $\frac{3}{2}$ because $\frac{2}{3} \cdot \frac{3}{2} = \frac{6}{6} = 1$

(c) The reciprocal of $\frac{3}{5}$ is $\frac{5}{3}$ because $\frac{3}{5} \cdot \frac{5}{3} = \frac{15}{15} = 1$

(d) The reciprocal of 8 is $\frac{1}{8}$ because $\frac{8}{1} \cdot \frac{1}{8} = \frac{8}{8} = 1$ Think of 8 as $\frac{8}{1}$.

Work Problem **1** *at the Side.* ▶

> **Note**
> Every number has a reciprocal except 0. The number 0 has no reciprocal because there is no number that can be multiplied by 0 to get 1.
>
> $$0 \cdot (\text{reciprocal}) = 1$$
>
> ↑
> There is no number to use here that will give an answer of 1. When you multiply by 0, you always get 0.

1 Find the reciprocal of each fraction.

(a) $\frac{3}{4}$

(b) $\frac{3}{8}$

(c) $\frac{9}{4}$

(d) 16

ANSWERS

1. **(a)** $\frac{4}{3}$ **(b)** $\frac{8}{3}$ **(c)** $\frac{4}{9}$ **(d)** $\frac{1}{16}$

In **Chapter 1,** we saw that the division problem $12 \div 3$ asks how many 3s are in 12. In the same way, the division problem $\frac{2}{3} \div \frac{1}{6}$ asks how many $\frac{1}{6}$ s are in $\frac{2}{3}$. The figure illustrates $\frac{2}{3} \div \frac{1}{6}$.

The figure shows that there are 4 of the $\frac{1}{6}$ pieces in $\frac{2}{3}$, or

$$\frac{2}{3} \div \frac{1}{6} = 4$$

OBJECTIVE 2 Divide fractions. We will use reciprocals to divide fractions.

> **Dividing Fractions**
>
> To divide two fractions, multiply the first fraction by the reciprocal of the second fraction.

EXAMPLE 2 **Dividing One Fraction by Another**

Divide. Write answers in lowest terms and as mixed numbers where possible.

(a) $\frac{7}{8} \div \frac{15}{16}$

The reciprocal of $\frac{15}{16}$ is $\frac{16}{15}$

Reciprocals

$$\frac{7}{8} \div \frac{15}{16} = \frac{7}{8} \cdot \frac{16}{15}$$

Use $\frac{16}{15}$, the reciprocal of $\frac{15}{16}$, and multiply.

Change division to multiplication.

$$= \frac{7}{\overset{}{\underset{1}{8}}} \cdot \frac{\overset{2}{16}}{15}$$ Divide the numerator and denominator by 8.

$$= \frac{7 \cdot 2}{1 \cdot 15}$$ Multiply.

$$= \frac{14}{15}$$ ← Lowest terms

Continued on Next Page

(b) $\dfrac{\frac{4}{5}}{\frac{3}{10}}$

$$\frac{\frac{4}{5}}{\frac{3}{10}} = \frac{4}{5} \div \frac{3}{10}$$ Rewrite by using the ÷ symbol for division.

$$= \frac{4}{\underset{1}{\cancel{5}}} \cdot \frac{\overset{2}{\cancel{10}}}{3}$$ The reciprocal of $\frac{3}{10}$ is $\frac{10}{3}$. Change "÷" to " • ". Divide the numerator and denominator by 5.

$$= \frac{4 \cdot 2}{1 \cdot 3}$$ Multiply.

$$= \frac{8}{3} = 2\frac{2}{3}$$ Rewrite the answer as a mixed number.

CAUTION
Be certain that the divisor fraction is changed to its reciprocal *before* you divide numerators and denominators by common factors.

Work Problem **2** *at the Side.* ▶

EXAMPLE 3 Dividing with a Whole Number

Divide. Write all answers in lowest terms and as whole or mixed numbers where possible.

(a) $5 \div \dfrac{1}{4}$

Write 5 as $\frac{5}{1}$. Next, use the reciprocal of $\frac{1}{4}$, which is $\frac{4}{1}$.

$$5 \div \frac{1}{4} = \frac{5}{1} \cdot \frac{4}{1}$$ Reciprocal of $\frac{1}{4}$ is $\frac{4}{1}$.

Reciprocals

$$= \frac{5 \cdot 4}{1 \cdot 1}$$ Multiply.

$$= \frac{20}{1} = 20$$ Rewrite the answer as a whole number.

Continued on Next Page

2 Divide. Write answers in lowest terms and as mixed numbers where possible.

(a) $\dfrac{1}{4} \div \dfrac{2}{3}$

(b) $\dfrac{3}{8} \div \dfrac{5}{8}$

(c) $\dfrac{\frac{2}{3}}{\frac{4}{5}}$

(d) $\dfrac{\frac{5}{6}}{\frac{7}{12}}$

3 Divide. Write answers in lowest terms and as whole or mixed numbers where possible.

(a) $10 \div \dfrac{1}{2}$

(b) $6 \div \dfrac{6}{7}$

(c) $\dfrac{4}{5} \div 6$

(d) $\dfrac{3}{8} \div 4$

(b) $\dfrac{2}{3} \div 6$

Write 6 as $\frac{6}{1}$. The reciprocal of $\frac{6}{1}$ is $\frac{1}{6}$.

$$\dfrac{2}{3} \div \dfrac{6}{1} = \dfrac{2}{3} \cdot \dfrac{1}{6}$$

Reciprocals

> Careful! Divide out common factors **after** changing to the reciprocal and to multiplication.

$$= \dfrac{\overset{1}{2}}{3} \cdot \dfrac{1}{\underset{3}{6}}$$

Divide the numerator and denominator by 2, then multiply.

$$= \dfrac{1 \cdot 1}{3 \cdot 3} = \dfrac{1}{9}$$

Lowest terms

◀ Work Problem **3** at the Side.

OBJECTIVE 3 Solve application problems in which fractions are divided. Many application problems require division of fractions. Recall that typical indicator words for division are *goes into, per, divide, divided by, divided equally,* and *divided into.*

EXAMPLE 4 Applying Fraction Skills

Goldie, the manager of the Burnside Deli, must fill a 10-gallon kosher dill pickle crock with salt brine. She has only a $\frac{2}{3}$-gallon container to use. How many times must she fill the $\frac{2}{3}$-gallon container and empty it into the 10-gallon crock?

Step 1 **Read** the problem. We need to find the number of times Goldie needs to use a $\frac{2}{3}$-gallon container in order to fill a 10-gallon crock.

Step 2 **Work out a plan.** We can solve the problem by finding the number of times 10 can be divided by $\frac{2}{3}$.

Step 3 **Estimate** a reasonable answer. Round $\frac{2}{3}$ gallon to 1 gallon. In order to fill the 10-gallon container, she would have to use the 1-gallon container 10 times, so our estimate is 10.

Step 4 **Solve** the problem.

Reciprocals

$$10 \div \dfrac{2}{3} = \dfrac{10}{1} \cdot \dfrac{3}{2}$$

The reciprocal of $\frac{2}{3}$ is $\frac{3}{2}$. Change "÷" to " • ."

$$= \dfrac{\overset{5}{10}}{1} \cdot \dfrac{3}{\underset{1}{2}}$$

Divide the numerator and denominator by 2, and then multiply.

$$= \dfrac{15}{1} = 15$$

Step 5 **State the answer.** Goldie must fill the container 15 times.

Step 6 **Check.** The exact answer, 15 times, is reasonably close to our estimate of 10 times.

◀ Work Problem **4** at the Side.

4 Solve each problem using the six problem-solving steps.

(a) How many $\frac{3}{4}$-quart leafblower fuel tanks can be filled from 15 quarts of fuel?

(b) Find the number of $\frac{4}{5}$-quart bottles that can be filled from a 120-quart cask.

ANSWERS

3. **(a)** 20 **(b)** 7 **(c)** $\dfrac{2}{15}$ **(d)** $\dfrac{3}{32}$

4. **(a)** 20 tanks **(b)** 150 bottles

EXAMPLE 5 **Applying Fraction Skills**

At the Happi-Time Day Care Center, $\frac{6}{7}$ of the total budget goes to classroom operation. If there are 18 classrooms and each one receives the same amount, what fraction of the operating amount does each classroom receive?

Step 1 **Read** the problem. Since $\frac{6}{7}$ of the total budget must be split into 18 parts, we must find the fraction of the classroom operating amount received by each classroom.

Step 2 **Work out a plan.** We must divide the fraction of the total budget going to classroom operation $\left(\frac{6}{7}\right)$ by the number of classrooms (18).

Step 3 **Estimate** a reasonable answer. Round $\frac{6}{7}$ to 1. If all of the operating expenses (1 whole) were divided between 18 classrooms, each classroom would receive $\frac{1}{18}$ of the operating expenses, our estimate.

Step 4 **Solve** the problem. We solve by dividing $\frac{6}{7}$ by 18.

$$\frac{6}{7} \div 18 = \frac{6}{7} \div \frac{18}{1} \qquad \boxed{\text{Write 18 as } \frac{18}{1}}$$

$$= \frac{\overset{1}{\cancel{6}}}{7} \cdot \frac{1}{\underset{3}{\cancel{18}}} \qquad \text{The reciprocal of } \frac{18}{1} \text{ is } \frac{1}{18}. \text{ Change "}\div\text{" to "} \cdot \text{."}$$
$$\text{Divide the numerator and denominator by 6.}$$

$$= \frac{1}{21} \qquad \text{Multiply.}$$

Step 5 **State the answer.** Each classroom receives $\frac{1}{21}$ of the total budget.

Step 6 **Check.** The exact answer, $\frac{1}{21}$, is close to our estimate of $\frac{1}{18}$.

Work Problem ⑤ *at the Side.* ▶

⑤ Solve each problem using the six problem-solving steps.

(a) The top 12 employees at Mayfield Manufacturing will divide $\frac{3}{4}$ of the annual bonus money. What fraction of the bonus money will each employee receive?

(b) A winning lottery ticket was purchased by 8 employees of United States Marketing and Promotions (USMP). They will donate $\frac{1}{5}$ of the total winnings to pay the medical expenses of a fellow employee, and then divide the remaining winnings evenly. What fraction of the prize money will each receive?

ANSWERS

5. **(a)** $\frac{1}{16}$ of the bonus money

(b) $\frac{1}{10}$ of the lottery prize money

Math in the Media

Mathematics teachers attending conferences in New Orleans, Louisiana, and San Jose, California, found the following information about hotel rates on Internet Web sites. To find the hotel room rates in New Orleans, they went to www.hotel-rate.com/us/louisiana/neworleans/, while the hotel room rates in San Jose were found at www.sanjose.com.

Hotel New Orleans	Single	Double	Triple	Quad	Suites
Marriott	$229	$239	$249	$259	$806
Sheraton	$165	$174	$193	$223	$748
San Jose					
Crowne Plaza	$209	$209	$219	$219	—
Hilton Towers	$179	$179	$199	$219	—
Sainte Claire	$149	$149	$169	$189	$329

1. The double rate is for two people sharing a room. What fractional part does each person pay? $\frac{1}{2}$

2. **(a)** Multiply the double rate at the Hilton Towers in San Jose by $\frac{1}{2}$. What is the result? **$89\frac{1}{2}$**

   (b) Divide the double rate at the Hilton Towers in San Jose by 2. What is the result? **$89\frac{1}{2}$**

   (c) Explain what happened. How much money would one person owe if he or she shared a double room at the Hilton Towers in San Jose?
   Multiplying by $\frac{1}{2}$ has the same result as dividing by 2; $89\frac{1}{2}$

3. The triple rate is for three people sharing a room. What fractional part does each person pay? $\frac{1}{3}$

4. How much money would one person owe if he or she shared a triple room at the Sainte Claire in San Jose? How much money would each person save if he or she could book a triple room at the Sainte Claire instead of the Crowne Plaza? Find your answer using two different methods, based on your observations in Problem 2. **$56\frac{1}{3}$; $16\frac{2}{3}$; $\frac{1}{3}$ of $219 − $\frac{1}{3}$ of $169 or**
   ($219 ÷ 3) − ($169 ÷ 3)

5. The quad rate is for four people sharing a room. What fractional part does each person pay? $\frac{1}{4}$

6. How much money would one person owe if he or she shared a quad room at the Sheraton in New Orleans? Find your answer using two different methods, based on your observations in Problem 2.
   $55\frac{3}{4}$; $\frac{1}{4}$ of $223 or $223 ÷ 4

7. How many people would have to share a suite at the Sheraton in New Orleans for the cost per person to be less than sharing a quad room at the same hotel? Round the answer to the nearest whole number. (*Hint:* Estimate the cost per person for a quad room first. Then estimate the number of people needed to share the cost of the suite. Check your work using actual values.)
   14 people

DO NOT DISTURB

NO MOLESTE

PRIÈRE DE NE PAS DÉRANGER

BITTE NICHT STÖREN

Find the reciprocal of each number. See Example 1.

 1. $\dfrac{3}{8}$ $\dfrac{8}{3}$

2. $\dfrac{2}{5}$ $\dfrac{5}{2}$

3. $\dfrac{5}{6}$ $\dfrac{6}{5}$

4. $\dfrac{12}{7}$ $\dfrac{7}{12}$

 5. $\dfrac{8}{5}$ $\dfrac{5}{8}$

6. $\dfrac{13}{20}$ $\dfrac{20}{13}$

 7. 4 $\dfrac{1}{4}$

8. 10 $\dfrac{1}{10}$

Divide. Write answers in lowest terms and as whole or mixed numbers where possible. See Examples 2 and 3.

 9. $\dfrac{1}{2} \div \dfrac{3}{4}$ $\dfrac{2}{3}$

10. $\dfrac{5}{8} \div \dfrac{7}{8}$ $\dfrac{5}{7}$

11. $\dfrac{7}{8} \div \dfrac{1}{3}$ $2\dfrac{5}{8}$

12. $\dfrac{7}{8} \div \dfrac{3}{4}$ $1\dfrac{1}{6}$

13. $\dfrac{3}{4} \div \dfrac{5}{3}$ $\dfrac{9}{20}$

14. $\dfrac{4}{5} \div \dfrac{9}{4}$ $\dfrac{16}{45}$

15. $\dfrac{7}{9} \div \dfrac{7}{36}$ 4

16. $\dfrac{5}{8} \div \dfrac{5}{16}$ 2

17. $\dfrac{15}{32} \div \dfrac{5}{64}$ 6

18. $\dfrac{7}{12} \div \dfrac{14}{15}$ $\dfrac{5}{8}$

 19. $\dfrac{\frac{13}{20}}{\frac{4}{5}}$ $\dfrac{13}{16}$

20. $\dfrac{\frac{9}{10}}{\frac{3}{5}}$ $1\dfrac{1}{2}$

 21. $\dfrac{\frac{5}{6}}{\frac{25}{24}}$ $\dfrac{4}{5}$

22. $\dfrac{\frac{28}{15}}{\frac{21}{5}}$ $\dfrac{4}{9}$

23. $12 \div \dfrac{2}{3}$ 18

24. $7 \div \dfrac{1}{4}$ 28

25. $\dfrac{18}{\frac{3}{4}}$ 24

26. $\dfrac{12}{\frac{3}{4}}$ 16

27. $\dfrac{\frac{4}{7}}{8}$ $\dfrac{1}{14}$

28. $\dfrac{\frac{7}{10}}{3}$ $\dfrac{7}{30}$

Solve each application problem by using division. See Examples 4 and 5.

29. Veterinarian Jasmine Cato has $\frac{8}{9}$ quart of medication. She wishes to prescribe this medication for 4 cats in her pet hospital. If she divides the medication evenly, how much will each cat receive?

$\frac{2}{9}$ **quart**

30. Harold Pishke, barber, has 15 quarts of conditioning shampoo. If he wants to put this shampoo into $\frac{3}{8}$-quart containers, how many containers can be filled?

40 containers

31. Some college roommates want to make pancakes for their neighbors. They need 5 cups of flour, but have only a $\frac{1}{3}$-cup measuring cup. How many times will they need to fill their measuring cup?

15 times

32. A cross-country bike rider completes $\frac{5}{6}$ of her trip in 15 days. What fraction of her total trip does she complete each day?

$\frac{1}{18}$ **of total trip**

33. How many $\frac{1}{8}$-ounce eye drop dispensers can be filled with 11 ounces of eye drops?

88 dispensers

34. It is estimated that each guest at a party will eat $\frac{5}{16}$ pound of peanuts. How many guests may be served with 10 pounds of peanuts?

32 guests

35. Pam Trizlia had a small pickup truck that could carry $\frac{2}{3}$ cord of firewood. Find the number of trips needed to deliver 40 cords of wood.

60 trips

36. Manuel Servin has a 200-yard roll of weather stripping material. Find the number of pieces of weather stripping $\frac{5}{8}$ yard in length that may be cut from the roll.

320 pieces

✎ 37. Your classmate is confused on how to divide by a fraction. Write a short note telling him how this should be done.

You can divide two fractions by using the reciprocal of the second fraction (divisor) and then multiplying.

✎ 38. If you multiply positive proper fractions, the product is smaller than the fractions multiplied. When you divide by a proper fraction, is the quotient smaller than the numbers in the problem? Prove your answer with examples.

Sometimes the answer is smaller and sometimes it is larger.

$$\frac{1}{4} \div \frac{7}{8} = \frac{1}{\underset{1}{\cancel{4}}} \cdot \frac{\overset{2}{\cancel{8}}}{7} = \frac{2}{7} \text{ (smaller than } \tfrac{7}{8} \text{ but not smaller than } \tfrac{1}{4})$$

$$\frac{1}{2} \div \frac{1}{4} = \frac{1}{\underset{1}{\cancel{2}}} \cdot \frac{\overset{2}{\cancel{4}}}{1} = \frac{2}{1} = 2 \text{ (larger)}$$

Solve each application problem using multiplication or division.

39. The recipe for a Jelly Belly Express loafcake calls for $\frac{3}{4}$ pound of Jelly Belly jelly beans in assorted colors. If you want to make 16 cakes, how many pounds of Jelly Belly jelly beans will you need?

12 pounds

40. In a recent study, it was found that one month after leaving the hospital only $\frac{7}{8}$ of the 1520 heart attack patients were still taking the life-saving drugs prescribed for them. How many of these patients were still taking their drugs? (*Source:* Dr. Michael Ho, Denver Veterans Medical Center.)

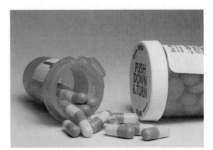

1330 patients

41. Broadly Plumbing finds that $\frac{3}{4}$ can of pipe joint compound is needed to plumb each new home. How many homes can be plumbed with 156 cans of compound?

208 homes

42. A mechanic uses an average of $\frac{2}{3}$ gallon of gear lube to service each tractor differential. Find the number of tractors that can be serviced with 28 gallons of gear lube.

42 tractors

43. In recordings of 186 patient visits, doctors failed to mention a new drug's side effects or how long to take the drug in $\frac{2}{3}$ of the visits. In how many visits did doctors fail to discuss these issues with patients? (*Source:* Dr. Deijieng Tam, UCLA.)

124 visits

44. Laura has been working on a job that will require 81 hours to complete. If she has completed $\frac{7}{9}$ of the job, how many hours has she worked?

63 hours

45. A dish towel manufacturer requires $\frac{3}{8}$ yard of cotton fabric for each towel. Find the number of dish towels that can be made from 912 yards of fabric.

2432 towels

46. Each patient will receive $\frac{7}{10}$ vial of medication. How many patients can be treated with 3150 vials of medication?

4500 patients

Relating Concepts (Exercises 47–52) For Individual or Group Work

Many application problems are solved using multiplication and division of fractions.
Work Exercises 47–52 in order.

47. Perhaps the most common indicator word for multiplication is the word *of.* Circle the words in the list below that are also indicator words for multiplication.

more than   per

(double)   (twice)

(times)   (product)

less than   difference

equals   (twice as much)

48. Circle the words in the list below that are indicator words for division.

fewer   sum of

(goes into)   (divide)

(per)   (quotient)

equals   double

loss of   (divided by)

49. To divide two fractions, multiply the first fraction by the __reciprocal__ of the second fraction.

50. Find the reciprocals for each number.

$$\frac{3}{4} \qquad \frac{7}{8} \qquad 5 \qquad \frac{12}{19} \qquad \frac{4}{3}, \frac{8}{7}, \frac{1}{5}, \frac{19}{12}$$

The size of a U.S.A. first-class Forever postage stamp is shown here. Use this to answer Exercises 51 and 52.

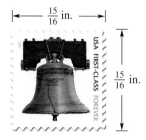

$\frac{15}{16}$ in.

$\frac{15}{16}$ in.

51.(a) Explain how to find the perimeter of any regular 3-, 4-, 5-, or 6-sided figure using multiplication.

Multiply the length of one side by 3, 4, 5, or 6.

(b) Find the perimeter of the stamp using multiplication.

$$\frac{15}{16} \times 4 = \frac{15}{\overset{}{\underset{4}{16}}} \times \frac{\overset{1}{4}}{1} = \frac{15}{4} = 3\frac{3}{4} \text{ in.}$$

52. Find the area of the postage stamp. Explain how to find the area of any rectangle.

$$\frac{225}{256} \text{ in.}^2$$

Multiply the length by the width.

2.8 ▶▶▶ Multiplying and Dividing Mixed Numbers

In **Section 2.2** we worked with mixed numbers—a whole number and a fraction written together. Many of the fraction problems you encounter in everyday life involve mixed numbers.

> **OBJECTIVE 1 Estimate the answer and multiply mixed numbers.** When multiplying mixed numbers, it is a good idea to estimate the answer first. Then multiply the mixed numbers by using the following steps.

OBJECTIVES

1 Estimate the answer and multiply mixed numbers.

2 Estimate the answer and divide mixed numbers.

3 Solve application problems with mixed numbers.

Multiplying Mixed Numbers

Step 1 *Change* each mixed number to an improper fraction.

Step 2 *Multiply* as fractions.

Step 3 *Simplify* the answer, which means to write it in *lowest terms,* and change it to a mixed number or whole number where possible.

To estimate the answer, round each mixed number to the nearest whole number. If the numerator is *half* of the denominator or *more,* round up the whole number part. If the numerator is *less* than half the denominator, leave the whole number as it is.

$$1\frac{5}{8} \quad \begin{matrix} \leftarrow 5 \text{ is more than } 4. \\ \leftarrow \text{Half of } 8 \text{ is } 4. \end{matrix} \Bigg\} \quad 1\frac{5}{8} \text{ rounds up to } 2$$

$$3\frac{2}{5} \quad \begin{matrix} \leftarrow 2 \text{ is less than } 2\frac{1}{2}. \\ \leftarrow \text{Half of } 5 \text{ is } 2\frac{1}{2}. \end{matrix} \Bigg\} \quad 3\frac{2}{5} \text{ rounds to } 3$$

> Round mixed numbers to the nearest whole number when estimating.

Work Problem **1** *at the Side.* ▶

EXAMPLE 1 **Multiplying Mixed Numbers**

First estimate the answer. Then multiply to get an exact answer. Simplify your answers.

(a) $2\frac{1}{2} \cdot 3\frac{1}{5}$

Estimate the answer by rounding the mixed numbers.

$$2\frac{1}{2} \text{ rounds to } 3 \quad \text{and} \quad 3\frac{1}{5} \text{ rounds to } 3$$
$$3 \cdot 3 = 9 \qquad \text{Estimated answer}$$

To find the exact answer, change each mixed number to an improper fraction.

$$\textit{Step 1} \qquad 2\frac{1}{2} = \frac{5}{2} \quad \text{and} \quad 3\frac{1}{5} = \frac{16}{5}$$

Continued on Next Page

---

1 Round each mixed number to the nearest whole number.

(a) $4\frac{2}{3}$

(b) $3\frac{2}{5}$

(c) $5\frac{3}{4}$

(d) $4\frac{7}{12}$

(e) $1\frac{1}{2}$

(f) $8\frac{4}{9}$

ANSWERS

1. **(a)** 5 **(b)** 3 **(c)** 6 **(d)** 5
(e) 2 **(f)** 8

2 First estimate the answer. Then multiply to find the exact answer. Simplify your answers.

(a) $3\dfrac{1}{2}$ • $6\dfrac{1}{3}$

_____ • _____

= _____ estimate

(b) $4\dfrac{2}{3}$ • $2\dfrac{3}{4}$

_____ • _____

= _____ estimate

(c) $3\dfrac{3}{5}$ • $4\dfrac{4}{9}$

_____ • _____

= _____ estimate

(d) $5\dfrac{1}{4}$ • $3\dfrac{3}{5}$

_____ • _____

= _____ estimate

ANSWERS

2. (a) *Estimate:* $4 • 6 = 24$; *Exact:* $22\dfrac{1}{6}$

(b) *Estimate:* $5 • 3 = 15$; *Exact:* $12\dfrac{5}{6}$

(c) *Estimate:* $4 • 4 = 16$; *Exact:* 16

(d) *Estimate:* $5 • 4 = 20$; *Exact:* $18\dfrac{9}{10}$

Next, multiply.

Step 1   *Step 2*   *Step 3*

$$2\dfrac{1}{2} • 3\dfrac{1}{5} = \dfrac{5}{2} • \dfrac{16}{5} = \dfrac{\overset{1}{\cancel{5}}}{\underset{1}{\cancel{2}}} • \dfrac{\overset{8}{\cancel{16}}}{\cancel{5}} = \dfrac{1 • 8}{1 • 1} = \dfrac{8}{1} = 8$$

Remember: Change to improper fractions, then divide out common factors, and finally multiply.

The estimated answer is 9 and the exact answer is 8. The exact answer is reasonable.

(b) $3\dfrac{5}{8} • 4\dfrac{4}{5}$

$3\dfrac{5}{8}$ rounds to 4   and $4\dfrac{4}{5}$ rounds to 5

$4 • 5 = 20$ Estimated answer

Now find the exact answer.

Step 1   *Step 2*

$$3\dfrac{5}{8} • 4\dfrac{4}{5} = \dfrac{29}{8} • \dfrac{24}{5} = \dfrac{29}{\underset{1}{\cancel{8}}} • \dfrac{\overset{3}{\cancel{24}}}{5} = \dfrac{29 • 3}{1 • 5} = \dfrac{87}{5}$$

Step 3

$$\dfrac{87}{5} = 17\dfrac{2}{5}$$

Simplify this answer by writing it as a mixed number.

The estimate was 20, so the exact answer of $17\dfrac{2}{5}$ is reasonable.

(c) $1\dfrac{3}{5} • 3\dfrac{1}{3}$

$1\dfrac{3}{5}$ rounds to 2   and $3\dfrac{1}{3}$ rounds to 3

$2 • 3 = 6$ Estimated answer

The exact answer is shown below.

$$1\dfrac{3}{5} • 3\dfrac{1}{3} = \dfrac{8}{\underset{1}{\cancel{5}}} • \dfrac{\overset{2}{\cancel{10}}}{3} = \dfrac{8 • 2}{1 • 3} = \dfrac{16}{3} = 5\dfrac{1}{3}$$

The estimate was 6, so the exact answer of $5\dfrac{1}{3}$ is reasonable.

◀ *Work Problem* **2** *at the Side.*

OBJECTIVE 2 Estimate the answer and divide mixed numbers.
Just as you did when multiplying mixed numbers, it is also a good idea to estimate the answer when dividing mixed numbers. To divide mixed numbers, use the following steps.

Dividing Mixed Numbers

Step 1 *Change* each mixed number to an improper fraction.

Step 2 Use the *reciprocal* of the second fraction (divisor).

Step 3 *Multiply.*

Step 4 *Simplify* the answer, which means to write it in *lowest terms,* and change it to a mixed number or whole number where possible.

Note
Recall that the reciprocal of a fraction is found by interchanging the numerator and the denominator.

EXAMPLE 2 **Dividing Mixed Numbers**

First estimate the answer. Then divide to find the exact answer. Simplify your answers.

(a) $2\frac{2}{5} \div 1\frac{1}{2}$

First estimate the answer by rounding each mixed number to the nearest whole number.

$$2\frac{2}{5} \quad \div \quad 1\frac{1}{2}$$

$$\downarrow \quad \text{Rounded} \quad \downarrow$$

$$2 \quad \div \quad 2 = 1 \qquad \text{Estimated answer}$$

To find the exact answer, first change each mixed number to an improper fraction.

$$\overset{\textit{Step 1}}{\overbrace{}}$$

$$2\frac{2}{5} \div 1\frac{1}{2} = \frac{12}{5} \div \frac{3}{2}$$

Next, use the reciprocal of the second fraction and multiply.

$$\overset{\textit{Step 2}}{\overbrace{}} \quad \overset{\textit{Step 3}}{\overbrace{}} \quad \overset{\textit{Step 4}}{\overbrace{}}$$

$$\frac{12}{5} \div \frac{3}{2} = \frac{\overset{4}{\cancel{12}}}{5} \cdot \frac{2}{\underset{1}{\cancel{3}}} = \frac{4 \cdot 2}{5 \cdot 1} = \frac{8}{5} = 1\frac{3}{5} \qquad \text{Exact answer simplified}$$

Reciprocals

Remember: Use the reciprocal of the *second* fraction.

The estimate was 1, so the exact answer of $1\frac{3}{5}$ is reasonable.

Continued on Next Page

3 First estimate the answer. Then divide to find the exact answer. Simplify all answers.

(a) $3\frac{1}{8} \div 6\frac{1}{4}$

$$\underline{} \div \underline{}$$

$$= \underline{} \; estimate$$

(b) $10\frac{1}{3} \div 2\frac{1}{2}$

$$\underline{} \div \underline{}$$

$$= \underline{} \; estimate$$

(c) $8 \div 5\frac{1}{3}$

$$\underline{} \div \underline{}$$

$$= \underline{} \; estimate$$

(d) $13\frac{1}{2} \div 18$

$$\underline{} \div \underline{}$$

$$= \underline{} \; estimate$$

ANSWERS

3. **(a)** *Estimate:* $3 \div 6 = \frac{1}{2}$; *Exact:* $\frac{1}{2}$

(b) *Estimate:* $10 \div 3 = 3\frac{1}{3}$; *Exact:* $4\frac{2}{15}$

(c) *Estimate:* $8 \div 5 = 1\frac{3}{5}$; *Exact:* $1\frac{1}{2}$

(d) *Estimate:* $14 \div 18 = \frac{7}{9}$; *Exact:* $\frac{3}{4}$

(b) $8 \div 3\frac{3}{5}$

$$8 \div 3\frac{3}{5}$$
$$\downarrow \; \text{Rounded} \; \downarrow$$
$$8 \div 4 = 2 \qquad \text{Estimate}$$

Now find the exact answer.

$$8 \div 3\frac{3}{5} = \frac{8}{1} \div \frac{18}{5} = \overset{4}{\frac{8}{1}} \cdot \frac{5}{\underset{9}{18}} = \frac{20}{9} = 2\frac{2}{9}$$

Write 8 as $\frac{8}{1}$.

> Divide out common factors *only after* you have changed to the reciprocal and are multiplying.

The estimate was 2, so the exact answer of $2\frac{2}{9}$ is reasonable.

(c) $4\frac{3}{8} \div 5$

$$4\frac{3}{8} \div 5$$
$$\downarrow \; \text{Rounded} \; \downarrow$$
$$4 \div 5 = \frac{4}{1} \div \frac{5}{1} = \frac{4}{1} \cdot \frac{1}{5} = \frac{4}{5} \qquad \text{Estimate}$$

The exact answer is shown below.

$$4\frac{3}{8} \div 5 = \frac{35}{8} \div \frac{5}{1} = \overset{7}{\frac{35}{8}} \cdot \frac{1}{\underset{1}{5}} = \frac{7}{8}$$

Write 5 as $\frac{5}{1}$.

The estimate was $\frac{4}{5}$, so the exact answer of $\frac{7}{8}$ is reasonable.

◀ **Work Problem 3** at the Side.

OBJECTIVE 3 Solve application problems with mixed numbers.
The next two examples show how to solve application problems involving mixed numbers.

EXAMPLE 3 **Applying Multiplication Skills**

The local Habitat for Humanity chapter is looking for 11 contractors who will each donate $3\frac{1}{4}$ days of labor to a community building project. How many days of labor will be donated in all?

Step 1 **Read** the problem. The problem asks for the total days of labor donated by the 11 contractors.

Step 2 **Work out a plan.** Multiply the number of contractors (11) and the amount of labor that each donates ($3\frac{1}{4}$ days).

Continued on Next Page

Step 3 **Estimate** a reasonable answer. Round $3\frac{1}{4}$ days to 3 days. Multiply 3 days by 11 contractors (3 • 11) to get an estimate of 33 days.

Step 4 **Solve** the problem. Find the exact answer.

$$11 \cdot 3\frac{1}{4} = 11 \cdot \frac{13}{4}$$

$$= \frac{11}{1} \cdot \frac{13}{4} = \frac{143}{4} = 35\frac{3}{4}$$

Step 5 **State the answer.** The community building project will receive $35\frac{3}{4}$ days of donated labor.

> Always check to see if the answer is close to the estimate.

Step 6 **Check.** The exact answer, $35\frac{3}{4}$ days, is close to our estimate of 33 days.

Work Problem **4** *at the Side.* ▶

EXAMPLE 4 **Applying Division Skills**

A dome tent for backpacking requires $7\frac{1}{4}$ yards of nylon material. How many tents can be made from $65\frac{1}{4}$ yards of material?

Step 1 **Read** the problem. The problem asks how many tents can be made from $65\frac{1}{4}$ yards of material.

Step 2 **Work out a plan.** Divide the number of yards of cloth ($65\frac{1}{4}$ yd) by the number of yards needed for one tent ($7\frac{1}{4}$ yd).

Step 3 **Estimate** a reasonable answer.

$$65\frac{1}{4} \quad \div \quad 7\frac{1}{4}$$

$$\downarrow \text{ Rounded } \downarrow$$

$$65 \quad \div \quad 7 \approx 9 \text{ tents} \qquad \text{Estimate}$$

Step 4 **Solve** the problem.

$$65\frac{1}{4} \div 7\frac{1}{4} = \frac{261}{4} \div \frac{29}{4}$$

$$= \frac{\overset{9}{\cancel{261}}}{\underset{1}{\cancel{4}}} \cdot \frac{\overset{1}{\cancel{4}}}{\underset{1}{\cancel{29}}} = \frac{9}{1} = 9 \qquad \text{Matches estimate}$$

Step 5 **State the answer.** 9 tents can be made from $65\frac{1}{4}$ yards of cloth.

Step 6 **Check.** The exact answer, 9, matches our estimate.

Work Problem **5** *at the Side.* ▶

Note

When rounding mixed numbers to estimate the answer to a problem, the estimated answer usually varies somewhat from the exact answer. However, the importance of the estimated answer is that it will show you whether your exact answer is reasonable or not.

---

4 Use the six problem-solving steps. Simplify all answers.

(a) If one automobile requires $2\frac{5}{8}$ quarts of paint, find the number of quarts needed to paint 15 cars.

(b) Clare earns $\$9\frac{1}{4}$ per hour. How much would she earn in $6\frac{1}{2}$ hours? Write the answer as a mixed number.

5 Use the six problem-solving steps. Simplify all answers.

(a) The manufacture of one outboard engine propeller requires $4\frac{3}{4}$ pounds of brass. How many propellers can be manufactured from 57 pounds of brass?

(b) Jack Armstrong Trucking uses $21\frac{3}{4}$ quarts of motor oil for each oil change on his diesel engine truck. Find the number of oil changes that can be made with 609 quarts of oil.

ANSWERS

4. (a) *Estimate:* 3 • 15 = 45; *Exact:* $39\frac{3}{8}$ quarts

(b) *Estimate:* 9 • 7 = 63; *Exact:* $\$60\frac{1}{8}$

5. (a) *Estimate:* 57 ÷ 5 ≈ 11; *Exact:* 12 propellers

(b) *Estimate:* 600 ÷ 22 ≈ 27; *Exact:* 28 oil changes

Math in the Media

Alaska Burgers

1 pound 93% lean ground beef

$\frac{1}{2}$ medium Spanish onion, minced or processed

4 shakes Worcestershire sauce

$\frac{1}{4}$ teaspoon allspice (1 good pinch)

$\frac{1}{2}$ teaspoon ground cumin (two good pinches)

Cracked black pepper

$\frac{1}{3}$ pound brick of smoked cheddar cheese, cut into $\frac{1}{2}$ inch slices

4 fresh, crusty onion rolls

Thick-sliced tomato and lettuce to top

Mix beef, onion, Worcestershire, all-spice, cumin, and black pepper in a bowl. Separate a quarter of the mixture. Take a slice of the smoked cheese and place it in the middle of the mixture. Form the pattie shape around the cheese filling. Patties should be no more than $\frac{3}{4}$ inch thick. Repeat with rest of mixture to have a total of 4 patties.

Heat a nonstick griddle or frying pan to medium hot. Cook burgers 5 to 6 minutes on each side. Meat should be cooked through and cheese melted. Check each burger with an instant-read thermometer for an internal temp of 170°F for well done if undercooking concerns you. Or cut into one and check the color of the meat.

Salt burgers after preparation to your taste. (Salting beef before cooking draws out juices and flavor.) Top with tomato slices and lettuce. Serves 4.

Rachael Ray is a Food Network television host, bestselling cookbook author, and the editor of her own lifestyle magazine. Rachael Ray's recipes can be found on her Web site, www.rachaelray.com, on her television show, or in one of her many cookbooks. The recipe at the side is from her book Rachael Ray: 30-Minute Meals.

1. Following the recipe, **(a)** what is the weight of one $\frac{1}{2}$-inch slice of cheese, and **(b)** what is the thickness of a $\frac{1}{3}$-pound brick of smoked cheddar cheese?

   (a) $\frac{1}{12}$ **pound**   **(b) 2 in.**

2. According to the recipe, **(a)** how many teaspoons of allspice are in 8 good pinches, and **(b)** how many teaspoons of ground cumin are in 7 good pinches?

   (a) 2 teaspoons   **(b)** $1\frac{3}{4}$ **teaspoons**

3. Suppose you are preparing Alaska Burgers for 18 guests. By what factor will you change the ingredient amounts?

   Multiply by $4\frac{1}{2}$

4. You know that of the 15 guests at your next party, 5 large eaters will eat $1\frac{1}{2}$ burgers each, 5 children will eat $\frac{1}{2}$ burger each, and the rest of the guests will each eat 1 burger. **(a)** How many burgers will you need, and **(b)** by what factor will you change the ingredient amounts?

   (a) 15 burgers   **(b) Multiply by** $3\frac{3}{4}$

5. Fill in the blanks with the ingredient amounts needed to make 9 servings of Alaska Burgers

Lean ground beef	$2\frac{1}{4}$ pounds
Spanish onions	$1\frac{1}{8}$ onions
Worcestershire sauce	9 shakes
Allspice	$\frac{9}{16}$ tsp
Ground cumin	$1\frac{1}{8}$ tsp
Cheddar cheese	$\frac{3}{4}$ pound

6. If you have $5\frac{3}{4}$ pounds of beef, how many servings of Alaska Burgers can be prepared?

   23 servings

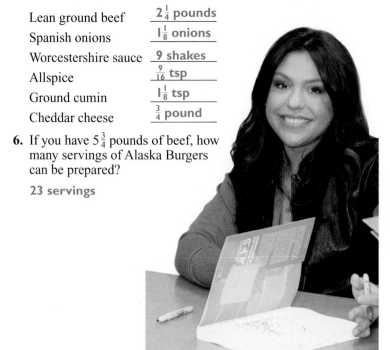

2.8 ▶▶▶ Exercises

First estimate the answer. Then multiply to find the exact answer. Simplify all answers.
See Example 1.

1. *Exact:*

$4\frac{1}{2} \cdot 1\frac{3}{4}$ $7\frac{7}{8}$

Estimate:

$\underline{\ 5\ } \cdot \underline{\ 2\ } = \underline{\ 10\ }$

2. *Exact:*

$2\frac{1}{2} \cdot 2\frac{1}{4}$ $5\frac{5}{8}$

Estimate:

$\underline{\ 3\ } \cdot \underline{\ 2\ } = \underline{\ 6\ }$

3. *Exact:*

$1\frac{2}{3} \cdot 2\frac{7}{10}$ $4\frac{1}{2}$

Estimate:

$\underline{\ 2\ } \cdot \underline{\ 3\ } = \underline{\ 6\ }$

4. *Exact:*

$4\frac{1}{2} \cdot 2\frac{1}{4}$ $10\frac{1}{8}$

Estimate:

$\underline{\ 5\ } \cdot \underline{\ 2\ } = \underline{\ 10\ }$

5. *Exact:*

$3\frac{1}{9} \cdot 1\frac{2}{7}$ 4

Estimate:

$\underline{\ 3\ } \cdot \underline{\ 1\ } = \underline{\ 3\ }$

6. *Exact:*

$6\frac{1}{4} \cdot 3\frac{1}{5}$ 20

Estimate:

$\underline{\ 6\ } \cdot \underline{\ 3\ } = \underline{\ 18\ }$

7. *Exact:*

$8 \cdot 6\frac{1}{4}$ 50

Estimate:

$\underline{\ 8\ } \cdot \underline{\ 6\ } = \underline{\ 48\ }$

8. *Exact:*

$6 \cdot 2\frac{1}{3}$ 14

Estimate:

$\underline{\ 6\ } \cdot \underline{\ 2\ } = \underline{\ 12\ }$

9. *Exact:*

$4\frac{1}{2} \cdot 2\frac{1}{5} \cdot 5$ $49\frac{1}{2}$

Estimate:

$\underline{\ 5\ } \cdot \underline{\ 2\ } \cdot \underline{\ 5\ } = \underline{\ 50\ }$

10. *Exact:*

$5\frac{1}{2} \cdot 1\frac{1}{3} \cdot 2\frac{1}{4}$ $16\frac{1}{2}$

Estimate:

$\underline{\ 6\ } \cdot \underline{\ 1\ } \cdot \underline{\ 2\ } = \underline{\ 12\ }$

11. *Exact:*

$3 \cdot 1\frac{1}{2} \cdot 2\frac{2}{3}$ 12

Estimate:

$\underline{\ 3\ } \cdot \underline{\ 2\ } \cdot \underline{\ 3\ } = \underline{\ 18\ }$

12. *Exact:*

$\frac{2}{3} \cdot 3\frac{2}{3} \cdot \frac{6}{11}$ $1\frac{1}{3}$

Estimate:

$\underline{\ 1\ } \cdot \underline{\ 4\ } \cdot \underline{\ 1\ } = \underline{\ 4\ }$

First estimate the answer. Then divide to find the exact answer. Simplify all answers.
See Example 2.

13. *Exact:*

$1\frac{1}{4} \div 3\frac{3}{4}$ $\frac{1}{3}$

Estimate:

$\underline{\ 1\ } \div \underline{\ 4\ } = \underline{\ \frac{1}{4}\ }$

14. *Exact:*

$1\frac{1}{8} \div 2\frac{1}{4}$ $\frac{1}{2}$

Estimate:

$\underline{\ 1\ } \div \underline{\ 2\ } = \underline{\ \frac{1}{2}\ }$

15. *Exact:*

$2\frac{1}{2} \div 3$ $\frac{5}{6}$

Estimate:

$\underline{\ 3\ } \div \underline{\ 3\ } = \underline{\ 1\ }$

16. *Exact:*

$2\frac{3}{4} \div 2$ $1\frac{3}{8}$

Estimate:

$\underline{\ 3\ } \div \underline{\ 2\ } = \underline{\ 1\frac{1}{2}\ }$

17. *Exact:*

$9 \div 2\frac{1}{2}$ $3\frac{3}{5}$

Estimate:

$\underline{\ 9\ } \div \underline{\ 3\ } = \underline{\ 3\ }$

18. *Exact:*

$5 \div 1\frac{7}{8}$ $2\frac{2}{3}$

Estimate:

$\underline{\ 5\ } \div \underline{\ 2\ } = \underline{\ 2\frac{1}{2}\ }$

19. *Exact:*

$\dfrac{5}{8} \div 1\dfrac{1}{2}$ $\dfrac{5}{12}$

Estimate:

$\underline{\quad 1 \quad} \div \underline{\quad 2 \quad} = \underline{\dfrac{1}{2}}$

20. *Exact:*

$\dfrac{3}{4} \div 2\dfrac{1}{2}$ $\dfrac{3}{10}$

Estimate:

$\underline{\quad 1 \quad} \div \underline{\quad 3 \quad} = \underline{\dfrac{1}{3}}$

21. *Exact:*

$1\dfrac{7}{8} \div 6\dfrac{1}{4}$ $\dfrac{3}{10}$

Estimate:

$\underline{\quad 2 \quad} \div \underline{\quad 6 \quad} = \underline{\dfrac{1}{3}}$

22. *Exact:*

$8\dfrac{2}{5} \div 3\dfrac{1}{2}$ $2\dfrac{2}{5}$

Estimate:

$\underline{\quad 8 \quad} \div \underline{\quad 4 \quad} = \underline{\quad 2 \quad}$

23. *Exact:*

$5\dfrac{2}{3} \div 6$ $\dfrac{17}{18}$

Estimate:

$\underline{\quad 6 \quad} \div \underline{\quad 6 \quad} = \underline{\quad 1 \quad}$

24. *Exact:*

$5\dfrac{3}{4} \div 2$ $2\dfrac{7}{8}$

Estimate:

$\underline{\quad 6 \quad} \div \underline{\quad 2 \quad} = \underline{\quad 3 \quad}$

For Exercises 25–42, first estimate the answer. Then solve each application problem by using the six problem-solving steps. Simplify all answers. See Examples 3 and 4. Use the recipe for Carrot Cake Cupcakes to work Exercises 25–28.

CARROT CAKE CUPCAKES

12 paper bake cups
$1\frac{3}{4}$ cups flour
1 cup packed brown sugar
1 tsp. baking powder
1 tsp. baking soda
1 tsp. ground cinnamon
$\frac{1}{2}$ tsp. salt
1 cup shredded carrots
$\frac{3}{4}$ cup applesauce

$\frac{1}{3}$ cup vegetable oil
1 large egg
$\frac{1}{2}$ tsp. vanilla extract
1 container (16 oz.) ready-to-
    spread cream cheese frosting
Shredded coconut,
    tinted green
3 oz. Jelly Belly jelly beans,
    Orange Sherbet flavor

Preheat oven to 350° F. Place 12 bake cups in a muffin pan; set aside. In a large bowl, using a wire whisk, stir together flour, brown sugar, baking powder, baking soda, cinnamon, and salt. In a medium bowl, combine carrots, applesauce, oil, egg, and vanilla until blended. Add carrot mixture to flour mixture, stir well. Spoon batter into bake cups, filling $\frac{2}{3}$ full. Bake until toothpick inserted in center comes out clean,

20–25 minutes. Cool cupcakes in pan 10 minutes; remove to wire rack and cool completely. Frost with cream cheese frosting. To make carrot design on cupcakes, place gourmet Jelly Belly jelly beans in a carrot shape on each cupcake, top with green coconut for carrot top. Makes 12 cupcakes.

---

25. If 30 cupcakes are baked ($2\frac{1}{2}$ times the recipe), find the amount of each ingredient.

(a) Applesauce

Estimate: $1 \cdot 3 = 3$ cups

Exact: $1\dfrac{7}{8}$ cups of applesauce

(b) Salt

Estimate: $1 \cdot 3 = 3$ teaspoons

Exact: $1\dfrac{1}{4}$ teaspoons of salt

(c) Flour

Estimate: $2 \cdot 3 = 6$ cups

Exact: $4\dfrac{3}{8}$ cups of flour

26. If 18 cupcakes are baked ($1\frac{1}{2}$ times the recipe), find the amount of each ingredient.

(a) Flour

Estimate: $2 \cdot 2 = 4$ cups

Exact: $2\dfrac{5}{8}$ cups of flour

(b) Applesauce

Estimate: $1 \cdot 2 = 2$ cups

Exact: $1\dfrac{1}{8}$ cups applesauce

(c) Vegetable oil

Estimate: $0 \cdot 2 = 0$ cup

Exact: $\dfrac{1}{2}$ cup vegetable oil

27. How much of each ingredient is needed if you bake one-half of the recipe?

  (a) Vanilla extract

  Estimate: $1 \div 2 = \dfrac{1}{2}$ teaspoon

  Exact: $\dfrac{1}{4}$ teaspoon vanilla extract

  (b) Applesauce

  Estimate: $1 \div 2 = \dfrac{1}{2}$ cup

  Exact: $\dfrac{3}{8}$ cup applesauce

  (c) Flour

  Estimate: $2 \div 2 = 1$ cup

  Exact: $\dfrac{7}{8}$ cup of flour

28. How much of each ingredient is needed if you bake one-third of the recipe?

  (a) Flour

  Estimate: $2 \div 3 = \dfrac{2}{3}$ cup

  Exact: $\dfrac{7}{12}$ cup of flour

  (b) Salt

  Estimate: $1 \div 3 = \dfrac{1}{3}$ teaspoon

  Exact: $\dfrac{1}{6}$ teaspoon of salt

  (c) Applesauce

  Estimate: $1 \div 3 = \dfrac{1}{3}$ cup

  Exact: $\dfrac{1}{4}$ cup of applesauce

29. A new condominium conversion project requires $11\frac{3}{4}$ gallons of paint for each unit. How many units can be painted with 1316 gallons of paint?

  Estimate: $1316 \div 12 \approx 110$

  Exact: **112 units**

30. According to an old English system of time units, a moment is one and one-half minutes. How many moments are there in an 8-hour work day? (8 hours = 480 minutes). (*Source:* hightechscience.org/funfacts.htm)

  Estimate: $480 \div 2 = 240$ moments

  Exact: **320 moments**

31. A manufacturer of floor jacks is ordering steel tubing to make the handles for the jack shown below. How much steel tubing is needed to make 135 of these jacks? The symbol for inches is ″, as in 5″ means 5 inches. (*Source:* Harbor Freight Tools.)

  Estimate: $135 \cdot 20 = 2700$ in.

  Exact: $2632\frac{1}{2}$ in.

32. A wheelbarrow manufacturer uses handles made of hardwood. Find the amount of wood that is necessary to make 182 handles. The longest dimension shown in the advertisement below is the handle length. (*Source:* Harbor Freight Tools.)

  Estimate: $182 \cdot 62 = 11,284$ in.

  Exact: **11,193 in.**

33. Write the three steps for multiplying mixed numbers. Use your own words.

The answer should include

Step 1 Change mixed numbers to improper fractions.

Step 2 Multiply the fractions.

Step 3 Write the answer in lowest terms, changing to mixed or whole numbers where possible.

34. Refer to Exercise 33. In your own words, write the additional step that must be added to the rule for multiplying mixed numbers to make it the rule for dividing mixed numbers.

The additional step is to use the reciprocal of the second fraction (divisor).

35. A tire manufacturer uses $20\frac{3}{4}$ pounds of rubber to make a tire. Find the number of tires that can be manufactured with 51,460 pounds of rubber.

Estimate: **51,460 ÷ 20 = 2573 tires**

Exact: **2480 tires**

36. The manager of the flooring department at The Home Depot determines that each apartment unit requires $62\frac{1}{2}$ square yards of carpet. Find the number of apartment units that can be carpeted with 6750 square yards of carpet.

Estimate: **6750 ÷ 63 ≈ 107 units**

Exact: **108 units**

37. Mother Nature, manufacturer of bird feeders, cuts spacers from a tube that is $9\frac{3}{4}$ inches long. How many spacers can be cut from the tube if each spacer must be $\frac{3}{4}$ inch thick?

Estimate: **10 ÷ 1 = 10 spacers**

Exact: **13 spacers**

38. A building contractor must move 12 tons of sand. If his truck can carry $\frac{3}{4}$ ton of sand, how many trips must he make to move the sand?

Estimate: **12 ÷ 1 = 12 trips**

Exact: **16 trips**

Use the information on bottle jacks in the advertisement to answer Exercises 39–40. The " symbol is for inches.

12 TON HEAVY DUTY INDUSTRIAL BOTTLE JACKS

These all-purpose hydraulic jacks easily handle lifting jobs up to 12 tons, even at 45 degree and 90 degree angles. Rugged steel construction. Jacks feature twist tops for even more height. Certified to meet ASME standards. Base dimensions: 5-1/4" x 5-3/8"

CENTRAL HYDRAULICS®

17-3/4" MAXIMUM HEIGHT

Item 93378

STANDARD JACK

15-1/16" MAXIMUM HEIGHT

Item 93376

LOW PROFILE

$**19**99 REGULAR PRICE $29.99 SAVE 33%

39. A mechanic needs a hydraulic lift that will raise a car 4 times as high as the standard jack shown.

   (a) How high must it lift?

   (b) Will a mechanic 6 feet tall be able to fit under the vehicle without bending down? *Hint:* 1 ft = 12 in.

   (a) *Estimate:* **18 • 4 = 72 in.** **(b)** **No, because**

   Exact: **71 in.** **6 ft = 72 in.**

40. A race car driver needs a jack that will raise a vehicle only $\frac{1}{3}$ as high as the low-profile jack pictured.

   (a) How high must it lift?

   (b) Will a 6-inch part fit under the car?

   (a) *Estimate:* **15 ÷ 3 = 5 in.** **(b)** **No, because 6 in. is**

   Exact: **$5\frac{1}{48}$ in.** **greater than $5\frac{1}{48}$ in.**

41. A grape grower uses $6\frac{3}{4}$ gallons of a chemical for each acre of grapes. If she has $25\frac{1}{2}$ acres of grapes, how many gallons of the chemical are needed?

Estimate: **7 • 26 = 182 gallons**

Exact: **$172\frac{1}{8}$ gallons**

42. A flooring contractor needs $24\frac{2}{7}$ boxes of tile to cover a kitchen floor. If there are 24 homes in a subdivision, how many boxes of tile are needed to cover all of the floors?

Estimate: **24 • 24 = 576 boxes**

Exact: **$582\frac{6}{7}$ boxes**

Study Skills

This is the second part of the Study Cards activity. As you get further into a chapter, you can choose particular problems that will serve as a good test review. Here are two more types of study cards that will help you.

When you are doing your homework and find yourself saying, "This is really hard," or "I'm having trouble with this," make a **tough problem** study card! On the front, write out the procedure to work the type of problem *in words*. If there are special notes (like what *not* to do), include them. On the back, work at least one example; make sure you label what you are doing.

OBJECTIVES

1 Create study cards for difficult problems.

2 Create study cards of quiz problems.

Tough Problems Card

Front of Card

> <u>Warning:</u> Division is NOT commutative. The order in which you write the numbers DOES matter.
>
> Example: $1\frac{1}{3} \div 4 = \frac{\cancel{4}}{3} \cdot \frac{1}{4} = \frac{1}{3}$ ← Very different answers!
>
> But $4 \div 1\frac{1}{3} = \frac{\cancel{4}}{1} \cdot \frac{3}{\cancel{4}_1} = \frac{3}{1} = 3$ ←
>
> In an application problem, do NOT assume the numbers are given in the correct order. Use estimation to check that the answer is <u>reasonable!</u>

Back of Card

> Maite painted 4 windows using $1\frac{1}{3}$ cans of paint. How much paint did she use on each window?
>
> Try $4 \div 1\frac{1}{3}$
> ↓ ↓ (round)
> $4 \div 1 = 4$ cans on each window ← Estimate
> <u>Not reasonable</u> — she only used $1\frac{1}{3}$ cans in all!
>
> Need to find: paint on each window
> ↓ ↓ ↓
> $1\frac{1}{3}$ cans $\div$ $4 = \frac{\cancel{4}}{3} \cdot \frac{1}{\cancel{4}_1} = \frac{1}{3}$ can
> Reasonable!

Choose three types of difficult problems, and work them out on study cards. Be sure to put the words for solving the problem on one side and the worked problem on the other side.

◀◀◀ **Now Try This**

Practice Quiz Cards

Make up a few **quiz cards** for each type of problem you learn, and use them to prepare for a test. Choose two or three problems from the different sections of the chapter. Be sure you don't just choose the easiest problems! Put the problem **with the direction words** (like *solve, simplify, estimate*) on the front, and work the problem on the back. If you like, put the page number from the text there, too. When you review, you work the problem on a separate paper, and check it by looking at the back.

Solve this application problem.

Tiffany's monthly income is $1275. She spends $\frac{2}{5}$ of her income on rent and utilities. How much does she pay for rent and utilities?

Front of Card

Proper fraction followed by "of" indicates multiplication.

She spends $\frac{2}{5}$ of her income

$$\frac{2}{5} \cdot 1275 = \frac{2}{\overset{1}{\cancel{5}}} \cdot \frac{\overset{255}{\cancel{1275}}}{1} = \frac{510}{1} = 510$$

Tiffany spends $510 on rent and utilities.

Back of Card

Why Are Study Cards Brain Friendly?

First, making the study cards is an **active technique** that really gets your dendrites growing. You have to make decisions about what is most important and how to put it on the card. This kind of thinking is more in depth than just memorizing, and as a result, you will understand the concepts better and remember them longer.

Second, the cards are **visually appealing** (if you write neatly and try some color). Your brain responds to pleasant visual images, and again, you will remember longer and may even be able to "picture in your mind" how your cards look. This will help you during tests.

Third, because study cards are small and portable, you can review them easily whenever you have a few minutes. Even while you're waiting for a bus or have a few minutes between classes you can take out your cards and read them to yourself. Your **brain really benefits from repetition;** each time you review your cards your dendrites are growing thicker and stronger. After a while, the information will become automatic and you will remember it for a long time.

Chapter 2 ▷▷▷ Summary

▶ Key Terms

2.1	**numerator**	The number above the fraction bar in a fraction is called the numerator. It shows how many of the equivalent parts are being considered.
	denominator	The number below the fraction bar in a fraction is called the denominator. It shows the number of equal parts in a whole.
	proper fraction	In a proper fraction, the numerator is smaller than the denominator. The fraction is less than 1.
	improper fraction	In an improper fraction, the numerator is greater than or equal to the denominator. The fraction is equal to or greater than 1.
2.2	**mixed number**	A mixed number includes a fraction and a whole number written together.
2.3	**factors**	Numbers that are multiplied to give a product are factors.
	composite number	A composite number has at least one factor other than itself and 1.
	prime number	A prime number is a whole number other than 0 and 1 that has exactly two factors, itself and 1.
	factorizations	The numbers that can be multiplied to give a specific number (product) are factorizations of that number.
	prime factorization	In a prime factorization, every factor is a prime number.
2.4	**equivalent fractions**	Two fractions are equivalent when they represent the same portion of a whole.
	common factor	A common factor is a number that can be divided evenly into two or more whole numbers.
	lowest terms	A fraction is written in lowest terms when its numerator and denominator have no common factor other than 1.
2.5	**multiplication shortcut**	When multiplying or dividing fractions, the process of dividing a numerator and denominator by a common factor can be used as a shortcut.
2.6	**reciprocal**	Two numbers are reciprocals of each other if their product is 1. To find the reciprocal of a fraction, interchange the numerator and the denominator.

▶ New Formula

Area of a rectangle: Area = length • width

▶ Test Your Word Power

See how well you have learned the vocabulary in this chapter. Answers follow the Quick Review.

1. A **numerator** is
   A. a number greater than 5
   B. the number above the fraction bar in a fraction
   C. any number
   D. the number below the fraction bar in a fraction.

2. A **proper fraction**
   A. has a value less than 1
   B. has a whole number and a fraction
   C. has a value greater than 1
   D. is equal to 1.

3. A **mixed number** is
   A. equal to 1
   B. less than 1
   C. a whole number and a fraction written together
   D. a number multiplied by another number.

4. A **factor** is
   A. one of two or more numbers that are added to get another number
   B. the answer in division
   C. one of two or more numbers that are multiplied to get another number
   D. the answer in multiplication.

5. A whole number greater than 1 is **prime** if
   A. it cannot be factored
   B. it has just one factor
   C. it has only itself and 1 as factors
   D. it has more than two different factors.

6. A **common factor** can
   A. only be divided by itself and 1
   B. be divided evenly into two or more whole numbers
   C. never be divided by 2
   D. only be divided by the numbers 5 and 10.

7. A fraction is in **lowest terms** when
   A. it cannot be divided
   B. it is a common fraction
   C. its numerator and denominator have no common factor other than 1
   D. it has a value less than 1.

8. To find the **reciprocal** of a fraction.
   A. multiply it by itself
   B. interchange the numerator and the denominator
   C. change it to an improper fraction
   D. change it to lowest terms.

▶ Quick Review

Concepts	Examples

2.1 Types of Fractions

Proper
Numerator smaller than denominator; a value less than 1

$$\frac{2}{3} \quad \frac{3}{4} \quad \frac{15}{16} \quad \frac{1}{8} \qquad \text{Proper fractions}$$

Improper
Numerator equal to or greater than denominator; a value equal to or greater than 1

$$\frac{17}{8} \quad \frac{19}{12} \quad \frac{11}{2} \quad \frac{5}{3} \quad \frac{7}{7} \qquad \text{Improper fractions}$$

2.2 Converting Fractions

Mixed to Improper
Multiply denominator by whole number, add numerator, and place over denominator.

$$7\frac{2}{3} = \frac{23}{3} \leftarrow 3 \times 7 + 2$$

Same denominator

Improper to Mixed
Divide numerator by denominator and place remainder over denominator.

$$\frac{17}{5} = 3\frac{2}{5}$$

Same denominator

$$\begin{array}{r} 3\frac{2}{5} \\ 5\overline{)17} \leftarrow \text{Divide numerator} \\ \underline{15} \quad \text{by denominator.} \\ 2 \end{array}$$

Concepts	Examples

(2.3) Prime Numbers

Determine whether a whole number is evenly divisible only by itself and 1. (By definition, 0 and 1 are not prime.)

The prime numbers less than 100 are 2, 3, 5, 7, 11, 13, 17, 19, 23, 29, 31, 37, 41, 43, 47, 53, 59, 61, 67, 71, 73, 79, 83, 89, and 97.

(2.3) Finding the Prime Factorization of a Number

Divide each factor by a prime number using a diagram that forms the shape of tree branches.

Find the prime factorization of 30. Use a factor tree.

Prime factors are circled.
$$30 = 2 \cdot 3 \cdot 5$$

(2.4) Writing Fractions in Lowest Terms

Divide the numerator and denominator by the greatest common factor.

Write $\frac{30}{42}$ in lowest terms.
$$\frac{30}{42} = \frac{30 \div 6}{42 \div 6} = \frac{5}{7}$$

(2.5) Multiplying Fractions

1. Multiply the numerators and multiply the denominators.
2. Write answers in lowest terms if the multiplication shortcut was not used.

Multiply.
$$\frac{6}{11} \cdot \frac{7}{8} = \frac{\overset{3}{\cancel{6}}}{11} \cdot \frac{7}{\underset{4}{\cancel{8}}} = \frac{3 \cdot 7}{11 \cdot 4} = \frac{21}{44}$$

(2.7) Finding the Reciprocal

To find the reciprocal of a fraction, interchange the numerator and denominator.

Find the reciprocal of each fraction.

$\frac{3}{4}$ The reciprocal of $\frac{3}{4}$ is $\frac{4}{3}$.

$\frac{8}{5}$ The reciprocal of $\frac{8}{5}$ is $\frac{5}{8}$.

9 The reciprocal of 9 is $\frac{1}{9}$.

(2.7) Dividing Fractions

Use the reciprocal of the second fraction (divisor) and multiply as fractions.

Divide.
$$\frac{25}{36} \div \frac{15}{18} = \frac{25}{\underset{2}{\cancel{36}}} \cdot \frac{\overset{1}{\cancel{18}}}{\underset{3}{\cancel{15}}} = \frac{5 \cdot 1}{2 \cdot 3} = \frac{5}{6}$$

Reciprocals

Concepts	Examples

(2.8) Multiplying Mixed Numbers

First estimate the answer. Then follow these steps.

Step 1 *Change* each mixed number to an improper fraction.

Step 2 *Multiply.*

Step 3 *Simplify* the answer, which means to write it in *lowest terms,* and change it to a mixed number or whole number where possible.

First estimate the answer. Then multiply to get the exact answer.

Estimate: *Exact:*

$$1\frac{3}{5} \quad \cdot \quad 3\frac{1}{3} \qquad 1\frac{3}{5} \cdot 3\frac{1}{3} = \frac{8}{\overset{1}{\cancel{5}}} \cdot \frac{\overset{2}{\cancel{10}}}{3}$$

Rounded

$$2 \quad \cdot \quad 3 = 6 \qquad\qquad = \frac{8 \cdot 2}{1 \cdot 3}$$

$$= \frac{16}{3} = 5\frac{1}{3}$$

Close to estimate

(2.8) Dividing Mixed Numbers

First estimate the answer. Then follow these steps.

Step 1 *Change* each mixed number to an improper fraction.

Step 2 Use the *reciprocal* of the second fraction (divisor).

Step 3 *Multiply.*

Step 4 *Simplify* the answer, which means to write it in *lowest terms,* and change it to a mixed number or whole number where possible.

First estimate the answer. Then divide to get the exact answer.

Estimate: *Exact:*

$$3\frac{5}{9} \quad \div \quad 2\frac{2}{5} \qquad 3\frac{5}{9} \div 2\frac{2}{5} = \frac{32}{9} \div \frac{12}{5}$$

Rounded

$$4 \quad \div \quad 2 = 2 \qquad\qquad = \frac{\overset{8}{\cancel{32}}}{9} \cdot \frac{5}{\underset{3}{\cancel{12}}} = \frac{40}{27}$$

$$= 1\frac{13}{27}$$

Close to estimate

ANSWERS TO TEST YOUR WORD POWER

1. B; *Example:* In $\frac{3}{8}$, the numerator is 3.

2. A; *Example:* $\frac{1}{2}, \frac{3}{4},$ and $\frac{7}{8}$ are all proper fractions with a value less than 1.

3. C; *Example:* $2\frac{3}{8}$ and $5\frac{3}{4}$ are mixed numbers.

4. C; *Example:* Since $3 \cdot 5 = 15$, the numbers 3 and 5 are factors of 15.

5. C; *Example:* 3, 5, and 11 are prime numbers; 4, 8, and 12 are composite numbers.

6. B; *Example:* 3 is a common factor of both 6 and 9 because it can be evenly divided into each of them.

7. C; *Example:* $\frac{3}{8}, \frac{4}{5},$ and $\frac{5}{6}$ are in lowest terms but $\frac{6}{8}, \frac{3}{6},$ and $\frac{2}{4}$ are not.

8. B; *Example:* The reciprocal of $\frac{3}{8}$ is $\frac{8}{3}$, the reciprocal of $\frac{25}{4}$ is $\frac{4}{25}$, and the reciprocal of 6 or $\frac{6}{1}$ is $\frac{1}{6}$.

Chapter 2 ▶▶▶ Review Exercises

[2.1] *Write the fraction that represents each shaded portion.*

1. $\dfrac{1}{3}$

2. $\dfrac{5}{8}$

3. $\dfrac{2}{4}$

List the proper and improper fractions in each group.

	Proper	**Improper**
4. $\dfrac{1}{8}\ \dfrac{4}{3}\ \dfrac{5}{5}\ \dfrac{3}{4}\ \dfrac{2}{3}$	$\dfrac{1}{8},\dfrac{3}{4},\dfrac{2}{3}$	$\dfrac{4}{3},\dfrac{5}{5}$
5. $\dfrac{6}{5}\ \dfrac{15}{16}\ \dfrac{16}{13}\ \dfrac{1}{8}\ \dfrac{5}{3}$	$\dfrac{15}{16},\dfrac{1}{8}$	$\dfrac{6}{5},\dfrac{16}{13},\dfrac{5}{3}$

[2.2] *Write each mixed number as an improper fraction. Write each improper fraction as a mixed number.*

6. $4\dfrac{3}{4}$ $\dfrac{19}{4}$ **7.** $9\dfrac{5}{6}$ $\dfrac{59}{6}$ **8.** $\dfrac{27}{8}$ $3\dfrac{3}{8}$ **9.** $\dfrac{63}{5}$ $12\dfrac{3}{5}$

[2.3] *Find all factors of each number.*

10. 6   **11.** 24   **12.** 55   **13.** 90

1, 2, 3, 6   1, 2, 3, 4, 6, 8, 12, 24   1, 5, 11, 55   1, 2, 3, 5, 6, 9, 10, 15, 18, 30, 45, 90

Write the prime factorization of each number by using exponents.

14. 27 3^3 **15.** 150 $2 \cdot 3 \cdot 5^2$ **16.** 420 $2^2 \cdot 3 \cdot 5 \cdot 7$

Simplify each expression.

17. 5^2 25   **18.** $6^2 \cdot 2^3$ 288   **19.** $8^2 \cdot 3^3$ 1728   **20.** $4^3 \cdot 2^5$ 2048

[2.4] *The fineness (purity) of gold is regulated by law and is the same in all parts of the world. Exercises 21–24 show the fineness of 24-kt, 18-kt, 14-kt, and 10-kt gold by comparing the parts of gold to the parts of alloy (metals other than gold). Write a fraction in lowest terms to show the portion that is gold. (Source: Costco Wholesale.)*

21. 24 kt. = (24 parts gold, 0 parts alloy) $\dfrac{24}{24} = 1$

22. 18 kt. = (18 parts gold, 6 parts alloy) $\dfrac{18}{24} = \dfrac{3}{4}$

23. 14 kt. = (14 parts gold, 10 parts alloy) $\dfrac{14}{24} = \dfrac{7}{12}$

24. 10 kt. = (10 parts gold, 14 parts alloy) $\dfrac{10}{24} = \dfrac{5}{12}$

Write the numerator and denominator of each fraction as a product of prime factors. Then, write the fraction in lowest terms.

25. $\dfrac{25}{60}$ $\dfrac{\overset{1}{\cancel{5}} \cdot 5}{2 \cdot 2 \cdot 3 \cdot \underset{1}{\cancel{5}}}$; $\dfrac{5}{12}$

26. $\dfrac{384}{96}$ $\dfrac{\overset{1}{\cancel{2}} \cdot \overset{1}{\cancel{2}} \cdot \overset{1}{\cancel{2}} \cdot \overset{1}{\cancel{2}} \cdot \overset{1}{\cancel{2}} \cdot 2 \cdot 2 \cdot \overset{1}{\cancel{3}}}{\underset{1}{\cancel{2}} \cdot \underset{1}{\cancel{2}} \cdot \underset{1}{\cancel{2}} \cdot \underset{1}{\cancel{2}} \cdot \underset{1}{\cancel{2}} \cdot \underset{1}{\cancel{3}}}$; 4

Decide whether each pair of fractions is equivalent or not equivalent, using the method of prime factors.

27. $\dfrac{3}{4}$ and $\dfrac{48}{64}$ **equivalent**

28. $\dfrac{5}{8}$ and $\dfrac{70}{120}$ **not equivalent**

29. $\dfrac{2}{3}$ and $\dfrac{360}{540}$ **equivalent**

[2.5–2.8] *Multiply. Write answers in lowest terms, and as mixed numbers or whole numbers where possible.*

30. $\dfrac{4}{5} \cdot \dfrac{3}{4}$ $\dfrac{3}{5}$

31. $\dfrac{3}{10} \cdot \dfrac{5}{8}$ $\dfrac{3}{16}$

32. $\dfrac{70}{175} \cdot \dfrac{5}{14}$ $\dfrac{1}{7}$

33. $\dfrac{44}{63} \cdot \dfrac{3}{11}$ $\dfrac{4}{21}$

34. $\dfrac{5}{16} \cdot 48$ 15

35. $\dfrac{5}{8} \cdot 1000$ 625

Divide. Write answers in lowest terms, and as mixed numbers or whole numbers where possible.

36. $\dfrac{2}{3} \div \dfrac{1}{2}$ $1\dfrac{1}{3}$

37. $\dfrac{5}{6} \div \dfrac{1}{2}$ $\dfrac{5}{3} = 1\dfrac{2}{3}$

38. $\dfrac{\frac{15}{18}}{\frac{10}{30}}$ $\dfrac{5}{2} = 2\dfrac{1}{2}$

39. $\dfrac{\frac{3}{4}}{\frac{3}{8}}$ 2

40. $7 \div \dfrac{7}{8}$ 8

41. $18 \div \dfrac{3}{4}$ 24

42. $\dfrac{5}{8} \div 3$ $\dfrac{5}{24}$

43. $\dfrac{2}{3} \div 5$ $\dfrac{2}{15}$

44. $\dfrac{\frac{12}{13}}{3}$ $\dfrac{4}{13}$

Find the area of each rectangle.

45.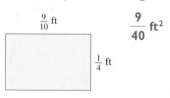
$\dfrac{9}{10}$ ft $\dfrac{9}{40}$ ft^2
$\dfrac{1}{4}$ ft

46.
$\dfrac{2}{3}$ in. $\dfrac{7}{12}$ in.^{2}
$\dfrac{7}{8}$ in.

47. Ceramic tile is being installed on the floor of a meeting hall that is 108 ft long and $72\dfrac{3}{4}$ ft wide. Find the area.

7857 ft^2

48. Find the area of a display shelf that is a rectangle measuring 6 ft long and $\dfrac{11}{12}$ ft wide.

$5\dfrac{1}{2}$ ft^2

First estimate the answer. Then multiply or divide to find the exact answer. Simplify all answers.

49. *Exact:*

$5\frac{1}{2} \cdot 1\frac{1}{4}$ $6\frac{7}{8}$

Estimate:

$\underline{\quad 6 \quad} \cdot \underline{\quad 1 \quad} = \underline{\quad 6 \quad}$

50. *Exact:*

$2\frac{1}{4} \cdot 7\frac{1}{8} \cdot 1\frac{1}{3}$ $21\frac{3}{8}$

Estimate:

$\underline{\quad 2 \quad} \cdot \underline{\quad 7 \quad} \cdot \underline{\quad 1 \quad} = \underline{\quad 14 \quad}$

51. *Exact:*

$15\frac{1}{2} \div 3$ $5\frac{1}{6}$

Estimate:

$\underline{\quad 16 \quad} \div \underline{\quad 3 \quad} = \underline{\quad 5\frac{1}{3} \quad}$

52. *Exact:*

$4\frac{3}{4} \div 6\frac{1}{3}$ $\frac{3}{4}$

Estimate:

$\underline{\quad 5 \quad} \div \underline{\quad 6 \quad} = \underline{\quad \frac{5}{6} \quad}$

Solve each application problem by using the six problem-solving steps.

53. Blue Diamond Almonds has 320 tons of almonds. How many $\frac{5}{8}$ ton bins will be needed to store the almonds?

512 bins

54. An estate is divided so that each of 5 children receives equal shares of $\frac{2}{3}$ of the estate. What fraction of the total estate will each receive?

$\dfrac{2}{15}$ **of the estate**

55. How many window-blind pull cords can be made from $157\frac{1}{2}$ yards of cord if $4\frac{3}{8}$ yards of cord are needed for each blind? First estimate, and then find the exact answer.

Estimate: **158 ÷ 4 ≈ 40 pull cords**

Exact: **36 pull cords**

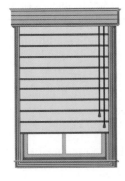

56. A gallon of water weighs $8\frac{1}{3}$ pounds. Find the weight of the water in two 50-gallon aquariums. First estimate, and then find the exact answer.

Estimate: **8 · 2 · 50 = 800 pounds**

Exact: $833\frac{1}{3}$ **pounds**

57. Ebony Wilson purchased 100 pounds of rice at the food co-op. After selling $\frac{1}{4}$ of this to her neighbor, she gives $\frac{2}{3}$ of the remaining rice to her parents. How many pounds of rice does she have left?

25 pounds

58. Sheila Spinney, a recent college graduate, receives a salary of $2976 each month. She pays $\frac{3}{8}$ of this amount in taxes, social security, and a retirement plan. Of the remainder, $\frac{9}{10}$ goes for basic living expenses. How much money remains?

$186

59. The Citrus Heights Park District will divide $\frac{3}{4}$ of its budget among 8 local parks. What fraction of the total budget will each park receive?

$\dfrac{3}{32}$ **of the budget**

60. In a morning of deep-sea fishing, 5 fishermen catch $\frac{4}{5}$ ton of salmon. If they divide the fish evenly, how much will each receive?

$\dfrac{4}{25}$ **ton**

>>> **Mixed Review Exercises**

Multiply or divide as indicated. Simplify all answers.

61. $\frac{1}{2} \cdot \frac{3}{4}$ $\frac{3}{8}$

62. $\frac{2}{3} \cdot \frac{3}{5}$ $\frac{2}{5}$

63. $12\frac{1}{2} \cdot 2\frac{1}{2}$ $31\frac{1}{4}$

64. $12\frac{1}{2} \cdot 2\frac{1}{4}$ $28\frac{1}{8}$

65. $\dfrac{\frac{4}{5}}{8}$ $\frac{1}{10}$

66. $\dfrac{\frac{5}{8}}{4}$ $\frac{5}{32}$

67. $\frac{15}{31} \cdot 62$ 30

68. $3\frac{1}{4} \div 1\frac{1}{4}$ $2\frac{3}{5}$

Write each mixed number as an improper fraction. Write each improper fraction as a mixed number.

69. $\frac{8}{5}$ $1\frac{3}{5}$

70. $\frac{153}{4}$ $38\frac{1}{4}$

71. $5\frac{2}{3}$ $\frac{17}{3}$

72. $38\frac{3}{8}$ $\frac{307}{8}$

Write the numerator and denominator of each fraction as a product of prime factors; then write the fraction in lowest terms.

73. $\frac{8}{12}$ $\frac{\overset{1}{\cancel{2}} \cdot \overset{1}{\cancel{2}} \cdot 2}{\underset{1}{\cancel{2}} \cdot \underset{1}{\cancel{2}} \cdot 3} = \frac{2}{3}$

74. $\frac{108}{210}$ $\frac{\overset{1}{\cancel{2}} \cdot 2 \cdot \overset{1}{\cancel{3}} \cdot 3 \cdot 3}{\underset{1}{\cancel{2}} \cdot \underset{1}{\cancel{3}} \cdot 5 \cdot 7} = \frac{18}{35}$

Write each fraction in lowest terms.

75. $\frac{75}{90}$ $\frac{5}{6}$

76. $\frac{48}{72}$ $\frac{2}{3}$

77. $\frac{44}{110}$ $\frac{2}{5}$

78. $\frac{87}{261}$ $\frac{1}{3}$

Solve each application problem.

79. The directions on a can of fabric glue say to apply $3\frac{3}{5}$ ounces of glue to each square yard of fabric. How many ounces are needed for $43\frac{3}{4}$ square yards? First estimate, and then find the exact answer.

Estimate: $4 \cdot 44 = 176$ **ounces**

Exact: $157\frac{1}{2}$ **ounces**

80. Valley Farms purchased some diesel fuel additive. The instructions say to use $7\frac{1}{4}$ quarts of additive for each tank of fuel. How many quarts are needed for $25\frac{1}{2}$ tanks? First estimate, and then find the exact answer.

Estimate: $7 \cdot 26 = 182$ **quarts**

Exact: $184\frac{7}{8}$ **quarts**

81. The U.S.A. Breast Cancer stamp is $1\frac{3}{4}$ in. by $\frac{7}{8}$ in. Find its area.

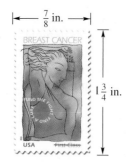

$\longleftarrow \frac{7}{8}$ in. $\longrightarrow$

$1\frac{3}{4}$ in.

$1\frac{17}{32}$ **in.²**

82. A section of marble countertop is $\frac{1}{2}$ yard by $\frac{5}{8}$ yard. What is its area?

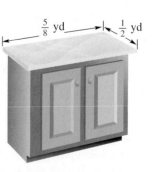

$\frac{5}{8}$ yd $\frac{1}{2}$ yd

$\frac{5}{16}$ **yd²**

Study Skills

▶▶▶ REVIEWING A CHAPTER

This activity is really about **preparing for tests.** Some of the suggestions are ideas that you will learn to use a little later in the term, but get started trying them out now. Often, the first chapters in your math textbook will be review, so it is good to practice some of the study techniques on material that is not too challenging.

Use these **chapter reviewing techniques.**

▶ **Make a study card for each vocabulary word and concept.** Include a definition, an example, a sketch, and a page reference. Include the symbol or formula if there is one. See the *Using Study Cards* activity for a quick look at some sample study cards.

▶ **Go back to the section** to find more explanations or information about any new vocabulary, formulas, or symbols.

▶ **Use the Chapter Summary** to practice each type of problem. Do not expect the Summary to substitute for reading and working through the whole chapter! First, take the "Test Your Word Power" quiz to check your understanding of new vocabulary. The answers are at the end of the Quick Review. Then read the Quick Review. **Pay special attention to the red headings.** Check the explanations for the solutions to problems given. Try to think about how all the topics in the **whole chapter** are related.

▶ **Study your lecture notes** to see what your instructor has emphasized in class. Then review that material in your text.

▶ **Do the Review Exercises.**
  ✓ Check your answers **after** you're done with each **section of exercises.**
  ✓ If you get stuck on a problem, **first** check the Chapter Summary. If that doesn't clear up your confusion, then check the section and your lecture notes.
  ✓ Pay attention to **direction words** for the problems, such as *simplify, round, solve,* and *estimate.*
  ✓ Make **study cards for especially difficult problems.**

▶ **Do the Mixed Review exercises.** This is a good check to see if you can still do the problems when they are in mixed-up order. **Check your answers carefully** in the Answers section in the back of your book. Are your answers **exact** and **complete?** Make sure you are **labeling** answers correctly, using the right **units.** For example, does your answer need to include *$, cm², ft,* and so on?

OBJECTIVES

1 Use the Chapter Summary to practice every type of problem.

2 Create study cards for vocabulary.

3 Practice by doing review and mixed review exercises.

4 Take the Chapter Test as a practice test.

Chapter Reviewing Techniques

You have already become familiar with the features of your textbook. This activity requires you to make good use of them. Your **brain needs repetition** to strengthen dendrites and the connections between them. By following the steps outlined here, you will be reinforcing the concepts, procedures, and skills you need to use for tests (and for the next chapters).

The combination of techniques provides repetition in different ways. That **promotes good branching of dendrites** instead of just relying on one branch, or route, to connect to the other dendrites. A thorough review of each chapter will **solidify your dendrite connections.** It will help you be sure that you understand the concepts **completely and accurately.** Also, taking the Chapter Test will **simulate the testing situation,** which gives you practice in test taking conditions.

Now Try This ▶▶▶

▶ **Take the Chapter Test as if it is a real test.** If your instructor has skipped sections in the chapter, figure out which problems to skip on the test before you start.

   ✓ **Time yourself** just as you would for a real test.

   ✓ **Use a calculator or notes** just as you would be permitted to (or not) on a real test.

   ✓ **Take the test in one sitting,** just like a real test is given in one sitting.

   ✓ **Show all your work.** Practice showing your work just the way your instructor has asked you to show it.

   ✓ **Practice neatness.** Can someone else follow your steps?

   ✓ **Check your answers** in the back of the book.

Notice that reviewing a chapter will take some time. Remember that it takes time for dendrites to grow! You cannot grow a good network of dendrites by rushing through a review in one night. But if you use the suggestions over a few days or evenings, you will notice that you understand the material more thoroughly and remember it longer.

Follow the reviewing techniques listed above for your next test. For each technique, write a comment *about how it worked for you in the spaces below.*

1. **Make a study card for each vocabulary word and concept.**

2. **Go back to the section** to find more explanations or information.

3. **Take the Test Your Word Power quiz and use the Quick Review** to review each concept in the chapter.

4. **Study your lecture notes** to see what your instructor has emphasized in class.

5. **Do the Review Exercises,** following the specific suggestions on the previous page.

6. **Do the Mixed Review exercises.**

7. **Take the Chapter Test** as if it is a real test.

Chapter 2 ▶▶▶ Test

Use the Chapter Test Prep Video CD to see fully worked-out solutions to any of the exercises you want to review.

Write a fraction to represent each shaded portion.

1.

2.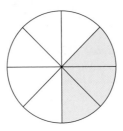

3. Identify all the proper fractions in this list:

$$\frac{2}{3} \quad \frac{4}{4} \quad \frac{6}{7} \quad \frac{5}{2} \quad \frac{1}{4} \quad \frac{5}{8} \quad \frac{30}{18}$$

4. Write $3\frac{3}{8}$ as an improper fraction.

5. Write $\frac{123}{4}$ as a mixed number.

6. Find all factors of 18.

Find the prime factorization of each number. Write the answers using exponents.

7. 45

8. 144

9. 500

Write each fraction in lowest terms.

10. $\frac{36}{48}$

11. $\frac{60}{72}$

✎ 12. The method of prime factors is used to write a fraction in lowest terms. Briefly explain how this is done. Use the fraction $\frac{56}{84}$ to show how this works.

**Write the prime factorization of both numerator and denominator.
Divide the numerator and denominator by any common factors.
Multiply the remaining factors in the numerator and denominator.**

1. $\dfrac{5}{6}$ _____

2. $\dfrac{3}{8}$ _____

3. $\dfrac{2}{3}, \dfrac{6}{7}, \dfrac{1}{4}, \dfrac{5}{8}$ _____

4. $\dfrac{27}{8}$ _____

5. $30\dfrac{3}{4}$ _____

6. 1, 2, 3, 6, 9, 18 _____

7. $3^2 \cdot 5$ _____

8. $2^4 \cdot 3^2$ _____

9. $2^2 \cdot 5^3$ _____

10. $\dfrac{3}{4}$ _____

11. $\dfrac{5}{6}$ _____

12. $\dfrac{56}{84} = \dfrac{\cancel{2} \cdot \cancel{2} \cdot 2 \cdot \cancel{7}}{\cancel{2} \cdot \cancel{2} \cdot 3 \cdot \cancel{7}} = \dfrac{2}{3}$

13. Answers will vary.
See one possibility
at the right.

14. $\dfrac{1}{3}$

15. 36

16. $\dfrac{5}{12}$ yd²

17. 5475 seedlings

18. $\dfrac{9}{10}$

19. $15\dfrac{3}{4}$

20. 24 pieces

21. *Estimate:* $4 \cdot 4 = 16$

Exact: $14\dfrac{7}{16}$

22. *Estimate:* $2 \cdot 4 = 8$

Exact: $7\dfrac{17}{18}$

23. *Estimate:* $10 \div 2 = 5$

Exact: $4\dfrac{4}{15}$

24. *Estimate:* $9 \div 2 = 4\dfrac{1}{2}$

Exact: $5\dfrac{1}{10}$

25. *Estimate:* $3 \cdot 12 = 36$

Exact: $30\dfrac{5}{8}$ grams

✏ **13.** Explain how to multiply fractions. What additional step must be taken when dividing fractions?

Multiply fractions by multiplying the numerators and multiplying the denominators. Divide two fractions by using the reciprocal of the second fraction (divisor) and then multiplying.

Multiply or divide. Write answers in lowest terms, and as mixed numbers or whole numbers where possible.

14. $\dfrac{3}{4} \cdot \dfrac{4}{9}$ **15.** $54 \cdot \dfrac{2}{3}$

16. A rectangular barbecue grill is $\frac{15}{16}$ yard by $\frac{4}{9}$ yard. Find the area of the grill.

17. The Sierra College Conservation Club planted 8760 Douglass Fir seedlings. If $\frac{3}{8}$ of these seedlings are not expected to survive, find the number of seedlings that do survive.

18. $\dfrac{3}{4} \div \dfrac{5}{6}$ **19.** $\dfrac{\dfrac{7}{4}}{\dfrac{9}{}}$

20. To complete a custom-designed cabinet, oak trim pieces must be cut exactly $2\frac{1}{4}$ inches long so that they can be used as dividers in a spice rack. Find the number of pieces that can be cut from a piece of oak that is 54 inches in length.

First estimate the answer. Then find the exact answer. Simplify all answers.

21. $4\dfrac{1}{8} \cdot 3\dfrac{1}{2}$ **22.** $1\dfrac{5}{6} \cdot 4\dfrac{1}{3}$

23. $9\dfrac{3}{5} \div 2\dfrac{1}{4}$ **24.** $\dfrac{8\dfrac{1}{2}}{1\dfrac{2}{3}}$

25. A new vaccine is synthesized at the rate of $2\frac{1}{2}$ grams per day. How many grams can be synthesized in $12\frac{1}{4}$ days?

Study Skills

►►► PREPARING FOR TESTS

Many things besides studying can improve your test scores. You may not realize that eating the right foods, and getting enough exercise and sleep, can also improve your scores. Your brain (and therefore your ability to think) is affected by the condition of your whole body. So, part of your preparation for tests includes keeping yourself in good physical shape as well as spending time on the actual course material. Try these suggestions and see the difference.

OBJECTIVES

1 Restate the importance of sleep and good nutrition as it affects learning.

2 Explain the effect of anxiety and stress on learning.

Performance Health Tips to Improve Your Test Score	Explanation
Get **seven to eight hours of sleep** the night before the exam. (It's helpful to get that much sleep *every* night.)	*Fatigue and exhaustion* reduce efficiency. They also cause poor memory and recall. If you didn't sleep much the night before a test, 20 minutes of relaxation or meditation can help.
Eat a **small, high-energy meal** about two hours before the test. Start the meal with a small amount of protein such as fish, chicken, or nonfat yogurt. Include carbohydrates if you like, but no high-fat foods.	Just 3 to 4 ounces of protein increases the amount of a chemical in the brain called tyrosine, which **improves your alertness, accuracy, and motivation.** High-fat foods dull your mind and slow down your brain.
Drink plenty of water. Don't wait until you feel thirsty; your body is already dehydrated by the time you feel it.	Research suggests that staying well hydrated improves the electrochemical communications in your brain.
Give your brain the time it needs to grow dendrites!	**Cramming doesn't work;** your brain cannot grow dendrites that quickly. **Studying every day** is the way to give your brain the time it needs.

To Prevent Anxiety	Explanation
Practice slow, deep breathing for five minutes each day. Then do a minute or two of deep breathing right before the test. Also, if you feel your anxiety building during the test, stop for a minute, close your eyes, and do some deep breathing.	When **test anxiety** hits, you breathe more quickly and shallowly, which causes hyperventilation. Symptoms may be confusion, inability to concentrate, shaking, dizziness, and more. Slow, deep breathing will **calm you and prevent panic.**

Performance Health Tips

Anxiety Prevention Tips

195

To Prevent Anxiety	Explanation
Do 15 to 20 minutes of *moderate exercise* (like walking) shortly before the test. Daily exercise is even better!	*Exercise reduces stress* and will help prevent "blanking out" on a test. Exercise also increases your alertness, clear thinking, and energy.
To help you sleep the night before the test, or any time you need to calm down, *eat high carbohydrate foods* such as popcorn, bread, rice, crackers, muffins, bagels, pasta, corn, baked potatoes (not fries or chips), and cereals.	Carbohydrates increase the level of a chemical in the brain called serotonin, which has a *calming effect on the mind*. It reduces feelings of tension and stress and improves your ability to concentrate. You only need to eat a small amount, like half a bagel, to get this effect.
Before the test, *go easy on caffeinated beverages* such as coffee, tea, and soft drinks. Do not eat candy bars or other sugary snacks.	Extra caffeine can *make you jittery,* "hyper," and shaky for the test. It can increase the tendency to panic. Too much sugar causes negative emotional reactions in some people.

Now Try This ▶▶▶

What will you do to improve your next test score? List the three or four tips you think will help you the most.

1. _____

2. _____

3. _____

4. _____

What changes will you have to make in order to try the tips you chose?

_____

_____

_____

_____

See *Tips for Taking Math Tests* **and** *Preparing for Your Final Exam* **for more ideas about managing anxiety.** (Check the Table of Contents to find their locations.)

Cumulative Review Exercises ▷▷▷ Chapters 1–2

Name the digit that has the given place value in each number.

1. 783
hundreds **7**
tens **8**

2. 8,621,785
millions **8**
ten thousands **2**

Add, subtract, multiply, or divide as indicated.

3. $\begin{array}{r} 71 \\ 23 \\ 47 \\ +\ 36 \\ \hline \textbf{177} \end{array}$

4. $\begin{array}{r} 82,121 \\ 5\ 468 \\ 316 \\ +\ 61,294 \\ \hline \textbf{149,199} \end{array}$

5. $\begin{array}{r} 6537 \\ -\ 2085 \\ \hline \textbf{4452} \end{array}$

6. $\begin{array}{r} 4,819,604 \\ -\ 1,597,783 \\ \hline \textbf{3,221,821} \end{array}$

7. $\begin{array}{r} 83 \\ \times\ 9 \\ \hline \textbf{747} \end{array}$

8. $9 \cdot 4 \cdot 2$ **72**

9. $\begin{array}{r} 3784 \\ \times\ 573 \\ \hline \textbf{2,168,232} \end{array}$

10. $\begin{array}{r} 563 \\ \times\ 800 \\ \hline \textbf{450,400} \end{array}$

11. $\dfrac{63}{7}$ **9**

12. $18\overline{)136,458}$ **7 581**

13. $33,886 \div 4$
8471 R2

14. $492\overline{)10,850}$ **22 R26**

Round each number to the nearest ten, nearest hundred, and nearest thousand.

	Ten	Hundred	Thousand
15. 6583	6580	6600	7000
16. 76,271	76,270	76,300	76,000

Simplify.

17. $2^5 - 6(4)$ **8**

18. $\sqrt{36} - 2 \cdot 3 + 5$ **5**

Solve each application problem using the six problem-solving steps.

19. A Tour of Gettysburg van uses 9 gallons of fuel on a half-day tour and 17 gallons of fuel on a full-day tour. Find the total number of gallons of fuel used in 26 half-day tours and 18 full-day tours.
540 gallons

20. Sounds louder than 80 decibels may harm a person's hearing. A loud rock concert produces 150 decibels, while a lawn mower, shop tool, truck traffic, or subway produces 90 decibels. How many more decibels does a rock concert produce than a lawn mower? (*Source:* Ear Foundation and Plantronics.)
60 decibels

21. A typical adult loses 100 hairs a day out of approximately 120,000 hairs. If the lost hairs were not replaced, find the number of hairs remaining after two years. (1 year = 365 days.)   **47,000 hairs**

22. A group of 22 health care workers will divide 4136 work hours evenly this month. Find the number of hours each will work.   **188 hours**

23. George Ann Horner worked $38\frac{5}{6}$ hours and is paid $18 per hour. How much money did she earn?
$699

24. A welder needs angle iron pieces that are $3\frac{1}{3}$ inches long. Find the number of pieces that can be cut from a piece of angle iron that is 70 inches in length.
21 pieces

Write proper *or* improper *for each fraction.*

25. $\frac{2}{3}$ proper

26. $\frac{6}{6}$ improper

27. $\frac{9}{18}$ proper

Write each mixed number as an improper fraction. Write each improper fraction as a whole or mixed number.

28. $3\frac{3}{8}$ $\frac{27}{8}$

29. $6\frac{2}{5}$ $\frac{32}{5}$

30. $\frac{14}{7}$ 2

31. $\frac{103}{8}$ $12\frac{7}{8}$

Find the prime factorization of each number. Write answers using exponents.

32. 72 $2^3 \cdot 3^2$

33. 126 $2 \cdot 3^2 \cdot 7$

34. 350 $2 \cdot 5^2 \cdot 7$

Simplify each expression.

35. $4^2 \cdot 2^2$ 64

36. $2^3 \cdot 6^2$ 288

37. $2^3 \cdot 4^2 \cdot 5$ 640

Write each fraction in lowest terms.

38. $\frac{42}{48}$ $\frac{7}{8}$

39. $\frac{24}{36}$ $\frac{2}{3}$

40. $\frac{30}{54}$ $\frac{5}{9}$

Multiply or divide as indicated. Simplify all answers.

41. $\frac{1}{2} \cdot \frac{3}{4}$ $\frac{3}{8}$

42. $30 \cdot \frac{2}{3} \cdot \frac{3}{5}$ 12

43. $7\frac{1}{2} \cdot 3\frac{1}{3}$ 25

44. $\frac{3}{5} \div \frac{5}{8}$ $\frac{24}{25}$

45. $\frac{7}{8} \div 1\frac{1}{2}$ $\frac{7}{12}$

46. $3 \div 1\frac{1}{4}$ $2\frac{2}{5}$

Adding and Subtracting Fractions

When Bryan Berg was a small boy, his grandfather taught him how to build a simple house of cards. Berg has set 8 Guinness world records for the tallest freestanding card structures—no glue, no tape, no hidden supports, just the force of gravity and a very steady hand. While a high school student, he set his first record of 14 feet, 6 inches tall. That house of cards required 4 days, 208 decks of cards, and a scaffold to build. Berg is a graduate of the Harvard Graduate School of Design and makes his living "stacking cards." At the 2007 State of Texas Fair, he set a new world record of 25 feet, $9\frac{7}{16}$ inches. As an architect and when *Stacking the Deck* (the title of his book), Berg will be adding and subtracting fractions. (See Exercises 49 and 66 in **Section 3.4.**) (*Source: Reader's Digest.*)

3.1 ▶▶▶ Adding and Subtracting Like Fractions

OBJECTIVES

1 Define like and unlike fractions.

2 Add like fractions.

3 Subtract like fractions.

In **Chapter 2** we looked at the basics of fractions and then practiced with multiplication and division of fractions and mixed numbers. In this chapter we will work with addition and subtraction of fractions and mixed numbers.

OBJECTIVE 1 Define like and unlike fractions. Fractions with the same denominators are **like fractions.** Fractions with different denominators are **unlike fractions.**

EXAMPLE 1 Identifying Like and Unlike Fractions

(a) $\dfrac{3}{4}, \dfrac{1}{4}, \dfrac{5}{4}, \dfrac{6}{4},$ and $\dfrac{4}{4}$ are **like** fractions.

⇡ ⇡ ⇡ ⇡   ⇡ —— All denominators are the same.

(b) $\dfrac{7}{12}$ and $\dfrac{12}{7}$ are **unlike** fractions.

⇡   ⇡ —— Denominators are different.

> **Note**
> Like fractions have the *same* denominator.

◀ *Work Problem* **1** *at the Side.*

1 Next to each pair of fractions write *like* or *unlike*.

(a) $\dfrac{2}{5}$ $\dfrac{3}{5}$ _____

(b) $\dfrac{2}{3}$ $\dfrac{3}{4}$ _____

(c) $\dfrac{7}{12}$ $\dfrac{11}{12}$ _____

(d) $\dfrac{3}{8}$ $\dfrac{3}{16}$ _____

OBJECTIVE 2 Add like fractions. The figures below show you how to add the fractions $\frac{2}{7}$ and $\frac{4}{7}$.

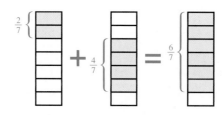

As the figures show,

$$\frac{2}{7} + \frac{4}{7} = \frac{6}{7}$$

Add like fractions as follows.

> **Adding Like Fractions**
> *Step 1* Add the numerators to find the numerator of the sum.
> *Step 2* Write the denominator of the like fractions as the denominator of the sum.
> *Step 3* Write the sum in lowest terms.

ANSWERS

1. **(a)** like **(b)** unlike **(c)** like **(d)** unlike

EXAMPLE 2 **Adding Like Fractions**

Add and write the sum in lowest terms.

(a) $\dfrac{1}{5} + \dfrac{2}{5}$

$$\dfrac{1}{5} + \dfrac{2}{5} = \dfrac{\overbrace{1 + 2}^{\text{Add numerators.}}}{5} = \dfrac{3}{5} \leftarrow \text{Same denominator}$$

(b) $\dfrac{1}{12} + \dfrac{7}{12} + \dfrac{1}{12}$ *Fractions are ready to be added if they are like fractions.*

Step 1 $\dfrac{\overbrace{1 + 7 + 1}^{\text{Add numerators.}}}{12}$

Step 2 $= \dfrac{9}{12} \leftarrow \text{Sum of numerators}$ $\leftarrow \text{Same denominator}$

Step 3 $= \dfrac{9 \div 3}{12 \div 3} = \dfrac{3}{4}$ In lowest terms

CAUTION
Fractions may be added ***only*** if they have like denominators.

────────────────── *Work Problem* **2** *at the Side.* ▶

OBJECTIVE 3 Subtract like fractions. The figures below show $\frac{7}{8}$ broken into $\frac{4}{8}$ and $\frac{3}{8}$.

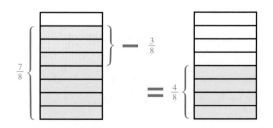

Subtracting $\frac{3}{8}$ from $\frac{7}{8}$ gives the answer $\frac{4}{8}$, or

$$\dfrac{7}{8} - \dfrac{3}{8} = \dfrac{4}{8}$$

2 Add and write the sums in lowest terms.

(a) $\dfrac{3}{8} + \dfrac{1}{8}$

(b) $\dfrac{2}{9}$
$+ \dfrac{5}{9}$
_____

(c) $\dfrac{3}{16} + \dfrac{1}{16}$

(d) $\dfrac{3}{10} + \dfrac{1}{10} + \dfrac{4}{10}$

3 Find the difference and simplify.

(a) $\dfrac{5}{6} - \dfrac{1}{6}$

(b) $\dfrac{16}{10}$
$-\dfrac{7}{10}$

(c) $\dfrac{15}{3} - \dfrac{5}{3}$

(d) $\dfrac{25}{32}$
$-\dfrac{6}{32}$

Write $\frac{4}{8}$ in lowest terms.

$$\frac{7}{8} - \frac{3}{8} = \frac{4 \div 4}{8 \div 4} = \frac{1}{2}$$

The steps for subtracting like fractions are very similar to those for adding like fractions.

> **Subtracting Like Fractions**
>
> **Step 1** Subtract the numerators to find the numerator of the difference.
>
> **Step 2** Write the denominator of the like fractions as the denominator of the difference.
>
> **Step 3** Write the answer in lowest terms.

EXAMPLE 3 **Subtracting Like Fractions**

Find the difference and simplify the answer.

(a) $\dfrac{15}{16} - \dfrac{3}{16}$

Step 1
$$\frac{15}{16} - \frac{3}{16} = \frac{\overbrace{15 - 3}^{\text{Subtract numerators.}}}{16}$$

Step 2
$$= \frac{12}{16} \begin{array}{l} \leftarrow \text{Difference of numerators} \\ \leftarrow \text{Same denominator} \end{array}$$

Step 3
$$= \frac{12 \div 4}{16 \div 4} = \frac{3}{4} \quad \text{In lowest terms}$$

(b) $\dfrac{13}{4} - \dfrac{6}{4}$ ◁ Fractions are ready to be subtracted if they are *like* fractions.

$$\frac{13}{4} - \frac{6}{4} = \frac{\overbrace{13 - 6}^{\text{Subtract numerators.}}}{4} \begin{array}{l} \leftarrow \text{Difference of numerators} \\ \leftarrow \text{Same denominator} \end{array}$$

$$= \frac{7}{4}$$

To simplify the answer, write $\frac{7}{4}$ as a mixed number.

$$\frac{7}{4} = 1\frac{3}{4}$$

> **CAUTION**
> Fractions may be subtracted **only** if they have like denominators.

◁ *Work Problem* 3 *at the Side.*

ANSWERS

3. (a) $\dfrac{2}{3}$ (b) $\dfrac{9}{10}$ (c) $3\dfrac{1}{3}$ (d) $\dfrac{19}{32}$

3.1 ▶▶▶ Exercises

Find the sum and simplify the answer. See Example 2.

1. $\dfrac{3}{8} + \dfrac{2}{8}$ **$\dfrac{5}{8}$**

2. $\dfrac{1}{5} + \dfrac{3}{5}$ **$\dfrac{4}{5}$**

3. $\dfrac{2}{6} + \dfrac{3}{6}$ **$\dfrac{5}{6}$**

4. $\dfrac{9}{11} + \dfrac{1}{11}$ **$\dfrac{10}{11}$**

5. $\dfrac{1}{4} + \dfrac{1}{4}$ **$\dfrac{1}{2}$**

6. $\dfrac{1}{14} + \dfrac{1}{14}$ **$\dfrac{1}{7}$**

7. $\begin{array}{r} \dfrac{9}{10} \\ + \dfrac{3}{10} \\ \hline \end{array}$ **$1\dfrac{1}{5}$**

8. $\begin{array}{r} \dfrac{13}{12} \\ + \dfrac{5}{12} \\ \hline \end{array}$ **$1\dfrac{1}{2}$**

9. $\begin{array}{r} \dfrac{2}{9} \\ + \dfrac{1}{9} \\ \hline \end{array}$ **$\dfrac{1}{3}$**

10. $\dfrac{7}{12} + \dfrac{3}{12}$ **$\dfrac{5}{6}$**

11. $\dfrac{6}{20} + \dfrac{4}{20} + \dfrac{3}{20}$ **$\dfrac{13}{20}$**

12. $\dfrac{1}{7} + \dfrac{2}{7} + \dfrac{3}{7}$ **$\dfrac{6}{7}$**

13. $\dfrac{4}{15} + \dfrac{2}{15} + \dfrac{5}{15}$ **$\dfrac{11}{15}$**

14. $\dfrac{5}{11} + \dfrac{1}{11} + \dfrac{4}{11}$ **$\dfrac{10}{11}$**

15. $\dfrac{3}{8} + \dfrac{7}{8} + \dfrac{2}{8}$ **$1\dfrac{1}{2}$**

16. $\dfrac{4}{9} + \dfrac{1}{9} + \dfrac{7}{9}$ **$1\dfrac{1}{3}$**

17. $\dfrac{2}{54} + \dfrac{8}{54} + \dfrac{12}{54}$ **$\dfrac{11}{27}$**

18. $\dfrac{7}{64} + \dfrac{15}{64} + \dfrac{20}{64}$ **$\dfrac{21}{32}$**

Find the difference and simplify the answer. See Example 3.

19. $\dfrac{7}{8} - \dfrac{4}{8}$ **$\dfrac{3}{8}$**

20. $\dfrac{2}{3} - \dfrac{1}{3}$ **$\dfrac{1}{3}$**

21. $\dfrac{10}{11} - \dfrac{4}{11}$ **$\dfrac{6}{11}$**

22. $\dfrac{4}{5} - \dfrac{3}{5}$ **$\dfrac{1}{5}$**

23. $\dfrac{9}{10} - \dfrac{3}{10}$ **$\dfrac{3}{5}$**

24. $\dfrac{7}{14} - \dfrac{3}{14}$ **$\dfrac{2}{7}$**

25. $\begin{array}{r} \dfrac{31}{21} \\ - \dfrac{7}{21} \\ \hline \end{array}$ **$1\dfrac{1}{7}$**

26. $\begin{array}{r} \dfrac{43}{24} \\ - \dfrac{13}{24} \\ \hline \end{array}$ **$1\dfrac{1}{4}$**

27. $\begin{array}{r} \dfrac{27}{40} \\ - \dfrac{19}{40} \\ \hline \end{array}$ **$\dfrac{1}{5}$**

28. $\begin{array}{r} \dfrac{38}{55} \\ - \dfrac{16}{55} \\ \hline \end{array}$ **$\dfrac{2}{5}$**

29. $\dfrac{47}{36} - \dfrac{5}{36}$ **$1\dfrac{1}{6}$**

30. $\dfrac{76}{45} - \dfrac{21}{45}$ **$1\dfrac{2}{9}$**

31. $\dfrac{73}{60} - \dfrac{7}{60}$ **$1\dfrac{1}{10}$**

32. $\dfrac{181}{100} - \dfrac{31}{100}$ **$1\dfrac{1}{2}$**

33. In your own words, write an explanation of how to add like fractions. Consider using three steps in your explanation.

Three steps to add like fractions are:

1. **Add the numerators of the fractions to find the numerator of the sum (the answer).**

2. **Use the denominator of the fractions as the denominator of the sum.**

3. **Write the answer in lowest terms.**

34. Describe in your own words the difference between *like* fractions and *unlike* fractions. Give three examples of each type.

Like fractions have the same denominator. Unlike fractions have denominators that are different. Some examples are:

like: $\dfrac{3}{8}, \dfrac{1}{8}, \dfrac{7}{8},$ unlike: $\dfrac{1}{2}, \dfrac{3}{4}, \dfrac{5}{8}$

Solve each application problem. Write answers in lowest terms.

35. The Fair Oaks Save the Bluffs Committee raised $\frac{2}{9}$ of their target goal last year and another $\frac{5}{9}$ of the goal this year. What fraction of their goal has been raised?

$\dfrac{7}{9}$

36. After an initial payment to a winner at an arcade fundraiser, the organization still owed the winner $\frac{7}{10}$ of her total winnings. If the organization pays the winner another $\frac{3}{10}$ of the winnings, what fraction is still owed?

$\dfrac{2}{5}$ **of the winnings**

37. Robert Hernandez has saved $\frac{3}{7}$ of the amount needed for his children's college education fund. In the next five years he plans to save another $\frac{2}{7}$ of the total amount needed. What fraction of the amount needed will he have saved?

$\dfrac{5}{7}$

38. Phil Fravesi, general contractor, completes $\frac{3}{12}$ of a hobby and toy train addition to his home in April. In May, he completes another $\frac{5}{12}$ of the project. What portion of the project has he completed?

$\dfrac{2}{3}$

39. An organic farmer purchased $\frac{9}{10}$ acre of land one year and $\frac{3}{10}$ acre the next year. She then planted carrots on $\frac{7}{10}$ acre of the land and squash on the remainder. How much land is planted in squash?

$\dfrac{1}{2}$ **acre**

40. A forester planted $\frac{5}{12}$ acre in seedlings in the morning and $\frac{11}{12}$ acre in the afternoon. If $\frac{7}{12}$ acre of seedlings were destroyed by frost, how many acres of seedlings remained?

$\dfrac{3}{4}$ **acre**

Study Skills

Many college students find themselves juggling a difficult schedule and multiple responsibilities. Perhaps you are going to school, working part time, and managing family demands. Here are some tips to help you develop good time management skills and habits.

▶ **Read the syllabus for each class.** Check on class policies, such as attendance, late homework, and make-up tests. Find out how you are graded. Keep the syllabus in your notebook.

▶ **Make a semester or quarter calendar.** Put test dates and major due dates for *all* your classes on the same calendar. That way you will see which weeks are the really busy ones. Try using a different color pen for each class. Your brain responds well to the use of color. A semester calendar is on the next page.

▶ **Make a weekly schedule.** After you fill in your classes and other regular responsibilities (such as work, picking up kids from school, etc.), block off some study periods during the day that you can guarantee you will use for studying. Aim for 2 hours of study for each 1 hour you are in class.

▶ **Make "To Do" lists.** Then use them by crossing off the tasks as you complete them. You might even number them in the order they need to be done (most important ones first).

▶ **Break big assignments into smaller chunks.** They won't seem so big that way. Make deadlines for each small part so you stay on schedule.

▶ **Give yourself small breaks in your studying.** Do not try to study for hours at a time! Your brain needs rest between periods of learning. Try to give yourself a 10 minute break each hour or so. You will learn more and remember it longer.

▶ **If you get off schedule, just try to get back on schedule tomorrow.** We all slip from time to time. All is not lost! Make a new "To Do" list and start doing the most important things first.

▶ **Get help when you need it.** Talk with your instructor during office hours. Also, most colleges have some kind of Learning Center, tutoring center, or counseling office. If you feel lost and overwhelmed, ask for help. Someone can help you decide what to do first and what to spend your time on right away.

Which two or three of the suggestions above will you try this week? How do you think they will help you?

1. _____

2. _____

3. _____

OBJECTIVES

1 Create a semester schedule.

2 Create a "to do" list.

Why Are These Techniques Brain Friendly?

Your brain appreciates some order. It enjoys a little routine, for example, choosing the same study time and place each day. You will find that you quickly settle in to your reading or homework.

Also, your brain **functions better when you are calm.** Too much rushing around at the last minute to get your homework and studying done sends hostile chemicals to your brain and makes it more difficult for you to learn and remember. So, a little planning can really pay off.

Building rest into your schedule is good for your brain. Remember, it takes time for dendrites to grow.

We've suggested using color on your calendars. This too, is brain friendly. Remember, your brain **likes pleasant colors and visual material** that are nice to look at. Messy and hard to read calendars will not be helpful, and you probably won't look at them often.

SEMESTER CALENDAR

WEEK	MON	TUES	WED	THUR	FRI	SAT	SUN
1							
2							
3							
4							
5							
6							
7							
8							
9							
10							
11							
12							
13							
14							
15							
16							

3.2 ▶▶▶ Least Common Multiples

Only *like* fractions can be added or subtracted. Because of this, we must rewrite *unlike* fractions as *like* fractions before we can add or subtract them.

OBJECTIVE 1 Find the least common multiple. We can rewrite unlike fractions as like fractions by finding the *least common multiple* of the denominators.

> **Least Common Multiple (LCM)**
> The **least common multiple (LCM)** of two whole numbers is the smallest whole number divisible by both of those numbers.

EXAMPLE 1 Finding the Least Common Multiple

Find the least common multiple of 6 and 9.
First, find the multiples of 6.

$$\underbrace{6 \cdot 1}_{6,} \quad \underbrace{6 \cdot 2}_{12,} \quad \underbrace{6 \cdot 3}_{18,} \quad \underbrace{6 \cdot 4}_{24,} \quad \underbrace{6 \cdot 5}_{30,} \quad \underbrace{6 \cdot 6}_{36,} \quad \underbrace{6 \cdot 7}_{42,} \quad \underbrace{6 \cdot 8}_{48, \ldots}$$

(The three dots at the end of the list show that the list continues in the same pattern without stopping.) Now, find the multiples of 9.

$$\underbrace{9 \cdot 1}_{9,} \quad \underbrace{9 \cdot 2}_{18,} \quad \underbrace{9 \cdot 3}_{27,} \quad \underbrace{9 \cdot 4}_{36,} \quad \underbrace{9 \cdot 5}_{45,} \quad \underbrace{9 \cdot 6}_{54,} \quad \underbrace{9 \cdot 7}_{63,} \quad \underbrace{9 \cdot 8}_{72, \ldots}$$

The smallest number found in *both* lists is 18, so 18 is the **least common multiple** of 6 and 9; the number 18 is the smallest whole number divisible by both 6 and 9.

Multiples of 6:   6, 12, **18**, 24, 30, 36, 42, 48, . . .

Multiples of 9:   9, **18**, 27, 36, 45, 54, 63, 72, . . .

> The *smallest* number in *both* lists is the LCM.

18 is the smallest number found in both lists. **18** is the least common multiple (LCM) of 6 and 9.

Work Problem **1** *at the Side.* ▶

OBJECTIVE 2 Find the least common multiple using multiples of the largest number. There are several ways to find the least common multiple. If the numbers are small, the least common multiple can often be found by inspection. Can you think of a number that can be divided evenly by both 3 and 4? The number 12 will work; it is the least common multiple of the numbers 3 and 4. A method that works well to find the least common multiple is to write multiples of the larger number.

In this case, 4 is larger than 3, so write the multiples of 4.

$$4, 8, 12, 16, 20, \ldots$$

Now, check each multiple of 4 to see if it is divisible by 3.

4 is *not* divisible by 3
8 is *not* divisible by 3
12 *is* divisible by 3

The first multiple of 4 that is divisible by 3 is 12, so 12 is the least common multiple of 3 and 4.

OBJECTIVES

1. **Find the least common multiple.**

2. **Find the least common multiple using multiples of the largest number.**

3. **Find the least common multiple using prime factorization.**

4. **Find the least common multiple using an alternative method.**

5. **Write a fraction with an indicated denominator.**

---

1 **(a)** List the multiples of 5.

5, _____, _____, _____,

_____, _____, _____,

_____, ...

(b) List the multiples of 8.

8, _____, _____, _____,

_____, _____, _____, ...

(c) Find the least common multiple of 5 and 8.

ANSWERS

1. **(a)** 10, 15, 20, 25, 30, 35, 40, . . .
   (b) 16, 24, 32, 40, 48, 56, . . .
   (c) 40

2 Use multiples of the larger number to find the least common multiple in each set of numbers.

(a) 2 and 5

(b) 3 and 9

(c) 6 and 8

(d) 4 and 7

3 Use prime factorization to find the LCM for each pair of numbers.

(a) 15 and 18

(b) 12 and 20

2. (a) 10 (b) 9 (c) 24 (d) 28

3. (a)

15
$\wedge$
15 = 3 • 5   LCM = 2 • 3 • 3 • 5 = 90
18 = 2 • 3 • 3
$\swarrow$$\searrow$
18

(b)

12
$\wedge$
12 = 2 • 2 • 3   LCM = 2 • 2 • 3 • 5 = 60
20 = 2 • 2 • 5
$\nwarrow$$\nearrow$
20

EXAMPLE 2 **Finding the Least Common Multiple**

Use multiples of the larger number to find the least common multiple of 6 and 9.

We start by writing the first few multiples of 9.

Multiples of 9
$\overbrace{\qquad\qquad\qquad}$
9, 18, 27, 36, 45, 54, . . .

Now, we check each multiple of 9 to see if it is divisible by 6. The first multiple of 9 that is divisible by 6 is 18.

9, **18**, 27, 36, 45, 54, . . .

First multiple divisible $\longrightarrow$ by 6, because $18 \div 6 = 3$

The least common multiple of the numbers 6 and 9 is 18.

◀ Work Problem **2** at the Side.

OBJECTIVE **3** **Find the least common multiple using prime factorization.** Example 2 shows how to find the least common multiple of two numbers by making a list of the multiples of the *larger* number. Although this method works well if both numbers are fairly small, it is usually easier to find the least common multiple for larger numbers by using *prime factorization,* as shown in the next example.

EXAMPLE 3 **Applying Prime Factorization Knowledge**

Use prime factorization to find the least common multiple of 9 and 12.
We start by finding the prime factorization of each number.

Factors of 9
$\wedge$
$9 = 3 \cdot 3$
$12 = 2 \cdot 2 \cdot 3$ $LCM = 3 \cdot 3 \cdot 2 \cdot 2 = 36$
$\searrow$$\nearrow$
Factors of 12

Check to see that 36 is divisible by 9 (yes) and by 12 (yes). The smallest whole number divisible by both 9 and 12 is 36.

> **CAUTION**
> Notice that we did ***not*** have to repeat the factors that 9 and 12 have in common. In this case, the **3** in 2 • 2 • **3** = 12 was ***not*** used because 3 is already included in 3 • 3 = 9.

◀ Work Problem **3** at the Side.

EXAMPLE 4 **Using Prime Factorization**

Find the least common multiple of 12, 18, and 20.
　　Find the prime factorization of each number. Then use the prime factors to build the LCM.

$$12 = 2 \cdot 2 \cdot 3$$
$$18 = 2 \cdot 3 \cdot 3 \qquad \text{LCM} = 2 \cdot 2 \cdot 3 \cdot 3 \cdot 5 = 180$$
$$20 = 2 \cdot 2 \cdot 5$$

Check to see that 180 is divisible by 12 (yes) and by 18 (yes) and by 20 (yes). This smallest whole number divisible by 12, 18, and 20 is 180.

> **Note**
> The LCM did *not* repeat the factors that 12, 18, and 20 have in common.

―――――――――――――――― Work Problem **4** at the Side. ▶

EXAMPLE 5 **Finding the Least Common Multiple**

Find the least common multiple for each set of numbers.

(a) 5, 6, 35
　　Find the prime factorization for each number.

$$5 = 5$$
$$6 = 2 \cdot 3 \qquad \text{LCM} = 2 \cdot 3 \cdot 5 \cdot 7 = 210$$
$$35 = 5 \cdot 7$$

　　The least common multiple of 5, 6, and 35 is 210.

(b) 10, 20, 24
　　Find the prime factorization for each number.

$$10 = 2 \cdot 5$$
$$20 = 2 \cdot 2 \cdot 5 \qquad \text{LCM} = 2 \cdot 2 \cdot 2 \cdot 3 \cdot 5 = 120$$
$$24 = 2 \cdot 2 \cdot 2 \cdot 3$$

The least common multiple of 10, 20, and 24 is 120.

―――――――――――――――― Work Problem **5** at the Side. ▶

OBJECTIVE 4 Find the least common multiple using an alternative method. Some people like the following *alternative method* for finding the least common multiple for larger numbers. Try both methods, and *use the one you prefer.* As a review, a list of the first few prime numbers follows.

$$2, 3, 5, 7, 11, 13, 17$$

4 Find the least common multiple of the denominators in each set of fractions.

(a) $\dfrac{3}{8}$ and $\dfrac{6}{5}$

(b) $\dfrac{5}{6}$ and $\dfrac{1}{14}$

(c) $\dfrac{4}{9}, \dfrac{5}{18},$ and $\dfrac{7}{24}$

5 Find the least common multiple for each set of numbers.

(a) 4, 8, 9

(b) 3, 6, 8

(c) 9, 36, 48

(d) 15, 20, 30, 40

ANSWERS

4. (a)
$$8 = 2 \cdot 2 \cdot 2 \qquad \text{LCM} = 2 \cdot 2 \cdot 2 \cdot 5 = 40$$
$$5 = 5$$

(b)
$$6 = 2 \cdot 3 \qquad \text{LCM} = 2 \cdot 3 \cdot 7 = 42$$
$$14 = 2 \cdot 7$$

(c)
$$9 = 3 \cdot 3 \qquad \text{LCM} = 2 \cdot 2 \cdot 2 \cdot 3 \cdot 3 = 72$$
$$18 = 2 \cdot 3 \cdot 3$$
$$24 = 2 \cdot 2 \cdot 2 \cdot 3$$

5. (a) 72　**(b)** 24　**(c)** 144　**(d)** 120

6 In the following problems, the divisions have already been worked out. Multiply the prime numbers on the left to find the least common multiple.

(a) 2 |6   1̸5̸
    3 |3   15
    5 |1̸   5
       1   1

(b) 2 |20   36
    2 |10   18
    3 |5̸   9
    3 |5̸   3
    5 |5   1̸
       1   1

EXAMPLE 6 **Alternative Method for Finding the Least Common Multiple**

Find the least common multiple for each set of numbers.

(a) 14 and 21

Start by trying to divide 14 and 21 by the first prime number in the list of prime numbers: 2, 3, 5, 7, 11, 13, and 17. Use the following shortcut. Divide by 2, the first prime.

$$2\ \underline{|14\quad 2\!\!\!/1}$$
$$7\quad 21$$

Because 21 cannot be divided evenly by 2, cross out 21 and bring it down. Divide by 3, the second prime.

$$2\ \underline{|14\quad 2\!\!\!/1}$$
$$3\ \underline{|\ 7\quad 21}$$
$$7\quad 7$$

Since 7 cannot be divided evenly by the third prime, 5, skip 5 and divide by the next prime, 7. Divide by 7, the fourth prime.

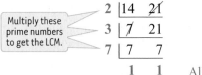

Multiply these prime numbers to get the LCM.

$$2\ \underline{|14\quad 2\!\!\!/1}$$
$$3\ \underline{|\ 7\quad 21}$$
$$7\ \underline{|\ 7\quad 7}$$
$$1\quad 1$$ All quotients are 1.

When all quotients are 1, multiply the prime numbers on the left side.

least common multiple = 2 • 3 • 7 = 42

The least common multiple of 14 and 21 is 42.

(b) 6, 15, 18

Divide by 2.

$$2\ \underline{|6\quad 1\!\!\!/5\quad 18}$$
$$3\quad 15\quad 9$$

Cross out 15 and bring it down.

Divide by 3.

$$2\ \underline{|6\quad 1\!\!\!/5\quad 18}$$
$$3\ \underline{|3\quad 15\quad 9}$$
$$1\quad 5\quad 3$$

Divide by 3 again, since the remaining 3 can be divided.

$$2\ \underline{|6\quad 1\!\!\!/5\quad 18}$$
$$3\ \underline{|3\quad 15\quad 9}$$
$$3\ \underline{|1\!\!\!/\quad 5\!\!\!/\quad 3}$$
$$1\quad 5\quad 1$$

Finally, divide by 5.

$$2\ \underline{|6\quad 1\!\!\!/5\quad 18}$$
$$3\ \underline{|3\quad 15\quad 9}$$
$$3\ \underline{|1\!\!\!/\quad 5\!\!\!/\quad 3}$$
$$5\ \underline{|1\!\!\!/\quad 5\quad 1\!\!\!/}$$
$$1\quad 1\quad 1$$ All quotients are 1.

Multiply the prime numbers on the left side.

2 • 3 • 3 • 5 = 90 ← Least common multiple

◀ *Work Problem* **6** *at the Side.*

Work Problem **7** *at the Side.* ▶

OBJECTIVE 5 Write a fraction with an indicated denominator.
Before we can add or subtract *unlike* fractions, we must find the least common multiple, which is then used as the denominator of the fractions.

EXAMPLE 7 Writing a Fraction with an Indicated Denominator

Write the fraction $\frac{2}{3}$ with a denominator of 15
  Find a numerator, so that these fractions are equivalent.

$$\frac{2}{3} = \frac{?}{15}$$

To find the new numerator, first divide **15** by **3.**

$$\frac{2}{3} = \frac{?}{15} \qquad 15 \div 3 = 5$$

Multiply both numerator and denominator of the fraction $\frac{2}{3}$ by 5

$$\frac{2}{3} = \frac{2 \cdot 5}{3 \cdot 5} = \frac{10}{15}$$

  This process is just the opposite of writing a fraction in lowest terms. Check the answer by writing $\frac{10}{15}$ in lowest terms; you should get $\frac{2}{3}$ again.

EXAMPLE 8 Writing Fractions with a New Denominator

Rewrite each fraction with the indicated denominator.

(a) $\dfrac{3}{8} = \dfrac{?}{48}$

  Divide 48 by 8, getting 6. Now multiply both the numerator and the denominator of $\frac{3}{8}$ by 6.

$$\frac{3}{8} = \frac{3 \cdot 6}{8 \cdot 6} = \frac{18}{48}$$ Multiply numerator and denominator by 6.

> Recall that multiplying a number by 1 does *not* change the number, and $\frac{6}{6} = 1$.

That is, $\frac{3}{8} = \frac{18}{48}$. As a check, write $\frac{18}{48}$ in lowest terms; you should get $\frac{3}{8}$ again.

(b) $\dfrac{5}{6} = \dfrac{?}{42}$

  Divide 42 by 6, getting 7. Next, multiply both the numerator and the denominator of $\frac{5}{6}$ by 7.

$$\frac{5}{6} = \frac{5 \cdot 7}{6 \cdot 7} = \frac{35}{42}$$ Multiply numerator and denominator by 7.

> Notice that $\frac{7}{7} = 1$.

This shows that $\frac{5}{6} = \frac{35}{42}$. As a check, write $\frac{35}{42}$ in lowest terms. Did you get $\frac{5}{6}$ again?

Continued on Next Page

---

7 Find the least common multiple of each set of numbers. Use whichever method you prefer.

(a) 3, 6, 10

(b) 15 and 40

(c) 9, 24

(d) 8, 21, 24

ANSWERS

7. (a) 30 **(b)** 120 **(c)** 72 **(d)** 168

8 Rewrite each fraction with the indicated denominator.

(a) $\dfrac{1}{4} = \dfrac{?}{16}$

> **Note**
>
> In Example 7, on the previous page, the fraction $\frac{2}{3}$ was multiplied by $\frac{5}{5}$. In Example 8, the fraction $\frac{3}{8}$ was multiplied by $\frac{6}{6}$ and the fraction $\frac{5}{6}$ was multiplied by $\frac{7}{7}$. The fractions, $\frac{5}{5}$, $\frac{6}{6}$, and $\frac{7}{7}$ are all equal to 1.
>
> $$\frac{5}{5} = 1 \qquad \frac{6}{6} = 1 \qquad \frac{7}{7} = 1$$
>
> Recall that any number multiplied by 1 is the number itself.

◀ Work Problem **8** at the Side.

(b) $\dfrac{2}{3} = \dfrac{?}{15}$

(c) $\dfrac{7}{16} = \dfrac{?}{32}$

(d) $\dfrac{6}{11} = \dfrac{?}{33}$

Answers

8. (a) $\dfrac{4}{16}$ (b) $\dfrac{10}{15}$ (c) $\dfrac{14}{32}$ (d) $\dfrac{18}{33}$

3.2 ▶▶▶ Exercises

Use multiples of the larger number to find the least common multiple in each set of numbers. See Examples 1 and 2.

1. 3 and 6   6

2. 2 and 4   4

3. 3 and 5   15

4. 3 and 7   21

5. 4 and 9   36

6. 4 and 10   20

7. 2 and 7   14

8. 6 and 8   24

9. 6 and 10   30

10. 12 and 16   48

11. 20 and 50   100

12. 25 and 75   75

Find the least common multiple of each set of numbers. Use any method. See Examples 2–6.

13. 4, 10   20

14. 8, 10   40

15. 12, 20   60

16. 9 and 15   45

17. 6, 9, 12   36

18. 20, 24, 30   120

19. 4, 6, 8, 10   120

20. 8, 9, 12, 18   72

21. 12, 15, 18, 20   180

22. 6, 8, 9, 27, 36   216

23. 8, 10, 12, 16, 36   720

24. 5, 6, 8, 25, 30   600

Rewrite each fraction with a denominator of 24. See Examples 7 and 8.

25. $\dfrac{2}{3} = \dfrac{16}{24}$

26. $\dfrac{3}{8} = \dfrac{9}{24}$

27. $\dfrac{3}{4} = \dfrac{18}{24}$

28. $\dfrac{5}{12} = \dfrac{10}{24}$

29. $\dfrac{5}{6} = \dfrac{20}{24}$

30. $\dfrac{7}{8} = \dfrac{21}{24}$

Rewrite each fraction with the indicated denominator.

31. $\dfrac{1}{2} = \dfrac{3}{6}$

32. $\dfrac{2}{3} = \dfrac{6}{9}$

33. $\dfrac{3}{4} = \dfrac{12}{16}$

34. $\dfrac{7}{10} = \dfrac{21}{30}$

35. $\dfrac{7}{8} = \dfrac{28}{32}$

36. $\dfrac{5}{12} = \dfrac{20}{48}$

37. $\dfrac{3}{16} = \dfrac{12}{64}$

38. $\dfrac{7}{8} = \dfrac{84}{96}$

39. $\dfrac{8}{5} = \dfrac{32}{20}$

40. $\dfrac{5}{8} = \dfrac{25}{40}$

41. $\dfrac{9}{7} = \dfrac{72}{56}$

42. $\dfrac{3}{2} = \dfrac{96}{64}$

43. $\dfrac{7}{4} = \dfrac{84}{48}$

44. $\dfrac{7}{6} = \dfrac{140}{120}$

45. $\dfrac{8}{11} = \dfrac{96}{132}$

46. $\dfrac{4}{15} = \dfrac{44}{165}$

47. $\dfrac{3}{16} = \dfrac{27}{144}$

48. $\dfrac{7}{16} = \dfrac{49}{112}$

49. There are several methods for finding the least common multiple (LCM). Do you prefer the method using multiples of the largest number or the method using prime factorizations? Why? Would you ever use the other method?

It probably depends on how large the numbers are. If the numbers are small, the method using multiples of the largest number seems best. If the numbers are larger, or if there are more than two numbers, then the factorization method will be better.

50. Explain in your own words how to write a fraction with an indicated denominator. As part of your explanation, show how to change $\frac{3}{4}$ to a fraction having 12 as a denominator.

Divide the desired denominator by the given denominator. Next, multiply both the given numerator and denominator by this number.

Divide 12 by 4 to get 3. Then multiply.

$$\frac{3}{4} = \frac{3 \cdot 3}{4 \cdot 3} = \frac{9}{12}$$

Find the least common multiple of the denominators of each pair of fractions.

51. $\dfrac{25}{400}, \dfrac{38}{1800}$ **3600**

52. $\dfrac{53}{600}, \dfrac{115}{4000}$ **12,000**

53. $\dfrac{109}{1512}, \dfrac{23}{392}$ **10,584**

54. $\dfrac{61}{810}, \dfrac{37}{1170}$ **10,530**

Relating Concepts (Exercises 55—62) For Individual or Group Work

Most people think that addition and subtraction of fractions is more difficult than multiplication and division of fractions. This is probably because a common denominator must be used. **Work Exercises 55–62 in order.**

55. Fractions with the same denominators are <u>like</u> fractions and fractions with different denominators are <u>unlike</u> fractions.

56. To subtract like fractions, first subtract the <u>numerators</u> to find the numerator of the difference. Write the denominator of the like fractions as the <u>denominator</u> of the difference. Finally, write the answer in <u>lowest</u> terms.

57. The <u>least</u> common multiple (LCM) of two numbers is the <u>smallest</u> whole number
(smallest/largest)
divisible by both those numbers.

58. The following shows the common multiples for both 8 and 10. What is the least common multiple for these two numbers?

Multiples of 8:   8, 16, 24, 32, 40, 48, 56, 64, 72, 80, 88, . . .

Multiples of 10:   10, 20, 30, 40, 50, 60, 70, 80, 90, . . .

40 is the least common multiple.

Find the least common multiple for each set of numbers.

59. 5, 7, 14, 10

70

60. 25, 18, 30, 5

450

61. Explain why the least common multiple for 8, 3, 5, 4, and 10 is not 240. Find the least common multiple.

240 is a common multiple but twice as large as the least common multiple; 120 is the LCM.

62. Demonstrate that the least common multiple of 55 and 1760 is 1760.

The least common multiple can be no smaller than the largest number in a group and the number 1760 is a multiple of 55.

3.3 ▶▶▶ Adding and Subtracting Unlike Fractions

OBJECTIVE 1 Add unlike fractions. In this section, we add and subtract unlike fractions. To add unlike fractions, we must first change them to like fractions (fractions with the same denominator). For example, the figures below show $\frac{3}{8}$ and $\frac{1}{4}$.

OBJECTIVES

1 Add unlike fractions.

2 Add unlike fractions vertically.

3 Subtract unlike fractions.

4 Subtract unlike fractions vertically.

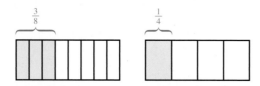

These fractions can be added by changing them to like fractions. Make like fractions by changing $\frac{1}{4}$ to the equivalent fraction $\frac{2}{8}$.

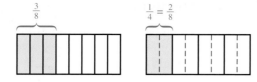

Now you can add the fractions.

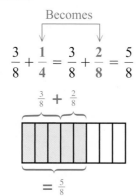

$$\frac{3}{8} + \frac{1}{4} = \frac{3}{8} + \frac{2}{8} = \frac{5}{8}$$

Use the following steps to add or subtract unlike fractions.

Adding or Subtracting Unlike Fractions

Step 1 Rewrite the *unlike fractions* as *like fractions* with the least common multiple as their new denominator. This new denominator is called the **least common denominator (LCD).**

Step 2 Add or subtract as with like fractions.

Step 3 Simplify the answer by writing it in lowest terms and as a whole or mixed number where possible.

EXAMPLE 1 Adding Unlike Fractions

Add $\frac{2}{3}$ and $\frac{1}{9}$.

The least common multiple of 3 and 9 is 9, so first write the fractions as like fractions with a denominator of 9. This is the *least common denominator (LCD)* of 3 and 9.

Continued on Next Page

1 Add.

(a) $\dfrac{1}{2} + \dfrac{3}{8}$

(b) $\dfrac{3}{4} + \dfrac{1}{8}$

(c) $\dfrac{3}{5} + \dfrac{3}{10}$

(d) $\dfrac{1}{12} + \dfrac{5}{6}$

2 Add. Simplify all answers.

(a) $\dfrac{3}{10} + \dfrac{1}{5}$

(b) $\dfrac{5}{8} + \dfrac{1}{3}$

(c) $\dfrac{1}{10} + \dfrac{1}{3} + \dfrac{1}{6}$

Step 1
$$\dfrac{2}{3} = \dfrac{?}{9}$$
Divide 9 by 3, getting 3. Next, multiply numerator and denominator by 3.
$$\dfrac{2}{3} = \dfrac{2 \cdot 3}{3 \cdot 3} = \dfrac{6}{9}$$

Now, add the like fractions $\frac{6}{9}$ and $\frac{1}{9}$.

Becomes

> Both fractions must have the *same* denominator **before** you add them.

Step 2
$$\dfrac{2}{3} + \dfrac{1}{9} = \dfrac{6}{9} + \dfrac{1}{9} = \dfrac{6+1}{9} = \dfrac{7}{9}$$

Step 3   Step 3 is not needed because $\frac{7}{9}$ is already in lowest terms.

◀ *Work Problem* **1** *at the Side.*

EXAMPLE 2   **Adding Fractions**

Add each pair of fractions using the three steps. Simplify all answers.

(a) $\dfrac{1}{3} + \dfrac{1}{6}$

The least common multiple of 3 and 6 is 6. Rewrite both fractions as fractions with a least common denominator of 6.

Rewritten as like fractions

Step 1 $\dfrac{1}{3} + \dfrac{1}{6} = \dfrac{2}{6} + \dfrac{1}{6}$
> 6 is the LCD (least common denominator).

Add numerators.

Step 2 $\dfrac{2}{6} + \dfrac{1}{6} = \dfrac{2+1}{6} = \dfrac{3}{6}$
> Only *like* fractions can be added.

Step 3 $\dfrac{3}{6} = \dfrac{1}{2}$ ← In lowest terms

(b) $\dfrac{6}{15} + \dfrac{3}{10}$

The least common multiple of 15 and 10 is 30, so rewrite both fractions with a least common denominator of 30.

Rewritten as like fractions

Step 1 $\dfrac{6}{15} + \dfrac{3}{10} = \dfrac{12}{30} + \dfrac{9}{30}$
> The least common multiple of the denominators is the LCD.

Add numerators.

Step 2 $\dfrac{12}{30} + \dfrac{9}{30} = \dfrac{12+9}{30} = \dfrac{21}{30}$

Step 3 $\dfrac{21}{30} = \dfrac{7}{10}$ ← In lowest terms

◀ *Work Problem* **2** *at the Side.*

ANSWERS

1. (a) $\frac{7}{8}$ (b) $\frac{7}{8}$ (c) $\frac{9}{10}$ (d) $\frac{11}{12}$

2. (a) $\frac{1}{2}$ (b) $\frac{23}{24}$ (c) $\frac{3}{5}$

OBJECTIVE 2 Add unlike fractions vertically. Fractions can also be added vertically (one fraction written below the other).

EXAMPLE 3 **Vertical Addition of Fractions**

Add the following fractions vertically.

(a)
$$\frac{3}{8} = \frac{3 \cdot 3}{8 \cdot 3} = \frac{9}{24}$$

24 is the LCD.

Rewritten as like fractions

$$+\frac{7}{12} = \frac{7 \cdot 2}{12 \cdot 2} = \frac{14}{24}$$

$$\frac{23}{24} \leftarrow \text{Add the numerators.}$$
$$\leftarrow \text{Denominator is 24, the LCD.}$$

(b)
$$\frac{2}{9} = \frac{2 \cdot 4}{9 \cdot 4} = \frac{8}{36}$$

$$+\frac{1}{4} = \frac{1 \cdot 9}{4 \cdot 9} = \frac{9}{36}$$

Rewritten as like fractions

$$\frac{17}{36} \leftarrow \text{Add the numerators.}$$
$$\leftarrow \text{Denominator is 36, the LCD.}$$

Work Problem **3** *at the Side.* ▶

OBJECTIVE 3 Subtract unlike fractions. The next example shows subtraction of unlike fractions.

EXAMPLE 4 **Subtracting Unlike Fractions**

Subtract. Simplify all answers.

As with addition, rewrite unlike fractions with a least common denominator.

(a) $\frac{3}{4} - \frac{3}{8}$

Rewritten as like fractions

Step 1
$$\frac{3}{4} - \frac{3}{8} = \frac{6}{8} - \frac{3}{8}$$

The LCD is 8.

Subtract numerators.

Step 2
$$\frac{6}{8} - \frac{3}{8} = \frac{6 - 3}{8} = \frac{3}{8}$$

Step 3 Not needed because $\frac{3}{8}$ is in lowest terms.

Continued on Next Page

---

3 Add the following fractions vertically.

(a)
$$\frac{5}{8}$$
$$+\frac{1}{12}$$

(b)
$$\frac{7}{16}$$
$$+\frac{1}{4}$$

4 Subtract. Simplify all answers.

(a) $\dfrac{5}{8} - \dfrac{1}{4}$

(b) $\dfrac{4}{5} - \dfrac{3}{4}$

5 Subtract vertically. Simplify all answers.

(a) $\dfrac{7}{8}$

$-\dfrac{2}{3}$

(b) $\dfrac{5}{6}$

$-\dfrac{1}{12}$

(b) $\dfrac{3}{4} - \dfrac{7}{12}$

Rewritten as like fractions

Step 1 $\dfrac{3}{4} - \dfrac{7}{12} = \dfrac{9}{12} - \dfrac{7}{12}$

Subtract numerators.

Step 2 $\dfrac{9}{12} - \dfrac{7}{12} = \dfrac{9-7}{12} = \dfrac{2}{12}$ ← Subtract the numerators.
← Denominator is 12, the LCD.

Step 3 $\dfrac{2}{12} = \dfrac{1}{6}$ ← Lowest terms

Always write the final answer in lowest terms.

◀ *Work Problem* **4** *at the Side.*

OBJECTIVE 4 Subtract unlike fractions vertically.

EXAMPLE 5 **Vertical Subtraction of Fractions**

Subtract the following fractions vertically.

(a) $\dfrac{4}{5} = \dfrac{4 \cdot 8}{5 \cdot 8} = \dfrac{32}{40}$

$-\dfrac{3}{8} = \dfrac{3 \cdot 5}{8 \cdot 5} = \dfrac{15}{40}$

Rewritten as like fractions

$\dfrac{17}{40}$ ← Subtract numerators.
← Denominator is 40, the LCD.

(b) $\dfrac{3}{7} = \dfrac{3 \cdot 12}{7 \cdot 12} = \dfrac{36}{84}$

$-\dfrac{5}{12} = \dfrac{5 \cdot 7}{12 \cdot 7} = \dfrac{35}{84}$

Rewritten as like fractions

$\dfrac{1}{84}$ ← Subtract numerators.
← Denominator is 84, the LCD.

◀ *Work Problem* **5** *at the Side.*

ANSWERS

4. (a) $\dfrac{3}{8}$ (b) $\dfrac{1}{20}$

5. (a) $\dfrac{5}{24}$ (b) $\dfrac{3}{4}$

Add the following fractions. Simplify all answers. See Examples 1–3.

1. $\dfrac{3}{4} + \dfrac{1}{8}$ $\dfrac{7}{8}$

2. $\dfrac{1}{6} + \dfrac{2}{3}$ $\dfrac{5}{6}$

3. $\dfrac{2}{3} + \dfrac{2}{9}$ $\dfrac{8}{9}$

4. $\dfrac{3}{7} + \dfrac{1}{14}$ $\dfrac{1}{2}$

5. $\dfrac{9}{20} + \dfrac{3}{10}$ $\dfrac{3}{4}$

6. $\dfrac{5}{8} + \dfrac{1}{4}$ $\dfrac{7}{8}$

7. $\dfrac{3}{5} + \dfrac{3}{8}$ $\dfrac{39}{40}$

8. $\dfrac{5}{7} + \dfrac{3}{14}$ $\dfrac{13}{14}$

 9. $\dfrac{2}{9} + \dfrac{5}{12}$ $\dfrac{23}{36}$

10. $\dfrac{5}{8} + \dfrac{1}{12}$ $\dfrac{17}{24}$

11. $\dfrac{1}{3} + \dfrac{3}{5}$ $\dfrac{14}{15}$

12. $\dfrac{2}{5} + \dfrac{3}{7}$ $\dfrac{29}{35}$

13. $\dfrac{1}{4} + \dfrac{2}{9} + \dfrac{1}{3}$ $\dfrac{29}{36}$

14. $\dfrac{3}{7} + \dfrac{2}{5} + \dfrac{1}{10}$ $\dfrac{13}{14}$

15. $\dfrac{3}{10} + \dfrac{2}{5} + \dfrac{3}{20}$ $\dfrac{17}{20}$

16. $\dfrac{1}{3} + \dfrac{3}{8} + \dfrac{1}{4}$ $\dfrac{23}{24}$

17. $\dfrac{4}{15} + \dfrac{1}{6} + \dfrac{1}{3}$ $\dfrac{23}{30}$

18. $\dfrac{5}{12} + \dfrac{2}{9} + \dfrac{1}{6}$ $\dfrac{29}{36}$

19. $\begin{array}{r} \dfrac{1}{4} \\ + \dfrac{1}{8} \\ \hline \dfrac{3}{8} \end{array}$

20. $\begin{array}{r} \dfrac{7}{12} \\ + \dfrac{1}{8} \\ \hline \dfrac{17}{24} \end{array}$

21. $\begin{array}{r} \dfrac{5}{12} \\ + \dfrac{1}{16} \\ \hline \dfrac{23}{48} \end{array}$

22. $\begin{array}{r} \dfrac{3}{7} \\ + \dfrac{1}{3} \\ \hline \dfrac{16}{21} \end{array}$

Subtract the following fractions. Simplify all answers. See Example 4.

23. $\dfrac{5}{6} - \dfrac{1}{3}$ $\dfrac{1}{2}$

24. $\dfrac{3}{4} - \dfrac{5}{8}$ $\dfrac{1}{8}$

25. $\dfrac{2}{3} - \dfrac{1}{6}$ $\dfrac{1}{2}$

26. $\dfrac{3}{4} - \dfrac{5}{8}$ $\dfrac{1}{8}$

27. $\dfrac{2}{3} - \dfrac{1}{5}$ $\dfrac{7}{15}$

28. $\dfrac{5}{6} - \dfrac{7}{9}$ $\dfrac{1}{18}$

29. $\dfrac{5}{12} - \dfrac{1}{4}$ $\dfrac{1}{6}$

30. $\dfrac{5}{7} - \dfrac{1}{3}$ $\dfrac{8}{21}$

31. $\dfrac{8}{9} - \dfrac{7}{15}$ $\dfrac{19}{45}$

32.
$$\begin{array}{r} \dfrac{4}{5} \\ -\,\dfrac{1}{3} \\ \hline \dfrac{7}{15} \end{array}$$

33.
$$\begin{array}{r} \dfrac{7}{8} \\ -\,\dfrac{4}{5} \\ \hline \dfrac{3}{40} \end{array}$$

34.
$$\begin{array}{r} \dfrac{5}{8} \\ -\,\dfrac{1}{3} \\ \hline \dfrac{7}{24} \end{array}$$

35.
$$\begin{array}{r} \dfrac{5}{12} \\ -\,\dfrac{1}{16} \\ \hline \dfrac{17}{48} \end{array}$$

36.
$$\begin{array}{r} \dfrac{7}{12} \\ -\,\dfrac{1}{3} \\ \hline \dfrac{1}{4} \end{array}$$

Solve each application problem.

Use the newspaper advertisement for this 4-piece chisel set to answer Exercises 37–38. (*Source:* Harbor Freight Tools.)

37. Find the difference in the cutting-edge width of the two chisels with the widest blades. The symbol ″ is for inches.

$\dfrac{1}{4}$ in.

38. Find the difference in the cutting-edge width of the two chisels with the narrowest blades. The symbol ″ is for inches.

$\dfrac{1}{4}$ in.

39. A sports and entertainment center has $\frac{4}{5}$ of its total area devoted to seating of fans and guests. If $\frac{3}{8}$ of the seating area is used for general admission seating and the rest for reserved seating, find the fraction of the total area used for reserved seating.

$\frac{17}{40}$ for reserved seating

40. Della Daniel wants to open a day care center and has saved $\frac{2}{5}$ of the amount needed for start-up costs. If she saves another $\frac{1}{8}$ of the amount needed and then $\frac{1}{6}$ more, find the total portion of the start-up costs she has saved.

$\frac{83}{120}$ of the start-up costs

41. When installing cabinets for The Home Depot, Sarah Bryn must be certain that the proper type and size of mounting screw is used. Find the total length of the screw shown.

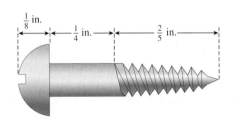

$\frac{31}{40}$ in.

42. When installing a computer chassis, Bonnie Bottorff must be certain that the proper type and size of bolt is used. Find the total length of the bolt shown.

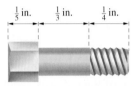

$\frac{47}{60}$ in.

43. Bill Newton is a general contractor and begins a job with $\frac{3}{4}$ of a tank of fuel in his back-hoe. If he uses $\frac{1}{3}$ of the tank in the morning and $\frac{3}{8}$ of the tank in the afternoon, what fraction of the tank of fuel remains?

$\frac{1}{24}$ of the tank

44. Adrian Ortega drives a tanker for the British Petroleum Company. He leaves the refinery with his tanker filled to $\frac{7}{8}$ of capacity. If he delivers $\frac{1}{4}$ of the tanker's capacity at the first stop and $\frac{1}{3}$ of the tanker's capacity at the second stop, find the fraction of the tanker's capacity remaining.

$\frac{7}{24}$ of the tanker's capacity

✎ 45. Step 1 in adding or subtracting unlike fractions is to rewrite the fractions so they have the least common multiple as a denominator. Explain in your own words why this is necessary.

You cannot add or subtract until all the fractional pieces are the same size. For example, halves are larger than fourths, so you cannot add $\frac{1}{2} + \frac{1}{4}$ until you rewrite $\frac{1}{2}$ as $\frac{2}{4}$.

✎ 46. Briefly list the three steps used for addition and subtraction of unlike fractions.

Step 1 **Rewrite the unlike fractions as like fractions.**

Step 2 **Add or subtract the like fractions.**

Step 3 **Simplify the answer.**

Refer to the circle graph to answer Exercises 47–50.

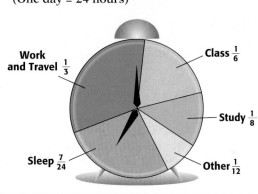

THE DAY OF THE STUDENT
(One day = 24 hours)

Work and Travel $\frac{1}{3}$

Class $\frac{1}{6}$

Study $\frac{1}{8}$

Sleep $\frac{7}{24}$

Other $\frac{1}{12}$

47. What fraction of the day was spent in class and study?

$$\frac{7}{24}$$

48. What fraction of the day was spent in work and travel and other?

$$\frac{5}{12}$$

49. In which activity was the greatest amount of time spent? How many hours did this activity take? What fraction of the day was spent on this activity and class time?

work and travel; 8 hr; $\frac{1}{2}$

50. In which activity was the least amount of time spent? How many hours did this activity take? What fraction of the day was spent on this activity and studying?

other; 2 hr; $\frac{5}{24}$

51. Find the diameter of the hole in the mounting bracket shown. (The diameter is the distance across the center of the hole.)

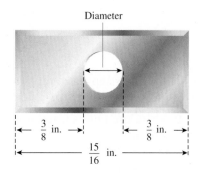

Diameter

$\frac{3}{8}$ in. $\frac{3}{8}$ in.

$\frac{15}{16}$ in.

$$\frac{3}{16} \text{ in.}$$

52. Chakotay is fitting a turquoise stone into a bear claw pendant. Find the diameter of the hole in the pendant. (The diameter is the distance across the center of the hole.)

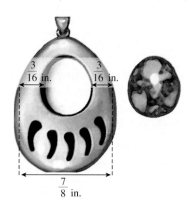

$\frac{3}{16}$ in. $\frac{3}{16}$ in.

$\frac{7}{8}$ in.

$$\frac{1}{2} \text{ in.}$$

Study Skills

Mind mapping is a visual way to show information that you have learned. It is an excellent way to review. Mapping is flexible and can be personalized, which is helpful for your memory. Your brain likes to see things that are **pleasing** to look at, **colorful,** and that **show connections** between ideas. Take advantage of that by creating maps that

▶ are easy to read,

▶ use color in a systematic way, and

▶ clearly show you how different concepts are related (using arrows or dotted lines, for example).

Here are some general directions for making a map. After you read them, work on completing the map that has been started for you on the next page. It is from Chapters 2 and 3: Fractions.

▶ To begin a mind map, write the concept in the center of a piece of paper and either circle it or draw a box around it.

▶ Make a line out from the center concept, and draw a box large enough to write the definition of the concept.

▶ Think of the other aspects (subpoints) of the concept that you have learned, such as procedures to follow or formulas. Make a separate line and box connecting each subpoint to the center.

▶ From each of the new boxes, add the information you've learned. You can continue making new lines and boxes or circles, or you can list items below the new information.

▶ Use color to highlight the major points. For example, everything related to one subpoint might be the same color. That way you can easily see related ideas.

▶ You may also use arrows, underlining, or small drawings to help yourself remember.

OBJECTIVES

1 **Create mind maps for appropriate concepts.**

2 **Visually show how concepts relate to each other using arrows or lines.**

Directions for Making a Mind Map

Why Is Mapping Brain Friendly?

Remember that your brain grows dendrites when you are **actively thinking** about and working with information. Making a map requires you to think hard about **how to place the information, how to show connections** between parts of the map, and **how color will be useful.** It also takes a lot of thinking to fill in all related details and **show how those details connect to the larger concept.** All that thinking will let your brain grow a complex, many-branched neural network of interconnected dendrites. It is time well spent.

Try This Fractions Mind Map Using Sections 2.5, 2.7, and 3.3

On a separate paper, make a map that summarizes Computations with Fractions. Follow the directions and use the starter map below.

▶ The longest rectangles are *instructions* for all four operations. (The first one starts "Rewrite all numbers as fractions..." and the second one is at the bottom of the map.)

▶ Notice the wavy dividing lines that separate the map into two sides.

▶ Your job is to complete the map by writing the steps used in multiplying and dividing fractions (from **Sections 2.5 and 2.7**) and the steps used in adding and subtracting fractions (from **Section 3.3**).

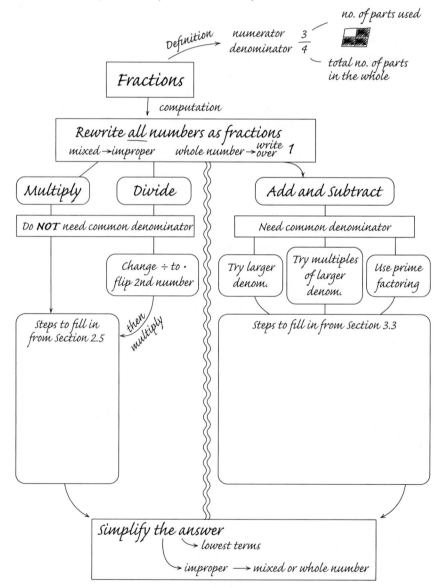

3.4 ▶▶▶ Adding and Subtracting Mixed Numbers

Recall that a mixed number is the sum of a whole number and a fraction. For example,

$$3\frac{2}{5} \quad \text{means} \quad 3 + \frac{2}{5}$$

OBJECTIVE 1 Estimate an answer, then add or subtract mixed numbers. Add or subtract mixed numbers by adding or subtracting the fraction parts and then the whole number parts. It is a good idea to estimate the answer first, as we did when multiplying and dividing mixed numbers in **Section 2.8.**

Work Problem **1** *at the Side.* ▶

1 Estimate an answer, then add or subtract mixed numbers.

2 Estimate an answer, then subtract mixed numbers by regrouping.

3 Add or subtract mixed numbers using an alternate method.

EXAMPLE 1 **Adding and Subtracting Mixed Numbers**

First estimate the answer. Then add or subtract to find the exact answer.

(a) $16\frac{1}{8} + 5\frac{5}{8}$

Estimate: *Exact:*

$$16 \xleftarrow{\text{Rounds to}} \left\{ 16\frac{1}{8} \right.$$

$$+ \;\; 6 \xleftarrow{\text{Rounds to}} \left\{ + \;\; 5\frac{5}{8} \right.$$

First, estimate the answer. → 22 $21\frac{6}{8} = 21\frac{3}{4}$ ← Lowest terms

Sum of ──┘ └── Sum of fractions
whole numbers

In lowest terms $\frac{6}{8}$ is $\frac{3}{4}$, so the exact answer of $21\frac{3}{4}$ is in lowest terms. The exact answer is *reasonable* because it is close to the estimate of 22.

(b) $8\frac{5}{8} - 3\frac{1}{12}$

Estimate: *Exact:*

$$9 \xleftarrow{\text{Rounds to}} \left\{ \; 8\frac{5}{8} = \;\; 8\frac{15}{24} \right.$$

$$- \; 3 \xleftarrow{\text{Rounds to}} \left\{ -3\frac{1}{12} = -3\frac{2}{24} \right.$$

24 is the least common denominator.

$$6 \qquad\qquad 5\frac{13}{24}$$

Subtract ──┘ └── Subtract fractions.
whole numbers. ──┘

The exact answer of $5\frac{13}{24}$ is *reasonable* because it is close to the estimated answer of 6. Check by adding $5\frac{13}{24}$ and $3\frac{1}{12}$; the sum should be $8\frac{5}{8}$.

Continued on Next Page

1 As a review of mixed numbers, write each mixed number as an improper fraction and each improper fraction as a mixed number.

(a) $\dfrac{9}{2}$

(b) $\dfrac{8}{3}$

(c) $4\dfrac{3}{4}$

(d) $3\dfrac{7}{8}$

ANSWERS

1. **(a)** $4\frac{1}{2}$ **(b)** $2\frac{2}{3}$ **(c)** $\frac{19}{4}$ **(d)** $\frac{31}{8}$

2 First estimate, and then add or subtract to find the exact answer.

(a) *Estimate:*   *Exact:*

$$7 \xleftarrow{\text{Rounds to}} \begin{cases} 6\frac{7}{8} = 6\frac{7}{8} \\ \\ +2\frac{1}{4} = 2\frac{2}{8} \end{cases}$$
$$+ 2 \xleftarrow{\text{Rounds to}}$$

(b) $25\frac{3}{5} + 12\frac{3}{10}$

Estimate:

_____ + _____ = _____

Exact: _____

(c) *Estimate:*   *Exact:*

$$\xleftarrow{\text{Rounds to}} \begin{cases} 4\frac{7}{9} \end{cases}$$

$$- \qquad \xleftarrow{\text{Rounds to}} \begin{cases} -2\frac{2}{3} \end{cases}$$

3 First estimate, and then add to find the exact answer.

(a) *Estimate:*   *Exact:*

$$\xleftarrow{\text{Rounds to}} \begin{cases} 9\frac{3}{4} \end{cases}$$

$$+ \qquad \xleftarrow{\text{Rounds to}} \begin{cases} +7\frac{1}{2} \end{cases}$$

(b) *Estimate:*   *Exact:*

$$\xleftarrow{\text{Rounds to}} \begin{cases} 15\frac{4}{5} \end{cases}$$

$$+ \qquad \xleftarrow{\text{Rounds to}} \begin{cases} +12\frac{2}{3} \end{cases}$$

ANSWERS

2. (a) $7 + 2 = 9; 9\frac{1}{8}$

(b) $26 + 12 = 38; 37\frac{9}{10}$

(c) $5 - 3 = 2; 2\frac{1}{9}$

3. (a) $10 + 8 = 18; 17\frac{1}{4}$

(b) $16 + 13 = 29; 28\frac{7}{15}$

> **Note**
> When estimating, if the numerator is *half* of the denominator or *more*, round up the whole number part. If the numerator is *less* than *half* the denominator, leave the whole number part as it is.

◀ *Work Problem* **2** *at the Side.*

When you add the fraction parts of mixed numbers, the sum may be greater than 1. If this happens, simplify the fraction and regroup in the whole number column. (You cannot leave a whole number along with an improper fraction as the answer.)

EXAMPLE 2 **Simplify and Regroup When Adding Mixed Numbers**

First estimate, and then add $9\frac{5}{8} + 13\frac{7}{8}$.

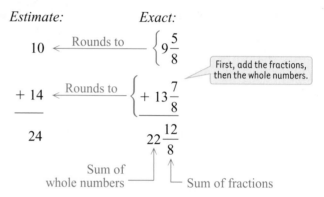

Estimate: *Exact:*

$$10 \xleftarrow{\text{Rounds to}} \begin{cases} 9\frac{5}{8} \end{cases}$$

First, add the fractions, then the whole numbers.

$$+ 14 \xleftarrow{\text{Rounds to}} \begin{cases} +13\frac{7}{8} \end{cases}$$

$$24 \qquad\qquad 22\frac{12}{8}$$

Sum of whole numbers ———— ┘   └ —— Sum of fractions

The improper fraction $\frac{12}{8}$ can be written in lowest terms as $\frac{3}{2}$. Because $\frac{3}{2} = 1\frac{1}{2}$, the simplified sum is

Becomes

$$22\frac{12}{8} = 22 + \frac{12}{8} = 22 + 1\frac{1}{2} = 23\frac{1}{2}.$$

The estimate was 24, so the exact answer of $23\frac{1}{2}$ is reasonable.

> **Note**
> When adding mixed numbers, first add the fraction parts, then add the whole number parts. Then combine the two answers, simplifying the fraction part when necessary.

◀ *Work Problem* **3** *at the Side.*

OBJECTIVE 2 Estimate an answer, then subtract mixed numbers by regrouping. When subtracting mixed numbers, **regrouping** is necessary when the fraction part of the first number is less than the fraction part of the second number.

EXAMPLE 3 **Regroup When Subtracting Mixed Numbers**

First estimate, and then subtract to find the exact answer.

(a) $7 - 2\frac{5}{6}$

Continued on Next Page

Estimate: *Exact:*

$$7 \quad \xleftarrow{\text{Rounds to}} \quad \begin{cases} 7 \end{cases}$$

> There is no fraction here from which to subtract $\frac{5}{6}$.

$$\frac{-3}{4} \quad \xleftarrow{\text{Rounds to}} \quad \begin{cases} -2\dfrac{5}{6} \end{cases}$$

It is **not** possible to subtract $\frac{5}{6}$ without regrouping the whole number **7** first.

> Regroup 7 as 6 + 1

$$7 = \overbrace{6 + 1}$$

> $1 = \frac{6}{6}$

$$= 6 + \frac{6}{6}$$

$$= 6\frac{6}{6}$$

Now you can subtract.

$$7 = \quad 6\frac{6}{6}$$

> 7 was written as $6\frac{6}{6}$.

$$-2\frac{5}{6} = \quad -2\frac{5}{6}$$

$$\overline{\qquad\qquad} \quad \overline{4\frac{1}{6}}$$

The estimate was 4, so the exact answer of $4\frac{1}{6}$ is reasonable.

(b) $8\dfrac{1}{3} - 4\dfrac{3}{5}$

Estimate: *Exact:*

$$8 \xleftarrow{\text{Rounds to}} \begin{cases} 8\dfrac{1}{3} = & 8\dfrac{5}{15} \end{cases}$$

$$\frac{-5}{3} \xleftarrow{\text{Rounds to}} \begin{cases} -4\dfrac{3}{5} = & -4\dfrac{9}{15} \end{cases}$$

> 15 is the least common denominator

It is **not** possible to subtract $\frac{9}{15}$ from $\frac{5}{15}$, so regroup the whole number **8.**

> Regroup 8 as 7 + 1

$$8\frac{5}{15} = 8 + \frac{5}{15} = \overbrace{7 + 1} + \frac{5}{15}$$

> $1 = \frac{15}{15}$

$$= 7 + \frac{15}{15} + \frac{5}{15}$$

$$= 7 + \frac{20}{15} \quad \leftarrow \frac{15}{15} + \frac{5}{15}$$

$$= 7\frac{20}{15}$$

Continued on Next Page

4 First estimate and then subtract to find the exact answer.

(a) *Estimate:*   *Exact:*

Rounds to $\left\{ 7\dfrac{1}{3} \right.$

$-$ Rounds to $\left\{ -4\dfrac{5}{6} \right.$

(b) *Estimate:*   *Exact:*

$\left\{ 4\dfrac{5}{8} \right.$

$-$ $\left\{ -2\dfrac{15}{16} \right.$

(c) *Estimate:*   *Exact:*

$\left\{ 15 \right.$

$-$ $\left\{ -6\dfrac{4}{9} \right.$

5 Add or subtract by changing mixed numbers to improper fractions. Simplify answers.

(a) $3\dfrac{3}{8}$

   $+\ 2\dfrac{1}{2}$

(b) $6\dfrac{3}{4}$

   $-\ 4\dfrac{2}{3}$

Now you can subtract.

$$8\dfrac{1}{3} = 8\dfrac{5}{15} = 7\dfrac{20}{15}$$
$$-\ 4\dfrac{3}{5} = 4\dfrac{9}{15} = 4\dfrac{9}{15}$$
$$3\dfrac{11}{15}$$

The exact answer is $3\dfrac{11}{15}$ (lowest terms), which is reasonable because it is close to the estimate of 3.

◀ *Work Problem* **4** *at the Side.*

OBJECTIVE **3** **Add or subtract mixed numbers using an alternate method.** An alternate method for adding or subtracting mixed numbers is to first change the mixed numbers to improper fractions. Then rewrite the unlike fractions as like fractions. Finally, add or subtract the numerators and write the answer in lowest terms.

EXAMPLE 4 **Adding or Subtracting Mixed Numbers**

Add or subtract.

Rewrite $2\dfrac{3}{8}$ as $\dfrac{19}{8}$ and $3\dfrac{3}{4}$ as $\dfrac{15}{4}$.

(a) $2\dfrac{3}{8} = \dfrac{19}{8} = \dfrac{19}{8}$

   $+\ 3\dfrac{3}{4} = \dfrac{15}{4} = \dfrac{30}{8}$ — 8 is the least common denominator.

$$\dfrac{49}{8} = 6\dfrac{1}{8} \quad \text{Answer as mixed number}$$

Change to improper fractions.

(b) $4\dfrac{2}{3} = \dfrac{14}{3} = \dfrac{70}{15}$

   $-\ 2\dfrac{1}{5} = \dfrac{11}{5} = \dfrac{33}{15}$ — 15 is the least common denominator.

$$\dfrac{37}{15} = 2\dfrac{7}{15} \quad \text{Simplify the answer by writing it as a mixed number.}$$

Improper fractions

◀ *Work Problem* **5** *at the Side.*

ANSWERS

4. **(a)** $7 - 5 = 2; 2\dfrac{1}{2}$

    (b) $5 - 3 = 2; 1\dfrac{11}{16}$

    (c) $15 - 6 = 9; 8\dfrac{5}{9}$

5. **(a)** $\dfrac{47}{8} = 5\dfrac{7}{8}$

    (b) $\dfrac{25}{12} = 2\dfrac{1}{12}$

Note

The advantage of this alternate method of adding or subtracting mixed numbers is that it eliminates the need to regroup. It is also the most useful method for working with algebraic fractions. (See **Section 9.7.**) However, if the mixed numbers are large, then the numerators of the improper fractions may become so large that they are difficult to work with. In such cases, you may want to keep the numbers as mixed numbers.

3.4 ▶▶▶ Exercises

First estimate the answer. Then add to find the exact answer. Write answers as mixed numbers. See Examples 1 and 2.

1. Estimate: Exact:

$6 \xleftarrow{\text{Rounds to}} \begin{cases} 5\dfrac{1}{2} \end{cases}$

$+3 \xleftarrow{\text{Rounds to}} \begin{cases} +3\dfrac{1}{3} \end{cases}$

9 $8\dfrac{5}{6}$

2. Estimate: Exact:

7 $6\dfrac{3}{5}$

$+7$ $+7\dfrac{1}{10}$

14 $13\dfrac{7}{10}$

3. Estimate: Exact:

7 $7\dfrac{1}{3}$

$+4$ $+4\dfrac{1}{6}$

11 $11\dfrac{1}{2}$

4. Estimate: Exact:

10 $10\dfrac{1}{4}$

$+6$ $+5\dfrac{5}{8}$

16 $15\dfrac{7}{8}$

5. Estimate: Exact:

1 $\dfrac{5}{8}$

$+4$ $+3\dfrac{7}{12}$

5 $4\dfrac{5}{24}$

6. Estimate: Exact:

13 $12\dfrac{4}{5}$

$+1$ $+\dfrac{7}{10}$

14 $13\dfrac{1}{2}$

7. Estimate: Exact:

25 $24\dfrac{5}{6}$

$+19$ $+18\dfrac{5}{6}$

44 $43\dfrac{2}{3}$

8. Estimate: Exact:

15 $14\dfrac{6}{7}$

$+16$ $+15\dfrac{1}{2}$

31 $30\dfrac{5}{14}$

9. Estimate: Exact:

34 $33\dfrac{3}{5}$

$+19$ $+18\dfrac{1}{2}$

53 $52\dfrac{1}{10}$

10. Estimate: Exact:

19 $18\dfrac{5}{8}$

$+7$ $+6\dfrac{2}{3}$

26 $25\dfrac{7}{24}$

11. Estimate: Exact:

23 $22\dfrac{3}{4}$

$+15$ $+15\dfrac{3}{7}$

38 $38\dfrac{5}{28}$

12. Estimate: Exact:

7 $7\dfrac{1}{4}$

$+26$ $+25\dfrac{7}{8}$

33 $33\dfrac{1}{8}$

13. *Estimate:*   *Exact:*

$$13 \qquad 12\frac{8}{15}$$

$$19 \qquad 18\frac{3}{5}$$

$$+\,15 \qquad +\,14\frac{7}{10}$$

$$\overline{\rule{1cm}{0.4pt}} \qquad \overline{\rule{2cm}{0.4pt}}$$

$$47 \qquad 45\frac{5}{6}$$

14. *Estimate:*   *Exact:*

$$15 \qquad 14\frac{9}{10}$$

$$8 \qquad 8\frac{1}{4}$$

$$+\,14 \qquad +\,13\frac{3}{5}$$

$$\overline{\rule{1cm}{0.4pt}} \qquad \overline{\rule{2cm}{0.4pt}}$$

$$37 \qquad 36\frac{3}{4}$$

First estimate the answer. Then subtract to find the exact answer. Simplify all answers. See Examples 1 and 3.

15. *Estimate:*   *Exact:*

$$15 \qquad 14\frac{7}{8}$$

$$-\,12 \qquad -\,12\frac{1}{4}$$

$$\overline{\rule{1cm}{0.4pt}} \qquad \overline{\rule{2cm}{0.4pt}}$$

$$3 \qquad 2\frac{5}{8}$$

16. *Estimate:*   *Exact:*

$$15 \qquad 14\frac{3}{4}$$

$$-\,11 \qquad -\,11\frac{3}{8}$$

$$\overline{\rule{1cm}{0.4pt}} \qquad \overline{\rule{2cm}{0.4pt}}$$

$$4 \qquad 3\frac{3}{8}$$

17. *Estimate:*   *Exact:*

$$13 \qquad 12\frac{2}{3}$$

$$-\,1 \qquad -\,1\frac{1}{5}$$

$$\overline{\rule{1cm}{0.4pt}} \qquad \overline{\rule{2cm}{0.4pt}}$$

$$12 \qquad 11\frac{7}{15}$$

18. *Estimate:*   *Exact:*

$$11 \qquad 11\frac{9}{20}$$

$$-\,5 \qquad -\,4\frac{3}{5}$$

$$\overline{\rule{1cm}{0.4pt}} \qquad \overline{\rule{2cm}{0.4pt}}$$

$$6 \qquad 6\frac{17}{20}$$

19. *Estimate:*   *Exact:*

$$28 \qquad 28\frac{3}{10}$$

$$-\,6 \qquad -\,6\frac{1}{15}$$

$$\overline{\rule{1cm}{0.4pt}} \qquad \overline{\rule{2cm}{0.4pt}}$$

$$22 \qquad 22\frac{7}{30}$$

20. *Estimate:*   *Exact:*

$$15 \qquad 15\frac{7}{20}$$

$$-\,6 \qquad -\,6\frac{1}{8}$$

$$\overline{\rule{1cm}{0.4pt}} \qquad \overline{\rule{2cm}{0.4pt}}$$

$$9 \qquad 9\frac{9}{40}$$

21. *Estimate:*   *Exact:*

$$17 \qquad 17$$

$$-\,7 \qquad -\,6\frac{5}{8}$$

$$\overline{\rule{1cm}{0.4pt}} \qquad \overline{\rule{2cm}{0.4pt}}$$

$$10 \qquad 10\frac{3}{8}$$

22. *Estimate:*   *Exact:*

$$22 \qquad 22$$

$$-\,5 \qquad -\,4\frac{5}{8}$$

$$\overline{\rule{1cm}{0.4pt}} \qquad \overline{\rule{2cm}{0.4pt}}$$

$$17 \qquad 17\frac{3}{8}$$

23. *Estimate:*   *Exact:*

$$19 \qquad 18\frac{3}{4}$$

$$-\,6 \qquad -\,5\frac{4}{5}$$

$$\overline{\rule{1cm}{0.4pt}} \qquad \overline{\rule{2cm}{0.4pt}}$$

$$13 \qquad 12\frac{19}{20}$$

24. *Estimate:*   *Exact:*

$$15 \qquad 14\frac{5}{8}$$

$$\underline{-\ 4} \qquad \underline{-\ 3\frac{2}{3}}$$

$$11 \qquad 10\frac{23}{24}$$

☢ 25. *Estimate:*   *Exact:*

$$20 \qquad 19\frac{2}{3}$$

$$\underline{-\ 12} \qquad \underline{-\ 11\frac{3}{4}}$$

$$8 \qquad 7\frac{11}{12}$$

26. *Estimate:*   *Exact:*

$$21 \qquad 20\frac{3}{5}$$

$$\underline{-\ 12} \qquad \underline{-\ 12\frac{7}{15}}$$

$$9 \qquad 8\frac{2}{15}$$

Add or subtract by changing mixed numbers to improper fractions. Write answers as mixed numbers when possible. See Example 4.

27.

$$7\frac{5}{8}$$

$$\underline{+\ 1\frac{1}{2}}$$

$$9\frac{1}{8}$$

28.

$$8\frac{3}{4}$$

$$\underline{+\ 1\frac{5}{8}}$$

$$10\frac{3}{8}$$

☢ 29.

$$4\frac{2}{3}$$

$$\underline{+\ 6\frac{5}{6}}$$

$$11\frac{1}{2}$$

30.

$$7\frac{5}{12}$$

$$\underline{+\ 6\frac{2}{3}}$$

$$14\frac{1}{12}$$

31.

$$2\frac{2}{3}$$

$$\underline{+\ 1\frac{1}{6}}$$

$$3\frac{5}{6}$$

32.

$$4\frac{1}{2}$$

$$\underline{+\ 2\frac{3}{4}}$$

$$7\frac{1}{4}$$

33.

$$3\frac{1}{4}$$

$$\underline{+\ 3\frac{2}{3}}$$

$$6\frac{11}{12}$$

34.

$$2\frac{4}{5}$$

$$\underline{+\ 5\frac{1}{3}}$$

$$8\frac{2}{15}$$

35.

$$1\frac{3}{8}$$

$$\underline{+\ 6\frac{3}{4}}$$

$$8\frac{1}{8}$$

36.

$$1\frac{5}{12}$$

$$\underline{+\ 1\frac{7}{8}}$$

$$3\frac{7}{24}$$

37.

$$3\frac{1}{2}$$

$$\underline{-\ 2\frac{2}{3}}$$

$$\frac{5}{6}$$

38.

$$4\frac{1}{4}$$

$$\underline{-\ 3\frac{7}{12}}$$

$$\frac{2}{3}$$

39.
$$8\frac{3}{4}$$
$$-5\frac{7}{8}$$
$$\overline{2\frac{7}{8}}$$

40.
$$12\frac{2}{3}$$
$$-7\frac{11}{12}$$
$$\overline{4\frac{3}{4}}$$

41.
$$7\frac{1}{4}$$
$$-4\frac{2}{3}$$
$$\overline{2\frac{7}{12}}$$

42.
$$4\frac{1}{10}$$
$$-3\frac{7}{8}$$
$$\overline{\frac{9}{40}}$$

43.
$$9\frac{1}{5}$$
$$-3\frac{3}{4}$$
$$\overline{5\frac{9}{20}}$$

44.
$$10\frac{2}{7}$$
$$-5\frac{5}{14}$$
$$\overline{4\frac{13}{14}}$$

45.
$$6\frac{3}{7}$$
$$-2\frac{2}{3}$$
$$\overline{3\frac{16}{21}}$$

46.
$$8\frac{2}{15}$$
$$-6\frac{1}{2}$$
$$\overline{1\frac{19}{30}}$$

47. In your own words, explain the steps you would take to add two large mixed numbers.

Find the least common denominator. Change the fraction parts so that they have the same denominator. Add the fraction parts. Add the whole number parts. Write the answer as a mixed number.

48. When subtracting mixed numbers, explain when you need to regroup. Explain how to regroup using your own example.

You need to regroup when the fraction in the minuend (top number) is smaller than the fraction in the subtrahend (bottom number). One example is shown below.

$$5\frac{1}{4} = 5\frac{1}{4} = 4\frac{5}{4}$$
$$-3\frac{1}{2} = 3\frac{2}{4} = 3\frac{2}{4}$$
$$\overline{1\frac{3}{4}}$$

First estimate the answer. Then solve each application problem.

49. At the beginning of this chapter you read about Bryan Berg, who builds houses of cards. While in high school, Bryan built a house of cards $14\frac{1}{2}$ ft tall, setting his first world record. Today his current world record is $25\frac{3}{4}$ ft tall. How much taller is his current world record than his first world record? (*Source: Guinness World Records.*)

Estimate: **26 − 15 = 11 ft**

Exact: **11$\frac{1}{4}$ ft**

50. The heaviest marine mammal in the world is the blue whale at $143\frac{3}{10}$ tons. The sixth heaviest is the humpback whale at $29\frac{1}{5}$ tons. Find the difference in their weights. (*Source: Top 10 of Everything.*)

Estimate: **143 − 29 = 114 tons**

Exact: **114$\frac{1}{10}$ tons**

Use the newspaper advertisement for this 6-piece jumbo wrench set to answer Exercises 51–54. The symbol " is for inches. (*Source:* Harbor Freight Tools.)

6-Piece Jumbo Wrench Set

• Sizes: $1\frac{3}{8}$" to 2"

• Length: $17\frac{3}{4}$" to $22\frac{3}{4}$"

$22\frac{3}{4}$"

$22\frac{1}{4}$"

21"

$20\frac{3}{4}$"

$18\frac{5}{8}$"

$17\frac{3}{4}$"

SALE!

$27⁹⁹

REGULAR PRICE $39.99

51. How much longer is the longest wrench than the second-to-shortest wrench?

Estimate: 23 − 19 = 4 in.

Exact: $4\frac{1}{8}$ in.

52. Find the difference in length between the shortest wrench and the second-to-longest wrench.

Estimate: 22 − 18 = 4 in.

Exact: $4\frac{1}{2}$ in.

53. What is the total length of the three longest wrenches?

Estimate: 23 + 22 + 21 = 66 in.

Exact: 66 in.

54. What is the total length of the three shortest wrenches?

Estimate: 21 + 19 + 18 = 58 in.

Exact: $57\frac{1}{8}$ in.

Storehouse 34 PC. GEAR HOSE CLAMP ASSORTMENT

LOT NO. 1420

Includes: two 2-3/4", two 2-1/2", four 2-1/4", four 2", six 1-3/4", six 1-1/2", six 1-1/4", four 9/16" through 1-1/16"

SALE!

$5⁹⁷

SAVE 40%

REGULAR PRICE $9.99

A mechanic buys the 34-piece hose clamp assortment shown at the left. Use the advertisement to answer Exercises 55–56. The symbol " is for inches. (*Source:* Harbor Freight Tools.)

55. Find the difference in size between the largest hose clamp and the smallest hose clamp.

Estimate: 3 − 1 = 2 in.

Exact: $2\frac{3}{16}$ in.

56. Find the difference in size between the second to largest hose clamp and the smallest hose clamp.

Estimate: 3 − 1 = 2 in.

Exact: $1\frac{15}{16}$ in.

57. The four sides of Andre Herrebout's vegetable garden are $15\frac{1}{2}$ feet, $18\frac{3}{4}$ feet, $24\frac{1}{4}$ feet, and $30\frac{1}{2}$ feet. How many feet of fencing are needed to go around the garden?

Estimate: 16 + 19 + 24 + 31 = 90 ft

Exact: $87\frac{8}{4}$ = 89 ft

58. On a recent vacation to Canada, Erin Gavin drove for $7\frac{3}{4}$ hours on the first day, $5\frac{1}{4}$ hours on the second day, $6\frac{1}{2}$ hours on the third day, and 9 hours on the fourth day. How many hours did she drive altogether?

Estimate: 8 + 5 + 7 + 9 = 29 hr

Exact: $28\frac{1}{2}$ hr

59. A craftsperson must attach a lead strip around all four sides of a stained glass window before it is installed. Find the length of lead stripping needed.

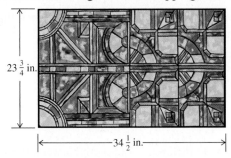

$23\frac{3}{4}$ in.

$34\frac{1}{2}$ in.

Estimate: **24 + 35 + 24 + 35 = 118 in.**

Exact: **116$\frac{1}{2}$ in.**

60. To complete a custom order, Zak Morten of Home Depot must find the number of inches of brass trim needed to go around the four sides of the lamp base plate shown. Find the length of brass trim needed.

$5\frac{1}{8}$ in.

$9\frac{7}{8}$ in.

Estimate: **10 + 5 + 10 + 5 = 30 in.**

Exact: **30 in.**

61. A museum humidifier contains 100 gallons of water. The system uses $10\frac{1}{4}$ gallons of water on Monday, $13\frac{1}{2}$ gallons on Tuesday, $8\frac{7}{8}$ gallons on Wednesday, $18\frac{3}{4}$ gallons on Thursday, $12\frac{3}{8}$ gallons on Friday, $9\frac{1}{2}$ gallons on Saturday, and $14\frac{1}{8}$ gallons on Sunday. Find the total number of gallons of water remaining.

Estimate: **100 − 10 − 14 − 9 − 19 − 12 − 10 − 14 = 12 gallons**

Exact: **12$\frac{5}{8}$ gallons**

62. Marv Levinson bought 15 yards of Italian silk fabric. He made two shirts, needing a total of $3\frac{3}{4}$ yards of the material, a suit for his wife with $4\frac{1}{8}$ yards, and a jacket with $3\frac{7}{8}$ yards. Find the number of yards of material remaining.

Estimate: **15 − 4 − 4 − 4 = 3 yards**

Exact: **3$\frac{1}{4}$ yards**

63. The exercise yard at the correction center has four sides and is surrounded by $527\frac{1}{24}$ ft of security fencing. If three sides of the yard measure $107\frac{2}{3}$ ft, $150\frac{3}{4}$ ft, and $138\frac{5}{8}$ ft, find the length of the fourth side.

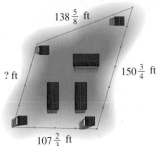

$138\frac{5}{8}$ ft

? ft

$150\frac{3}{4}$ ft

$107\frac{2}{3}$ ft

Estimate: **527 − 108 − 151 − 139 = 129 ft**

Exact: **130 ft**

64. Three sides of a parking lot are $108\frac{1}{4}$ ft, $162\frac{3}{8}$ ft, and $143\frac{1}{2}$ ft. If the distance around the lot is $518\frac{3}{4}$ ft, find the length of the fourth side.

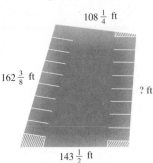

$108\frac{1}{4}$ ft

$162\frac{3}{8}$ ft

? ft

$143\frac{1}{2}$ ft

Estimate: **519 − 108 − 162 − 144 = 105 ft**

Exact: **104$\frac{5}{8}$ ft**

65. A freight car is loaded with Morton Salt products consisting of $58\frac{1}{2}$ tons of coarse rock salt, $23\frac{5}{8}$ tons of medium rock salt, $16\frac{5}{6}$ tons of table salt, and $29\frac{1}{4}$ tons of animal salt lick blocks. If the weight of the freight car is $58\frac{1}{3}$ tons, find the weight of the loaded freight car.

Estimate: **59 + 24 + 17 + 29 + 58 = 187 tons**

Exact: $186\dfrac{13}{24}$ **tons**

66. Bryan Berg, from the opening page of this chapter, built houses of cards reaching heights of $14\frac{1}{2}$ ft, $19\frac{3}{8}$ ft, $23\frac{5}{12}$ ft, and $25\frac{3}{4}$ ft (his current world record). Find the total height of these four houses of cards. (*Source: Reader's Digest.*)

Estimate: **15 + 19 + 23 + 26 = 83 ft**

Exact: $83\dfrac{1}{24}$ **ft**

Find the unknown length, labeled with a question mark, in each figure.

67.

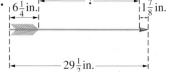

$4\dfrac{11}{16}$ in.

$2\frac{3}{8}$ in. $2\frac{3}{8}$ in.

$9\frac{7}{16}$ in.

68.

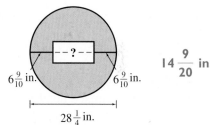

$14\dfrac{9}{20}$ in.

$6\frac{9}{10}$ in. $6\frac{9}{10}$ in.

$28\frac{1}{4}$ in.

69.

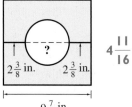

$6\frac{1}{4}$ in.   ? $1\frac{7}{8}$ in.

$29\frac{1}{2}$ in.

$21\dfrac{3}{8}$ in.

70.

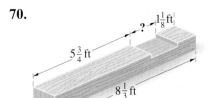

$5\frac{3}{4}$ ft   ? $1\frac{1}{8}$ ft

$8\frac{1}{3}$ ft

$1\dfrac{11}{24}$ ft

Relating Concepts (Exercises 71–76) For Individual or Group Work

Most fraction problems include fractions with different denominators.
Work Exercises 71–76 in order.

71. To add or subtract fractions, we must first rewrite them as like fractions. Rewrite each fraction with the indicated denominator.

(a) $\dfrac{5}{9} = \dfrac{30}{54}$ (b) $\dfrac{7}{12} = \dfrac{28}{48}$

(c) $\dfrac{5}{8} = \dfrac{25}{40}$ (d) $\dfrac{11}{5} = \dfrac{264}{120}$

72. When rewriting unlike fractions as like fractions with the least common multiple as a denominator, the new denominator is called the __least__ __common__ __denominator__, or LCD.

73. Add or subtract as indicated. Write answers in lowest terms.

(a) $\dfrac{5}{8} + \dfrac{1}{3}$ $\dfrac{23}{24}$ (b) $\dfrac{19}{20} - \dfrac{5}{12}$ $\dfrac{8}{15}$

(c) $\begin{array}{r} \dfrac{7}{12} \\ \dfrac{3}{16} \\ +\dfrac{3}{24} \\ \hline \dfrac{43}{48} \end{array}$ (d) $\begin{array}{r} \dfrac{6}{7} \\ -\dfrac{2}{3} \\ \hline \dfrac{4}{21} \end{array}$

74. A common method for adding or subtracting mixed numbers is to add or subtract the __fraction__ __parts__ and then add or subtract the whole number parts

75. Another method for adding or subtracting mixed numbers is to first change the mixed numbers to __improper__ fractions. After adding or subtracting, write the answer in lowest terms and as a mixed number when possible. This method is difficult to use if the mixed numbers are __large__.
(large/small)

76. Add or subtract these fractions as indicated. First use the method where you add or subtract fraction parts and then whole number parts. Then use the method where you change each mixed number to an improper fraction before adding or subtracting. Do you get the same answer using both methods? Which method do you prefer?

(a) $\begin{array}{r} 4\dfrac{5}{8} \\ +3\dfrac{3}{4} \\ \hline \end{array}$ (b) $\begin{array}{r} 12\dfrac{2}{5} \\ -8\dfrac{7}{8} \\ \hline \end{array}$

(a) $4\dfrac{5}{8} + 3\dfrac{6}{8} = 7\dfrac{11}{8} = 8\dfrac{3}{8}$; $\dfrac{37}{8} + \dfrac{30}{8} = \dfrac{67}{8} = 8\dfrac{3}{8}$

(b) $11\dfrac{56}{40} - 8\dfrac{35}{40} = 3\dfrac{21}{40}$; $\dfrac{496}{40} - \dfrac{355}{40} = \dfrac{141}{40} = 3\dfrac{21}{40}$

3.5 ▶▶▶ Order Relations and the Order of Operations

There are times when we want to compare the size of two numbers. For example, we might want to know which is the greater amount, the larger size, or the longer distance.

Fractions, like whole numbers, can be graphed on a number line. For fractions, divide the space between whole numbers into equal parts.

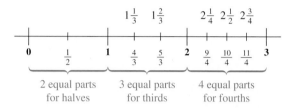

2 equal parts for halves   3 equal parts for thirds   4 equal parts for fourths

OBJECTIVES

1 Identify the greater of two fractions.

2 Use exponents with fractions.

3 Use the order of operations with fractions.

OBJECTIVE 1 Identify the greater of two fractions. To compare the size of two numbers, place the two numbers on a number line and use the following rule.

Comparing the Size of Two Numbers

The number farther to the *left* on the number line is always *less*, and the number farther to the *right* on the number line is always *greater*.

For example, on the number line above, $\frac{1}{2}$ is to the *left* of $\frac{4}{3}$, so $\frac{1}{2}$ is *less than* $\frac{4}{3}$.

Work Problem **1** *at the Side.* ▶

Write *order relations* using the symbols shown below.

Symbols Used to Show Order Relations

$<$ is less than $>$ is greater than

EXAMPLE 1 Using Less-Than and Greater-Than Symbols

Rewrite the following using $<$ and $>$ symbols.

(a) $\frac{1}{2}$ is less than $\frac{4}{3}$

$\frac{1}{2}$ is less than $\frac{4}{3}$ is written as $\frac{1}{2} < \frac{4}{3}$ ◀ The number farther to the *left* on the number line is *less*.

(b) $\frac{9}{4}$ is greater than 1

$\frac{9}{4}$ is greater than 1 is written as $\frac{9}{4} > 1$ ◀ The number farther to the *right* on the number line is *greater*.

(c) $\frac{5}{3}$ is less than $\frac{11}{4}$

$\frac{5}{3}$ is less than $\frac{11}{4}$ is written as $\frac{5}{3} < \frac{11}{4}$

Continued on Next Page

1 Locate each fraction on the number line.

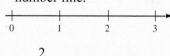

(a) $\frac{2}{3}$

(b) $1\frac{1}{2}$

(c) $2\frac{3}{4}$

ANSWER

1.

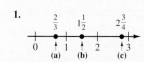

2 Use the number line on the previous page to help you write < or > in each blank to make a true statement.

(a) 1 _____ $\dfrac{5}{4}$

(b) $\dfrac{8}{3}$ _____ $\dfrac{3}{2}$

(c) 0 _____ 1

(d) $\dfrac{17}{8}$ _____ $\dfrac{8}{4}$

3 Write < or > in each blank to make a true statement.

(a) $\dfrac{7}{8}$ _____ $\dfrac{3}{4}$

(b) $\dfrac{13}{8}$ _____ $\dfrac{15}{9}$

(c) $\dfrac{9}{4}$ _____ $\dfrac{7}{3}$

(d) $\dfrac{9}{10}$ _____ $\dfrac{14}{15}$

Note

A number line is a very useful tool when working with order relations.

◀ *Work Problem* **2** *at the Side.*

The fraction $\frac{7}{8}$ represents 7 of 8 equivalent parts, while $\frac{3}{8}$ means 3 of 8 equivalent parts. Because $\frac{7}{8}$ represents more of the equivalent parts, $\frac{7}{8}$ is greater than $\frac{3}{8}$.

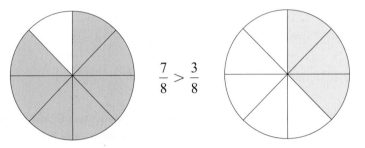

$$\dfrac{7}{8} > \dfrac{3}{8}$$

To identify the greater fraction, use the following steps.

Identifying the Greater Fraction

Step 1 Write the fractions as like fractions (same denominators).

Step 2 Compare the numerators. The fraction with the greater numerator is the greater fraction.

EXAMPLE 2 **Identifying the Greater Fraction**

Determine which fraction in each pair is greater.

(a) $\dfrac{7}{8}, \dfrac{9}{10}$

First, write the fractions as like fractions. The least common multiple for 8 and 10 is 40.

$$\dfrac{7}{8} = \dfrac{7 \cdot 5}{8 \cdot 5} = \dfrac{35}{40} \quad \text{and} \quad \dfrac{9}{10} = \dfrac{9 \cdot 4}{10 \cdot 4} = \dfrac{36}{40}$$

Rewrite both fractions with 40 as the denominator.

Look at the numerators. Because 36 is greater than 35, $\frac{36}{40}$ is greater than $\frac{35}{40}$. Then, because $\frac{36}{40}$ is equivalent to $\frac{9}{10}$,

$$\dfrac{9}{10} > \dfrac{7}{8} \quad \text{or} \quad \dfrac{7}{8} < \dfrac{9}{10}$$

The greater fraction is $\frac{9}{10}$.

(b) $\dfrac{8}{5}, \dfrac{23}{15}$

The least common multiple of 5 and 15 is 15.

$$\dfrac{8}{5} = \dfrac{8 \cdot 3}{5 \cdot 3} = \dfrac{24}{15} \quad \text{and} \quad \dfrac{23}{15} = \dfrac{23}{15}$$

This shows that $\frac{8}{5}$ is greater than $\frac{23}{15}$, or

$$\dfrac{8}{5} > \dfrac{23}{15}$$

◀ *Work Problem* **3** *at the Side.*

ANSWERS

2. (a) < (b) > (c) < (d) >
3. (a) > (b) < (c) < (d) <

OBJECTIVE 2 Use exponents with fractions. Exponents were used in **Section 1.8** to write repeated multiplication. For example,

$$3^2 = \underbrace{3 \cdot 3}_{\text{Two factors of 3}} = 9 \quad \text{and} \quad 5^3 = \underbrace{5 \cdot 5 \cdot 5}_{\text{Three factors of 5}} = 125$$

where the overbraces point to the **Exponent**.

The next example shows exponents used with fractions.

EXAMPLE 3 Using Exponents with Fractions

Simplify.

(a) $\left(\dfrac{1}{2}\right)^3$

$$\left(\dfrac{1}{2}\right)^3 = \overbrace{\dfrac{1}{2} \cdot \dfrac{1}{2} \cdot \dfrac{1}{2}}^{\text{Three factors of } \frac{1}{2}} = \dfrac{1}{8}$$

(b) $\left(\dfrac{5}{8}\right)^2$

$$\left(\dfrac{5}{8}\right)^2 = \overbrace{\dfrac{5}{8} \cdot \dfrac{5}{8}}^{\text{Two factors of } \frac{5}{8}} = \dfrac{25}{64}$$

(c) $\left(\dfrac{3}{4}\right)^2 \cdot \left(\dfrac{2}{3}\right)^3$

$$\left(\dfrac{3}{4}\right)^2 \cdot \left(\dfrac{2}{3}\right)^3 = \left(\dfrac{3}{4} \cdot \dfrac{3}{4}\right) \cdot \left(\dfrac{2}{3} \cdot \dfrac{2}{3} \cdot \dfrac{2}{3}\right)$$

$$= \dfrac{\overset{1}{\cancel{3}} \cdot \overset{1}{\cancel{3}} \cdot \overset{1}{\cancel{2}} \cdot \overset{1}{\cancel{2}} \cdot \overset{1}{\cancel{2}}}{\underset{2}{\cancel{4}} \cdot \underset{2}{\cancel{4}} \cdot \underset{1}{\cancel{3}} \cdot \underset{1}{\cancel{3}} \cdot 3} \quad \text{Divide out all the common factors}$$

$$= \dfrac{1}{6} \quad \boxed{\text{The fraction is in lowest terms after all common factors are divided out.}}$$

Work Problem **4** *at the Side.* ▶

OBJECTIVE 3 Use the order of operations with fractions. Recall the *order of operations* from **Section 1.8.**

Order of Operations

1. Do all operations inside *parentheses or other grouping symbols.*
2. Simplify any expressions with *exponents* and find any *square roots.*
3. *Multiply* or *divide,* proceeding from left to right.
4. *Add* or *subtract,* proceeding from left to right.

---

4 Simplify.

(a) $\left(\dfrac{1}{2}\right)^4$

(b) $\left(\dfrac{3}{4}\right)^2$

(c) $\left(\dfrac{1}{2}\right)^3 \cdot \left(\dfrac{2}{3}\right)^2$

(d) $\left(\dfrac{1}{5}\right)^2 \cdot \left(\dfrac{5}{3}\right)^2$

The next example shows how to apply the order of operations with fractions.

5 Simplify by using the order of operations.

(a) $\dfrac{5}{9} - \dfrac{3}{4}\left(\dfrac{2}{3}\right)$

EXAMPLE 4 **Using the Order of Operations with Fractions**

Simplify by using the order of operations.

(a) $\dfrac{1}{3} + \dfrac{1}{2}\left(\dfrac{4}{5}\right)$

Multiply $\dfrac{1}{2}\left(\dfrac{4}{5}\right)$ first because multiplication and division are done *before* adding.

> Do **not** add $\dfrac{1}{3} + \dfrac{1}{2}$ as the first step.

$$\dfrac{1}{3} + \dfrac{1}{\underset{1}{2}}\left(\dfrac{\overset{2}{4}}{5}\right) = \dfrac{1}{3} + \dfrac{2}{5}$$

Next, add. The least common denominator of 3 and 5 is 15.

$$\dfrac{1}{3} + \dfrac{2}{5} = \dfrac{5}{15} + \dfrac{6}{15} = \dfrac{11}{15}$$

(b) $\dfrac{3}{4}\left(\dfrac{2}{3} \cdot \dfrac{3}{5}\right)$

(b) $\dfrac{3}{8}\left(\dfrac{1}{2} + \dfrac{1}{3}\right)$

$$\dfrac{3}{8}\left(\dfrac{1}{2} + \dfrac{1}{3}\right) = \dfrac{3}{8}\left(\underbrace{\dfrac{3}{6} + \dfrac{2}{6}}\right)$$

Work inside parentheses first.

$$= \dfrac{3}{8}\left(\dfrac{5}{6}\right)$$

$$= \dfrac{\overset{1}{\cancel{3}}}{8}\left(\dfrac{5}{\underset{2}{\cancel{6}}}\right) \qquad \text{Divide numerator and denominator by 3.}$$

$$= \dfrac{5}{16} \qquad \text{Multiply.}$$

(c) $\dfrac{7}{8}\left(\dfrac{2}{3}\right) - \left(\dfrac{1}{2}\right)^2$

(c) $\left(\dfrac{2}{3}\right)^2 - \dfrac{4}{5}\left(\dfrac{1}{2}\right)$

Simplify the expression with the exponent. $\dfrac{2}{3} \cdot \dfrac{2}{3}$ is $\dfrac{4}{9}$.

> No work inside parentheses, so operations with exponents are next.

$$\left(\dfrac{2}{3}\right)^2 - \dfrac{4}{5}\left(\dfrac{1}{2}\right) = \dfrac{4}{9} - \dfrac{4}{5}\left(\dfrac{1}{2}\right)$$

$$= \dfrac{4}{9} - \dfrac{\overset{2}{\cancel{4}}}{5}\left(\dfrac{1}{\underset{1}{\cancel{2}}}\right) \qquad \text{Now, multiply.}$$

$$= \dfrac{4}{9} - \dfrac{2}{5}$$

$$= \dfrac{20}{45} - \dfrac{18}{45} \qquad \text{Subtract last. (Least common denominator is 45.)}$$

$$= \dfrac{2}{45}$$

(d) $\dfrac{\left(\dfrac{5}{6}\right)^2}{\dfrac{4}{3}}$

◀ *Work Problem* **5** *at the Side.*

Locate each fraction in Exercises 1–12 on the following number line. See Margin Problem 1.

1. $\dfrac{1}{2}$

2. $\dfrac{1}{4}$

3. $\dfrac{3}{2}$

4. $\dfrac{5}{4}$

5. $\dfrac{7}{3}$

6. $\dfrac{11}{4}$

7. $2\dfrac{1}{6}$

8. $3\dfrac{4}{5}$

9. $\dfrac{7}{2}$

10. $\dfrac{7}{8}$

11. $3\dfrac{1}{4}$

12. $1\dfrac{7}{8}$

Write < or > to make a true statement. See Examples 1 and 2.

13. $\dfrac{1}{2}$ __>__ $\dfrac{3}{8}$

14. $\dfrac{5}{8}$ __<__ $\dfrac{3}{4}$

15. $\dfrac{5}{6}$ __<__ $\dfrac{11}{12}$

16. $\dfrac{13}{18}$ __<__ $\dfrac{5}{6}$

17. $\dfrac{5}{12}$ __>__ $\dfrac{3}{8}$

18. $\dfrac{7}{15}$ __>__ $\dfrac{9}{20}$

19. $\dfrac{7}{12}$ __<__ $\dfrac{11}{18}$

20. $\dfrac{17}{24}$ __<__ $\dfrac{5}{6}$

21. $\dfrac{11}{18}$ __>__ $\dfrac{5}{9}$

22. $\dfrac{13}{15}$ __<__ $\dfrac{8}{9}$

23. $\dfrac{37}{50}$ __>__ $\dfrac{13}{20}$

24. $\dfrac{7}{12}$ __>__ $\dfrac{11}{20}$

Simplify. See Example 3.

25. $\left(\dfrac{1}{3}\right)^2$ $\dfrac{1}{9}$

26. $\left(\dfrac{2}{3}\right)^2$ $\dfrac{4}{9}$

27. $\left(\dfrac{5}{8}\right)^2$ $\dfrac{25}{64}$

28. $\left(\dfrac{7}{8}\right)^2$ $\dfrac{49}{64}$

29. $\left(\dfrac{3}{4}\right)^2$ $\dfrac{9}{16}$

30. $\left(\dfrac{3}{5}\right)^3$ $\dfrac{27}{125}$

31. $\left(\dfrac{4}{5}\right)^3$ $\dfrac{64}{125}$

32. $\left(\dfrac{4}{7}\right)^3$ $\dfrac{64}{343}$

33. $\left(\dfrac{3}{2}\right)^4$ $\dfrac{81}{16} = 5\dfrac{1}{16}$

34. $\left(\dfrac{4}{3}\right)^4$ $\dfrac{256}{81} = 3\dfrac{13}{81}$

35. $\left(\dfrac{3}{4}\right)^4$ $\dfrac{81}{256}$

36. $\left(\dfrac{2}{3}\right)^5$ $\dfrac{32}{243}$

✎ **37.** Describe in your own words what a number line is, and draw a picture of one. Be sure to include how it works and how it can be used.

A number line is a horizontal line with a range of equally spaced whole numbers placed on it. The lowest number is on the left and the greatest number is on the right. It can be used to compare the size or value of numbers.

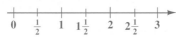

$0 \quad \frac{1}{2} \quad 1 \quad 1\frac{1}{2} \quad 2 \quad 2\frac{1}{2} \quad 3$

✎ **38.** You have used the order of operations with whole numbers and again with fractions. List from memory the steps in the order of operations.

1. **Do all operations inside parentheses or other grouping symbols.**
2. **Simplify any expressions with exponents or square roots.**
3. **Multiply or divide, proceeding from left to right.**
4. **Add or subtract, proceeding from left to right.**

Use the order of operations to simplify each expression. See Example 4.

39. $2^4 - 4\,(3)$

4

40. $3^2 + 4\,(1)$

13

41. $3 \cdot 2^2 - \dfrac{6}{3}$

10

42. $5 \cdot 2^3 - \dfrac{6}{2}$

37

43. $\left(\dfrac{1}{2}\right)^2 \cdot 4$

1

44. $\left(\dfrac{1}{4}\right)^2 \cdot 4$

$\dfrac{1}{4}$

45. $\left(\dfrac{3}{4}\right)^2 \cdot \left(\dfrac{1}{3}\right)$

$\dfrac{3}{16}$

46. $\left(\dfrac{2}{3}\right)^3 \cdot \left(\dfrac{1}{2}\right)$

$\dfrac{4}{27}$

47. $\left(\dfrac{4}{5}\right)^2 \cdot \left(\dfrac{5}{6}\right)^2$

$\dfrac{4}{9}$

48. $\left(\dfrac{5}{8}\right)^2 \cdot \left(\dfrac{4}{25}\right)^2$

$\dfrac{1}{100}$

49. $6\left(\dfrac{2}{3}\right)^2 \left(\dfrac{1}{2}\right)^3$

$\dfrac{1}{3}$

50. $9\left(\dfrac{1}{3}\right)^3 \left(\dfrac{4}{3}\right)^2$

$\dfrac{16}{27}$

51. $\dfrac{3}{5}\left(\dfrac{1}{3}\right) + \dfrac{2}{5}\left(\dfrac{3}{4}\right)$

$\dfrac{1}{2}$

52. $\dfrac{1}{4}\left(\dfrac{3}{4}\right) + \dfrac{3}{8}\left(\dfrac{4}{3}\right)$

$\dfrac{11}{16}$

53. $\dfrac{1}{2} + \left(\dfrac{1}{2}\right)^2 - \dfrac{3}{8}$

$\dfrac{3}{8}$

54. $\dfrac{2}{3} + \left(\dfrac{1}{3}\right)^2 - \dfrac{5}{9}$

$\dfrac{2}{9}$

55. $\left(\dfrac{1}{3} + \dfrac{1}{6}\right) \cdot \dfrac{1}{2}$

$\dfrac{1}{4}$

56. $\left(\dfrac{3}{5} - \dfrac{3}{20}\right) \cdot \dfrac{4}{3}$

$\dfrac{3}{5}$

57. $\dfrac{9}{8} \div \left(\dfrac{2}{3} + \dfrac{1}{12}\right)$

$1\dfrac{1}{2}$

58. $\dfrac{6}{5} \div \left(\dfrac{3}{5} - \dfrac{3}{10}\right)$

4

59. $\left(\dfrac{7}{8} - \dfrac{3}{4}\right) \div \dfrac{3}{2}$

$\dfrac{1}{12}$

60. $\left(\dfrac{4}{5} - \dfrac{3}{10}\right) \div \dfrac{4}{5}$

$\dfrac{5}{8}$

61. $\dfrac{3}{8}\left(\dfrac{1}{4} + \dfrac{1}{2}\right) \cdot \dfrac{32}{3}$

3

62. $\dfrac{1}{3}\left(\dfrac{4}{5} - \dfrac{3}{10}\right) \cdot \dfrac{4}{2}$

$\dfrac{1}{3}$

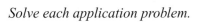 **63.** $\left(\dfrac{3}{4}\right)^2 - \left(\dfrac{1}{2} - \dfrac{1}{6}\right) \div \dfrac{4}{3}$

$\dfrac{5}{16}$

64. $\left(\dfrac{2}{3}\right)^2 - \left(\dfrac{5}{8} - \dfrac{1}{2}\right) \div \dfrac{3}{2}$

$\dfrac{13}{36}$

65. $\left(\dfrac{7}{8} - \dfrac{1}{4}\right) - \dfrac{2}{3}\left(\dfrac{3}{4}\right)^2$

$\dfrac{1}{4}$

66. $\left(\dfrac{5}{6} - \dfrac{7}{12}\right) - \dfrac{3}{4}\left(\dfrac{1}{3}\right)^2$

$\dfrac{1}{6}$

67. $\left(\dfrac{3}{4}\right)^2\left(\dfrac{2}{3} - \dfrac{5}{9}\right) - \dfrac{1}{4}\left(\dfrac{1}{8}\right)$

$\dfrac{1}{32}$

68. $\left(\dfrac{2}{3}\right)^2\left(\dfrac{1}{2} - \dfrac{1}{8}\right) - \dfrac{2}{3}\left(\dfrac{1}{8}\right)$

$\dfrac{1}{12}$

Solve each application problem.

69. The population of Las Vegas, Nevada has had an increase of $\dfrac{4}{25}$ since the turn of the century. During this same period, the population in Atlanta, Georgia has had an increase of $\dfrac{5}{30}$. Which city has had a higher rate of population growth? (*Source:* Census Bureau estimates.)

70. Two of the cities with the most expensive average home prices are Newport Beach, California ($\$1\dfrac{5}{8}$ million) and Santa Barbara, California ($\$1\dfrac{21}{32}$ million). Which city has the highest average home price? (*Source:* Coldwell Banker Home Price Comparison index.)

$\dfrac{5}{30}$ **in Atlanta is greater.**

Santa Barbara with $\$1\dfrac{21}{32}$ **million is higher.**

Relating Concepts (Exercises 71–80) For Individual or Group Work

You often need to use order relations and the order of operations when solving problems.
Work Exercises 71–80 in order.

71. When comparing the size of two numbers, the symbol $\leq$ means **is less than** and the symbol $\geq$ means **is greater than**.

72. (a) To identify the greater of two or more fractions, we must first write the fractions as **like** fractions and then compare the **numerators**. The fraction with the greater **numerator** is the greater fraction.

(b) Write four pairs of fractions, all with different denominators. Write the symbol for **less than** or for **greater than** between each pair.

Answers will vary.

73. Fill in the blanks to complete the order of operations.

1. Do all operations inside **parentheses** or other grouping symbols.
2. Simplify any expressions with **exponents** and find any **square** roots.
3. **Multiply** or **divide** proceeding from left to right.
4. **Add** or **subtract** proceeding from left to right.

74. Use the order of operations to simplify the following.

$$\left(\frac{2}{3}\right)^2 - \left(\frac{4}{5} - \frac{3}{10}\right) \div \frac{5}{4}$$

$\dfrac{2}{45}$

Simplify, then place the results on the number line.

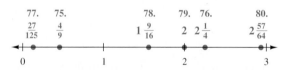

75. $\left(\dfrac{2}{3}\right)^2 \quad \dfrac{4}{9}$

76. $\left(\dfrac{3}{2}\right)^2 \quad 2\dfrac{1}{4}$

77. $\left(\dfrac{3}{5}\right)^3$

$\dfrac{27}{125}$; Use $\dfrac{25}{125} = \dfrac{1}{5}$ to help place the answer on the number line.

78. $\left(\dfrac{5}{4}\right)^2$

$1\dfrac{9}{16}$; Use $1\dfrac{8}{16} = 1\dfrac{1}{2}$ to help place the answer on the number line.

79. $4 + 2 - 2^2$

2

80. $\left(\dfrac{5}{8}\right)^2 + \left(\dfrac{7}{8} - \dfrac{1}{4}\right) \div \dfrac{1}{4}$

$2\dfrac{57}{64}$; Use $2\dfrac{56}{64} = 2\dfrac{7}{8}$ to help place the answer on the number line.

Summary Exercises on Fractions

Write proper *or* improper *for each fraction.*

1. $\dfrac{3}{4}$ proper

2. $\dfrac{4}{3}$ improper

3. $\dfrac{10}{10}$ improper

4. $\dfrac{11}{12}$ proper

Write each fraction in lowest terms.

5. $\dfrac{30}{36}$ $\dfrac{5}{6}$

6. $\dfrac{175}{200}$ $\dfrac{7}{8}$

7. $\dfrac{15}{35}$ $\dfrac{3}{7}$

8. $\dfrac{115}{235}$ $\dfrac{23}{47}$

Add, subtract, multiply, or divide as indicated. Simplify all answers.

9. $\dfrac{3}{4} \cdot \dfrac{2}{3}$ $\dfrac{1}{2}$

10. $\dfrac{7}{12} \cdot \dfrac{9}{14}$ $\dfrac{3}{8}$

11. $56 \cdot \dfrac{5}{8}$ 35

12. $\dfrac{5}{8} \div \dfrac{3}{4}$ $\dfrac{5}{6}$

13. $\dfrac{35}{45} \div \dfrac{10}{15}$ $1\dfrac{1}{6}$

14. $21 \div \dfrac{3}{8}$ 56

15. $\dfrac{7}{8} + \dfrac{2}{3}$ $1\dfrac{13}{24}$

16. $\dfrac{5}{8} + \dfrac{3}{4} + \dfrac{7}{16}$ $1\dfrac{13}{16}$

17. $\dfrac{7}{12} + \dfrac{5}{6} + \dfrac{2}{3}$ $2\dfrac{1}{12}$

18. $\dfrac{5}{6} - \dfrac{3}{4}$ $\dfrac{1}{12}$

19. $\dfrac{7}{8} - \dfrac{5}{12}$ $\dfrac{11}{24}$

20. $\dfrac{4}{5} - \dfrac{2}{3}$ $\dfrac{2}{15}$

First estimate the answer. Then add, subtract, multiply, or divide to find the exact answer.

21. *Exact:*

$3\dfrac{1}{2} \cdot 2\dfrac{1}{4}$ $7\dfrac{7}{8}$

Estimate:

$\underline{4} \cdot \underline{2} = \underline{8}$

22. *Exact:*

$5\dfrac{3}{8} \cdot 3\dfrac{1}{4}$ $17\dfrac{15}{32}$

Estimate:

$\underline{5} \cdot \underline{3} = \underline{15}$

23. *Exact:*

$8 \cdot 5\dfrac{2}{3} \cdot 2\dfrac{3}{8}$ $107\dfrac{2}{3}$

Estimate:

$\underline{8} \cdot \underline{6} \cdot \underline{2} = \underline{96}$

24. *Exact:*

$4\dfrac{3}{8} \div 3\dfrac{3}{4}$ $1\dfrac{1}{6}$

Estimate:

$\underline{4} \div \underline{4} = \underline{1}$

25. *Exact:*

$6\dfrac{7}{8} \div 2$ $3\dfrac{7}{16}$

Estimate:

$\underline{7} \div \underline{2} = \underline{3\dfrac{1}{2}}$

26. *Exact:*

$4\dfrac{5}{8} \div \dfrac{3}{4}$ $6\dfrac{1}{6}$

Estimate:

$\underline{5} \div \underline{1} = \underline{5}$

27. *Estimate:*   *Exact:*

$$6 \xleftarrow{\text{Rounds to}} \left\{ 5\frac{2}{3} \right.$$

$$+4 \xleftarrow{\text{Rounds to}} \left\{ +4\frac{1}{4} \right.$$

$$\overline{10} \qquad \overline{9\frac{11}{12}}$$

28. *Estimate:*   *Exact:*

$$18 \qquad 18\frac{5}{12}$$

$$+10 \qquad +9\frac{3}{4}$$

$$\overline{28} \qquad \overline{28\frac{1}{6}}$$

29. *Estimate:*   *Exact:*

$$15 \qquad 14\frac{3}{5}$$

$$+11 \qquad +10\frac{2}{3}$$

$$\overline{26} \qquad \overline{25\frac{4}{15}}$$

30. *Estimate:*   *Exact:*

$$9 \qquad 8\frac{3}{4}$$

$$-4 \qquad -3\frac{4}{5}$$

$$\overline{5} \qquad \overline{4\frac{19}{20}}$$

31. *Estimate:*   *Exact:*

$$14 \qquad 14$$

$$-7 \qquad -7\frac{3}{8}$$

$$\overline{7} \qquad \overline{6\frac{5}{8}}$$

32. *Estimate:*   *Exact:*

$$32 \qquad 31\frac{5}{6}$$

$$-23 \qquad -22\frac{7}{12}$$

$$\overline{9} \qquad \overline{9\frac{1}{4}}$$

Simplify by using the order of operations.

33. $\dfrac{1}{5}\left(\dfrac{2}{3} - \dfrac{1}{4}\right)$ **$\dfrac{1}{12}$**

34. $\dfrac{3}{4} \div \left(\dfrac{1}{2} + \dfrac{1}{3}\right)$ **$\dfrac{9}{10}$**

35. $\dfrac{2}{3} + \left(\dfrac{2}{3}\right)^2 - \dfrac{5}{6}$ **$\dfrac{5}{18}$**

Find the least common multiple of each set of numbers.

36. 8, 10   **40**

37. 9, 18, 24   **72**

38. 4, 12, 21   **84**

Rewrite each fraction with the indicated denominator.

39. $\dfrac{5}{6} = \dfrac{\mathbf{35}}{42}$

40. $\dfrac{3}{7} = \dfrac{\mathbf{12}}{28}$

41. $\dfrac{11}{12} = \dfrac{\mathbf{55}}{60}$

Write $<$ or $>$ to make a true statement.

42. $\dfrac{7}{8} \underline{\;\mathbf{<}\;} \dfrac{15}{16}$

43. $\dfrac{16}{20} \underline{\;\mathbf{>}\;} \dfrac{23}{30}$

44. $\dfrac{11}{15} \underline{\;\mathbf{>}\;} \dfrac{7}{10}$

Chapter 3 ▶▷▷ Summary

▶ Key Terms

3.1	**like fractions**	Fractions with the same denominator are called *like fractions.*
	unlike fractions	Fractions with different denominators are called *unlike fractions.*
3.2	**least common multiple**	Given two or more whole numbers, the least common multiple is the smallest whole number that is divisible by all the numbers.
	LCM	The abbreviation for *least common multiple* is LCM.
3.3	**least common denominator**	When unlike fractions are rewritten as like fractions having the least common multiple as the denominator, the new denominator is the least common denominator.
	LCD	The abbreviation for *least common denominator* is LCD.
3.4	**regrouping when adding fractions**	Regrouping is used in the addition of mixed numbers when the sum of the fractions is greater than 1.
	regrouping when subtracting fractions	Regrouping is used in the subtraction of mixed numbers when the fraction part of the minuend is less than the fraction part of the subtrahend.

▶ New Symbols

$<$ is less than ($2 < 5$) $>$ is greater than ($4 > 2$)

▶ Test Your Word Power

See how well you have learned the vocabulary in this chapter. Answers follow the Quick Review.

1. **Like fractions** are
   A. fractions that are equivalent
   B. fractions that are not equivalent
   C. fractions that have the same numerator
   D. fractions that have the same denominator.

2. Two or more fractions are **unlike fractions** if
   A. they are not equivalent
   B. they have different numerators
   C. they have different denominators
   D. they are improper fractions.

3. The abbreviation **LCM** stands for
   A. the largest common multiple
   B. the longest common multiple
   C. the most likely common multiplier
   D. the least common multiple.

4. The **least common multiple** is
   A. the smallest whole number that is divisible by each of two or more numbers
   B. the smallest numerator
   C. the smallest denominator
   D. the smallest whole number that is not divisible by a group of numbers.

5. The abbreviation **LCD** stands for
   A. the largest common denominator
   B. the least common denominator
   C. the least common divisor
   D. the most likely common denominator.

6. The **least common denominator** is
   A. needed when multiplying fractions
   B. needed when dividing fractions
   C. the least common multiple of the denominators in a fraction problem
   D. any denominator that is common to a group of fractions.

▶ Quick Review

Concepts	Examples

3.1 Adding Like Fractions

Add numerators and keep the same denominator. Write the result in lowest terms.

$$\frac{3}{4} + \frac{1}{4} + \frac{5}{4} = \frac{3+1+5}{4} = \frac{9}{4} = 2\frac{1}{4}$$

3.1 Subtracting Like Fractions

Subtract numerators and keep the same denominator. Write the result in lowest terms.

$$\frac{7}{8} - \frac{5}{8} = \frac{7-5}{8} = \frac{2}{8} = \frac{2 \div 2}{8 \div 2} = \frac{1}{4}$$

3.2 Finding the Least Common Multiple (LCM)

Method of using multiples of the larger number: List the first few multiples of the larger number. Check each one until you find the multiple that is divisible by the smaller number.

$$\frac{1}{3} + \frac{1}{4}$$

$4, 8, 12, 16, \ldots \longleftarrow$ Multiples of 4

First multiple divisible by 3 $(12 \div 3 = 4)$

The least common multiple (LCM) of 3 and 4 is 12.

3.2 Finding the Least Common Multiple (LCM)

Method of prime numbers: First find the prime factorization of each number. Then use the prime factors to build the least common multiple.

Factors of 9

$9 = 3 \cdot 3$

$15 = 3 \cdot 5$

$\text{LCM} = 3 \cdot 3 \cdot 5 = 45$

Factors of 15

The least common multiple (LCM) of 9 and 15 is 45.

3.3 Adding Unlike Fractions

Step 1 Find the least common multiple (LCM).
Step 2 Rewrite the fractions with the least common multiple as the denominator.
Step 3 Add the numerators, placing the sum over the common denominator, and simplify the answer.

$$\frac{1}{3} + \frac{1}{4} + \frac{1}{10} \qquad \text{LCM} = 60$$

$$\frac{1}{3} = \frac{20}{60} \quad \frac{1}{4} = \frac{15}{60} \quad \frac{1}{10} = \frac{6}{60}$$

$$\frac{20}{60} + \frac{15}{60} + \frac{6}{60} = \frac{41}{60} \quad \text{Lowest terms}$$

3.3 Subtracting Unlike Fractions

Step 1 Find the least common multiple (LCM).
Step 2 Rewrite the fractions with the least common multiple as the denominator.
Step 3 Subtract the numerators, place the difference over the common denominator, and simplify the answer.

$$\frac{5}{8} - \frac{1}{3} \qquad \text{LCM} = 24$$

$$\frac{5}{8} = \frac{15}{24} \quad \frac{1}{3} = \frac{8}{24}$$

$$\frac{15}{24} - \frac{8}{24} = \frac{7}{24} \quad \text{Lowest Terms}$$

Concepts	Examples

3.4 Adding Mixed Numbers

Round the numbers and estimate the answer. Then find the exact answer using these steps.

Step 1   Add the fractions using a common denominator.
Step 2   Add the whole numbers.
Step 3   Combine the sums of the whole numbers and the fractions, simplifying the fraction part when necessary.

Compare the exact answer to the estimate to see if it is reasonable.

Estimate: *Exact:*

$$10 \xleftarrow{\text{Rounds to}} \begin{cases} 9\frac{2}{3} = 9\frac{8}{12} \end{cases}$$

$$+\ 7 \xleftarrow{\text{Rounds to}} \begin{cases} +6\frac{3}{4} = 6\frac{9}{12} \end{cases}$$

$$17 \qquad\qquad 15\frac{17}{12} = 16\frac{5}{12}$$

The exact answer of $16\frac{5}{12}$ is reasonable because it is close to the estimate of 17.

3.4 Subtracting Mixed Numbers

Round the numbers and estimate the answer. Then find the exact answer using these steps.

Step 1   Subtract the fractions, regrouping if necessary.
Step 2   Subtract the whole numbers.
Step 3   Combine the differences of the whole numbers and the fractions, simplifying the fraction part when necessary.

Compare the exact answer to the estimate to see if it is reasonable.

Estimate: *Exact:*

$$9 \xleftarrow{\text{Rounds to}} \begin{cases} 8\frac{5}{8} = 8\frac{15}{24} = 7\frac{39}{24} \end{cases}$$

$$-\ 4 \xleftarrow{\text{Rounds to}} \begin{cases} -3\frac{11}{12} = 3\frac{22}{24} = 3\frac{22}{24} \end{cases}$$

$$5 \qquad\qquad 4\frac{17}{24}$$

The exact answer of $4\frac{17}{24}$ is reasonable because it is close to the estimate of 5.

3.4 Adding or Subtracting Mixed Numbers Using an Alternate Method

Step 1   Change the mixed numbers to improper fractions.
Step 2   Rewrite the unlike fractions as like fractions.
Step 3   Add or subtract the numerators and simplify the answer.

Add.

$$2\frac{2}{3} = \frac{8}{3} = \frac{64}{24}$$

$$+1\frac{3}{8} = \frac{11}{8} = +\frac{33}{24}$$

24 is the least common denominator.

$$\frac{97}{24} = 4\frac{1}{24} \quad \text{Answer as mixed number}$$

Improper fractions

Subtract.

$$8\frac{2}{3} = \frac{26}{3} = \frac{104}{12}$$

$$-5\frac{3}{4} = \frac{23}{4} = -\frac{69}{12}$$

12 is the least common denominator.

$$\frac{35}{12} = 2\frac{11}{12} \quad \text{Answer as mixed number}$$

Improper fractions

Concepts	Examples

(3.5) Identifying the Greater of Two Fractions

With unlike fractions, change to like fractions first. The fraction with the greater numerator is the greater fraction. Use these symbols:

$<$ is less than

$>$ is greater than

Identify the greater fraction.

$$\frac{7}{8}, \frac{9}{10}$$

$$\frac{7}{8} = \frac{7 \cdot 5}{8 \cdot 5} = \frac{35}{40}$$

$$\frac{9}{10} = \frac{9 \cdot 4}{10 \cdot 4} = \frac{36}{40}$$

$\frac{35}{40}$ is smaller than $\frac{36}{40}$, so $\frac{7}{8} < \frac{9}{10}$ or $\frac{9}{10} > \frac{7}{8}$. $\frac{9}{10}$ is greater.

(3.5) Using the Order of Operations with Fractions

Follow the order of operations.
1. Do all operations inside parentheses or other grouping symbols.
2. Simplify any expressions with exponents and find any square roots.
3. Multiply or divide, proceeding from left to right.
4. Add or subtract, proceeding from left to right.

Simplify by using the order of operations.

$$\frac{1}{2}\left(\frac{2}{3}\right) - \left(\frac{1}{4}\right)^2 \quad \text{Simplify fraction with exponent.}$$

$$= \frac{1}{\underset{1}{2}}\left(\frac{\overset{1}{2}}{3}\right) - \frac{1}{16} \quad \text{Next, multiply.}$$

$$= \frac{1}{3} - \frac{1}{16}$$

$$= \frac{16}{48} - \frac{3}{48} \quad \text{Change to common denominator and subtract.}$$

$$= \frac{13}{48}$$

ANSWERS TO TEST YOUR WORD POWER

1. D; *Example:* Because the fractions $\frac{3}{8}$ and $\frac{10}{8}$ both have 8 as a denominator, they are like fractions.

2. C; *Example:* The fractions $\frac{2}{3}$ and $\frac{3}{4}$ are unlike fractions because they have different denominators.

3. D; *Example:* LCM is the abbreviation for least common multiple.

4. A; *Example:* The least common multiple of 4 and 5 is 20 because 20 is the smallest number into which 4 and 5 will divide evenly.

5. B; *Example:* LCD is the abbreviation for least common denominator.

6. C; *Example:* The least common denominator of the fractions $\frac{2}{3}$ and $\frac{1}{2}$ is 6 because 6 is the least common multiple of 3 and 2.

When written using the least common denominator, $\frac{2}{3}$ and $\frac{1}{2}$ become $\frac{4}{6}$ and $\frac{3}{6}$, respectively.

Chapter 3 ▶▶▶ Review Exercises

[3.1] *Add or subtract. Write answers in lowest terms.*

1. $\dfrac{5}{7} + \dfrac{1}{7}$

   $\dfrac{6}{7}$

2. $\dfrac{4}{9} + \dfrac{3}{9}$

   $\dfrac{7}{9}$

3. $\dfrac{1}{8} + \dfrac{3}{8} + \dfrac{2}{8}$

   $\dfrac{3}{4}$

4. $\dfrac{5}{16} - \dfrac{3}{16}$

   $\dfrac{1}{8}$

5. $\dfrac{5}{10} + \dfrac{3}{10}$

   $\dfrac{4}{5}$

6. $\dfrac{5}{12} - \dfrac{3}{12}$

   $\dfrac{1}{6}$

7. $\dfrac{36}{62} - \dfrac{10}{62}$

   $\dfrac{13}{31}$

8. $\dfrac{68}{75} - \dfrac{43}{75}$

   $\dfrac{1}{3}$

Solve each application problem. Write answers in lowest terms.

9. Jaime Villagranna earns $\frac{7}{12}$ of his income installing kitchen cabinets for the Home Depot and $\frac{4}{12}$ of his income by operating his own cabinet business. What fraction of his total income comes from the two activities?

$\dfrac{11}{12}$ of his total income

10. The Koats for Kids committee members completed $\frac{5}{8}$ of their Web-page design in the morning and $\frac{3}{8}$ in the afternoon. How much less did they complete in the afternoon than in the morning?

$\dfrac{1}{4}$ Web page less

[3.2] *Find the least common multiple of each set of numbers.*

11. 5, 2

    10

12. 3, 4

    12

13. 10, 12, 20

    60

14. 3, 8, 4

    24

15. 6, 8, 5, 15

    120

16. 15, 9, 20

    180

Rewrite each fraction using the indicated denominator.

17. $\dfrac{2}{3} = \dfrac{8}{12}$

18. $\dfrac{3}{8} = \dfrac{21}{56}$

19. $\dfrac{2}{5} = \dfrac{10}{25}$

20. $\dfrac{5}{9} = \dfrac{45}{81}$

21. $\dfrac{4}{5} = \dfrac{32}{40}$

22. $\dfrac{5}{16} = \dfrac{20}{64}$

[3.1–3.3] *Add or subtract. Write answers in lowest terms.*

23. $\dfrac{1}{2} + \dfrac{1}{3}\quad \dfrac{5}{6}$

24. $\dfrac{1}{5} + \dfrac{3}{10} + \dfrac{3}{8}\quad \dfrac{7}{8}$

25. $\begin{array}{r} \dfrac{5}{12} \\ + \dfrac{5}{24} \\ \hline \dfrac{5}{8} \end{array}$

26. $\dfrac{2}{3} - \dfrac{1}{4}\quad \dfrac{5}{12}$

27. $\begin{array}{r} \dfrac{7}{8} \\ - \dfrac{1}{3} \\ \hline \dfrac{13}{24} \end{array}$

28. $\begin{array}{r} \dfrac{11}{12} \\ - \dfrac{4}{9} \\ \hline \dfrac{17}{36} \end{array}$

Solve each application problem.

29. The San Juan School District operates an after school program for students. This year $\frac{2}{5}$ of the students played after school sports, $\frac{1}{6}$ participated in arts and crafts, and $\frac{1}{3}$ spent their time in tutoring and study hall. What fraction of the total students participated in these activities?

30. Lynn Couch Catering serves food and beverages when and where you like it. She finds that $\frac{1}{3}$ of her business is for company events, $\frac{3}{8}$ for wedding parties, and $\frac{1}{4}$ for club and membership events. What portion of her business comes from these three categories?

$\dfrac{23}{24}$ **of her business**

$\dfrac{9}{10}$ **of the students**

[3.4] *First estimate the answer. Then add or subtract to find the exact answer. Write exact answers as mixed numbers.*

31. *Estimate:*   *Exact:*

$19 \xleftarrow{\text{Rounds to}} \begin{cases} 18\dfrac{5}{8} \\[2mm] \end{cases}$

$+14 \xleftarrow{\text{Rounds to}} \begin{cases} +13\dfrac{3}{4} \end{cases}$

$\overline{33} \qquad\qquad \overline{32\dfrac{3}{8}}$

32. *Estimate:*   *Exact:*

$23 \qquad\qquad 22\dfrac{2}{3}$

$+15 \qquad\qquad +15\dfrac{4}{9}$

$\overline{38} \qquad\qquad \overline{38\dfrac{1}{9}}$

33. *Estimate:*   *Exact:*

$13 \qquad\qquad 12\dfrac{3}{5}$

$9 \qquad\qquad 8\dfrac{5}{8}$

$+10 \qquad\qquad +10\dfrac{5}{16}$

$\overline{32} \qquad\qquad \overline{31\dfrac{43}{80}}$

34. *Estimate:*   *Exact:*

$32 \qquad\qquad 31\dfrac{3}{4}$

$-15 \qquad\qquad -14\dfrac{2}{3}$

$\overline{17} \qquad\qquad \overline{17\dfrac{1}{12}}$

35. *Estimate:*   *Exact:*

$34 \qquad\qquad 34$

$-16 \qquad\qquad -15\dfrac{2}{3}$

$\overline{18} \qquad\qquad \overline{18\dfrac{1}{3}}$

36. *Estimate:*   *Exact:*

$215 \qquad\qquad 215\dfrac{7}{16}$

$-136 \qquad\qquad -136$

$\overline{79} \qquad\qquad \overline{79\dfrac{7}{16}}$

Add or subtract by changing mixed numbers to improper fractions. Simplify all answers.

37. $5\dfrac{2}{5}$

$+3\dfrac{7}{10}$

$\overline{9\dfrac{1}{10}}$

38. $4\dfrac{3}{4}$

$+5\dfrac{2}{3}$

$\overline{10\dfrac{5}{12}}$

39. 5

$-1\dfrac{3}{4}$

$\overline{3\dfrac{1}{4}}$

40. $6\dfrac{1}{2}$

$-4\dfrac{5}{6}$

$\overline{1\dfrac{2}{3}}$

41. $8\dfrac{1}{3}$

$-2\dfrac{5}{6}$

$\overline{5\dfrac{1}{2}}$

42. $5\dfrac{5}{12}$

$-2\dfrac{5}{8}$

$\overline{2\dfrac{19}{24}}$

First estimate the answer and then find the exact answer for each application problem.

43. Two long-distance runners began an $18\frac{3}{4}$ mile run. They ran uphill $5\frac{5}{8}$ miles, downhill $7\frac{1}{3}$ miles, and the rest of the course was level. Find the distance of the level portion of the course.

Estimate: $19 - 6 - 7 = 6$ miles

Exact: $5\dfrac{19}{24}$ miles

44. The Boys and Girls Clubs of America collected $28\frac{2}{3}$ tons of newspapers on Saturday and $24\frac{3}{4}$ tons on Sunday. Find the total weight of the newspapers collected.

Estimate: $29 + 25 = 54$ tons

Exact: $53\dfrac{5}{12}$ tons

45. In a recent bass-fishing derby, Darrel Holmes caught three largemouth bass weighing $8\frac{7}{8}$ pounds, $9\frac{1}{2}$ pounds, and $6\frac{3}{4}$ pounds. Find their total weight.

Estimate: $9 + 10 + 7 = 26$ pounds

Exact: $25\dfrac{1}{8}$ pounds

46. A developer wants to build a shopping center. She bought two parcels of land, one, $1\frac{11}{16}$ acres, and the other, $2\frac{3}{4}$ acres. If she needs a total of $8\frac{1}{2}$ acres for the center, how much additional land does she need to buy?

Estimate: $9 - 2 - 3 = 4$ acres

Exact: $4\dfrac{1}{16}$ acres

[3.5] *Locate each fraction in Exercises 47–50 on the number line.*

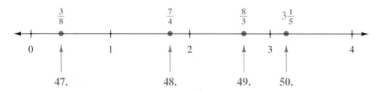

47. $\dfrac{3}{8}$
48. $\dfrac{7}{4}$
49. $\dfrac{8}{3}$
50. $3\dfrac{1}{5}$

Write < or > in each blank to make a true statement.

51. $\dfrac{2}{3}$ _<_ $\dfrac{3}{4}$
52. $\dfrac{3}{4}$ _<_ $\dfrac{7}{8}$
53. $\dfrac{1}{2}$ _>_ $\dfrac{7}{15}$
54. $\dfrac{7}{10}$ _>_ $\dfrac{8}{15}$

55. $\dfrac{9}{16}$ _<_ $\dfrac{5}{8}$
56. $\dfrac{7}{20}$ _>_ $\dfrac{8}{25}$
57. $\dfrac{19}{36}$ _<_ $\dfrac{29}{54}$
58. $\dfrac{19}{132}$ _>_ $\dfrac{7}{55}$

Simplify each expression.

59. $\left(\dfrac{1}{2}\right)^{2}$ $\dfrac{1}{4}$
60. $\left(\dfrac{2}{3}\right)^{2}$ $\dfrac{4}{9}$
61. $\left(\dfrac{3}{10}\right)^{3}$ $\dfrac{27}{1000}$
62. $\left(\dfrac{3}{8}\right)^{4}$ $\dfrac{81}{4096}$

Simplify by using the order of operations.

63. $8\left(\dfrac{1}{4}\right)^2$ $\dfrac{1}{2}$

64. $12\left(\dfrac{3}{4}\right)^2$ $6\dfrac{3}{4}$

65. $\left(\dfrac{2}{3}\right)^2 \cdot \left(\dfrac{3}{8}\right)^2$ $\dfrac{1}{16}$

66. $\dfrac{7}{8} \div \left(\dfrac{1}{8} + \dfrac{3}{4}\right)$ 1

67. $\left(\dfrac{1}{2}\right)^2 \cdot \left(\dfrac{1}{4} + \dfrac{1}{2}\right)$ $\dfrac{3}{16}$

68. $\left(\dfrac{1}{4}\right)^3 + \left(\dfrac{5}{8} + \dfrac{3}{4}\right)$ $1\dfrac{25}{64}$

▶▶▶ Mixed Review Exercises

Simplify. Use the order of operations as necessary.

69. $\dfrac{7}{8} - \dfrac{1}{8}$ $\dfrac{3}{4}$

70. $\dfrac{7}{10} - \dfrac{3}{10}$ $\dfrac{2}{5}$

71. $\dfrac{29}{32} - \dfrac{5}{16}$ $\dfrac{19}{32}$

72. $\dfrac{1}{4} + \dfrac{1}{8} + \dfrac{5}{16}$ $\dfrac{11}{16}$

73. $6\dfrac{2}{3}$
$\underline{\;-\;4\dfrac{1}{2}\;}$
   $2\dfrac{1}{6}$

74. $9\dfrac{1}{2}$
$\underline{\;+\,16\dfrac{3}{4}\;}$
   $26\dfrac{1}{4}$

75. 7
$\underline{\;-\;1\dfrac{5}{8}\;}$
   $5\dfrac{3}{8}$

76. $2\dfrac{3}{5}$
$8\dfrac{5}{8}$
$\underline{\;+\;\dfrac{5}{16}\;}$
   $11\dfrac{43}{80}$

77. $32\dfrac{5}{12}$
$\underline{\;-17\;}$
   $15\dfrac{5}{12}$

78. $\dfrac{7}{22} + \dfrac{3}{22} + \dfrac{3}{11}$ $\dfrac{8}{11}$

79. $\left(\dfrac{1}{4}\right)^2 \cdot \left(\dfrac{2}{5}\right)^3$ $\dfrac{1}{250}$

80. $\dfrac{3}{8} \div \left(\dfrac{1}{2} + \dfrac{1}{4}\right)$ $\dfrac{1}{2}$

81. $\left(\dfrac{2}{3}\right)^2 \cdot \left(\dfrac{1}{3} + \dfrac{1}{6}\right)$ $\dfrac{2}{9}$

82. $\left(\dfrac{2}{3}\right)^3 + \left(\dfrac{2}{3} - \dfrac{5}{9}\right)$ $\dfrac{11}{27}$

Write < or > in each blank to make a true statement.

83. $\dfrac{2}{3}$ __>__ $\dfrac{7}{12}$

84. $\dfrac{8}{9}$ __<__ $\dfrac{15}{8}$

85. $\dfrac{17}{30}$ __<__ $\dfrac{36}{60}$

86. $\dfrac{5}{8}$ __>__ $\dfrac{17}{30}$

Find the least common multiple of each set of numbers.

87. 12, 18 **36**

88. 6, 8, 10, 12 **120**

89. 9, 14, 21 **126**

Rewrite each fraction using the indicated denominator.

90. $\dfrac{2}{3} = \dfrac{18}{27}$

91. $\dfrac{9}{12} = \dfrac{108}{144}$

92. $\dfrac{4}{5} = \dfrac{60}{75}$

First estimate the answer and then find the exact answer for each application problem.

93. A cement contractor needs $13\frac{1}{2}$ ft of wire mesh for a concrete walkway and $22\frac{3}{8}$ ft of wire mesh for a driveway. If the contractor starts with a roll of wire that is $92\frac{3}{4}$ ft long, find the number of feet remaining after the two jobs have been completed.

Estimate: **93 − 14 − 22 = 57 ft**

Exact: **$56\dfrac{7}{8}$ ft**

94. A baker had four 50-pound bags of sugar. If she used $68\frac{1}{2}$ pounds of sugar to bake cakes, $76\frac{5}{8}$ pounds for baking pies, and $33\frac{1}{4}$ pounds for baking cookies, how many pounds of sugar remain?

Estimate: **4 · 50 = 200 pounds of sugar**
200 − 69 − 77 − 33 = 21 pounds

Exact: **$21\dfrac{5}{8}$ pounds**

Add or subtract. Write answers in lowest terms.

1. $\dfrac{5}{8} + \dfrac{1}{8}$

2. $\dfrac{1}{16} + \dfrac{7}{16}$

3. $\dfrac{7}{10} - \dfrac{3}{10}$

4. $\dfrac{7}{12} - \dfrac{5}{12}$

Find the least common multiple of each set of numbers.

5. 2, 3, 4

6. 6, 3, 5, 15

7. 6, 9, 27, 36

Add or subtract. Write answers in lowest terms.

8. $\dfrac{3}{8} + \dfrac{1}{4}$

9. $\dfrac{2}{9} + \dfrac{5}{12}$

10. $\dfrac{7}{8} - \dfrac{2}{3}$

11. $\dfrac{2}{5} - \dfrac{3}{8}$

First estimate the answer. Then add or subtract to find the exact answer; simplify exact answers.

12. $7\dfrac{2}{3} + 4\dfrac{5}{6}$

13. $16\dfrac{2}{5} - 11\dfrac{2}{3}$

14. $18\dfrac{3}{4} + 9\dfrac{2}{5} + 12\dfrac{1}{3}$

15. $24 - 18\dfrac{3}{8}$

---

1. $\dfrac{3}{4}$

2. $\dfrac{1}{2}$

3. $\dfrac{2}{5}$

4. $\dfrac{1}{6}$

5. 12

6. 30

7. 108

8. $\dfrac{5}{8}$

9. $\dfrac{23}{36}$

10. $\dfrac{5}{24}$

11. $\dfrac{1}{40}$

12. *Estimate:* $8 + 5 = 13$

   Exact: $12\dfrac{1}{2}$

13. *Estimate:* $16 - 12 = 4$

   Exact: $4\dfrac{11}{15}$

14. *Estimate:* $19 + 9 + 12 = 40$

   Exact: $40\dfrac{29}{60}$

15. *Estimate:* $24 - 18 = 6$

   Exact: $5\dfrac{5}{8}$

16. Answers will vary; see sample at the right.

✎ **16.** Most students say that "addition and subtraction of fractions is more difficult than multiplication and division of fractions." Why do you think they say this? Do you agree with these students?

Probably addition and subtraction of fractions is more difficult because you have to find the least common denominator and then change the fractions to the same denominator.

17. Answers will vary; see sample at the right.

✎ **17.** Devise and explain a method of estimating an answer to addition and subtraction problems involving mixed numbers. Might your estimated answer vary from the exact answer? If it did, what would the estimation accomplish?

Round mixed numbers to the nearest whole number. Then add or subtract to estimate the answer. The estimate may vary from the exact answer but it lets you know if your answer is reasonable.

First estimate the answer and then find the exact answer for each application problem.

18. *Estimate:* $10 + 85 + 37 + 8 = 140$ pounds

Exact: $140\dfrac{1}{24}$ pounds

18. In one week, a kennel owner used $10\frac{3}{8}$ pounds of puppy chow, $84\frac{1}{2}$ pounds of dry kibble, $36\frac{5}{6}$ pounds of high-protein mature dog mix, and $8\frac{1}{3}$ pounds of fresh ground meat products. Find the total number of pounds used.

19. *Estimate:* $148 - 69 - 37 - 6 = 36$ gallons

Exact: $35\dfrac{7}{8}$ gallons

19. A painting contractor arrived at a 6-unit apartment complex with $147\frac{1}{2}$ gallons of exterior paint. If his crew sprayed $68\frac{1}{2}$ gallons on the wood siding, rolled $37\frac{3}{8}$ gallons on the masonry exterior, and brushed $5\frac{3}{4}$ gallons on the trim, find the number of gallons of paint remaining.

Write $<$ or $>$ between each pair of fractions to make a true statement.

20. $>$

20. $\dfrac{3}{4}$ ____ $\dfrac{17}{24}$

21. $>$

21. $\dfrac{19}{24}$ ____ $\dfrac{17}{36}$

Simplify. Use the order of operations as needed.

22. $\dfrac{2}{13}$

22. $\left(\dfrac{1}{3}\right)^3 \cdot 54$

23. $\dfrac{48}{?}$

23. $\left(\dfrac{3}{4}\right)^2 - \left(\dfrac{7}{8} \cdot \dfrac{1}{3}\right)$

24. $1\dfrac{3}{4}$

24. $4\left(\dfrac{7}{8} - \dfrac{7}{16}\right)$

25. $1\dfrac{1}{3}$

25. $\dfrac{5}{6} + \dfrac{4}{3}\left(\dfrac{3}{8}\right)$

Study Skills

TIPS FOR TAKING MATH TESTS

OBJECTIVES

1 Apply suggestions to tests and quizzes.

2 Develop a set of "best practices" to apply while testing.

To Improve Your Test Score	Comments
Come prepared with a pencil, eraser, and calculator, if allowed. If you are easily distracted, sit in the corner farthest from the door.	**Working in pencil lets you erase,** keeping your work neat and readable.
Scan the entire test, note the point value of different problems, and plan your time accordingly. Allow at least five minutes to check your work at the end of the testing time.	If you have 50 minutes to do 20 problems, $50 \div 20 = 2.5$ minutes per problem. **Spend less time on easy ones,** more time on problems with higher point values.
Read directions carefully, and circle any significant words. When you finish a problem, read the directions again to make sure you did what was asked.	**Pay attention to announcements** written on the board or made by your instructor. Ask if you don't understand. You don't want to get problems wrong because you misread the directions!
Show your work. Most math teachers give partial credit if some of the steps in your work are correct, even if the final answer is wrong. **Write neatly.** If you like to scribble when first working or checking a problem, do it on scratch paper.	**If your teacher can't read your writing, you won't get credit for it.** If you need more space to work, ask if you can use extra pieces of paper that you hand in with your test paper.
Check that the **answer to an application problem is reasonable** and makes sense. Read the problem again to make sure you've answered the question.	**Use common sense.** Can the father really be seven years old? Would a month's rent be $32,140? Label your answer: $, years, inches, etc.
To check for careless errors, you need to **rework the problem again, without looking at your previous work.** Cover up your work with a piece of scratch paper, and pretend you are doing the problem for the first time. Then compare the two answers.	If you just "look over" your work, your mind can make the same mistake again without noticing it. Reworking the problem from the beginning **forces you to rethink it.** If possible, use a different method to solve the problem the second time.

Improving Your Test Score

To Reduce Anxiety	Comments
Do not try to review up until the last minute before the test. Instead, go for a walk, do some deep breathing, and arrive just in time for the test. Ignore other students.	Listening to anxious classmates before the test **may cause you to panic.** Moderate exercise and deep breathing will calm your mind.

Reducing Anxiety

Reducing Anxiety
(continued)

To Reduce Anxiety	Comments
Do a "knowledge dump" as soon as you get the test. Write important notes to yourself in a corner of the test paper: formulas, or common errors you want to watch out for.	Writing down tips and things that you've memorized ***lets you relax;*** you won't have to worry about forgetting those things and can refer to them as needed.
Do the easy problems first in order to build confidence. If you feel your anxiety starting to build, *immediately* stop for a minute, close your eyes, and take several slow, deep breaths.	Greater confidence helps you ***get the easier problems correct.*** Anxiety causes shallow breathing, which leads to confusion and reduced concentration. Deep breathing calms you.
As you work on more difficult problems, ***notice your "inner voice."*** You may have negative thoughts such as, "I can't do it," or "who cares about this test anyway." In your mind, yell, "STOP" and take several deep, slow breaths. Or, replace the negative thought with a positive one.	Here are ***examples of positive statements.*** Try writing one of them on the top of your test paper. • I know I can do it. • I can do this one step at a time. • I've studied hard, and I'll do the best I can.
If you still can't solve a difficult problem when you come back to it the second time, ***make a guess and do not change it.*** In this situation, your first guess is your best bet. Do not change an answer just because you're a little unsure. ***Change it only if you find an obvious mistake.***	If you are thinking about changing an answer, be sure you have a good reason for changing it. If you cannot find a specific error, leave your first answer alone. ***When the tests are returned, check to see if changing answers helped or hurt you.***
Read the harder problems twice. Write down *anything* that might help solve the problem: a formula, a picture, etc. If you still can't get it, circle the problem and ***come back to it later***. Do *not* erase any of the things you wrote down.	If you know even a *little* bit about the problem, write it down. The ***answer may come to you*** as you work on it, or you may get partial credit. Don't spend too long on any one problem. Your subconscious mind will work on the tough problem while you go on with the test.
Ignore students who finish early. Use the entire test time. *You do not get extra credit for finishing early.* Use the extra time to rework problems and correct careless errors.	Students who leave early are often the ones who didn't study or who are too anxious to continue working. If they bother you, ***sit as far from the door as possible.***

Why Are These Suggestions Brain Friendly?

Several suggestions address anxiety. **Reducing anxiety allows your brain to make the connections between dendrites;** in other words, you can think clearly.

Remember that **your brain continues to work on a difficult problem** even if you skip it and go on to the next one. Your subconscious mind will come through for you if you are open to the idea!

Some of the suggestions ask you to **use your common sense.** Follow the directions, show your work, write neatly, and pay attention to whether your answers really make sense.

Cumulative Review Exercises ▷▷▷ Chapters 1–3

1. In this number, name the digit that has the given place value. 5,639,428

millions **5**   ten-thousands **3**

thousands **9**   hundreds **2**

2. Round 59,803 to the nearest ten, then to the nearest hundred, and finally to the nearest thousand.

59,800; 59,800; 60,000

Use front end rounding to estimate each answer. Then find the exact answer.

3. *Estimate:*   *Exact:*

$$
\begin{array}{r}
20,000 \\
-\,10,000 \\
\hline
10,000
\end{array}
\qquad
\begin{array}{r}
24,276 \\
-\,9,887 \\
\hline
14,389
\end{array}
$$

4. *Estimate:*   *Exact:*

$$
\begin{array}{r}
2\,500 \\
40\overline{)100,000}
\end{array}
\qquad
\begin{array}{r}
3\,211 \\
35\overline{)112,385}
\end{array}
$$

Add, subtract, multiply, or divide as indicated.

5.
$$
\begin{array}{r}
375,899 \\
521,742 \\
+\,357,968 \\
\hline
1,255,609
\end{array}
$$

6.
$$
\begin{array}{r}
3,896,502 \\
-\,1,094,807 \\
\hline
2,801,695
\end{array}
$$

7. 5(8)(4) **160**

8.
$$
\begin{array}{r}
962 \\
\times\,384 \\
\hline
369,408
\end{array}
$$

9.
$$
\begin{array}{r}
135 \\
8\overline{)1080}
\end{array}
$$

10. 13,467 ÷ 5

2693 R2

Use front end rounding to estimate the answer to each application problem. Then find the exact answer.

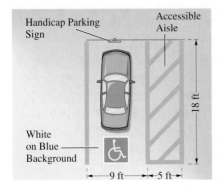

Handicap Parking Sign

Accessible Aisle

White on Blue Background

18 ft

9 ft   5 ft

11. The Americans with Disabilities Act provides the single parking space design shown at the left. Find the perimeter of (distance around) this parking space, including the accessible aisle.

Estimate: **20 + 9 + 5 + 20 + 9 + 5 = 68 ft**

Exact: **64 ft**

12. The single parking space design in Exercise 22 measures 18 ft by 14 ft. Find its area.

Estimate: **20 · 10 = 200 ft²**

Exact: **252 ft²**

Round the mixed numbers in each problem to the nearest whole number and estimate the answer. Then find the exact answer.

13. The top of a rectangular pool table is $1\frac{3}{4}$ yards by $2\frac{2}{3}$ yards. Find its area.

Estimate: **2 • 3 = 6 yd²**

Exact: $4\frac{2}{3}$ **yd²**

$2\frac{2}{3}$ yards

$1\frac{3}{4}$ yards

14. Fleas can jump 130 times higher than their own height. In human terms, how far could a $70\frac{1}{4}$ inch person jump? (*Source:* High Tech Productions Science and Technology Center.)

Estimate: **70 • 130 = 9100 in.**

Exact: **9132$\frac{1}{2}$ in.**

Simplify.

15. $2^4 \cdot 3^2$ **144**

16. $5^2 - 2\,(8)$ **9**

17. $\sqrt{25} + 5 \cdot 9 - 6$ **44**

18. $\frac{2}{3}\left(\frac{4}{5} - \frac{2}{3}\right)$ $\frac{4}{45}$

19. $\frac{3}{4} \div \left(\frac{1}{3} + \frac{1}{2}\right)$ $\frac{9}{10}$

20. $\frac{7}{8} + \left(\frac{3}{4}\right)^2 - \frac{3}{8}$ $1\frac{1}{16}$

21. $\frac{3}{4} \cdot \frac{2}{3}$ $\frac{1}{2}$

22. $42 \cdot \frac{7}{8}$ $36\frac{3}{4}$

23. $\frac{25}{40} \div \frac{10}{35}$ $2\frac{3}{16}$

24. $9 \div \frac{2}{3}$ $13\frac{1}{2}$

First estimate the answer and then add or subtract to find the exact answer. Write exact answers as mixed numbers.

25. *Estimate:*   *Exact:*

$3 \xleftarrow{\text{Rounds to}} \begin{cases} 3\frac{3}{8} \\ \end{cases}$

$+\ 5 \xleftarrow{\text{Rounds to}} \begin{cases} +\ 4\frac{1}{2} \end{cases}$

$\overline{8} \qquad \overline{7\frac{7}{8}}$

26. *Estimate:*   *Exact:*

$22 \qquad 21\frac{7}{8}$

$+\ 4 \qquad +\ 4\frac{5}{12}$

$\overline{26} \qquad \overline{26\frac{7}{24}}$

27. *Estimate:*   *Exact:*

$5 \qquad 5$

$-\ 2 \qquad -\ 2\frac{3}{8}$

$\overline{3} \qquad \overline{2\frac{5}{8}}$

Locate each fraction in Exercises 28–31 on the number line at the right.

28. $2\frac{3}{4}$

29. $\frac{1}{9}$

30. $\frac{5}{3}$

31. $\frac{10}{3}$

$\frac{1}{9} \qquad\qquad \frac{5}{3} \qquad\qquad 2\frac{3}{4} \qquad \frac{10}{3}$

0   1   2   3   4

29.   30.   28.   31.

Write < or > in each blank to make a true statement.

32. $\frac{3}{5}$ __<__ $\frac{5}{8}$

33. $\frac{17}{20}$ __>__ $\frac{3}{4}$

34. $\frac{7}{12}$ __<__ $\frac{11}{18}$

Decimals

Over 41 million Americans go fishing at least once a year, making it America's sixth most popular recreational activity. (*Source:* National Sporting Goods Association.) Record-size fish caught include a 67.5-pound muskie in Wisconsin, a 58-pound channel catfish in South Carolina, and a 97.25-pound chinook salmon in Alaska.

In **Section 4.3,** Exercises 47–50, this father and daughter will use decimal numbers when paying for new fishing equipment. But will decimals help them catch their limit? (See **Section 4.1,** Exercises 59–62, and **Section 4.6,** Exercises 65–68.)

4.1 ▶▶▶ Reading and Writing Decimals

OBJECTIVES

 Write parts of a whole using decimals.

2 Identify the place value of a digit.

3 Read and write decimals in words.

4 Write decimals as fractions or mixed numbers.

Fractions are used to represent parts of a whole. In this chapter, **decimals** are used as another way to show parts of a whole. For example, our money system is based on decimals. One dollar is divided into 100 equivalent parts. One cent ($0.01) is one of the parts, and a dime ($0.10) is 10 of the parts. Metric measurement (see **Chapter 7**) is also based on decimals.

OBJECTIVE 1 Write parts of a whole using decimals. Decimals are used when a whole is divided into 10 equivalent parts, or into 100 or 1000 or 10,000 equivalent parts. In other words, decimals are fractions with denominators that are a power of 10. For example, the square below is cut into 10 equivalent parts. Written as a fraction, each part is $\frac{1}{10}$ of the whole. Written as a decimal, each part is 0.1. Both $\frac{1}{10}$ and 0.1 are read as "*one tenth.*"

$\frac{1}{10}$ ◁————————▷ 0.1

One-tenth of the square is shaded.

The dot in 0.1 is called the **decimal point.**

$$0.1$$
Decimal point ⤴

1 There are 10 dimes in one dollar. Each dime is $\frac{1}{10}$ of a dollar. Write the yellow shaded portion of each dollar as a fraction, as a decimal, and in words.

(a)

(b)

(c)

The square at the right has 7 of its 10 parts shaded.

Written as a *fraction*, $\frac{7}{10}$ of the square is shaded.

Written as a *decimal*, 0.7 of the square is shaded.

Both $\frac{7}{10}$ and 0.7 are read as "*seven tenths.*"

◀ *Work Problem* **1** *at the Side.*

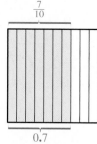

$\frac{7}{10}$

0.7

Seven-tenths of the square is shaded.

The square below is cut into 100 equivalent parts. Written as a *fraction*, each part is $\dfrac{1}{100}$ of the whole.

Written as a decimal, each part is **0.01** of the whole. Both $\frac{1}{100}$ and 0.01 are read as "one hundredth."

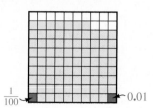

$\frac{1}{100}$ ————0.01

The square above has 87 parts shaded.

Written as a fraction, $\dfrac{87}{100}$ of the total area is shaded.

Written as a decimal, **0.87** of the total area is shaded.
Both $\frac{87}{100}$ and 0.87 are read as "*eighty-seven hundredths.*"

Work Problem 2 *at the Side.* ▶

Example 1 below shows several numbers written as fractions, as decimals, and in words.

| EXAMPLE 1 | **Using the Decimal Forms of Fractions** |

Fraction	Decimal	Read As
(a) $\dfrac{4}{10}$	0.4	four tenths
(b) $\dfrac{9}{100}$	0.09	nine hundredths
(c) $\dfrac{71}{100}$	0.71	seventy-one hundredths
(d) $\dfrac{8}{1000}$	0.008	eight thousandths
(e) $\dfrac{45}{1000}$	0.045	forty-five thousandths
(f) $\dfrac{832}{1000}$	0.832	eight hundred thirty-two thousandths

Work Problem 3 *at the Side.* ▶

OBJECTIVE 2 Identify the place value of a digit. The decimal point separates the *whole number part* from the *fractional part* in a decimal number. In the chart below, you see that the **place value** names for fractional parts are similar to those on the whole number side, but end in "*ths*."

Decimal Place Value Chart

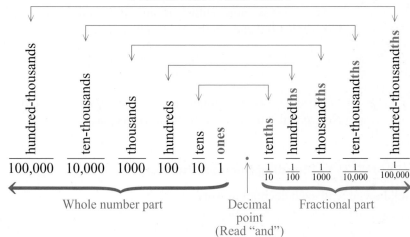

hundred-thousands	ten-thousands	thousands	hundreds	tens	ones	tenths	hundredths	thousandths	ten-thousandths	hundred-thousandths
100,000	10,000	1000	100	10	1	$\frac{1}{10}$	$\frac{1}{100}$	$\frac{1}{1000}$	$\frac{1}{10,000}$	$\frac{1}{100,000}$

◄──── Whole number part ──── Decimal point (Read "and") Fractional part ────►

Note

Notice that the **ones** place is at the center of the place value chart. There is no "oneths" place.

Also notice that each place is 10 times the value of the place to its right.

Finally, be sure to write a hyphen (dash) in ten-thousand**ths** and hundred-thousand**ths**.

2 Write the portion of each square that is shaded as a fraction, as a decimal, and in words.

(a)

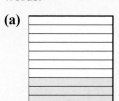

(b)

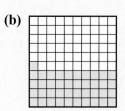

3 Write each decimal as a fraction.

(a) 0.7

(b) 0.2

(c) 0.03

(d) 0.69

(e) 0.047

(f) 0.351

ANSWERS

2. (a) $\dfrac{3}{10}$; 0.3; three tenths

 (b) $\dfrac{41}{100}$; 0.41; forty-one hundredths

3. (a) $\dfrac{7}{10}$ (b) $\dfrac{2}{10}$ (c) $\dfrac{3}{100}$

 (d) $\dfrac{69}{100}$ (e) $\dfrac{47}{1000}$ (f) $\dfrac{351}{1000}$

4 Identify the place value of each digit.

(a) 971.54

(b) 0.4

(c) 5.60

(d) 0.0835

5 Tell how to read each decimal in words.

(a) 0.6

(b) 0.46

(c) 0.05

(d) 0.409

(e) 0.0003

(f) 0.2703

(g) 0.088

ANSWERS

4. (a) 9 7 1 . 5 4   (b) 0 . 4
  (hundreds tens ones . tenths hundredths)   (ones . tenths)

  (c) 5 . 6 0   (d) 0 . 0 8 3 5
  (ones . tenths hundredths)   (ones . tenths hundredths thousandths ten-thousandths)

4. (a) six tenths
  (b) forty-six hundredths
  (c) five hundredths
  (d) four hundred nine thousandths
  (e) three ten-thousandths
  (f) two thousand seven hundred three ten-thousandths
  (g) eighty-eight thousandths

CAUTION
If a number does *not* have a decimal point, it is a *whole number*. A whole number has no fractional part. If you want to show the decimal point in a whole number, it is just to the *right* of the digit in the ones place. Here are two examples.

$$8 = 8. \qquad 306 = 306.$$

↑ Decimal point   ↑ Decimal point

EXAMPLE 2 **Identifying the Place Value of a Digit**

Identify the place value of each digit.

(a) 178.36   **(b)** 0.00935

hundreds | tens | ones . tenths | hundredths
1 7 8 . 3 6

ones . tenths | hundredths | thousandths | ten-thousandths | hundred-thousandths
0 . 0 0 9 3 5

Notice in Example 2(b) that we do *not* use commas on the right side of the decimal point.

◀ *Work Problem* **4** *at the Side.*

OBJECTIVE 3 **Read and write decimals in words.** A decimal is read according to its form as a fraction.

ones . tenths
0 . 9

We read 0.9 as "nine tenths" because 0.9 is the same as $\frac{9}{10}$. Notice that 0.9 ends in the tenths place.

ones . tenths hundredths
0 . 0 2

We read 0.02 as "two hundredths" because 0.02 is the same as $\frac{2}{100}$. Notice that 0.02 ends in the hundredths place.

EXAMPLE 3 **Reading Decimal Numbers**

Tell how to read each decimal in words.

(a) 0.3

Because $0.3 = \frac{3}{10}$, read the decimal as three <u>tenths</u>.

(b) 0.49   Read it as: forty-nine <u>hundredths</u>.

(c) 0.08   Read it as: eight <u>hundredths</u>.   *Think:* $0.08 = \frac{8}{100}$ so write *hundredths*.

(d) 0.918   Read it as: nine hundred eighteen <u>thousandths</u>.

(e) 0.0106   Read it as: one hundred six <u>ten-thousandths</u>.   *Think:* $0.0106 = \frac{106}{10,000}$

◀ *Work Problem* **5** *at the Side.*

Reading Decimal Numbers

Step 1 Read any whole number part to the *left* of the decimal point as you normally would.

Step 2 Read the decimal point as "*and.*"

Step 3 Read the part of the number to the *right* of the decimal point as if it were an ordinary whole number.

Step 4 Finish with the place value name of the rightmost digit; these names all end in "*ths.*"

Note

If there is *no whole number part,* you will use only Steps 3 and 4.

EXAMPLE 4 **Reading Decimals**

Read each decimal.

(a)

9 is in tenths place.

.9

sixteen **and** nine **tenths**

Remember to say or write "and" *only* when you see a decimal point.

16**.**9 is read "sixteen and nine tenths."

(b)

5 is in hundredths place.

482**.**35

four hundred eighty-two **and** thirty-five **hundredths**

482**.**35 is read "four hundred eighty-two and thirty-five hundredths."

3 is in thousandths place.

(c) 0.063 is "sixty-three **thousandths**." (No whole number part.)

(d) 11**.**1085 is "eleven **and** one thousand eighty-five **ten-thousandths**."

CAUTION

Use "and" *only* when reading a decimal point. A common mistake is to read the whole number 405 as "four hundred *and* five." But there is *no decimal point* shown in 405, so it is read "four hundred five."

Work Problem **6** *at the Side.* ▶

OBJECTIVE **4** **Write decimals as fractions or mixed numbers.**
Knowing how to read decimals will help you when writing decimals as fractions.

Writing Decimals as Fractions or Mixed Numbers

Step 1 The digits to the right of the decimal point are the numerator of the fraction.

Step 2 The denominator is 10 for tenths, 100 for hundredths, 1000 for thousandths, 10,000 for ten-thousandths, and so on.

Step 3 If the decimal has a whole number part, it will be written as a mixed number with the same whole number part.

6 Tell how to read each decimal in words.

(a) 3**.**8

(b) 15**.**001

(c) 0.0073

(d) 64**.**309

ANSWERS

6. **(a)** three and eight tenths
   (b) fifteen and one thousandth
   (c) seventy-three ten-thousandths
   (d) sixty-four and three hundred nine thousandths

7 Write each decimal as a fraction or mixed number.

(a) 0.7

(b) 12.21

(c) 0.101

(d) 0.007

(e) 1.3717

8 Write each decimal as a fraction or mixed number in lowest terms.

(a) 0.5

(b) 12.6

(c) 0.85

(d) 3.05

(e) 0.225

(f) 420.0802

EXAMPLE 5 Writing Decimals as Fractions or Mixed Numbers

Write each decimal as a fraction or mixed number.

(a) 0.19

The digits to the right of the decimal point, 19, are the numerator of the fraction. The denominator is 100 for hundredths because the rightmost digit is in the hundredths place.

$$0.19 = \frac{19}{100} \leftarrow 100 \text{ for hundredths}$$

Hundredths place

(b) 0.863

$$0.863 = \frac{863}{1000} \leftarrow 1000 \text{ for thousandths}$$

Thousandths place

(c) 4.0099

The whole number part stays the same.

$$4.0099 = 4\frac{99}{10,000} \leftarrow 10,000 \text{ for ten-thousandths}$$

Ten-thousandths place

◄ Work Problem **7** at the Side.

CAUTION

After you write a decimal as a fraction or a mixed number, make sure the fraction is in lowest terms.

EXAMPLE 6 Writing Decimals as Fractions or Mixed Numbers in Lowest Terms

Write each decimal as a fraction or mixed number in lowest terms.

(a) $0.4 = \frac{4}{10} \leftarrow 10 \text{ for tenths}$ Write $\frac{4}{10}$ in lowest terms.

$$\frac{4}{10} = \frac{4 \div 2}{10 \div 2} = \frac{2}{5} \leftarrow \text{Lowest terms}$$

(b) $0.75 = \frac{75}{100} = \frac{75 \div 25}{100 \div 25} = \frac{3}{4} \leftarrow \text{Lowest terms}$

The whole number part stays the same.

(c) $18.105 = 18\frac{105}{1000} = 18\frac{105 \div 5}{1000 \div 5} = 18\frac{21}{200} \leftarrow \text{Lowest terms}$

(d) $42.8085 = 42\frac{8085}{10,000} = 42\frac{8085 \div 5}{10,000 \div 5} = 42\frac{1617}{2000} \leftarrow \text{Lowest terms}$

◄ Work Problem **8** at the Side.

Calculator Tip   In this book we will write a 0 in the ones place for decimal fractions. We write **0**.45 instead of just .45, to emphasize that there is no whole number. Many calculators show these zeros also. Try entering ⊙ ④ ⑤; the display probably shows 0.45 even though you did not press 0. For comparison, enter the whole number 45 by pressing ④ ⑤ ⊕ and notice where the decimal point is shown in the display. (It automatically appears to the *right* of the 5.)

ANSWERS

7. (a) $\frac{7}{10}$ (b) $12\frac{21}{100}$ (c) $\frac{101}{1000}$
   (d) $\frac{7}{1000}$ (e) $1\frac{3717}{10,000}$

8. (a) $\frac{1}{2}$ (b) $12\frac{3}{5}$ (c) $\frac{17}{20}$ (d) $3\frac{1}{20}$
   (e) $\frac{9}{40}$ (f) $420\frac{401}{5000}$

4.1 ▶▶▶ Exercises

Identify the digit that has the given place value. See Example 2.

1. 70.489
tens 7
ones 0
tenths 4

2. 135.296
ones 5
tenths 2
tens 3

3. 0.2518
hundredths 5
thousandths 1
ten-thousandths 8

4. 0.9347
hundredths 3
thousandths 4
ten-thousandths 7

5. 93.01472
thousandths 4
ten-thousandths 7
tenths 0

6. 0.51968
tenths 5
ten-thousandths 6
hundredths 1

7. 314.658
tens 1
tenths 6
hundreds 3

8. 51.325
tens 5
tenths 3
hundredths 2

9. 149.0832
hundreds 1
hundredths 8
ones 9

10. 3458.712
hundreds 4
hundredths 1
tenths 7

⊙ 11. 6285.7125
thousands 6
thousandths 2
hundredths 1

12. 5417.6832
thousands 5
thousandths 3
ones 7

Write the decimal number that has the specified place values. See Example 2.

13. 0 ones, 5 hundredths, 1 ten, 4 hundreds, 2 tenths
410.25

14. 7 tens, 9 tenths, 3 ones, 6 hundredths, 8 hundreds
873.96

15. 3 thousandths, 4 hundredths, 6 ones, 2 ten-thousandths, 5 tenths
6.5432

16. 8 ten-thousandths, 4 hundredths, 0 ones, 2 tenths, 6 thousandths
0.2468

17. 4 hundredths, 4 hundreds, 0 tens, 0 tenths, 5 thousandths, 5 thousands, 6 ones
5406.045

18. 7 tens, 7 tenths, 6 thousands, 6 thousandths, 3 hundreds, 3 hundredths, 2 ones
6372.736

Write each decimal as a fraction or mixed number in lowest terms. See Examples 5 and 6.

19. 0.7

$\frac{7}{10}$

20. 0.1

$\frac{1}{10}$

 21. 13.4

$13\frac{2}{5}$

22. 9.8

$9\frac{4}{5}$

23. 0.25

$\frac{1}{4}$

24. 0.55

$\frac{11}{20}$

25. 0.66

$\frac{33}{50}$

26. 0.33

$\frac{33}{100}$

27. 10.17

$10\frac{17}{100}$

28. 31.99

$31\frac{99}{100}$

29. 0.06

$\frac{3}{50}$

30. 0.08

$\frac{2}{25}$

31. 0.205

$\frac{41}{200}$

32. 0.805

$\frac{161}{200}$

33. 5.002

$5\frac{1}{500}$

34. 4.008

$4\frac{1}{125}$

35. 0.686

$\frac{343}{500}$

36. 0.492

$\frac{123}{250}$

Tell how to read each decimal in words. See Examples 3 and 4.

37. 0.5

five tenths

38. 0.2

two tenths

39. 0.78

seventy-eight hundredths

40. 0.55

fifty-five hundredths

41. 0.105

one hundred five thousandths

42. 0.609

six hundred nine thousandths

43. 12.04

twelve and four hundredths

44. 86.09

eighty-six and nine hundredths

45. 1.075

one and seventy-five thousandths

46. 4.025

four and twenty-five thousandths

Write each decimal in numbers. See Examples 3 and 4.

47. six and seven tenths

6.7

48. eight and twelve hundredths

8.12

49. thirty-two hundredths

0.32

50. one hundred eleven thousandths

0.111

51. four hundred twenty and eight thousandths

420.008

52. two hundred and twenty-four thousandths

200.024

53. seven hundred three ten-thousandths

0.0703

54. eight hundred and six hundredths

800.06

55. seventy-five and thirty thousandths

75.030

56. sixty and fifty hundredths

60.50

57. Anne read the number 4302 as "four thousand three hundred and two." Explain what is wrong with the way Anne read the number.

Anne should not say "and" because that denotes a decimal point.

58. Jerry read the number 9.0106 as "nine and one hundred and six ten-thousandths." Explain the error he made.

Jerry used "and" twice; only the first "and" is correct.

The father on the first page of this chapter needs to select the correct fishing line for his daughter's reel. Fishing line is sold according to how many pounds of "pull" the line can withstand before breaking. Use the table to answer Exercises 59–62. Write all fractions in lowest terms. (Note: The diameter of the fishing line is its thickness.)

RELATING FISHING LINE DIAMETER TO TEST STRENGTH

Test Strength (pounds)	Average Diameter (inches)
4	0.008
8	0.010
12	0.013
14	0.014
17	0.015
20	0.016

Source: Berkley Outdoor Technologies Group.

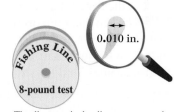

The diameter is the distance across the end of the line (or its thickness).

59. Write the diameter of 8-pound test line in words and as a fraction.

ten thousandths inch; $\frac{10}{1000} = \frac{1}{100}$ inch

60. Write the diameter of 17-pound test line in words and as a fraction.

fifteen thousandths inch; $\frac{15}{1000} = \frac{3}{200}$ inch

61. What is the test strength of the line with a diameter of $\frac{13}{1000}$ inch?

12 pounds

62. What is the test strength of the line with a diameter of sixteen thousandths inch?

20 pounds

Suppose your job is to take phone orders for precision parts. Use the table, and in Exercises 63–66, write the correct part number that matches what you hear the customer say over the phone. In Exercises 67–68, write the words you would say to the customer.

Part Number	Size in Centimeters
3-A	0.06
3-B	0.26
3-C	0.6
3-D	0.86
4-A	1.006
4-B	1.026
4-C	1.06
4-D	1.6
4-E	1.602

63. "Please send the six-tenths centimeter bolt."

Part number ____3-C____.

64. "The part missing from our order was the one and six hundredths size."

Part number ____4-C____.

65. "The size we need is one and six thousandths centimeters."

Part number ____4-A____.

66. "Do you still stock the twenty-six hundredths centimeter bolt?"

Part number ____3-B____.

67. "What size is part number 4-E?" Write your answer in words.

One and six hundred two thousandths centimeters

68. "What size is part number 4-B?" Write your answer in words.

One and twenty-six thousandths centimeters

Relating Concepts (Exercises 69–76) For Individual or Group Work

*Use your knowledge of place value to **work Exercises 69–76 in order.***

69. Look back at the decimal place value chart on page 267. What do you think would be the names of the next four places to the *right* of hundred-thousandths? What information did you use to come up with these names?

millionths, ten-millionths, hundred-millionths, billionths; these match the words on the left side of the chart with "ths" attached.

70. A common mistake is to think that the first place to the right of the decimal point is "oneths" and the second place is "tenths." Why might someone make that mistake? How would you explain why there is no "oneths" place?

The first place to the left of the decimal point is ones, so the first place to the right could be one*ths*, like tens and ten*ths*. But anything that is 1 or more is to the *left* of the decimal point.

71. Use your answer to Exercise 69 to write 0.72436955 in words.

seventy-two million four hundred thirty-six thousand nine hundred fifty-five hundred-millionths

72. Use your answer to Exercise 69 to write 0.000678554 in words.

six hundred seventy-eight thousand five hundred fifty-four billionths

73. Write 8006.500001 in words.

eight thousand six and five hundred thousand one millionths

74. Write 20,060.000505 in words.

twenty thousand sixty and five hundred five millionths

75. Write this decimal in numbers.

three hundred two thousand forty ten-millionths

0.0302040

76. Write this decimal in numbers.

nine billion, eight hundred seventy-six million, five hundred forty-three thousand, two hundred ten and one hundred million two hundred thousand three hundred billionths

9,876,543,210.100200300

4.2 ▶▶▶ Rounding Decimals

Section 1.7 showed how to round whole numbers. For example, 89 rounded to the nearest ten is 90, and 8512 rounded to the nearest hundred is 8500.

OBJECTIVE 1 Learn the rules for rounding decimals. It is also important to be able to **round** decimals. For example, a store is selling 2 candy mints for $0.75 but you want only one mint. The price of each mint is $0.75 ÷ 2, which is $0.375, but you cannot pay part of a cent. Is $0.375 closer to $0.37 or to $0.38? Actually, it's exactly halfway between. When this happens in everyday situations, the rule is to round *up*. The store will charge you $0.38 for the mint.

OBJECTIVES

1 Learn the rules for rounding decimals.

2 Round decimals to any given place.

3 Round money amounts to the nearest cent or nearest dollar.

Rounding Decimals

Step 1   Find the place to which the rounding is being done. Draw a "cut off" line *after* that place to show that you are cutting off and dropping the rest of the digits.

Step 2   Look *only* at the *first* digit you are cutting off.

Step 3(a) If this digit is *4 or less,* the part of the number you are keeping *stays the same.*

Step 3(b) If this digit is *5 or more,* you must *round up* the part of the number you are keeping.

Step 4   You can use the ≈ symbol or the ≐ symbol to indicate that the rounded number is now an approximation (close, but not exact). Both symbols mean "is approximately equal to." In this book we will use the ≈ symbol.

CAUTION
Do *not* move the decimal point when rounding.

OBJECTIVE 2 Round decimals to any given place. The following examples show you how to round decimals.

EXAMPLE 1 Rounding a Decimal Number

Round 14.39652 to the nearest thousandth.

Step 1 Draw a "cut-off" line after the thousandths place.

$$1\ 4\ .\ 3\ 9\ 6\ |\ 5\ 2$$
Thousandths

You are cutting off the 5 and 2.
They will be dropped.

Step 2 Look *only* at the *first* digit you are cutting off. Ignore the other digits you are cutting off.

$$1\ 4\ .\ 3\ 9\ 6\ |\ 5\ 2$$

Look *only* at the 5.
Ignore the 2.

Continued on Next Page

1 Round to the nearest thousandth.

(a) 0.33492

(b) 8.00851

(c) 265.42038

(d) 10.70180

Step 3 If the first digit you are cutting off is *5 or more,* round up the part of the number you are keeping.

First digit cut is *5 or more,* so round up by adding 1 thousandth to the part you are keeping.

$$14.396 \text{ | } 5\,2$$
$$+0.001$$
$$\overline{14.397}$$

Think: Rounding to *thousandths* means *three* decimal places.

So, 14.39652 rounded to the nearest thousandth is 14.397
We can write 14.39652 ≈ 14.397

CAUTION
When rounding whole numbers in **Section 1.7,** you kept all the digits but changed some to zeros. With decimals, you cut off and *drop the extra digits.* In Example 1 above, 14.39652 rounds to 14.397, ***not*** 14.39700.

◀ *Work Problem* **1** *at the Side.*

In Example 1, the rounded number 14.397 had *three decimal places.* **Decimal places** are the number of digits to the *right* of the decimal point. The first decimal place is tenths, the second is hundredths, the third is thousandths, and so on.

EXAMPLE 2 **Rounding Decimals to Different Places**

Round to the place indicated.

(a) 5.3496 to the nearest tenth

Tenths is one decimal place.

Step 1 Draw a cut-off line after the tenths place.

$$5 . 3 \text{ | } 4\,9\,6$$

Tenths

You are cutting off the 4, 9, and 6. They will be dropped.

Step 2
$$5 . 3 \text{ | } 4\,\underline{9\,6}$$

Look *only* at the 4.

Ignore these digits.

Step 3
$$\underline{5 . 3} \text{ | } 4\,9\,6$$
$$5 . 3 \leftarrow \text{Stays the same}$$

First digit cut is *4 or less,* so the part you are keeping stays the same.

5.3496 rounded to the nearest tenth is 5.3 (*one* decimal place for *tenths*).
We can write 5.3496 ≈ 5.3
Notice: 5.3496 does ***not*** round to 5.3000, which would be ten-thousandths.

(b) 0.69738 to the nearest hundredth

Step 1
$$0 . 6\,9 \text{ | } 7\,3\,8$$

Hundredths

Draw a cut-off line after the hundredths place.

Step 2
$$0 . 6\,9 \text{ | } 7\,3\,8$$

Look *only* at the 7.

Continued on Next Page

Step 3   0 . 6 9 | 7 3 8   First digit cut is *5 or more,* so round up by adding 1 hundredth to the part you are keeping.

     1
  0 . 6 9 ← Keep this part.
+ 0 . 0 1 ← To round up, add 1 hundredth.
  0 . 7 0 ← 9 + 1 is 10; write 0 and regroup 1 to the tenths place.

0.69738 rounded to the nearest hundredth is 0.70. Hundredths is *two* decimal places so you *must* write the 0 in the hundredths place.
We can write 0.69738 ≈ 0.70

Think: Rounding to *hundredths* means *two* decimal places.

CAUTION
If a *rounded* number has a 0 in the rightmost place, you *must* keep the 0. As shown above, 0.69738 rounded to the nearest hundredth is 0.70. Do *not* write 0.7, which is rounded to tenths instead of hundredths.

(c) 0.01806 to the nearest thousandth

First digit cut is *4 or less,* so the part you are keeping stays the same.

0 . 0 1 8 | 0 6

0 . 0 1 8 ← Stays the same
0.01806 rounded to the nearest thousandth is 0.018 (*three* decimal places for *thousandths*). We can write 0.01806 ≈ 0.018

(d) 57.976 to the nearest tenth

First digit cut is *5 or more,* so round up by adding 1 tenth to the part you are keeping.

57.9 | 76
57.9
+ 0.1
58.0 ← 9 + 1 is 10; write the 0 and regroup the 1 to the ones place.

Be sure to write the 0 in the tenths place.

57.976 rounded to the nearest tenth is 58.0. We can write 57.976 ≈ 58.0
You *must* write the 0 in the tenths place to show that the number was rounded to the nearest tenth.

CAUTION
Check that your rounded answer shows *exactly* the number of decimal places asked for in the problem. Be sure your answer shows *one* decimal place if you rounded to *tenths,* *two* decimal places for *hundredths,* *three* decimal places for *thousandths,* and so on.

Work Problem ② *at the Side.* ▶

OBJECTIVE 3 Round money amounts to the nearest cent or nearest dollar.   In many everyday situations, such as shopping in a store, money amounts are rounded to the nearest cent. There are 100 cents in a dollar.

Each cent is $\frac{1}{100}$ of a dollar.

Another way to write $\frac{1}{100}$ is 0.01. So rounding to the *nearest cent* is the same as rounding to the *nearest hundredth of a dollar.*

② Round to the place indicated.
(a) 0.8988 to the nearest hundredth

(b) 5.8903 to the nearest hundredth

(c) 11.0299 to the nearest thousandth

(d) 0.545 to the nearest tenth

ANSWERS
2. (a) 0.90 **(b)** 5.89 **(c)** 11.030 **(d)** 0.5

3 Round each money amount to the nearest cent.

(a) $14.595

(b) $578.0663

(c) $0.849

(d) $0.0548

EXAMPLE 3 **Rounding to the Nearest Cent**

How much will you pay in each shopping situation? Round each money amount to the nearest cent.

(a) $2.4238 (Is it closer to $2.42 or to $2.43?)

First digit cut is *4 or less,* so the part you are keeping stays the same.

$2.42 | 38

$2.42 ⟵ You pay.

> Rounding to the *nearest cent* is rounding to *hundredths.*

You pay $2.42 because $2.4238 is closer to $2.42 than to $2.43.

(b) $0.695 (Is it closer to $0.69 or to $0.70?)

5 or more; round up

$0.69 | 5

$0.69
+ $0.01 ⟵ To round up, add 1 hundredth (1 cent).
———
$0.70 ⟵ You pay.

◀ Work Problem **3** at the Side.

Note

Some stores round *all* money amounts up to the next higher cent, even if the next digit is *4 or less.* In Example 3(a) above, some stores would round $2.4238 *up* to $2.43, even though it is closer to $2.42.

It is also common to round money amounts to the nearest dollar. For example, you can do that on your federal and state income tax returns to make the calculations easier.

EXAMPLE 4 **Rounding to the Nearest Dollar**

Round to the nearest dollar.

(a) $48.69 (Is it closer to $48 or to $49?)

First digit cut is *5 or more,* so round up by adding $1.

$48. | 69

$48
+ 1
———
$49

> Write $49 *not* $49.00

$48.69 is closer to $49 than to $48.
So $48.69 rounded to the nearest dollar is $49

CAUTION

$48.69 rounded to the nearest dollar is $49. Write the answer as $49 to show that the rounding is to the *nearest dollar.* Writing $49.00 would show rounding to the *nearest cent.*

Continued on Next Page

(b) $594.36 (Is it closer to $594 or $595?)

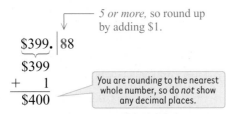

First digit cut is *4 or less,* so the part you keep stays the same.

$594. | 36

$594

> Careful!
> Write $594,
> *not* $594.00

$594.36 rounded to the nearest dollar is $594

(c) $399.88 (Is it closer to $399 or to $400?)

5 or more, so round up by adding $1.

$399. | 88

$399

+   1

$400

> You are rounding to the nearest whole number, so do *not* show any decimal places.

$399.88 rounded to the nearest dollar is $400

(d) $2689.50 (Is it closer to $2689 or $2690?)

5 or more, so round up by adding $1.

$2689. | 50

$2689

+   1

$2690

$2689.50 rounded to the nearest dollar is $2690

Note

When rounding $2689.50 to the nearest dollar, above, notice that it is exactly halfway between $2689 and $2690. When this happens in everyday situations, the rule is to round *up.* (Scientists working with technical data may use a more complicated rule when rounding numbers that are exactly in the middle.)

(e) $0.61 (Is it closer to $0 or to $1?)

5 or more, so round up.

$0. | 61

> Write the rounded amount as $1, *not* $1.00

$0.61 rounded to the nearest dollar is $1

⌨ **Calculator Tip**   Accountants and other people who work with money amounts often set their calculators to automatically round to two decimal places (nearest cent) or to round to zero decimal places (nearest dollar). Your calculator may have this feature.

Work Problem **4** *at the Side.* ▶

4 Round to the nearest dollar.

(a) $29.10

(b) $136.49

(c) $990.91

(d) $5949.88

(e) $49.60

(f) $0.55

(g) $1.08

Math in the Media

LAWN FERTILIZER

Gotta Be Green

A lot's being written about personal responsibility these days, and the idea seems to be ending up on the front lawn—literally! Each spring, homeowners across the country gear up to green up their lawns, and the increased use of fertilizer has a lot of environmentalists concerned about the potential effects of chemical runoff into nearby rivers and streams.

Every year, according to a study conducted by the University of Minnesota's Department of Agriculture, each household in the Minneapolis/St. Paul metro area uses an average of 36 pounds of lawn fertilizer. That adds up to 25,529,295 pounds, or 12,765 tons. Add to that another 193,000 pounds of weed killer and you're looking at the total picture for keeping it green in the Twin Cities.

Source: Minneapolis Star Tribune.

1. According to the article,

   (a) How many pounds of lawn fertilizer are used each year in the *entire metro area?*
   25,529,295 pounds

   (b) Do a division on your calculator to find the number of *households* in the metro area.
   709,147.0833 households

   (c) Why does it make sense to round your answer to part (b)? How would you round it?
   It makes sense because you can't have part of a household. Round to 709,147 (nearest whole number) or better choice, 709,000 (nearest thousand).

2. There are 2000 pounds in one ton.

   (a) Find the number of tons equivalent to 25,529,295 pounds of fertilizer.   **12,764.6475 tons**

   (b) Does your answer match the figure given in the article? If not, what did the author of the article do to get 12,765 tons?
   No; the author rounded to the nearest whole number.

   (c) Is the author's figure accurate? Why or why not?
   Answers vary; but it's probably accurate enough for newspaper reporting.

3. **(a)** When the average amount of lawn fertilizer per household was calculated, the answer was probably not *exactly* 36 pounds. List six different values that are *less than* 36 that would round to 36. List two values with one decimal place; two values with two decimal places; and two values with three decimal places.   **There are many correct answers. Some possibilities are 35.5 and 35.8; 35.52 and 35.65; 35.902 and 35.565.**

   (b) List six different values that are *greater than* 36 that would round to 36. List two values each with one, two, and three decimal places.   **There are many correct answers. Some possibilities are 36.1 and 36.4; 36.01 and 36.49; 36.002 and 36.465.**

4.2 ▶▶▶ Exercises

Round each number to the place indicated. See Examples 1 and 2.

1. 16.8974 to the nearest tenth

16.9

2. 193.845 to the nearest hundredth

193.85

3. 0.95647 to the nearest thousandth

0.956

4. 96.81584 to the nearest ten-thousandth

96.8158

5. 0.799 to the nearest hundredth

0.80

6. 0.952 to the nearest tenth

1.0

7. 3.66062 to the nearest thousandth

3.661

8. 1.5074 to the nearest hundredth

1.51

9. 793.988 to the nearest tenth

794.0

10. 476.1196 to the nearest thousandth

476.120

11. 0.09804 to the nearest ten-thousandth

0.0980

12. 176.004 to the nearest tenth

176.0

13. 48.512 to the nearest one

49

14. 3.385 to the nearest one

3

15. 9.0906 to the nearest hundredth

9.09

16. 30.1290 to the nearest thousandth

30.129

17. 82.000151 to the nearest ten-thousandth

82.0002

18. 0.400594 to the nearest ten-thousandth

0.4006

Nardos is grocery shopping. The store will round the amount she pays for each item to the nearest cent. Write the rounded amounts. See Example 3.

19. Soup is three cans for $2.45, so one can is $0.81666. Nardos pays **$0.82**.

20. Orange juice is two cartons for $3.89, so one carton is $1.945. Nardos pays **$1.95**.

21. Facial tissue is four boxes for $4.89, so one box is $1.2225. Nardos pays **$1.22**.

22. Muffin mix is three packages for $1.75, so one package is $0.58333. Nardos pays **$0.58**.

23. Candy bars are six for $2.99, so one bar is $0.4983. Nardos pays **$0.50**.

24. Spaghetti is four boxes for $4.39, so one box is $1.0975. Nardos pays **$1.10**.

As she gets ready to do her income tax return, Ms. Chen rounds each amount to the nearest dollar. Write the rounded amounts. See Example 4.

25. Income from job, $48,649.60

$48,650

26. Income from interest on bank account, $69.58

$70

27. Union dues, $310.08

$310

28. Federal withholding, $6064.49

$6064

29. Donations to charity, $848.91

$849

30. Medical expenses, $609.38

$609

Round each money amount as indicated.

31. $499.98 to the nearest dollar.

$500

32. $9899.59 to the nearest dollar

$9900

33. $0.996 to the nearest cent

$1.00

34. $0.09929 to the nearest cent.

$0.10

35. $999.73 to the nearest dollar.

$1000

36. $9999.80 to the nearest dollar.

$10,000

The table lists speed records for various types of transportation. Use the table to answer Exercises 37–40.

Record	Speed (miles per hour)
Land speed record (specially built car)	763.04
Motorcycle speed record (specially adapted motorcycle)	322.16
Fastest roller coaster	106.9
Fastest military jet	2193.167
Boeing 737-300 airplane (regular passenger service)	495
Indianapolis 500 auto race (fastest average winning speed)	185.981
Daytona 500 auto race (fastest average winning speed)	177.602

Sources: Guinness World Records and The World Almanac.

37. Round these speed records to the nearest whole number.

(a) Motorcycle

322 miles per hour

(b) Roller coaster

107 miles per hour

38. Round these speed records to the nearest hundredth.

(a) Daytona 500 average winning speed

177.60 miles per hour

(b) Indianapolis 500 average winning speed

185.98 miles per hour

39. Round these speed records to the nearest tenth.

(a) Indianapolis 500 average winning speed

186.0 miles per hour

(b) Land speed record

763.0 miles per hour

40. Round these speed records to the nearest hundred.

(a) military jet

2200 miles per hour

(b) Boeing 737-300 airplane

500 miles per hour

Relating Concepts (Exercises 41–44) For Individual or Group Work

Use your knowledge about rounding money amounts to **work Exercises 41–44 in order.**

41. Explain what happens when you round $0.499 to the nearest dollar. Why does this happen?

rounds to $0 (zero dollars) because $0.499 is closer to $0 than to $1

42. Look again at Exercise 41. How else could you round $0.499 that would be more helpful? What kind of guideline does this suggest about rounding to the nearest dollar?

Round amounts less than $1.00 to the nearest cent instead of the nearest dollar.

43. Explain what happens when you round $0.0015 to the nearest cent. Why does this happen?

rounds to $0.00 (zero cents) because $0.0015 is closer to $0.00 than to $0.01

44. Suppose you want to know which of these amounts is less, so you round them both to the nearest cent.

$0.5968 $0.6014

Explain what happens. Describe what you could do instead of rounding to the nearest cent.

Both round to $0.60. Rounding to the nearest thousandth (tenth of a cent) would allow you to identify $0.597 as less than $0.601.

4.3 ▶▶▶ Adding and Subtracting Decimals

OBJECTIVE 1 Add decimals. When adding or subtracting *whole* numbers **(Sections 1.2** and **1.3),** you lined up the numbers in columns so that you were adding ones to ones, tens to tens, and so on. A similar idea applies to adding or subtracting *decimal* numbers. With decimals, you line up the decimal points to make sure you are adding tenths to tenths, hundredths to hundredths, and so on.

Adding and Subtracting Decimals

Step 1 Write the numbers in columns with the decimal points lined up.

Step 2 If necessary, write in zeros so both numbers have the same number of decimal places. Then add or subtract as if they were whole numbers.

Step 3 Line up the decimal point in the answer directly below the decimal points in the problem.

EXAMPLE 1 Adding Decimal Numbers

Find each sum.

(a) 16.92 and 48.34

Step 1 Write the numbers in columns with the decimal points lined up.

$$
\begin{array}{r}
\text{tens}\ \text{ones}\ .\ \text{tenths}\ \text{hundredths} \\
1\ 6\ .\ 9\ 2 \\
+\ 4\ 8\ .\ 3\ 4 \\
\end{array}
$$

—— Decimal points are lined up.

Step 2 Add as if these were whole numbers.

$$
\begin{array}{r}
\overset{1\ 1}{16}.92 \\
+\ 48.34 \\
\hline
65.26
\end{array}
$$

Step 3

Decimal point in answer is lined up under decimal points in problem.

(b) 5.897 + 4.632 + 12.174
Write the numbers vertically with decimal points lined up. Then add.

$$
\begin{array}{r}
\overset{1\,1\ 2\,1}{5}.897 \\
4.632 \\
+\ 12.174 \\
\hline
22.703
\end{array}
$$

—— Decimal points are lined up.

Work Problem **1** *at the Side.* ▶

In Example 1(a) above, both numbers had *two* decimal places (two digits to the right of the decimal point). In Example 1(b), all the numbers had *three decimal places* (three digits to the right of the decimal point). That made it easy to add tenths to tenths, hundredths to hundredths, and so on.

1 Find each sum.

(a) 2.86 + 7.09

(b) 13.761 + 8.325

(c) 0.319 + 56.007 + 8.252

(d) 39.4 + 0.4 + 177.2

2 Find each sum.

(a) $6.54 + 9.8$

(b) $0.831 + 222.2 + 10$

(c) $8.64 + 39.115 + 3.0076$

(d) $5 + 429.823 + 0.76$

If the number of decimal places does *not* match, you can write in zeros as placeholders to make them match. This is shown in Example 2.

EXAMPLE 2 | **Writing Zeros as Placeholders before Adding**

Find each sum.

(a) $7.3 + 0.85$

There are two decimal places in 0.85 (tenths and hundredths), so write a 0 in the hundredths place in 7.3 so that it has two decimal places also.

$$\begin{array}{r} 7.30 \\ + \ 0.85 \\ \hline 8.15 \end{array}$$ ← One 0 is written in.

7.30 is equivalent to 7.3 because

$7\dfrac{30}{100}$ in lowest terms is $7\dfrac{3}{10}$

(b) $6.42 + 9 + 2.576$

Write in zeros so that all the addends have three decimal places. Notice how the whole number 9 is written with the decimal point at the *far right* side. (If you put the decimal point on the *left* side of the 9, you would turn it into the decimal fraction 0.9.)

Write the decimal point in 9 on the *right* side.

$$\begin{array}{r} 6.420 \\ 9.000 \\ + \ 2.576 \\ \hline 17.996 \end{array}$$

← One 0 is written in.
← 9 is a whole number; decimal point and three zeros are written in.
← No zeros are needed.

Decimal points are lined up.

> **Note**
>
> Writing zeros to the right of a *decimal* number does *not* change the value of the number, as shown in Example 2(a) above.

◀ Work Problem **2** at the Side.

OBJECTIVE **2** **Subtract decimals.** Subtraction of decimals is done in much the same way as addition of decimals. You can check the answers to subtraction problems using addition, as you did with whole numbers (see **Section 1.3**).

EXAMPLE 3 | **Subtracting Decimal Numbers**

Find each difference. Check your answers using addition.

(a) 15.82 from 28.93

Step 1

$$\begin{array}{r} 28.93 \\ - \ 15.82 \end{array}$$

Line up decimal points. Then you will be subtracting hundredths from hundredths and tenths from tenths.

Continued on Next Page

Step 2

$$\begin{array}{r} 28\;.\;93 \\ -\;15\;.\;82 \\ \hline 13\;.\;11 \end{array}$$

Both numbers have two decimal places; no need to write in zeros.

← Subtract as if they were whole numbers.

└── Decimal point in answer is lined up.

Step 3

Check the answer by adding 13.11 and 15.82. If the subtraction is done correctly, the sum will be 28.93.

(b) 146.35 minus 58.98
Regrouping is needed here.

$$\begin{array}{r} \overset{0\;\;13\;\;15}{\cancel{1}\;\cancel{4}\;\cancel{6}}\;.\;\overset{12\;\;15}{\cancel{3}\;\cancel{5}} \\ -\;\;5\;8\;.\;9\;8 \\ \hline 8\;7\;.\;3\;7 \end{array}$$

└── Line up decimal points.

Check the answer by adding 87.37 and 58.98. If you did the subtraction correctly, the sum will be 146.35. (If it *isn't*, you need to rework the problem.)

Work Problem **3** *at the Side.* ▶

| **EXAMPLE 4** | **Writing Zeros as Placeholders before Subtracting** |

Find each difference.

(a) 16.5 from 28.362

Use the same steps as in Example 3 above. Remember to write in zeros so both numbers have three decimal places.

┌── Line up decimal points.

16.500 is equivalent to 16.5

$$\begin{array}{r} 28.362 \\ -\;16.500 \\ \hline 11.862 \end{array}$$

← Write two zeros.

← Subtract as usual.

Check the answer by adding.

$$\begin{array}{r} 16.500 \\ +\;11.862 \\ \hline 28.362 \end{array}$$

← Matches minuend in original problem.

(b) 59.7 − 38.914

$$\begin{array}{r} 59.700 \\ -\;38.914 \\ \hline 20.786 \end{array}$$

← Write two zeros.

← Subtract as usual.

(c) 12 less 5.83

12.00 is equivalent to 12

$$\begin{array}{r} 12.00 \\ -\;5.83 \\ \hline 6.17 \end{array}$$

← Write a decimal point and two zeros.

← Subtract as usual.

Work Problem **4** *at the Side.* ▶

OBJECTIVE 3 Estimate the answer when adding or subtracting decimals. A common error in working decimal problems by hand is to misplace the decimal point in the answer. Or, when using a calculator, you may accidentally press the wrong key. **Estimating** the answer will help you avoid these mistakes. Start by using *front end rounding* on each number (as you did in **Section 1.7**). Here are several examples. Notice that in the rounded numbers, only the leftmost digit is something other than 0.

3.25	rounds to	3	6.812	rounds to	7
532.6	rounds to	500	26.397	rounds to	30
7094.2	rounds to	7000	351.24	rounds to	400

---

3 Find each difference. Check your answers using addition.

(a) 22.7 from 72.9

(b) 6.425 from 11.813

(c) 20.15 − 19.67

4 Find each difference. Check your answers using addition.

(a) 18.651 from 25.3

(b) 5.816 − 4.98

(c) 40 less 3.66

(d) 1 − 0.325

ANSWERS

3. **(a)** 50.2; 50.2 + 22.7 = 72.9
   (b) 5.388; 5.388 + 6.425 = 11.813
   (c) 0.48; 0.48 + 19.67 = 20.15
4. **(a)** 6.649; 6.649 + 18.651 = 25.3
   (b) 0.836; 0.836 + 4.98 = 5.816
   (c) 36.34; 36.34 + 3.66 = 40
   (d) 0.675; 0.675 + 0.325 = 1

5 First, use front end rounding and estimate each answer. Then add or subtract to find the exact answer.

(a) 2.83 + 5.009 + 76.1

Estimate:

Exact:

(b) 11.365 from 58

Estimate:

Exact:

(c) 398.81 + 47.658 + 4158.7

Estimate:

Exact:

(d) Find the difference between 12.837 meters and 46.091 meters.

Estimate:

Exact:

(e) $19.28 plus $1.53

Estimate:

Exact:

EXAMPLE 5   **Estimating Decimal Answers**

Use front end rounding to round each number. Then add or subtract the rounded numbers to get an estimated answer. Finally, find the exact answer.

(a) Find the sum of 194.2 and 6.825.

Estimate: Exact:

$$200 \xleftarrow{\text{Rounds to}} \quad 194.200$$
$$+ \quad 7 \xleftarrow{\text{Rounds to}} \quad + \quad 6.825$$
$$\overline{207} \qquad\qquad \overline{201.025}$$

The estimate goes out to the hundreds place (three places to the *left* of the decimal point), and so does the exact answer. Therefore, the decimal point is probably in the correct place in the exact answer.

(b) $69.42 + $13.78

Estimate: Exact:

$$\$70 \xleftarrow{\text{Rounds to}} \quad \$69.42$$
$$+ \quad 10 \xleftarrow{\text{Rounds to}} \quad + \quad 13.78$$
$$\overline{\$80} \qquad\qquad \overline{\$83.20} \leftarrow \text{Exact answer is close to estimate, so it is probably correct.}$$

(c) Find the difference between 0.92 ft and 8 ft.
   Use subtraction to find the difference between two numbers. The larger number, 8, is written on top.

Estimate: Exact:

$$8 \xleftarrow{\text{Rounds to}} \quad 8.00 \leftarrow \text{Write a decimal point and two zeros.}$$
$$- \quad 1 \xleftarrow{\text{Rounds to}} \quad - \quad 0.92$$
$$\overline{7} \qquad\qquad \overline{7.08} \text{ ft} \leftarrow \text{Exact answer is close to estimate.}$$

(d) Subtract 1.8614 from 7.3.

Estimate: Exact:

$$7 \xleftarrow{\text{Rounds to}} \quad 7.3000 \leftarrow \text{Write three zeros.}$$
$$- \quad 2 \xleftarrow{\text{Rounds to}} \quad - \quad 1.8614$$
$$\overline{5} \qquad\qquad \overline{5.4386} \leftarrow \text{Exact answer is close to estimate.}$$

◀ Work Problem **5** at the Side.

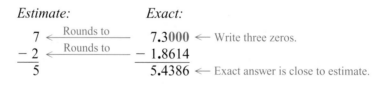

Calculator Tip   If you are *adding* numbers, you can enter them in any order on your calculator. Try these; jot down the answers.

9.82 ⊕ 1.86 ⊜   1.86 ⊕ 9.82 ⊜

The answers are the same because addition is *commutative*. (See **Section 1.2**.) But subtraction is *not* commutative. It *does* matter which number you enter first. Try these:

9.82 ⊖ 1.86 ⊜   1.86 ⊖ 9.82 ⊜

The second answer has a negative sign (−) next to it. A negative number is *less* than 0. If it was in your checkbook, you'd be "in the hole" by $7.96. (See **Section 9.1** for more about negative numbers.)

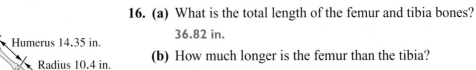

4.3 ▶▶▶ **Exercises**

FOR EXTRA HELP

Find each sum or difference. See Examples 1–4.

🌐 **1.** 5.69 + 11.79
   17.48

2. 372.1 − 33.7
   338.4

3. 24.008 − 0.995
   23.013

4. 0.7759 + 9.8883
   10.6642

5. 8.263 − 0.5
   7.763

6. 47.658 − 20.9
   26.758

7. 76.5 + 0.506
   77.006

8. 1.87 + 9.749
   11.619

9. 21 − 0.896
   20.104

10. 9 − 1.183
   7.817

🌐 **11.** Subtract 0.291 from 0.4
   0.109

12. Subtract 0.088 from 0.35
   0.262

13. 39.76005 + 182 + 4.799 + 98.31 + 5.9999
   330.86895

14. 489.76 + 0.9993 + 38 + 8.55087 + 80.697
   618.00717

This drawing of a human skeleton shows the average length of the longest bones, in inches.
Use the drawing to answer Exercises 15–18. (Source: Top Ten of Everything.)

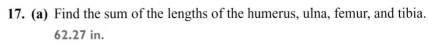

7th rib 9.45 in.

Humerus 14.35 in.

Radius 10.4 in.

8th rib 9.06 in.

Ulna 11.1 in.

Femur 19.88 in.

Tibia 16.94 in.   Fibula 15.94 in.

15. (a) What is the combined length of the humerus and radius bones?
   24.75 in.

   (b) What is the difference in the lengths of these two bones?
   3.95 in.

16. (a) What is the total length of the femur and tibia bones?
   36.82 in.

   (b) How much longer is the femur than the tibia?
   2.94 in.

17. (a) Find the sum of the lengths of the humerus, ulna, femur, and tibia.
   62.27 in.

   (b) How much shorter is the 8th rib than the 7th rib?
   0.39 in.

18. (a) What is the difference in the lengths of the two bones in the lower arm?
   0.7 in.

   (b) What is the difference in the lengths of the two bones in the lower leg?
   1.00 or 1 in.

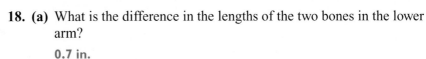

19. Explain and correct
the error that a student
made when he added
$0.72 + 6 + 39.5$ this way:

$$\begin{array}{r} 0.72 \\ 6 \\ +\ 39.50 \\ \hline 40.28 \end{array}$$

6 should be written 6.00; the sum is 46.22.

20. Explain the difference between saying "subtract 2.9 from 8" and saying "2.9 minus 8."

The two problems are done in a different order: 8 − 2.9 is not the same as 2.9 − 8 because subtraction is not commutative.

Use front end rounding to round each number. Then add or subtract the rounded numbers to get an estimated answer. Finally, find the exact answer. See Example 5.

21. *Estimate:*   *Exact:*

$$\begin{array}{r} \$20 \\ -\ 7 \\ \hline \$13 \end{array} \qquad \begin{array}{r} \$19.74 \\ -\ 6.58 \\ \hline \$13.16 \end{array}$$

22. *Estimate:*   *Exact:*

$$\begin{array}{r} \$30 \\ -\ 8 \\ \hline \$22 \end{array} \qquad \begin{array}{r} \$27.96 \\ -\ 8.39 \\ \hline \$19.57 \end{array}$$

23. *Estimate:*   *Exact:*

$$\begin{array}{r} 400 \\ 1 \\ +\ 20 \\ \hline 421 \end{array} \qquad \begin{array}{r} 392.7 \\ 0.865 \\ +\ 21.08 \\ \hline 414.645 \end{array}$$

24. *Estimate:*   *Exact:*

$$\begin{array}{r} 40 \\ 8 \\ +\ 1 \\ \hline 49 \end{array} \qquad \begin{array}{r} 38.55 \\ 7.716 \\ +\ 0.6 \\ \hline 46.866 \end{array}$$

25. What is 8.6 less 3.751?

Estimate: *Exact:*

$9 - 4 = 5$ 4.849

26. What is 31.7 less 4.271?

Estimate: *Exact:*

$30 - 4 = 26$ 27.429

27. *Estimate:*   *Exact:*

$$\begin{array}{r} 60 \\ 500 \\ +\ 6 \\ \hline 566 \end{array} \qquad \begin{array}{r} 62.8173 \\ 539.99 \\ +\ 5.629 \\ \hline 608.4363 \end{array}$$

28. *Estimate:*   *Exact:*

$$\begin{array}{r} 300 \\ 10 \\ +\ 800 \\ \hline 1110 \end{array} \qquad \begin{array}{r} 332.607 \\ 12.5 \\ +\ 823.3949 \\ \hline 1168.5019 \end{array}$$

Use your estimation skills to pick the most reasonable answer for each example.
*Do **not** solve the problems. Circle your choice.*

29. $12 - 11.725$

2.75   (0.275)   27.5

30. $20 - 1.37$

0.1863   1.863   (18.63)

31. $6.5 + 0.007$

(6.507)   0.6507   65.07

32. $9.67 + 0.09$

0.976   (9.76)   0.00976

33. $456.71 - 454.9$

18.1   181   (1.81)

34. $803.25 - 0.6$

(802.65)   0.80265   8.0265

35. $6004.003 + 52.7172$

60.567202   605.67202   (6056.7202)

36. $128.35 + 97.0093$

2253.593   (225.3593)   0.2253593

First use front end rounding to round each number and estimate the answer. Then find the exact answer. Use the information in the table below for Exercises 37–40.

INTERNET USERS IN SELECTED COUNTRIES

Country	Number of Users
United States	197.8 million
China	119.5 million
Japan	86.3 million
India	50.6 million
South Korea	34 million
Canada	21.9 million
Mexico	16.9 million
World total	**1081.1 million**

Source: Computer Industry Almanac.

39. How many Internet users are there in all the countries listed in the table?

Estimate: 200 + 100 + 90 + 50 + 30 + 20 + 20 = 510 million people

Exact: **527.0 million people**

41. The tallest known land mammal is a prehistoric ancestor of the rhino measuring 6.4 meters. Find the combined heights of these NBA basketball stars: Allen Iverson at 1.83 meters, Shaquille O'Neal at 2.16 meters, and Kevin Garnett at 2.11 meters. Is their combined height greater or less than the prehistoric rhino? By how much? (*Source:* www.NBA.com/players)

6.4 meters

Estimate: 2 + 2 + 2 = 6 meters

Exact: **6.1 meters, which is less than the rhino by 0.3 meter**

37. How many fewer Internet users are there in Canada compared to South Korea?

Estimate: 30 − 20 = 10 million people

Exact: **12.1 million people**

38. How many more users are there in China compared to India?

Estimate: 100 − 50 = 50 million people

Exact: **68.9 million people**

40. Using the answer from Exercise 39, calculate the number of worldwide Internet users in countries other than the ones in the table.

Estimate: 1000 − 500 = 500 million people

Exact: **554.1 million people**

42. At a bakery, Sue Chee bought $7.42 worth of muffins and $10.09 worth of croissants for a staff party and a $0.69 cookie for herself. How much change did she receive from two $10 bills?

Estimate: $20 − ($7 + $10 + $1) = $20 − $18 = $2

Exact: **$1.80**

43. Namiko is comparing two boxes of chicken nuggets. One box weighs 9.85 ounces and the other weighs 10.5 ounces. What is the difference in the weight of the two boxes?

Estimate: 11 − 10 = 1 ounce

Exact: **0.65 ounce**

44. Sammy works in a veterinarian's office. He weighed two young kittens. One was 3.9 ounces and the other was 4.05 ounces. What was the difference in the weight of the two kittens?

Estimate: 4 − 4 = 0

Exact: **0.15 ounce**

Find the perimeter of (distance around) each figure by adding the lengths of the sides.

45.

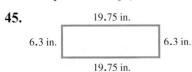

19.75 in.

6.3 in.   6.3 in.

19.75 in.

Estimate: 20 + 6 + 20 + 6 = 52 in.

Exact: **52.1 in.**

46.

2 meters   1 meter

0.9 meter

1.7 meters

1.18 meters

0.86 meter

2.095 meters

Estimate: 2 + 1 + 2 + 1 + 2 + 1 + 1 = 10 meters

Exact: **9.735 meters**

The father and daughter on the first page of this chapter are buying fishing equipment. They brought along the store's sale insert from the Sunday paper. Use the information below on sale prices to answer Exercises 47–50. When estimating, round to the nearest whole number.

☆ **Fishing Opener Sale** ☆
Catch your limit of savings!

Bobbers 3 for 87¢

8-pound test fishing line
regular $4.84
invisible $7.47
fluorescent $5.14
No-See Line

Environmentally safe
tin split shot
$2.07
Leaded split shot
94¢

Tackle boxes
Two trays $7.96
Three trays $9.96

Spinning reels: $9.88, $12.54, $18.84, $24.96
Spinning rods: $9.97, $18.97, $22.96, $28.94

Source: Wal-Mart.

47. What is the difference in price between the fluorescent and regular fishing line?

Estimate: **$5 − 5 = 0**

Exact: **$0.30**

48. How much more does the least expensive spinning rod cost than the least expensive spinning reel?

Estimate: **$10 − 10 = 0**

Exact: **$0.09**

49. Find the total cost of the second highest priced spinning reel, two packages of tin split shot, and a three-tray tackle box. Sales tax for all the items was $2.31.

Estimate: **$19 + 2 + 2 + 10 + 2 = $35**

Exact: **$35.25**

50. The father bought three bobbers on sale. He also bought some SPF15 sunscreen for $7.53 and a flotation vest for $44.96. Sales tax was $3.74. How much did he spend in all?

Estimate: **$1 + 8 + 45 + 4 = $58**

Exact: **$57.10**

Olivia Sanchez kept track of her expenses for one month. Use her list to answer Exercises 51–56.

MONTHLY EXPENSES	
Rent	$994
Car payment	$190.78
Car repairs, gas	$205
Cable TV	$39.95
Internet access	$19.95
Electricity	$40.80
Cell phone	$57.32
Groceries	$186.81
Entertainment	$97.75
Clothing, laundry	$107

51. What were Olivia's total expenses for the month?

$1939.36

52. How much did Olivia pay for cell phone, cable TV, and Internet access?

$117.22

53. What was the difference in the amounts spent for groceries and for the car payment?

$3.97

54. Compare the amount Olivia spent on entertainment to the amount spent on car repairs and gas. What is the difference?

$107.25

55. How much more did Olivia spend on rent than on all her car expenses?

$598.22

56. How much less did Olivia spend on clothing and laundry than on all her car expenses?

$288.78

Find the length of the dashed line in each rectangle or circle.

57.

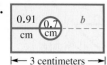

0.91 cm   0.7 cm   *b*

|← 3 centimeters →|

b = 1.39 centimeters

58.

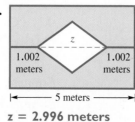

z

1.002 meters   1.002 meters

|← 5 meters →|

z = 2.996 meters

59.

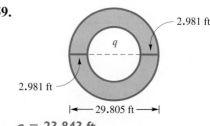

2.981 ft

q

2.981 ft

|← 29.805 ft →|

q = 23.843 ft

4.4 ▶▶▶ Multiplying Decimals

OBJECTIVE 1 Multiply decimals.   The decimals 0.3 and 0.07 can be multiplied by writing them as fractions.

$$0.\underset{\uparrow}{3} \quad \times \quad 0.0\underset{\uparrow}{7} = \frac{3}{10} \times \frac{7}{100} = \frac{3 \times 7}{10 \times 100} = \frac{21}{1000} = 0.0\underset{\uparrow\uparrow\uparrow}{21}$$

1 decimal place + 2 decimal places $\longrightarrow$ 3 decimal places

Can you see a way to multiply decimals without writing them as fractions? Try these steps. Remember that each number in a multiplication problem is called a *factor,* and the answer is called the *product.*

> **Multiplying Decimals**
>
> **Step 1** Multiply the numbers (the factors) as if they were whole numbers.
>
> **Step 2** Find the *total* number of decimal places in *both* factors.
>
> **Step 3** Write the decimal point in the product (the answer) so it has the same number of decimal places as the total from Step 2. You may need to write in extra zeros on the left side of the product to get the correct number of decimal places.

> **Note**
> When multiplying decimals, you do ***not*** need to line up decimal points. (You ***do*** need to line up decimal points when adding or subtracting.)

EXAMPLE 1 Multiplying Decimal Numbers

Find the product of 8.34 and 4.2

Step 1 Multiply the numbers as if they were whole numbers.

```
      8.3 4      You do not have to
   ×    4.2      line up decimal points
   ───────       when multiplying.
   1 6 6 8
 3 3 3 6
 ─────────
 3 5 0 2 8
```

Step 2 Count the total number of decimal places in both factors.

```
      8.3 4  ← 2 decimal places
   ×    4.2  ← 1 decimal place
   ─────────
   1 6 6 8       3 total decimal places
 3 3 3 6
 ─────────
 3 5 0 2 8
```

Step 3 Count over 3 places in the product and write the decimal point. Count from *right to left.*

```
      8.3 4  ← 2 decimal places
   ×    4.2  ← 1 decimal place
   ─────────
   1 6 6 8       3 total decimal places
 3 3 3 6
 ─────────
 3 5.0 2 8  ← 3 decimal places in product
```

Count over 3 places from right to left to position the decimal point.

Work Problem 1 at the Side. ▶

---

1 Find each product.

(a)   2.6
    × 0.4

(b)   45.2
    × 0.25

(c)   0.104 ← 3 decimal places
    ×   7 ← 0 decimal places
           ← 3 decimal places
             in the product

(d)   3.18
    × 2.23

(e)   611
    × 3.7

ANSWERS

1. **(a)** 1.04 **(b)** 11.300 **(c)** 0.728
   (d) 7.0914 **(e)** 2260.7

2 Find each product.

(a) 0.04×0.09

(b) $(0.2)(0.008)$

(c) $(0.003)^2$ *Hint:* Recall that the 2 is an exponent. See **Section 1.8**.

(d) $(0.0081)(0.003)$

(e) $(0.11)(0.0005)$

3 First use front end rounding and estimate the answer. Then find the exact answer.

(a) $(11.62)(4.01)$

(b) $(5.986)(33)$

(c) $8.31(4.2)$

(d) 58.6×17.4

EXAMPLE 2 **Writing Zeros as Placeholders in the Product**

Find the product: $(0.042)(0.03)$

Start by multiplying, then count decimal places.

$$
\begin{array}{r}
0.0\,4\,2 \leftarrow \text{3 decimal places} \\
\times \quad 0.0\,3 \leftarrow \text{2 decimal places} \\
\hline
1\,2\,6 \leftarrow \text{5 decimal places needed in product}
\end{array}
$$

After multiplying, the answer has only three decimal places, but five are needed, so write two zeros on the *left* side of the answer.

$$
\begin{array}{r}
0.0\,4\,2 \\
\times \quad 0.0\,3 \\
\hline
0\,0\,1\,2\,6 \\
\uparrow\uparrow
\end{array}
\qquad
\begin{array}{r}
0.0\,4\,2 \leftarrow \text{3 decimal places} \\
\times \quad 0.0\,3 \leftarrow \text{2 decimal places} \\
\hline
.0\,0\,1\,2\,6 \leftarrow \text{5 decimal places}
\end{array}
$$

Write two zeros on *left* side of answer.

Now count over 5 places and write in the decimal point.

The final product is 0.00126, which has five decimal places.

◀ *Work Problem* **2** *at the Side.*

OBJECTIVE **2** **Estimate the answer when multiplying decimals.** If you are doing multiplication problems by hand, estimating the answer helps you check that the decimal point is in the right place. When you are using a calculator, estimating helps you catch an error like pressing the ÷ key instead of the × key.

EXAMPLE 3 **Estimating before Multiplying**

First estimate the answer to $(76.34)(12.5)$ using front end rounding. Then find the exact answer.

Estimate:

$$
\begin{array}{r}
80 \xleftarrow{\text{Rounds to}} \\
\times \ 10 \xleftarrow{\text{Rounds to}} \\
\hline
800
\end{array}
$$

Exact:

$$
\begin{array}{r}
7\,6.3\,4 \leftarrow \text{2 decimal places} \\
\times \quad 1\,2.5 \leftarrow \text{1 decimal place} \\
\hline
3\,8\,1\,7\,0 \quad \text{3 decimal places needed in product} \\
1\,5\,2\,6\,8 \\
7\,6\,3\,4 \\
\hline
9\,5\,4.2\,5\,0
\end{array}
$$

Both the estimate and the exact answer go out to the hundreds place, so the decimal point in 954.250 is probably in the correct place.

◀ *Work Problem* **3** *at the Side.*

Calculator Tip When working with money amounts, you may need to write a 0 in your answer. For example, try multiplying 3.54×5 on your calculator. Write down the result.

$$3.54 \ \text{×} \ 5 \ \text{=}$$

Notice that the result is 17.7, which is *not* the way to write a money amount. You have to write the 0 in the hundredths place: $17.70 is correct. The calculator does not show the "extra" 0 because:

$$17.70 \text{ or } 17\frac{70}{100} \quad \text{simplifies to} \quad 17\frac{7}{10} \text{ or } 17.7$$

So keep an eye on your calculator—it doesn't know when you're working with money amounts.

4.4 ▶▶▶ Exercises

Find each product. See Example 1.

1. 0.042
× 3.2
0.1344

2. 0.571
× 2.9
1.6559

3. 21.5
× 7.4
159.10

4. 85.4
× 3.5
298.90

5. (0.666)(23.4)
15.5844

6. (0.799)(0.896)
0.715904

7. $51.88
× 665
$34,500.20

8. $736.75
× 118
$86,936.50

Use the fact that 72 × 6 = 432 to solve Exercises 9–16 by simply counting decimal places and writing the decimal point in the correct location. See Examples 1 and 2.

9. 72 × 0.6 = 4 3 2
43.2

10. 7.2 × 6 = 4 3 2
43.2

11. (7.2)(0.06) = 4 3 2
0.432

12. (0.72)(0.6) = 4 3 2
0.432

13. 0.72(0.06) = 4 3 2
0.0432

14. 72(0.0006) = 4 3 2
0.0432

15. 0.0072(0.6) = 4 3 2
0.00432

16. 0.072(0.006) = 4 3 2
0.000432

Find each product. See Example 2.

17. (0.006)(0.0052)
0.0000312

18. (0.0052)(0.009)
0.0000468

19. (0.005)^2
0.000025

20. (0.03)^2
0.0009

Relating Concepts (Exercises 21–22) For Individual or Group Work

Look for patterns in the multiplications as you **work Exercises 21 and 22 in order.**

21. Do these multiplications:

(5.96)(10) = _59.6_   (3.2)(10) = _32_

(0.476)(10) = _4.76_   (80.35)(10) = _803.5_

(722.6)(10) = _7226_   (0.9)(10) = _9_

What pattern do you see? Write a "rule" for multiplying by 10. What do you think the rule is for multiplying by 100? by 1000? Write the rules and try them out on the numbers above.

multiplying by 10, decimal point moves one place to the right; by 100, two places to the right; by 1000, three places to the right

22. Do these multiplications:

(59.6)(0.1) = _5.96_   (3.2)(0.1) = _0.32_

(0.476)(0.1) = _0.0476_   (80.35)(0.1) = _8.035_

(65)(0.1) = _6.5_   (523)(0.1) = _52.3_

What pattern do you see? Write a "rule" for multiplying by 0.1. What do you think the rule is for multiplying by 0.01? by 0.001? Write the rules and try them out on the numbers above.

multiplying by 0.1, decimal point moves one place to the left; by 0.01, two places to the left; by 0.001, three places to the left

First use front end rounding to round each number and estimate the answer. Then find the exact answer. See Example 3.

23. Estimate: Exact:

    40 ←Rounds to— 39.6
  × 5 ←Rounds to— × 4.8
   200   190.08

24. Estimate: Exact:

    20   18.7
  × 2   × 2.3
   40   43.01

25. Estimate: Exact:

    40   37.1
  × 40   × 42
  1600   1558.2

26. Estimate: Exact:

     5   5.08
  × 70   × 71
   350   360.68

27. Estimate: Exact:

     7   6.53
  × 5   × 4.6
   35   30.038

28. Estimate: Exact:

     8   7.51
  × 8   × 8.2
   64   61.582

29. Estimate: Exact:

     3   2.809
  × 7   × 6.85
   21   19.24165

30. Estimate: Exact:

    70   73.52
  × 20   × 22.34
  1400   1642.4368

Even with most of the problem missing, you can tell whether or not these answers are reasonable. Circle reasonable or unreasonable. If the answer is unreasonable, move the decimal point, or insert a decimal point, to make the answer reasonable.

31. How much was his car payment? $28.90

reasonable

(unreasonable,) should be **$289.00**

32. How many hours did she work today? 25 hours

reasonable

(unreasonable,) should be **2.5 hours**

33. How tall is her son? 60.5 in.

(reasonable)

unreasonable, should be _____

34. How much does he pay for rent now? $6.92

reasonable

(unreasonable,) should be **$692.00**

35. What is the price of one gallon of milk? $419

reasonable

(unreasonable,) should be **$4.19**

36. How long is the living room? 16.8 feet

(reasonable)

unreasonable, should be _____

37. How much did Mrs. Brown's baby weigh? 0.095 pound

reasonable

(unreasonable,) should be **9.5 pounds**

38. What was the sale price of the jacket? $1.49

reasonable

(unreasonable,) should be **$14.90 or $149**

Solve each application problem. Round money answers to the nearest cent when necessary.

39. LaTasha worked 50.5 hours over the last two weeks. She earns $18.73 per hour. How much did she make?

$945.87 (rounded)

40. Michael's time card shows 42.2 hours at $10.03 per hour. What are his earnings?

$423.27 (rounded)

41. Sid needs 0.6 meter of canvas material to make a carry-all bag that fits on his wheelchair. If canvas is $4.09 per meter, how much will Sid spend? (*Note:* $4.09 *per* meter means $4.09 for *one* meter.)

$2.45 (rounded)

42. How much will Mrs. Nguyen pay for 3.5 yards of lace trim that costs $0.87 per yard?

$3.05 (rounded)

43. Michelle filled the tank of her pickup truck with regular unleaded gas. Use the information shown on the pump to find how much she paid for gas.

GALLONS	PRICE PER GALLON	GALLONS	PRICE PER GALLON
10.329	$ 4.679	20.510	$ 4.389
SUPRA UNLEADED Minimum Octane Rating 90		UNLEADED REGULAR Minimum Octane Rating 87	

Source: Holiday.

$90.02 (rounded)

44. Ground beef and chicken legs are on sale. Juma bought 1.7 pounds of legs. Use the information in the ad to find the amount she paid.

BIG ONE FOODS Sale

Ground Beef
$2.09
per pound
For juicy burgers

Chicken Legs
$0.98
per pound

PRICES GOOD THROUGH SUNDAY!

$1.67 (rounded)

45. Ms. Rolack is a real estate broker who helps people sell their homes. Her fee is 0.07 times the price of the home. What was her fee for selling a $289,500 home?

$20,265

46. Manny Ramirez of the Boston Red Sox had a batting average of 0.296 in the 2007 season. He went to bat 483 times. How many hits did he make? (*Hint:* Multiply the number of times at bat by his batting average.) Round to the nearest whole number. (*Source: World Almanac.*)

143 hits (rounded)

Paper money in the United States has not always been the same size. Shown below are the measurements of bills printed before 1929 and the measurements from 1929 on. Use this information to answer Exercises 47–50. Recall that perimeter is the total distance around the edges of a figure (see **Section 1.2**). *The area of a rectangle is found by multiplying the length by the width (see* **Section 2.5**). (*Source:* www.moneyfactory.com)

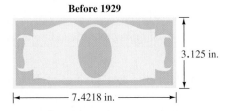

Before 1929

3.125 in.

7.4218 in.

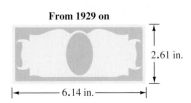

From 1929 on

2.61 in.

6.14 in.

47. (a) Find the area of each bill, rounded to the nearest tenth.

Area before 1929 ≈ 23.2 in.²
Area today ≈ 16.0 in.²

(b) What is the difference in the rounded areas?

7.2 in.²

48. (a) Find the perimeter of each bill, to the nearest hundredth.

Perimeter before 1929 ≈ 21.09 in.
Perimeter today = 17.50 in.

(b) How much less is the perimeter of today's bills than the bills printed before 1929?

3.59 in.

49. The thickness of one piece of today's paper money is 0.0043 inch.

(a) If you had a pile of 100 bills, how high would the pile be?

0.43 in.

(b) How high would a pile of 1000 bills be?

4.3 in.

50. (a) Use your answers from Exercise 49 to find the number of bills in a pile that is 43 inches high.

10,000 bills

(b) How much money would you have if the pile is all $20 bills?

$200,000

51. Judy Lewis pays $38.96 per month for basic cable TV. The one-time installation fee was $49. How much will she pay for cable over two years? How much would she pay in two years for the deluxe cable package that costs $89.95 per month?

$984.04; $2207.80

52. Chuck's car payment is $420.27 per month for four years. He also made a down payment of $5000 at the time he bought the car. How much will he pay altogether?

$25,172.96

53. Paper for the copy machine at the library costs $0.015 per sheet. How much will the library pay for 5100 sheets?

$76.50

54. A student group collected 2200 pounds of plastic as a fund-raiser. How much will they make if the recycling center pays $0.142 per pound?

$312.40

55. Barry bought 16.5 meters of rope at $0.47 per meter and three meters of wire at $1.05 per meter. How much change did he get from three $5 bills?

$4.09 (rounded)

56. Susan bought a 42-inch plasma HDTV that cost $1999.99. She paid $68.83 per month for 36 months. How much could she have saved by paying for the HDTV when she bought it?

$477.89

Use the information from the Look Smart mail-order catalog to answer Exercises 57–60.

Knit Shirt Ordering Information		
43-2A	Short sleeved, solid colors	$14.75 each
43-2B	Short sleeved, stripes	$16.75 each
43-3A	Long sleeved, solid colors	$18.95 each
43-3B	Long sleeved, stripes	$21.95 each
XXL size, add $2 per shirt.		
Monogram, $4.95 each. Gift box, $5 each.		

Total Price of All Items (excluding monograms and gift boxes)	Shipping, Packing, and Handling
$0–25.00	$3.50
$25.01–75.00	$5.95
$75.01–125.00	$7.95
$125.01 +	$9.95
Shipping to each additional address, add $4.25.	

57. Find the total cost of ordering four long-sleeved, solid-color shirts and two short-sleeved, striped shirts, all in the XXL size, and all shipped to your home.

$129.25

58. What is the total cost of eight long-sleeved shirts, size L, five in solid colors and three striped? Include the cost of shipping the solid shirts to your home and the striped shirts to your brother's home.

$174.80

59. (a) What is the total cost, including shipping, of sending three short-sleeved, solid-color shirts, size M, with monograms, in a gift box to your aunt for her birthday?

$70.05

(b) How much did the monograms, gift box, and shipping add to the cost of your gift?

$25.80

60. (a) Suppose you order one of each type of shirt for yourself, adding a monogram to each of the solid-color shirts. At the same time, you order three long-sleeved striped shirts, in the XXL size, shipped to your dad in a gift box. Find the total cost of your order.

$173.35

(b) What is the difference in total cost (excluding shipping) between the shirts for yourself and the gift for your dad?

$5.45

Summary Exercises on Decimals

Write each decimal as a fraction or mixed number in lowest terms.

1. 0.8 $\dfrac{4}{5}$

2. 6.004 $6\dfrac{1}{250}$

3. 0.35 $\dfrac{7}{20}$

Write each decimal in words.

4. 94.5

   ninety-four and five tenths

5. 2.0003

   two and three
   ten-thousandths

6. 0.706

   seven hundred six
   thousandths

Write each decimal in numbers.

7. five hundredths

   0.05

8. three hundred nine
ten-thousandths

   0.0309

9. ten and seven tenths

   10.7

Round to the place indicated.

10. 6.1873 to the nearest hundredth

   6.19

11. 0.95 to the nearest tenth

   1.0

12. 0.42025 to the nearest
thousandth

   0.420

13. $0.893 to the nearest cent

   $0.89

14. $3.0017 to the nearest cent

   $3.00

15. $99.64 to the nearest dollar

   $100

Simplify.

16. 0.27 (3.5)

   0.945

17. 50 − 0.3801

   49.6199

18. $\begin{array}{r} 0.205 \\ \times \quad 9 \\ \hline 1.845 \end{array}$

19. 25 ($3.74)

   $93.50

20. (0.004) (1.22)

   0.00488

21. 3.7 − 1.55

   2.15

22. 0.95 + 10.005

   10.955

23. 3.6 + 0.718 + 9 + 5.0829

   18.4009

24. 32.305 − 28 − 0.0007

   $4.3043

25. 8.9 − 0.4 − 0.03 − 7.6

   0.87

26. $18 + $2.09 + $90 + $0.75

   $110.84

27. (0.99) (83.672)   Round your answer to the
nearest hundredth.

   82.84 (rounded)

28. Find the perimeter and area of this rectangular computer chip. Round your answers to the nearest tenth.

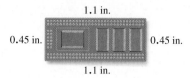

1.1 in.

0.45 in.   0.45 in.

1.1 in.

$P = 3.1$ in.; $A \approx 0.5$ in.^{2}

29. Find the perimeter and area of this rectangular watch face. Round your answers to the nearest hundredth.

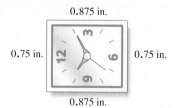

0.875 in.

0.75 in.   0.75 in.

0.875 in.

$P = 3.25$ in.; $A \approx 0.66$ in.^{2}

Use the table on heaviest fruits and vegetables to answer Exercises 30–34.

HEAVIEST FRUITS AND VEGETABLES

Fruit/Vegetable	Weight (in pounds)	Location
Apple	4.0625	Japan
Grapefruit	6.75	Australia
Onion	15.94	England
Peach	1.6	Michigan, USA
Radish	68.563	Japan
Squash	962	Canada
Strawberry	0.5	England
Tomato	7.75	Oklahoma, USA

Source: Guinness World Records.

30. (a) Identify the heaviest and lightest weight items in the table.

Squash is heaviest; strawberry is lightest.

(b) Find the difference in weight between the heaviest and lightest items.

961.5 pounds

31. If you used the radish, tomato, and onion in a salad, what would be their combined weight?

92.253 pounds

32. (a) How much would six grapefruits weigh?

40.5 pounds

(b) Find the total weight of a dozen strawberries.

6.0 or 6 pounds

33. How much heavier is the tomato than the apple?

3.6875 pounds

34. (a) Round the weight of each fruit or vegetable to the nearest whole number.

4; 7; 16; 2; 69; 962; 1; 8 (all pounds)

(b) Use the rounded numbers to find the approximate total weight if you had one of each item in the table.

1069 pounds

35. A craft cooperative paid $40.32 for enough fabric to make eight baby blankets for their store. They sold the blankets for $12.95 each. How much profit was made on the blankets?

$63.28

4.5 ▶▶▶ Dividing Decimals

There are two kinds of decimal division problems: those in which a decimal is divided by a whole number, and those in which a number is divided by a decimal. First recall the parts of a division problem from **Section 1.5**.

$$\text{Divisor} \rightarrow 4\overline{)33} \begin{array}{l} 8 \leftarrow \text{Quotient} \\ \leftarrow \text{Dividend} \\ \underline{32} \\ 1 \leftarrow \text{Remainder} \end{array}$$

OBJECTIVE 1 Divide a decimal by a whole number. When the divisor is a whole number, use these steps.

> **Dividing Decimals by Whole Numbers**
>
> **Step 1** Write the decimal point in the quotient (answer) directly above the decimal point in the dividend.
>
> **Step 2** Divide as if both numbers were whole numbers.

EXAMPLE 1 **Dividing Decimals by Whole Numbers**

Find each quotient. Check the quotients by multiplying.

(a) 21.93 by 3
   Dividend └ Divisor

Rewrite the division problem. $3\overline{)21.93}$

Step 1 Write the decimal point in the quotient directly above the decimal point in the dividend.

— Decimal points lined up
$$3\overline{)21\overset{.}{.}93}$$

Step 2 Divide as if the numbers were whole numbers.

Check by multiplying the quotient times the divisor.

$$\begin{array}{r} 7.31 \\ 3\overline{)21.93} \end{array}$$

Matches, so 7.31 is correct.

Check:
$$\begin{array}{r} 7.31 \\ \times 3 \\ \hline 21.93 \end{array}$$

The quotient (answer) is 7.31.

(b) $9\overline{)470.7}$
   Divisor ┘ └ Dividend

Write the decimal point in the quotient above the decimal point in the dividend. Then divide as if the numbers were whole numbers.

Decimal points lined up
$$\begin{array}{r} 52.3 \\ 9\overline{)470.7} \\ \underline{45} \\ 20 \\ \underline{18} \\ 27 \\ \underline{27} \\ 0 \end{array}$$

Matches

Check:
$$\begin{array}{r} 52.3 \\ \times 9 \\ \hline 470.7 \end{array}$$

Multiply the quotient by the divisor. The result should match the dividend.

The quotient is 52.3.

Work Problem **1** *at the Side.* ▶

OBJECTIVES

1 **Divide a decimal by a whole number.**

2 **Divide a number by a decimal.**

3 **Estimate the answer when dividing decimals.**

4 **Use the order of operations with decimals.**

1 Find each quotient. Check the quotients by multiplying.

(a) $4\overline{)93.6}$

(b) $6\overline{)6.804}$

(c) $11\overline{)278.3}$

(d) $0.51835 \div 5$

(e) $213.45 \div 15$

ANSWERS

1. **(a)** 23.4; (23.4)(4) = 93.6
   (b) 1.134; (1.134)(6) = 6.804
   (c) 25.3; (25.3)(11) = 278.3
   (d) 0.10367; (0.10367)(5) = 0.51835
   (e) 14.23; (14.23)(15) = 213.45

2 Divide. Check each quotient by multiplying.

(a) $5\overline{)6.4}$

(b) $30.87 \div 14$

(c) $\dfrac{259.5}{30}$

(d) $0.3 \div 8$

EXAMPLE 2   **Writing Extra Zeros to Complete a Division**

Divide 1.5 by 8. Check the quotient by multiplying.

   Keep dividing until the remainder is 0, or until the digits in the quotient begin to repeat in a pattern. In Example 1(b), you ended up with a remainder of 0. But sometimes you run out of digits in the dividend before that happens. If so, write extra zeros on the right side of the dividend so you can continue dividing.

$$
\begin{array}{r}
0.1 \\
8\overline{)1.5} \leftarrow \text{All digits have been used.} \\
\underline{8} \\
7 \leftarrow \text{Remainder is not yet 0.}
\end{array}
$$

Write a 0 after the 5 in the dividend so you can continue dividing. Keep writing more zeros in the dividend, if needed. Recall that writing zeros to the *right* of a decimal number does *not* change its value.

Check:

$$
\begin{array}{r}
0.1\ 8\ 7\ 5 \\
8\overline{)1.5\ 0\ 0\ 0} \leftarrow \begin{array}{l}\text{Three zeros needed} \\ \text{to complete the} \\ \text{division}\end{array} \\
\underline{8} \downarrow \\
7\ 0 \\
\underline{6\ 4} \downarrow \\
6\ 0 \\
\underline{5\ 6} \downarrow \\
4\ 0 \\
\underline{4\ 0} \\
0 \leftarrow \text{Stop dividing when the remainder is 0.}
\end{array}
$$

$$
\begin{array}{r}
0.1875 \\
\times \quad\quad 8 \\
\hline
1.5000
\end{array}
$$

Matches dividend, so 0.1875 is correct.

CAUTION

When dividing decimals, notice that the dividend might *not* be the larger number. In Example 2 above, the dividend is 1.5, which is *smaller* than the divisor 8.

◀ *Work Problem* **2** *at the Side.*

▦ **Calculator Tip**   When *multiplying* numbers, you can enter them in any order because multiplication is commutative (see **Section 1.4**). But division is *not* commutative. It *does* matter which number you enter first. Try Example 2 both ways; jot down your answers.

1.5 ÷ 8 = ____   8 ÷ 1.5 = ____

Notice that the first answer, 0.1875, matches the result from Example 2. But the second answer is much different: 5.333333333. Be careful to enter the dividend first.

   The next example shows a quotient (answer) that must be rounded because you will never get a remainder of 0.

EXAMPLE 3 **Rounding a Decimal Quotient**

Divide 4.7 by 3. Round the quotient to the nearest thousandth. Write extra zeros in the dividend so you can continue dividing.

$$
\begin{array}{r}
1.5\,6\,6\,6 \\
3\overline{)4.7\,0\,0\,0} \\
\underline{3} \\
1\,7 \\
\underline{1\,5} \\
2\,0 \\
\underline{1\,8} \\
2\,0 \\
\underline{1\,8} \\
2\,0 \\
\underline{1\,8} \\
2
\end{array}
$$

← Three zeros added so far

← Remainder is still not 0.

Notice that the digit 6 in the answer is repeating. It will continue to do so. The remainder will *never be 0.* There are two ways to show that the answer is a **repeating decimal** that goes on forever. You can write three dots after the answer, or you can write a bar above the digits that repeat (in this case, the 6).

$$1.5666\ldots \quad \text{or} \quad 1.5\overline{6}$$

← Bar above repeating digit

Three dots

When repeating decimals occur, round the quotient according to the directions in the problem. In this example, to round to thousandths, divide out one *more* place, to ten-thousandths.

$$4.7 \div 3 = 1.5666\ldots \quad \text{rounds to} \quad 1.567$$

Nearest thousandth is *three* decimal places.

Check the answer by multiplying 1.567 by 3. Because 1.567 is a rounded answer, the check will not give exactly 4.7, but it should be very close.

$$(1.567)(3) = 4.701 \leftarrow \text{Does not equal exactly 4.7}$$
because 1.567 was rounded

CAUTION
When checking quotients that you've rounded, the check will *not* match the dividend exactly, but it should be very close.

Work Problem **3** *at the Side.* ▶

OBJECTIVE 2 Divide a number by a decimal.   To divide by a *decimal* divisor, first change the divisor to a whole number. Then divide as before. To see how this is done, write the problem in fraction form. Here is an example.

$$1.2\overline{)6.36} \quad \text{can be written} \quad \frac{6.36}{1.2}$$

In **Section 3.2** you learned that multiplying the numerator and denominator by the same number gives an equivalent fraction. We want the divisor (1.2) to be a whole number. Multiplying by 10 will accomplish that.

$$\underset{\substack{\text{Decimal} \\ \text{divisor}}}{\frac{6.36}{1.2}} = \frac{(6.36)(10)}{(1.2)(10)} = \underset{\substack{\text{Whole number} \\ \text{divisor}}}{\frac{63.6}{12}}$$

3 Divide. Round quotients to the nearest thousandth. If it is a repeating decimal, also write the answer using a bar. Check your quotients by multiplying.

(a) $13\overline{)267.01}$

(b) $6\overline{)20.5}$

(c) $\dfrac{10.22}{9}$

(d) $16.15 \div 3$

(e) $116.3 \div 11$

ANSWERS

3. **(a)** 20.539 (rounded); no repeating digits visible on calculator; $(20.539)(13) = 267.007$
   (b) 3.417 (rounded); $3.41\overline{6}$; $(3.417)(6) = 20.502$
   (c) 1.136 (rounded); $1.13\overline{5}$; $(1.136)(9) = 10.224$
   (d) 5.383 (rounded); $5.38\overline{3}$; $(5.383)(3) = 16.149$
   (e) 10.573 (rounded); $10.5\overline{72}$; $(10.573)(11) = 116.303$

4 Divide. If the quotient does not come out even, round to the nearest hundredth.

(a) $0.2\overline{)1.04}$

(b) $0.06\overline{)1.8072}$

(c) $0.005\overline{)32}$

(d) $8.1 \div 0.025$

(e) $\dfrac{7}{1.3}$

(f) $5.3091 \div 6.2$

The short way to multiply by 10 is to move the decimal point *one place* to the *right* in both the divisor and the dividend.

$$1.2\overline{)6.3\,6} \quad \text{is equivalent to} \quad 12\overline{)63.6}$$

> **Note**
> Moving the decimal points the **same** number of places in **both** the divisor and dividend will **not** change the answer.

> **Dividing by Decimals**
> **Step 1** Count the number of decimal places in the divisor and move the decimal point that many places to the *right*. (This changes the divisor to a whole number.)
>
> **Step 2** Move the decimal point in the dividend the *same* number of places to the *right*. (Write in extra zeros if needed.)
>
> **Step 3** Write the decimal point in the quotient directly above the decimal point in the dividend. Then divide as usual.

> **EXAMPLE 4** Dividing by Decimals

(a) $0.003\overline{)27.69}$

Move the decimal point in the divisor *three* places to the *right* so 0.003 becomes the whole number 3. To move the decimal point in the dividend the same number of places, write in an extra 0.

$$0.003\overline{)27.690}$$

Moving decimal point three places to the right is the same as multiplying by 1000.

Move decimal points in divisor and dividend. Then line up the decimal point in the quotient.

$$\begin{array}{r} 9230. \\ 3\overline{)27690.} \end{array}$$

Divide as usual.

(b) Divide 5 by 4.2. Round to the nearest hundredth.

Move the decimal point in the divisor one place to the right so 4.2 becomes the whole number 42. The decimal point in the dividend starts on the right side of 5 and is also moved one place to the right.

$$\begin{array}{r} 1.1\,9\,0 \\ 4.2\overline{)5.0\,0\,0\,0} \\ 4\,2 \\ \hline 8\,0 \\ 4\,2 \\ \hline 3\,8\,0 \\ 3\,7\,8 \\ \hline 2\,0 \end{array}$$

← In order to round to hundredths, divide out one *more* place, to thousandths.

Move the decimal points the *same* number of places.

Round the quotient. It is 1.19 (rounded to the nearest hundredth).

◀ Work Problem **4** at the Side.

ANSWERS
4. **(a)** 5.2 **(b)** 30.12 **(c)** 6400 **(d)** 324 **(e)** 5.38 (rounded) **(f)** 0.86 (rounded)

OBJECTIVE **3** **Estimate the answer when dividing decimals.**
Estimating answers helps you catch errors. Compare the estimate to your exact answer. If they are very different, work the problem again.

EXAMPLE 5 **Estimating before Dividing**

First use front end rounding to round each number and estimate the answer. Then divide to find the exact answer.

$$580.44 \div 2.8$$

Here is how one student solved this problem. She rounded 580.44 to 600 and rounded 2.8 to 3 to estimate the answer.

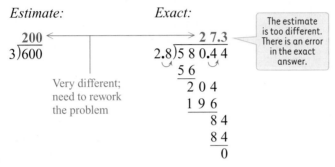

Notice that the estimate, which is in the hundreds, is very different from the exact answer, which is only in the tens. This tells the student that she needs to rework the problem. Can you find the error?
(The exact answer should be 207.3, which fits with the estimate of 200.)

Work Problem **5** *at the Side.* ▶

OBJECTIVE **4** **Use the order of operations with decimals.** Use the order of operations when a decimal problem involves more than one operation, as you did with whole numbers in **Section 1.8.**

> **Order of Operations**
> 1. Do all operations inside *parentheses* or *other grouping symbols.*
> 2. Simplify any expressions with *exponents* and find any *square roots.*
> 3. *Multiply* or *divide,* proceeding from left to right.
> 4. *Add* or *subtract,* proceeding from left to right.

EXAMPLE 6 **Using the Order of Operations**

Use the order of operations to simplify each expression.

(a) $2.5 + \underbrace{6.3^2} + 9.62$ Apply the exponent: $(6.3)(6.3)$ is 39.69

$\underbrace{2.5 + 39.69} + 9.62$ Add from left to right.

$\underbrace{42.19 + 9.62}$

51.81

(b) $1.82 + \underbrace{(6.7 - 5.2)}(5.8)$ Work inside parentheses.

$1.82 + \underbrace{(1.5)(5.8)}$ Multiply next.

$\underbrace{1.82 + 8.7}$ Add last.

10.52

Continued on Next Page

---

5 Decide whether each answer is reasonable by using front end rounding to estimate the answer. If the exact answer is *not* reasonable, find and correct the error.

(a) $42.75 \div 3.8 = 1.125$

Estimate:

(b) $807.1 \div 1.76 = 458.580$
to nearest thousandth

Estimate:

(c) $48.63 \div 52 = 93.519$
to nearest thousandth

Estimate:

(d) $9.0584 \div 2.68 = 0.338$

Estimate:

ANSWERS

5. **(a)** Estimate is $40 \div 4 = 10$; answer is not reasonable; should be 11.25
   (b) Estimate is $800 \div 2 = 400$; answer is reasonable.
   (c) Estimate is $50 \div 50 = 1$; answer is not reasonable, should be 0.935 (rounded)
   (d) Estimate is $9 \div 3 = 3$; answer is not reasonable, should be 3.38

6 Use the order of operations to simplify each expression.

(a) $4.6 - 0.79 + 1.5^2$

(c)

$$3.7^2 - 1.8 \div 5(1.5)$$ Apply the exponent.

$$13.69 - \underbrace{1.8 \div 5}(1.5)$$ Multiply and divide from left to right, so first divide 1.8 by 5

$$13.69 - \underbrace{0.36(1.5)}$$ Then multiply 0.36 by 1.5

$$\underbrace{13.69 - \quad 0.54}$$ Subtract last.

$$13.15$$

◀ *Work Problem* **6** *at the Side.*

⊞ **Calculator Tip** Most scientific calculators that have parentheses keys ⓛ ⓘ can handle calculations like those in Example 6 if you just enter the numbers in the order given. For example, the keystrokes for Example 6(b) on the previous page are:

┌─── Parentheses ───┐

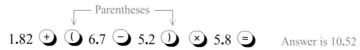

1.82 ⊕ ⓘ 6.7 ⊖ 5.2 ⓘ ⊗ 5.8 ⊜   Answer is 10.52

Standard, four-function calculators generally do not have parentheses keys and will *not* give the correct answer if you simply enter the numbers in the order given.

Check the instruction manual that came with your calculator for information on "order of calculations" to see if your model has the rules for order of operations built into it. For a quick check, try entering this problem.

$$2 ⊕ 2 ⊗ 2 ⊜$$

If the result is 6, the calculator follows the order of operations. If the result is 8, it does *not* have the rules built into it. To see why this test works, do the calculations by hand.

(b) $3.64 \div 1.3 \times 3.6$

(c) $0.08 + 0.6(3 - 2.99)$

Follow the order of operations.	Work from left to right.
$2 + \underbrace{2 \times 2}$ Multiply before adding.	$\underbrace{2 + 2} \times 2$
$\underbrace{2 + \quad 4}$	$\underbrace{4 \quad \times 2}$
$6 \longleftarrow$ Correct	✗ $8 \longleftarrow$ Incorrect

(d) $10.85 - 2.3(5.2) \div 3.2$

4.5 ▶▶▶ Exercises

Find each quotient. See Examples 1 and 4.

🌐 **1.** $7)\overline{27.3}$ quotient 3.9

2. $8)\overline{50.4}$ quotient 6.3

🌐 **3.** $\dfrac{4.23}{9}$ 0.47

4. $\dfrac{1.62}{6}$ 0.27

5. $0.05)\overline{20.01}$ quotient 400.2

6. $0.08)\overline{16.04}$ quotient 200.5

7. $1.5)\overline{54}$ quotient 36

8. $2.4)\overline{132}$ quotient 55

Use the fact that $108 \div 18 = 6$ to work Exercises 9–12 simply by moving decimal points.

9. $0.108 \div 1.8$ 0.06

10. $10.8 \div 18$ 0.6

11. $0.018)\overline{108}$ quotient 6000

12. $0.18)\overline{1.08}$ quotient 6

Divide. Round quotients to the nearest hundredth if necessary. See Examples 3 and 4.

13. $4.6)\overline{116.38}$ quotient 25.3

14. $2.6)\overline{4.992}$ quotient 1.92

15. $\dfrac{3.1}{0.006}$
516.67 (rounded)

16. $\dfrac{1.7}{0.09}$
18.89 (rounded)

Divide. Round quotients to the nearest thousandth.

17. $240.8 \div 9$
26.756 (rounded)

18. $76.43 \div 7$
10.919 (rounded)

🖩 **19.** $0.034)\overline{342.81}$
10,082.647 (rounded)

🖩 **20.** $0.043)\overline{1748.4}$
40,660.465 (rounded)

Relating Concepts (Exercises 21–22) For Individual or Group Work

*Look for patterns as you **work Exercises 21 and 22 in order.***

21. Do these division problems:

$3.77 \div 10 = \underline{0.377}$ $9.1 \div 10 = \underline{0.91}$

$0.886 \div 10 = \underline{0.0886}$ $30.19 \div 10 = \underline{3.019}$

$406.5 \div 10 = \underline{40.65}$ $6625.7 \div 10 = \underline{662.57}$

(a) What pattern do you see? Write a "rule" for dividing by 10. What do you think the rule is for dividing by 100? by 1000? Write the rules and try them out on the numbers above.

dividing by 10, decimal point moves one place to the left; by 100, two places to the left; by 1000, three places to the left

(b) Compare your rules to the ones you wrote in **Section 4.4,** Exercise 21. How are they different?

The decimal point moved to the *right* when *multiplying* by 10 or 100 or 1000. Here the decimal point moves to the *left* when *dividing* by those numbers.

22. Do these division problems:

$40.2 \div 0.1 = \underline{402}$ $7.1 \div 0.1 = \underline{71}$

$0.339 \div 0.1 = \underline{3.39}$ $15.77 \div 0.1 = \underline{157.7}$

$46 \div 0.1 = \underline{460}$ $873 \div 0.1 = \underline{8730}$

(a) What pattern do you see? Write a "rule" for dividing by 0.1. What do you think the rule is for dividing by 0.01? by 0.001? Write the rules and try them out on the numbers above.

dividing by 0.1, decimal point moves one place to the right; by 0.01, two places to the right; by 0.001, three places to the right.

(b) Compare your rules to the ones you wrote in **Section 4.4,** Exercise 22. How are they different?

The decimal point moved to the *left* when *multiplying* by 0.1 or 0.01 or 0.001. Here the decimal point moves to the *right* when *dividing* by those numbers.

Decide whether each answer is reasonable or unreasonable by rounding the numbers and estimating the answer. If the exact answer is not reasonable, find the correct answer. See Example 5.

23. 37.8 ÷ 8 = 47.25   unreasonable

Estimate: 4.725
40 ÷ 8 = 5   8)37.8

24. 345.6 ÷ 3 = 11.52   unreasonable

Estimate: 115.2
300 ÷ 3 = 100   3)345.6

25. 54.6 ÷ 48.1 = 1.135   reasonable

Estimate:
50 ÷ 50 = 1

26. 2428.8 ÷ 4.8 = 56   unreasonable

Estimate: 506
2000 ÷ 5 = 400   4.8)2428.8

27. 307.02 ÷ 5.1 = 6.2   unreasonable

Estimate: 60.2
300 ÷ 5 = 60   5.1)307.02

28. 395.415 ÷ 5.05 = 78.3   reasonable

Estimate:
400 ÷ 5 = 80

29. 9.3 ÷ 1.25 = 0.744   unreasonable

Estimate: 7.44
9 ÷ 1 = 9   1.25)9.3

30. 78 ÷ 14.2 = 5493   unreasonable

Estimate: 5.493
80 ÷ 10 = 8   14.2)78

Solve each application problem. Round money answers to the nearest cent, if necessary.

31. Rob discovered that his daughter's favorite brand of tights are on sale. He decided to buy one pair as a surprise for her. How much did he pay?

$4.00 (rounded)

32. The bookstore has a special price on notepads. How much did Randall pay for one notepad?

$0.42 (rounded)

33. It will take 21 equal monthly payments for Aimee to pay off her credit card balance of $1408.68. How much is she paying each month?

$67.08

34. Marcella Anderson bought 2.6 meters of microfiber woven suede fabric for $33.77. How much did she pay per meter?

$12.99 per meter (rounded)

35. Adrian Webb bought 619 bricks to build a barbecue pit, paying $185.70. Find the cost per brick. (*Hint:* Cost *per* brick means the cost for *one* brick.)

$0.30

36. Lupe Wilson is a newspaper distributor. Last week she paid the newspaper $130.51 for 842 copies. Find the cost per copy.

$0.16 (rounded)

37. Darren Jackson earned $476.80 for 40 hours of work. Find his earnings per hour.

$11.92 per hour

38. At a CD manufacturing company, 400 CDs cost $289. Find the cost per CD.

$0.72 (rounded)

39. It took 16.35 gallons of gas to fill the gas tank of Kim's car. She had driven 346.2 miles since her last fill-up. How many miles per gallon did her car get? Round to the nearest tenth.

21.2 miles per gallon (rounded)

40. Mr. Rodriquez pays $53.19 each month to Household Finance. How many months will it take him to pay off $1436.13?

27 months

Use the table of women's longest long jumps (through the year 2007) to answer Exercises 41–46. To find an average, add up the values you are interested in and then divide the sum by the number of values. Round your answers to the nearest hundredth.

Athlete	Country	Year	Length (meters)
Galina Christyakova	USSR	1988	7.52
Jackie Joyner-Kersee	U.S.	1994	7.49
Heike Drechsler	Germany	1992	7.48
Jackie Joyner-Kersee	U.S.	1987	7.45
Jackie Joyner-Kersee	U.S.	1988	7.40
Jackie Joyner-Kersee	U.S.	1991	7.32
Jackie Joyner-Kersee	U.S.	1996	7.20
Chioma Ajunwa	Nigeria	1996	7.12
Fiona May	Italy	2000	7.09
Tatyana Lebedeva	Russia	2004	7.07

Source: CNNSI.com

41. Find the average length of the long jumps made by Jackie Joyner-Kersee.

7.37 meters (rounded)

42. Find the average length of all the long jumps listed in the table.

7.31 meters (rounded)

43. How much longer was the fifth-longest jump than the sixth-longest jump?

0.08 meter

44. If the first-place athlete made five jumps of the same length, what would be the total distance jumped?

37.60 meters

45. What was the total length jumped by the top three athletes in the table?

22.49 meters

46. How much less was the last-place jump than the next-to-last-place jump?

0.02 meter

Use the order of operations to simplify each expression. See Example 6.

47. $7.2 - 5.2 + 3.5^2$
14.25

48. $6.2 + 4.3^2 - 9.72$
14.97

49. $38.6 + 11.6(13.4 - 10.4)$
73.4

50. $2.25 - 1.06(4.85 - 3.95)$
1.296

51. $8.68 - 4.6(10.4) \div 6.4$
1.205

52. $25.1 + 11.4 \div 7.5(3.75)$
30.8

53. 🌐 $33 - 3.2(0.68 + 9) - 1.3^2$
0.334

54. $0.6 + (1.89 + 0.11) \div 0.004(0.5)$
250.6

Solve each application problem.

55. Soup is on sale at six cans for $3.25, or you may purchase individual cans for $0.57. How much will you save per can if you buy six cans? Round to the nearest cent.

$0.03 per can (rounded)

56. Nadia's diet says she can eat 3.5 ounces of chicken nuggets. The package weighs 10.5 ounces and contains 15 nuggets. How many nuggets can Nadia eat?

5 nuggets

57. The U.S. Treasury prints about 38,000,000 pieces of paper money each day. The printing presses run 24 hours a day. How many pieces of money are printed, to the nearest whole number:

(a) each hour?

1,583,333 pieces (rounded)

(b) each minute?

26,389 pieces (rounded)

(c) each second?

440 pieces (rounded)

(*Source:* www.moneyfactory.com)

38,000,000 pieces of paper money are printed each day.

58. Mach 1 is the speed of sound. Dividing a vehicle's speed by the speed of sound gives its speed on the Mach scale. In 1997, a specially built car with two 110,000-horsepower engines broke the world land speed record by traveling 763.035 miles per hour. The speed of sound that day was 748.11 miles per hour. What was the car's Mach speed, to the nearest hundredth? (*Source:* Associated Press.)

Mach 1.02 (rounded)

General Mills will give a school 10¢ for each box top logo from its cereals and other products. A school can earn up to $10,000 per year. Use this information to answer Exercises 59–62. Round your answers to the nearest whole number when necessary. (Source: General Mills.)

59. How many box tops would a school need to collect in one year to earn the maximum amount?

100,000 box tops

60. (Complete Exercise 59 first.) If a school has 550 children, how many box tops would each child need to collect in one year to reach the maximum?

182 box tops (rounded)

61. How many box tops would need to be collected during each of the 38 weeks in the school year to reach the maximum amount?

2632 box tops (rounded)

62. How many box tops would each of the 550 children need to collect during each of the 38 weeks of school to reach the maximum amount?

5 box tops (rounded)

4.6 ▶▶▶ Writing Fractions as Decimals

Writing fractions as equivalent decimals can help you do calculations more easily or compare the size of two numbers.

OBJECTIVE 1 Write fractions as equivalent decimals.   Recall that a fraction is one way to show division (see **Section 1.5**). For example, $\frac{3}{4}$ means $3 \div 4$. If you are doing the division by hand, write it as $4\overline{)3}$. When you do the division, the result is 0.75, the decimal equivalent of $\frac{3}{4}$.

> **Writing a Fraction as an Equivalent Decimal**
>
> **Step 1**   Divide the numerator of the fraction by the denominator.
>
> **Step 2**   If necessary, round the answer to the place indicated.

Work Problem ① *at the Side.* ▶

EXAMPLE 1   Writing Fractions or Mixed Numbers as Decimals

(a) Write $\frac{1}{8}$ as a decimal.

$\frac{1}{8}$ means $1 \div 8$. Write it as $8\overline{)1}$. The decimal point in the dividend is on the *right* side of the 1. Write extra zeros in the dividend so you can continue dividing until the remainder is 0.

Decimal points lined up

$$\frac{1}{8} \rightarrow 1 \div 8 \rightarrow 8\overline{)1} \rightarrow \begin{array}{r} 0.125 \\ 8\overline{)1.000} \\ \underline{8} \\ 20 \\ \underline{16} \\ 40 \\ \underline{40} \\ 0 \end{array}$$

← Three extra zeros needed

← Remainder is 0.

Therefore, $\frac{1}{8} = 0.125$.
To check this, write 0.125 as a fraction, then change it to lowest terms.

$$0.125 = \frac{125}{1000} \qquad \text{In lowest terms} \qquad \frac{125 \div 125}{1000 \div 125} = \frac{1}{8} \leftarrow \begin{array}{l}\text{Original}\\\text{fraction}\end{array}$$

> **⊞ Calculator Tip**   When using your calculator to write fractions as decimals, enter the numbers from the top down. Remember that the order in which you enter the numbers *does* matter in division. Example 1(a) above works like this:
>
> $\frac{1}{8}$ ↓ Top down   Enter 1 ⊝ 8 ⊜   Answer is 0.125
>
> What happens if you enter 8 ⊝ 1 ⊜ ? Do you see why that cannot possibly be correct? (Answer: $8 \div 1 = 8$. A proper fraction like $\frac{1}{8}$ *cannot* be equivalent to a whole number.)

Continued on Next Page

OBJECTIVES

1 Write fractions as equivalent decimals.

2 Compare the size of fractions and decimals.

① Rewrite each fraction so you could do the division by hand. Do *not* complete the division.

(a) $\frac{1}{9}$ is written $9\overline{)}$

(b) $\frac{2}{3}$ is written $\overline{)}$

(c) $\frac{5}{4}$ is written $\overline{)}$

(d) $\frac{3}{10}$ is written $\overline{)}$

(e) $\frac{21}{16}$ is written $\overline{)}$

(f) $\frac{1}{50}$ is written $\overline{)}$

ANSWERS

1. **(a)** $9\overline{)1}$ **(b)** $3\overline{)2}$ **(c)** $4\overline{)5}$
   (d) $10\overline{)3}$ **(e)** $16\overline{)21}$ **(f)** $50\overline{)1}$

2 Write each fraction or mixed number as a decimal.

(a) $\dfrac{1}{4}$

(b) $2\dfrac{1}{2}$

(c) $\dfrac{5}{8}$

(d) $4\dfrac{3}{5}$

(e) $\dfrac{7}{8}$

(b) Write $2\frac{3}{4}$ as a decimal.

One method is to divide 3 by 4 to get 0.75 for the fraction part. Then add the whole number part to 0.75.

$$\frac{3}{4} \rightarrow 4\overline{)3.00}$$

0.75 ← Fraction part

$$\begin{array}{r} 2.00 \leftarrow \text{Whole number part}\\ +\ 0.75\\ \hline 2.75 \end{array}$$

$$\frac{2\ 8}{20}\quad\frac{20}{0}$$

So, $2\frac{3}{4} = 2.75$ *Check:* $2.75 = 2\frac{75}{100} = 2\frac{3}{4}$ ← Lowest terms

Whole number parts match.

A second method is to first write $2\frac{3}{4}$ as an improper fraction and then divide numerator by denominator.

$$2\frac{3}{4} = \frac{11}{4}$$

$$\frac{11}{4} \rightarrow 11 \div 4 \rightarrow 4\overline{)11} \rightarrow 4\overline{)11.00}$$ ← Two extra zeros needed

2.75

$$\frac{8}{3\ 0}\quad\frac{2\ 8}{20}\quad\frac{20}{0}$$

Whole number parts match.

So, $2\frac{3}{4} = 2.75$

$\frac{3}{4}$ is equivalent to $\frac{75}{100}$ or 0.75

◀ *Work Problem* **2** *at the Side.*

EXAMPLE 2 **Writing a Fraction as a Decimal with Rounding**

Write $\frac{2}{3}$ as a decimal and round to the nearest thousandth.

$\frac{2}{3}$ means $2 \div 3$. To round to thousandths, divide out one *more* place, to ten-thousandths.

$$\frac{2}{3} \rightarrow 2 \div 3 \rightarrow 3\overline{)2} \rightarrow 3\overline{)2.0000}$$ ← Four zeros needed for ten-thousandths

0.6666

Be careful to divide in the correct order!

$$\frac{1\ 8}{20}\quad\frac{18}{20}\quad\frac{18}{20}\quad\frac{18}{2}$$

Written as a repeating decimal, $\frac{2}{3} = 0.\overline{6}$ ← Bar above repeating digit

Rounded to the nearest thousandth, $\frac{2}{3} \approx 0.667$

— **Continued on Next Page**

▦ **Calculator Tip**   Try Example 2 on your calculator. Enter 2 ⊝ 3. Which answer do you get?

| **0.666666667** | or | **0.6666666** |

Many scientific calculators will show a 7 as the last digit. Because the sixes keep on repeating forever, the calculator automatically rounds in the last decimal place it has room to show. If you have a 10-digit display space, the calculator is rounding as shown below.

0.6666666666 (11 digits)   rounds to   0.666666667

Next digit is 5 or more, so 6 rounds to 7.

Other calculators, especially standard, four-function ones, may *not* round. They just cut off, or *truncate,* the extra digits. Such a calculator would show 0.6666666 in the display.

   Would this difference in calculators show up when changing $\frac{1}{3}$ to a decimal? Why not? (Answer: The repeating digit is 3, which is *4 or less,* so it stays as 3 whether it's rounded or not.)

——————— *Work Problem* ③ *at the Side.* ▶

OBJECTIVE ② **Compare the size of fractions and decimals.** You can use a number line to compare fractions and decimals. For example, the number line below shows the space between 0 and 1. The locations of some commonly used fractions are marked, along with their decimal equivalents.

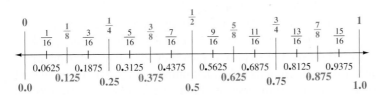

   The next number line shows the locations of some commonly used fractions between 0 and 1 that are equivalent to *repeating* decimals. The decimal equivalents use a bar above repeating digits.

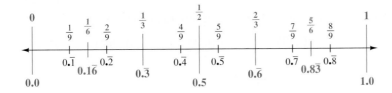

┌─ **EXAMPLE 3**   **Using a Number Line to Compare Numbers**

Use the number lines above to decide whether to write >, <, or = in the blank between each pair of numbers.

(a) 0.6875 ———— 0.625
   You learned in **Section 3.5** that the number farther to the right on the number line is the greater number. Look at the first number line above. Because 0.6875 is to the *right* of 0.625, use the > symbol.

0.6875 is greater than 0.625   0.6875 > 0.625

── **Continued on Next Page**

③ Write as decimals. Round to ▦ the nearest thousandth.

(a) $\frac{1}{3}$

(b) $2\frac{7}{9}$

(c) $\frac{10}{11}$

(d) $\frac{3}{7}$

(e) $3\frac{5}{6}$

4 Use the number lines on the previous page to help you decide whether to write $<$, $>$, or $=$ in each blank.

(a) 0.4375 _____ 0.5

(b) 0.75 _____ 0.6875

(c) 0.625 _____ 0.0625

(d) $\dfrac{2}{8}$ _____ 0.375

(e) $0.8\overline{3}$ _____ $\dfrac{5}{6}$

(f) $\dfrac{1}{2}$ _____ $0.\overline{5}$

(g) $0.\overline{1}$ _____ $0.1\overline{6}$

(h) $\dfrac{8}{9}$ _____ $0.\overline{8}$

(i) $0.\overline{7}$ _____ $\dfrac{4}{6}$

(j) $\dfrac{1}{4}$ _____ 0.25

5 Arrange each group in order from least to greatest.

(a) 0.7, 0.703, 0.7029

(b) 6.39, 6.309, 6.401, 6.4

(c) 1.085, $1\dfrac{3}{4}$, 0.9

(d) $\dfrac{1}{4}, \dfrac{2}{5}, \dfrac{3}{7}$, 0.428

ANSWERS

4. (a) < (b) > (c) > (d) < (e) =
   (f) < (g) < (h) = (i) > (j) =
5. (a) 0.7, 0.7029, 0.703
   (b) 6.309, 6.39, 6.4, 6.401
   (c) 0.9, 1.085, $1\dfrac{3}{4}$
   (d) $\dfrac{1}{4}, \dfrac{2}{5}$, 0.428, $\dfrac{3}{7}$

(b) $\dfrac{3}{4}$ _____ 0.75

On the first number line, $\frac{3}{4}$ and 0.75 are at the same point on the number line. They are equivalent.

$$\tfrac{3}{4} = 0.75$$

(c) 0.5 _____ $0.\overline{5}$

On the second number line, 0.5 is to the *left* of $0.\overline{5}$ (which is actually 0.555 . . .) so use the $<$ symbol.

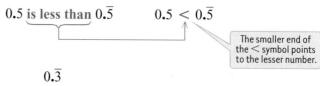

0.5 is less than $0.\overline{5}$ $0.5 < 0.\overline{5}$

> The smaller end of the $<$ symbol points to the lesser number.

(d) $\dfrac{2}{6}$ _____ $0.\overline{3}$

Write $\frac{2}{6}$ in lowest terms as $\frac{1}{3}$.
On the second number line you can see that $\frac{1}{3} = 0.\overline{3}$.

◀ *Work Problem* **4** *at the Side.*

Fractions can also be compared by first writing each one as a decimal. The decimals can then be compared by writing each one with the same number of decimal places.

EXAMPLE 4 **Arranging Numbers in Order**

Write each group of numbers in order, from least to greatest.

(a) 0.49   0.487   0.4903

It is easier to compare decimals if they are all tenths, or all hundredths, and so on. Because 0.4903 has four decimal places (ten-thousandths), write zeros to the right of 0.49 and 0.487 so they also have four decimal places. Writing zeros to the right of a decimal number does *not* change its value (see **Section 4.3**). Then find the least and greatest number of ten-thousandths.

$0.49 = 0.4900 = $ **4900** ten-thousandths ← 4900 is in the middle.

$0.487 = 0.4870 = $ **4870** ten-thousandths ← 4870 is the least.

$0.4903 = $ **4903** ten-thousandths ← 4903 is the greatest.

From least to greatest, the correct order is shown below.

0.487   0.49   0.4903

(b) $2\dfrac{5}{8}$ 2.63   2.6

Write $2\frac{5}{8}$ as $\frac{21}{8}$ and divide $8\overline{)21}$ to get the decimal form, 2.625. Then, because 2.625 has three decimal places, write zeros so all the numbers have three decimal places.

$2\dfrac{5}{8} = 2.625 = 2$ and **625** thousandths ← 625 is in the middle.

$2.63 = 2.630 = 2$ and **630** thousandths ← 630 is the greatest.

$2.6 = 2.600 = 2$ and **600** thousandths ← 600 is the least.

From least to greatest, the correct order is shown below.

2.6 $2\dfrac{5}{8}$ 2.63

◀ *Work Problem* **5** *at the Side.*

4.6 ▶▶▶ Exercises

Write each fraction or mixed number as a decimal. Round to the nearest thousandth if necessary. See Examples 1 and 2.

1. $\frac{1}{2}$ 0.5

2. $\frac{1}{4}$ 0.25

🌐 **3.** $\frac{3}{4}$ 0.75

4. $\frac{1}{10}$ 0.1

5. $\frac{3}{10}$ 0.3

6. $\frac{7}{10}$ 0.7

7. $\frac{9}{10}$ 0.9

8. $\frac{4}{5}$ 0.8

9. $\frac{3}{5}$ 0.6

10. $\frac{2}{5}$ 0.4

11. $\frac{7}{8}$ 0.875

12. $\frac{3}{8}$ 0.375

13. $2\frac{1}{4}$ 2.25

14. $1\frac{1}{2}$ 1.5

15. $14\frac{7}{10}$ 14.7

16. $23\frac{3}{5}$ 23.6

🌐 **17.** $3\frac{5}{8}$ 3.625

18. $2\frac{7}{8}$ 2.875

▦ **19.** $6\frac{1}{3}$ 6.333 (rounded)

▦ **20.** $5\frac{2}{3}$ 5.667 (rounded)

▦ **21.** $\frac{5}{6}$ 0.833 (rounded)

▦ **22.** $\frac{1}{6}$ 0.167 (rounded) 🌐

▦ **23.** $1\frac{8}{9}$ 1.889 (rounded)

▦ **24.** $5\frac{4}{7}$ 5.571 (rounded)

Relating Concepts (Exercises 25–28) For Individual or Group Work

*Use your knowledge of fractions and decimals to **work Exercises 25–28 in order.***

25. (a) Explain how you can tell that Keith made an error *just by looking at his final answer.* Here is his work.

$$\frac{5}{9} = 5\overline{)9.0}^{\,1.8} \quad \text{so} \quad \frac{5}{9} = 1.8$$

A proper fraction like $\frac{5}{9}$ is less than 1, so it cannot be equivalent to a decimal number that is greater than 1.

(b) Show the correct way to change $\frac{5}{9}$ to a decimal. Explain why your answer makes sense.

$\frac{5}{9}$ means $5 \div 9$ or $9\overline{)5}$ so the correct answer is 0.556 (rounded). This makes sense because both the fraction and decimal are less than 1.

26. (a) How can you prove to Sandra that $2\frac{7}{20}$ is *not* equivalent to 2.035? Here is her work.

$$2\frac{7}{20} = 20\overline{)7.00}^{\,0.35} \quad \text{so} \quad 2\frac{7}{20} = 2.035$$

$$2.035 = 2\frac{35}{1000} = 2\frac{7}{200}, \text{not } 2\frac{7}{20}$$

(b) What is the correct answer? Show how to prove that it is correct.

Adding the whole number part gives 2 + 0.35, which is 2.35, not 2.035. To check,

$$2.35 = 2\frac{35}{100} = 2\frac{7}{20}$$

27. Ving knows that $\frac{3}{8} = 0.375$. How can he write $1\frac{3}{8}$ as a decimal *without* having to do a division? How can he write $3\frac{3}{8}$ as a decimal? $295\frac{3}{8}$? Explain your answer.

Just add the whole number part to 0.375.

So $1\frac{3}{8} = 1.375$; $3\frac{3}{8} = 3.375$; $295\frac{3}{8} = 295.375$.

28. Iris has found a shortcut for writing mixed numbers as decimals.

$$2\frac{7}{10} = 2.7 \qquad 1\frac{13}{100} = 1.13$$

Does her shortcut work for all mixed numbers? Explain when it works and why it works.

It works only when the fraction part has a one-digit numerator and a denominator of 10, a two-digit numerator and a denominator of 100, and so on.

Find each decimal or fraction equivalent. Write fractions in lowest terms.

	Fraction	Decimal		Fraction	Decimal
29.	$\frac{2}{5}$ _____	0.4	**30.**	$\frac{3}{4}$ _____	0.75
31.	$\frac{5}{8}$ _____	0.625	**32.**	$\frac{111}{1000}$ _____	0.111
33.	$\frac{7}{20}$ _____	0.35	**34.**	$\frac{9}{10}$ _____	0.9
35.	$\frac{7}{20}$	_0.35_	**36.**	$\frac{1}{40}$	_0.025_
37.	$\frac{1}{25}$ _____	0.04	**38.**	$\frac{13}{25}$ _____	0.52
39.	$\frac{3}{20}$ _____	0.15	**40.**	$\frac{17}{20}$ _____	0.85
41.	$\frac{1}{5}$	_0.2_	**42.**	$\frac{1}{8}$	_0.125_
43.	$\frac{9}{100}$ _____	0.09	**44.**	$\frac{1}{50}$ _____	0.02

Solve each application problem.

45. The average length of a newborn baby is 20.8 inches. Charlene's baby is 20.08 inches long. Is her baby longer or shorter than the average? By how much?

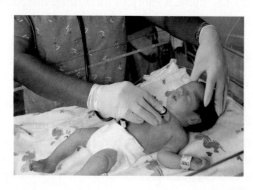

shorter; 0.72 inch

46. The patient in room 830 is supposed to get 8.3 milligrams of medicine. She was actually given 8.03 milligrams. Did she get too much or too little medicine? What was the difference?

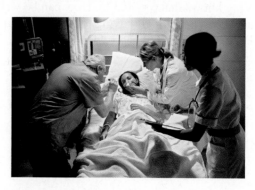

too little; 0.27 milligram

47. The label on the bottle of vitamins says that each capsule contains 0.5 gram of calcium. When checked, each capsule had 0.505 gram of calcium. Was there too much or too little calcium? What was the difference?

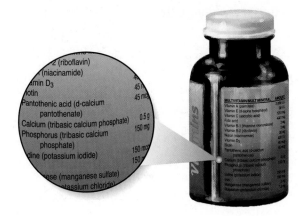

too much; 0.005 gram

48. The glass mirror of the Hubble telescope had to be repaired in space because it would not focus properly. The problem was that the mirror's outer edge had a thickness of 0.6248 centimeter when it was supposed to be 0.625 centimeter. Was the edge too thick or too thin? By how much? (*Source:* NASA.)

too thin; 0.0002 centimeter.

49. Precision Medical Parts makes an artificial heart valve that must measure between 0.998 centimeter and 1.002 centimeters. Circle the lengths that are acceptable:

1.01 cm (0.9991 cm) (1.0007 cm) 0.99 cm

50. The white rats in a medical experiment must start out weighing between 2.95 ounces and 3.05 ounces. Circle the weights that can be used:

(3.0 ounces)   (2.995 ounces)   3.055 ounces
(3.005 ounces)

51. Ginny Brown hoped her crops would get $3\frac{3}{4}$ inches of rain this month. The newspaper said the area received 3.8 inches of rain. Was that more or less than Ginny had hoped for? By how much?

more; 0.05 inch

52. The rats in the experiment in Exercise 50 gained $\frac{3}{8}$ ounce. They were expected to gain 0.3 ounce. Was their actual gain more or less than expected? By how much?

more; 0.075 ounce

Arrange each group of numbers in order, from least to greatest. See Example 4.

53. 0.54, 0.5455, 0.5399

0.5399, 0.54, 0.5455

54. 0.76, 0.7, 0.7006

0.7, 0.7006, 0.76

55. 5.8, 5.79, 5.0079, 5.804

5.0079, 5.79, 5.8, 5.804

56. 12.99, 12.5, 13.0001, 12.77

12.5, 12.77, 12.99, 13.0001

57. 0.628, 0.62812, 0.609, 0.6009

    0.6009, 0.609, 0.628, 0.62812

58. 0.27, 0.281, 0.296, 0.3

    0.27, 0.281, 0.296, 0.3

59. 5.8751, 4.876, 2.8902, 3.88

    2.8902, 3.88, 4.876, 5.8751

60. 0.98, 0.89, 0.904, 0.9

    0.89, 0.9, 0.904, 0.98

61. $0.043, 0.051, 0.006, \dfrac{1}{20}$

    $\mathbf{0.006, 0.043, \dfrac{1}{20}, 0.051}$

62. $0.629, \dfrac{5}{8}, 0.65, \dfrac{7}{10}$

    $\mathbf{\dfrac{5}{8}, 0.629, 0.65, \dfrac{7}{10}}$

63. $\dfrac{3}{8}, \dfrac{2}{5}, 0.37, 0.4001$

    $\mathbf{0.37, \dfrac{3}{8}, \dfrac{2}{5}, 0.4001}$

64. $0.1501, 0.25, \dfrac{1}{10}, \dfrac{1}{5}$

    $\mathbf{\dfrac{1}{10}, 0.1501, \dfrac{1}{5}, 0.25}$

Four boxes of fishing line are in the sale bin. The thicker the line, the stronger it is. The diameter of the fishing line is its thickness. Use the information on the boxes to answer Exercises 65–68.

65. Which color box has the strongest line?

    red box

66. Which color box has the line with the least strength?

    green box

67. Which color box has the line that is $\frac{1}{125}$ inch in diameter?

    green box

68. What is the difference in line diameter between the blue and purple boxes?

    No difference; they are equivalent.

Some rulers for technical occupations show each inch divided into tenths. Use this scale drawing for Exercises 69–74. Change the measurements on the drawing to decimals and round them to the nearest tenth of an inch.

69. Length **(a)** is   **1.4 in. (rounded)**

70. Length **(b)** is   **1.1 in. (rounded)**

71. Length **(c)** is   **0.3 in. (rounded)**

72. Length **(d)** is   **0.5 in.**

73. Length **(e)** is   **0.4 in. (rounded)**

74. Length **(f)** is   **0.7 in. (rounded)**

Chapter 4  Summary

▶ Key Terms

4.1	**decimals**	Decimals, like fractions, are used to show parts of a whole.
	decimal point	A decimal point is the dot that is used to separate the whole number part from the fractional part of a decimal number.
	place value	A place value is assigned to each place to the right or left of the decimal point. Whole numbers such as ones and tens, are to the *left* of the decimal point. Fractional parts, such as tenths and hundredths, are to the *right* of the decimal point.
4.2	**rounding**	Rounding is "cutting off" a number after a certain place, such as rounding to the nearest hundredth. The rounded number is less accurate than the original number. You can use the symbol "≈" to mean "is approximately equal to."
	decimal places	Decimal places are the number of digits to the *right* of the decimal point. For example, 6.37 has two decimal places, and 4.706 has three decimal places.
4.3	**estimating**	Estimating is the process of rounding the numbers in a problem and getting an approximate answer. This helps you check that the decimal point is in the correct place in the exact answer.
4.5	**repeating decimal**	A repeating decimal (like the 6 in 0.1666 . . .) is a decimal number with one or more digits that repeat forever; it never ends. Use three dots to indicate that it is a repeating decimal. Or, write the number with a bar above the repeating digits, as in $0.1\overline{6}$. (Use the dots or the bar, but not both.)

▶ New Symbols

$3.8\overline{6}$ ◀── Bar above repeating digit(s) in a decimal number 3.866 . . .   Three dots indicate a repeating decimal

▶ Test Your Word Power

See how well you have learned the vocabulary in this chapter. Answers follow the Quick Review.

1. **Decimal numbers** are like fractions in that they both
   A. must be written in lowest terms
   B. need common denominators
   C. have decimal points
   D. represent parts of a whole.

2. **Decimal places** refer to
   A. the digits from 0 to 9
   B. digits to the left of the decimal point
   C. digits to the right of the decimal point
   D. the number of zeros in a decimal number.

3. When a decimal number is **rounded,** it
   A. always ends in 0
   B. has the same number of decimal places as the original number
   C. is less accurate than the original number
   D. is less than one whole.

4. The **decimal point**
   A. separates the whole number part from the fractional part
   B. is always moved when finding a quotient
   C. separates tenths from hundredths
   D. is at the far left side of a whole number.

5. The number $0.\overline{3}$ is an example of
   A. an estimate
   B. a repeating decimal
   C. a rounded number
   D. a truncated number.

6. The **place value** names on the right side of the decimal point are
   A. ones, tens, hundreds, and so on
   B. ones, tenths, hundredths, and so on
   C. zero, one, two, three, four, and so on
   D. tenths, hundredths, thousandths, and so on.

▶ Quick Review

4.1 Reading and Writing Decimals

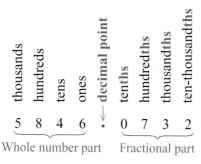

Write each decimal in words.

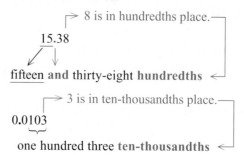

fifteen **and** thirty-eight **hundredths**

> 3 is in ten-thousandths place.

0.0103

one hundred three **ten-thousandths**

4.1 Writing Decimals as Fractions

The digits to the right of the decimal point are the numerator. The place value of the rightmost digit determines the denominator.

Always write the fractions in lowest terms.

Write 0.45 as a fraction in lowest terms.

The numerator is 45. The rightmost digit, 5, is in the hundredths place, so the denominator is 100. Then write the fraction in lowest terms.

$$\frac{45}{100} = \frac{45 \div 5}{100 \div 5} = \frac{9}{20} \leftarrow \text{Lowest terms}$$

4.2 Rounding Decimals

Find the place to which you are rounding. Draw a cut-off line to the right of that place; the rest of the digits will be dropped. Look *only* at the first digit being cut. If it is *4 or less,* the part you are keeping stays the same. If it is *5 or more,* the part you are keeping rounds up. Do not move the decimal point when rounding. Write "≈" to mean "is approximately equal to."

Round 0.17952 to the nearest thousandth.

> First digit cut is *5 or more,* so round up.

$$0.179 | 52$$

$$
\begin{array}{l}
0.179 \leftarrow \text{Keep this part.}\\
+\ 0.001 \leftarrow \text{To round up, add 1 thousandth.}\\
\hline
0.180
\end{array}
$$

0.17952 rounds to 0.180. Write $0.17952 \approx 0.180$

4.3 Adding and Subtracting Decimals

Use front end rounding to round each number and estimate the answer.

To find the exact answer, line up the decimal points. If needed, write in zeros as placeholders. Add or subtract as if they were whole numbers. Line up the decimal point in the answer directly below the decimal points in the problem.

Add $5.68 + 785.3 + 12 + 2.007$.

Estimate:	*Exact:*
$6 \longleftarrow$	5.680
$800 \longleftarrow$	785.300
$10 \longleftarrow$	12.000
$+\ \ \ 2 \longleftarrow$	$+\ \ \ 2.007$
818	804.987

Use zeros as place-holders so that all numbers have three decimal places.

Line up decimal points.

The estimate and exact answer are both in the hundreds, so the decimal point is probably in the correct place.

4.4 Multiplying Decimals

Step 1 Multiply as you would for whole numbers.
Step 2 Count the total number of decimal places in both factors.
Step 3 Write the decimal point in the product so it has the same number of decimal places as the total from Step 2. You may need to write extra zeros on the left side of the product to get enough decimal places.

Multiply 0.169×0.21.

$$
\begin{array}{r}
0.169 \leftarrow \text{3 decimal places}\\
\times\ \ 0.21 \leftarrow \text{2 decimal places}\\
\hline
169 \quad\quad \text{5 total decimal places}\\
338\quad\ \ \\
\hline
.03549 \leftarrow \text{5 decimal places in product}
\end{array}
$$

Write a 0 in the product so you can count over 5 decimal places. The final answer is 0.03549.

Concepts	Examples

(4.5) Dividing by Decimals

Step 1   Change the divisor to a whole number by moving the decimal point to the right.

Step 2   Move the decimal point in the dividend the same number of places to the right.

Step 3   Write the decimal point in the quotient directly above the decimal point in the dividend.

Step 4   Divide as with whole numbers.

Divide 52.8 by 0.75.

$$
\begin{array}{r}
7\,0.4 \\
0.75\overline{)5\,2.8\,0\,0} \\
5\,2\,5 \\
\hline
3\,0\,0 \\
3\,0\,0 \\
\hline
0
\end{array}
$$

Move the decimal point two places to the right in the divisor and dividend. Write zeros in the dividend so you can move the decimal point and continue dividing until the remainder is 0.

To check your answer, multiply 70.4 times 0.75. If the result matches the dividend (52.8), you solved the problem correctly.

(4.6) Writing Fractions as Decimals

Divide the numerator by the denominator. If necessary, round to the place indicated.

Write $\frac{1}{8}$ as a decimal.

$\frac{1}{8}$ means $1 \div 8$. Write it as $8\overline{)1}$.

The decimal point is on the right side of 1.

$$
\begin{array}{r}
0.125 \\
8\overline{)1.000} \\
8 \\
\hline
20 \\
16 \\
\hline
40 \\
40 \\
\hline
0
\end{array}
$$

← Write a decimal point and three zeros so you can continue dividing.

Therefore, $\frac{1}{8}$ is equivalent to 0.125

(4.6) Comparing the Size of Fractions and Decimals

Step 1   Write any fractions as decimals.

Step 2   Write zeros so that all the numbers being compared have the same number of decimal places.

Step 3   Use < to mean "is less than," > to mean "is greater than," or list the numbers from least to greatest.

Arrange in order from least to greatest.

$$0.505 \qquad \frac{1}{2} \qquad 0.55$$

$0.505 = 505$ thousandths ← 505 is in the middle.

$\frac{1}{2} = 0.5 = 0.500 = 500$ thousandths ← 500 is least.

$0.55 = 0.550 = 550$ thousandths ← 550 is greatest.

$$(\text{least}) \ \frac{1}{2} \qquad 0.505 \qquad 0.55 \ (\text{greatest})$$

ANSWERS TO TEST YOUR WORD POWER

1. D; *Example:* For 0.7, the whole is cut into ten parts, and you are interested in 7 of the parts.
2. C; *Examples:* The number 6.87 has two decimal places; 0.309 has three decimal places.
3. C; *Example:* When 0.815 is rounded to 0.8 it is accurate only to the nearest tenth, while the original number was accurate to the nearest thousandth.
4. A; *Example:* In 5.42, the whole number part is 5 ones, and the decimal part is 42 hundredths.
5. B; *Example:* The bar above the 3 in $0.\overline{3}$ indicates that the 3 repeats forever.
6. D; *Example:* In 6.219, the 2 is in the tenths place, the 1 is in the hundredths place, and the 9 is in the thousandths place.

Math in the Media

DOLLAR-COST AVERAGING

Making money in the stock market can be difficult. You want to buy shares of stock when the price is low and sell them when the price is high. Predicting the right time to buy is tricky. One strategy is dollar-cost averaging. You invest the same amount at regular intervals, like the first of each month. Over time you will usually buy more shares at lower prices, though there is no guarantee of making a profit.

The table below shows Microsoft's closing share price on the first day of each month in 2007. See what happens if you invest $100 each month rather than buying a lot of shares at one time.

Microsoft Corp. (MSFT)	Amount Invested	Price Per Share	Number of Shares Bought
January	$100	30.86	($100 ÷ 30.86) ≈ 3.2404 shares
February	$100	28.17	3.5499
March	$100	27.87	3.5881
April	$100	29.94	3.3400
May	$100	30.69	3.2584
June	$100	29.47	3.3933
July	$100	28.99	3.4495
August	$100	28.73	3.4807
September	$100	29.46	3.3944
October	$100	36.81	2.7167
November	$100	34.74	2.8785
December	$100	33.60	2.9762
Total investment	**$1200**		**39.2661**

Source: http://finance.yahoo.com

▦ Use a calculator to help answer these questions.

1. Calculate the total amount invested and enter the value in the table. **$1200**

2. Calculate the number of shares bought each month. Round the number of shares to the nearest ten-thousandth. The calculation for January is shown. **See table.**

3. Calculate the average market price per share. (*Hint:* Add the monthly prices per share and divide by 12.) **$30.78**

4. Calculate the average price based on *dollar-cost averaging*. (*Hint:* Divide the total amount invested by the total number of shares.) **$30.56**

5. Rank the following scenarios in order of which was the best investment (most profit or least loss). Show the value of each investment in December 2007 as a basis for your answer.

   (a) $1200 invested in Microsoft in January 2007 **rank 2; $1306.55 profit**

   (b) $1200 invested in Microsoft using *dollar-cost averaging* **rank 1; $1319.34 profit**

   (c) $1200 invested in Microsoft in October 2007 **rank 3; $1095.35 profit**

320

Chapter 4 Review Exercises

[4.1] *Name the digit that has the given place value.*

1. 243.059
tenths   **0**
hundredths   **5**

2. 0.6817
ones   **0**
tenths   **6**

3. $5824.39
hundreds   **8**
hundredths   **9**

4. 896.503
tenths   **5**
tens   **9**

5. 20.73861
tenths   **7**
ten-thousandths   **6**

Write each decimal as a fraction or mixed number in lowest terms.

6. 0.5 $\dfrac{1}{2}$

7. 0.75 $\dfrac{3}{4}$

8. 4.05 $4\dfrac{1}{20}$

9. 0.875 $\dfrac{7}{8}$

10. 0.027 $\dfrac{27}{1000}$

11. 27.8 $27\dfrac{4}{5}$

Write each decimal in words.

12. 0.8   **eight tenths**

13. 400.29   **four hundred and twenty-nine hundredths**

14. 12.007   **twelve and seven thousandths**

15. 0.0306   **three hundred six ten-thousandths**

Write each decimal in numbers.

16. eight and three tenths   **8.3**

17. two hundred five thousandths   **0.205**

18. seventy and sixty-six ten-thousandths   **70.0066**

19. thirty hundredths   **0.30**

[4.2] *Round each decimal to the place indicated.*

20. 275.635 to the nearest tenth   **275.6**

21. 72.789 to the nearest hundredth   **72.79**

22. 0.1604 to the nearest thousandth   **0.160**

23. 0.0905 to the nearest thousandth   **0.091**

24. 0.98 to the nearest tenth   **1.0**

Round each money amount to the nearest cent.

25. $15.8333   **$15.83**

26. $0.698   **$0.70**

27. $17,625.7906   **$17,625.79**

Round each income or expense item to the nearest dollar.

28. Income from pancake breakfast was $350.48.   **$350**

29. Members paid $129.50 in dues.   **$130**

30. Refreshments cost $99.61.   **$100**

31. Bank charges were $29.37.   **$29**

[4.3] *First use front end rounding to round each number and estimate the answer. Then find the exact answer.*

32. *Estimate:*   *Exact:*

$$
\begin{array}{r}
6 \\
400 \\
+\ 20 \\
\hline
426
\end{array}
\qquad
\begin{array}{r}
5.81 \\
423.96 \\
+\ 15.09 \\
\hline
444.86
\end{array}
$$

33. *Estimate:*   *Exact:*

$$
\begin{array}{r}
80 \\
1 \\
100 \\
1 \\
+\ 30 \\
\hline
212
\end{array}
\qquad
\begin{array}{r}
75.6 \\
1.29 \\
122.045 \\
0.88 \\
+\ 33.7 \\
\hline
233.515
\end{array}
$$

34. *Estimate:*   *Exact:*

$$
\begin{array}{r}
300 \\
-\ 20 \\
\hline
280
\end{array}
\qquad
\begin{array}{r}
308.5 \\
-\ 17.8 \\
\hline
290.7
\end{array}
$$

35. *Estimate:*   *Exact:*

$$
\begin{array}{r}
9 \\
-\ 8 \\
\hline
1
\end{array}
\qquad
\begin{array}{r}
9.2 \\
-\ 7.9316 \\
\hline
1.2684
\end{array}
$$

36. American's favorite household pet is a cat. There are about 90.5 million pet cats, 73.9 million pet dogs, and 16.6 million pet birds. How many more pet cats are there than pet birds? (*Source:* American Pet Products Manufacturers Association.)

Estimate: **90 million − 20 million = 70 million**

Exact: **73.9 million more cats**

37. Today, Jasmin had $306 in her checking account. She wrote a check to the day care center for $215.53 and a check for $44.67 at the grocery store. What is the new balance in her account?

Estimate: **$300 − $200 − $40 = $60**

Exact: **$45.80**

38. Joey spent $1.59 for toothpaste, $5.33 for a gift, and $18.94 for a toaster. He gave the clerk three $10 bills. How much change did he get?

Estimate: **$2 + $5 + $20 = $27;**
                   $30 − $27 = $3

Exact: **$4.14**

39. Roseanne is training for a wheelchair race. She raced 2.3 kilometers on Monday, 4 kilometers on Wednesday, and 5.25 kilometers on Friday. How far did she race altogether?

Estimate: **2 + 4 + 5 = 11 kilometers**

Exact: **11.55 kilometers**

[4.4] *First use front end rounding to round each number and estimate the answer. Then find the exact answer.*

40. *Estimate:*   *Exact:*

$$
\begin{array}{r}
6 \\
\times\ 4 \\
\hline
24
\end{array}
\qquad
\begin{array}{r}
6.138 \\
\times\ 3.7 \\
\hline
22.7106
\end{array}
$$

41. *Estimate:*   *Exact:*

$$
\begin{array}{r}
40 \\
\times\ 3 \\
\hline
120
\end{array}
\qquad
\begin{array}{r}
42.9 \\
\times\ 3.3 \\
\hline
141.57
\end{array}
$$

Find each product.

42. (5.6) (0.002)

0.0112

43. 0.071 (0.005)

0.000355

[4.5] *Decide whether each answer is reasonable by rounding the numbers and estimating the answer. If the exact answer is not reasonable, find and correct the error.*

44. $706.2 \div 12 = 58.85$ **reasonable**

Estimate:
700 ÷ 10 = 70

45. $26.6 \div 2.8 = 0.95$ **unreasonable**

Estimate:
30 ÷ 3 = 10

$$\begin{array}{r} 9.5 \\ 2.8\overline{)26.6} \end{array}$$

Divide. Round each quotient to the nearest thousandth if necessary.

46. $3\overline{)43.4}$

14.467 (rounded)

47. $\dfrac{72}{0.06}$

1200

48. $0.00048 \div 0.0012$

0.4

[4.4–4.5] *Solve each application problem.*

49. Adrienne worked 46.5 hours this week. Her hourly wage is $14.24 for the first 40 hours and 1.5 times that rate over 40 hours. Find her total earnings to the nearest dollar.

$708 (rounded)

50. A book of 12 tickets costs $35.89 at the State Fair midway. What is the cost per ticket, to the nearest cent?

$2.99 (rounded)

51. Stock in MathTronic sells for $3.75 per share. Kenneth is thinking of investing $500. How many whole shares could he buy?

133 shares (rounded)

52. Grapes are on sale at $0.99 per pound. How much will Ms. Lee pay for 3.5 pounds of grapes, to the nearest cent?

$3.47 (rounded)

Simplify each expression.

53. $3.5^2 + 8.7(1.95)$

29.215

54. $11 - 3.06 \div (3.95 - 0.35)$

10.15

[4.6] *Write each fraction or mixed number as a decimal. Round to the nearest thousandth when necessary.*

55. $3\dfrac{4}{5}$

3.8

56. $\dfrac{16}{25}$

0.64

57. $1\dfrac{7}{8}$

1.875

58. $\dfrac{1}{9}$

0.111 (rounded)

Arrange each group of numbers in order from least to greatest.

59. 3.68, 3.806, 3.6008

3.6008, 3.68, 3.806

60. 0.215, 0.22, 0.209, 0.2102

0.209, 0.2102, 0.215, 0.22

61. $0.17, \dfrac{3}{20}, \dfrac{1}{8}, 0.159$

$\dfrac{1}{8}, \dfrac{3}{20}, 0.159, 0.17$

▶▶▶ Mixed Review Exercises

Add, subtract, multiply, or divide as indicated.

62. $89.19 + 0.075 + 310.6 + 5$

404.865

63. 72.8×3.5

254.8

64. $1648.3 \div 0.46$ Round to the nearest thousandth.

3583.261 (rounded)

65. $30 - 0.9102$

29.0898

66. $4.38(0.007)$

0.03066

67. $0.005\overline{)0.047}$

9.4

68. $72.105 + 8.2 + 95.37$

175.675

69. $81.36 \div 9$

9.04

70. $(5.6 - 1.22) + 4.8\,(3.15)$

19.50

71. 0.455×18

8.19

72. $(1.6)\,(0.58)$

0.928

73. $0.218\overline{)7.63}$

35

74. $21.059 - 20.8$

0.259

75. $18.3 - 3^2 \div 0.5$

0.3

Use the information in the ad to answer Exercises 76–80. Round money answers to the nearest cent. (Disregard any sales tax.)

Grand Opening Sale!
Save on Clothing for the Entire Family

Jeans for Teens
only $19.95 each
women's sizes $24.99

Athletic Shoes
regularly priced
$89.99 to $149.50
NOW just $71 to $119.60

Men's socks *NOW* *3 pairs for $8.99*
Children's socks *6 pairs for $5*

Hurry in — *TWO DAYS ONLY*

76. How much would one pair of men's socks cost?

$3.00 (rounded)

77. How much more would one pair of men's socks cost than one pair of children's socks?

$2.17 (rounded)

78. How much would Fernando pay for a dozen pair of men's socks?

$35.96

79. How much would Akiko pay for five pairs of teen jeans and four pairs of women's jeans?

$199.71

80. What is the difference between the cheapest sale price for athletic shoes and the highest regular price?

$78.50

To decrease your risk of clogged arteries, it is recommended that you get at least 2 milligrams of vitamin B-6 each day. Use the information in the table to answer Exercises 81–82.

**SOURCES OF VITAMIN B-6
(AMOUNTS IN MILLIGRAMS)**

1 cup orange juice	0.11
1 banana	0.66
1 baked potato with skin	0.7
$\frac{1}{2}$ cup strawberries	0.45
3 ounces skinless chicken	0.5
3 ounces water-packed tuna	0.3
$\frac{1}{2}$ cup chickpeas	0.57

Source: Harvard Health Letter.

81. (a) Which food item has the highest amount of vitamin B-6?

baked potato with skin

(b) Which food item has the lowest amount?

cup of orange juice

(c) What is the difference in the amount of vitamin B-6 between the food items with the highest and lowest amounts?

0.59 milligram

82. (a) Suppose you ate a banana, 3 ounces of skinless chicken, and 1 cup of strawberries. How many milligrams of vitamin B-6 would you get?

2.06 milligrams

(b) Did you get more or less than the recommended daily amount? By how much?

more, by 0.06 milligram

Chapter 4 ▸▸▸▸ Test

 **Test Prep** VIDEO CD

Use the Chapter Test Prep Video CD to see fully worked-out solutions to any of the exercises you want to review.

Write each decimal as a fraction or mixed number in lowest terms.

1. 18.4

2. 0.075

Write each decimal in words.

3. 60.007

4. 0.0208

Round each decimal to the place indicated.

5. 725.6089 to the nearest tenth

6. 0.62951 to the nearest thousandth

7. $1.4945 to the nearest cent

8. $7859.51 to the nearest dollar

First use front end rounding to round each number and estimate the answer. Then find the exact answer.

9. 7.6 + 82.0128 + 39.59

10. 79.1 − 3.602

11. 5.79 (1.2)

12. 20.04 ÷ 4.8

Find the exact answer.

13. 53.1 + 4.631 + 782 + 0.031

14. 670 − 0.996

15. (0.0069) (0.007)

16. 0.15)‾72‾

1. $18\frac{2}{5}$

2. $\dfrac{3}{40}$

3. sixty and seven thousandths

4. two hundred eight ten-thousandths

5. 725.6

6. 0.630

7. $1.49

8. $7860

9. Estimate: 8 + 80 + 40 = 128
Exact: 129.2028

10. Estimate: 80 − 4 = 76
Exact: 75.498

11. Estimate: 6(1) = 6
Exact: 6.948

12. Estimate: 20 ÷ 5 = 4
Exact: 4.175

13. 839.762

14. 669.004

15. 0.0000483

16. 480

17. _2.625_ _____

17. Write $2\dfrac{5}{8}$ as a decimal. Round to the nearest thousandth, if necessary.

Arrange in order from least to greatest.

18. _0.44, $\dfrac{9}{20}$, 0.4506, 0.451_ _____

18. 0.44, 0.451, $\dfrac{9}{20}$, 0.4506

Simplify this expression.

19. _35.49_ _____

19. $6.3^2 - 5.9 + 3.4\,(0.5)$

Solve each application problem.

20. _$446.87_ _____

20. Jennifer had $71.15 in her checking account. Yesterday her account earned $0.95 interest for the month, and she deposited a paycheck for $390.77. The bank charged her $16 for new checks. What is the new balance in her account?

21. _pintails, wigeons, gadwalls_ _____

21. Three types of ducks that are hunted in the United States are gadwalls, wigeons, and pintails. The estimated populations of these ducks are 2.5 million gadwalls, 2.551 million wigeons, and 2.56 million pintails. List the ducks in order from the greatest number to the least. (*Source:* U.S. Fish and Wildlife Service.)

22. _$5.35 (rounded)_ _____

22. Mr. Yamamoto bought 1.85 pounds of cheese at $2.89 per pound. What was the total amount he paid for the cheese, to the nearest cent?

23. _2.8 degrees_ _____

23. Loren's baby had a temperature of 102.7 degrees. Later in the day it was 99.9 degrees. How much had the baby's temperature dropped?

24. _$1.79 per foot (rounded)_ _____

24. Pat bought 6.5 ft of decorative gold chain to hang a light over her dining table. She paid $11.64. What was the cost per foot, to the nearest cent?

25. _Answers will vary._ _____

✐ **25.** Write your own application problem using decimals. Make it different from Problems 20–24. Then show how to solve your problem.

Study Skills

▶▶▶ ANALYZING YOUR TEST RESULTS

OBJECTIVES

1 Determine the reason for errors.

2 Develop a plan to avoid test taking errors.

3 Review material to correct misunderstandings.

Immediately After the Test

After the Test Is Returned

Find Out Why You Made the Errors You Made

After taking a test, many students heave a big sigh of relief and try to forget it ever happened. Don't fall into this trap! An exam is a learning opportunity. It gives you clues about *what your instructor thinks is important,* what *concepts and skills are valued* in mathematics, and *if you are on the right track.*

Jot down problems that caused you trouble. Find out how to solve them by checking your textbook, notes, or asking your instructor or tutor (if available). You might see those same problems again on a final exam.

Find out what you got wrong and why you had points deducted. Write down the problem so you can learn how to do it correctly. Sometimes you only have a short time in class to review your test. *If you need more time,* ask your instructor if you can look at the test in his or her office.

Here is a list of typical reasons for making errors on math tests.

1. You read the directions wrong.
2. You read the question wrong or skipped over something.
3. You made a computation error (maybe even an easy one).
4. Your answer is not accurate.
5. Your answer is not complete.
6. You labeled your answer wrong. For example, you labeled it "feet" and it should have been "feet^2."
7. You didn't show your work.
8. *You didn't understand the concept.
9. *You were unable to go from words (in a word problem) to setting up the problem.
10. *You were unable to apply a procedure to a new situation.
11. You were so anxious that you made errors even when you knew the material.

The first seven errors are **test-taking errors.** They are easy to correct if you decide to carefully read test questions and directions, proofread or rework your problems, show all your work, and double check units and labels every time.

The three starred errors (*) are **test preparation errors.** Remember that to grow a complex neural network, you need to practice the kinds of problems that you will see on the tests. So, for example, if application problems are difficult for you, you must *do more application problems!* If you have practiced the study skills techniques, however, you are less likely to make these kinds of errors on tests because you will have a deeper understanding of course concepts and you will be able to remember them better.

The last error isn't really an error. **Anxiety** can play a big part in your test results. Go back to the *Preparing for Tests* activity and read the suggestions about exercise and deep breathing. Recall from the *Your Brain Can Learn Mathematics* activity that when you are anxious, your body produces adrenaline. The presence of *adrenaline in the brain blocks connections* between dendrites. If you can *reduce* the adrenaline in your system, you will be able to *think more clearly* during your test. Just five minutes of brisk walking right before your test can help do that. Also, *practicing a relaxation technique while you do your homework* will make it more likely that you can benefit from using the technique during a test. *Deep breathing* is helpful because it gets *oxygen into your brain*. When you are anxious you tend to breathe more shallowly, which can make you feel confused and easily distracted.

Make a Plan for the Next Test

Make a plan for your next test based on your results from this test. You might review the Chapter Summary and work the problems in the Chapter Review Exercises or the Chapter Test. Ask your instructor or a tutor (if available) for more help if you are confused about any of the problems.

Now Try This ▶▶▶

Below is a record sheet to track your progress in test taking. Use it to find out if you make particular kinds of errors. Then you can work specifically on correcting them. Just check in the box when you made one of the errors. If you take more than five tests, make your own grid on separate paper.

What will you do to avoid Test Taking Errors?

_____

_____

_____

Test Taking Errors

Test #	Read directions wrong	Read question wrong	Computation error	Not exact or accurate	Not complete	Labeled wrong	Didn't show work
1							
2							
3							
4							
5							

What will you do to avoid Test Preparation Errors?

_____

_____

_____

Test Preparation Errors

Test #	Didn't understand concept	Didn't set up problem correctly	Couldn't apply concept to new situation
1			
2			
3			
4			
5			

What will you do to reduce anxiety?

_____

_____

_____

Anxiety

Test #	Felt anxious *before* the exam	Felt anxious *during* the exam	Blanked out on questions	Got questions wrong that I knew how to do
1				
2				
3				
4				
5				

Cumulative Review Exercises ▷▷▷ Chapters 1–4

Round each number as indicated.

1. 499,501 to the nearest thousand
500,000

2. 602.4937 to the nearest hundredth
602.49

3. $709.60 to the nearest dollar
$710

4. $0.0528 to the nearest cent
$0.05

Simplify each expression in Exercises 5–21.

5. $10 - 0.329$
9.671

6. $2\frac{3}{5} \cdot \frac{5}{9}$
$1\frac{4}{9}$

7. $11\frac{1}{5} \div 8$
$1\frac{2}{5}$

8. $5006 - 92$
4914

9. $0.7 + 85 + 7.903$
93.603

10. Write your answer using R for the remainder.
$$\overset{404 \text{ R}3}{7\overline{)2831}}$$

11. $\frac{5}{6} + \frac{7}{8}$
$1\frac{17}{24}$

12. 332×704
233,728

13. $(0.006)(5.44)$
0.03264

14. $3.2(2.5)$
8

15. $25.2 \div 0.56$
45

16. $\frac{2}{3} \div 5\frac{1}{6}$
$\frac{4}{31}$

17. $5\frac{1}{4} - 4\frac{7}{12}$
$\frac{2}{3}$

18. $4.7 \div 9.3$
Round to the nearest hundredth.
0.51 (rounded)

19. $10 - 4 \div 2 \cdot 3$
4

20. $\sqrt{36} + 3(8) - 4^2$
14

21. Write 40.035 in words.
forty and thirty-five thousandths

22. Write three hundred six ten-thousandths in numbers.
0.0306

Arrange each group of numbers in order, from least to greatest.

23. 7.005, 7.5005, 7.5, 7.505
7.005, 7.5, 7.5005, 7.505

24. $\frac{7}{8}$, 0.8, $\frac{21}{25}$, 0.8015
0.8, 0.8015, $\frac{21}{25}$, $\frac{7}{8}$

First round the numbers and estimate the answer to each application problem. Then find the exact answer.

25. Lameck had two $20 bills. He spent $17.96 on gasoline and $0.87 for a candy bar at the convenience store. How much money does he have left?

Estimate: $40 − $18 − $1 = $21

Exact: $21.17

26. Manuela's daughter is 50 inches tall. Last year she was $46\frac{5}{8}$ inches tall. How much has she grown? (When estimating, round to the nearest whole number.)

Estimate: 50 − 47 = 3 inches

Exact: $3\frac{3}{8}$ inches

27. Sharon records textbooks on tape for students who are blind. Her hourly wage is $11.63. How much did she earn working 16.5 hours last week, to the nearest cent?

Estimate: $10 × 20 = $200

Exact: $191.90 (rounded)

28. The Farnsworth Elementary School has eight classrooms with 22 students in each one and 12 classrooms with 26 students in each one. How many students attend the school?

Estimate: (8 × 20) + (10 × 30) = 160 + 300 = 460 students

Exact: 488 students

29. Toshihiro bought $2\frac{1}{3}$ yd of cotton fabric and $3\frac{7}{8}$ yd of wool fabric. How many yards did he buy in all?

Estimate: 2 + 4 = 6 yd

Exact: $6\frac{5}{24}$ yd

30. Kimberly had $29.44 in her checking account. She wrote a check for $40 and deposited a $220.06 paycheck into her account, but not in time to prevent an $18 overdraft charge. What is the new balance in her account?

Estimate: $30 + $200 − $40 − $20 = $170

Exact: $191.50

31. Paulette bought 2.7 pounds of apples for $6.18. What was the cost per pound, to the nearest cent?

Estimate: $6 ÷ 3 = $2 per pound

Exact: $2.29 per pound (rounded)

32. Carter Community College received a $78,000 grant from a local hospital to help students pay tuition for nursing classes. How much money could be given to each of 107 students? Round to the nearest dollar.

Estimate: $80,000 ÷ 100 = $800

Exact: $729 (rounded)

Use the information in the table to answer Exercises 33–36. All measurements are in inches.

Children's Hats	Head Size	Order Size:
Measure around head above eyebrow ridges.	16½"–18"	XXS
	18¼"–19"	XS
	19¼"–20"	S
	20¼"–21⅛"	M
	21½"–22¼"	L

Source: Lands' End.

33. If the distance around your child's head is $21\frac{1}{16}$ in., which hat size should you order?

size M (medium)

34. Find the difference between the smaller and larger measurements for each of the hat sizes.

XXS $1\frac{1}{2}$ in.; **XS** $\frac{3}{4}$ in.; **S** $\frac{3}{4}$ in.; **M** $\frac{7}{8}$ in.; **L** $\frac{3}{4}$ in.

35. Change the measurements for the medium (M) size to decimals. Then find the difference in the measurements. Show why this answer is equivalent to your answer from Exercise 34.

21.125 − 20.25 = 0.875 in.; 7 ÷ 8 = 0.875 in.

or $\frac{875}{1000} = \frac{875 ÷ 125}{1000 ÷ 125} = \frac{7}{8}$ in.

36. Did you prefer doing the subtraction using fractions (as in Exercise 34) or using decimals (as in Exercise 35)? Explain your reasoning.

Answers will vary. Many people prefer using decimals because you do not need to find a common denominator or rewrite answers in lowest terms.

5

Ratio and Proportion

Three out of every four Americans have cellular phones. Worldwide, about 1 out of 3 people use a cell phone! (*Source:* International Telecommunication Union.)

You can use unit rates to find the best deal on cell phone service plans. But most plans do not cover long-distance calls to certain U.S. locations or to other countries. Then you can use unit rates to find the cheapest long-distance calling card. (See **Section 5.2**, Exercises 29–32 for calling cards and Exercises 37–40 for cell phone plans; see Cumulative Review, Exercises 30–33 for international calling cards.)

5.1 ▶▶▶ Ratios

OBJECTIVES

1. Write ratios as fractions.

2. Solve ratio problems involving decimals or mixed numbers.

3. Solve ratio problems after converting units.

A **ratio** compares two quantities. You can compare two numbers, such as 8 and 4, or two measurements that have the *same* type of units, such as 3 *days* and 12 *days*. (*Rates* compare measurements with different types of units and are covered in the next section.)

Ratios can help you see important relationships. For example, if the ratio of your monthly expenses to your monthly income is 10 to 9, then you are spending $10 for every $9 you earn and going deeper into debt.

OBJECTIVE 1 Write ratios as fractions. A ratio can be written in three ways.

Writing a Ratio

The ratio of $7 **to** $3 can be written:

$$7 \text{ to } 3 \qquad \text{or} \quad 7{:}3 \quad \text{or} \quad \frac{7}{3} \quad \leftarrow \text{Fraction bar indicates "to."}$$

"**:**" indicates "**to**"

Writing a ratio as a fraction is the most common method, and the one we will use here. All three ways are read, "the ratio of 7 to 3." The word **to** separates the quantities being compared.

Writing a Ratio as a Fraction

Order is important when writing a ratio. The quantity mentioned **first** is the **numerator**. The quantity mentioned **second** is the **denominator**. For example:

$$\text{The ratio of } 5 \text{ to } 12 \text{ is written } \frac{5}{12}$$

EXAMPLE 1 **Writing Ratios**

Ancestors of the Pueblo Indians built multistory apartment towns in New Mexico about 1100 years ago. A room might measure 14 ft long, 11 ft wide, and 15 ft high.

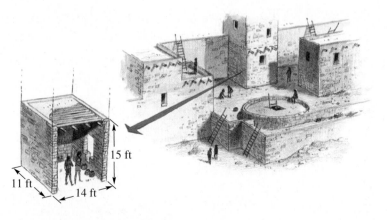

Continued on Next Page

Write each ratio as a fraction, using the room measurements.

(a) Ratio of length to width

The ratio of **length to width** is $\dfrac{14 \,\cancel{ft}}{11 \,\cancel{ft}} = \dfrac{14}{11}$

Do not rewrite the ratio as $1\frac{3}{11}$.

Numerator (mentioned first) — Denominator (mentioned second)

You can divide out common *units* just like you divided out common *factors* when writing fractions in lowest terms. (See **Section 2.4.**) However, do *not* rewrite the fraction as a mixed number. Keep it as the ratio of 14 to 11.

(b) Ratio of width to height

Divide out the common units (ft).

The ratio of width **to** height is $\dfrac{11 \,\cancel{ft}}{15 \,\cancel{ft}} = \dfrac{11}{15}$

> **CAUTION**
> Remember, the order of the numbers is important in a ratio. Look for the words "ratio of a to b." Write the ratio as $\dfrac{a}{b}$, ***not*** $\dfrac{b}{a}$. The quantity mentioned first is the numerator.

Work Problem *at the Side.* ▶

Any ratio can be written as a fraction. Therefore, you can write a ratio in *lowest terms,* just as you do with any fraction.

EXAMPLE 2 **Writing Ratios in Lowest Terms**

Write each ratio in lowest terms.

(a) 60 days to 20 days

The ratio is $\frac{60}{20}$. Write this ratio in lowest terms by dividing the numerator and denominator by 20.

$$\frac{60}{20} = \frac{60 \div 20}{20 \div 20} = \frac{3}{1} \quad \left\{ \begin{array}{l} \text{Ratio in} \\ \text{lowest terms} \end{array} \right.$$

So, the ratio of 60 days to 20 days is 3 to 1 or, written as a fraction, $\frac{3}{1}$.

> **CAUTION**
> In the fractions chapters you would have rewritten $\frac{3}{1}$ as 3. But a *ratio* compares *two* quantities, so you need to keep both parts of the ratio and write it as $\frac{3}{1}$.

(b) 50 ounces of medicine to 120 ounces of medicine

The ratio is $\frac{50}{120}$. Divide the numerator and denominator by 10.

$$\frac{50}{120} = \frac{50 \div 10}{120 \div 10} = \frac{5}{12} \quad \left\{ \begin{array}{l} \text{Ratio in} \\ \text{lowest terms} \end{array} \right.$$

So, the ratio of 50 ounces to 120 ounces is $\frac{5}{12}$.

—— Continued on Next Page

---

1 Shane spent \$14 on meat, \$5 on milk, and \$7 on fresh fruit. Write each ratio as a fraction.

(a) The ratio of amount spent on fruit to amount spent on milk.

(b) The ratio of amount spent on milk to amount spent on meat.

(c) The ratio of amount spent on meat to amount spent on milk.

ANSWERS

1. **(a)** $\dfrac{7}{5}$ **(b)** $\dfrac{5}{14}$ **(c)** $\dfrac{14}{5}$

2 Write each ratio as a fraction in lowest terms.

(a) 9 hours to 12 hours

(b) 100 meters to 50 meters

(c) Write the ratio of width to length for this rectangle.

Length
48 ft

Width
24 ft

3 Write each ratio as a ratio of whole numbers in lowest terms.

(a) The price of Tamar's favorite brand of lipstick increased from $5.50 to $7.00. Find the ratio of the increase in price to the original price.

(b) Last week, Lance worked 4.5 hours each day. This week he cut back to 3 hours each day. Find the ratio of the decrease in hours to the original number of hours.

(c) 15 people in a large van to 6 people in a small van

$$\text{The ratio is } \frac{15}{6} = \frac{15 \div 3}{6 \div 3} = \frac{5}{2} \quad \left\{ \begin{array}{l} \text{Ratio in} \\ \text{lowest terms} \end{array} \right.$$

Note
Although $\frac{5}{2} = 2\frac{1}{2}$, ratios are *not* written as mixed numbers. Nevertheless, in Example 2(c) above, the ratio $\frac{5}{2}$ does mean the large van holds $2\frac{1}{2}$ times as many people as the small van.

◀ *Work Problem* **2** *at the Side.*

OBJECTIVE **2** **Solve ratio problems involving decimals or mixed numbers.** Sometimes a ratio compares two decimal numbers or two fractions. It is easier to understand if we rewrite the ratio as a ratio of two whole numbers.

EXAMPLE 3 **Using Decimal Numbers in a Ratio**

The price of a Sunday newspaper increased from $1.50 to $1.75. Find the ratio of the increase in price to the original price.
  The words increase in price are mentioned first, so the increase will be the numerator. How much did the price go up? Use subtraction.

$$\text{new price} - \text{original price} = \text{increase}$$
$$\$1.75 \quad - \quad \$1.50 \quad = \quad \$0.25$$

The words the original price are mentioned second, so the original price of $1.50 is the denominator.
  The ratio of increase in price to original price is shown below.

$$\frac{0.25}{1.50} \begin{array}{l} \leftarrow \text{increase in price} \\ \leftarrow \text{original price} \end{array}$$

Now rewrite the ratio as a ratio of whole numbers. Recall that if you multiply both the numerator and denominator of a fraction by the same number, you get an equivalent fraction. The decimals in this example are hundredths, so multiply by 100 to get whole numbers. (If the decimals are tenths, multiply by 10. If thousandths, multiply by 1000.) Then write the ratio in lowest terms.

$$\frac{0.25}{1.50} = \underbrace{\frac{0.25 \times 100}{1.50 \times 100} = \frac{25}{150}}_{\substack{\text{Ratio as two} \\ \text{whole numbers}}} = \frac{25 \div 25}{150 \div 25} = \frac{1}{6} \quad \left\{ \begin{array}{l} \text{Ratio in} \\ \text{lowest terms} \end{array} \right.$$

◀ *Work Problem* **3** *at the Side.*

EXAMPLE 4 **Using Mixed Numbers in Ratios**

Write each ratio as a comparison of whole numbers in lowest terms.

(a) 2 days to $2\frac{1}{4}$ days

  Write the ratio as follows. Divide out the common units.

$$\frac{2 \text{ days}}{2\frac{1}{4} \text{ days}} = \frac{2}{2\frac{1}{4}}$$

Continued on Next Page

ANSWERS
2. (a) $\frac{3}{4}$ (b) $\frac{2}{1}$ (c) $\frac{1}{2}$

3. (a) $\frac{1.50 \times 100}{5.50 \times 100} = \frac{150 \div 50}{550 \div 50} = \frac{3}{11}$
   (b) $\frac{1.5 \times 10}{4.5 \times 10} = \frac{15 \div 15}{45 \div 15} = \frac{1}{3}$

Next, write 2 as $\frac{2}{1}$ and $2\frac{1}{4}$ as the improper fraction $\frac{9}{4}$.

> Think: $4 \cdot 2 = 8$
> and $8 + 1 = 9$
> so $2\frac{1}{4} = \frac{9}{4}$

$$\frac{\frac{2}{1}}{2\frac{1}{4}} = \frac{\frac{2}{1}}{\frac{9}{4}}$$

Now rewrite the problem in horizontal format, using the "÷" symbol for division. Finally, multiply by the reciprocal of the divisor, as you did in **Section 2.7.**

$$\frac{\frac{2}{1}}{\frac{9}{4}} = \frac{2}{1} \div \frac{9}{4} = \frac{2}{1} \cdot \frac{4}{9} = \frac{8}{9}$$

Reciprocals

The ratio, in lowest terms, is $\frac{8}{9}$.

(b) $3\frac{1}{4}$ to $1\frac{1}{2}$

Write the ratio as $\dfrac{3\frac{1}{4}}{1\frac{1}{2}}$. Then write $3\frac{1}{4}$ and $1\frac{1}{2}$ as improper fractions.

$$3\frac{1}{4} = \frac{13}{4} \quad \text{and} \quad 1\frac{1}{2} = \frac{3}{2}$$

The ratio is shown below.

$$\frac{3\frac{1}{4}}{1\frac{1}{2}} = \frac{\frac{13}{4}}{\frac{3}{2}}$$

Rewrite as a division problem in horizontal format, using the "÷" symbol. Then multiply by the reciprocal of the divisor.

$$\frac{13}{4} \div \frac{3}{2} = \frac{13}{\overset{}{\underset{2}{4}}} \cdot \frac{\overset{1}{2}}{3} = \frac{13}{6} \quad \left\{ \begin{array}{l} \text{Ratio in} \\ \text{lowest terms} \end{array} \right.$$

Work Problem **4** at the Side. ▶

OBJECTIVE 3 Solve ratio problems after converting units.
When a ratio compares measurements, both measurements must be in the *same* units. For example, *feet* must be compared to *feet, hours* to *hours, pints* to *pints,* and *inches* to *inches.*

EXAMPLE 5 **Ratio Applications Using Measurement**

(a) Write the ratio of the length of the shorter board on the left to the length of the longer board on the right. Compare in inches.

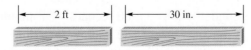

First, express 2 ft in inches. Because 1 ft has 12 in., 2 ft is

$$2 \cdot 12 \text{ in.} = 24 \text{ in.}$$

Continued on Next Page

---

4 Write each ratio as a ratio of whole numbers in lowest terms.

(a) $3\frac{1}{2}$ to 4

(b) $5\frac{5}{8}$ pounds to $3\frac{3}{4}$ pounds

(c) $3\frac{1}{2}$ in. to $\frac{7}{8}$ in.

5 Write each ratio as a fraction in lowest terms. (*Hint:* Recall that it is usually easier to write the ratio using the smaller measurement unit.)

(a) 9 in. to 6 ft

(b) 2 days to 8 hours

(c) 7 yd to 14 ft

(d) 3 quarts to 3 gallons

(e) 25 minutes to 2 hours

(f) 4 pounds to 12 ounces

On the previous page, the length of the board on the left is 24 in., so the ratio of the lengths is

$$\frac{2 \text{ ft}}{30 \text{ in.}} = \frac{24 \text{ in.}}{30 \text{ in.}} = \frac{24}{30}$$

Once the units match, you can divide them out.

Write the ratio in lowest terms.

$$\frac{24}{30} = \frac{24 \div 6}{30 \div 6} = \frac{4}{5} \quad \leftarrow \left\{ \begin{array}{l} \text{Ratio in} \\ \text{lowest terms} \end{array} \right.$$

The shorter board is $\frac{4}{5}$ the length of the longer board.

> **Note**
>
> Notice in the example above that we wrote the ratio using the smaller unit (inches are smaller than feet). Using the smaller unit will help you avoid working with fractions. If we wrote the ratio using feet, then
>
> $$30 \text{ in.} = 2\frac{1}{2} \text{ ft}$$
>
> So the ratio in feet is shown below.
>
> $$\frac{2 \text{ ft}}{2\frac{1}{2} \text{ ft}} = \frac{2}{1} \div \frac{5}{2} = \frac{2}{1} \cdot \frac{2}{5} = \frac{4}{5} \quad \leftarrow \text{Same result}$$
>
> The ratio is the same, but it takes more steps to get the answer. Using the smaller unit is usually easier.

(b) Write the ratio of 28 days to 3 weeks.

Since it is easier to write the ratio using the smaller measurement unit, compare in *days* because days are shorter than weeks.

First express 3 weeks in days. Because 1 week has 7 days, 3 weeks is

$$3 \cdot 7 \text{ days} = 21 \text{ days.}$$

So the ratio in days is shown below.

$$\frac{28 \text{ days}}{3 \text{ weeks}} = \frac{28 \text{ days}}{21 \text{ days}} = \frac{28}{21} = \frac{28 \div 7}{21 \div 7} = \frac{4}{3} \quad \leftarrow \left\{ \begin{array}{l} \text{Ratio in} \\ \text{lowest terms} \end{array} \right.$$

The following table will help you set up ratios that compare measurements. You will work with these measurements again in **Chapter 7.**

Measurement Comparisons	
Length	**Capacity (Volume)**
12 inches = 1 foot	2 cups = 1 pint
3 feet = 1 yard	2 pints = 1 quart
5280 feet = 1 mile	4 quarts = 1 gallon
Weight	**Time**
16 ounces = 1 pound	60 seconds = 1 minute
2000 pounds = 1 ton	60 minutes = 1 hour
	24 hours = 1 day
	7 days = 1 week

◀ **Work Problem** **5** *at the Side.*

5.1 ▶▶▶ Exercises

Write each ratio as a fraction in lowest terms. See Examples 1 and 2.

1. 8 days to 9 days $\dfrac{8}{9}$
   2. $11 to $15 $\dfrac{11}{15}$
   3. $100 to $50 $\dfrac{2}{1}$
   4. 35¢ to 7¢ $\dfrac{5}{1}$

5. 30 minutes to 90 minutes $\dfrac{1}{3}$
   6. 9 pounds to 36 pounds $\dfrac{1}{4}$

7. 80 miles to 50 miles $\dfrac{8}{5}$
   8. 300 people to 450 people $\dfrac{2}{3}$

9. 6 hours to 16 hours $\dfrac{3}{8}$
   10. 45 books to 35 books $\dfrac{9}{7}$

Write each ratio as a ratio of whole numbers in lowest terms. See Examples 3 and 4.

11. $4.50 to $3.50 $\dfrac{9}{7}$
   12. $0.08 to $0.06 $\dfrac{4}{3}$
   13. 15 to $2\dfrac{1}{2}$ $\dfrac{6}{1}$

14. 5 to $1\dfrac{1}{4}$ $\dfrac{4}{1}$
   15. $1\dfrac{1}{4}$ to $1\dfrac{1}{2}$ $\dfrac{5}{6}$
   16. $2\dfrac{1}{3}$ to $2\dfrac{2}{3}$ $\dfrac{7}{8}$

Write each ratio as a fraction in lowest terms. For help, use the table of measurement relationships on the previous page. See Example 5.

17. 4 ft to 30 in. $\dfrac{8}{5}$
   18. 8 ft to 4 yd $\dfrac{2}{3}$

19. 5 minutes to 1 hour $\dfrac{1}{12}$
   20. 8 quarts to 5 pints $\dfrac{16}{5}$

21. 15 hours to 2 days $\dfrac{5}{16}$
   22. 3 pounds to 6 ounces $\dfrac{8}{1}$

23. 5 gallons to 5 quarts $\dfrac{4}{1}$
   24. 3 cups to 3 pints $\dfrac{1}{2}$

The table shows the number of greeting cards that Americans buy for various occasions. Use the information to answer Exercises 25–30. Write each ratio as a fraction in lowest terms.

Holiday/Event	Cards Sold
Valentine's Day	900 million
Mother's Day	150 million
Father's Day	95 million
Graduation	60 million
Thanksgiving	30 million
Halloween	25 million

Source: Hallmark Cards.

25. Find the ratio of Thanksgiving cards to graduation cards.

$$\frac{1}{2}$$

26. Find the ratio of Halloween cards to Mother's Day cards.

$$\frac{1}{6}$$

27. Find the ratio of Valentine's Day cards to Halloween cards.

$$\frac{36}{1}$$

28. Find the ratio of Mother's Day cards to Father's Day cards.

$$\frac{30}{19}$$

29. Explain how you might use the information in the table if you owned a shop selling gifts and greeting cards.

Answers will vary. One possibility is stocking cards of various types in the same ratios as those in the table.

30. Why is the ratio of Valentine's Day cards to graduation cards $\frac{15}{1}$? Give two possible reasons.

Answers will vary. Possibilities include: a person may send Valentine's Day cards every year but graduation cards only once every few years; Valentine's Day may have more advertising than graduations.

The bar graph shows worldwide sales of the most popular songs of all time. Use the graph to complete Exercises 31–34. Write each ratio as a fraction in lowest terms.

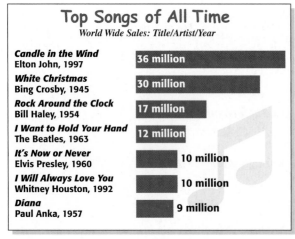

Top Songs of All Time
World Wide Sales: Title/Artist/Year

Candle in the Wind
Elton John, 1997 — 36 million

White Christmas
Bing Crosby, 1945 — 30 million

Rock Around the Clock
Bill Haley, 1954 — 17 million

I Want to Hold Your Hand
The Beatles, 1963 — 12 million

It's Now or Never
Elvis Presley, 1960 — 10 million

I Will Always Love You
Whitney Houston, 1992 — 10 million

Diana
Paul Anka, 1957 — 9 million

Source: The Music Information Database.

31. Write a ratio that compares the top-selling song to the second best seller, and a ratio that compares the top-selling song to the third best seller.

$$\frac{6}{5}, \frac{36}{17}$$

32. Write two ratios that compare sales of Elvis Presley's song with sales of the songs just ahead and just behind it in the graph.

$$\frac{5}{6}, \frac{1}{1}$$

33. Sales of which two songs give a ratio of $\frac{3}{1}$? There may be more than one correct answer.

White Christmas to It's Now or Never; White Christmas to I Will Always Love You; Candle in the Wind to I Want to Hold Your Hand.

34. Sales of which two songs give a ratio of $\frac{5}{6}$? There may be more than one correct answer.

I Will Always Love You to I Want to Hold Your Hand; It's Now or Never to I Want to Hold Your Hand; White Christmas to Candle in the Wind

*Use the circle graph of one person's monthly budget to complete Exercises 35–36.
Write each ratio as a fraction in lowest terms.*

35. Find the ratio of

(a) rent to utilities

(b) rent to food

(c) rent and utilities to total budget.

(a) $\frac{6}{1}$ (b) $\frac{5}{2}$ (c) $\frac{7}{16}$

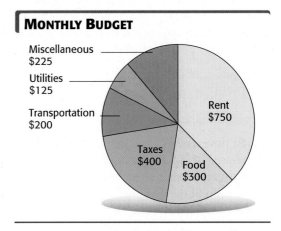

MONTHLY BUDGET

Miscellaneous
$225

Utilities
$125

Transportation
$200

Rent
$750

Taxes
$400

Food
$300

36. Find the ratio of

(a) taxes to rent

(b) food to transportation

(c) taxes and food to rent and utilities.

(a) $\frac{8}{15}$ (b) $\frac{3}{2}$ (c) $\frac{4}{5}$

*For each figure, find the ratio of the length of the longest side to the length of the shortest
side. Write each ratio as a fraction in lowest terms.*

37.

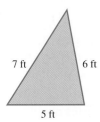

7 ft   6 ft

5 ft

$\frac{7}{5}$

38.

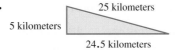

25 kilometers

5 kilometers

24.5 kilometers

$\frac{5}{1}$

39.

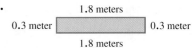

1.8 meters

0.3 meter   0.3 meter

1.8 meters

$\frac{6}{1}$

40.

0.09 in.

0.12 in.   0.12 in.

0.09 in.

$\frac{4}{3}$

41.

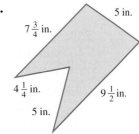

5 in.

$7\frac{3}{4}$ in.

$4\frac{1}{4}$ in. $9\frac{1}{2}$ in.

5 in.

$\frac{38}{17}$

42.

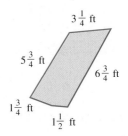

$3\frac{1}{4}$ ft

$5\frac{3}{4}$ ft $6\frac{3}{4}$ ft

$1\frac{3}{4}$ ft

$1\frac{1}{2}$ ft

$\frac{9}{2}$

Write each ratio as a fraction in lowest terms.

43. The price of automobile engine oil has gone from $10 to $12.50 for the 5 quarts needed for an oil change. Find the ratio of the increase in price to the original price.

$\dfrac{1}{4}$

44. The price that a pharmacy pays for an antibiotic decreased from $8.80 to $5.60 for 10 tablets. Find the ratio of the decrease in price to the original price.

$\dfrac{4}{11}$

45. The first time a movie was made in Minnesota, the cast and crew spent $59\frac{1}{2}$ days filming winter scenes. The next year, another movie was filmed in $8\frac{3}{4}$ weeks. Find the ratio of the first movie's filming time to the second movie's time. Compare in weeks.

$\dfrac{34}{35}$

46. The percheron, a large draft horse, measures about $5\frac{3}{4}$ ft at the shoulder. The prehistoric ancestor of the horse measured only $15\frac{3}{4}$ in. at the shoulder. Find the ratio of the percheron's height to its prehistoric ancestor's height. Compare in inches. (*Source: Eyewitness Books: Horse.*)

$\dfrac{92}{21}$

$5\frac{3}{4}$ ft $15\frac{3}{4}$ in.

Relating Concepts (Exercises 47–50) For Individual or Group Work

*Use your knowledge of ratios to **work Exercises 47–50 in order.***

47. In this painting, what is the ratio of the length of the longest side to the length of the shortest side? What other measurements could the painting have and still maintain the same ratio?

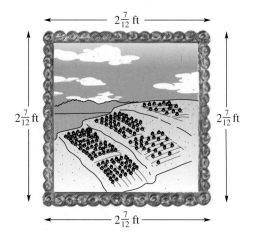

$2\frac{7}{12}$ ft

$2\frac{7}{12}$ ft $2\frac{7}{12}$ ft

$2\frac{7}{12}$ ft

$\dfrac{1}{1}$; as long as the sides all have the same length, any measurement you choose will maintain the ratio.

48. The ratio of my son's age to my daughter's age is 4 to 5. One possibility is that my son is 4 years old and my daughter is 5 years old. Find six other possibilities that fit the 4 to 5 ratio.

Answers will vary. Some possibilities are:

$$\dfrac{4}{5} = \dfrac{8}{10} = \dfrac{12}{15} = \dfrac{16}{20} = \dfrac{20}{25} = \dfrac{24}{30} = \dfrac{28}{35}.$$

49. Amelia said that the ratio of her age to her mother's age is 5 to 3. Is this possible? Explain your answer.

It is not possible. Amelia would have to be older than her mother to have a ratio of 5 to 3.

50. Would you prefer that the ratio of your income to your friend's income be 1 to 3 or 3 to 1? Explain your answers.

Answers will vary, but a ratio of 3 to 1 means your income is 3 times your friend's income.

5.2 ▶▶▶ Rates

A *ratio* compares two measurements with the same type of units, such as 9 <u>feet</u> to 12 <u>feet</u> (both length measurements). But many of the comparisons we make use measurements with different types of units, such as the following.

160 <u>dollars</u> **for** 8 <u>hours</u>   (money to time)

450 <u>miles</u> **on** 15 <u>gallons</u>   (distance to capacity)

This type of comparison is called a **rate**.

OBJECTIVE 1 Write rates as fractions. Suppose you hiked 18 <u>miles</u> **in** 4 <u>hours</u>. The *rate* at which you hiked can be written as a fraction in lowest terms.

$$\frac{18 \text{ miles}}{4 \text{ hours}} = \frac{18 \text{ miles} \div 2}{4 \text{ hours} \div 2} = \frac{9 \text{ miles}}{2 \text{ hours}} \left.\begin{cases} \\ \\ \end{cases}\right\} \begin{array}{l} \text{Rate in} \\ \text{lowest terms} \end{array}$$

In a rate, you often find these words separating the quantities you are comparing.

in   for   on   per   from

> **CAUTION**
> When writing a rate, always include the units, such as miles, hours, dollars, and so on. Because the units in a rate are different, the units do *not* divide out.

EXAMPLE 1 Writing Rates in Lowest Terms

Write each rate as a fraction in lowest terms.

(a) 5 gallons of chemical **for** $60.

$$\frac{5 \text{ gallons} \div 5}{60 \text{ dollars} \div 5} = \frac{1 \text{ gallon}}{12 \text{ dollars}} \quad \begin{array}{l}\leftarrow \\ \leftarrow\end{array} \begin{array}{l}\text{Write the units:} \\ \text{gallons and dollars.}\end{array}$$

(b) $1500 wages **in** 10 weeks

$$\frac{1500 \text{ dollars} \div 10}{10 \text{ weeks} \div 10} = \frac{150 \text{ dollars}}{1 \text{ week}}$$

> Be sure to write the units in a *rate*: dollars, miles, gallons, and so on.

(c) 2225 miles **on** 75 gallons of gas

$$\frac{2225 \text{ miles} \div 25}{75 \text{ gallons} \div 25} = \frac{89 \text{ miles}}{3 \text{ gallons}}$$

Work Problem **1** *at the Side.* ▶

OBJECTIVE 2 Find unit rates. When the *denominator* of a rate is 1, it is called a **unit rate.** We use unit rates frequently. For example, you earn $12.75 for *1 hour* of work. This unit rate is written:

$12.75 **per** hour   or $12.75/hour.

Or, you drive 28 miles on *1 gallon* of gas. This unit rate is written

28 miles **per** gallon   or   28 miles/gallon.

Use **per** or a slash mark (/) when writing unit rates.

---

1 Write each rate as a fraction in lowest terms.

(a) $6 for 30 packages

(b) 500 miles in 10 hours

(c) 4 teachers for 90 students

(d) 1270 bushels from 30 acres

ANSWERS

1. **(a)** $\dfrac{1 \text{ dollar}}{5 \text{ packages}}$ **(b)** $\dfrac{50 \text{ miles}}{1 \text{ hour}}$

   (c) $\dfrac{2 \text{ teachers}}{45 \text{ students}}$ **(d)** $\dfrac{127 \text{ bushels}}{3 \text{ acres}}$

2 Find each unit rate.

(a) $4.35 for 3 pounds of cheese

(b) 304 miles on 9.5 gallons of gas

(c) $850 in 5 days

(d) 24-pound turkey for 15 people

EXAMPLE 2   Finding Unit Rates

Find each unit rate.

(a) 337.5 miles on 13.5 gallons of gas
Write the rate as a fraction.

$$\frac{337.5 \text{ miles}}{13.5 \text{ gallons}} \;\longleftarrow \text{ The fraction bar indicates division.}$$

Divide 337.5 by 13.5 to find the unit rate.

$$13.5\overline{)337.5} \;\; \overset{2\;5.}{}$$

$$\frac{337.5 \text{ miles} \;\div\; 13.5}{13.5 \text{ gallons} \;\div\; 13.5} = \frac{25 \text{ miles}}{1 \text{ gallon}}$$

The unit rate is 25 miles **per** gallon, or 25 miles/gallon.

(b) 549 miles in 18 hours.

$$\frac{549 \text{ miles}}{18 \text{ hours}} \qquad \text{Divide: } 18\overline{)549.0} \;\; \overset{30.5}{}$$

The unit rate is 30.5 miles/hour

(c) $810 in 6 days

$$\frac{810 \text{ dollars}}{6 \text{ days}} \qquad \text{Divide: } 6\overline{)810} \;\; \overset{135}{}$$

> Use *per* or a slash mark to write unit rates.

The unit rate is $135/day.

◀ *Work Problem* **2** *at the Side.*

OBJECTIVE 3 Find the best buy based on cost per unit.   When shopping for groceries, household supplies, and health and beauty items, you will find many different brands and package sizes. You can save money by finding the lowest *cost per unit.*

> **Cost per Unit**
> **Cost per unit** is a rate that tells how much you pay for *one* item or *one* unit. Examples are $3.25 per gallon, $47 per shirt, and $2.98 per pound.

EXAMPLE 3   Determining the Best Buy

The local store charges the following prices for pancake syrup. Find the best buy.

Continued on Next Page

The best buy is the container with the *lowest* cost per unit. All the containers are measured in *ounces* (oz), so you first need to find the *cost per ounce* for each one. Divide the price of the container by the number of ounces in it. Round to the nearest thousandth, if necessary.

Let the *order* of the *words* help you set up the rate.

$$
\begin{array}{ccc}
\text{cost} & \longrightarrow & \$1.28 \\
\text{per (means divide)} & \longrightarrow & \overline{} \\
\text{ounce} & \longrightarrow & 12 \text{ ounces}
\end{array}
$$

Size	Cost per Unit (rounded)
12 ounces	$\dfrac{\$1.28}{12 \text{ ounces}} \approx \$0.107$ per ounce (highest)
24 ounces	$\dfrac{\$1.81}{24 \text{ ounces}} \approx \$0.075$ per ounce (lowest)
36 ounces	$\dfrac{\$2.73}{36 \text{ ounces}} \approx \$0.076$ per ounce

The lowest cost per ounce is $0.075, so the 24-ounce container is the best buy.

Note

Earlier we rounded money amounts to the nearest hundredth (nearest cent). But when comparing unit costs, rounding to the nearest thousandth will help you see the difference between very similar unit costs. Notice that the 24-ounce and 36-ounce syrup containers above would both have rounded to $0.08 per ounce if we had rounded to hundredths.

Work Problem **3** *at the Side.* ▶

🖩 **Calculator Tip**   When using a calculator to find unit prices, remember that division is *not* commutative. In Example 3 you wanted to find cost per ounce. Let the *order* of the *words* help you enter the numbers in the correct order.

cost	**per**	**ounce**
↓	↓	↓
Enter the cost.	*Per* means divide.	Enter number of ounces.
↓	↓	↓
$2.73	÷	36 = 0.076 (rounded)

If you entered 36 ÷ 2.73 = , you'd get the number of *ounces* per *dollar.* How could you use that information to find the best buy? (*Answer:* The best buy would be to get the greatest number of ounces per dollar.)

Finding the best buy is sometimes a complicated process. Things that affect the cost per unit can include "cents off" coupons and differences in how much use you'll get out of each unit.

3 Find the best buy (lowest cost per unit) for each purchase.

(a) 2 quarts for $3.25
3 quarts for $4.95
4 quarts for $6.48

(b) 6 cans of cola for $1.99
12 cans of cola for $3.49
24 cans of cola for $7

4 Solve each problem.

(a) Some batteries claim to last longer than others. If you believe these claims, which brand is the best buy?

Four-pack of AA-size batteries for $2.79

One AA-size battery for $1.19; lasts twice as long

(b) Which tube of toothpaste is the best buy? You have a coupon for 85¢ off Brand C and a coupon for 20¢ off Brand D.

Brand C is $3.89 for 6 ounces.

Brand D is $1.59 for 2.5 ounces.

EXAMPLE 4 **Solving Best Buy Applications**

Solve each application problem.

(a) There are many brands of liquid laundry detergent. If you feel they all do a good job of cleaning your clothes, you can base your purchase on cost per unit. But some brands are "concentrated" so you can use less detergent for each load of clothes. Which of the choices shown below is the best buy?

50 fluid ounces for $3.99
Does same number of washloads as the old 64-ounce bottle!

One gallon (128 ounces) for $9.89
Does twice the washloads of the old gallon bottle!

To find Sudzy's unit cost, divide $3.99 by 64 ounces, not 50 ounces. You're getting as many clothes washed as if you bought 64 ounces. Similarly, to find White-O's unit cost, divide $9.89 by 256 ounces (twice 128 ounces, or 2 • 128 ounces = 256 ounces).

Sudzy $\dfrac{\$3.99}{64 \text{ ounces}} \approx \$0.062$ per ounce

White-O $\dfrac{\$9.89}{256 \text{ ounces}} \approx \$0.039$ per ounce

> The best buy is the *lower* cost per ounce.

White-O has the lower cost per ounce and is the better buy. (However, if you try it and it really doesn't get out all the stains, Sudzy may be worth the extra cost.)

(b) "Cents-off" coupons also affect the best buy. Suppose you are looking at these choices for "extra-strength" pain reliever. Both brands have the same amount of pain reliever in each tablet.

Brand X is $2.29 for 50 tablets.

Brand Y is $10.75 for 200 tablets.

You have a 40¢ coupon for Brand X and a 75¢ coupon for Brand Y. Which choice is the best buy?

To find the best buy, first subtract the coupon amounts, then divide to find the lower cost per ounce.

Brand X costs $2.29 − $0.40 = $1.89

$\dfrac{\$1.89}{50 \text{ tablets}} \approx \$0.038$ per tablet

Brand Y costs $10.75 − $0.75 = $10.00

> Look for the *lower* cost per tablet.

$\dfrac{\$10.00}{200 \text{ tablets}} = \$0.05$ per tablet

Brand X has the lower cost per tablet and is the better buy.

◀ *Work Problem* **4** *at the Side.*

ANSWERS

4. (a) One battery that lasts twice as long (like getting two) is the better buy. The cost per unit is $0.595 per battery. The four-pack is $0.698 per battery (rounded).
(b) Brand C with the 85¢ coupon is the better buy at $0.507 per ounce (rounded). Brand D with the 20¢ coupon is $0.556 per ounce.

5.2 ▶▶▶ Exercises

Write each rate as a fraction in lowest terms. See Example 1.

1. 10 cups for 6 people $\dfrac{5 \text{ cups}}{3 \text{ people}}$

2. $12 for 30 pens $\dfrac{2 \text{ dollars}}{5 \text{ pens}}$

3. 15 feet in 35 seconds $\dfrac{3 \text{ feet}}{7 \text{ seconds}}$

4. 100 miles in 30 hours $\dfrac{10 \text{ miles}}{3 \text{ hours}}$

5. 72 miles on 4 gallons $\dfrac{18 \text{ miles}}{1 \text{ gallon}}$

6. 132 miles on 8 gallons $\dfrac{33 \text{ miles}}{2 \text{ gallons}}$

Find each unit rate. See Example 2.

7. $60 in 5 hours
  $12 per hour or $12/hour

8. $2500 in 20 days
  $125 per day or $125/day

9. 7.5 pounds for 6 people
  1.25 pounds/person

10. 44 bushels from 8 trees
  5.5 bushels/tree

11. $413.20 for 4 days
  $103.30/day

12. $74.25 for 9 hours
  $8.25/hour

Earl kept the following record of the gas he bought for his car. For each entry, find the number of miles he traveled and the unit rate. Round your answers to the nearest tenth.

	Date	Odometer at Start	Odometer at End	Miles Traveled	Gallons Purchased	Miles per Gallon
13.	2/4	27,432.3	27,758.2	**325.9**	15.5	**21.0 (rounded)**
14.	2/9	27,758.2	28,058.1	**299.9**	13.4	**22.4 (rounded)**
15.	2/16	28,058.1	28,396.7	**338.6**	16.2	**20.9 (rounded)**
16.	2/20	28,396.7	28,704.5	**307.8**	13.3	**23.1 (rounded)**

Source: Author's car records.

Find the best buy (based on the cost per unit) for each item. See Example 3.
(*Sources:* Cub Foods, Target, Rainbow Foods.)

17. Black pepper

**4 ounces for $3.65,
about $0.913/ounce**

18. Shampoo

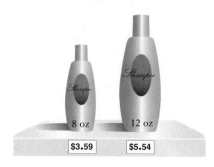

**8 ounces for $3.59,
about $0.449/ounce**

19. Cereal
12 ounces for $2.49
14 ounces for $2.89
18 ounces for $3.96

**14 ounces for $2.89,
about $0.206/ounce**

20. Soup (same size cans)
2 cans for $2.18
3 cans for $3.57
5 cans for $5.29

**5 cans for $5.29,
or $1.058/can**

21. Chunky peanut butter
12 ounces for $1.29
18 ounces for $1.79
28 ounces for $3.39
40 ounces for $4.39

**18 ounces for $1.79,
about $0.099/ounce**

22. Baked beans
8 ounces for $0.59
16 ounces for $0.99
21 ounces for $1.29
28 ounces for $1.89

**21 ounces for $1.29,
about $0.061/ounce**

23. Suppose you are choosing between two brands of chicken noodle soup. Brand A is $0.88 per can and Brand B is $0.98 per can. The cans are the same size but Brand B has more chunks of chicken in it. Which soup is the better buy? Explain your choice.

Answers will vary. For example, you might choose Brand B because you like more chicken, so the cost per chicken chunk may actually be the same or less than Brand A.

24. A small bag of potatoes costs $0.19 per pound. A large bag costs $0.15 per pound. But there are only two people in your family, so half the large bag would probably rot before you used it up. Which bag is the better buy? Explain.

Answers will vary. For example, if you use only half of the larger bag, you really pay $0.30 per pound, so the smaller bag is the better buy.

Solve each application problem. See Examples 2–4.

25. Makesha lost 10.5 pounds in six weeks. What was her rate of loss in pounds per week?

1.75 pounds/week

26. Enrique's taco recipe uses three pounds of meat to feed 10 people. Give the rate in pounds per person.

0.3 pound/person

27. Russ works 7 hours to earn $85.82. What is his pay rate per hour?

$12.26/hour

28. Find the cost of 1 gallon of Hawaiian Punch beverage if 18 gallons for a graduation party cost $55.62.

$3.09/gallon

The table lists information about three long-distance calling cards. The connection fee is charged each time you make a call, no matter how long the call lasts. Use the table to answer Exercises 29–32. Round answers to the nearest thousandth when necessary.

LONG-DISTANCE CALLING CARDS (U.S.)

Card Name	Cost per Minute	Connection Fee
Radiant Penny	$0.01	$0.39
IDT Special	0.022	0.14
Access America	0.047	0.00

Source: www.1callcard.com

31. Find the *actual* total cost per minute for a 15-minute call and a 25-minute call using each card. Then select the best buy for each call.

For a 15-minute call: Radiant, $0.036/min; IDT, $0.031/min; Access, $0.047/min; IDT is the best buy. For a 25-minute call: Radiant, $0.026/min; IDT, $0.028/min; Access, $0.047/min; Radiant is the best buy.

33. If you believe the claims that some batteries last longer, which is the better buy?

one battery for 1.79; like getting 3 batteries so $1.79 ÷ 3 ≈ $0.597 per battery

35. Three brands of cornflakes are available. Brand G is priced at $2.39 for 10 ounces. Brand K is $3.99 for 20.3 ounces and Brand P is $3.39 for 16.5 ounces. You have a coupon for 50¢ off Brand P and a coupon for 60¢ off Brand G. Which cereal is the best buy based on cost per unit?

Brand P with the 50¢ coupon is the best buy. ($3.39 − $0.50 = $2.89; $2.89 ÷ 16.5 ounces ≈ $0.175 per ounce)

29. (a) Find the *actual* total cost, including the connection charge, for a five-minute call using each card.

(b) Find the cost per minute for this call using each card and select the best buy.

(a) Radiant, $0.44; IDT, $0.25; Access, $0.235

(b) Radiant, $0.088/min; IDT, $0.05/min; Access, $0.047/min; Access America is the best buy.

30. (a) Find the *actual* total cost, including the connection charge, for a 30-minute call using each card.

(b) Find the cost per minute for this call using each card and select the best buy.

(a) Radiant, $0.69; IDT, $0.80; Access, $1.41

(b) Radiant, $0.023/min; IDT, $0.027/min; Access, $0.047/min; Radiant is the best buy.

32. All the cards round calls up to the next full minute.

(a) Suppose you call the wrong number. How much would you pay for this 40-*second* call on each card?

(b) How much would you save on this call by using Access America instead of Radiant Penny?

(a) Radiant, $0.40; IDT, $0.162; Access, $0.047

(b) $0.353

34. Which is the better buy, assuming these laundry detergents both clean equally well?

Dazzle is the better buy; 2 · 64 ounces = 128 ounces; $5.99 ÷ 128 ounces ≈ $0.047 per ounce

36. Two brands of facial tissue are available. Brand K is specially priced at three boxes of 175 tissues each for $5. Brand S is priced at $1.29 per box of 125 tissues. You have a coupon for 20¢ off one box of Brand S and a coupon for 45¢ off one box of Brand K. How can you get the best buy on one box of tissue?

Best buy is 1 box of Brand K with a 45¢ coupon. ($5 ÷ 3 boxes ≈ $1.67 per box; $1.67 − $0.45 = $1.22; $1.22 ÷ 175 tissues ≈ $0.007 per tissue)

Relating Concepts (Exercises 37–40) For Individual or Group Work

On the first page of this chapter, we said that unit rates can help you get the best deal on cell phone service. Use the information in the table to **work Exercises 37–40 in order.**

CELL PHONE SERVICE PLANS

Company	Anytime Minutes	One-time Activation Fee	Monthly Charge	Termination Fee
Verizon	400	$35	$59.99	$175
T-Mobile	600	$35	$39.99	$200
Nextel	500	$35	$45.99	$200
Sprint	500	$36	$55	$150

Source: Advertisements appearing in Minneapolis Star Tribune.

Notes:

1. All companies require that you sign a contract for one year of service and charge a one-time termination fee if you quit early.
2. Unused minutes cannot be carried over to the next month.

37. How much will each company's activation fee cost you on a *monthly* basis during the one-year contract?

   $2.92 (rounded) per month for Verizon, T-Mobile, and Nextel; $3 per month for Sprint.

38. All the plans allow unlimited calls on nights and weekends, so you would be using the "anytime minutes" on weekdays. Figure out the average number of weekdays per month. Then, for each plan, how many minutes could you use per weekday? Round to the nearest whole minute.

   Average number of weekdays per month is about 21.7; Verizon ≈ 18 minutes/weekday; T-Mobile ≈ 28 minutes/weekday; Nextel and Sprint ≈ 23 minutes/weekday.

39. Find the actual average cost per "anytime minute" during the one-year contract for each company, including the activation fee. Assume you use all the minutes and no more. Decide how to round your answers so you can find the best buy.

   Round to hundredths (nearest cent) to see that T-Mobile is the best buy at $0.07 per "anytime minute"; Verizon ≈ $0.16; Sprint ≈ $0.12; Nextel ≈ $0.10.

40. Suppose that after two months you canceled your service because you found that you only used 100 "anytime minutes" per month. Under those conditions, find the actual cost per "anytime minute" for each company, to the nearest cent.

   Verizon ≈ $1.65 per "anytime minute"; T-Mobile ≈ $1.57; Nextel ≈ $1.63; Sprint = $1.48

5.3 ▶▶▶ Proportions

OBJECTIVE **1** **Write proportions.** A **proportion** states that two ratios (or rates) are equivalent. For example,

$$\frac{\$20}{4 \text{ hours}} = \frac{\$40}{8 \text{ hours}}$$

is a proportion that says the rate $\frac{\$20}{4 \text{ hours}}$ is equivalent to the rate $\frac{\$40}{8 \text{ hours}}$. As the amount of money doubles, the number of hours also doubles. This proportion is read:

20 dollars **is to** 4 hours **as** 40 dollars **is to** 8 hours.

EXAMPLE 1 **Writing Proportions**

Write each proportion.

(a) 6 ft is to 11 ft **as** 18 ft is to 33 ft.

$$\frac{6 \text{ ft}}{11 \text{ ft}} = \frac{18 \text{ ft}}{33 \text{ ft}} \quad \text{so} \quad \frac{6}{11} = \frac{18}{33} \qquad \text{The common units (ft) divide out and are not written.}$$

(b) \$9 is to 6 liters **as** \$3 is to 2 liters.

$$\frac{\$9}{6 \text{ liters}} = \frac{\$3}{2 \text{ liters}}$$

The units do *not* match so you must write them in the proportion.

Work Problem **1** *at the Side.* ▶

OBJECTIVE **2** **Determine whether proportions are true or false.** There are two ways to see whether a proportion is true. One way is to *write both of the ratios in lowest terms.*

EXAMPLE 2 **Writing Both Ratios in Lowest Terms**

Determine whether each proportion is true or false by writing both ratios in lowest terms.

(a) $\dfrac{5}{9} = \dfrac{18}{27}$

Write each ratio in lowest terms.

$$\frac{5}{9} \leftarrow \begin{array}{l}\text{Already in} \\ \text{lowest terms}\end{array} \qquad \frac{18 \div 9}{27 \div 9} = \frac{2}{3} \leftarrow \begin{array}{l}\text{Lowest} \\ \text{terms}\end{array}$$

Because $\frac{5}{9}$ is *not* equivalent to $\frac{2}{3}$, the proportion is *false.*

(b) $\dfrac{16}{12} = \dfrac{28}{21}$

Write each ratio in lowest terms.

$$\frac{16 \div 4}{12 \div 4} = \frac{4}{3} \quad \text{and} \quad \frac{28 \div 7}{21 \div 7} = \frac{4}{3}$$

Both ratios are equivalent to $\frac{4}{3}$, so the proportion is *true.*

Work Problem **2** *at the Side.* ▶

---

1 Write each proportion.

(a) \$7 is to 3 cans as \$28 is to 12 cans

(b) 9 meters is to 16 meters as 18 meters is to 32 meters

(c) 5 is to 7 as 35 is to 49

(d) 10 is to 30 as 60 is to 180

2 Determine whether each proportion is true or false by writing both ratios in lowest terms.

(a) $\dfrac{6}{12} = \dfrac{15}{30}$

(b) $\dfrac{20}{24} = \dfrac{3}{4}$

(c) $\dfrac{25}{40} = \dfrac{30}{48}$

(d) $\dfrac{35}{45} = \dfrac{12}{18}$

ANSWERS

1. **(a)** $\dfrac{\$7}{3 \text{ cans}} = \dfrac{\$28}{12 \text{ cans}}$ **(b)** $\dfrac{9}{16} = \dfrac{18}{32}$

   (c) $\dfrac{5}{7} = \dfrac{35}{49}$ **(d)** $\dfrac{10}{30} = \dfrac{60}{180}$

2. **(a)** $\dfrac{1}{2} = \dfrac{1}{2}$; true **(b)** $\dfrac{5}{6} \neq \dfrac{3}{4}$; false

   (c) $\dfrac{5}{8} = \dfrac{5}{8}$; true **(d)** $\dfrac{7}{9} \neq \dfrac{2}{3}$; false

OBJECTIVE **3** **Find cross products.** Another way to test whether the ratios in a proportion are equivalent is to compare *cross products*.

Using Cross Products to Determine Whether a Proportion Is True

To see whether a proportion is true, first multiply along one diagonal, then multiply along the other diagonal, as shown here.

$$5 \cdot 4 = 20$$

$$\frac{2}{5} = \frac{4}{10}$$

Cross products are equal.

$$2 \cdot 10 = 20$$

In this case the **cross products** are both 20. When cross products are *equal*, the proportion is *true*. If the cross products are *unequal*, the proportion is *false*.

Note

The cross products test is based on rewriting both fractions with a common denominator of 5 · 10, or 50.

$$\frac{2 \cdot 10}{5 \cdot 10} = \frac{20}{50} \quad \text{and} \quad \frac{4 \cdot 5}{10 \cdot 5} = \frac{20}{50}$$

We see that $\frac{2}{5}$ and $\frac{4}{10}$ are equivalent because both can be rewritten as $\frac{20}{50}$. The cross product test takes a shortcut by comparing only the two numerators (20 = 20).

EXAMPLE 3 **Using Cross Products**

Use cross products to see whether each proportion is true or false.

(a) $\dfrac{3}{5} = \dfrac{12}{20}$

Multiply along one diagonal, then multiply along the other diagonal.

$$5 \cdot 12 = 60$$

$$\frac{3}{5} = \frac{12}{20}$$

Equal cross products; proportion is *true*.

$$3 \cdot 20 = 60$$

The cross products are *equal,* so the proportion is *true*.

CAUTION

Use cross products *only* when working with *proportions*. Do ***not*** use cross products when multiplying fractions, adding fractions, or writing fractions in lowest terms.

Continued on Next Page

(b) $\dfrac{2\frac{1}{3}}{3\frac{1}{3}} = \dfrac{9}{16}$

Find the cross products.

> Write $3\frac{1}{3}$ as $\frac{10}{3}$ and 9 as $\frac{9}{1}$.

$$3\frac{1}{3} \cdot 9 = \frac{10}{\overset{1}{\cancel{3}}} \cdot \frac{\overset{3}{\cancel{9}}}{1} = \frac{30}{1} = 30$$

$$\dfrac{2\frac{1}{3}}{3\frac{1}{3}} \underset{\times}{=} \dfrac{9}{16}$$

Unequal cross products; proportion is *false*.

$$2\frac{1}{3} \cdot 16 = \frac{7}{3} \cdot \frac{16}{1} = \frac{112}{3} = 37\frac{1}{3}$$

> Write $2\frac{1}{3}$ as $\frac{7}{3}$ and 16 as $\frac{16}{1}$.

The cross products are *unequal,* so the proportion is *false.*

Note
The numbers in a proportion do *not* have to be whole numbers. They can be fractions, mixed numbers, decimal numbers, and so on.

Work Problem **3** *at the Side.* ▶

---

3 Find the cross products to see whether each proportion is true or false.

(a) $\dfrac{5}{9} = \dfrac{10}{18}$

(b) $\dfrac{32}{15} = \dfrac{16}{8}$

(c) $\dfrac{10}{17} = \dfrac{20}{34}$

(d) $\dfrac{2.4}{6} \underset{\times}{=} \dfrac{5}{12}$ $(6)(5) =$ $(2.4)(12) =$

(e) $\dfrac{3}{4.25} = \dfrac{24}{34}$

(f) $\dfrac{1\frac{1}{6}}{2\frac{1}{3}} = \dfrac{4}{8}$

Answers

3. **(a)** $90 = 90$; true **(b)** $240 \neq 256$; false
   (c) $340 = 340$; true
   (d) $(6)(5) = 30$; $(2.4)(12) = 28.8$; false
   (e) $102 = 102$; true **(f)** $9\frac{1}{3} = 9\frac{1}{3}$; true

Math in the Media

MOVIES MAKE BIG BUCKS

Americans love movies and spend millions of dollars each year to see them. The table below lists some of the movies that have made the most money (gross earnings).

SELECTED ALL-TIME TOP GROSSING AMERICAN MOVIES

Title (Year Released)	Gross Earnings (nearest $10 million dollars)
Titanic (1997)	$600 million
Star Wars: Episode IV (1977)	$460 million
Spider-Man (2002)	$400 million
Jurassic Park (1993)	$360 million
Harry Potter and the Sorcerer's Stone (2001)	$320 million
Pirates of the Caribbean (2003)	$300 million

Source: The World Almanac.

Write all ratios as fractions in lowest terms.

1. **(a)** Write a ratio to compare *Titanic's* earnings to the earnings from *Pirates of the Caribbean*. $\dfrac{2}{1}$

   (b) Write a ratio to compare the earnings from *Pirates of the Caribbean* to *Titanic's* earnings. $\dfrac{1}{2}$

   (c) Look at the two ratios you wrote. What is special about them? What is the mathematical word that describes the relationship between these two ratios?
   The ratios are reciprocals of each other.

2. **(a)** Write a ratio to compare *Jurassic Park's* earnings to those of the movie just below it. $\dfrac{9}{8}$

   (b) One way to use the ratio you wrote in part (a) is to say, "For every $9 earned by *Jurassic Park,* the *Harry Potter* movie earned $8." Now use the ratios you wrote in Problem 1 above to talk about those movies in the same way. **For every $2 earned by *Titanic, Pirates* earned $1. For every $1 earned by *Pirates, Titanic* earned $2.**

   (c) Find a pair of movies where one of them earned $4 to every $3 earned by the other. ***Spider-Man* to *Pirates of the Caribbean***

   (d) Find a pair of movies where one of them earned $4 to every $5 earned by the other. ***Harry Potter and the Sorcerer's Stone* to *Spider-Man***

5.3 ▶▶▶ **Exercises**

FOR EXTRA HELP

 MyMathLab

 Math XL
PRACTICE

WATCH

DOWNLOAD

READ

 REVIEW

Write each proportion. See Example 1.

1. $9 is to 12 cans as $18 is to 24 cans.

$$\frac{\$9}{12 \text{ cans}} = \frac{\$18}{24 \text{ cans}}$$

2. 28 people is to 7 cars as 16 people is to 4 cars.

$$\frac{28 \text{ people}}{7 \text{ cars}} = \frac{16 \text{ people}}{4 \text{ cars}}$$

3. 200 adults is to 450 children as 4 adults is to 9 children.

$$\frac{200 \text{ adults}}{450 \text{ children}} = \frac{4 \text{ adults}}{9 \text{ children}}$$

4. 150 trees is to 1 acre as 1500 trees is to 10 acres.

$$\frac{150 \text{ trees}}{1 \text{ acre}} = \frac{1500 \text{ trees}}{10 \text{ acres}}$$

5. 120 ft is to 150 ft as 8 ft is to 10 ft.

$$\frac{120}{150} = \frac{8}{10}$$

6. $6 is to $9 as $10 is to $15.

$$\frac{6}{9} = \frac{10}{15}$$

Determine whether each proportion is true or false by writing the ratios in lowest terms. Show the simplified ratios and then write true *or* false. *See Example 2.*

7. $\frac{6}{10} = \frac{3}{5}$ $\frac{3}{5} = \frac{3}{5}$; true

8. $\frac{1}{4} = \frac{9}{36}$ $\frac{1}{4} = \frac{1}{4}$; true

9. $\frac{5}{8} = \frac{25}{40}$ $\frac{5}{8} = \frac{5}{8}$; true

10. $\frac{2}{3} = \frac{20}{27}$ $\frac{2}{3} \neq \frac{20}{27}$; false

11. $\frac{150}{200} = \frac{200}{300}$ $\frac{3}{4} \neq \frac{2}{3}$; false

12. $\frac{100}{120} = \frac{75}{100}$ $\frac{5}{6} \neq \frac{3}{4}$; false

13. $\frac{42}{15} = \frac{28}{10}$ $\frac{14}{5} = \frac{14}{5}$; true

14. $\frac{18}{16} = \frac{36}{32}$ $\frac{9}{8} = \frac{9}{8}$; true

15. $\frac{32}{18} = \frac{48}{27}$ $\frac{16}{9} = \frac{16}{9}$; true

16. $\frac{15}{48} = \frac{10}{24}$ $\frac{5}{16} \neq \frac{5}{12}$; false

17. $\frac{7}{6} = \frac{54}{48}$ $\frac{7}{6} \neq \frac{9}{8}$; false

18. $\frac{28}{21} = \frac{44}{33}$ $\frac{4}{3} = \frac{4}{3}$; true

Use cross products to determine whether each proportion is true or false. Show the cross products and then circle true *or* false. *See Example 3.*

19. $\frac{2}{9} = \frac{6}{27}$ $54 = 54$

(True)   False

20. $\frac{20}{25} = \frac{4}{5}$ $100 = 100$

(True)   False

21. $\frac{20}{28} = \frac{12}{16}$ $336 \neq 320$

True   (False)

22. $\dfrac{16}{40} = \dfrac{22}{55}$ $880 = 880$

(True)   False

23. $\dfrac{110}{18} = \dfrac{160}{27}$ $2880 \neq 2970$

True   (False)

24. $\dfrac{600}{420} = \dfrac{20}{14}$ $8400 = 8400$

(True)   False

25. $\dfrac{3.5}{4} = \dfrac{7}{8}$ $28 = 28$

(True)   False

26. $\dfrac{36}{23} = \dfrac{9}{5.75}$ $207 = 207$

(True)   False

27. $\dfrac{18}{16} = \dfrac{2.8}{2.5}$ $44.8 \neq 45$

True   (False)

28. $\dfrac{0.26}{0.39} = \dfrac{1.3}{1.9}$ $0.507 \neq 0.494$

True   (False)

29. $\dfrac{6}{3\frac{2}{3}} = \dfrac{18}{11}$ $66 = 66$

(True)   False

30. $\dfrac{16}{13} = \dfrac{2}{1\frac{5}{8}}$ $26 = 26$

(True)   False

31. $\dfrac{2\frac{5}{8}}{3\frac{1}{4}} = \dfrac{21}{26}$ $68\frac{1}{4} = 68\frac{1}{4}$

(True)   False

32. $\dfrac{28}{17} = \dfrac{9\frac{1}{3}}{5\frac{2}{3}}$ $158\frac{2}{3} = 158\frac{2}{3}$

(True)   False

33. $\dfrac{\frac{2}{3}}{2} = \dfrac{2.7}{8}$ $5\frac{2}{5} \neq 5\frac{1}{3}$

True   (False)

34. $\dfrac{3.75}{1\frac{1}{4}} = \dfrac{7.5}{2\frac{1}{2}}$ $9\frac{3}{8} = 9\frac{3}{8}$ or $9.375 = 9.375$

(True)   False

35. $\dfrac{2\frac{3}{10}}{8.05} = \dfrac{\frac{1}{4}}{0.9}$ $2\frac{1}{80} \neq 2\frac{7}{100}$ or $2.0125 \neq 2.07$

True   (False)

36. $\dfrac{3}{\frac{5}{6}} = \dfrac{1.5}{\frac{7}{12}}$ $1\frac{1}{4} \neq 1\frac{3}{4}$

True   (False)

37. Suppose Joe Mauer of the Minnesota Twins had 17 hits in 50 times at bat and Freddy Sanchez of the Pittsburgh Pirates was at bat 450 times and got 153 hits. Paul is trying to convince Jamie that the two men hit equally well. Show how you could use a proportion and cross products to see whether Paul is correct.

$$\dfrac{17 \text{ hits}}{50 \text{ at bats}} = \dfrac{153 \text{ hits}}{450 \text{ at bats}}$$

$50 \cdot 153 = 7650$
$17 \cdot 450 = 7650$

Cross products are *equal* so the proportion is *true*; they hit equally well.

38. Jay worked 3.5 hours and packed 91 cartons. Craig packed 126 cartons in 5.25 hours. To see whether the men worked equally fast, Barry set up this proportion:

$$\dfrac{3.5}{91} = \dfrac{126}{5.25}$$

Explain what is wrong with Barry's proportion and write a correct one. Is the correct proportion true or false?

Left-hand ratio compares hours to cartons but right-hand ratio compares cartons to hours. Correct proportion is shown.

$$\dfrac{3.5 \text{ hours}}{91 \text{ cartons}} = \dfrac{5.25 \text{ hours}}{126 \text{ cartons}}$$

$91 \cdot 5.25 = 477.75$
$3.5 \cdot 126 = 441$

Cross products are *not* equal so the proportion is *false*; the men do not work equally fast.

5.4 ▶▶▶ Solving Proportions

OBJECTIVES

1 Find the unknown number in a proportion.

2 Find the unknown number in a proportion with mixed numbers or decimals.

OBJECTIVE 1 Find the unknown number in a proportion. Four numbers are used in a proportion. If any three of these numbers are known, the fourth can be found. For example, find the unknown number that will make this proportion true.

$$\frac{3}{5} = \frac{x}{40}$$

The x represents the unknown number. Start by finding the cross products.

$$\frac{3}{5} = \frac{x}{40}$$

$$\left. \begin{array}{c} 5 \cdot x \\ 3 \cdot 40 \end{array} \right\} \text{Cross products}$$

To make the proportion true, the cross products must be equal.

$$5 \cdot x = 3 \cdot 40$$
$$5 \cdot x = 120$$

The equal sign says that $5 \cdot x$ and 120 are equivalent. If $5 \cdot x$ and 120 are *both* divided by 5, the results will still be equivalent.

$$\frac{5 \cdot x}{5} = \frac{120}{5} \quad \leftarrow \text{Divide both sides by 5.}$$

On the left side, divide out the common factor of 5; slashes indicate the divisions.

$$\frac{\overset{1}{\cancel{5}} \cdot x}{\underset{1}{\cancel{5}}} = 24$$

On the right side, divide 120 by 5 to get 24.

Multiplying by 1 does *not* change a number, so in the numerator on the left side, $1 \cdot x$ is the same as x.

$$\frac{x}{1} = 24$$

Dividing by 1 does *not* change a number, so on the left side, $\frac{x}{1}$ is the same as x.

$$x = 24$$

The unknown number in the proportion is 24. The complete proportion is shown below.

$$\frac{3}{5} = \frac{24}{40} \quad \leftarrow x \text{ is 24.}$$

Check by finding the cross products. If they are equal, you solved the problem correctly. If they are unequal, rework the problem.

$$\frac{3}{5} = \frac{24}{40}$$

$$\left. \begin{array}{c} 5 \cdot 24 = 120 \\ 3 \cdot 40 = 120 \end{array} \right\} \text{Equal; proportion is true.}$$

The cross products are equal, so the solution, $x = 24$, is correct.

> **CAUTION**
> The solution is 24, which is the unknown number in the proportion. 120 is *not* the solution; it is the cross product you get when *checking* the solution.

Solve a proportion for an unknown number by using the following steps.

Finding an Unknown Number in a Proportion

Step 1 Find the cross products.

Step 2 Show that the cross products are equivalent.

Step 3 Divide both products by the number multiplied by x (the number next to x).

Step 4 Check by writing the solution in the *original* proportion and finding the cross products.

EXAMPLE 1 **Solving Proportions for Unknown Numbers**

Find the unknown number in each proportion. Round answers to the nearest hundredth when necessary.

(a) $\dfrac{16}{x} = \dfrac{32}{20}$

Recall that ratios can be rewritten in lowest terms. If desired, you can do that *before* finding the cross products. In this example, write $\frac{32}{20}$ in lowest terms as $\frac{8}{5}$, which gives the proportion $\dfrac{16}{x} = \dfrac{8}{5}$.

Step 1
$$\frac{16}{x} = \frac{8}{5}$$
$x \cdot 8$
$16 \cdot 5$
Find the cross products.

Step 2
$$x \cdot 8 = 16 \cdot 5 \quad \leftarrow \text{Show that cross products are equivalent.}$$
$$x \cdot 8 = \quad 80$$

Step 3
$$\frac{x \cdot \overset{1}{\cancel{8}}}{\cancel{8}_1} = \frac{80}{8} \quad \leftarrow \text{Divide both sides by 8.}$$
$$x = 10 \quad \leftarrow \text{Find } x. \text{ (No rounding necessary)}$$

The unknown number in the proportion is 10.

Step 4 Write the solution in the *original* proportion and check by finding cross products.

$10 \cdot 32 = 320$

$$x \text{ is } 10 \rightarrow \frac{16}{10} = \frac{32}{20}$$

Equal: proportion is true.

$16 \cdot 20 = 320$

> The solution is 10, **not** 320.

The cross products are equal, so 10 is the correct solution.

Note

It is not necessary to write the ratios in lowest terms before solving. However, if you do, you will work with smaller numbers.

Continued on Next Page

(b) $\dfrac{7}{12} = \dfrac{15}{x}$

Step 1

$12 \cdot 15 = 180 \leftarrow$

$\dfrac{7}{12} \bowtie \dfrac{15}{x}$ ← Find the cross products.

$7 \cdot x \leftarrow$

Step 2 $7 \cdot x = 180 \leftarrow$ Show that cross products are equal.

Step 3 $\dfrac{\overset{1}{\cancel{7}} \cdot x}{\underset{1}{\cancel{7}}} = \dfrac{180}{7} \leftarrow$ Divide both sides by 7.

$x \approx 25.71 \leftarrow$ Rounded to nearest hundredth

When the division does not come out even, check for directions on how to round your answer. Divide out one more place, then round.

$$\begin{array}{r} 25.714 \\ 7\overline{)180.000} \end{array} \begin{array}{l} \leftarrow \text{Divide out to thousandths} \\ \text{so you can round to hundredths.} \end{array}$$

The unknown number in the proportion is 25.71 (rounded).

Step 4 Write the solution in the original proportion and check by finding the cross products.

$12 \cdot 15 = \mathbf{180} \leftarrow$

$\dfrac{\mathbf{7}}{\mathbf{12}} \bowtie \dfrac{\mathbf{15}}{\mathbf{25.71}}$ Very close, but *not* equal due to rounding the solution.

$7 \cdot \mathbf{25.71} = \mathbf{179.97} \leftarrow$

The cross products are slightly different because you rounded the value of x. However, they are close enough to see that the problem was done correctly and that 25.71 is the approximate solution.

Work Problem **1** at the Side. ▶

OBJECTIVE 2 Find the unknown number in a proportion with mixed numbers or decimals. The next example shows how to work with mixed numbers or decimals in a proportion.

EXAMPLE 2 Solving Proportions with Mixed Numbers and Decimals

Find the unknown number in each proportion.

(a) $\dfrac{2\frac{1}{5}}{6} = \dfrac{x}{10}$ $\dfrac{2\frac{1}{5}}{6} \bowtie \dfrac{x}{10}$

$6 \cdot x \leftarrow$ Find the cross products.

$2\frac{1}{5} \cdot 10 \leftarrow$

Find $2\frac{1}{5} \cdot 10$.

$$2\frac{1}{5} \cdot 10 = \dfrac{11}{5} \cdot \dfrac{10}{1} = \dfrac{11}{\cancel{5}} \cdot \dfrac{\overset{2}{\cancel{10}}}{1} = \dfrac{22}{1} = 22$$

Changed to improper fraction

Continued on Next Page

---

1 Find the unknown numbers. Round to hundredths when necessary. Check your solutions by finding the cross products.

(a) $\dfrac{1}{2} = \dfrac{x}{12}$

(b) $\dfrac{6}{10} = \dfrac{15}{x}$

(c) $\dfrac{28}{x} = \dfrac{21}{9}$

(d) $\dfrac{x}{8} = \dfrac{3}{5}$

(e) $\dfrac{14}{11} = \dfrac{x}{3}$

ANSWERS

1. **(a)** $x = 6$ **(b)** $x = 25$
   (c) $x = 12$ **(d)** $x = 4.8$
   (e) $x \approx 3.82$ (rounded to nearest hundredth)

2 Find the unknown numbers. Round to hundredths on the decimal problems, if necessary. Check your solutions by finding the cross products.

(a) $\dfrac{3\frac{1}{4}}{2} = \dfrac{x}{8}$

(b) $\dfrac{x}{3} = \dfrac{1\frac{2}{3}}{5}$

(c) $\dfrac{0.06}{x} = \dfrac{0.3}{0.4}$

(d) $\dfrac{2.2}{5} = \dfrac{13}{x}$

(e) $\dfrac{x}{6} = \dfrac{0.5}{1.2}$

(f) $\dfrac{0}{2} = \dfrac{x}{7.092}$

Show that the cross products are equivalent.

$$6 \cdot x = 22$$

Divide both sides by 6.

$$\dfrac{\overset{1}{\cancel{6}} \cdot x}{\cancel{6}} = \dfrac{22}{6}$$

Write the solution as a mixed number in lowest terms.

$$x = \dfrac{22 \div 2}{6 \div 2} = \dfrac{11}{3} = 3\dfrac{2}{3}$$

The unknown number is $3\frac{2}{3}$.

Write the solution in the proportion and check by finding the cross products.

$$6 \cdot 3\dfrac{2}{3} = \dfrac{\overset{2}{\cancel{6}}}{1} \cdot \dfrac{11}{\cancel{3}} = \dfrac{22}{1} = 22$$

$$\dfrac{2\frac{1}{5}}{6} = \dfrac{3\frac{2}{3}}{10}$$

Equal

$$2\dfrac{1}{5} \cdot 10 = \dfrac{11}{\cancel{5}} \cdot \dfrac{\overset{2}{\cancel{10}}}{1} = \dfrac{22}{1} = 22$$

The solution is $3\frac{2}{3}$, **not** 22.

The cross products are equal, so $3\frac{2}{3}$ is the correct solution.

(b) $\dfrac{1.5}{0.6} = \dfrac{2}{x}$

Show that cross products are equivalent.

$$(1.5)(x) = \underbrace{(0.6)(2)}$$
$$(1.5)(x) = \quad 1.2$$

Divide both sides by 1.5

$$\dfrac{\overset{1}{\cancel{(1.5)}}(x)}{\cancel{1.5}} = \dfrac{1.2}{1.5}$$

$$x = \dfrac{1.2}{1.5}$$

Complete the division.

$$x = 0.8 \qquad 1.5\overline{)1.20}^{\,.8}$$

So the unknown number is 0.8. Write the solution in the original proportion and check it by finding the cross products.

$$(0.6)(2) = 1.2$$

$$\dfrac{1.5}{0.6} = \dfrac{2}{0.8}$$

Equal

$$(1.5)(0.8) = 1.2$$

The solution is 0.8, **not** 1.2.

The cross products are equal, so 0.8 is the correct solution.

◀ *Work Problem* **2** *at the Side.*

5.4 ▶▶▶ Exercises

FOR EXTRA HELP

MyMathLab

Math XL
PRACTICE

WATCH

DOWNLOAD

READ

REVIEW

Find the unknown number in each proportion. Round your answers to hundredths, if necessary. Check your answers by finding the cross products. See Examples 1 and 2.

1. $\dfrac{1}{3} = \dfrac{x}{12}$ **x = 4**

2. $\dfrac{x}{6} = \dfrac{15}{18}$ **x = 5**

3. $\dfrac{15}{10} = \dfrac{3}{x}$ **x = 2**

4. $\dfrac{5}{x} = \dfrac{20}{8}$ **x = 2**

5. $\dfrac{x}{11} = \dfrac{32}{4}$ **x = 88**

6. $\dfrac{12}{9} = \dfrac{8}{x}$ **x = 6**

7. $\dfrac{42}{x} = \dfrac{18}{39}$ **x = 91**

8. $\dfrac{49}{x} = \dfrac{14}{18}$ **x = 63**

9. $\dfrac{x}{25} = \dfrac{4}{20}$ **x = 5**

10. $\dfrac{6}{x} = \dfrac{4}{8}$ **x = 12**

11. $\dfrac{8}{x} = \dfrac{24}{30}$ **x = 10**

12. $\dfrac{32}{5} = \dfrac{x}{10}$ **x = 64**

13. $\dfrac{99}{55} = \dfrac{44}{x}$

   x ≈ 24.44 (rounded)

14. $\dfrac{x}{12} = \dfrac{101}{147}$

   x ≈ 8.24 (rounded)

15. $\dfrac{0.7}{9.8} = \dfrac{3.6}{x}$

   x = 50.4

16. $\dfrac{x}{3.6} = \dfrac{4.5}{6}$

   x = 2.7

17. $\dfrac{250}{24.8} = \dfrac{x}{1.75}$

   x ≈ 17.64 (rounded)

18. $\dfrac{4.75}{17} = \dfrac{43}{x}$

   x ≈ 153.89 (rounded)

Find the unknown number in each proportion. Write your answers as whole or mixed numbers when possible. See Example 2.

19. $\dfrac{15}{1\frac{2}{3}} = \dfrac{9}{x}$ $x = 1$

20. $\dfrac{x}{\frac{3}{10}} = \dfrac{2\frac{2}{9}}{1}$ $x = \dfrac{2}{3}$

21. 🌐 $\dfrac{2\frac{1}{3}}{1\frac{1}{2}} = \dfrac{x}{2\frac{1}{4}}$ $x = 3\frac{1}{2}$

22. $\dfrac{1\frac{5}{6}}{x} = \dfrac{\frac{3}{14}}{\frac{6}{7}}$ $x = 7\frac{1}{3}$

Solve each proportion two different ways. First change all the numbers to decimal form and solve. Then change all the numbers to fraction form and solve; write your answers in lowest terms.

23. $\dfrac{\frac{1}{2}}{x} = \dfrac{2}{0.8}$ $x = 0.2$ or $x = \dfrac{1}{5}$

24. $\dfrac{\frac{3}{20}}{0.1} = \dfrac{0.03}{x}$ $x = 0.02$ or $x = \dfrac{1}{50}$

25. $\dfrac{x}{\frac{3}{50}} = \dfrac{0.15}{1\frac{4}{5}}$ $x = 0.005$ or $x = \dfrac{1}{200}$

26. $\dfrac{8\frac{4}{5}}{1\frac{1}{10}} = \dfrac{x}{0.4}$ $x = 3.2$ or $x = 3\frac{1}{5}$

Relating Concepts (Exercises 27–28) For Individual or Group Work

Work Exercises 27–28 in order. *First prove that the proportions are **not** true. Then create four true proportions for each exercise by changing one number at a time.*

27. $\dfrac{10}{4} = \dfrac{5}{3}$ Find cross products: $20 \neq 30$, so the proportion is false.

$\dfrac{6\frac{2}{3}}{4} = \dfrac{5}{3}$ or $\dfrac{10}{6} = \dfrac{5}{3}$ or $\dfrac{10}{4} = \dfrac{7.5}{3}$ or $\dfrac{10}{4} = \dfrac{5}{2}$

28. $\dfrac{6}{8} = \dfrac{24}{30}$ Find cross products: $192 \neq 180$, so the proportion is false.

$\dfrac{6.4}{8} = \dfrac{24}{30}$ or $\dfrac{6}{7.5} = \dfrac{24}{30}$ or $\dfrac{6}{8} = \dfrac{22.5}{30}$ or $\dfrac{6}{8} = \dfrac{24}{32}$

Summary Exercises on Ratios, Rates, and Proportions

Use the circle graph of one college's enrollment to complete Exercises 1–4. Write each ratio as a fraction in lowest terms.

1. Write the ratio of freshmen to juniors.

   $\frac{6}{5}$

2. What is the ratio of freshmen to the total college enrollment?

   $\frac{2}{7}$

3. Find the ratio of seniors and sophomores to juniors.

   $\frac{2}{1}$

4. Write the ratio of freshmen and sophomores to juniors and seniors.

   $\frac{13}{11}$

COLLEGE ENROLLMENT

Seniors 1850 | Freshmen 2400 | Juniors 2000 | Sophomores 2150

The bar graph shows the number of Americans who play various instruments. Use the graph to complete Exercises 5 and 6. Write each ratio as a fraction in lowest terms.

AMERICANS MAKE MUSIC BY THE MILLIONS

How many people play each type of instrument?

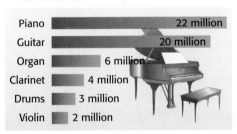

Piano — 22 million
Guitar — 20 million
Organ — 6 million
Clarinet — 4 million
Drums — 3 million
Violin — 2 million

Source: *America by the Numbers.*

5. Write six ratios that compare the least popular instrument to each of the other instruments.

   Comparing the violin to piano, guitar, organ, clarinet, and drums gives ratios of $\frac{1}{11}, \frac{1}{10}, \frac{1}{3}, \frac{1}{2}$ and $\frac{2}{3}$, respectively.

6. Which two instruments give each of these ratios: **(a)** $\frac{5}{1}$; **(b)** $\frac{2}{1}$? There may be more than one correct answer.

   (a) guitar to clarinet; (b) organ to drums, or clarinet to violin

The table lists data on the top three individual scoring NBA basketball games of all time. Use the data to answer Exercises 7 and 8. Round answers to the nearest tenth.

7. What was Wilt Chamberlain's scoring rate in points per minute and in minutes per point?

   2.1 points/min; 0.5 min/point

8. Find David Robinson's scoring rate in points per minute and minutes per point.

   1.6 points/min; 0.6 min/point

Player	Date	Points	Min.
Wilt Chamberlain	3/2/62	100	48
David Thompson	4/9/78	73	43
David Robinson	4/24/94	71	44

Source: www.NBA.com

Wilt Chamberlain (#13)

9. Lucinda's paycheck showed gross pay of $652.80 for 40 hours of work and $195.84 for 8 hours of overtime work. Find her regular hourly pay rate and her overtime rate.

$16.32/hour; $24.48/hour for overtime

10. Satellite TV is being offered to new subscribers at $25 per month for 50 channels, $39 per month for 100 channels, or $48 per month for 150 channels. What is the monthly cost per channel under each plan? (*Source:* Dish1Up Satellites.)

$0.50/channel; $0.39/channel; $0.32/channel

11. Find the best buy on gourmet coffee beans.

    11 ounces for $6.79

    12 ounces for $7.24

    16 ounces for $10.99

(*Source:* Cub Foods.)

12 ounces, at about $0.60 per ounce

12. Which brand of cat food is the best buy? You have a coupon for $2 off Brand P, and another for $1 off Brand N.

    Brand N is $3.75 for 3.5 pounds.

    Brand P is $5.99 for 7 pounds.

    Brand R is $10.79 for 18 pounds.

Brand P with the $2 coupon is the best buy at $0.57 per pound.

Use either the method of writing in lowest terms or the method of finding cross products to decide whether each proportion is true or false. Show your work and then write true *or* false.

13. $\dfrac{28}{21} = \dfrac{44}{33}$

$\dfrac{4}{3} = \dfrac{4}{3}$ **or** **924 = 924; true**

14. $\dfrac{2.3}{8.05} = \dfrac{0.25}{0.9}$

2.0125 ≠ 2.07; false

15. $\dfrac{2\frac{5}{8}}{3\frac{1}{4}} = \dfrac{21}{26}$

$68\frac{1}{4} = 68\frac{1}{4}$**; true**

Solve each proportion to find the unknown number. Round your answers to hundredths when necessary.

16. $\dfrac{7}{x} = \dfrac{25}{100}$

x = 28

17. $\dfrac{15}{8} = \dfrac{6}{x}$

x = 3.2

18. $\dfrac{x}{84} = \dfrac{78}{36}$

x = 182

19. $\dfrac{10}{11} = \dfrac{x}{4}$

x ≈ 3.64 (rounded)

20. $\dfrac{x}{17} = \dfrac{3}{55}$

x ≈ 0.93 (rounded)

21. $\dfrac{2.6}{x} = \dfrac{13}{7.8}$

x = 1.56

22. $\dfrac{0.14}{1.8} = \dfrac{x}{0.63}$

x ≈ 0.05 (rounded)

23. $\dfrac{\frac{1}{3}}{8} = \dfrac{x}{24}$

x = 1

24. $\dfrac{6\frac{2}{3}}{4\frac{1}{6}} = \dfrac{6\frac{5}{}}{x}$

x = $\dfrac{3}{4}$

5.5 ▶▶▶ Solving Application Problems with Proportions

OBJECTIVE 1 Use proportions to solve application problems.
Proportions can be used to solve a wide variety of problems. Watch for problems in which you are given a ratio or rate and then are asked to find part of a corresponding ratio or rate. Remember that a ratio or rate compares two quantities and often includes one of the following indicator words.

OBJECTIVE

1 Use proportions to solve application problems.

in   for   on   per   from   to

Use the six problem-solving steps you learned in **Section 1.10.**

Step 1   **Read** the problem.

Step 2   **Work out a plan.**

Step 3   **Estimate** a reasonable answer.

Step 4   **Solve** the problem.

Step 5   **State the answer.**

Step 6   **Check** your work.

EXAMPLE 1   Solving a Proportion Application

Mike's car can travel 163 **miles on** 6.4 **gallons** of gas. How far can it travel on a full tank of 14 **gallons** of gas? Round to the nearest whole mile.

Step 1   **Read** the problem. The problem asks for the number of miles the car can travel on 14 gallons of gas.

Step 2   **Work out a plan.** Decide what is being compared. This example compares **miles** to **gallons**. Write a proportion using the two rates. Be sure that *both* rates compare miles to gallons in the same order. In other words, miles is in both numerators and gallons is in both denominators. Use a letter to represent the unknown number.

Matching units

This rate compares **miles** to **gallons**. } ⟶ $\dfrac{163 \text{ miles}}{6.4 \text{ gallons}} = \dfrac{x \text{ miles}}{14 \text{ gallons}}$ ⟵ { This rate compares **miles** to **gallons**.

Matching units

Step 3   **Estimate** a reasonable answer. To estimate the answer, notice that 14 gallons is a little more than *twice as much* as 6.4 gallons, so the car should travel a little more than *twice as far.* So use 2 • 163 miles = 326 miles as the estimate.

Step 4   **Solve** the problem. Ignore the units while solving for *x.*

$$\frac{163 \text{ miles}}{6.4 \text{ gallons}} = \frac{x \text{ miles}}{14 \text{ gallons}}$$

$(6.4)\,(x) = \underbrace{(163)\,(14)}$ Show that cross products are equivalent.

$(6.4)\,(x) = 2282$

$\dfrac{(6.4)(x)}{6.4} = \dfrac{2282}{6.4}$ Divide both sides by 6.4.

$x = 356.5625$ Round to 357.   Check the problem for rounding directions; this one asks for nearest whole number.

Continued on Next Page

1 Set up and solve a proportion for each problem.

(a) If 2 pounds of fertilizer will cover 50 square feet of garden, how many pounds are needed for 225 square feet?

(b) A U.S. map has a scale of 1 inch to 75 miles. Lake Superior is 4.75 inches long on the map. What is the lake's actual length to the nearest whole mile?

(c) Cough syrup is to be given at the rate of 30 milliliters for each 100 pounds of body weight. How much should be given to a 34-pound child? Round to the nearest whole milliliter.

ANSWERS

1. **(a)** $\dfrac{2 \text{ pounds}}{50 \text{ square feet}} = \dfrac{x \text{ pounds}}{225 \text{ square feet}}$
$x = 9$ pounds

(b) $\dfrac{1 \text{ inch}}{75 \text{ miles}} = \dfrac{4.75 \text{ inches}}{x \text{ miles}}$
$x = 356.25$ miles, rounds to 356 miles

(c) $\dfrac{30 \text{ milliliters}}{100 \text{ pounds}} = \dfrac{x \text{ milliliters}}{34 \text{ pounds}}$
$x \approx 10$ milliliters (rounded)

Step 5 **State the answer.** Rounded to the nearest mile, the car can travel about 357 miles on a full tank of gas.

Step 6 **Check** your work. The answer, 357 miles, is a little more than the estimate of 326 miles, so it is reasonable.

> **CAUTION**
> When setting up a proportion do *not* mix up the units in the rates.
>
> $\left. \text{compares } \mathbf{miles} \atop \text{to gallons} \right\} \quad \dfrac{163 \text{ miles}}{6.4 \text{ gallons}} \neq \dfrac{14 \text{ gallons}}{x \text{ miles}} \quad \left\{ \text{compares } \mathbf{gallons} \atop \text{to } \mathbf{miles} \right.$
>
> These rates do *not* compare things in the same order and *cannot* be set up as a proportion.

◀ *Work Problem* **1** *at the Side.*

EXAMPLE 2 **Solving a Proportion Application**

A newspaper report says that 7 out of 10 people surveyed watch the news on TV. At that rate, how many of the 3200 people in town would you expect to watch the news?

Step 1 **Read** the problem. The problem asks how many of the 3200 people in town would be expected to watch TV news.

Step 2 **Work out a plan.** You are comparing people who watch the news to people surveyed. Set up a proportion using the two rates described in the example. Be sure that both rates make the same comparison. "People who watch the news" is mentioned first, so it should be in the numerator of *both* rates.

People who watch news $\rightarrow \dfrac{7}{10} = \dfrac{x}{3200} \leftarrow$ People who watch news
Total group $\rightarrow$ $\quad\quad\quad\quad \leftarrow$ Total group
(people surveyed) $\quad\quad\quad\quad\quad$ (people in town)

Step 3 **Estimate** a reasonable answer. To estimate the answer, notice that 7 out of 10 people is more than half the people, but less than all the people. Half of 3200 people is $3200 \div 2 = 1600$, so our estimate is between 1600 and 3200 people.

Step 4 **Solve** the problem. Solve for the unknown number in the proportion.

$$\frac{7}{10} = \frac{x}{3200}$$

$10 \cdot x = 7 \cdot 3200$ Show that cross products are equivalent.

$10 \cdot x = 22{,}400$

$\dfrac{\overset{1}{\cancel{10}} \cdot x}{\underset{1}{\cancel{10}}} = \dfrac{22{,}400}{10}$ Divide both sides by 10.

$x = 2240$ No rounding is needed here.

Continued on Next Page

Step 5 **State the answer.** You would expect 2240 people in town to watch the news on TV.

Step 6 **Check** your work. The answer, 2240 people, is between 1600 and 3200, as called for in the estimate.

CAUTION

Always check that your answer is reasonable. If it is not, look at the way your proportion is set up. Be sure you have matching units in the numerators and matching units in the denominators.

For example, suppose you had set up the last proportion **incorrectly** as shown here.

$$\frac{7}{10} = \frac{3200}{x} \quad \leftarrow \text{ Incorrect setup}$$

$$7 \cdot x = 10 \cdot 3200$$

$$\frac{\overset{1}{\cancel{7}} \cdot x}{\underset{1}{\cancel{7}}} = \frac{32,000}{7}$$

$$x \approx 4571 \text{ people} \quad \leftarrow \text{ Unreasonable answer}$$

This answer is **unreasonable** because there are only 3200 people in the town; it is **not** possible for 4571 people to watch the news.

Work Problem (**2**) *at the Side.* ▶

Work Problem (2) at the Side. ▶

(**2**) Solve each problem to find a reasonable answer. Then flip one side of your proportion to see what answer you get with an *incorrect* setup. Explain why the second answer is *unreasonable.*

(a) A survey showed that 2 out of 3 people would like to lose weight. At this rate, how many people in a group of 150 want to lose weight?

(b) In one state, 3 out of 5 college students receive financial aid. At this rate, how many of the 4500 students at Central Community College receive financial aid?

(c) An advertisement says that 9 out of 10 dentists recommend sugarless gum. If the ad is true, how many of the 60 dentists in our city would recommend sugarless gum?

ANSWERS

2. (a) 100 people (reasonable); incorrect setup gives 225 people (only 150 people in the group).
(b) 2700 students (reasonable); incorrect setup gives 7500 students (only 4500 students at the college).
(c) 54 dentists (reasonable); incorrect setup gives about 67 dentists (only 60 dentists in the city).

Math in the Media

FEEDING HUMMINGBIRDS

After getting a hummingbird feeder, the next step is to fill it! You have two choices at this point: you can either buy one of the commercial mixtures or you can make your own solution. See the recipe at the right.

The concentration of the sugar is important. The 1 to 4 ratio of sugar to water is recommended because it approximates the ratio of sugar to water found in the nectar of many hummingbird flowers.

Boiling the solution helps retard fermentation. Sugar-and-water solutions are subject to rapid spoiling, especially in hot weather.

Source: The Hummingbird Book.

> **Recipe for Homemade Mixture:**
> 1 part sugar (not honey)
> 4 parts water
> Boil for 1 to 2 minutes. Cool.
> Store extra in refrigerator.

A recipe can be used to make as much of a mixture as you need as long as the ingredients are kept proportional. Use the recipe for a homemade mixture of sugar water for hummingbird feeders to answer these problems.

1. What is the ratio of sugar to water in the recipe?

   1 to 4

   What is the ratio of water to sugar in the recipe?

   4 to 1

2. Complete each table.

Sugar	Water
1 cup	4 cups
$1\frac{1}{4}$ cups	5 cups
$1\frac{1}{2}$ cups	6 cups
$1\frac{3}{4}$ cups	7 cups
2 cups	8 cups

Sugar	Water
1 cup	4 cups
$\frac{3}{4}$ cup	3 cups
$\frac{1}{2}$ cup	2 cups
$\frac{1}{4}$ cup	1 cup

3. How much water would you need if you used

   (a) 3 cups of sugar? **12 cups**

   (b) 4 cups of sugar? **16 cups**

   (c) $\frac{1}{3}$ cup of sugar? $\frac{4}{3}$ **cup**

4. As you change the amounts of water and sugar, should you change the length of time that you boil the mixture? Explain your answer. **No. If you boil the mixture for a longer time, the water will evaporate and the ratio of sugar to water will increase.**

FOR EXTRA HELP

MyMathLab   Math XL PRACTICE   WATCH   DOWNLOAD   READ   REVIEW

Set up and solve a proportion for each application problem. See Example 1.

1. Caroline can sketch four cartoon strips in five hours. How long will it take her to sketch 18 strips?

22.5 hours

2. The Cosmic Toads recorded eight songs on their first CD in 26 hours. How long will it take them to record 14 songs for their second CD?

45.5 hours

3. Sixty newspapers cost $27. Find the cost of 16 newspapers.

$7.20

4. Twenty-two guitar lessons cost $528. Find the cost of 12 lessons.

$288

5. If three pounds of fescue grass seed cover about 350 square feet of ground, how many pounds are needed for 4900 square feet?

42 pounds

6. Anna earns $1242.08 in 14 days. How much does she earn in 260 days?

$23,067.20

7. Tom makes $672.80 in 5 days. How much does he make in 3 days?

$403.68

8. If 5 ounces of a medicine must be mixed with 8 ounces of water, how many ounces of medicine would be mixed with 20 ounces of water?

12.5 ounces

9. The bag of rice noodles below makes 7 servings. At that rate, how many ounces of noodles do you need for 12 servings, to the nearest ounce?

10 ounces (rounded)

10. This can of sweet potatoes is enough for 4 servings. How many ounces are needed for 9 servings, to the nearest ounce?

52 ounces (rounded)

11. Three quarts of a latex enamel paint will cover about 270 square feet of wall surface. How many quarts will you need to cover 350 square feet of wall surface in your kitchen and 100 square feet of wall surface in your bathroom?

5 quarts

12. One gallon of clear gloss wood finish covers about 550 square feet of surface. If you need to apply three coats of finish to 400 square feet of surface, how many gallons do you need, to the nearest tenth?

2.2 gallons (rounded)

Use the floor plan shown to complete Exercises 13–16. On the plan, one inch represents four feet.

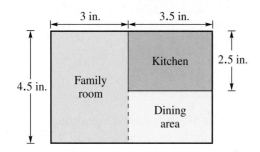

13. What is the actual length and width of the kitchen?

14 ft, 10 ft

14. What is the actual length and width of the family room?

18 ft, 12 ft

15. What is the actual length and width of the dining area?

14 ft, 8 ft

16. What is the actual length and width of the entire floor plan?

26 ft, 18 ft

The table below lists recommended amounts of food to order for 25 party guests. Use the table to answer Exercises 17 and 18. (Source: CubFoods.)

FOOD FOR 25 GUESTS

Item	Amount
Fried chicken	40 pieces
Lasagna	14 pounds
Deli meats	4.5 pounds
Sliced cheese	$2\frac{1}{3}$ pounds
Bakery buns	3 dozen
Potato salad	6 pounds

17. How much of each food item should Nathan and Amanda order for a graduation party with 60 guests?

96 pieces chicken; 33.6 pounds lasagna;

10.8 pounds deli meats; $5\frac{3}{5}$ pounds cheese;

7.2 dozen (about 86) buns; 14.4 pounds salad

18. Taisha is having 20 neighbors over for a Fourth of July picnic. How much food should she buy?

32 pieces chicken; 11.2 pounds lasagna;

3.6 pounds deli meats; $1\frac{13}{15}$ pounds cheese;

2.4 dozen (about 29) buns; 4.8 pounds salad

Set up a proportion to solve each problem. Check to see whether your answer is reasonable. Then flip one side of your proportion to see what answer you get with an incorrect setup. Explain why the second answer is unreasonable. See Example 2.

19. About 7 out of 10 people entering a community college need to take a refresher math course. If there are 2950 entering students, how many will probably need refresher math? (*Source:* Minneapolis Community and Technical College.)

2065 students (reasonable); about 4214 students with incorrect setup (only 2950 students in the group)

20. In a survey, only 3 out of 100 people like their eggs poached. At that rate, how many of the 60 customers who ordered eggs at Soon-Won's restaurant this morning asked to have them poached? Round to the nearest whole person.

about 2 people (reasonable); 2000 people with incorrect setup (only 60 people at the restaurant)

21. About 1 out of 3 people choose vanilla as their favorite ice cream flavor. If 250 people attend an ice cream social, how many would you expect to choose vanilla? Round to the nearest whole person.

about 83 people (reasonable); 750 people with incorrect setup (only 250 people attended)

22. In a test of 200 sewing machines, only one had a defect. At that rate, how many of the 5600 machines shipped from the factory have defects?

28 sewing machines (reasonable); 1,120,000 machines with incorrect setup (only 5600 machines shipped)

23. About 98 out of 100 U.S. households have at least one TV set. There were 113,100,000 U.S. households in 2005. How many households have one or more TVs? (*Source:* Nielsen Media Research.)

110,838,000 households (reasonable); about 115,408,163 households with incorrect setup (only 113,100,000 U.S. households)

24. In a survey, 3 out of 100 dog owners wash their pets by having the dogs go into the shower with them. If the survey is accurate, how many of the 31,200,000 dog owners in the United States use this method? (*Source:* Teledyne Water Pik, American Veterinary Medical Association.)

936,000 dog owners (reasonable); 1,040,000,000 dog owners with incorrect setup (only 31,200,000 U.S. dog owners)

Set up and solve a proportion for each problem.

25. The stock market report says that 5 stocks went up for every 6 stocks that went down. If 750 stocks went down yesterday, how many went up?

625 stocks

26. The human body contains 90 pounds of water for every 100 pounds of body weight. How many pounds of water are in a child who weighs 80 pounds?

72 pounds

27. The ratio of the length of an airplane wing to its width is 8 to 1. If the length of a wing is 32.5 meters, how wide must it be? Round to the nearest hundredth.

4.06 meters (rounded)

28. The Rosebud School District wants a student-to-teacher ratio of 19 to 1. How many teachers are needed for 1850 students? Round to the nearest whole number.

97 teachers (rounded)

29. The number of calories you burn is proportional to your weight. A 150-pound person burns 222 calories during 30 minutes of tennis. How many calories would a 210-pound person burn, to the nearest whole number? (*Source: Wellness Encyclopedia.*)

311 calories (rounded)

30. (Complete Exercise 29 first.) A 150-pound person burns 189 calories during 45 minutes of grocery shopping. How many calories would a 115-pound person burn, to the nearest whole number? (*Source: Wellness Encyclopedia.*)

145 calories (rounded)

31. At 3 P.M., Coretta's shadow is 1.05 meters long. Her height is 1.68 meters. At the same time, a tree's shadow is 6.58 meters long. How tall is the tree? Round to the nearest hundredth.

10.53 meters (rounded)

32. Refer to Exercise 31. Later in the day, Coretta's shadow was 2.95 meters long. How long a shadow did the tree have at that time? Round to the nearest hundredth.

18.49 meters (rounded)

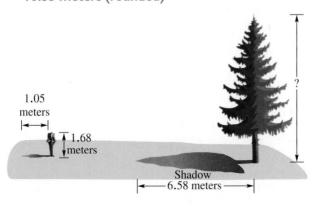

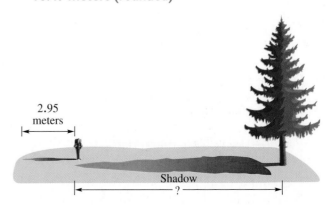

33. Can you set up a proportion to solve this problem? Explain why or why not. Jim is 25 years old and weighs 180 pounds. How much will he weigh when he is 50 years old?

You cannot solve this problem using a proportion because the ratio of age to weight is not constant. As Jim's age increases, his weight may decrease, stay the same, or increase.

34. Write your own application problem that can be solved by setting up a proportion. Also show the proportion and the steps needed to solve your problem.

Answers will vary; Exercises 1-32 are all examples of application problems.

35. A survey of college students shows that 4 out of 5 drink coffee. Of the students who drink coffee, 1 out of 8 adds cream to it. How many of the 50,500 students at Ohio State University would be expected to use cream in their coffee?

5050 students use cream

36. About 9 out of 10 adults think it is a good idea to exercise regularly. But of the ones who think it is a good idea, only 1 in 6 actually exercises at least three times a week. At this rate, how many of the 300 employees in our company exercise regularly?

45 employees

37. The nutrition information on a bran cereal box says that a $\frac{1}{3}$-cup serving provides 80 calories and 8 grams of dietary fiber. At that rate, how many calories and grams of fiber are in a $\frac{1}{2}$-cup serving? (*Source:* Kraft Foods, Inc.)

120 calories and 12 grams of fiber

38. A $\frac{2}{3}$-cup serving of penne pasta has 210 calories and 2 grams of dietary fiber. How many calories and grams of fiber would be in a 1-cup serving? (*Source:* Borden Foods.)

315 calories; 3 grams of fiber

Relating Concepts (Exercises 39–42) For Individual or Group Work

A box of instant mashed potatoes has the list of ingredients shown in the table. Use this information to **work Exercises 39–42 in order.**

Ingredient	For 12 Servings
Water	$3\frac{1}{2}$ cups
Margarine	6 Tbsp
Milk	$1\frac{1}{2}$ cups
Potato flakes	4 cups

Source: General Mills.

39. Find the amount of each ingredient needed for 6 servings. Show *two* different methods for finding the amounts. One method should use proportions.

$1\frac{3}{4}$ **cups water, 3 Tbsp margarine,**

$\frac{3}{4}$ **cup milk, 2 cups flakes**

40. Find the amount of each ingredient needed for 18 servings. Show *two* different methods for finding the amounts, one using proportions and one using your answers from Exercise 39.

$5\frac{1}{4}$ **cups water, 9 Tbsp margarine,**

$2\frac{1}{4}$ **cups milk, 6 cups flakes**

41. Find the amount of each ingredient needed for 3 servings, using your answers from either Exercise 39 or Exercise 40.

$\frac{7}{8}$ **cup water, $1\frac{1}{2}$ Tbsp margarine,**

$\frac{3}{8}$ **cup milk, 1 cup flakes**

42. Find the amount of each ingredient needed for 9 servings, using your answers from either Exercise 40 or Exercise 41.

$2\frac{5}{8}$ **cups water, $4\frac{1}{2}$ Tbsp margarine,**

$1\frac{1}{8}$ **cups milk, 3 cups flakes**

Chapter 5 ▶▶▶ Summary

▶ Key Terms

5.1	**ratio**	A ratio compares two quantities having the same type of units. For example, the ratio of 6 apples to 11 apples is written in fraction form as $\frac{6}{11}$. The common units (apples) divide out.
5.2	**rate**	A rate compares two measurements with different types of units. Examples are 96 dollars for 8 hours, or 450 miles on 18 gallons.
	unit rate	A unit rate has 1 in the denominator.
	cost per unit	Cost per unit is a rate that tells how much you pay for one item or one unit. The lowest cost per unit is the best buy.
5.3	**proportion**	A proportion states that two ratios or rates are equivalent.
	cross products	Multiply along one diagonal and then multiply along the other diagonal to find the cross products of a proportion. If the cross products are equal, the proportion is true.

▶ Test Your Word Power

See how well you have learned the vocabulary in this chapter. Answers follow the Quick Review.

1. **A ratio**
   A. can be written only as a fraction
   B. compares two quantities that have the same type of units
   C. compares two quantities that have different types of units
   D. is the reciprocal of a rate.

2. **A rate**
   A. can be written only as a decimal
   B. compares two quantities that have the same type of units
   C. compares two quantities that have different types of units
   D. is the reciprocal of a ratio.

3. **A unit rate**
   A. has a numerator of 1
   B. has a denominator of 1
   C. is found by cross multiplying
   D. is usually written in fraction form.

4. **Cost per unit is**
   A. the best buy
   B. a ratio written in lowest terms
   C. found by comparing cross products
   D. the price of one item or one unit.

5. **A proportion**
   A. shows that two ratios or rates are equivalent
   B. contains only whole numbers or decimals
   C. always has one unknown number
   D. states that two improper fractions are equivalent.

6. **Cross products are**
   A. used only with ratios, not with rates
   B. equal when a proportion is false.
   C. used to find the best buy
   D. equal when a proportion is true.

▶ Quick Review

Concepts

5.1 Writing a Ratio

A ratio compares two quantities that have the same type of units. A ratio is usually written as a fraction with the number that is mentioned first in the numerator. The common units divide out and are not written in the answer. Check that the fraction is in lowest terms.

Examples

Write this ratio as a fraction in lowest terms.

60 ounces of medicine **to** 160 ounces of medicine

$$\frac{60 \text{ ounces}}{160 \text{ ounces}} = \frac{60 \div 20}{160 \div 20} = \frac{3}{8} \left\{ \begin{array}{l} \text{Ratio in lowest} \\ \text{terms} \end{array} \right.$$

Divide out common units.

Concepts	Examples

5.1 Using Mixed Numbers in a Ratio

If a ratio has mixed numbers, change the mixed numbers to improper fractions. Rewrite the problem in horizontal format using the "÷" symbol for division. Finally, multiply by the reciprocal of the divisor.

Write as a ratio of whole numbers in lowest terms.

$$2\frac{1}{2} \text{ to } 3\frac{3}{4}$$

$$\frac{2\frac{1}{2}}{3\frac{3}{4}} = \frac{\frac{5}{2}}{\frac{15}{4}}$$

Reciprocal

$$= \frac{5}{2} \div \frac{15}{4} = \frac{5}{2} \cdot \frac{4}{15}$$

$$= \frac{\overset{1}{\cancel{5}}}{\underset{1}{\cancel{2}}} \cdot \frac{\overset{2}{\cancel{4}}}{\underset{3}{\cancel{15}}} = \frac{2}{3} \quad \leftarrow \text{Ratio in lowest terms}$$

5.1 Using Measurements in Ratios

When a ratio compares measurements, both measurements must be in the *same* units. It is usually easier to compare the measurements using the smaller unit, for example, inches instead of feet.

Write as a ratio in lowest terms.

8 in. to 6 ft

Compare using the smaller unit, inches. Because 1 ft has 12 in., 6 ft is

$$6 \cdot \textbf{12 in.} = 72 \text{ in.}$$

The ratio is shown below.

$$\frac{8 \text{ in.}}{72 \text{ in.}} = \frac{8 \div 8}{72 \div 8} = \frac{1}{9}$$

↑
Divide out common units.

5.2 Writing Rates

A rate compares two measurements with different types of units. The units do *not* divide out, so you must write them as part of the rate.

Write the rate as a fraction in lowest terms.

475 miles in 10 hours

$$\frac{475 \text{ miles} \div 5}{10 \text{ hours} \div 5} = \frac{95 \text{ miles}}{2 \text{ hours}} \quad \begin{array}{l}\leftarrow \text{Must write units:} \\ \leftarrow \text{miles and hours}\end{array}$$

5.2 Finding a Unit Rate

A unit rate has 1 in the denominator. To find the unit rate, divide the numerator by the denominator. Write unit rates using the word **per** or a / mark.

Write as a unit rate: $1278 in 9 days.

$$\frac{\$1278}{9 \text{ days}} \quad \leftarrow \text{The fraction bar indicates division.}$$

$$9\overline{)1278}^{\,142} \quad \text{so} \quad \frac{\$1278 \div 9}{9 \text{ days} \div 9} = \frac{\$142}{1 \text{ day}}$$

Write the answer as $142 **per** day or $142/day.

Concepts	Examples

5.2 Finding the Best Buy

The best buy is the item with the lowest cost per unit. Divide the price by the number of units. Round to thousandths, if necessary. Then compare to find the lowest cost per unit.

Find the best buy on cheese. You have a coupon for 50¢ off on 2 pounds or 75¢ off on 3 pounds.

$$2 \text{ pounds for } \$2.75$$

$$3 \text{ pounds for } \$4.15$$

Find the cost per unit (cost per pound) after subtracting the coupon.

$$2 \text{ pounds cost } \$2.75 - \$0.50 = \$2.25$$

$$\frac{\$2.25}{2} = \$1.125 \text{ per pound}$$

$$3 \text{ pounds cost } \$4.15 - \$0.75 = \$3.40$$

$$\frac{\$3.40}{3} \approx \$1.133 \text{ per pound}$$

The lower cost per pound is $1.125, so 2 pounds of cheese is the best buy.

5.3 Writing Proportions

A proportion states that two ratios or rates are equivalent. The proportion "5 is to 6 as 25 is to 30" is written as shown below.

$$\frac{5}{6} = \frac{25}{30}$$

To see whether a proportion is true or false, multiply along one diagonal, then multiply along the other diagonal. If the two cross products are equal, the proportion is true. If the two cross products are unequal, the proportion is false.

Write as a proportion: 8 is to 40 as 32 is to 160

$$\frac{8}{40} = \frac{32}{160}$$

Is this proportion true or false?

$$\frac{6}{8\frac{1}{2}} = \frac{24}{34}$$

Find the cross products.

$$8\frac{1}{2} \cdot 24 = \frac{17}{2} \cdot \frac{\overset{12}{24}}{1} = 204$$

$$\frac{6}{8\frac{1}{2}} = \frac{24}{34} \qquad 6 \cdot 34 = 204 \longleftarrow \text{Equal}$$

The cross products are equal, so the proportion is true.

5.4 Solving Proportions

Solve for an unknown number in a proportion by using the steps shown on the next page.

Find the unknown number.

$$\frac{12}{x} = \frac{6}{8} \quad \left.\begin{array}{l} \\ \\ \end{array}\right\} \text{Write } \frac{6}{8} \text{ in}$$

$$\frac{12}{x} = \frac{3}{4} \quad \left.\begin{array}{l} \\ \\ \end{array}\right\} \text{lowest terms as } \frac{3}{4}$$

(continued)

Concepts	Examples

5.4 Solving Proportions (*continued*)

Step 1 Find the cross products. (If desired, you can rewrite the ratios in lowest terms before finding the cross products.)

Step 1
$$\frac{12}{x} = \frac{3}{4}$$
$x \cdot 3$ — Find cross products
$12 \cdot 4$

Step 2 Show that the cross products are equivalent.

Step 2 $x \cdot 3 = \underbrace{12 \cdot 4}$ Show that cross products are equivalent.

$x \cdot 3 = \ \ 48$

Step 3 Divide both products by the number multiplied by x (the number next to x).

Step 3
$$\frac{x \cdot \overset{1}{\cancel{3}}}{\underset{1}{\cancel{3}}} = \frac{48}{3}$$ Divide both sides by 3.

$x = 16$

Step 4 Check by writing the solution in the original proportion and finding the cross products.

Step 4

x is 16. →
$$\frac{12}{16} = \frac{6}{8}$$
$16 \cdot 6 = 96$
$12 \cdot 8 = 96$ Equal

The cross products are equal, so 16 is the correct solution (**not** 96).

5.5 Solving Application Problems with Proportions

Decide what is being compared. Set up and solve a proportion using the two rates described in the problem. Be sure that *both* rates compare things in the *same order*. Use a letter, like x, to represent the unknown number.

Use the six problem-solving steps.

If 3 pounds of grass seed cover 450 square feet of lawn, how much seed is needed for 1500 square feet of lawn?

Step 1 **Read** the problem carefully.

Step 1 The problem asks for the pounds of grass seed needed for 1500 square feet of lawn.

Step 2 **Work out a plan.**

Step 2 Pounds of seed is compared to square feet of lawn. Set up and solve a proportion using the two given rates. Be sure that pounds of seed is in both numerators and square feet of lawn is in both denominators.

Step 3 **Estimate** a reasonable answer.

Step 3 Because 1500 square feet is about three times as much lawn as 450 square feet, about three times as much seed is needed. So, $3 \cdot 3$ pounds $= 9$ pounds as our estimate.

Step 4 **Solve** the problem.

Step 4 With the proportion set up correctly, solve for the unknown number.

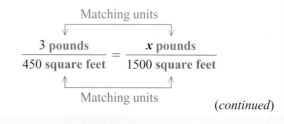

(continued)

Concepts	Examples
5.5 **Solving Application Problems with Proportions** (*continued*)	Both sides compare pounds to square feet. Ignore the units while finding the cross products and solving for *x*.

$$450 \cdot x = \underbrace{3 \cdot 1500}$$ Show that cross products are equivalent.

$$450 \cdot x = 4500$$

$$\frac{\overset{1}{\cancel{450}} \cdot x}{\cancel{450}_{1}} = \frac{4500}{450}$$ Divide both sides by 450.

$$x = 10$$

Step 5   State the answer.

Step 6   **Check** your work.

Step 5   10 pounds of grass seed are needed.

Step 6   The exact answer, 10 pounds of seed, is close to our estimate of 9 pounds, so it is reasonable.

ANSWERS TO TEST YOUR WORD POWER

1. B; *Example:*   The ratio of 3 miles to 4 miles is $\frac{3}{4}$; the common units (miles) divide out.

2. C; *Example:* $4.50 for 3 pounds is a rate comparing dollars to pounds.

3. B; *Example:* $\frac{\$1.79}{1 \text{ pound}}$ is a unit rate. We write it as $1.79 per pound or $1.79/pound.

4. D; *Example:* $3.95 per gallon tells the price of one gallon (one unit).

5. A; *Example:*   The proportion $\frac{5}{6} = \frac{25}{30}$ says that $\frac{5}{6}$ is equivalent to $\frac{25}{30}$.

6. D; *Example:*   The cross products for $\frac{5}{6} = \frac{25}{30}$ are $6 \cdot 25 = 150$ and $5 \cdot 30 = 150$.

Math in the Media

CURRENCY EXCHANGE

When you travel between countries, you will exchange U.S. dollars for the local currency. The exchange rate between currencies changes daily, and you can easily find the updated rates using the Internet or any major newspaper. The table shown below has been extracted from the Oanda Web page, www.oanda.com. It shows how much of each country's currency was equivalent to 1 U.S. dollar on April 16, 2008.

NORTH AMERICA/CARIBBEAN CURRENCY RATES (APRIL 16, 2008)

Currency	Symbol	Value
Canadian dollar	CAD	1.0198
Cayman Islands dollar	KYD	0.833
Jamaican dollar	JMD	74.75
Mexican peso	MXN	10.5
United States dollar	USD	1.00

From the table, $1.00 U.S. was equivalent to 10.5 Mexican pesos. You can set up a proportion to convert dollars to pesos. For example, suppose you want to determine the number of pesos that is equivalent to $50.00.

$$\frac{\$1}{10.5 \text{ pesos}} = \frac{\$50}{x \text{ pesos}} \quad \text{or} \quad \frac{1}{10.5} = \frac{50}{x}$$

$$(1)(x) = (10.5)(50)$$

$$x = 525.0 \text{ pesos}$$

So $50 buys 525 pesos.

1. Based on the currency exchange rates for April 16, 2008, find the amount of each local currency that is equivalent to $50 U.S. and find the number of U.S. dollars that is equivalent to 200 units of each local currency. Round your answers to the nearest hundredth.

   (a) $50 = __50.99__ Canadian dollars, and 200 Canadian dollars = __$196.12__ U.S. dollars.

   (b) $50 = __41.65__ Cayman Islands dollars, and 200 Cayman Islands dollars = __240.10__ U.S. dollars.

   (c) $50 = __3737.5__ Jamaican dollars, and 200 Jamaican dollars = __2.68__ U.S. dollars.

2. Set up a proportion to find the number of U.S. dollars that was equivalent to 1 Mexican Peso. Round your answer to the nearest cent. 1 Mexican peso was equivalent to $__0.10__ (U.S.).

3. From Problem 2, you should recognize the conversion rate based on 1 Mexican peso as the expression $\frac{1}{10.5}$. What is the mathematical word that describes the relationship between the conversion rates 10.5 and $\frac{1}{10.5}$?

   The rates 10.5 and $\frac{1}{10.5}$ are reciprocals.

Chapter 5 ▶▶▶ Review Exercises

[5.1] *Write each ratio as a fraction in lowest terms. Change to the same units when necessary, using the table of measurement comparisons in* **Section 5.1.** *Use the information in the graph to answer Exercises 1–3.*

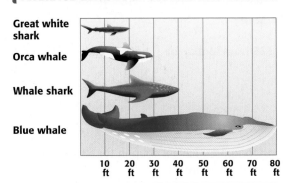

Average Length of Sharks and Whales

Great white shark

Orca whale

Whale shark

Blue whale

10 ft 20 ft 30 ft 40 ft 50 ft 60 ft 70 ft 80 ft

Source: *Grolier Multimedia Encyclopedia.*

1. Ratio of orca whale's length to whale shark's length

$\dfrac{3}{4}$

2. Ratio of blue whale's length to great white shark's length

$\dfrac{4}{1}$

3. Which two animals' lengths give a ratio of $\frac{1}{2}$? There are several answers.

great white shark to whale shark; whale shark to blue whale

4. $2.50 to $1.25

$\dfrac{2}{1}$

5. $0.30 to $0.45

$\dfrac{2}{3}$

6. $1\dfrac{2}{3}$ cups to $\dfrac{2}{3}$ cup

$\dfrac{5}{2}$

7. $2\dfrac{3}{4}$ miles to $16\dfrac{1}{2}$ miles

$\dfrac{1}{6}$

8. 5 hours to 100 minutes

$\dfrac{3}{1}$

9. 9 in. to 2 ft

$\dfrac{3}{8}$

10. 1 ton to 1500 pounds

$\dfrac{4}{3}$

11. 8 hours to 3 days

$\dfrac{1}{9}$

12. Jake sold $350 worth of his kachina figures. Ramona sold $500 worth of her pottery. What is the ratio of Ramona's sales to Jake's sales?

$\dfrac{10}{7}$

13. Ms. Wei's new car gets 35 miles per gallon. Her old car got 25 miles per gallon. Find the ratio of the new car's mileage to the old car's mileage.

$\dfrac{7}{5}$

14. This fall, 6000 students are taking math courses and 7200 students are taking English courses. Find the ratio of math students to English students.

$\dfrac{5}{6}$

[5.2] *Write each rate as a fraction in lowest terms.*

15. $88 for 8 dozen

$$\frac{\$11}{1 \text{ dozen}}$$

16. 96 children in 40 families

$$\frac{12 \text{ children}}{5 \text{ families}}$$

17. When entering data into his computer, Patrick can type four pages in 20 minutes. Give his rate in pages per minute and minutes per page.

0.2 page/minute or $\frac{1}{5}$ page/minute;

5 minutes/page

18. Elena made $24 in three hours. Give her earnings in dollars per hour and hours per dollar.

$8/hour;

0.125 hour/dollar or $\frac{1}{8}$ hour/dollar

Find the best buy.

19. Minced onion

8 ounces for $4.98

3 ounces for $2.49

2 ounces for $1.89

8 ounces for $4.98, about $0.623/ounce

20. Dog food; you have a coupon for $1 off on 8 pounds or more.

35.2 pounds for $36.96

17.6 pounds for $18.69

3.5 pounds for $4.25

17.6 pounds for $18.69 − $1 coupon, about $1.005/pound

[5.3] *Use either the method of writing in lowest terms or the method of finding cross products to decide whether each proportion is true or false. Show your work and then write* true *or* false.

21. $\dfrac{6}{10} = \dfrac{9}{15}$ $\dfrac{3}{5} = \dfrac{3}{5}$ or

true   90 = 90

22. $\dfrac{6}{48} = \dfrac{9}{36}$ $\dfrac{1}{8} \neq \dfrac{1}{4}$ or

false   432 ≠ 216

23. $\dfrac{47}{10} = \dfrac{98}{20}$ $\dfrac{47}{10} \neq \dfrac{49}{10}$ or

false   980 ≠ 940

24. $\dfrac{64}{36} = \dfrac{96}{54}$ $\dfrac{16}{9} = \dfrac{16}{9}$ or

true   3456 = 3456

25. $\dfrac{1.5}{2.4} = \dfrac{2}{3.2}$

true   4.8 = 4.8

26. $\dfrac{3\frac{1}{2}}{2\frac{1}{3}} = \dfrac{6}{4}$

true   14 = 14

[5.4] *Find the unknown number in each proportion. Round answers to the nearest hundredth, if necessary.*

27. $\dfrac{4}{42} = \dfrac{150}{x}$ x = 1575

28. $\dfrac{16}{x} = \dfrac{12}{15}$ x = 20

29. $\dfrac{100}{14} = \dfrac{x}{56}$ x = 400

30. $\dfrac{5}{8} = \dfrac{x}{20}$ x = 12.5

31. $\dfrac{x}{24} = \dfrac{11}{18}$

x ≈ 14.67 (rounded)

32. $\dfrac{7}{x} = \dfrac{18}{21}$

x ≈ 8.17 (rounded)

33. $\dfrac{x}{3.6} = \dfrac{9.8}{0.7}$ x = 50.4

34. $\dfrac{13.5}{1.7} = \dfrac{4.5}{x}$

x ≈ 0.57 (rounded)

35. $\dfrac{0.82}{1.89} = \dfrac{x}{5.7}$

x ≈ 2.47 (rounded)

[5.5] *Set up and solve a proportion for each application problem.*

36. The ratio of cats to dogs at the animal shelter is 3 to 5. If there are 45 dogs, how many cats are there?

27 cats

37. Danielle had 8 hits in 28 times at bat during last week's games. If she continues to hit at the same rate, how many hits will she gets in 161 times at bat?

46 hits

38. If 3.5 pounds of ground beef cost $9.77, what will 5.6 pounds cost? Round to the nearest cent.

$15.63 (rounded)

39. About 4 out of 10 students are expected to vote in campus elections. There are 8247 students. How many are expected to vote? Round to the nearest whole number.

3299 students (rounded)

40. The scale on Brian's model railroad is 1 in. to 16 ft. One of the scale model boxcars is 4.25 in. long. What is the length of a real boxcar in feet?

68 feet

41. Marvette makes necklaces to sell at a local gift shop. She made 2 dozen necklaces in $16\frac{1}{2}$ hours. How long will it take her to make 40 necklaces?

$27\frac{1}{2}$ hours or 27.5 hours

42. A 180-pound person burns 284 calories playing basketball for 25 minutes. At this rate, how many calories would the person burn in 45 minutes, to the nearest whole number? (*Source: Wellness Encyclopedia.*)

511 calories (rounded)

43. In the hospital pharmacy, Michiko sees that a medicine is to be given at the rate of 3.5 milligrams for every 50 pounds of body weight. How much medicine should be given to a patient who weighs 210 pounds?

14.7 milligrams

▶▶▶ Mixed Review Exercises

Find the unknown number in each proportion. Round answers to the nearest hundredth, if necessary.

44. $\dfrac{x}{45} = \dfrac{70}{30}$ **x = 105**

45. $\dfrac{x}{52} = \dfrac{0}{20}$ **x = 0**

46. $\dfrac{64}{10} = \dfrac{x}{20}$ **x = 128**

47. $\dfrac{15}{x} = \dfrac{65}{100}$

x ≈ 23.08 (rounded)

48. $\dfrac{7.8}{3.9} = \dfrac{13}{x}$

x = 6.5

49. $\dfrac{34.1}{x} = \dfrac{0.77}{2.65}$

x ≈ 117.36 (rounded)

Find cross products to decide whether each proportion is true or false. Show the cross products and then circle true *or* false.

50. $\dfrac{55}{18} = \dfrac{80}{27}$ **1440 ≠ 1485**

True   (False)

51. $\dfrac{5.6}{0.6} = \dfrac{18}{1.94}$ **10.8 ≠ 10.864**

True   (False)

52. $\dfrac{\frac{1}{5}}{2} = \dfrac{1\frac{1}{6}}{11\frac{2}{3}}$ $2\frac{1}{3} = 2\frac{1}{3}$

(True)   False

Write each ratio as a fraction in lowest terms. Change to the same units when necessary.

53. 4 dollars to 10 quarters $\dfrac{8}{5}$

54. $4\frac{1}{8}$ in. to 10 in. $\dfrac{33}{80}$

55. 10 yd to 8 ft $\dfrac{15}{4}$

56. $3.60 to $0.90 $\dfrac{4}{1}$

57. 12 eggs to 15 eggs $\dfrac{4}{5}$

58. 37 meters to 7 meters $\dfrac{37}{7}$

59. 3 pints to 4 quarts $\dfrac{3}{8}$

60. 15 minutes to 3 hours $\dfrac{1}{12}$

61. $4\dfrac{1}{2}$ miles to $1\dfrac{3}{10}$ miles $\dfrac{45}{13}$

62. Nearly 7 out of 8 fans buy something to drink at rock concerts. How many of the 28,500 fans at today's concert would be expected to buy a beverage? Round to the nearest hundred fans.

24,900 fans (rounded)

63. Emily spent $150 on car repairs and $400 on car insurance. What is the ratio of the amount spent on insurance to the amount spent on repairs?

$\dfrac{8}{3}$

64. Antonio is choosing among three packages of plastic wrap. Is the best buy 25 ft for $0.78; 75 ft for $1.99; or 100 ft for $2.59? He has a coupon for 50¢ off either of the larger two packages.

75 ft for $1.99 − $0.50 coupon, about $0.020/ft

65. On this scale drawing of a backyard patio, 0.5 in. represents 6 ft. If the patio measures 1.75 in. long and 1.25 in. wide on the drawing, what will be the actual length and width of the patio when it is built?

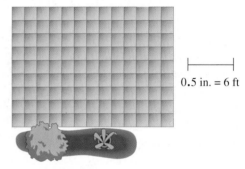

0.5 in. = 6 ft

21 ft long; 15 ft wide

66. A lawn mower uses 0.8 gallon of gas every 3 hours. How long can the mower run on 2 gallons of gas?

7.5 hours or $7\dfrac{1}{2}$ hours

67. An antibiotic is to be given at the rate of $1\dfrac{1}{2}$ teaspoons for every 24 pounds of body weight. How much should be given to an infant who weighs 8 pounds?

$\dfrac{1}{2}$ **teaspoon or 0.5 teaspoon**

68. Charles made 251 points during 169 minutes of playing time last year. At that same rate, how many points would you expect him to make if he plays 14 minutes in tonight's game? Round to the nearest whole number.

21 points (rounded)

69. Refer to Exercise 67. Explain each step you took in solving the problem. Be sure to tell how you decided which way to set up the proportion and how you checked your answer.

Set up the proportion to compare teaspoons to pounds on both sides.

$$\dfrac{1.5 \text{ teaspoons}}{24 \text{ pounds}} = \dfrac{x \text{ teaspoons}}{8 \text{ pounds}}$$

Show that cross products are equal.

$$(24)(x) = (1.5)(8)$$

Divide both sides by 24.

$$\dfrac{(\overset{1}{\cancel{24}})(x)}{\underset{1}{\cancel{24}}} = \dfrac{12}{24} \text{ so } x = \dfrac{1}{2} \text{ teaspoon or 0.5 teaspoon}$$

70. A vitamin supplement for cats is to be given at the rate of 1000 milligrams for a 5-pound cat. (*Source:* St. Jon Pet Care Products.)

(a) How much should be given to a 7-pound cat?

1400 milligrams

(b) How much should be given to an 8-ounce kitten?   **100 milligrams**

Chapter 5 ▷▷▷ Test

Use the Chapter Test Prep Video CD to see fully worked-out solutions to any of the exercises you want to review.

Write each rate or ratio as a fraction in lowest terms. Change to the same units when necessary.

1. 16 fish to 20 fish

2. 300 miles on 15 gallons

3. $15 for 75 minutes

4. 3 hours to 40 minutes

5. The little theater at our college has 320 seats. The auditorium has 1200 seats. Find the ratio of auditorium seats to theater seats.

6. Use the information in the table about Quiznos chicken with bacon sub sandwich to find the best buy.

Size	Length of Sub	Price
Small	5 inches	$4.39
Regular	8 inches	$5.79
Large	11 inches	$8

▦ **7.** Find the best buy on spaghetti sauce. You have a coupon for 75¢ off Brand X and a coupon for 50¢ off Brand Y.

26 ounces of Brand X for $3.89

16 ounces of Brand Y for $1.89

14 ounces of Brand Z for $1.29

✎ **8.** Suppose the ratio of your income last year to your income this year is 3 to 2. Explain what this means. Give an example of the dollars earned last year and this year that fits the 3 to 2 ratio.

Determine whether each proportion is true or false. Show your work and then write true *or* false.

9. $\dfrac{6}{14} = \dfrac{18}{45}$

10. $\dfrac{8.4}{2.8} = \dfrac{2.1}{0.7}$

---

1. $\dfrac{4}{5}$

2. $\dfrac{20 \text{ miles}}{1 \text{ gallon}}$

3. $\dfrac{\$1}{5 \text{ minutes}}$

4. $\dfrac{9}{2}$

5. $\dfrac{15}{4}$

6. Best buy is 8-inch sub, about $0.724/inch.

7. 16 ounces for $1.89 − $0.50 coupon, about $0.087/ounce

8. You earned less this year. An example is:
Last year→ $\dfrac{\$30,000}{}$ This year→ $\dfrac{}{\$20,000} = \dfrac{3}{2}$

9. $\dfrac{3}{7} \neq \dfrac{2}{5}$ or $252 \neq 270$; false

10. $5.88 = 5.88$; true

Find the unknown number in each proportion. In Problems 11–13, round the answers to the nearest hundredth, if necessary.

11. $\dfrac{5}{9} = \dfrac{x}{45}$

12. $\dfrac{3}{1} = \dfrac{8}{x}$

13. $\dfrac{x}{20} = \dfrac{6.5}{0.4}$

14. $\dfrac{2\frac{1}{3}}{x} = \dfrac{\frac{8}{9}}{4}$

Set up and solve a proportion for each application problem.

15. Pedro entered 18 orders into his computer in thirty minutes at his job. At that rate, how many orders could he enter in forty minutes?

16. Just 0.8 ounce of wildflower seeds is enough for 50 square feet of ground. What weight of seeds is needed for a garden with 225 square feet? (*Source:* White Swan Ltd.)

17. About 2 out of every 15 people are left-handed. How many of the 650 students in our school would you expect to be left-handed? Round to the nearest whole number.

18. A student set up the proportion for Problem 17 this way and arrived at an answer of 4875.

$$\frac{2}{15} = \frac{650}{x} \qquad Check: \qquad \frac{2}{15} \bowtie \frac{650}{4875}$$

$$15 \cdot 650 = 9750$$

$$2 \cdot 4875 = 9750$$

Because the cross products are equal, the student said the answer is correct. Is the student right? Explain why or why not.

19. A medication is given at the rate of 8.2 grams for every 50 pounds of body weight. How much should be given to a 145-pound person? Round to the nearest tenth.

20. On a scale model, 1 in. represents 8 ft. If a building in the model is 7.5 in. tall, what is the actual height of the building in feet?

---

11. $x = 25$

12. $x \approx 2.67$ (rounded)

13. $x = 325$

14. $x = 10\frac{1}{2}$

15. 24 orders

16. 3.6 ounces

17. 87 students (rounded)

18. No, 4875 cannot be correct because there are only 650 students in the whole school.

19. 23.8 grams (rounded)

20. 60 ft

Cumulative Review Exercises ▶▶▶ Chapters 1–5

Round each number as indicated.

1. 9903 to the nearest hundred

9900

2. 617.0519 to the nearest tenth

617.1

3. $99.81 to the nearest dollar

$100

4. $3.0555 to the nearest cent

$3.06

Simplify.

5. $30 - 0.66$

29.34

6. Write the answer using R for the remainder.

$$33\overline{)20{,}157} \quad \text{610 R27}$$

7. $(1.9)(0.004)$

0.0076

8. $3020 - 708$ **2312**

9. $0.401 + 62.98 + 5$ **68.381**

10. $1.39 \div 0.025$ **55.6**

11. $36 + 18 \div 6$ **39**

12. $8 \div 4 + (10 - 3^2) \cdot 4^2$ **18**

13. $88 \div \sqrt{121} \cdot 2^3$ **64**

14. $(16.2 - 5.85) - 2.35(4)$

0.95

15. Write 0.0105 in words.

one hundred five ten-thousandths

16. Write sixty and seventy-one thousandths in numbers.

60.071

In a survey of 1000 adults, 550 drank coffee daily, 250 drank coffee occasionally, and 200 never drank coffee. Use this information for Exercises 17–18. Write each ratio as a fraction in lowest terms.

17. What is the ratio of those who do not drink coffee to the total number of people in the survey?

$$\frac{1}{5}$$

18. Find the ratio of all the adults who drink coffee to those who never drink it.

$$\frac{4}{1}$$

Nest boxes for birds are made with different sizes of entry holes. Then, only certain types of birds can use the nest box and it helps prevent entry by predators. Use the information in the table to answer Exercises 19–20. The diameter is the distance across the opening at its widest point.

Eastern bluebird nest box with 1.5-inch opening

Bird	Diameter of Opening
Eastern bluebird	1.5 inches
Western bluebird	$1\frac{9}{16}$ inches
Chickadee	1.25 inches
Swallow	$1\frac{3}{8}$ inches
Wren	$1\frac{1}{8}$ inches

Source: Duncraft.

19. List the nest box openings in order from smallest to largest.

$1\frac{1}{8}$, 1.25, $1\frac{3}{8}$, 1.5, $1\frac{9}{16}$ (all inches)

20. What is the difference in diameter between the openings for an eastern bluebird and a western bluebird? Write your answer as a fraction and as a decimal.

$\frac{1}{16}$ inch or 0.063 inch (rounded)

Find the unknown number in each proportion. Round your answers to the nearest hundredth, when necessary.

21. $\dfrac{9}{12} = \dfrac{x}{28}$

$x = 21$

22. $\dfrac{7}{12} = \dfrac{10}{x}$

$x \approx 17.14$ (rounded)

▦ **23.** $\dfrac{6.7}{x} = \dfrac{62.8}{9.15}$

$x \approx 0.98$ (rounded)

Solve each application problem.

24. The college honor society has a goal of collecting 1500 pounds of food to fill Thanksgiving baskets. So far they've collected $\frac{5}{6}$ of their goal. How many more pounds do they need?

250 pounds

25. Tara has a photo that is 10 centimeters wide by 15 centimeters long. If the photo is enlarged to a length of 40 centimeters, find the new width, to the nearest tenth.

26.7 centimeters (rounded)

26. The distance around Dunning Pond is $1\frac{1}{10}$ miles. Norma ran around the pond four times in the morning and $2\frac{1}{2}$ times in the afternoon. How far did she run in all?

$7\dfrac{3}{20}$ **miles**

▦ **27.** Find the best buy on instant mashed potatoes. You have a coupon for 50¢ off either the 34-serving or 42-serving box.

A box that makes 17 servings for $1.75

A box that makes 34 servings for $3.85

A box that makes 42 servings for $4.75

34 servings for $3.85 − $0.50 coupon, about $0.099/serving

28. In a survey, 5 out of 8 apartment residents said they are sometimes bothered by noise from their neighbors. How many of the 224 residents at Harris Towers would you expect to be bothered by noise?

140 residents

29. The directions on a bottle of plant food call for $\frac{1}{2}$ teaspoon in two quarts of water. How much plant food is needed for five quarts? (*Source:* Schultz Company.)

$1\dfrac{1}{4}$ **teaspoons**

Use the information in the table on international long-distance calling card rates to answer Exercises 30–33. Round answers to the nearest whole minute when necessary.

$10 International Phone Cards *No Connnection Fee!*	
Place a call to	**Cost per minute**
Mexico City 🇲🇽	**$0.05**
Philippines 🇵🇭	**$0.12**
Japan 🇯🇵	**$0.043**
Hong Kong 🇭🇰	**$0.028**
Canada 🇨🇦	**$0.052**
Ghana 🇬🇭	**$0.106**
South Korea 🇰🇷	**$0.04**

Sources: www.1callcard.com; Access America; Original Gold Card.

30. List the per-minute rates from least to greatest.

$0.028, $0.04, $0.043, $0.05, $0.052, $0.106, $0.12

31. If you buy a $10 calling card, how long a call could you make to Japan?

233 minutes (rounded)

32. How many $10 cards would you have to buy to make four hours' worth of calls to the Philippines?

83 minutes (rounded) per $10 card, so buy 3 cards to cover 240 minutes (4 hours)

33. What is the ratio of minutes you can call South Korea for $10 to minutes you can call Mexico City for $10?

$\dfrac{250}{200} = \dfrac{5}{4}$

Percent

Percents are widely used in our everyday lives. For example, interest rates on savings and investments, automobile loans, home loans, and other installment loans are almost always given as percents. Stores often advertise sale prices as being a certain percent off the regular price. Sales tax, commission rates, and current government figures about inflation, recession, and unemployment are also reported as percents. An everyday example of the importance of understanding percent is knowing how to calculate the sales tax on items that you purchase. This allows you to know the true cost of anything you buy. (See Examples 1 and 2, Exercises 27–30, 47, and 48 in **Section 6.6.**)

6

6.1 ▶▶▶ Basics of Percent

OBJECTIVES

1. **Learn the meaning of percent.**

2. **Write percents as decimals.**

3. **Write decimals as percents.**

4. **Understand 100%, 200%, and 300%.**

5. **Use 50%, 10%, and 1%.**

Notice that the figure below has one hundred squares of equal size. Eleven of the squares are shaded. The shaded portion is $\frac{11}{100}$, or 0.11, of the total figure.

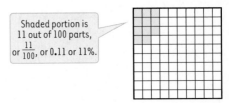

Shaded portion is 11 out of 100 parts, or $\frac{11}{100}$, or 0.11 or 11%.

The shaded portion is also 11% of the total, or "eleven parts out of 100 parts." Read **11%** as "eleven percent."

OBJECTIVE 1 Learn the meaning of percent. As we just saw, a percent is a ratio with a denominator of 100.

> **The Meaning of Percent**
> **Percent** means *per one hundred*. The "%" sign is used to show the number of parts out of one hundred parts.

EXAMPLE 1 Understanding Percent

(a) If *43 out of 100* students are men, then *43 per 100* or $\frac{43}{100}$ or **43%** of the students are men.

(b) If a person pays a tax of $7 on every $100 of purchases, then the tax rate is $7 per $100. The ratio is $\frac{7}{100}$ and the percent of tax is **7%**.

◀ *Work Problem* **1** *at the Side.*

OBJECTIVE 2 Write percents as decimals. If 8% means 8 parts out of 100 parts or $\frac{8}{100}$, then $p\%$ means p parts out of 100 parts or $\frac{p}{100}$. Because $\frac{p}{100}$ is another way to write the division $p \div 100$, we have

$$p\% = \frac{p}{100} = p \div 100$$

> **Writing a Percent as a Decimal**
> $$p\% = \frac{p}{100} \qquad \text{or} \qquad p\% = p \div 100$$
> As a fraction   As a decimal

EXAMPLE 2 Writing Percents as Decimals

Write each percent as a decimal.

(a) 47%

$$p\% = p \div 100$$
$$47\% = 47 \div 100 = 0.47 \leftarrow \text{Decimal form}$$

0.47 is $\frac{47}{100}$ which is equivalent to 47%

Continued on Next Page

1 Write as percents.

(a) In a group of 100 people, 46 are homeowners. What percent are homeowners?

(b) The sales tax is $6 per $100. What percent is this?

(c) Out of 100 students, 68 are working full- or part-time. What percent are working?

ANSWERS

1. (a) 46% (b) 6% (c) 68%

(b) 76%   76% = 76 ÷ 100 = 0.76 ⟵ Decimal form

(c) 28.2%   28.2% = 28.2 ÷ 100 = 0.282 ⟵ Decimal form

(d) 100%   100% = 100 ÷ 100 = 1.00 ⟵ Decimal form

> **CAUTION**
> In Example 2(d) above, notice that 100% is 1.00, or 1, which is a whole number. Whenever you have a percent that is *100% or greater*, the equivalent decimal number will be *1 or greater than 1.*

─────────────── *Work Problem* **2** *at the Side.* ▶

The answers in Example 2 above suggest these steps for writing a percent as a decimal.

> **Writing a Percent as a Decimal**
> **Step 1**   Drop the percent sign.
> **Step 2**   Divide by 100.

> **Note**
> Recall from **Section 4.6** that a quick way to divide a number by 100 is to move the decimal point **two places to the left.**

EXAMPLE 3   **Writing Percents as Decimals by Moving the Decimal Point**

Write each percent as a decimal by moving the decimal point two places to the left.

(a) 17%

17% = 17.%   Decimal point starts at far right side.

0.17 ⟵ Percent sign is dropped. (Step 1)

──── Decimal point is moved two places to the left. (Step 2)

17% = 0.17

(b) 160%

> 1.60 is equivalent to 1.6
> because $1\frac{60}{100}$ simplifies to $1\frac{6}{10}$

160% = 160.% = 1.60 or 1.6   Decimal point starts at far right side.

(c) 4.9%

.049   0 is attached so the decimal point can be moved two places to the left.

4.9% = 0.049

──── **Continued on Next Page**

2 Write each percent as a decimal.

(a) 68%

(b) 34%

(c) 58.5%

(d) 175%

(e) 200%

3 Write each percent as a decimal.

(a) 96%

(b) 6%

(c) 24.8%

(d) 0.9%

(d) 0.6%

$$0.6\% = 0.006$$ Two zeros are attached so the decimal point can be moved two places to the left.

> **CAUTION**
> Look at Example 3(d) above, where 0.6% is less than 1%. Because 0.6% is $\frac{6}{10}$ of 1%, it is *less than 1%*. Any fraction of a percent is *less than 1%*.

◀ *Work Problem* **3** *at the Side.*

OBJECTIVE 3 Write decimals as percents. You can write any decimal as a percent. For example, the decimal 0.78 is the same as the fraction

$$\frac{78}{100}$$

This fraction means 78 out of 100 parts, or 78%. The following steps give the same result.

> **Writing a Decimal as a Percent**
> *Step 1* Multiply by 100.
> *Step 2* Attach a percent sign.

> **Note**
> A quick way to divide or multiply a number by 100 is to move the decimal point two places to the left or two places to the right, respectively.
>
> Multiply decimal by 100; move decimal point two places to the *right*.
> **Decimal** ⟶ **Percent**
> Divide percent by 100; move decimal point two places to the *left*.

EXAMPLE 4 **Writing Decimals as Percents by Moving the Decimal Point**

Write each decimal as a percent by moving the decimal point two places to the right.

(a) 0.21

0.21

Decimal point is moved two places to the right. (Step 1)

$$0.21 = 21\%$$ ← Percent sign is attached. (Step 2)

Remember to attach the percent (%) sign.

Decimal point is not written with whole number percents.

Continued on Next Page

(b) $0.529 = 52.9\%$

(c) $1.92 = 192\%$

(d) 2.5

$\underset{\curvearrowright}{2.5\underline{0}}$ 0 is attached so the decimal point can be moved two places to the right.

$2.5 = 250\%$ Attach % sign.

> When necessary, attach zeros so you can move the decimal point.

(e) 3

$3. = 3.\underset{\curvearrowright}{\underline{00}}$ Two zeros are attached so the decimal point can be moved two places to the right.

so $3 = 300\%$ Attach % sign.

> **CAUTION**
> Look at Examples 4(c), 4(d), and 4(e) above, where 1.92, 2.5, and 3 are greater than 1. Because the number 1 is equivalent to 100%, all numbers greater than 1 will be *greater than 100%*.

Work Problem **4** *at the Side.* ▶

OBJECTIVE 4 Understand 100%, 200%, and 300%. When working with percents, it is helpful to have several reference points. 100%, 200%, and 300% are three such helpful reference points.

100% means 100 parts out of 100 parts. That's *all* of the parts. If 100% of the 18 people attending last week's meeting attended this week's meeting, then 18 people (*all* of them) attended this week.

If attendance at the meeting this week is 200% of last week's attendance of 18 people, then this week's attendance is 36 people, or *two* times as many people $(2 \cdot 18 = 36)$. Likewise, if attendance is 300% of last week's attendance, then *three* times as many people, or 54 people, attended $(3 \cdot 18 = 54)$.

EXAMPLE 5 **Finding 100%, 200%, and 300% of a Number**

Fill in the blanks.

> Notice that 100% of something is all of it (the whole thing).

(a) 100% of 82 people is _____ .
100% is *all* of the people. So, 100% of 82 people is <u>82 people</u> .

(b) 200% of $63 is _____ .
200% is twice (2 times) as much money.
So, 200% of $63 is $2 \cdot \$63 = \underline{\$126}$.

(c) 300% of 32 employees is _____ .
300% is 3 times as many employees.
So, 300% of 32 employees is $3 \cdot 32 = \underline{96 \text{ employees}}$.

Work Problem **5** *at the Side.* ▶

---

4 Write each decimal as a percent.

(a) 0.74   **(b)** 0.15

(c) 0.09   **(d)** 0.617

(e) 0.834   **(f)** 5.34

(g) 2.8   **(h)** 4

5 Fill in the blanks.

(a) 100% of $7.80 is.

_____ .

(b) 100% of 1850 workers is.

_____ .

(c) 200% of 24 photographs is _____ .

(d) 300% of 8 miles is

_____ .

ANSWERS

4. (a) 74%   **(b)** 15%   **(c)** 9%   **(d)** 61.7%
    (e) 83.4%   **(f)** 534%   **(g)** 280%
    (h) 400%
5. (a) $7.80   **(b)** 1850 workers
    (c) 48 photographs   **(d)** 24 miles

6 Fill in the blanks.

(a) 50% of 200 patients is

_____.

(b) 50% of 64 e-mails is

_____.

(c) 10% of 3850 elm trees
is _____.

(d) 10% of 7 pounds
is _____.

(e) 1% of 240 ft is

_____.

(f) 1% of $3000
is _____.

OBJECTIVE **5** **Use 50%, 10%, and 1%.**   50% means 50 parts out of 100 parts, which is *half* of the parts ($\frac{50}{100} = \frac{1}{2}$). So, 50% of $18 is $9 (*half* of the money).

When using 10%, we have 10 parts out of 100 parts, which is $\frac{1}{10}$ of the parts ($\frac{10}{100} = \frac{1}{10}$). To find 10% or $\frac{1}{10}$ of a number, we move the decimal point **one** place to the left. 10% of $285 is $28.50 (because $28\underset{\frown}{5}. = $28.50).

To find 1% of a number ($\frac{1}{100}$), we move the decimal point **two** places to the left. 1% of $198 is $1.98 (because $1\underset{\frown}{9\,8}. = $1.98).

EXAMPLE 6 **Finding 50%, 10%, and 1% of a Number**

Fill in the blanks.

> Think: 50% of something is $\frac{50}{100}$ or $\frac{1}{2}$ of it.

(a) 50% of 24 hours is _____.
50% is half of the hours. So, 50% of 24 hours is 12 hours .

(b) 10% of 280 pages is _____.
10% is $\frac{1}{10}$ of the pages. Move the decimal point *one* place to the left.
So, 10% of 28$\underset{\frown}{0}$. pages is 28 pages .

(c) 1% of $540 is _____.
1% is $\frac{1}{100}$ of the money. Move the decimal point *two* places to the left.
So, 1% of 5\underset{\frown}{4\,0}$. is $5.40 .

◀ *Work Problem* **6** *at the Side.*

6.1 ▶▶▶ Exercises

FOR EXTRA HELP

MyMathLab

Math XL
PRACTICE

WATCH

DOWNLOAD

READ

REVIEW

Write each percent as a decimal. See Examples 2 and 3.

🌐 **1.** 12% **0.12**   **2.** 57% **0.57**   **3.** 70% **0.70 or 0.7**   **4.** 40% **0.40 or 0.4**

5. 25% **0.25**   **6.** 35% **0.35**   **7.** 140% **1.40 or 1.4**   **8.** 250% **2.50 or 2.5**

🌐 **9.** 5.5% **0.055**   **10.** 6.7% **0.067**   🌐 **11.** 100% **1.00 or 1**   **12.** 600% **6.00 or 6**

🌐 **13.** 0.5% **0.005**   **14.** 0.25% **0.0025**   🌐 **15.** 0.35% **0.0035**   **16.** 0.75% **0.0075**

Write each decimal as a percent. See Example 4.

17. 0.6 **60%**   **18.** 0.9 **90%**   **19.** 0.58 **58%**   **20.** 0.25 **25%**

21. 0.01 **1%**   **22.** 0.07 **7%**   **23.** 0.125 **12.5%**   **24.** 0.875 **87.5%**

25. 0.375 **37.5%**   **26.** 0.625 **62.5%**   **27.** 2 **200%**   **28.** 5 **500%**

29. 3.7 **370%**   **30.** 2.2 **220%**   **31.** 0.0312 **3.12%**   **32.** 0.0625 **6.25%**

33. 4.162 **416.2%**   **34.** 8.715 **871.5%**   **35.** 0.0028 **0.28%**   **36.** 0.0064 **0.64%**

37. Fractions, decimals, and percents are all used to describe a part of something. The use of percents is much more common than fractions and decimals. Why do you suppose this is true?

Answers will vary. Some possibilities are: No common denominators are needed with percents. The denominator is always 100 with percent, which makes comparisons easier to understand.

38. List five uses of percent that are or will be part of your life. Consider the activities of working, shopping, saving, and planning for the future.

Answers will vary. Some answers might be: When using discounts on purchases, calculating sales tax, figuring interest on loans, examining investments, finding tips in restaurants, calculating interest on savings, and doing math problems in this book.

Write each percent as a decimal and each decimal as a percent. See Examples 2–4.

39. When asked, 38% of those 50 years of age or older don't think they need a flu shot. **0.38**

40. At College of DuPage, 82% of the students work part-time. **0.82**

41. Lack of parking spaces bothers 47% of holiday shoppers. **0.47**

42. There was a 43.2% voter turnout at the election. **0.432**

43. The property tax rate in Alpine County is 0.035. **3.5%**

44. A church building fund has 0.89 of the money needed. **89%**

45. The number of people successfully completing CPR training this session is 2 times that of the last session. **200%**

46. The number of newspaper subscribers was 4 times as great as last quarter. **400%**

47. Only 0.005 of the total population has this genetic defect. **0.5%**

48. The return rate of defective keyboards is 0.0075 of total output. **0.75%**

49. The patient's blood pressure was 153.6% of normal. **1.536**

50. Success with the diet was 248.7% greater than anticipated. **2.487**

Fill in the blanks. Remember that 100% is all of something, 200% is two times as many, and 300% is three times as many. See Example 5.

51. There are 20 children in the preschool class. 100% of the children are served breakfast and lunch. How many children are served both meals?

  20 children

52. When 500 adults were asked, "Do you think your taxes are too high," 100% said yes. How many said yes?

  500 adults

53. Last year we had 210 employees. This year we have 200% of that number. How many employees do we have this year?   **420 employees**

54. This week's expenses are 200% of last week's $380. This week's expenses are   **$760**   .

55. Last week 90 chairs were used for the meeting. This week we need 300% of that number of chairs. We'll need   **270 chairs**   .

56. Wayman's new hybrid car gets 300% of the 12 miles per gallon that his old car got. His new car gets   **36 miles per gallon**   .

Fill in the blanks. Remember that 50% is half of something, 10% is found by moving the decimal point one place to the left, and 1% is found by moving the decimal point two places to the left. See Example 6.

57. Jacob owes $755 for tuition. Financial aid will pay 50% of the cost. Financial aid will pay   **$377.50**   .

58. Elizabeth gets 3200 "off-peak" minutes of calling time on her cell phone plan. Last month she used 50% of her allowed minutes. The number of minutes she used was   **1600 minutes**   .

59. Only 10% of 8200 commuters are carpooling to work. How many commuters carpool?   **820 commuters**

60. Sarah expects that 10% of the 240 dozen plants in her greenhouse will not be sold. The expected number of unsold plants is   **24 dozen**   .

61. The naturalist said that 1% of the 2600 plants in the park are poisonous. How many plants are poisonous?   **26 plants**

62. Of the 4800 accidents, only 1% were caused by mechanical failure. How many accidents were caused by mechanical failure?   **48 accidents**   .

63. (a) Describe a shortcut method of finding 100% of a number.

   Since 100% means 100 parts out of 100 parts, 100% is all of the number.

   (b) Show an example using your shortcut.

   Answers will vary. For example, 100% of $72 is $72.

64. (a) Describe a shortcut method of finding 50% of a number.

   50% means 50 parts out of 100 parts. That's half of the number. A shortcut for finding 50% of a number is to divide the number by 2.

   (b) Show an example using your shortcut.

   Answers will vary. For example, 50% of $14 is $14 ÷ 2 = $7.

65. (a) Describe a shortcut method of finding 200% of a number.

Since 200% is two times a number, find 200% of the number by multiplying the number by 2 (double it).

(b) Show an example using your shortcut.

Answers will vary. For example, 200% of $20 is 2 · $20 = $40.

66. (a) Describe a shortcut method of finding 300% of a number.

Since 300% is three times a number, find 300% of the number by multiplying the number by 3 (triple it).

(b) Show an example using your shortcut.

Answers will vary. For example, 300% of $10 is 3 · $10 = $30.

67. (a) Describe a shortcut method of finding 10% of a number.

Since 10% means 10 parts out of 100 parts or $\frac{1}{10}$, the shortcut for finding 10% of a number is to move the decimal point in the number one place to the left.

(b) Show an example using your shortcut.

Answers will vary. For example, 10% of $90 is $9.

68. (a) Describe a shortcut method of finding 1% of a number.

Since 1% means 1 part out of 100 parts or $\frac{1}{100}$, the shortcut for finding 1% of a number is to move the decimal point in the number two places to the left.

(b) Show an example using your shortcut.

Answers will vary. For example, 1% of $500 is $5.

More than 7.4 million households will dress up their pets (dogs and cats) in Halloween costumes this year. The bar graph shows the ranking of the top pet costumes and the percent of pet owners selecting each costume. Use this graph to answer Exercises 69–72. Write each answer as a percent and as a decimal. (Source: BIGresearch survey of 8877 adult pet owners.)

69. What portion of the pet owners selected the devil costume for their pet?

12%; 0.12

70. What portion of the pet owners selected the pirate costume for their pet?

2.8%; 0.028

71. (a) What was the third-most-popular costume?

Witch costume

(b) Write the portion of the pet owners who selected this costume.

4.5%; 0.045

72. (a) What was the second-most-popular costume?

Pumpkin costume

(b) Write the portion of the pet owners who selected this costume.

9.2%; 0.092

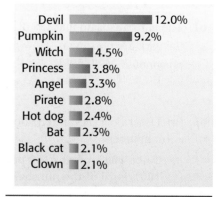

Pets get dressed up

This year, 7.4 million households plan to put their furry friends into a Halloween costume. Top outfits:

Costume	Percent
Devil	12.0%
Pumpkin	9.2%
Witch	4.5%
Princess	3.8%
Angel	3.3%
Pirate	2.8%
Hot dog	2.4%
Bat	2.3%
Black cat	2.1%
Clown	2.1%

Source: BIGresearch survey of 8,877 adults, Sept. 4–11.

There are more than 200,000 women serving in the U.S. military. The circle graph shows the percent of these women serving in each branch of the armed forces. Use this graph to answer Exercises 73–76. Write each answer as a percent and as a decimal.

73. What portion of the women in the U.S. military are in the Air Force?

26%; 0.26

74. What portion of the women in the U.S. military are in the Navy?

24%; 0.24

75. (a) Which branch of the U.S. military has the lowest portion of women?

Marine Corps

(b) What portion is this?

5%; 0.05

76. (a) Which branch of the U.S. military has the highest portion of women?

Army

(b) What portion is this?

45%; 0.45

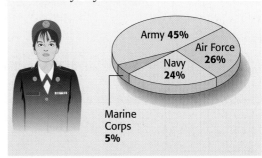

ENLISTED WOMEN

Over 200,000 women serve in the U.S. military. By branch:

Source: U.S. Department of Defense.

In the United States 12.7% of the population is 65 years of age or older. The bar graph shows the countries with the highest percent of people 65 years of age or older. Use this graph to answer Exercises 77–80. Write each answer as a percent and as a decimal.

77. What portion of the population of Spain is 65 or older? **16.8%; 0.168**

78. What portion of the population of Greece is 65 or older? **17.2%; 0.172**

79. (a) Which two countries in the graph have the lowest portion of the population 65 or older?

Germany; Bulgaria

(b) What portion is this?

16.5%; 0.165

80. (a) Which country has the highest portion of the population 65 or older?

Italy

(b) What portion is this?

18.2%; 0.182

65 AND UP

Countries with the highest percentage of seniors:

Country	Percentage
Italy	18.2%
Sweden	17.2%
Greece	17.2%
Belgium	17.1%
Japan	17%
Spain	16.8%
Germany	16.5%
Bulgaria	16.5%

Source: U.S. Bureau of the Census International Programs Center.

Write a percent for both the shaded and unshaded parts of each figure.

81. 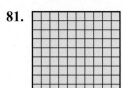 95% shaded; 5% unshaded

82. 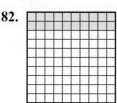 20% shaded; 80% unshaded

83. 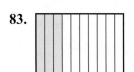 30% shaded; 70% unshaded

84. 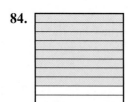 80% shaded; 20% unshaded

85. 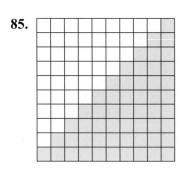 55% shaded; 45% unshaded

86. 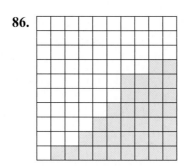 37% shaded; 63% unshaded

87. 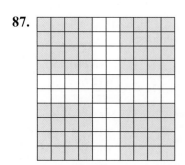 64% shaded; 36% unshaded

88. 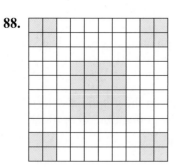 32% shaded; 68% unshaded

6.2 ▷▷▷ Percents and Fractions

OBJECTIVE 1 Write percents as fractions. Percents can be written as fractions by using what we learned in the previous section.

Writing a Percent as a Fraction

$$p\% = \frac{p}{100}, \quad \text{as a fraction.}$$

EXAMPLE 1 Writing Percents as Fractions

Write each percent as a fraction or mixed number in lowest terms.

(a) 25%

As we saw in the last section, 25% can be written as a decimal.

$$25\% = 25 \div 100 = 0.25 \qquad \text{Percent sign dropped}$$

Because 0.25 means 25 hundredths,

$$0.25 = \frac{25}{100} = \frac{25 \div 25}{100 \div 25} = \frac{1}{4} \qquad \text{Lowest terms}$$

It is not necessary, however, to write 25% as a decimal first. Just write

$$25\% = \frac{25}{100} \qquad \text{25 per 100}$$

$$= \frac{1}{4} \quad \longleftarrow \text{Lowest terms}$$

(b) 76%

The percent becomes the numerator.

Write 76% as $\frac{76}{100}$

The *denominator* is always 100 because percent means *parts per 100.*

Write $\frac{76}{100}$ in lowest terms.

> To write a fraction in lowest terms, divide numerator and denominator by the *same number.*

$$\frac{76 \div 4}{100 \div 4} = \frac{19}{25} \quad \longleftarrow \text{Lowest terms}$$

(c) 150%

$$150\% = \frac{150}{100} = \frac{150 \div 50}{100 \div 50} = \frac{3}{2} = 1\frac{1}{2} \quad \longleftarrow \text{Mixed number}$$

Note
Remember that percent means *per 100.*

Work Problem ① *at the Side.* ▶

---

① Write each percent as a fraction or mixed number in lowest terms.

(a) 50%

(b) 75%

(c) 48%

(d) 23%

(e) 125%

(f) 250%

ANSWERS

1. **(a)** $\frac{1}{2}$ **(b)** $\frac{3}{4}$ **(c)** $\frac{12}{25}$ **(d)** $\frac{23}{100}$

   (e) $1\frac{1}{4}$ **(f)** $2\frac{1}{2}$

The next example shows how to write decimal and fraction percents as fractions.

2 Write each percent as a fraction in lowest terms.

(a) 37.5%

EXAMPLE 2 **Writing Decimal or Fraction Percents as Fractions**

Write each percent as a fraction in lowest terms.

(a) 15.5%

Write 15.5 over 100

$$15.5\% = \frac{15.5}{100}$$

(b) 62.5%

To get a whole number in the numerator, multiply the numerator and denominator by 10. (Recall that multiplying by $\frac{10}{10}$ is the same as multiplying by 1.)

$$\frac{15.5}{100} = \frac{15.5\,(10)}{100\,(10)} = \frac{155}{1000}$$

Move the decimal point 1 place to the *right* to multiply by 10.

Now write the fraction in lowest terms.

$$\frac{155 \div 5}{1000 \div 5} = \frac{31}{200}$$

(c) 4.5%

(b) $33\frac{1}{3}\%$

Write $33\frac{1}{3}$ over 100

$$33\frac{1}{3}\% = \frac{33\frac{1}{3}}{100}$$

When we have a mixed number in the numerator, we must write the mixed number as an improper fraction.

(d) $66\frac{2}{3}\%$

$$\frac{33\frac{1}{3}}{100} = \frac{\frac{100}{3}}{100}$$

Write $33\frac{1}{3}$ as $\frac{100}{3}$

Next, rewrite the division problem in a horizontal form. Finally, multiply by the reciprocal of the divisor.

(e) $10\frac{1}{3}\%$

Reciprocals

$$\frac{\frac{100}{3}}{100} = \frac{100}{3} \div 100 = \frac{100}{3} \div \frac{100}{1} = \frac{\overset{1}{\cancel{100}}}{3} \cdot \frac{1}{\underset{1}{\cancel{100}}} = \frac{1}{3}$$

(f) $87\frac{1}{2}\%$

Note

In Example 2(a) at the top of the page we could have changed 15.5% to $15\frac{1}{2}\%$ and then written it as the improper fraction $\frac{31}{2}$ over 100. But it is usually easier *not* to change decimals to fractions but to leave decimal percents as they are.

◀ Work Problem **2** at the Side.

ANSWERS
2. (a) $\frac{3}{8}$ (b) $\frac{5}{8}$ (c) $\frac{9}{200}$ (d) $\frac{2}{3}$
   (e) $\frac{31}{300}$ (f) $\frac{7}{8}$

OBJECTIVE 2 Write fractions as percents. We will use the formula from the beginning of this section to write fractions as percents.

$$p\% = \frac{p}{100}$$

EXAMPLE 3 **Writing Fractions as Percents**

Write each fraction as a percent. Round to the nearest tenth if necessary.

(a) $\frac{3}{5}$

Write $\frac{3}{5}$ as a percent by solving for p in the proportion below.

$$\frac{3}{5} = \frac{p}{100}$$

Find cross products and show that they are equivalent.

$$5 \cdot p = 3 \cdot 100$$
$$5 \cdot p = 300$$
$$\frac{\cancel{5}^1 \cdot p}{\cancel{5}_1} = \frac{300}{5}$$

Divide both sides by 5.

$$p = 60$$

This result means that $\frac{3}{5} = \frac{60}{100}$ or 60%.

Note

Solving proportions can be reviewed in **Section 5.4.**

(b) $\frac{7}{8}$

Write a proportion.

$$\frac{7}{8} = \frac{p}{100}$$
$$8 \cdot p = 7 \cdot 100 \quad \text{Show that cross products are equivalent.}$$
$$8 \cdot p = 700$$
$$\frac{\cancel{8}^1 \cdot p}{\cancel{8}_1} = \frac{700}{8} \quad \text{Divide both sides by 8.}$$
$$p = 87.5$$

So, $\frac{7}{8} = 87.5\%$.

Note

If you think of $\frac{700}{8}$ as an improper fraction, changing it to a mixed number gives an answer of $87\frac{1}{2}$. So $\frac{7}{8} = 87.5\%$ or $87\frac{1}{2}\%$.

Continued on Next Page

3 Write as percents. Round to the nearest tenth if necessary.

(a) $\dfrac{1}{4}$

(b) $\dfrac{3}{10}$

(c) $\dfrac{6}{25}$

(d) $\dfrac{5}{8}$

(e) $\dfrac{1}{6}$

(f) $\dfrac{2}{9}$

(c) $\dfrac{5}{6}$

Start with a proportion.

$$\frac{5}{6} = \frac{p}{100}$$

$$6 \cdot p = 5 \cdot 100 \qquad \text{Show that cross products are equivalent.}$$

$$6 \cdot p = 500$$

$$\frac{\overset{1}{\cancel{6}} \cdot p}{\underset{1}{\cancel{6}}} = \frac{500}{6} \qquad \text{Divide both sides by 6.}$$

$$p = 83.\overline{3} \qquad \boxed{\text{A bar over the 3 indicates that the decimal keeps repeating 83.3333 forever.}}$$

$$p \approx 83.3 \qquad \text{Round to the nearest tenth.}$$

So, $\dfrac{5}{6} = 83.\overline{3}\% \approx 83.3\%$ (rounded) $\qquad \boxed{\text{The "} \approx \text{" symbol shows that 83.3\% is rounded.}}$

> **Note**
> You can change $\frac{500}{6}$ to a mixed number to get an exact answer of $83\frac{1}{3}\%$.

◀ *Work Problem* **3** *at the Side.*

OBJECTIVE 3 Use the table of percent equivalents. The table on the next two pages shows common and not so common fractions and mixed numbers and their decimal and percent equivalents. The more you work with them, the more familiar they will become.

EXAMPLE 4 Using the Table of Percent Equivalents

Read the following from the table.

(a) $\frac{1}{12}$ as a percent
Find $\frac{1}{12}$ in the "fraction" column. The equivalent percent is 8.3% (rounded) or $8\frac{1}{3}\%$ (exact).

(b) 0.375 as a fraction
Look in the "decimal" column for 0.375. The equivalent fraction is $\frac{3}{8}$.

(c) $\frac{13}{16}$ as a percent
Find $\frac{13}{16}$ in the "fraction" column. The equivalent percent is 81.25% or $81\frac{1}{4}\%$.

🖩 **Calculator Tip** Example 4 (c) above can be solved on a calculator as shown below.

$$13 \;\div\; 16 \;=\; 0.8125 \;\times\; 100 \;=\; \mathbf{81.25}$$

Multiplying by 100 changes the decimal to a percent.
Or, if your calculator has a percent key, follow these steps.

$$13 \;\div\; 16 \;\%\; \mathbf{81.25}$$

↑ —— Press % key instead of = key.

On scientific calculators, you may need to press the (2nd) key to access the % function.

Note

When a fraction like $\frac{1}{12}$ is changed to a decimal, it is a *repeating decimal* that goes on forever, $0.083333\ldots$. In the table below these decimals are rounded to the nearest thousandth. When the decimal is then changed to a percent, it will be to the nearest tenth of a percent. Decimals that do not repeat are usually not rounded.

Work Problem ④ *at the Side.* ▶

Percent, Decimal, and Fraction Equivalents

Percent (rounded to tenths when necessary)	Decimal	Fraction
1%	0.01	$\frac{1}{100}$
2%	0.02	$\frac{1}{50}$
4%	0.04	$\frac{1}{25}$
5%	0.05	$\frac{1}{20}$
6.25% or $6\frac{1}{4}$%	0.0625	$\frac{1}{16}$
8.3% (rounded) or $8\frac{1}{3}$% (exact)	$0.08\overline{3}$ rounds to 0.083	$\frac{1}{12}$
10%	0.1	$\frac{1}{10}$
12.5% or $12\frac{1}{2}$%	0.125	$\frac{1}{8}$
16.7% (rounded) or $16\frac{2}{3}$% (exact)	$0.1\overline{6}$ rounds to 0.167	$\frac{1}{6}$
18.75% or $18\frac{3}{4}$%	0.1875	$\frac{3}{16}$
20%	0.2	$\frac{1}{5}$
25%	0.25	$\frac{1}{4}$
30%	0.3	$\frac{3}{10}$
31.25% or $31\frac{1}{4}$%	0.3125	$\frac{5}{16}$
33.3% (rounded) or $33\frac{1}{3}$% (exact)	$0.\overline{3}$ rounds to 0.333	$\frac{1}{3}$
37.5% or $37\frac{1}{2}$%	0.375	$\frac{3}{8}$
40%	0.4	$\frac{2}{5}$

(continued)

④ Read the following fractions, mixed numbers, decimals, and percents from the table on this or the next page. If you already know the answer or can solve for the answer quickly, don't use the table.

(a) $\frac{3}{4}$ as a percent

(b) 10% as a fraction

(c) $0.\overline{6}$ as a fraction

(d) $37\frac{1}{2}$% as a fraction

(e) $\frac{7}{8}$ as a percent

(f) $\frac{1}{2}$ as a percent

(g) $33\frac{1}{3}$% as a fraction

(h) $1\frac{3}{4}$ as a percent

Answers

4. **(a)** 75% **(b)** $\frac{1}{10}$ **(c)** $\frac{2}{3}$ **(d)** $\frac{3}{8}$

(e) 87.5% **(f)** 50% **(g)** $\frac{1}{3}$ **(h)** 175%

Percent, Decimal, and Fraction Equivalents (*continued*)

Percent	Decimal	Fraction
43.75% or $43\frac{3}{4}$%	0.4375	$\frac{7}{16}$
50%	0.5	$\frac{1}{2}$
56.25% or $56\frac{1}{4}$%	0.5625	$\frac{9}{16}$
60%	0.6	$\frac{3}{5}$
62.5% or $62\frac{1}{2}$%	0.625	$\frac{5}{8}$
66.7% (rounded) or $66\frac{2}{3}$% (exact)	$0.\overline{6}$ rounds to 0.667	$\frac{2}{3}$
68.75% or $68\frac{3}{4}$%	0.6875	$\frac{11}{16}$
70%	0.7	$\frac{7}{10}$
75%	0.75	$\frac{3}{4}$
80%	0.8	$\frac{4}{5}$
81.25% or $81\frac{1}{4}$%	0.8125	$\frac{13}{16}$
83.3% (rounded) or $83\frac{1}{3}$% (exact)	$0.8\overline{3}$ rounds to 0.833	$\frac{5}{6}$
87.5% or $87\frac{1}{2}$%	0.875	$\frac{7}{8}$
90%	0.9	$\frac{9}{10}$
93.75% or $93\frac{3}{4}$%	0.9375	$\frac{15}{16}$
100%	1.0	1
110%	1.1	$1\frac{1}{10}$
125%	1.25	$1\frac{1}{4}$
133.3% (rounded) or $133\frac{1}{3}$% (exact)	$1.\overline{3}$ rounds to 1.333	$1\frac{1}{3}$
150%	1.5	$1\frac{1}{2}$
166.7% (rounded) or $166\frac{2}{3}$% (exact)	$1.\overline{6}$ rounds to 1.667	$1\frac{2}{3}$
175%	1.75	$1\frac{3}{4}$
200%	2.0	2

6.2 ▶▶▶ Exercises

Write each percent as a fraction or mixed number in lowest terms. See Examples 1 and 2.

1. 25% $\dfrac{1}{4}$

2. 30% $\dfrac{3}{10}$

3. 75% $\dfrac{3}{4}$

4. 80% $\dfrac{4}{5}$

 5. 85% $\dfrac{17}{20}$

6. 45% $\dfrac{9}{20}$

7. 62.5% $\dfrac{5}{8}$

8. 87.5% $\dfrac{7}{8}$

 9. 6.25% $\dfrac{1}{16}$

10. 43.75% $\dfrac{7}{16}$

11. $16\dfrac{2}{3}\%$ $\dfrac{1}{6}$

12. $66\dfrac{2}{3}\%$ $\dfrac{2}{3}$

13. $6\dfrac{2}{3}\%$ $\dfrac{1}{15}$

14. $46\dfrac{2}{3}\%$ $\dfrac{7}{15}$

15. 0.5% $\dfrac{1}{200}$

16. 0.8% $\dfrac{1}{125}$

 17. 180% $1\dfrac{4}{5}$

18. 140% $1\dfrac{2}{5}$

19. 375% $3\dfrac{3}{4}$

20. 225% $2\dfrac{1}{4}$

Write each fraction as a percent. Round percents to the nearest tenth if necessary. See Example 3.

21. $\dfrac{1}{2}$ 50%

22. $\dfrac{4}{10}$ 40%

23. $\dfrac{4}{5}$ 80%

24. $\dfrac{3}{10}$ 30%

25. $\dfrac{7}{10}$ 70%

26. $\dfrac{3}{4}$ 75%

27. $\dfrac{37}{100}$ 37%

28. $\dfrac{63}{100}$ 63%

29. $\dfrac{5}{8}$ 62.5%

30. $\dfrac{1}{8}$ 12.5%

 31. $\dfrac{7}{8}$ 87.5%

32. $\dfrac{3}{8}$ 37.5%

33. $\dfrac{12}{25}$ 48%

34. $\dfrac{15}{25}$ 60%

35. $\dfrac{23}{50}$ 46%

36. $\dfrac{18}{50}$ 36%

37. $\dfrac{7}{20}$ 35%   **38.** $\dfrac{9}{20}$ 45%   **39.** $\dfrac{5}{6}$ 83.3% (rounded)   **40.** $\dfrac{1}{6}$ 16.7% (rounded)

41. $\dfrac{5}{9}$ 55.6% (rounded)   **42.** $\dfrac{7}{9}$ 77.8% (rounded)   **43.** $\dfrac{1}{7}$ 14.3% (rounded)   **44.** $\dfrac{5}{7}$ 71.4% (rounded)

Complete the chart. Round decimals to the nearest thousandth and percents to the nearest tenth if necessary. See Examples 3 and 4.

Fraction	Decimal	Percent
45. $\dfrac{1}{2}$	0.5	50%
46. $\dfrac{1}{4}$	0.25	25%
47. $\dfrac{7}{8}$	0.875	87.5%
48. $\dfrac{3}{4}$	0.75	75%
49. $\dfrac{4}{5}$	0.8	80%
50. $\dfrac{3}{5}$	0.6	60%
51. $\dfrac{1}{6}$	0.167 (rounded)	16.7% (rounded)
52. $\dfrac{1}{3}$	0.333 (rounded)	33.3% (rounded)
53. $\dfrac{7}{10}$	0.7	70%

Fraction	Decimal	Percent
54. $\frac{3}{8}$	0.375	37.5%
55. $\frac{1}{8}$	0.125	12.5%
56. $\frac{5}{8}$	0.625	62.5%
57. $\frac{2}{3}$	0.667 (rounded)	66.7% (rounded)
58. $\frac{5}{6}$	0.833 (rounded)	83.3% (rounded)
59. $\frac{3}{50}$	0.06	6%
60. $\frac{3}{10}$	0.3	30%
61. $\frac{8}{100}$	0.08	8%
62. 1	1.00 or 1	100%
63. $\frac{1}{200}$	0.005	0.5%
64. $\frac{1}{400}$	0.0025	0.25%

Fraction	Decimal	Percent
65. $2\frac{1}{2}$	2.5	250%
66. $1\frac{7}{10}$	1.7	170%
67. $3\frac{1}{4}$	3.25	325%
68. $2\frac{4}{5}$	2.8	280%

69. Select a decimal percent and write it as a fraction. Select a different fraction and write it as a percent. Write an explanation of each step of your work.

There are many possible answers. Examples 2 and 3 show the steps that students should include in their answers.

70. Prepare a table showing fraction, decimal, and percent equivalents for five fractions and mixed numbers of your choice.

There are many correct answers. The table of percent equivalents shows some of the possibilities.

In the following application problems, write the answer as a fraction in lowest terms, as a decimal, and as a percent.

71. Many pet owners say they have used the Internet to find pet information. Of 500 people who used the Internet for this purpose, 90 said they used it when buying a pet. What portion used the Internet when buying a pet? (*Source:* American Animal Hospital Association.)

$\frac{9}{50}$; 0.18; 18%

72. About $\frac{1}{3}$ of all books purchased last year were for children. Of these children's books, 27 of every 100 purchased included a coloring activity. What portion of the children's books included a coloring activity? (*Source:* Consumer Research Study on Book Publishing, the American Booksellers, and the Book Industry Study Group.)

$\frac{27}{100}$; 0.27; 27%

73. Only 13 out of every 100 adults consume the recommended 1000 milligrams of calcium daily. What portion consumes the recommended daily amount? (*Source:* Market Facts for Milk Mustache.)

$\frac{13}{100}$; 0.13; 13%

74. In a survey on how people learn to parent, 360 parents out of 800 said they were most influenced by relatives, friends, and spouses. What portion learned to parent this way? (*Source:* Bama Research.)

$\frac{9}{20}$; 0.45; 45%

75. In a recent survey, "Attitudes in the American Workplace," 750 workers were asked if they would hire their own boss if they were in charge. A total of 150 workers said no, they would not. What portion of the workers said no? (*Source:* Marlin's 13th annual Attitudes in the Workplace Survey.)

$\frac{1}{5}$; 0.2; 20%

76. When 1500 adults were asked what was the most important factor to consider when relocating after retirement, 675 said that climate was most important. What portion consider climate to be most important? (*Source:* Longevity Alliance Retirement and Relocation Survey.)

$\frac{9}{20}$; 0.45; 45%

77. An insurance office has 80 employees. If 64 of the employees have cell phones, what portion of the employees do *not* have cell phones?

$\frac{1}{5}$; 0.20; 20%

78. A zoo has 125 types of animals, including 25 that are endangered. What portion is *not* endangered?

$\frac{4}{5}$; 0.8; 80%

79. An antibiotic is used to treat 380 people. If 342 people do not have side effects from the antibiotic, find the portion that do have side effects.

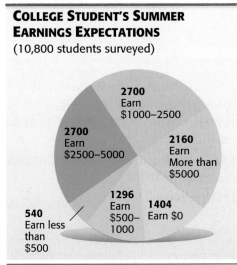

$\frac{1}{10}$; 0.1; 10%

80. A medical group includes 340 doctors. While electronic recordkeeping was used most often by the younger doctors, overall, only 85 of the doctors used it. What portion of the doctors did not use it? (*Source:* National Center for Health Statistics.)

$\frac{3}{4}$; 0.75; 75%

The circle graph shows how much money the 10,800 students at a college expect to earn this summer. Use this graph to answer Exercises 81–84, giving each answer as a fraction, as a decimal, and as a percent.

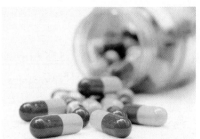

COLLEGE STUDENT'S SUMMER EARNINGS EXPECTATIONS
(10,800 students surveyed)

2700 Earn $1000–2500

2700 Earn $2500–5000

2160 Earn More than $5000

1296 Earn $500–1000

1404 Earn $0

540 Earn less than $500

Source: U.S. Collegeclub.com

81. What portion of the students expect to earn $2500 to $5000?

$\frac{1}{4}$; 0.25; 25%

82. What portion of the students expect to earn $500 to $1000?

$\frac{3}{25}$; 0.12; 12%

83. Find the portion who expect to earn less than $500.

$\frac{1}{20}$; 0.05; 5%

84. Find the portion who expect to earn more than $5000.

$\frac{1}{5}$; 0.20; 20%

Relating Concepts (Exercises 85–94) For Individual or Group Work

Work Exercises 85–94 in order to review the basics of percent.

85. 100% of a number means all of the parts or
<u>100</u> parts out of <u>100</u> parts.
200% means two times as many parts and 300%
means three times as many parts.

86. Fill in the blanks.

(a) 100% of 765 workers is <u>765 workers</u>.

(b) 200% of 48 letters is <u>96 letters</u>.

(c) 300% of 7 videos is <u>21 videos</u>.

87. 50% of a number is <u>50</u> parts out of
<u>100</u> parts, which is <u>half or $\frac{1}{2}$</u> of the parts.

88. 10% of a number is <u>10</u> parts out of <u>100</u>
parts and can be found quickly by moving the
decimal point <u>1</u> place to the <u>left</u>.

89. 1% of a number is <u>1</u> part out of
<u>100</u> parts and can be found quickly by moving
the decimal point <u>2</u> places to the <u>left</u>.

90. Fill in the blanks.

(a) 50% of 1050 homes is <u>525 homes</u>.

(b) 10% of 370 printers is <u>37 printers</u>.

(c) 1% of $8 is <u>$0.08</u>.

In Exercises 91–94, use the shortcut methods for finding 1%, 10%, 50%, 100%, 200%, *and* 300%.

91. Devise a shortcut method for finding 15% of a
number. Use your method to find 15% of $160.

Find 10% of $160, then add $\frac{1}{2}$ of the 10% amount.

$$10\% + 5\% = 15\%$$
$$\downarrow \quad \downarrow \quad \downarrow$$
$$\$16 + \$8 = \$24$$

92. Devise a shortcut method for finding 150% of a
number. Use your method to find 150% of $160.

Find 100% of $160, then add 50% of $160.

$$100\% + 50\% = 150\%$$
$$\downarrow \quad \downarrow \quad \downarrow$$
$$\$160 + \$80 = \$240$$

93. Explain a shortcut method for finding 90% of a
number. Show how to find 90% of $450 using your
method.

From 100% of $450, subtract 10% of $450.

$$100\% - 10\% = 90\%$$
$$\downarrow \quad \downarrow \quad \downarrow$$
$$\$450 - \$45 = \$405$$

94. Explain a shortcut method for finding 210% of a
number. Show how to find 210% of $800 using
your method.

To 100% of $800, add another 100% of $800, and
then add 10% of $800.

$$100\% + 100\% + 10\% = 210\%$$
$$\downarrow \quad \downarrow \quad \downarrow \quad \downarrow$$
$$\$800 + \$800 + \$80 = \$1680$$

6.3 ▶▶▶ Using the Percent Proportion and Identifying the Components in a Percent Problem

There are two ways to solve percent problems. One method uses proportions and is discussed in this and the next section. The other method uses the percent equation and is explained in **Section 6.5.**

OBJECTIVE 1 Learn the percent proportion. We have seen that a statement of two equivalent ratios is called a proportion.

$\frac{3}{5}$ or 3 out of 5 parts

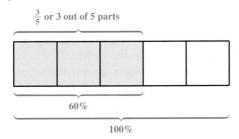

60%

100%

For example, the fraction $\frac{3}{5}$ is the same as the ratio 3 to 5, and 60% is the same as the ratio 60 to 100. As the figure above shows, these two ratios are equivalent and make a proportion.

Work Problem **1** *at the Side.* ▶

The percent proportion can be used to solve percent problems.

Percent Proportion

Part is to *whole* as *percent* is to *100.*

$$\frac{\text{part}}{\text{whole}} = \frac{\text{percent}}{100} \quad \leftarrow \text{Always 100 because percent means per 100}$$

In the figure at the top of the page, the **whole** is 5 (the entire quantity), the **part** is 3 (the part of the whole), and the **percent** is 60. Write the percent proportion as follows.

$$\text{part} \rightarrow \frac{3}{5} = \frac{60}{100} \leftarrow \text{percent} \qquad \boxed{\text{Remember: Percent means } per\ 100.}$$
$$\text{whole} \rightarrow 5 \qquad 100 \leftarrow 100$$

OBJECTIVE 2 Solve for an unknown value in a percent proportion. As shown in **Section 5.4,** if any three of the four values in a proportion are known, the fourth can be found by solving the proportion.

EXAMPLE 1 Using the Percent Proportion

Use the percent proportion and solve for the unknown value. Let x represent the unknown value.

(a) part = 12, percent = 25; find the whole.

$$\frac{\text{part}}{\text{whole}} = \frac{\text{percent}}{100} \qquad \text{Percent proportion}$$

Percent

$$\text{Part} \rightarrow \frac{12}{x} = \frac{25}{100} \qquad \text{or} \qquad \frac{12}{x} = \frac{1}{4} \qquad \frac{25}{100} \text{ is } \frac{1}{4} \text{ in lowest terms.}$$
$$\text{Whole (unknown)} \rightarrow x$$

── **Continued on Next Page**

OBJECTIVES

1 Learn the percent proportion.

2 Solve for an unknown value in a percent proportion.

3 Identify the percent.

4 Identify the whole.

5 Identify the part.

---

1 As a review of proportions, use the method of comparing cross products to decide whether each proportion is *true* or *false*. Show the cross products.

(a) $\dfrac{1}{2} = \dfrac{25}{50}$

(b) $\dfrac{3}{4} = \dfrac{150}{200}$

(c) $\dfrac{7}{8} = \dfrac{180}{200}$

(d) $\dfrac{32}{53} = \dfrac{160}{265}$

(e) $\dfrac{112}{41} = \dfrac{332}{123}$

ANSWERS

1. (a) 50 = 50; true **(b)** 600 = 600; true
 (c) 1440 ≠ 1400; false
 (d) 8480 = 8480; true
 (e) 13,612 ≠ 13,776; false

2 Use the percent proportion $\left(\dfrac{\text{part}}{\text{whole}} = \dfrac{\text{percent}}{100}\right)$ and solve for the unknown value.

(a) part = 12, percent = 16

(b) part = 30, whole = 120

(c) whole = 210, percent = 20

(d) whole = 4000, percent = 32

(e) part = 74, whole = 185

ANSWERS
2. **(a)** whole = 75
   (b) percent = 25
     (so, the percent is 25%)
   (c) part = 42
   (d) part = 1280
   (e) percent = 40
     (so, the percent is 40%)

First find the cross products.

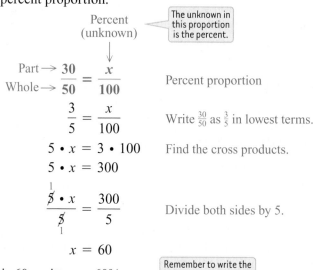

The unknown in this proportion is the *whole*.

$$\frac{12}{x} = \frac{1}{4}$$

$x \cdot 1$

$12 \cdot 4$

Show that the cross products are equivalent.

$$x \cdot 1 = 12 \cdot 4$$
$$x = 48$$

The whole is 48.

CAUTION
You **cannot** divide out a common factor from the numerator of one ratio and the denominator of the other ratio in **a proportion.** This can be done *only* when you are multiplying fractions.

(b) part = 30, whole = 50; find the percent.
  Use the percent proportion.

Percent (unknown) — The unknown in this proportion is the percent.

Part → $\dfrac{30}{50} = \dfrac{x}{100}$ ← Whole   Percent proportion

$$\frac{3}{5} = \frac{x}{100} \qquad \text{Write } \tfrac{30}{50} \text{ as } \tfrac{3}{5} \text{ in lowest terms.}$$

$$5 \cdot x = 3 \cdot 100 \qquad \text{Find the cross products.}$$
$$5 \cdot x = 300$$

$$\frac{\overset{1}{\cancel{5}} \cdot x}{\cancel{5}} = \frac{300}{5} \qquad \text{Divide both sides by 5.}$$

$$x = 60$$

The percent is 60, written as 60%. ← Remember to write the % sign when solving for an unknown percent.

(c) whole = 150, percent = 18; find the part.

Percent

Part (unknown) → $\dfrac{x}{150} = \dfrac{18}{100}$ or $\dfrac{x}{150} = \dfrac{9}{50}$ ← Whole

Write $\tfrac{18}{100}$ as $\tfrac{9}{50}$ in lowest terms.

$$x \cdot 50 = 150 \cdot 9 \qquad \text{Find the cross products.}$$
$$x \cdot 50 = 1350$$

$$\frac{x \cdot \overset{1}{\cancel{50}}}{\cancel{50}} = \frac{1350}{50} \qquad \text{Divide both sides by 50.}$$

$$x = 27$$

The part is 27.

◀ *Work Problem* **2** *at the Side.*

As a help in solving percent problems, keep in mind this basic idea.

> **Percent Problems**
>
> All percent problems involve a comparison between a part of something and the whole.

Solving these problems requires identifying the three components of a percent proportion: part, whole, and percent.

OBJECTIVE 3 Identify the percent. Look for the percent first. It is the easiest to identify.

> **Percent**
>
> The **percent** is the ratio of a part to a whole, with 100 as the denominator. In a problem, the percent appears with the word **percent** or with the symbol "**%**" after it.

EXAMPLE 2 **Finding the Percent in Percent Problems**

Find the percent in the following.

(a) 32% of the 900 men were too large for the imported car.

Percent — Look for the percent sign or the word percent.

The percent is 32. The number 32 appears with the symbol %.

(b) $150 is 25 percent of what number?

Percent

The percent is 25 because 25 appears with the word *percent*.

(c) What percent of 7000 pounds is 3500 pounds?

Percent (unknown)

The word *percent* has no number with it, so the percent is the unknown part of the problem.

—————— *Work Problem* ③ *at the Side.* ▶

OBJECTIVE 4 Identify the whole. Next, look for the whole.

> **Whole**
>
> The **whole** is the entire quantity. In a percent problem, the whole often appears after the word **of**.

③ Identify the percent.

(a) Of the 900 blood glucose tests, 25% will be completed by Adrian.

(b) Of the 620 preschool students, 65% will be served breakfast and lunch.

(c) Find the amount of sales tax by multiplying $590 and $6\frac{1}{2}$ percent.

(d) 8500 tons of recyclables is 42% of what number of tons?

(e) What percent of the 380 guests will return this year?

ANSWERS

3. (a) 25   **(b)** 65   **(c)** $6\frac{1}{2}$ **(d)** 42

(e) The percent is unknown.

4 Identify the whole.

(a) Of the 900 blood glucose tests, 25% will be completed by Adrian.

(b) Of the 620 preschool students, 65% will be served breakfast and lunch.

(c) Find the amount of sales tax by multiplying sales of $590 and $6\frac{1}{2}$ percent.

(d) 8500 tons of recyclables is 42% of what number of tons?

(e) What percent of the 380 guests will return this year?

5 Identify the part, then set up the percent proportion.

(a) Of the 900 blood glucose tests, 25% or 225 will be completed by Adrian.

(b) Of the 620 preschool students, 65% or 403 will be served breakfast and lunch.

(c) Find the sales tax by multiplying $590 and $6\frac{1}{2}$ percent.

(d) 8500 tons of recyclables is 42% of what number of tons?

(e) 80% of the 380 guests will return this year.

ANSWERS

4. (a) 900 (b) 620 (c) 590
   (d) what number (an unknown) (e) 380

5. (a) 225; $\dfrac{\text{Part} \to 225}{\text{Whole} \to 900} = \dfrac{25 \leftarrow \text{Percent}}{100 \leftarrow \text{Always 100}}$

   (b) 403; $\dfrac{\text{Part} \to 403}{\text{Whole} \to 620} = \dfrac{65 \leftarrow \text{Percent}}{100 \leftarrow \text{Always 100}}$

   (c) Unknown;

   $\dfrac{\text{Part} \to \text{unknown}}{\text{Whole} \to 590} = \dfrac{6\frac{1}{2} \leftarrow \text{Percent}}{100 \leftarrow \text{Always 100}}$

   (d) 8500;

   $\dfrac{\text{Part} \to 8500}{\text{Whole} \to \text{unknown}} = \dfrac{42 \leftarrow \text{Percent}}{100 \leftarrow \text{Always 100}}$

   (e) unknown;

   $\dfrac{\text{Part} \to \text{unknown}}{\text{Whole} \to 380} = \dfrac{80 \leftarrow \text{Percent}}{100 \leftarrow \text{Always 100}}$

EXAMPLE 3 **Finding the Whole in Percent Problems**

Identify the whole in the following.

(a) 32% **of** the 900 men were too large for the imported car.

Whole

The whole is 900. The number 900 appears after the word *of.*

(b) $150 is 25 percent **of** what number?

Whole   The whole is the unknown part of the problem.

(c) What percent **of** 7000 pounds is 3500 pounds?

Whole

◄ *Work Problem* **4** *at the Side.*

OBJECTIVE 5 Identify the part.   Finally, look for the part.

Part

The **part** is the portion being compared with the whole.

Note

If you have trouble identifying the part, find the percent and whole first. The remaining number is the part.

EXAMPLE 4 **Finding the Part in Percent Problems**

Identify the part. Then set up the percent proportion. (Do **not** solve the proportions.)

(a) 54% **of** 700 students is 378 students.
First find the percent and the whole.

54% **of** 700 students is 378 students.

Find the percent and then the whole. The remaining number, 378, is the part.

Percent; with % sign   Whole; follows "of"

The remaining number, 378, is the part.

54% **of** 700 students is 378 students. $\dfrac{\text{Part} \to 378}{\text{Whole} \to 700} = \dfrac{54 \leftarrow \text{Percent}}{100 \leftarrow \text{Always 100}}$

Percent Whole   Part

(b) $150 is 25% **of** what number? $\dfrac{\text{Part} \to 150}{\text{Whole} \to \text{unknown}} = \dfrac{25 \leftarrow \text{Percent}}{100 \leftarrow \text{Always 100}}$

Percent   Whole (unknown)

$150 is the remaining number, so the part is $150.

(c) 85% **of** 7000 is what number? $\dfrac{\text{Part} \to \text{unknown}}{\text{Whole} \to 7000} = \dfrac{85 \leftarrow \text{Percent}}{100 \leftarrow \text{Always 100}}$

Percent Whole   Part (unknown)

◄ *Work Problem* **5** *at the Side.*

Find the unknown value in the percent proportion $\dfrac{part}{whole} = \dfrac{percent}{100}$. *Round to the nearest tenth if necessary. If the answer is a percent, be sure to include a percent sign (%). See Example 1.*

1. part = 5, percent = 10
whole = 50

2. part = 20, percent = 25
whole = 80

3. part = 30, percent = 20
whole = 150

4. part = 25, percent = 25
whole = 100

5. part = 28, percent = 40
whole = 70

6. part = 11, percent = 5
whole = 220

🌐 **7.** part = 15, whole = 60
percent = 25%

8. part = 105, whole = 35
percent = 300%

9. part = 36, whole = 24
percent = 150%

10. part = 1.5, whole = 4.5
percent = 33.3% (rounded)

11. part = 9.25, whole = 27.75
percent = 33.3% (rounded)

12. part = 12.8, whole = 9.6
percent = 133.3% (rounded)

13. whole = 52, percent = 50
part = 26

14. whole = 160, percent = 35
part = 56

15. whole = 72, percent = 30
part = 21.6

16. whole = 115, percent = 38
part = 43.7

17. whole = 94.4, part = 25
percent = 26.5% (rounded)

18. whole = 89.6, part = 50
percent = 55.8% (rounded)

Solve each problem. If the answer is a percent, be sure to include a percent sign (%).
See Examples 2–4.

19. Find the whole if the part is 46 and the percent is 40.

115

20. The percent is 45 and the whole is 160. Find the part.

72

🌐 **21.** The whole is 5000 and the part is 20. Find the percent.

0.4%

22. Suppose the part is 15 and the whole is 2500. Find the percent.

0.6%

23. Find the percent if the whole is 4300 and the part is $107\dfrac{1}{2}$.

2.5%

24. What is the part, if the percent is $12\dfrac{3}{4}$ and the whole is 5600?

714

25. The whole is 6480 and the part is 19.44. Find the percent.

0.3%

26. Suppose the part is 281.25 and the percent is $1\dfrac{1}{4}$. Find the whole.

22,500

*In Exercises 27–42, set up the percent proportion, and write "unknown" for any value that is not given. Recall that the percent proportion is $\dfrac{part}{whole} = \dfrac{percent}{100}$. Do **not** try to solve for the unknowns. See Examples 2–4.*

27. 10% of how many bicycles is 60 bicycles?

$$\text{part} \rightarrow \frac{\mathbf{60}}{\mathbf{unknown}} \leftarrow \text{whole} = \frac{\mathbf{10}}{\mathbf{100}} \begin{array}{l} \leftarrow \text{percent} \\ \leftarrow \text{always 100} \end{array}$$

28. 58% of how many preschoolers is 203 preschoolers?

$$\frac{203}{\mathbf{unknown}} = \frac{58}{100}$$

29. 75% of $800 is $600.

$$\frac{600}{800} = \frac{75}{100}$$

30. 93% of $1500 is $1395.

$$\frac{1395}{1500} = \frac{93}{100}$$

31. What is 25% of $970?

$$\frac{\mathbf{unknown}}{970} = \frac{25}{100}$$

32. What is 61% of 830 homes?

$$\frac{\mathbf{unknown}}{830} = \frac{61}{100}$$

33. 12 injections is 20% of what number of injections?

$$\frac{12}{\mathbf{unknown}} = \frac{20}{100}$$

34. 92 servings is 26% of what number of servings?

$$\frac{92}{\mathbf{unknown}} = \frac{26}{100}$$

35. 34 trophies is 50% of 68 trophies.

$$\frac{34}{68} = \frac{50}{100}$$

36. 410 pallets is $33\frac{1}{3}$% of 1230 pallets.

$$\frac{410}{1230} = \frac{33\frac{1}{3}}{100}$$

37. What percent of $296 is $177?

$$\frac{177}{296} = \frac{\mathbf{unknown}}{100}$$

38. What percent of $121 is $30?

$$\frac{30}{121} = \frac{\mathbf{unknown}}{100}$$

39. 54.34 is 3.25% of what number?

$$\frac{54.34}{\mathbf{unknown}} = \frac{3.25}{100}$$

40. 16.74 is 11.9% of what number?

$$\frac{16.74}{\mathbf{unknown}} = \frac{11.9}{100}$$

41. 0.68% of $487 is what amount?

$$\frac{\mathbf{unknown}}{487} = \frac{0.68}{100}$$

42. What amount is 6.21% of $704.35?

$$\frac{\mathbf{unknown}}{704.35} = \frac{6.21}{100}$$

43. Identify the three components in a percent problem. In your own words, write a sentence telling how you will identify each of these three components.

Percent—the ratio of the part to the whole. It appears with the word *percent* or "%" after it. Whole—the entire quantity. Often appears after the word *of*. Part—the part being compared with the whole.

44. Write one short sentence using numbers and words. The sentence should include a percent, a whole, and a part. Identify each of these three components.

A possible sentence is: Of the 580 cars entering the parking lot, 464 cars, or 80%, had parking stickers on their windshield.

$$\text{percent} = 80; \text{whole} = 580; \text{part} = 464$$

*Set up the percent proportion for each application problem. Do **not** try to solve for any unknowns.*

45. Of the 1802 television sets sold at a Best Buy, 585 were high-definition sets. What percent of the televisions were high-definition sets?

$$\frac{585}{1082} = \frac{\text{unknown}}{100}$$

46. Ivory Soap is $99\frac{44}{100}\%$ pure. If a bar of Ivory Soap weighs 4 ounces, how many ounces are pure? (*Source:* Procter & Gamble.)

$$\frac{\text{unknown}}{4} = \frac{99\frac{44}{100}}{100}$$

47. Of the 142 people attending a movie theater, 86 bought buttered popcorn. What percent bought buttered popcorn?

$$\frac{86}{142} = \frac{\text{unknown}}{100}$$

48. On her first check from the Pizza Hut Restaurant, 15% is withheld from Maria's total earnings of $225. What amount is withheld?

$$\frac{\text{unknown}}{225} = \frac{15}{100}$$

49. Of the customers buying a salad at McDonald's, 23% prefer Newman's Own Light Salad Dressing. If the total number of salad customers is 610, find the number who prefer Newman's Own Light Dressing.

$$\frac{\text{unknown}}{610} = \frac{23}{100}$$

50. Of the total candy bars contained in a vending machine, 240 bars have been sold. If 25% of the bars have been sold, find the total number of candy bars that were in the machine.

$$\frac{240}{\text{unknown}} = \frac{25}{100}$$

51. There are 680 computer chips in a secured storage area designed to hold 2000 computer chips. What percent of the storage area is filled?

$$\frac{680}{2000} = \frac{\text{unknown}}{100}$$

52. There have been 36 cups of coffee served from a banquet-sized coffee pot. If this is 30% of the capacity of the pot, find the capacity of the pot.

$$\frac{36}{\text{unknown}} = \frac{30}{100}$$

53. In a recent survey of 480 adults, 55% said that they would prefer to have their wedding at a religious site. How many said they would prefer the religious site? (*Source:* National Family Opinion Research.)

$$\frac{\text{unknown}}{480} = \frac{55}{100}$$

54. Sue Ann needs 64 credits to graduate. If she has completed 48 of the credits needed, what percent of the credits has she already completed?

$$\frac{48}{64} = \frac{\text{unknown}}{100}$$

55. In a poll of 822 people, 49.5% said that they get their news from television. Find the number of people who said they get their news from television. (*Source: Brills Content.*)

$$\frac{\text{unknown}}{822} = \frac{49.5}{100}$$

56. The sales tax on a new car is $1575. If the sales tax rate is 7%, find the price of the car before the sales tax is added.

$$\frac{1575}{\text{unknown}} = \frac{7}{100}$$

57. A medical clinic found that 16.8% of the patients were late for their appointments. The number of patients who were late was 504. Find the total number of patients.

$$\frac{504}{\text{unknown}} = \frac{16.8}{100}$$

58. The state troopers tested 924 cars for safety. There were 231 cars that failed the safety test for one or more reasons. Find the percent of cars that failed the test.

$$\frac{231}{924} = \frac{\text{unknown}}{100}$$

59. Jerry Azzaro has listed 680 antique toys on eBay. If 45% of these were antique toy trains, find the number that were toy trains.

$$\frac{\text{unknown}}{680} = \frac{45}{100}$$

60. In a recent survey, 141 people said that they save money for up to two years to pay for a vacation that is at least a week long. If the total number of people in the survey was 1008 people, what percent of them save for up to two years to pay for their vacation? (*Source:* Capital One Direct Banking.)

$$\frac{141}{1008} = \frac{\text{unknown}}{100}$$

6.4 ▶▶▶ Using Proportions to Solve Percent Problems

This is the percent proportion that you learned about in the previous section.

$$\frac{\text{part}}{\text{whole}} = \frac{\text{percent}}{100}$$

Recall that if one of the values is unknown, you can find it by solving the percent proportion.

OBJECTIVE 1 Use the percent proportion to find the part. The first example shows how to use the percent proportion to find the part.

OBJECTIVES

1 Use the percent proportion to find the part.

2 Find the whole using the percent proportion.

3 Find the percent using the percent proportion.

EXAMPLE 1 Finding the Part with the Percent Proportion

Find 15% of $160.
    Here the percent is 15 and the whole is 160. (Recall that the whole often comes after the word *of.*) Now find the part. Let x represent the unknown part.

$$\frac{\text{part}}{\text{whole}} = \frac{\text{percent}}{100} \quad \text{so} \quad \frac{x}{160} = \frac{15}{100} \quad \text{or} \quad \frac{x}{160} = \frac{3}{20}$$

Write $\frac{15}{100}$ as $\frac{3}{20}$ in lowest terms.

Find the cross products in the proportion and show that they are equivalent.

$$x \cdot 20 = 160 \cdot 3 \qquad \text{Cross products}$$

$$x \cdot 20 = 480$$

$$\frac{x \cdot \overset{1}{\cancel{20}}}{\underset{1}{\cancel{20}}} = \frac{480}{20} \qquad \text{Divide both sides by 20.}$$

$$x = 24 \qquad \text{The unknown part is 24.}$$

15% of $160 is **$24**.

Work Problem ① *at the Side.* ▶

1 Use the percent proportion to find the part.

(a) 8% of 400 patients

(b) 15% of $3220

(c) 7% of 2700 miles

(d) 48% of 1580 kilowatts

    Just as with some of the fraction application problems in **Section 2.6,** the word *of* may be an indicator word meaning **_multiply._** Here is an example.

$$\text{15\% of 160}$$
$$\downarrow$$
$$\text{15\%} \cdot \text{160}$$

In this type of example, there is another way to find the part.

Finding the Part Using Multiplication

To find the part:

Step 1 Identify the percent. Write the percent as a decimal.

Step 2 Multiply this decimal by the whole.

2 Use multiplication to find the part.

(a) 55% of 10,000 injections

(b) 16% of 120 miles

(c) 135% of 60 dosages

(d) 0.5% of $238

EXAMPLE 2 Finding the Part Using Multiplication

Use multiplication to find the part.

(a) Find **42% of** 830 yards.

Step 1 Here, the percent is 42. Write 42% as the decimal 0.42.

Step 2 Multiply 0.42 and the whole, which is 830.

$$\text{part} = (0.42)(830)$$
$$= 348.6 \text{ yd}$$

When the percent and whole are given, the part must be found.

It is a good idea to estimate the answer, to make sure no mistakes were made with decimal points. Round 42% to 40% or 0.4, and round 830 as 800. Next, 40% of 800 is

$$(0.4)(800) = 320 \leftarrow \text{Estimate}$$

so the exact answer of 348.6 is reasonable.

(b) Find **25% of** 1680 cars.
Identify the percent as 25. Write 25% in decimal form as 0.25. Now, multiply 0.25 and 1680.

$$\text{part} = (0.25)(1680) = 420 \text{ cars} \quad \text{Multiply.}$$

You can also use a shortcut to find the answer. Since 25% means 25 parts out of 100 parts, this is the same as $\frac{1}{4}$ of the whole ($\frac{25}{100} = \frac{1}{4}$). Do you see a shortcut here? You can find $\frac{1}{4}$ of a number by dividing the number by 4. So, this shortcut gives us the exact answer, $1680 \div 4 = 420$.

(c) Find **140% of** 60 miles.
In this problem, the percent is 140. Write 140% as the decimal 1.40. Next, multiply 1.40 and 60.

$$\text{part} = (1.40)(60) = 84 \text{ miles} \quad \text{Multiply.}$$

You can estimate the answer by realizing that 140% is close to 150% (which is $1\frac{1}{2}$) and $1\frac{1}{2}$ times 60 is 90. So, 84 miles is a reasonable answer.

(d) Find **0.4% of** 50 kilometers.

$$\text{part} = (0.004)(50) = 0.2 \text{ kilometer} \quad \text{Multiply.}$$

Write 0.4% as a decimal.

Estimate the answer by realizing that 0.4% is less than 1%.

$$1\% \text{ of } 50 \text{ kilometers} = 50. = 0.5 \text{ kilometer}$$

So our exact answer should be *less than* 0.5 kilometer, and 0.2 kilometer fits this requirement.

◀ Work Problem **2** at the Side.

EXAMPLE 3 Solving for the Part in an Application Problem

Raley's Markets has 850 employees. Of these employees, 28% are students. How many of the employees are students? Use the six problem-solving steps.

Continued on Next Page

Step 1 **Read** the problem. The problem asks us to find the number of employees who are students.

Step 2 **Work out a plan.** Look for the word *of* as an indicator word for multiplication.

28% of the employees are students.
⌐ Indicator word

The total number of employees is 850, so the whole is 850. The percent is 28. To find the number of students, find the part.

Step 3 **Estimate** a reasonable answer. You can estimate the answer by rounding 28% to 25% and 850 to 900. Remember that 25% is 25 parts out of 100, which is equivalent to $\frac{1}{4}$. So divide 900 by 4.

$$900 \div 4 = 225 \text{ students} \leftarrow \text{Estimate}$$

Step 4 **Solve** the problem.

$$\text{part} = (0.28)(850) = 238 \quad \text{Multiply.}$$
⌐ Write 28% as a decimal.

> Notice that the decimal point was moved two places to the *left*.

Step 5 **State the answer.** Raley's Markets has 238 student employees.

Step 6 **Check.** The exact answer, 238 students, is close to our estimate of 225 students.

Work Problem ③ *at the Side.* ▶

🖩 **Calculator Tip**   If you are using a calculator, you could solve Example 3 above like this.

$$0.28 \; \boxed{\times} \; 850 \; \boxed{=} \; 238$$

Or, you can use this alternate approach on calculators with a % key.

$$850 \; \boxed{\times} \; 28 \; \boxed{\%} \; \boxed{=} \; 238$$

OBJECTIVE **2** **Find the whole using the percent proportion.**   The next example shows how to use the percent proportion to find the whole.

Note
Remember, the *whole* is the entire quantity.

EXAMPLE 4   **Finding the Whole with the Percent Proportion**

(a) 8 iPods is 4% of what number of iPods?
   Here the percent is 4, the whole is unknown, and the part is 8. Use the percent proportion to find the whole. Let *x* represent the unknown whole.

$$\frac{\text{part}}{\text{whole}} = \frac{\text{percent}}{100} \quad \text{so} \quad \frac{8}{x} = \frac{4}{100} \quad \text{or} \quad \frac{8}{x} = \frac{1}{25} \quad \begin{array}{l}\text{Write } \frac{4}{100} \text{ as } \frac{1}{25} \\ \text{in lowest terms.}\end{array}$$

$$x \cdot 1 = 8 \cdot 25 \quad \text{Cross products}$$

$$x = 200$$

8 iPods is 4% of **200 iPods**.

Continued on Next Page

---

③ Use the six problem-solving steps to solve each problem.

(a) One day on Jacob's mail route there were 2920 pieces of mail. If 45% of those were advertising pieces, find the number of advertising pieces.

(b) There are 9750 students at the college. If 12% of them wear glasses or contact lenses, how many students wear glasses or contact lenses?

4 Use the percent proportion to find the unknown whole.

(a) 750 Super Lotto Tickets is 25% of what number of tickets?

(b) 28 antiques is 35% of what number of antiques?

(c) 387 customers is 36% of what number of customers?

🖩 **(d)** 292.5 miles is 37.5% of what number of miles?

(b) 135 tourists is 15% of what number of tourists?
The percent is 15 and the part is 135.

$$\text{Part} \rightarrow \frac{135}{x} = \frac{15}{100} \begin{array}{l}\leftarrow \text{Percent} \\ \leftarrow \text{Always 100}\end{array}$$
Whole (unknown) $\rightarrow$

> If the part and percent are given, the *whole* must be found.

$$\frac{135}{x} = \frac{3}{20} \qquad \text{Write } \tfrac{15}{100} \text{ as } \tfrac{3}{20} \text{ in lowest terms.}$$

$$x \cdot 3 = 135 \cdot 20 \qquad \text{Cross products}$$

$$x \cdot 3 = 2700$$

$$\frac{x \cdot \overset{1}{\cancel{3}}}{\underset{1}{\cancel{3}}} = \frac{2700}{3} \qquad \text{Divide both sides by 3.}$$

$$x = 900$$

135 tourists is 15% of **900 tourists**.

◀ *Work Problem* **4** *at the Side.*

EXAMPLE 5 **Applying the Percent Proportion**

At Newark Salt Works, 78 employees are absent because of illness. If this is 5% of the total number of employees, how many employees does the company have? Use the six problem solving steps.

Step 1 **Read** the problem. The problem asks for the total number of employees.

Step 2 **Work out a plan.** From the information in the problem, the percent is 5 and the part of the total number of employees is 78. The total number of employees or entire quantity, which is the whole, is the unknown.

Step 3 **Estimate** a reasonable answer. Round the number of employees from 78 to 80. Then, 5% is equivalent to the fraction $\frac{1}{20}$, and 80 is $\frac{1}{20}$ of the total number of employees.

$$80 \cdot 20 = 1600 \text{ employees} \leftarrow \text{Estimate}$$

Step 4 **Solve** the problem. Use the percent proportion to find the whole (the total number of employees).

$$\text{Part} \rightarrow \frac{78}{x} = \frac{5}{100} \begin{array}{l}\leftarrow \text{Percent} \\ \leftarrow \text{Always 100}\end{array}$$
Whole (unknown) $\rightarrow$

$$\frac{78}{x} = \frac{1}{20} \qquad \text{Write } \tfrac{5}{100} \text{ as } \tfrac{1}{20} \text{ in lowest terms.}$$

$$x \cdot 1 = 78 \cdot 20 \qquad \text{Cross products}$$

$$x = 1560$$

Step 5 **State the answer.** The company has **1560 employees**.

Step 6 **Check.** The exact answer, 1560 employees, is close to our estimate of 1600 employees.

Continued on Next Page

Note

To estimate the answer to Example 5 on the previous page, the 5% was changed to its fraction equivalent, $\frac{1}{20}$. Because 80 (rounded) is $\frac{1}{20}$ of the total employees, 80 was multiplied by 20 to get 1600, the estimated answer.

Work Problem *at the Side.* ▶

OBJECTIVE 3 Find the percent using the percent proportion. If the part and the whole are known, the percent proportion can be used to find the percent.

EXAMPLE 6 **Using the Percent Proportion to Find the Percent**

(a) 13 coupons is what percent of 52 coupons?

The whole is 52 (follows *of*) and the part is 13. Next, find the percent.

$$\frac{\text{part}}{\text{whole}} = \frac{\text{percent}}{100}$$

$$\text{Part} \rightarrow \frac{13}{52} = \frac{x}{100} \begin{matrix} \leftarrow \text{Percent (unknown)} \\ \leftarrow \text{Always 100} \end{matrix}$$

Write $\frac{13}{52}$ as $\frac{1}{4}$ in lowest terms. $\frac{1}{4} = \frac{x}{100}$

Find the cross products.

$$4 \cdot x = 1 \cdot 100 \qquad \text{Cross products}$$

$$\frac{\overset{1}{\cancel{4}} \cdot x}{\underset{1}{\cancel{4}}} = \frac{100}{4} \qquad \text{Divide both sides by 4.}$$

$$x = 25$$

13 coupons is **25%** of 52 coupons.

(b) What percent of $500 is $100?

The whole is 500 (follows *of*) and the part is 100.

$$\frac{100}{500} = \frac{x}{100} \longleftarrow \text{Percent (unknown)}$$

$$\frac{1}{5} = \frac{x}{100} \qquad \begin{matrix}\text{Write } \frac{100}{500} \text{ as } \frac{1}{5} \text{ in} \\ \text{lowest terms.}\end{matrix}$$

$$5 \cdot x = 1 \cdot 100 \qquad \text{Cross products}$$

$$5 \cdot x = 100$$

$$\frac{\overset{1}{\cancel{5}} \cdot x}{\underset{1}{\cancel{5}}} = \frac{100}{5} \qquad \text{Divide both sides by 5.}$$

$$x = 20$$

20% of $500 is $100.

> Remember to write the % symbol in the answer.

Continued on Next Page

---

5 Use the six problem-solving steps and the percent proportion to solve each problem.

(a) A freeze resulted in a loss of 52% of an avocado crop. If the loss was 182 tons, find the total number of tons in the crop.

(b) A factory batch of cake mix contains 900 pounds of sugar, which is 18%, by weight, of the entire batch. What is the total weight of the batch?

6 Use the percent proportion to solve each problem.

(a) $21 is what percent of $105?

(b) What percent of 320 Internet companies is 48 Internet companies?

(c) What percent of 2280 court trials is 1026 trials?

(d) 432 snowboarders is what percent of 108 snowboarders?

7 Solve each problem.

(a) The bid price on an auction item is $289 while the minimum acceptable price is $425. The bid price is what percent of the minimum?

(b) A laboratory technician completes 80 tests in one day. If 52 of these tests were completed in the morning, what percent of the tests were completed in the morning?

> **CAUTION**
> When finding the percent, be sure to label your answer with the percent symbol (%).

◀ *Work Problem* **6** *at the Side.*

EXAMPLE 7 **Applying the Percent Proportion**

A roof is expected to last 20 years before needing replacement. If the roof is now 15 years old, what percent of the roof's life has been used?

Step 1 **Read** the problem. The problem asks for the percent of the roof's life that is already used.

Step 2 **Work out a plan.** The expected life of the roof is the entire quantity or *whole,* which is 20. The *part* of the roof's life that is already used is 15. Use the percent proportion to find the percent of the roof's life used.

Step 3 **Estimate** a reasonable answer. Since the roof is 15 years old, it is $\frac{15}{20}$ or $\frac{3}{4}$ used. Remember that $\frac{3}{4}$ is equivalent to 75%, so our estimate is 75%.

Step 4 **Solve** the problem. Let x represent the unknown percent.

$$\text{Part} \rightarrow \frac{15}{20} = \frac{x}{100} \quad \text{or} \quad \frac{3}{4} = \frac{x}{100} \qquad \text{Write } \tfrac{15}{20} \text{ as } \tfrac{3}{4} \text{ in lowest terms.}$$
$$\text{Whole} \rightarrow$$

$$4 \cdot x = 3 \cdot 100 \qquad \text{Cross products}$$
$$4 \cdot x = 300$$

$$\frac{\overset{1}{\cancel{4}} \cdot x}{\underset{1}{\cancel{4}}} = \frac{300}{4} \qquad \text{Divide both sides by 4.}$$

$$x = 75$$

Step 5 **State the answer.** 75% of the roof's life has been used.

Step 6 **Check.** The exact answer, 75%, matches our estimate of 75%.

◀ *Work Problem* **7** *at the Side.*

EXAMPLE 8 **Applying the Percent Proportion**

Rainfall this year was 33 inches, while normal rainfall is only 30 inches. What percent of normal rainfall is this year's rainfall?

Step 1 **Read** the problem. The problem asks us to find what percent this year's rainfall is of normal rainfall.

Step 2 **Work out a plan.** The normal rainfall is the *whole,* which is 30. This year's rainfall is *all of normal rainfall and more,* or 33 (part = 33). You need to find the percent that this year's rainfall is of normal rainfall.

Continued on Next Page

ANSWERS

6. (a) 20%   **(b)** 15%   **(c)** 45%   **(d)** 400%
7. (a) 68%   **(b)** 65%

Step 3 **Estimate** a reasonable answer. The increase in rainfall is 3 inches and the whole is 30 inches. The increase is $\frac{3}{30}$ or $\frac{1}{10}$ which is 10%. The whole is 100%, so 100% + 10% = 110%, our estimate.

Step 4 **Solve** the problem. Let x represent the unknown percent.

$$\frac{33}{30} = \frac{x}{100} \quad \text{or} \quad \frac{11}{10} = \frac{x}{100} \qquad \text{Write } \tfrac{33}{30} \text{ as } \tfrac{11}{10} \text{ in lowest terms.}$$

$$10 \cdot x = 11 \cdot 100 \qquad \text{Cross products}$$
$$10 \cdot x = 1100$$

$$\frac{\overset{1}{\cancel{10}} \cdot x}{\underset{1}{\cancel{10}}} = \frac{1100}{10} \qquad \text{Divide both sides by 10.}$$

$$x = 110$$

> If the part is *greater than* the whole, the percent is greater than 100.

Step 5 **State the answer.** This year's rainfall is **110%** of normal rainfall.

Step 6 **Check.** The exact answer, 110%, matches our estimate of 110%.

Work Problem **8** *at the Side.* ▶

8 Solve each problem.

(a) A new Toyota Prius Hybrid gets 32 miles per gallon on the highway and 48 miles per gallon around town. What percent of the highway mileage does the car get around town?

(b) The service department set a goal of 360 service calls this week. If they made 432 service calls, find the percent of their goal that they completed.

Math in the Media

EDUCATIONAL TAX INCENTIVES

The government sponsors tax incentive programs to make education more affordable. To qualify for the programs, you have to have an adjusted gross income below a certain level (most recently $47,000). You can find specific information at the Internal Revenue Service Web site.

- The Hope Scholarship offers 100% of the first $1200 spent for certain expenses, such as tuition and books, during the first year of college, plus 50% of the next $1200 incurred during the second year of college. The scholarship money is payable as a tax refund. The student cannot have completed the first two years of post-secondary education and must meet certain educational goals and workload criteria.

Suppose you are paying your own educational costs, and your adjusted gross income meets the guidelines to qualify for the Hope Scholarship. Your goals are to earn an Associate of Arts degree from a community college and then transfer to a state university to complete a Bachelor's degree. Tuition costs for resident students at North Harris Montgomery Community College District (NHMCCD) in Texas are used as an example of educational expenses.

Residents of NHMCCD pay a $12 registration fee for each semester enrolled plus tuition of $36 per semester hour. Assume that you must study a total of 15 semester hours in developmental work in mathematics, reading, and writing, and an additional 60 semester hours to complete an Associate of Arts degree. You decide to limit your course load to 15 credit hours each semester. Assume that one course is 3 semester hours, and you will have to purchase books at an approximate cost of $90 per course.

1. How many semesters and how many courses will it take you to finish the requirements for an Associate of Arts degree?   **5 semesters; 25 courses**

2. What is the total cost to complete the Associate of Arts degree for **(a)** books and **(b)** tuition and fees?   **$2250; $2760**

3. Calculate the total cost for tuition, fees, and books during the first two years (four semesters). What is the maximum tax incentive payable under the Hope Scholarship during **(a)** the first year and **(b)** the second year?

   $4008 in costs; $1200 first year; $600 second year

424

6.4 ▶▶▶ **Exercises**

FOR
EXTRA
HELP

MyMathLab

Math XL
PRACTICE

WATCH

DOWNLOAD

READ

REVIEW

Find the part using the multiplication shortcut. See Example 2.

1. 35% of 120 test tubes

42 test tubes

2. 20% of 1800 rentals

360 rentals

🌐 **3.** 45% of 4080 military personnel

1836 military personnel

4. 12% of 3650 Web sites

438 Web sites

5. 4% of 120 ft

4.8 ft

6. 9% of $150

$13.50

7. 150% of 210 files

315 files

8. 130% of 60 trees

78 trees

9. 52.5% of 1560 trucks

819 trucks

10. 38.2% of 4250 loads

1623.5 loads

11. 2% of $164

$3.28

12. 6% of $434

$26.04

13. 225% of 680 tables

1530 tables

14. 110% of 150 apartments

165 apartments

15. 17.5% of 1040 cell phones

182 cell phones

16. 46.1% of 843 kilograms

388.623 kilograms

17. 0.9% of $2400

$21.60

18. 0.3% of $1400

$4.20

Find the whole using the percent proportion. See Example 4.

🌐 **19.** 80 e-mails is 25% of what number of e-mails?

320 e-mails

20. 32 medical exams is 5% of what number of medical exams?

640 medical exams

21. 30% of what number of hay bales is 48 hay bales?

160 hay bales

22. 55% of what number of experiments is 209 experiments?

380 experiments

23. 495 successful students is 90% of what number of students?

550 students

24. 84 letters is 28% of what number of letters?

300 letters

25. 462 mountain bikes is 140% of what number of mountain bikes?

330 mountain bikes

26. 1496 graduates is 110% of what number of graduates?

1360 graduates

27. $12\frac{1}{2}\%$ of what number is 350?

$\left(\textit{Hint:} \text{ Write } 12\frac{1}{2}\% \text{ as } 12.5\%.\right)$

2800

28. $5\frac{1}{2}\%$ of what number is 176?

$\left(\textit{Hint:} \text{ Write } 5\frac{1}{2}\% \text{ as } 5.5\%.\right)$

3200

Find the percent using the percent proportion. Round your answers to the nearest tenth if necessary. See Example 6.

29. 18 bean burritos is what percent of 36 bean burritos?

50%

30. 62 hospital rooms is what percent of 248 hospital rooms?

25%

31. 390 SUVs is what percent of 750 SUVs?

52%

32. 650 liters is what percent of 1000 liters?

65%

33. 32 patients is what percent of 400 patients?

8%

34. 7 bridges is what percent of 350 bridges?

2%

35. 54 CDs is what percent of 3600 CDs?

1.5%

36. 60 cartons is what percent of 2400 cartons?

2.5%

37. What percent of $344 is $64?

18.6% (rounded)

38. What percent of $398 is $14?

3.5% (rounded)

39. What percent of 250 tires is 23 tires?

9.2%

40. What percent of 105 employees is 54 employees?

51.4% (rounded)

41. A student turned in the following answers on a test. You can tell that two of the answers are incorrect without even working the problems. Find the incorrect answers and explain how you identified them (without actually solving the problems).

50% of $84 is $42 .

150% of $30 is $20 .

25% of $16 is $32 .

100% of $217 is $217 .

150% of $30 cannot be less than $30 because 150% is greater than 1 (100%). The answer must be greater than $30.

25% of $16 cannot be greater than $16 because 25% is less than 1 (100%). The answer must be less than $16.

42. Write a percent problem on any topic you choose. Be sure to include only two of the three components so that you can solve for the third component. Identify each component of the problem and then solve it.

Answers will vary. One example is:

There are 600 vehicles in the parking lot and 45 of them are pickup trucks. What percent are pickup trucks?

Total vehicles, 600, is the whole, and the number of pickup trucks, 45, is the part.

$$\frac{45}{600} = \frac{x}{100}$$

$$600 \cdot x = 4500$$

$$\frac{600 \cdot x}{600} = \frac{4500}{600}$$

$$x = 7.5 \text{ or } 7.5\%$$

Solve each application problem. Round percent answers to the nearest tenth if necessary. See Examples 3, 5, 7, and 8.

43. Aimee Toit, who works part-time, earns $240 per week and has 22% of this amount withheld for taxes, Social Security, and Medicare. Find the amount withheld.

$52.80

44. An estimated 29.5% of automobile crashes are caused by driver distractions such as mobile communications devices. If there are 16,450 automobile crashes in a study, what number would be caused by driver distractions? Round to the nearest whole number. (*Source:* National Conference of State Legislatures.)

4853 accidents (rounded)

45. The guided-missile destroyer USS *Sullivans* has a 335-person crew, of which 13% are female. Find the number of female crew members. Round to the nearest whole number. (*Source:* U.S. Navy.)

44 females (rounded)

46. In a survey on where to hold their weddings, 45% of 480 adults preferred a nonreligious site. How many adults said they would prefer a religious site? (*Source:* National Family Opinion Research.)

264 adults

47. This year, there are 550 scholarship applicants. If 40% of the applicants will receive a scholarship, find the number of students who will receive a scholarship.

220 students

48. A U.S. Food and Drug Administration (FDA) biologist found that canned tuna is "relatively clean." Extraneous matter was found in 5% of the 1600 cans of tuna tested. How many cans of tuna contained extraneous matter?

80 cans

49. There are 1,094,751 active lawyers living in the United States. If 71.4% of these lawyers are male, find **(a)** the percent of the lawyers who are female and **(b)** the number of lawyers who are female. Round to the nearest whole number. (*Source:* American Bar Association.)

(a) 28.6% female lawyers

(b) 313,099 female lawyers (rounded)

50. According to the National Association of Realtors, 56% of first-time buyers make down payments of less than 5% of the purchase price. If 2.6 million first-time buyers bought homes, find **(a)** the percent who had down payments of 5% or more and **(b)** the number of buyers having down payments of less than 5% (*Source:* NAR Research.)

(a) 44% (5% or more)

(b) 1.456 million or
1,456,000 (less than 5%)

The bar graph below shows the percent of children 6–11 years of age who are neglecting dental hygiene. Use the information to answer Exercises 51–54.

DOWN IN THE MOUTH

The percentage[1] of children ages 6–11 who are neglecting dental hygiene by:

- Going to bed without brushing — 67%
- Brushing less than one minute — 61%
- Not flossing — 49%
- Brushing only once a day — 41%

[1]Respondents allowed to choose multiple answers.
Source: Services for Crest.

51. What percent of the children brush their teeth before going to bed?

33%

52. What percent of the children floss their teeth?

51%

53. If 3400 children answered the questions for this survey, how many of the children brush less than one minute?

2074 children

54. How many of the 3400 children in the survey brush only once a day?

1394 children

55. A recent study examined 48,000 military jobs, such as Army attack helicopter pilot or Navy gunner's mate. It was found that only 960 of these jobs are filled by women. What percent of these jobs are filled by women? (*Source:* Rand's National Defense Research Institute.)   2%

56. There are more than 55,000 words in *Webster's Dictionary,* but most educated people can identify only 20,000 of these words. What percent of the words in the dictionary can these people identify?

36.4% (rounded)

57. Ebony Durrant has 7.5% of her earnings deposited into her retirement plan. If $240 per month is deposited in the plan, find her monthly and yearly earnings.

$3200 monthly earnings; $38,400 yearly earnings

58. About 61% of the 43,000,000 people who receive Social Security benefits are paid with a direct deposit to their bank. How many of the people receiving benefits are paid with a direct deposit?

26,230,000 people

The circle graph shows the percent of various ice cream brands purchased by the
1582 Americans in a recent survey. Use this information to answer Exercises 59–62.

GET THE SCOOP

Last year, Americans spent a total of $4.5 billion
on ice cream. These are the brands they purchased.

37.7%
Other

22.3%
Private
label

14.7%
Breyers

10.6%
Dreyers/
Edy's Grand

4.5%
Ben &
Jerry's

4.8%
Häagen-
Dazs

5.4%
Blue Bell

Source: Information Resources Inc.; NPD Group.

59. Of the specific brands purchased, (not "Other" or
"Private Label,") which brand was purchased most
often?

Breyers

60. What percent of ice cream purchases were "Private
Label" or "Other" brands?

60%

61. Find the number of people in the survey who said
they purchase Häagen-Dazs. Round to the nearest
whole number.

76 people (rounded)

62. How many more people said they purchase Blue
Bell brand than Ben & Jerry's brand? Round to the
nearest whole number.

14 people (rounded)

The circle graph shows the sales at fast-food hamburger chains as a percent of total
fast-food hamburger sales. Use this information to answer Exercises 63–66.

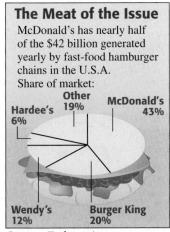

The Meat of the Issue

McDonald's has nearly half
of the $42 billion generated
yearly by fast-food hamburger
chains in the U.S.A.
Share of market:

Other
19%

McDonald's
43%

Hardee's
6%

Wendy's
12%

Burger King
20%

Source: Technomic.

63. Which of the hamburger chains had the lowest
sales?

Hardee's

64. What percent of the total hamburger sales were
made by the two companies having the lowest sales?

18%

65. Find the total annual sales for McDonald's.

$18.06 billion

66. Find the total annual sales for Burger King.

$8.4 billion

67. A collection agency, specializing in collecting
past-due child support, charges $25 as an
application fee plus 20% of the amount collected.
What is the total charge for collecting $3100 in
past-due child support?

$645

68. The income earned on an investment is 8.5% of the
amount invested. If the income is $12,750, find the
amount of the investment.

$150,000

69. Marketing Intelligence Service says that there were
15,401 new products introduced last year. If 86% of
the products introduced last year failed to reach
their business objectives, find the number of products that were successful. (Round to the nearest
whole number.)

2156 products

70. A family of four with a monthly income of $2900
spends 90% of its earnings and saves the balance.
Find **(a)** the monthly savings and **(b)** the annual
savings of this family.

(a) $290   (b) $3480

Relating Concepts (Exercises 71–77) For Individual or Group Work

Knowing and using the percent proportion is useful when solving percent problems.
Work Exercises 71–77 in order.

71. In the percent proportion, part is to ___whole___ as percent is to ___100___ .

72. All percent problems involve a comparison between a part of something and the ___whole___ .

Use this Ramen Noodles label of nutrition facts to answer Exercises 73–77. Read the label very carefully. Round to the nearest tenth of a percent.

Ramen Noodles

Nutrition Facts	Amount/serving	%DV*	Amount/serving	%DV*
Serving Size 1/2 Pkg. (15 oz/42.5 g)	Total Fat 8g	12%	Total Carbohydrates 27g	9%
	Saturated Fat 4g	20%	Dietary Fiber Less Than 1g	3%
Servings Per Package 2	Cholesterol 0mg	0%	Sugars Less Than 1g	
Calories 190	Sodium 670mg	28%	Protein 4g	
Calories from Fat 70				

Vitamin A 0% • Vitamin C 0% • Calcium 2% • Iron 4%

*Percent Daily Values (DV) are based on a 2,000 calorie diet.

Calories Per Gram
Fat 9 • Carbohydrates 4 • Protein 4

Source: Ramen Noodles package.

73. How many calories per serving are from total carbohydrates?

108 calories

74. The package label shows that the 27 grams (g) of carbohydrates in one serving are 9% of the recommended Daily Value (DV). What is the recommended daily value of carbohydrates?

300 grams

75. Find the number of grams of total fat that are needed to meet the recommended Daily Value (DV) of fat. Round to the nearest gram.

67 grams (rounded)

76. Will a person eating two packages of Ramen Noodles in one day exceed their recommended daily value of sodium? Explain your answer.

Yes, since they would eat 28% × 4 servings = 112% of the daily value.

77. How many packages of Ramen Noodles must be eaten in a day to meet the recommended daily value of fiber? Would this be possible? Would this result in good nutrition? (Round to the nearest whole package.)

17 packages (rounded). It may be possible but would result in a diet that is high in total fat, saturated fat, sodium, and total carbohydrates.

6.5 ▶▶▶ Using the Percent Equation

In the last section you used a proportion to solve percent problems. In this section we show another way to solve these problems by using the **percent equation.** The percent equation is just a rearrangement of the percent proportion.

> **Percent Equation**
>
> $$\text{part} = \text{percent} \cdot \text{whole}$$
>
> *Be sure to write the percent as a decimal before using the equation.*

OBJECTIVES

1 Use the percent equation to find the part.

2 Find the whole using the percent equation.

3 Find the percent using the percent equation.

When using the percent proportion, we did *not* have to write the percent as a decimal because 100 was used in the denominator of the proportion. However, because there is no 100 in the percent *equation, we must* first write the percent as a decimal by dividing by 100.

Some of the examples solved earlier will be reworked by using the percent equation. If you want to, you can look back at **Section 6.4** to see how some of these same problems were solved using proportions. This will give you a comparison of the two methods.

OBJECTIVE 1 Use the percent equation to find the part. The first example shows how to find the part.

EXAMPLE 1 **Finding the Part**

(a) Find 15% of $160.

Write 15% as the decimal 0.15. The whole, which comes after the word *of,* is 160. Next, use the percent equation. Let x represent the unknown part.

$$\text{part} = \text{percent} \cdot \text{whole}$$
$$x = (0.15)(160)$$

Multiply 0.15 and 160.　　　| 15% *must* be written as the decimal 0.15 |

$$x = 24$$

15% of $160 is **$24.**

(b) Find 110% of 80 cases.

Write 110% as the decimal 1.10. The whole is 80. Let x represent the unknown part.

$$\text{part} = \text{percent} \cdot \text{whole}$$
$$x = (1.10)(80)$$
$$x = 88$$

110% of 80 cases is **88 cases.**

(c) Find 0.4% of 250 patients.

Write 0.4% as the decimal 0.004. The whole is 250. Let x represent the unknown part.

$$\text{part} = \text{percent} \cdot \text{whole}$$
$$x = (0.004)(250) \quad \text{| Write 0.4\% as a decimal. } 0.4\% = 0.004 \text{ |}$$
$$x = 1$$

0.4% of 250 patients is **1 patient.**

Continued on Next Page

1 Use the percent equation to find the part.

(a) 15% of 880 policyholders

(b) 23% of 840 gallons

(c) 120% of $220

(d) 135% of $1080

(e) 0.5% of 1200 fruit cups

(f) 0.25% of 1600 lab tests

To estimate the answer, think of 0.4% as approximately 0.5% or $\frac{1}{2}$ of 1%. Because 1% is $\frac{1}{100}$, 1% of 250 is

$$250 \div 100 = 2.5$$

Since 1% of 250 is 2.5 then 0.5% of 250 is 1.25 (because $2.5 \div 2 = 1.25$). So, the exact answer of 1 patient is reasonable.

◀ *Work Problem* **1** *at the Side.*

> **CAUTION**
> When using the percent equation, the percent must always be *changed to a decimal* before multiplying.

OBJECTIVE **2** **Find the whole using the percent equation.** The next example shows how to use the percent equation to find the whole.

> **Note**
> When the word *of* follows a percent, it is an indicator word for *multiply.*

EXAMPLE 2 **Solving for the Whole**

(a) 8 tables is 4% of what number of tables?
  The part is 8 and the percent is 4% or the decimal 0.04. The whole is unknown.

8   is   4% of what number?
          └── Indicator word

Now, use the percent equation.

$$\textbf{part} = \textbf{percent} \cdot \textbf{whole}$$
$$8 = (0.04)(x) \qquad \text{Let } x \text{ represent the unknown whole.}$$

$$\frac{8}{0.04} = \frac{(0.04)(x)}{0.04} \qquad \text{Divide both sides by } 0.04.$$

$$200 = x \longleftarrow \text{Whole}$$

8 tables is 4% of **200 tables**.

(b) 135 tourists is 15% of what number of tourists?
  Write 15% as 0.15. The part is 135. Use the percent equation to find the whole.

$$\textbf{part} = \textbf{percent} \cdot \textbf{whole}$$
$$135 = (0.15)(x) \qquad \text{Let } x \text{ represent the unknown whole.}$$

$$\frac{135}{0.15} = \frac{(0.15)(x)}{0.15} \qquad \text{Divide both sides by } 0.15.$$

$$900 = x \longleftarrow \text{Whole}$$

135 tourists is 15% of **900 tourists**.

Continued on Next Page

(c) $8\frac{1}{2}\%$ of what number is 102?

Write $8\frac{1}{2}\%$ as 8.5%, or the decimal 0.085. The part is 102. Use the percent equation.

> Write 8.5% as a decimal
> 8.5% = 0.085

$$\textbf{part} = \textbf{percent} \cdot \textbf{whole}$$

$$102 = (0.085)(x) \qquad \text{Let } x \text{ represent the unknown whole.}$$

$$\frac{102}{0.085} = \frac{(0.085)(x)}{0.085} \qquad \text{Divide both sides by } 0.085.$$

$$1200 = x \longleftarrow \text{Whole}$$

102 is $8\frac{1}{2}\%$ of **1200**.

Estimate the answer. Notice that $8\frac{1}{2}\%$ is close to 10%. If 102 is 10% of a number, then the number is 10 times 102, or 1020. So the exact answer, 1200, is reasonable.

> **CAUTION**
> In Example 2(c) above, $8\frac{1}{2}\%$ was first written as 8.5%, which is the decimal form of $8\frac{1}{2}\%$ ($8\frac{1}{2} = 8.5$). **The percent sign still remained** in 8.5%. Then 8.5% was changed to the decimal 0.085 before solving the equation.

Work Problem **2** *at the Side.* ▶

OBJECTIVE 3 Find the percent using the percent equation.
The final example shows how to use the percent equation to find the percent.

EXAMPLE 3 Finding the Percent

(a) 13 auto mechanics is what percent of 52 auto mechanics?

Because 52 follows *of,* the whole is 52. The part is 13, and the percent is unknown. Use the percent equation.

$$\textbf{part} = \textbf{percent} \cdot \textbf{whole}$$

$$13 = x \cdot 52 \qquad \text{Let } x \text{ represent the unknown percent.}$$

$$\frac{13}{52} = \frac{x \cdot 52}{52} \qquad \text{Divide both sides by } 52.$$

$$0.25 = x$$

$$0.25 \text{ is } 25\%$$

> You *must* write the decimal answer as a percent. 0.25 = 25%

13 auto mechanics is **25%** of 52 auto mechanics.

The equation can also be set up using *of* as an indicator word for multiplication and *is* as an indicator word for "is equal to."

$$13 \text{ is what percent of } 52?$$

$$13 = \qquad x \qquad \cdot 52$$

$$13 = x \cdot 52 \qquad \text{Same equation as above}$$

Continued on Next Page

---

2 Find the whole using the percent equation.

(a) 18 supervisors is 45% of what number of supervisors?

(b) 67.5 containers is 27% of what number of containers?

(c) 666 inoculations is 45% of what number of inoculations?

(d) $5\frac{1}{2}\%$ of what number of policies is 66 policies?

3 Find the percent using the percent equation.

(a) What percent of 35 monitors is 7 monitors?

(b) 34 post office boxes is what percent of 85 post office boxes?

(c) What percent of 920 invitations is 1288 invitations?

(d) 9 world-class runners is what percent of 1125 runners?

(b) What percent of $500 is $100?
The whole is 500 and the part is 100. Let x represent the unknown percent.

part = percent • whole

$$100 = x \cdot 500 \qquad \text{Let } x \text{ represent the unknown percent.}$$

$$\frac{100}{500} = \frac{x \cdot \overset{1}{\cancel{500}}}{\underset{1}{\cancel{500}}} \qquad \text{Divide both sides by 500.}$$

$$0.20 = x$$

0.20 is 20% ◁ Write the decimal as a percent. $0.\underset{\curvearrowright}{20} = 20\%$

20% of $500 is $100.

(c) What percent of $300 is $390?
The whole is 300 and the part is 390. Let x represent the unknown percent.

part = percent • whole

$$390 = x \cdot 300 \qquad \text{Let } x \text{ represent the unknown percent.}$$

$$\frac{390}{300} = \frac{x \cdot \overset{1}{\cancel{300}}}{\underset{1}{\cancel{300}}} \qquad \text{Divide both sides by 300.}$$

$$1.3 = x$$

1.3 is 130% ◁ Write the decimal as a percent $1.3 = 1.\underset{\curvearrowright}{30} = 130\%$

130% of $300 is $390.

(d) 6 ladders is what percent of 1200 ladders?
Since 1200 follows *of,* the whole is 1200. The part is 6.

part = percent • whole

$$6 = x \cdot 1200 \qquad \text{Let } x \text{ represent the unknown percent.}$$

$$\frac{6}{1200} = \frac{x \cdot \overset{1}{\cancel{1200}}}{\underset{1}{\cancel{1200}}} \qquad \text{Divide both sides by 1200.}$$

$$0.005 = x$$

0.005 is 0.5% ◁ Write the decimal as a percent $0.\underset{\curvearrowright}{005} = 0.5\%$

6 ladders is **0.5%** of 1200 ladders.

You can estimate the answer because 1% of 1200 ladders is found by moving the decimal point two places to the left in 1200, resulting in 12. Since 6 ladders is half of 12 ladders, our answer should be $\frac{1}{2}$ of 1% or 0.5%. Our exact answer matches the estimate.

> **CAUTION**
> When you use the percent equation to solve for an unknown percent, the answer will always be in decimal form. Notice that in Example 3(a), (b), (c), and (d) above, **the decimal answer had to be changed to a percent** by multiplying by 100 and attaching the percent sign. The answers became: (a) $0.25 = 25\%$; (b) $0.20 = 20\%$; (c) $1.3 = 130\%$; and (d) $0.005 = 0.5\%$.

ANSWERS

3. (a) 20% (b) 40% (c) 140% (d) 0.8%

◁ *Work Problem* **3** *at the Side.*

6.5   ▶▶▶ Exercises

Find the part using the percent equation. See Example 1.

1. 25% of 1080 blood donors
270 donors

2. 19% of 700 MP3 players
133 MP3 players

 3. 45% of 3000 bath towels
1350 bath towels

4. 75% of 360 dosages
270 dosages

5. 32% of 260 quarts
83.2 quarts

6. 44% of 430 liters
189.2 liters

 7. 140% of 2500 air bags
3500 air bags

8. 145% of 580 hamburgers
841 hamburgers

9. 12.4% of 8300 meters
1029.2 meters

10. 26.4% of 4700 miles
1240.8 miles

 11. 0.8% of $520
$4.16

12. 0.3% of $480
$1.44

Find the whole using the percent equation. See Example 2.

13. 24 patients is 15% of what number of patients?
160 patients

14. 32 classrooms is 20% of what number of classrooms?
160 classrooms

15. 40% of what number of salads is 130 salads?
325 salads

16. 75% of what number of wrenches is 675 wrenches?
900 wrenches

17. 476 circuits is 70% of what number of circuits?
680 circuits

18. 270 lab tests is 45% of what number of lab tests?
600 lab tests

 19. $12\frac{1}{2}$% of what number of people is 135 people?
1080 people

20. $18\frac{1}{2}$% of what number of circuit breakers is 370 circuit breakers?
2000 circuit breakers

21. $1\frac{1}{4}$% of what number of gallons is 3.75 gallons?
300 gallons

22. $2\frac{1}{4}$% of what number of files is 9 files?
400 files

Find the percent using the percent equation. See Example 3.

23. 70 shipments is what percent of 140 shipments?

50%

24. 180 telemarketers is what percent of 450 telemarketers?

40%

25. 114 tuxedos is what percent of 150 tuxedos?

76%

26. 75 PDAs is what percent of 125 PDAs?

60%

27. What percent of $264 is $330?

125%

28. What percent of $480 is $696?

145%

29. What percent of 160 liters is 2.4 liters?

1.5%

30. What percent of 600 meters is 7.5 meters?

1.25%

31. 170 cartons is what percent of 68 cartons?

250%

32. 612 orders is what percent of 425 orders?

144%

33. When using the percent equation, the percent must always be changed to a decimal before doing any calculations. Show and explain how to change a fraction percent to a decimal. Use $2\frac{1}{2}\%$ in your explanation.

You must first write the fraction in the percent as a decimal, then divide the percent by 100 to change it to a decimal.

$$2\frac{1}{2}\% = 2.5\% = 0.025$$

2.5% as a decimal

Write $2\frac{1}{2}$ as 2.5

34. Suppose a problem on your homework assignment was, "Find $\frac{1}{2}\%$ of $1300." Your classmates got answers of $0.65, $6.50, $65, and $650. Which answer is correct? How and why are they getting all of these answers? Explain.

The correct answer is $6.50. The error is in changing $\frac{1}{2}\%$ to a decimal. $\frac{1}{2}\% = 0.5\% = 0.005$;

$$(0.005)(\$1300) = \$6.50 \quad \textbf{Correct}$$

Here are the incorrect answers and how your classmates got them.

$\frac{1}{2}\% = 0.0005; (0.0005)(\$1300) = \$0.65$

$\frac{1}{2}\% = 0.05; \quad (0.05) \quad (\$1300) = \$65$ **Incorrect**

$\frac{1}{2}\% = 0.5; \quad (0.5) \quad (\$1300) = \$650$

Solve each application problem.

35. A study of office workers found that 27% would like more storage space. If there are 14 million office workers, how many want more storage space? (*Source:* Steelcase Workplace Index.)

3.78 million or 3,780,000 office workers

36. Most shampoos contain 75% to 90% water. If a 16-ounce bottle of shampoo contains 78% water, find the number of ounces of water in the bottle. Round to the nearest tenth of an ounce.

12.5 ounces (rounded)

37. The household lubricant WD-40 is used in 79% of U.S. homes. If there are 115.8 million U.S. homes, find the number of homes in which WD-40 is used. Round to the nearest tenth of a million.

91.5 million homes (rounded)

38. Three Spam Mobiles travel throughout the country promoting Spam and Spam Lite. By using these kitchens on wheels, the annual goal is to give 1.5 million taste samples to the public. If 58.6% of the goal has been met, find the number of samples that have been given. (*Source:* Hormel Foods Corporation.)

0.879 million or 879,000 samples given

39. In the United States, 98% of all households have a refrigerator. Out of 18,000 households, **(a)** how many are expected to have a refrigerator and **(b)** how many are expected to not have a refrigerator? (*Source:* American Housing Authority.)

(a) 17,640 have a refrigerator

(b) 360 do not have a refrigerator

40. In the United States, 56% of the households have a dishwasher. Out of 214,500 households **(a)** how many are expected to have a dishwasher and **(b)** how many are expected to not have a dishwasher? (*Source:* American Housing Authority.)

(a) 120,120 have a dishwasher

(b) 94,380 do not have a dishwasher

41. In the United States, 2 of the 50 states (Alaska and Louisiana) do not have any drive-in movies. The remaining states do have drive-in movies. (*Source: USA Today*)

(a) What percent of the states do not have drive-in movies?   4%

(b) What percent have drive-in movies?   96%

42. Among the 50 companies receiving the greatest number of U.S. patents last year, 18 were Japanese companies. (*Source: Wall Street Journal.*)

(a) What percent of the companies were Japanese companies?   36%

(b) What percent of the companies were not Japanese companies?   64%

43. In a survey of 1250 Americans, 461 rated their health as excellent. What percent of these Americans rate their health as excellent? Round to the nearest tenth of a percent. (*Source:* National Health Interview Survey.)

36.9% (rounded)

44. General Nutrition Center now has 3200 stores and plans to add 450 more stores. Find the percent of additional stores that they have planned.

14.1% (rounded)

The graph shows the average time spent preparing weekday dinners. Assume that 5400 people were surveyed to gather this data. Use this graph to answer Exercises 45–48.

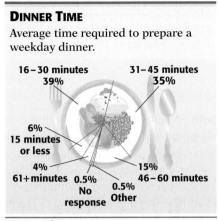

DINNER TIME
Average time required to prepare a weekday dinner.

16–30 minutes
39%

31–45 minutes
35%

6%
15 minutes
or less

4%
61+minutes

0.5%
No response

0.5%
Other

15%
46–60 minutes

Source: The NPD Group.

45. Find the number of people who said they spend "16–30 minutes" preparing weekday dinners.

2106 people

46. What total number of people answered "No response" and "Other"?

54 people

47. How many people said they spend over 30 minutes preparing weekday dinners?

2916 people

48. Find the number of people who said they spend 30 minutes or less preparing weekday dinners.

2430 people

49. Chemical Banking Corporation made $338 million worth of mortgage loans to minorities last year. If this represented 18.6% of all their mortgages, find the total value of all mortgages that they made last year. Round to the nearest tenth of a million.

$1817.2 million (rounded)

50. Rachel Williams has 8.5% of her monthly earnings deposited into the credit union. If this amounts to $131.75 per month, find her annual earnings.

12 ($1550) = $18,600 annual earnings

51. The Chevy Camaro was introduced in 1967. Sales that year were 220,917 Camaros, which was 46.2% of the number of Ford Mustangs sold in the same year. Find the number of Mustangs sold in 1967. Round to the nearest whole number.

478,175 Mustangs (rounded)

52. Chris Goodwin is a waiter and has sales of $822.25 on Saturday. If this is 28.6% of his sales for the week, find his weekly sales.

$2875

53. J & K Mustang has increased the sale of auto parts by $32\frac{1}{2}$% over last year. If the sale of parts last year amounted to $385,200, find the volume of sales this year.

$510,390

54. An ad for steel-belted radial tires promises 15% better mileage. If mileage had been 25.6 miles per gallon in the past, what mileage could be expected after these tires are installed? Round to the nearest tenth of a mile.

29.4 miles per gallon (rounded)

55. A Polaris Vac-Sweep is priced at $524 with an allowed trade-in of $125 on an old unit. If sales tax of $7\frac{3}{4}$% is charged on the price of the new Polaris unit before the trade-in, find the total cost to the customer after receiving the trade-in. (*Hint:* Trade-in is subtracted last.)

$439.61

56. General Motors car sales in China were 36.7% greater than last year's sales of 629,778 cars. Find this year's sales. Round to the nearest whole number. (*Source:* General Motors Corporation.)

860,907 cars

Summary Exercises on Percent

Write each percent as a decimal and each decimal as a percent.

1. 6.25%
   0.0625

2. 380%
   3.80 or 3.8

3. 0.375
   37.5%

4. 0.006
   0.6%

Write each percent as a fraction or mixed number in lowest terms and each fraction as a percent.

5. 87.5%
   $\dfrac{7}{8}$

6. 160%
   $1\dfrac{3}{5}$

7. $\dfrac{5}{8}$
   62.5%

8. $\dfrac{1}{125}$
   0.8%

The circle graph shows what 1100 adults dread the most about the Thanksgiving holiday. Use this graph to answer Exercises 9–12. Write each answer as a fraction, as a decimal, and as a percent. (*Source:* Opinion Research Corporation.)

THANKSGIVING TURNOFFS
What adults dread most about the holiday.

None of these/don't know 253
Fear of putting on weight 286
Too much family time 66
The expense 99
Traveling 187
Hassle of cooking 209

9. What portion of the adults have a fear of putting on weight?

   $\dfrac{13}{50}$; 0.26; 26%

10. What portion of the adults dread traveling?

   $\dfrac{17}{100}$; 0.17; 17%

11. Find the portion who dread too much family time.

   $\dfrac{3}{50}$; 0.06; 6%

12. Find the portion who dread the expense.

   $\dfrac{9}{100}$; 0.09; 9%

Find the part, whole, or percent as indicated. Round percent answers to the nearest tenth if necessary.

13. 6% of $780 is what amount?
   $46.80

14. 70 rolls of film is 14% of how many rolls of film?
   500 rolls

15. What percent of 320 policies is 176 policies?
   55%

16. 0.8% of 3500 screening exams is how many exams?
   28 screening exams

17. 1016.4 acres is 280% of what number of acres?
   363 acres

18. What percent of 658 circuits is 18 circuits?
   2.7% (rounded)

Solve each application problem. Round to the nearest cent or nearest tenth of a percent if necessary.

19. In a poll of 1582 people, 10.6% said that they prefer to purchase Dreyer's/Edy's brand ice cream. Find the number of people who said they prefer Dreyer's/Edy's brand ice cream. Round to the nearest whole number. (*Source:* Information Resources Inc; NPD Group.)

168 people (rounded)

20. A real estate broker wants to purchase a Palm i705 with a built-in radio modem priced at $399. If the sales tax rate is 7.75%, find the total price including the sales tax. (*Source:* Real Estate Technology.)

$429.92

21. The number of people who passed the real estate license exam was 832. If this is a 65% pass rate, how many took the exam?

1280 people

22. In a recent survey of dog owners, it was found that 901 or 34% of the owners take their dogs on vacation with them. Find the number of dog owners in the survey who do not take their dogs on vacation. (*Source:* American Animal Hospital Association.)

1749 owners

23. The distance around Lake Tahoe is 65 miles. If Lino Delgadillo has completed 39 miles of the Lake Tahoe run, what percent of the run remains?

40%

24. Beutler Heating and Air Conditioning plans to lay off 45 of its 1215 workers. What percent of the workers will be laid off? (*Source:* Beutler Heating and Air Conditioning.)

3.7% (rounded)

25. Netflix, a DVD-rental company, had 1,300,000 subscribers three years ago. Find the number of subscribers today after an increase of 338.5%. (*Source:* Netflix.)

5,700,500 subscribers

26. The number of ski-lift tickets sold last week was 3820. This week's sales are down 10.5%. Find the number of ski-lift tickets sold this week. Round to the nearest whole number.

3419 ski-lift tickets (rounded)

6.6 ▶▶▶ Solving Application Problems with Percent

Percent has many applications in our daily lives. This section discusses percent as it applies to sales tax, commissions, discounts, and the percent of change (increase and decrease).

OBJECTIVE 1 Find sales tax. States, countries, and cities often collect taxes on sales to customers. The **sales tax** is a percent of the total sale. The following formula for finding sales tax is based on the percent equation.

> **Sales Tax Formula**
>
> part   = percent • whole
>   ↓   ↓   ↓
> amount of sales tax = rate of tax • cost of item

OBJECTIVES

1 Find sales tax.

2 Find commissions.

3 Find the discount and sale price.

4 Find the percent of change.

EXAMPLE 1   Solving for Sales Tax

Office Max sells a flat panel computer monitor for $299. If the sales tax rate is 5%, how much tax is paid? What is the total cost of the flat panel computer monitor? Use the six problem-solving steps.

Step 1 **Read** the problem. The problem asks for the total cost of the flat panel display including the sales tax.

Step 2 **Work out a plan.** Use the sales tax formula to find the amount of sales tax. Write the tax rate (5%) as a decimal (0.05). The cost of the item is $299. Use the letter a to represent the unknown *amount* of tax. Add the sales tax to the cost of the item.

Step 3 **Estimate** a reasonable answer. Round $299 to $300. Recall that 5% is equivalent to $\frac{1}{20}$, so divide $300 by 20 to estimate the tax.

$$\$300 \div 20 = \$15 \text{ tax}$$

The total estimated cost is $300 + $15 = $315. ← Estimate

Step 4 **Solve** the problem.

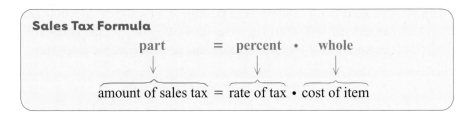

part   = percent • whole
  ↓   ↓   ↓
amount of sales tax = rate of tax • cost of item

$$a = (5\%)(\$299)$$
$$a = (0.05)(\$299)$$
$$a = \$14.95 \quad \text{Sales tax}$$

The tax paid on the flat panel computer monitor is **$14.95**. The customer would pay a total cost of $299 + **$14.95** = $313.95.

Step 5 **State the answer.** The total cost of the flat panel computer monitor is $313.95.

Step 6 **Check.** The exact answer, $313.95, is close to our estimate of $315.

Work Problem **1** *at the Side.* ▶

1 Suppose the sales tax rate in your state is 6%. Find the amount of the tax and the total you would pay for each item.

(a) $29 Little League bat

(b) $89 Guitar Hero: Legends of Rock bundle

(c) $1287 leather chair and ottoman

(d) $19,300 pickup truck

ANSWERS

1. (a) $1.74; $30.74 **(b)** $5.34; $94.34
  (c) $77.22; $1364.22 **(d)** $1158; $20,458

2 Find the rate of sales tax.

(a) The tax on a $320 patio set is $25.60.

(b) The tax on a $24 sweatshirt is $1.56.

(c) The tax on a $22,620 Ford Escape is $904.80.

EXAMPLE 2 Finding the Sales Tax Rate

The sales tax on a $14,800 Honda Civic is $962. Find the rate of the sales tax.

Step 1 **Read** the problem. This problem asks us to find the sales tax rate.

Step 2 **Work out a plan.** Use the sales tax formula.

$$\text{sales tax} = \text{rate of tax} \cdot \text{cost of item}$$

Solve for the rate of tax, which is the percent. The cost of the Honda Civic (the whole) is $14,800, and the amount of sales tax (the part) is $962. Use r to represent the unknown *rate* of tax (the percent).

Step 3 **Estimate** a reasonable answer. Round $14,800 to $15,000 and round $962 to $1000. The sales tax is $\frac{1000}{15,000}$ or $\frac{1}{15}$ of the cost of the car. So divide 1 by 15 to estimate the percent (rate) of sales tax.

$$\frac{1}{15} = 0.06\overline{6} \approx 7\% \leftarrow \text{Rounded estimate}$$

Step 4 **Solve** the problem.

$$\text{sales tax} = \text{rate of tax} \cdot \text{cost of item}$$
$$\$962 = r \cdot \$14,800$$

$$\frac{962}{14,800} = \frac{r \cdot \overset{1}{\cancel{14,800}}}{\underset{1}{\cancel{14,800}}} \quad \text{Divide both sides by 14,800.}$$

$$0.065 = r$$
$$0.065 \text{ is } 6.5\% \quad \boxed{\begin{array}{c}\text{Write the decimal}\\\text{as a percent.}\\0.065 = 6.5\%.\end{array}}$$

Step 5 **State the answer.** The sales tax rate is 6.5% or $6\frac{1}{2}\%$.

Step 6 **Check.** The exact answer, $6\frac{1}{2}\%$, is close to our estimate of 7%.

◀ *Work Problem* **2** *at the Side.*

Note

You can use the sales tax formula to find the amount of sales tax, the cost of an item, or the rate of sales tax (the percent).

OBJECTIVE **2** **Find commissions.** Many salespeople are paid by *commission* rather than an hourly wage. If you are paid by **commission,** you are paid a certain percent of your total sales dollars. The formula below for finding the commission is based on the percent equation.

Commission Formula

$$\text{part} = \text{percent} \cdot \text{whole}$$

$$\text{amount of commission} = \text{rate of commission} \cdot \text{amount of sales}$$

ANSWERS

2. (a) 8% **(b)** 6.5% or $6\frac{1}{2}\%$ **(c)** 4%

EXAMPLE 3   **Determining the Amount of Commission**

Scott Samuels had pharmaceutical sales of $42,500 last month. If his commission rate is 9%, find the amount of his commission.

Step 1   **Read** the problem. The problem asks for the amount of commission that Samuels earned.

Step 2   **Work out a plan.** Use the commission formula. Write the rate of commission (9%) as a decimal (0.09). The amount of Samuels' sales ($42,500) is the whole. Use c to represent the unknown *amount* of commission.

Step 3   **Estimate** a reasonable answer. Round the commission rate of 9% to 10%. Round the amount of sales from $42,500 to $40,000. Since 10% is equivalent to $\frac{1}{10}$, divide $40,000 by 10 to estimate the amount of commission.

$$\$40,000 \div 10 = \$4000 \longleftarrow \text{Estimate}$$

Step 4   **Solve** the problem.

amount of commission = rate of commission • amount of sales
$$c = (9\%)(\$42,500)$$
$$c = (0.09)(\$42,500)$$
$$c = \$3825 \quad \text{Amount of commission}$$

Step 5   **State the answer.** Samuels earned a commission of $3825 for selling the pharmaceuticals.

Step 6   **Check.** The exact answer, $3825, is close to our estimate of $4000.

Work Problem ③ *at the Side.* ▶

EXAMPLE 4   **Finding the Rate of Commission**

Chris Knudson earned a commission of $510 for selling $17,000 worth of shipping supplies. Find the rate of commission.

Step 1   **Read** the problem. In this problem, we must find the rate (percent) of commission.

Step 2   **Work out a plan.** You could use the commission formula. Another approach is to use the percent proportion. The *whole* is $17,000, the *part* is $510, and the *percent* is unknown. (The rate of commission is the percent.)

Step 3   **Estimate** a reasonable answer. Round the commission, $510, to $500, and round $17,000 to $20,000. The commission in fraction form is $\frac{\$500}{\$20,000}$, which simplifies to $\frac{1}{40}$. Changing $\frac{1}{40}$ to a percent gives $2\frac{1}{2}\%$ (rounded), as our estimate.

Step 4   **Solve** the problem.

$$\frac{\text{part}}{\text{whole}} = \frac{x}{100} \longleftarrow \text{Percent (unknown)} \longleftarrow \boxed{\text{Think: The rate of the commission is the percent.}}$$

$$\frac{510}{17,000} = \frac{x}{100}$$

$$17,000 \cdot x = 510 \cdot 100 \quad \text{Cross products.}$$

$$\frac{\overset{1}{\cancel{17,000}} \cdot x}{\underset{1}{\cancel{17,000}}} = \frac{51,000}{17,000} \quad \text{Divide both sides by 17,000.}$$

$$x = 3$$

Continued on Next Page

③ Find the amount of commission.

(a) Jill Buteo sells dental equipment at a commission rate of 12% and has sales for the month of $28,750.

(b) Last month Janis Cimperman sold a home for $220,500 for a client and earned a commission of 6%.

ANSWERS

3. (a) $3450 **(b)** $13,230

4 Find the rate of commission.

(a) A commission of $450 is earned on one sale of computer products worth $22,500.

(b) Jamal Story earns $2898 for selling office furniture worth $32,200.

5 Find the amount of the discount and the sale price.

(a) An Easy-Boy leather recliner originally priced at $950 is offered at a 42% discount.

(b) Wal-Mart has women's sweater sets on sale at 35% off. One sweater set was originally priced at $30.

Step 5   **State the answer.** The rate of commission is 3%.

Step 6   **Check.** The exact answer of 3% is close to our estimate of $2\frac{1}{2}$%.

◀ *Work Problem* **4** *at the Side.*

OBJECTIVE **3** **Find the discount and sale price.**   Most of us prefer buying things when they are on sale. A store will reduce prices, or **discount,** to attract additional customers. Use the following formula to find the discount and the sale price.

> **Discount Formula and Sale Price Formula**
> amount of discount = rate (or percent) of discount • original price
> sale price = original price − amount of discount

EXAMPLE 5   **Finding a Sale Price**

Whitings Oak Furniture Store has a home theater cabinet with an original price of $840 on sale at 15% off. Find the sale price of the cabinet.

Step 1   **Read** the problem. This problem asks for the price of a home theater cabinet after a discount of 15%.

Step 2   **Work out a plan.** The problem is solved in two steps. First, find the amount of the discount, that is, the amount that will be "taken off" (subtracted), by multiplying the original price ($840) by the rate of the discount (15%). The second step is to subtract the amount of discount from the original price. This gives you the sale price, which is what you will actually pay for the home theater cabinet.

Step 3   **Estimate** a reasonable answer. Round the original price from $840 to $800, and the rate of discount from 15% to 20%. Since 20% is equivalent to $\frac{1}{5}$, the estimated discount is $800 ÷ 5 = $160, so the estimated sale price is $800 − $160 = $640.

Step 4   **Solve** the problem. First find the exact amount of the discount.

amount of discount = rate of discount • original price

$$a = (0.15)(\$840) \quad \text{Write 15\% as a decimal.}$$
$$a = \$126 \quad \text{Amount of discount}$$

Now find the sale price of the home theater cabinet by subtracting the amount of the discount ($126) from the original price.

sale price = original price − amount of discount
$$= \$840 - \$126$$
$$= \$714 \quad \text{Sale price}$$

Step 5   **State the answer.** The sale price of the home theater cabinet is $714.

Step 6   **Check.** The exact answer, $714, is close to our estimate of $640.

◀ *Work Problem* **5** *at the Side.*

🔢 **Calculator Tip**   In Example 5 on the previous page, you can use a calculator to find the amount of discount and subtract the discount from the original price, all in one step.

840 ⊖ .15 ⊗ 840 ⊜ 714

↑   ︷   ↑
Original   Amount of   Sale
price   discount   price

A scientific calculator observes the order of operations, so it will automatically do the multiplication before the subtraction.

OBJECTIVE 4 Find the percent of change.   We are often interested in looking at increases or decreases in sales, production, population, and many other items. This type of problem involves finding the *percent of change.* Use the following steps to find the **percent of increase.**

Finding the Percent of Increase

Step 1   Use subtraction to find the amount of increase.

Step 2   Use the percent proportion to find the percent of increase.

$$\frac{\textbf{amount of increase (part)}}{\textbf{original value (whole)}} = \frac{\textbf{percent}}{\textbf{100}}$$

EXAMPLE 6   **Finding the Percent of Increase**

Attendance at county parks climbed from 18,300 last month to 56,730 this month. Find the percent of increase.

Step 1   **Read** the problem. The problem asks for the percent of increase.

Step 2   **Work out a plan.** Subtract the attendance last month (18,300) from the attendance this month (56,730) to find the amount of increase in attendance. Next, use the percent proportion. The whole is 18,300 (last month's original attendance), the part is 38,430 (amount of increase in attendance), and the percent is unknown.

Step 3   **Estimate** a reasonable answer. Round 18,300 to 20,000 and 56,730 to 60,000. The amount of increase is $60,000 - 20,000 = 40,000$. Since 40,000 (the increase) is *twice* as large as the original amount, the estimated percent of increase is 200%.

Step 4   **Solve** the problem.

$$56,730 - 18,300 = 38,430 \quad \leftarrow \begin{array}{c}\text{Subtract to find the}\\ \textbf{amount of increase}\\ \text{in attendance}\end{array}$$

Amount of increase → $\dfrac{38,430}{18,300} = \dfrac{x}{100}$ Percent proportion

$\boxed{\text{Use the } \textit{original} \text{ value of } 18,300 \text{ (not } 56,730\text{).}}$ ↗

Solve this proportion to find that $x = 210$.

Step 5   **State the answer.** The percent of increase is 210%.

Step 6   **Check.** The exact answer, 210%, is close to our estimate of 200%.

Work Problem **6** *at the Side.* ▶

---

6 Find the percent of increase.

(a) A manufacturer of snowboards increased production from 14,100 units last year to 19,035 this year.

(b) The number of flu cases rose from 496 cases last week to 620 this week.

ANSWERS

6. (a) 35% **(b)** 25%

Use the following steps to find the **percent of decrease.**

7 Find the percent of decrease.

(a) The number of service calls fell from 380 last month to 285 this month.

> **Finding the Percent of Decrease**
>
> *Step 1* Use subtraction to find the amount of decrease.
>
> *Step 2* Use the percent proportion to find the percent of decrease.
>
> $$\frac{\text{amount of decrease (part)}}{\text{original value (whole)}} = \frac{\text{percent}}{100}$$

EXAMPLE 7 **Finding the Percent of Decrease**

The number of production employees this week fell to 1406 people from 1480 people last week. Find the percent of decrease.

Step 1 **Read** the problem. The problem asks for the percent of decrease.

Step 2 **Work out a plan.** Subtract the number of employees this week (1406) from the number of employees last week (1480) to find the amount of decrease. Then, use the percent proportion. The whole is 1480 (last week's *original* number of employees), the part is 74 (amount of decrease in employees), and the percent is unknown.

Step 3 **Estimate** a reasonable answer. Estimate the answer by rounding 1406 to 1400 and 1480 to 1500. The decrease is $1500 - 1400 = 100$. Since 100 is $\frac{1}{15}$ of 1500, our estimate is $1 \div 15 \approx 0.07$ or 7%.

(b) The number of workers applying for unemployment fell from 4850 last month to 3977 this month.

Step 4 **Solve** the problem.

$$1480 - 1406 = 74 \quad \text{Subtract to find the } \textbf{amount of decrease} \text{ in number of employees.}$$

Amount of decrease $\rightarrow \dfrac{74}{1480} = \dfrac{x}{100}$ Percent proportion

Use the original value of 1480 (**not** 1406).

Solve this proportion to find that $x = 5$.

Step 5 **State the answer.** The percent of decrease is 5%.

Step 6 **Check.** The exact answer, 5%, is close to our estimate of 7%.

> **CAUTION**
> When solving for percent of increase or decrease, the **whole is always the original value** or **value before the change occurred.** The part is the change in values, that is, how much something went up or went down.

◀ *Work Problem* **7** *at the Side.*

FOR
EXTRA
HELP

Find the amount of sales tax or the tax rate and the total cost (amount of sale + amount of tax = total cost). Round money answers to the nearest cent if necessary. See Examples 1 and 2.

	Amount of Sale	Tax Rate	Amount of Tax	Total Cost
1.	$6	4%	$0.24	$6.24
2.	$45	5%	$2.25	$47.25
3.	$425	3%	$12.75	$437.75
4.	$322	6%	$19.32	$341.32
5.	$284	5%	$14.20	$298.20
6.	$84	7%	$5.88	$89.88
7.	$12,229	$5\frac{1}{2}\%$	$672.60 (rounded)	$12,901.60 (rounded)
8.	$11,789	$7\frac{1}{2}\%$	$884.18 (rounded)	$12,673.18 (rounded)

Find the commission earned or the rate of commission. Round money answers to the nearest cent if necessary. See Examples 3 and 4.

	Sales	Rate of Commission	Commission
9.	$280	8%	$22.40
10.	$660	10%	$66
11.	$3000	20%	$600
12.	$7800	15%	$1170
13.	$6183.50	3%	$185.51 (rounded)
14.	$4416.70	7%	$309.17 (rounded)
15.	$73,500	9%	$6615
16.	$55,800	6%	$3348

Find the amount or rate of discount and the sale price after the discount. Round money answers to the nearest cent if necessary. See Example 5.

	Original Price	Rate of Discount	Amount of Discount	Sale Price
17.	$199.99	10%	$20.00 (rounded)	$179.99 (rounded)
18.	$29.95	15%	$4.49 (rounded)	$25.46 (rounded)
19.	$180	30%	$54	$126
20.	$38	25%	$9.50	$28.50
21.	$17.50	25%	$4.38 (rounded)	$13.12
22.	$76	60%	$45.60	$30.40
23.	$58.40	15%	$8.76	$49.64
24.	$99.80	30%	$29.94	$69.86

25. You are trying to decide between Company A paying a 10% commission and Company B paying an 8% commission. For which company would you prefer to work? What considerations other than commission rate would be important to you?

On the basis of commission alone you would choose Company A. Other considerations might be: reputation of the company; expense allowances; other fringe benefits; travel; promotion and training; to name a few.

26. Give four examples of where you might use the percent of increase or the percent of decrease in your own personal activities. Think in terms of work, school, home, hobbies, and sports.

Some answers might be: Calculating percent pay increases or decreases; changes in the cost of utilities, groceries, gasoline, and insurance; changes in the value of investments; the economy (inflation or deflation); to name a few.

Solve each application problem. Round money answers to the nearest cent and rates to the nearest tenth of a percent if necessary. See Examples 1–7.

Country Store has a unique selection of merchandise that it sells by mail and over the Internet. Use the shipping and insurance delivery chart at the right and a sales tax rate of 5% to solve Exercises 27–30. There is no sales tax on shipping and insurance. (*Source:* Country Store Catalog.)

SHIPPING AND INSURANCE DELIVERY CHART

Up to $15.00	add $4.99
$15.01 to $25.00	add $6.99
$25.01 to $35.00	add $7.99
$35.01 to $50.00	add $8.99
$50.01 to $70.00	add $10.99
$70.01 to $99.99	add $12.99
$100.00 or more	add $14.99

27. Find the total cost of six Small Fry Handi-Pan electric skillets priced at $29.99 each.

$203.93 (rounded)

28. A customer ordered five sets of flour-sack towels priced at $12.99 per set. What is the total cost?

$79.19 (rounded)

29. Find the total cost of three pop-up hampers at $9.99 each and four nonstick mini doughnut pans at $10.99 each. **$90.62 (rounded)**

30. What is the total cost of five coach lamp bird feeders at $19.99 each and six garden weather centers at $14.99 each? **$214.37 (rounded)**

31. An Anderson wood-frame French door is priced at $1980 with a sales tax of $99. Find the rate of sales tax. **5%**

32. Textbooks for three classes cost $245 plus tax of $17.15. Find the sales tax rate. **7%**

33. Today there are 635,000 women motorcyclists in the United States, up from 467,400 just eight years ago. Find the percent of increase in the number of women motorcyclists. (*Source:* Motorcycle Industry Council.)

35.9% (rounded)

34. Americans are eating more fish. This year the average American will eat 15.5 pounds compared to only 12.5 pounds per year a decade ago. Find the percent of increase. (*Source: Consumer Reports.*)

24%

35. The number of industrial accidents this month fell to 989 accidents from 1276 accidents last month. Find the percent of decrease. **22.5% (rounded)**

36. The average number of hours worked in manufacturing jobs last week fell from 41.1 to 40.9. Find the percent of decrease. **0.5% (rounded)**

37. A "60% off sale" begins today at Wanda's Women's Wear. What is the sale price of women's wool coats normally priced at $335? **$134**

38. What is the sale price of a $769 Kenmore washer/dryer set with a discount of 25%? **$576.75**

A weekly sales report for the top four sales people at Active Sports is shown below. Use this information to answer Exercises 39–42.

Employee	Sales	Rate of Commission	Commission
Strong, A.	$18,960	3%	**$568.80**
Ferns, K.	$21,460	3%	**$643.80**
Keyes, B.	$17,680	4%	$707.20
Vargas, K.	$23,104	5.0% (rounded)	$1152.20

39. Find the commission for Strong. **$568.80**

40. Find the commission for Ferns. **$643.80**

41. What is the rate of commission for Keyes?
4%

42. What is the rate of commission for Vargas?
5.0% (rounded)

43. A Sony Micro Hi Fi Component System was priced at $390 and is on sale at 22% off. Find the discount and the sale price. **$85.80; $304.20**

44. A Honda Pilot is offered at 12% off the manufacturer's suggested retail price. Find the discount and the sale price of this SUV, originally priced at $32,500. **$3900; $28,600**

45. The price per share of Toys Я Us stock fell from $35.50 to $33.50. Find the percent of decrease in price.

5.6% (rounded)

46. In the past five years, the cost of generating electricity from the sun has been brought down from 24 cents per kilowatt hour to 8 cents (less than the newest nuclear power plants.) Find the percent of decrease.

66.7% (rounded)

47. College students are offered a 6% discount on a dictionary that sells for $18.50. If the sales tax is 6%, find the cost of the dictionary including the sales tax. **$18.43 (rounded)**

48. A fax machine priced at $398 is marked down 7% to promote the new model. If the sales tax is also 7%, find the cost of the fax machine including sales tax. **$396.05**

49. A real estate agent sells a condominium for $129,605. A sales commission of 6% is charged. The agent gets 55% of this commission. How much money does the agent get?   **$4276.97 (rounded)**

50. The local real estate agents' association collects a fee of 2% on all money received by its members. The members charge 6% of the selling price of a property as their fee. How much does the association get, if its members sell property worth a total of $8,680,000?   **$10,416**

51. What is the total price of a ski boat with an original price of $15,321, if it is sold at a 15% discount? The sales tax rate is $7\frac{3}{4}$%.

$14,032.12 (rounded)

52. A commercial security alarm system originally priced at $10,800 is discounted 22%. Find the total price of the system if the sales tax rate is $7\frac{1}{4}$%.

$9034.74

Relating Concepts (Exercises 53—58) For Individual or Group Work

Knowing how to use the percent equation is important when solving application problems involving sales tax. **Work Exercises 53–58 in order.**

53. The percent equation is

part = **percent** • **whole.**

54. The formula used to find sales tax is an application of the percent equation. The sales tax formula is

sales tax = **rate (percent) of tax** • **cost of item** .

*In the United States there are certain items on which an excise tax is charged in addition to a sales tax. A table of federal excise taxes is shown here. Use this table to answer Exercises 55–58. (Excise tax is calculated on the amount of the sale **before** sales tax is added.) Round answers to the nearest cent.*

FEDERAL EXCISE TAXES

Product or Service	Rate	Product or Service	Rate
Telephone service	3%	Tires (by weight)	
Teletypewriter service	3%	Under 40 pounds	No tax
Air transportation	7.5%	40–69 pounds	15¢/pound over 40 pounds
International air travel	$13.40/person	70–89 pounds	$4.50 plus 30¢/pound over 70 pounds
Air freight	6.25%		
		90 pounds and more	$10.50 plus 50¢/pound over 90 pounds
		Truck and trailer, chassis and bodies	12%
Fishing rods	10%	Inland waterways fuel	24.4¢/gallon
Bows and arrows	12.4%	Ship passenger tax	$3/passenger
Gasoline	18.4¢ gallon	Vaccines	75¢/dose
Diesel fuel	24.4¢ gallon		
Aviation fuel	21.9¢ gallon		

Source: Publication 510, I.R.S., Excise Taxes.

55. Some archery equipment (bows and arrows) is priced at $123. Use the federal excise tax table and a sales tax rate of $6\frac{1}{2}$% to find the cost of the equipment, including both taxes. (Round to the nearest cent.)   **$146.25 (rounded)**

56. Refer to Exercise 55. Calculate the two taxes separately and then add them together. Now, add the two tax rates together and then find the tax. Are your answers the same? Why or why not? (*Hint:* Recall the commutative and associative properties of multiplication.)   **Yes, the same;**
(12.4% • $123) + (6.5% • $123) = $23.25 (rounded)
and (12.4% + 6.5%) • ($123) = $23.25

57. The price of an international airline ticket is $1248. Use the federal excise tax table and a sales tax rate of $7\frac{3}{4}$% to find the total cost of one ticket. (Sales tax is not charged on the $13.40 federal excise tax.)

$1358.12

58. Refer to Exercise 57. Can the federal excise tax be added to the sales tax rate to find the total tax? Why or why not?   **No. In Exercise 57 the excise tax of $13.40 cannot be added to the sales tax rate of $7\frac{3}{4}$% because the excise tax is an amount, not a percent.**

6.7 ▶▶▶ Simple Interest

When we open a savings account, we are actually lending money to the bank or credit union. It will in turn lend this money to individuals and businesses. These people then become borrowers. The bank or credit union pays a fee to the savings account holders and charges a higher fee to its borrowers. These fees are called *interest*.

Interest is a fee paid or a charge made for lending or borrowing money. The amount of money borrowed is called the **principal.** The charge for interest is often given as a percent, called the interest rate or **rate of interest.** The rate of interest is assumed to be *per year,* unless stated otherwise. Time is always expressed in years or fractions of a year.

OBJECTIVE 1 Find the simple interest on a loan.   In most cases, interest on a loan is computed on the *original principal* and is called **simple interest.** We use the following **interest formula** to find simple interest.

> **Formula for Simple Interest**
>
> $$\text{Interest} = \text{principal} \cdot \text{rate} \cdot \text{time}$$
>
> The formula is usually written using letters.
>
> $$I = p \cdot r \cdot t$$

> **Note**
>
> Simple interest is used for most short-term business loans, most real estate loans, and many automobile and consumer loans.

EXAMPLE 1   **Finding Interest for a Year**

Find the interest on $5000 at 6% for 1 year.

The amount borrowed, or principal (p), is $5000. The interest rate (r) is 6%, which is 0.06 as a decimal, and the time of the loan (t) is 1 year. Use the formula.

$$I = p \cdot r \cdot t$$
$$I = (5000)(0.06)(1)$$
$$I = 300$$

Notice that "1" is used as the time for 1 year.

The interest is $300.

Work Problem **1** *at the Side.* ▶

EXAMPLE 2   **Finding Interest for More Than a Year**

Find the interest on $4200 at 8% for three and a half years.

The principal (p) is $4200. The rate ($r$) is 8%, or 0.08 as a decimal, and the time (t) is $3\frac{1}{2}$ or 3.5 years. Use the formula.

$$I = p \cdot r \cdot t$$
$$I = (4200)(0.08)(3.5)$$
$$I = 1176$$

3.5 years is equivalent to $3\frac{1}{2}$ years because 3.5 is $3\frac{5}{10}$ which simplifies to $3\frac{1}{2}$.

The interest is $1176.

Work Problem **2** *at the Side.* ▶

OBJECTIVES

1 Find the simple interest on a loan.

2 Find the total amount due on a loan.

1 Find the interest.

(a) $1000 at 5% for 1 year

(b) $3650 at 2% for 1 year

2 Find the interest.

(a) $820 at 6% for $3\frac{1}{2}$ years

(b) $4850 at 8% for $2\frac{1}{2}$ years

(c) $16,800 at 3% for $2\frac{3}{4}$ years

ANSWERS

1. (a) $50 (b) $73
2. (a) $172.20 (b) $970 (c) $1386

3 Find the interest.

(a) $1800 at 3% for 4 months

(b) $28,000 at $9\frac{1}{2}$% for 3 months

4 Find the total amount due on each loan.

(a) $3800 at $6\frac{1}{2}$% for 6 months

(b) $12,400 at 5% for 5 years

(c) $2400 at 11% for $2\frac{3}{4}$ years

Interest rates are given **per year.** For loan periods of less than one year, be careful to express time as a fraction of a year.

If time is given in months, for example, use a denominator of 12, because there are 12 months in a year. A loan for 9 months would be for $\frac{9}{12}$ of a year.

EXAMPLE 3 **Finding Interest for Less Than 1 Year**

Find the interest on $840 at $8\frac{1}{2}$% for 9 months.

The principal is $840. The rate is $8\frac{1}{2}$% or 0.085, and the time is $\frac{9}{12}$ of a year. Use the formula $I = p \cdot r \cdot t$.

$$I = \underbrace{(840)(0.085)}\left(\frac{9}{12}\right)$$

9 months $= \frac{9}{12}$ of a year

$$= \quad 71.4 \quad \left(\frac{3}{4}\right)$$

Write $\frac{9}{12}$ in lowest terms as $\frac{3}{4}$.

$$= \frac{(71.4)(3)}{4}$$

$$= \frac{214.2}{4} = 53.55$$

The interest is $53.55.

Calculator Tip The calculator solution to Example 3 above uses chain calculations.

840 ⊗ .085 ⊗ 9 ⊘ 12 ⊜ **53.55**

◄ *Work Problem* **3** *at the Side.*

OBJECTIVE **2** **Find the total amount due on a loan.** When a loan is repaid, the interest is added to the original principal to find the total amount due.

Formula for Amount Due

amount due = principal + interest

EXAMPLE 4 **Calculating the Total Amount Due**

A loan of $3240 was made at 12% for 3 months. Find the total amount due.

First find the interest. Then add the principal and the interest to find the total amount due.

$$I = (3240)(0.12)\left(\frac{3}{12}\right) \quad \text{3 months} = \tfrac{3}{12} \text{ of a year.}$$

$$I = \$97.20$$

The interest is $97.20.

Remember that the *total amount due* is the amount of the loan plus the interest.

$$\text{amount due} = \text{principal} + \text{interest}$$
$$= \$3240 + \$97.20 = \$3337.20$$

The total amount due is $3337.20.

◄ *Work Problem* **4** *at the Side.*

6.7 ▶▶▶ Exercises

Find the interest. See Examples 1 and 2.

	Principal	Rate	Time in Years	Interest
1.	$100	6%	1	$6
2.	$200	3%	1	$6
3.	$700	5%	3	$105
4.	$900	2%	4	$72
5.	$240	4%	3	$28.80
6.	$190	3%	2	$11.40
7.	$2300	$8\frac{1}{2}\%$	$2\frac{1}{2}$	$488.75
8.	$4700	$5\frac{1}{2}\%$	$1\frac{1}{2}$	$387.75
9.	$10,800	$7\frac{1}{2}\%$	$2\frac{3}{4}$	$2227.50
10.	$12,400	$6\frac{1}{2}\%$	$3\frac{3}{4}$	$3022.50

Find the interest. Round to the nearest cent if necessary. See Example 3.

	Principal	Rate	Time in Months	Interest
11.	$400	5%	6	$10
12.	$600	2%	5	$5
13.	$820	6%	12	$49.20

	Principal	Rate	Time in Months	Interest
14.	$780	8%	24	$124.80
15.	$940	3%	18	$42.30
16.	$178	4%	12	$7.12
17.	$1225	$5\frac{1}{2}\%$	3	$16.84 (rounded)
18.	$2660	$7\frac{1}{2}\%$	3	$49.88 (rounded)
19.	$15,300	$7\frac{1}{4}\%$	7	$647.06 (rounded)
20.	$13,700	$3\frac{3}{4}\%$	11	$470.94 (rounded)

Find the total amount due on the following loans. Round to the nearest cent if necessary. See Example 4.

	Principal	Rate	Time	Total Amount Due
21.	$200	5%	1 year	$210
22.	$400	2%	6 months	$404
23.	$740	6%	9 months	$773.30
24.	$1180	3%	2 years	$1250.80

	Principal	Rate	Time	Total Amount Due
25.	$1800	9%	18 months	$2043
26.	$9000	6%	7 months	$9315
27.	$3250	10%	6 months	$3412.50
28.	$7600	5%	1 year	$7980
29.	$16,850	$7\frac{1}{2}$%	9 months	$17,797.81 (rounded)
30.	$19,450	$5\frac{1}{2}$%	6 months	$19,984.88 (rounded)

31. The amount of interest paid on savings accounts and charged on loans can vary from one institution to another. However, when the amount of interest is calculated, three factors are used in the calculation. Name these three factors and describe them in your own words.

The answer should include:

amount of principal—This is the amount of money borrowed or loaned.

interest rate—This is the percent used to calculate the interest.

time of loan—The length of time that money is loaned or borrowed is an important factor in determining interest.

32. Interest rates are usually given as a rate per year (annual rate). Explain what must be done when time is given in months. Write your own problem where time is given in months and then show how to solve it.

When time is given in months, the number of months are placed over 12. This becomes a fraction of a year. Here is an example.

$$6 \text{ months} = \frac{6}{12} \text{ year} = \frac{1}{2} \text{ year} = 0.5 \text{ year}$$

Solve each application problem. Round to the nearest cent if necessary.

33. Reann Chang deposits $825 at 5% for 1 year. How much interest will she earn?

$41.25

34. The Jidobu family invests $18,000 at 9% for 6 months. What amount of interest will the family earn?

$810

35. CITI Bank loans $150,000 to Estelle Class to expand the size of her jewelry store. If the loan is for 30 months at 7%, how much interest will the bank earn?

$26,250

36. Esther Albert, a professional dancer, deposits $68,000 of her earnings at 5% for 5 years. How much interest will she earn?

$17,000

37. A student borrows $1200 at 8% for 5 months to pay for books and tuition. Find the total amount due.

$1240

38. Sarah Brynski borrows $2750 from her dad for a used car. The loan will be paid back with 8% interest at the end of 9 months. Find the total amount due.

$2915

39. Nicholas Thomas deposits $14,800 in his school credit union account for 10 months. If the credit union pays $2\frac{1}{4}$% interest, find the amount of interest he will earn.

$277.50

40. Sid and Shirley Kordell, owners of the Nut House, borrow $54,000 to update their store. If the loan is for 42 months at $7\frac{1}{4}$%, find the amount of interest they will owe.

$13,702.50

41. An investment fund pays $7\frac{1}{4}$% interest. If Beverly Habecker deposits $8800 in her account for $\frac{1}{4}$ year, find the amount of interest she will earn.

$159.50

42. Pat Carper owes $1900 in taxes. She is charged a penalty of $12\frac{1}{4}$% annual interest and pays the taxes and penalty after 6 months. Find the total amount she must pay.

$2016.38 (rounded)

43. A gift shop owner invests his profits of $11,500 at $8\frac{3}{4}$% interest for $\frac{3}{4}$ year. Find the total amount in his account at the end of this time.

$12,254.69 (rounded)

44. A pawn shop owner lends $35,400 to another business for $\frac{1}{2}$ year at an interest rate of 14.9%. How much interest will be earned on the loan?

$2637.30

45. The owners of Clear Lake Marina bought six canoes to use as rentals at a cost of $550 per canoe. If they borrowed 60% of the total cost for 9 months at $12\frac{1}{2}$% interest, find the amount due.

$2165.63 (rounded)

46. The owners of Baily and Daughters Excavating purchased four earth movers at a cost of $485,000 each. If they borrowed 80% of the total purchase price for $2\frac{1}{2}$ years at $10\frac{1}{2}$% interest, find the total amount due.

$1,959,400

6.8 ▶▶▶ Compound Interest

The interest we studied in **Section 6.7** was *simple interest* (interest only on the original principal). A common type of interest used with savings accounts and most investments is **compound interest** or interest paid on past interest as well as on the principal.

OBJECTIVE 1 Understand compound interest.   Suppose that you make a single deposit of $1000 in a savings account that earns 5% per year. What will happen to your savings over 3 years? At the end of the first year, 1 year's interest on the original deposit is found. Use the simple interest formula.

> Interest = principal • rate • time

Year 1   ($1000)(0.05)(1) = **$50**

> Add the interest to the $1000 to find the amount in your account at the end of the first year. $1000 + **$50** = **$1050** — In year 1, interest is calculated on principal.

The interest for the second year is found on $1050; that is, the interest is **compounded.**

Year 2   (**$1050**)(0.05)(1) = **$52.50**

> Add this interest to the $1050 to find the amount in your account at the end of the second year. **$1050** + **$52.50** = **$1102.50**.
> The interest for the third year is found on $1102.50.

In year 2 and thereafter, interest is calculated on the principal and all past interest.

Year 3   (**$1102.50**)(0.05)(1) ≈ **$55.13**

> Add this interest to the **$1102.50**. So, **$1102.50** + **$55.13** = **$1157.63**

At the end of 3 years, you will have **$1157.63** in your savings account. The $1157.63 that you have in your account is called the **compound amount.**

If you had earned only *simple* interest for 3 years, your interest would be as follows.

$$I = (\$1000)(0.05)(3)$$
$$= \$150 \leftarrow \text{Simple interest}$$

At the end of 3 years, you would have $1000 + $150 = $1150 in your account. Compounding the interest increased your earnings by $7.63 because $1157.63 − $1150 = $7.63.

With *compound* interest, the interest earned during the second year is greater than that earned during the first year, and the interest earned during the third year is greater than that earned during the second year. This happens because the interest earned each year is *added* to the principal, and the new total is used to find the amount of interest in the next year.

> **Compound Interest**
> Interest paid on principal plus past interest is called **compound interest.**

OBJECTIVES

1. **Understand compound interest.**
2. **Understand compound amount.**
3. **Find the compound amount.**
4. **Use a compound interest table.**
5. **Find the compound amount and the amount of compound interest.**

---

1 Find the compound amount given the following deposits. Round to the nearest cent if necessary.

(a) $500 at 4% for 2 years

(b) $2000 at 7% for 3 years

EXAMPLE 1 **Finding the Compound Amount**

Nancy Wegener deposits $3400 in an account that pays 6% interest compounded annually for 4 years. Find the compound amount. Round to the nearest cent when necessary.

Year	Interest	Compound Amount
1	($3400)(0.06)(1) = $204 $3400 + $204 =	$3604
2	($3604)(0.06)(1) = $216.24 $3604 + $216.24 =	$3820.24
3	($3820.24)(0.06)(1) ≈ $229.21 $3820.24 + $229.21 =	$4049.45
4	($4049.45)(0.06)(1) ≈ $242.97 $4049.45 + $242.97 =	$4292.42

The compound amount is $4292.42.

◀ *Work Problem* **1** *at the Side.*

---

2 Find the compound amount by multiplying the original deposit by 100% plus the compound interest rate. Round to the nearest cent if necessary.

(a) $1800 at 2% for 3 years

$(1800)(1.02)(1.02)(1.02) =$

_____

(b) $900 at 3% for 2 years

OBJECTIVE **3** **Find the compound amount.** A more efficient way of finding the compound amount is to add the interest rate to 100% and then multiply by the original deposit. Notice that in Example 1 above, at the end of the first year, you will have $3400 (100% of the original deposit) plus 6% (of the original deposit) or 106% (because 100% + 6% = 106%).

EXAMPLE 2 **Finding the Compound Amount**

Find the compound amount in Example 1 using multiplication.

Year 1 Year 2 Year 3 Year 4

($3400)(1.06)(1.06)(1.06)(1.06) ≈ $4292.42

This method works well with a calculator.

Original deposit   100% + 6% = 106% = 1.06   Compound amount

Our answer, $4292.42, is the same as in Example 1 above.

◀ *Work Problem* **2** *at the Side.*

Note

By adding the compound interest rate to 100%, we can then multiply by the original deposit. This will give us the compound amount at the end of each compound interest period.

(c) $2500 at 5% for 4 years

Calculator Tip If you use a calculator for Example 2 above, you can use the y^x key (exponent key).

3400 × 1.06 y^x 4 = **4292.42 (rounded)**

The 4 following the y^x key represents the number of compound interest periods.

OBJECTIVE **4** **Use a compound interest table.** The calculation of compound interest can be quite tedious. For this reason, compound interest tables have been developed.

Suppose you deposit $1 in a savings account today that earns 4% compounded annually and you allow the deposit to remain for 3 years. The diagram below shows the compound amount at the end of each of the 3 years.

Compound Amount

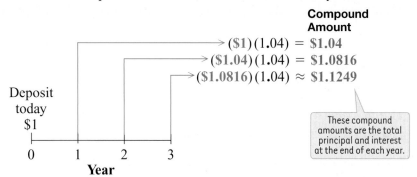

($1)(1.04) = **$1.04**
($1.04)(1.04) = **$1.0816**
($1.0816)(1.04) ≈ **$1.1249**

These compound amounts are the total principal and interest at the end of each year.

Deposit today $1

0   1   2   3
Year

Using the compound amounts for $1, a table can be formed. Look at the table below and find the column headed 4%. The first three numbers for years 1, 2, and 3 are the same as those we have calculated for $1 at 4% for 3 years. This table, giving the compound amounts on a $1 deposit for given lengths of time and interest rates, can be used for finding the compound amount on any amount of deposit.

COMPOUND INTEREST

Time Periods	3.00%	3.50%	4.00%	4.50%	5.00%	5.50%	6.00%	8.00%	Time Periods
1	1.0300	1.0350	1.0400	1.0450	1.0500	1.0550	1.0600	1.0800	1
2	1.0609	1.0712	1.0816	1.0920	1.1025	1.1130	1.1236	1.1664	2
3	1.0927	1.1087	1.1249	1.1412	1.1576	1.1742	1.1910	1.2597	3
4	1.1255	1.1475	1.1699	1.1925	1.2155	1.2388	1.2625	1.3605	4
5	1.1593	1.1877	1.2167	1.2462	1.2763	1.3070	1.3382	1.4693	5
6	1.1941	1.2293	1.2653	1.3023	1.3401	1.3788	1.4185	1.5869	6
7	1.2299	1.2723	1.3159	1.3609	1.4071	1.4547	1.5036	1.7138	7
8	1.2668	1.3168	1.3686	1.4221	1.4775	1.5347	1.5938	1.8509	8
9	1.3048	1.3629	1.4233	1.4861	1.5513	1.6191	1.6895	1.9990	9
10	1.3439	1.4106	1.4802	1.5530	1.6289	1.7081	1.7908	2.1589	10
11	1.3842	1.4600	1.5395	1.6229	1.7103	1.8021	1.8983	2.3316	11
12	1.4258	1.5111	1.6010	1.6959	1.7959	1.9012	2.0122	2.5182	12

EXAMPLE 3 **Using a Compound Interest Table**

Find each compound amount using the compound interest table. Round answers to the nearest cent.

(a) $1 is deposited at a 5% interest rate for 10 years.

Look down the column headed 5%, and across to row 10 (because 10 years = 10 time periods). At the intersection of the column and row, read the compound amount, **1.6289**, which can be rounded to $1.63.

(b) $1 is deposited at $3\frac{1}{2}$% for 6 years.

The intersection of the $3\frac{1}{2}$% (3.50%) column and row 6 shows **1.2293** as the compound amount. Round this to $1.23.

Work Problem **3** *at the Side.* ▶

---

3 Find the compound amount using the compound interest table. Round to the nearest cent.

(a) $1 at 3% for 6 years

(b) $1 at 6% for 8 years

(c) $1 at $4\frac{1}{2}$% for 12 years

ANSWERS

3. **(a)** $1.19 (rounded)   **(b)** $1.59 (rounded)
   (c) $1.70 (rounded)

4 Use the compound interest table to find the compound amount and the interest.

(a) $4000 at 3% for 10 years

(b) $12,600 at $3\frac{1}{2}$% for 8 years

(c) $32,700 at 6% for 12 years

OBJECTIVE 5 Find the compound amount and the amount of compound interest. Find the compound amount and interest as follows.

Finding the Compound Amount and the Interest

Compound Amount
Multiply the principal by the compound amount for $1 (from the table on the previous page).

Interest
Find the interest earned on a deposit by subtracting the original deposit from the compound amount.

EXAMPLE 4 **Finding Compound Amount and Interest**

Use the compound interest table to find the compound amount and the interest.

(a) $1000 at $5\frac{1}{2}$% interest for 12 years.

Look in the table on the previous page for $5\frac{1}{2}$% (5.50%) and 12 periods to find the number **1.9012** but do *not* round it. Multiply this number and the principal of $1000.

> Never round the numbers found in the table.

$$(\$1000)(1.9012) = \$1901.20$$

The account will contain $1901.20 after 12 years.

Find the interest by subtracting the original deposit from the compound amount.

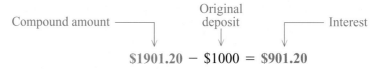

Compound amount — Original deposit — Interest

$$\$1901.20 - \$1000 = \$901.20$$

(b) $6400 at 8% for 7 years

Look in the table for 8% and 7 periods to find **1.7138**. Multiply.

$$(\$6400)(1.7138) = \$10,968.32 \quad \text{Compound amount}$$

Subtract the original deposit from the compound amount.

$$\$10,968.32 - \$6400 = \$4568.32 \quad \text{Interest}$$

> Remember:
> Compound amount −
> Principal = Interest.

A total of $4568.32 in interest was earned.

◀ *Work Problem* **4** *at the Side.*

6.8 ▸▸▸ **Exercises**

FOR EXTRA HELP

MyMathLab

Math XL
PRACTICE

WATCH

DOWNLOAD

READ

REVIEW

Find the compound amount given the following deposits. Calculate the interest each year, then add it to the previous year's amount. See Example 1.

1. $500 at 4% for 2 years **$540.80**

2. $1500 at 5% for 3 years **$1736.44 (rounded)**

3. $1800 at 3% for 3 years **$1966.91 (rounded)**

4. $2000 at 8% for 3 years **$2519.42 (rounded)**

5. $3500 at 7% for 4 years **$4587.79 (rounded)**

6. $5500 at 6% for 4 years **$6943.63 (rounded)**

Find each compound amount by multiplying the original deposit by 100% plus the compound rate given in the following. See Example 2. Round answers to the nearest cent if necessary.

7. $1000 at 5% for 2 years **$1102.50**

8. $500 at 4% for 3 years **$562.43 (rounded)**

9. $1400 at 6% for 5 years **$1873.52 (rounded)**

10. $2500 at 3% for 4 years **$2813.77 (rounded)**

11. $1180 at 7% for 8 years **$2027.46 (rounded)**

12. $12,800 at 6% for 7 years **$19,246.47 (rounded)**

13. $10,940 at 8% for 6 years **$17,360.41 (rounded)**

14. $15,710 at 10% for 8 years **$33,675.78 (rounded)**

Use the table on page 459 to find the compound amount and the interest. Interest is compounded annually. Round answers to the nearest cent if necessary. See Examples 3 and 4.

15. $1000 at 4% for 5 years

$1216.70; $216.70

16. $10,000 at 3% for 4 years

$11,255; $1255

17. $8000 at 6% for 10 years

$14,326.40; $6326.40

18. $7800 at 5% for 8 years

$11,524.50; $3724.50

19. $8428.17 at $4\frac{1}{2}$% for 6 years

$10,976.01 (rounded); $2547.84 (rounded)

20. $10,472.88 at $5\frac{1}{2}$% for 12 years

$19,911.04 (rounded); $9438.16 (rounded)

21. Write a definition for compound interest. Describe in your own words what compound interest means to you.

Interest paid on past interest as well as on the principal. Many people describe compound interest as "interest on interest."

22. What is the difference between the compound amount and compound interest?

Compound amount is the total amount, original deposit plus interest on deposit, at the end of the compound interest period. Compound interest is found by subtracting the original deposit from the compound amount.

🖩 *Use the table on page 459 to solve each application problem. Round answers to the nearest cent if necessary. See Examples 3 and 4.*

23. Jane Chavez deposited $8450 in an account that pays 6% interest compounded annually. Find the amount she will have (compound amount) at the end of 8 years.

$13,467.61

24. Yen Lee borrowed $32,800 from her family to open a small restaurant named Shanghai Winds. Her plan is to repay the loan at the end of 4 years at $3\frac{1}{2}$% interest compounded annually. Find the total amount that she must repay.

$37,638

25. Al Granard lends $76,000 to the owner of Rick's Limousine Service. He will be repaid at the end of 9 years at 6% interest compounded annually. Find **(a)** the total amount that he should be repaid and **(b)** the amount of interest earned.

(a) $128,402   **(b)** $52,402

26. Vincent and Shannon Zagorin have $48,500 in profit from the sale of their first home. They have invested all of it at 8% interest compounded annually for 6 years. Find **(a)** the total amount they will have at the end of 6 years and **(b)** the amount of interest earned.

(a) $76,964.65
(b) $28,464.65

27. Jennifer Barrister deposits $30,000 at 6% interest compounded annually. Two years after she makes the first deposit, she deposits another $40,000, also at 6% compounded annually.

(a) What total amount will she have five years after her first deposit?

$87,786.23

(b) What amount of interest will she have earned?

$17,786.23

28. Kara Ivee invests $25,000 at 8% interest compounded annually. Three years after she makes the first deposit, she deposits another $25,000, also at 8% compounded annually.

(a) What total amount will she have five years after her first deposit?

$65,892.85

(b) What amount of interest will she have earned?

$15,892.85

Relating Concepts (Exercises 29–34) For Individual or Group Work

Knowing how to solve interest problems is important to businesspeople and consumers alike. **Work Exercises 29–34 in order.**

29. Simple interest calculation is used for most short-term business loans, most real estate loans, and many automobile and consumer loans. The formula for simple interest is

Interest = ___principal___ · ___rate___ · ___time___

or I = ___p___ · ___r___ · ___t___.

30. When a loan is repaid, the interest is added to the original principal. The formula for amount due is

amount due = ___principal___ + ___interest___.

31. Compound interest is paid on most savings accounts and many other types of investments. Compound interest is interest calculated on ___principal___ plus past ___interest___.

32. The compound amount is the total amount in an account at the end of a period of time. Compound amount is the original ___principal___ + compound ___interest___.

Use the table on page 459 to answer Exercises 33 and 34.

33. Donna Gonsalves has two choices. She can invest $4350 at 6% simple interest for 6 years, or she can invest the same amount at 6% interest compounded annually for 6 years.

  (a) Find the difference in the amount of interest earned in these two accounts.

    $254.48

  (b) If the length of time is doubled from 6 to 12 years, will the difference in the interest earned also double?

    No; it has more than doubled (about five times).

  (c) Use your own example to determine that what you found in part (b) is true with a different interest rate and length of time.

    Examples will vary.

34. One account is opened with $20,000 at 8% simple interest for 12 years. Another account is opened with $16,000 at 8% interest compounded annually for 12 years.

  (a) At the end of the 12 years, which account has the higher balance and by how much?

    the compound interest account; $1091.20 more

  (b) What does this tell you about compound interest? Why?

    Greater amounts of interest are earned with compound interest than with simple interest because interest is earned on both principal and past interest.

Chapter 6 ▶▶▶ Summary

▶ Key Terms

6.1	**percent**	Percent means "per one hundred." A percent is a ratio with a denominator of 100.
6.3	**percent proportion**	The proportion $\dfrac{part}{whole} = \dfrac{percent}{100}$ is used to solve percent problems.
	whole	The whole in a percent problem is the entire quantity, the total, or the base.
	part	The part in a percent problem is the portion being compared with the whole.
6.5	**percent equation**	The percent equation is: part = percent • whole. It is another way to solve percent problems.
6.6	**sales tax**	Sales tax is a percent of the total sales charged as a tax.
	commission	Commission is a percent of the dollar value of total sales paid to a salesperson.
	discount	Discount is often expressed as a percent of the original price; it is then deducted from the original price, resulting in the sale price.
	percent of increase or decrease	Percent of increase or decrease is the amount of change (increase or decrease) expressed as a percent of the original amount.
6.7	**interest**	Interest is a fee paid or a charge made for lending or borrowing money
	interest formula	The interest formula is used to calculate interest. It is: Interest = principal • rate • time or $I = p \cdot r \cdot t$.
	simple interest	Interest that is computed only on the original principal is simple interest.
	principal	Principal is the amount of money on which interest is earned.
	rate of interest	Often referred to as "rate," it is the charge for interest and is given as a percent.
6.8	**compound interest**	Compound interest is interest paid on both the past interest and on the principal.
	compound amount	Compound amount is the total amount in an account including compound interest and the original principal.
	compounding	Interest that is compounded once each year is compounded annually.

▶ New Symbols

%   percent (per one hundred)

▶ New Formulas

To write percents as decimals: $p\% = p \div 100$ **To write percents as fractions:** $p\% = \dfrac{p}{100}$

Percent proportion: $\dfrac{part}{whole} = \dfrac{percent}{100}$ **Percent equation:**   part = percent • whole

Amount of sales tax: amount of sales tax = rate of tax • cost of item

Amount of commission: amount of commission = rate of commission • amount of sales

Amount of discount: amount of discount = rate of discount • original price

Sale price: sale price = original price − amount of discount

Percent of increase: $\dfrac{\text{amount of increase}}{\text{original value}} = \dfrac{percent}{100}$

Percent of decrease: $\dfrac{\text{amount of decrease}}{\text{original value}} = \dfrac{percent}{100}$

Interest: Interest = principal • rate • time   or $I = p \cdot r \cdot t$

Amount due: amount due = principal + interest

▶ Test Your Word Power

See how well you have learned the vocabulary in this chapter. Answers follow the Quick Review.

1. To write a **percent as a decimal,** you drop the percent sign
   A. after finding the decimal point
   B. and drop the decimal point
   C. and move the decimal point two places to the right
   D. after moving the decimal point two places to the left.

2. To write a **decimal as a percent,** you add the percent sign
   A. after finding the decimal point
   B. after removing the decimal point
   C. after moving the decimal point two places to the right
   D. after moving the decimal point two places to the left.

3. **Percent** means
   A. the same as interest
   B. per every ten
   C. per one thousand
   D. per one hundred.

4. When you use
   $$\frac{part}{whole} = \frac{percent}{100}$$ to solve
   percent problems, you are using the
   A. simple interest formula
   B. percent proportion
   C. percent equation
   D. percent sales tax formula.

5. The **percent equation** is
   A. part = percent • whole
   B. $I = p \cdot r \cdot t$
   C. $p\% = \dfrac{p}{100}$
   D. amount due = principal + interest.

6. In a **percent of increase** problem, the increase is a percent of
   A. the largest amount
   B. the original amount
   C. the new or most recent amount
   D. all the amounts.

7. In the formula $I = p \cdot r \cdot t,$ the p stands for
   A. proportion
   B. product
   C. principal
   D. percent.

8. The term **rate** in an interest problem represents the
   A. whole
   B. percent
   C. part
   D. amount of interest.

▶ Quick Review

Concepts	Examples

6.1 Basics of Percent

Writing a Percent as a Decimal
To write a percent as a decimal, drop the percent sign and move the decimal point two places to the left.

$$50\%°° (.50\%) = 0.50 \text{ or just } 0.5$$
$$3\%°° (.03\%) = 0.03$$
$$12.5\%°° (12.5\%) = 0.125$$

Writing a Decimal as a Percent
To write a decimal as a percent, move the decimal point two places to the right and attach a % sign.

$$0.75°° (0.75) = 75\%$$
$$0.875°° (0.875) = 87.5\%$$
$$3.6°° (3.60) = 360\%$$

6.2 Writing a Fraction as a Percent

Use a proportion and solve for p to change a fraction to percent.

$$\frac{2}{5} = \frac{p}{100} \qquad \text{Proportion}$$
$$5 \cdot p = 2 \cdot 100 \qquad \text{Cross products}$$
$$5 \cdot p = 200$$
$$\frac{\overset{1}{\cancel{5}} \cdot p}{\underset{1}{\cancel{5}}} = \frac{200}{5} \qquad \text{Divide both sides by 5.}$$
$$p = 40$$
$$\frac{2}{5} = 40\% \qquad \text{Attach \% sign.}$$

Concepts	Examples

6.3 Learning the Percent Proportion

Part is to whole as percent is to 100.

$$\frac{\text{part}}{\text{whole}} = \frac{\text{percent}}{100} \leftarrow \text{Always } 100$$

Use the percent proportion to solve for the unknown value. part = 30, whole = 50; find the percent.

Part → $\frac{30}{50} = \frac{x}{100}$ ← Always 100, Percent (unknown)

$$\frac{3}{5} = \frac{x}{100} \quad \text{Write } \tfrac{30}{50} \text{ as } \tfrac{3}{5} \text{ in lowest terms.}$$

$$5 \cdot x = 3 \cdot 100 \quad \text{Cross products}$$
$$5 \cdot x = 300$$

$$\frac{\overset{1}{\cancel{5}} \cdot x}{\cancel{5}} = \frac{300}{5} \quad \text{Divide both sides by 5.}$$

$$x = 60$$

The percent is 60, which is written as 60%.

6.3 Identifying Percent, Whole, and Part in a Percent Problem

The percent appears with the word **percent** or with the symbol %.

The whole often appears after the word **of**. The whole is the entire quantity or total.

The part is the portion of the total. If the percent and the whole are found first, the remaining number is the part.

Find the percent, whole, and part in the following.

10% of the 500 pies is how many pies?
Percent Whole Part (unknown)

20 cats is 5% of what number of cats?
Part Percent Whole (unknown)

What percent of $220 is $33?
Percent (unknown) Whole Part

6.4 Applying the Percent Proportion

Read the problem and identify the percent, whole, and part. Use the percent proportion to solve for the unknown quantity.

A liquid mixture in a tank contains 35% distilled water. If 28 gallons of distilled water are in the tank when it is full, find the capacity of the tank.

percent = 35 and part = 28

Use the percent proportion to find the whole.

$$\frac{\text{part}}{\text{whole} \; x} = \frac{\text{percent}}{100}$$

$$\frac{28}{x} = \frac{35}{100}$$

$$\frac{28}{x} = \frac{7}{20} \quad \text{Write } \tfrac{35}{100} \text{ as } \tfrac{7}{20} \text{ in lowest terms.}$$

$$x \cdot 7 = 560 \quad \text{Cross products}$$

$$\frac{x \cdot \overset{1}{\cancel{7}}}{\cancel{7}} = \frac{560}{7} \quad \text{Divide both sides by 7.}$$

$$x = 80$$

The capacity of the tank is 80 gallons.

Concepts	Examples

6.5 Using the Percent Equation

The percent equation is part = percent • whole. Identify the percent, whole, and part and solve for the unknown quantity. Always write the percent as a decimal before using the equation.

Solve each problem.

(a) Find 20% of 220 applicants.

$$\textbf{part (unknown)} = \text{percent} \cdot \text{whole}$$
$$x = (0.2)(220)$$
$$x = 44$$

20% of 220 applicants is 44 applicants.

(b) 8 balls is 4% of what number of balls?

$$\text{part} = \text{percent} \cdot \textbf{whole (unknown)}$$
$$8 = (0.04)(x)$$

$$\frac{8}{0.04} = \frac{\overset{1}{\cancel{(0.04)}}(x)}{\underset{1}{\cancel{0.04}}}$$

$$x = 200$$

8 balls is 4% of 200 balls.

(c) $13 is what percent of $52?

$$\text{part} = \textbf{percent (unknown)} \cdot \text{whole}$$
$$13 = x \cdot 52$$

$$\frac{13}{52} = \frac{x \cdot \overset{1}{\cancel{52}}}{\underset{1}{\cancel{52}}}$$

$$x = 0.25 = 25\%$$

$13 is 25% of $52.

6.6 Solving Application Problems with Proportions

To solve for **sales tax,** use this formula.

> **amount of sales tax = rate of tax • cost of item**

The price of a 37-inch plasma HD television is $699, and the sales tax is 5%. Find the sales tax.

$$\text{amount of sales tax} = (5\%)(\$699)$$
$$= (0.05)(\$699) = \$34.95$$

To find **commissions,** use this formula.

> **amount of commission =**
> **rate of commission • amount of sales**

The sales are $92,000 with a commission rate of 3%. Find the commission.

$$\text{amount of commission} = (3\%)(\$92,000)$$
$$= (0.03)(\$92,000)$$
$$= \$2760$$

To find the **discount** and the **sale price,** use these formulas.

> **amount of discount = rate of discount • original price**
>
> **sale price = original price − amount of discount**

A gas oven originally priced at $480 is offered at a 25% discount. Find the amount of the discount and the sale price.

$$\text{discount} = (0.25)(\$480) = \$120$$
$$\text{sale price} = \$480 - \$120 = \$360$$

Concepts	Examples

6.6 Solving Application Problems with Proportions (continued)

To find the **percent of change,** subtract to find the amount of change (increase or decrease), which is the part. The whole is the original value or value before the change.

The number of parking violations rose from 1980 violations to 2277. Find the percent of increase.

$$2277 - 1980 = 297 \quad \text{Increase}$$

$$\text{Original value} \rightarrow \frac{297}{1980} = \frac{\text{percent}}{100}$$

Solve the proportion to find that the percent = 15, so the percent of increase is 15%.

6.7 Finding Simple Interest

Use the formula $\quad I = p \cdot r \cdot t$

$$\text{Interest} = \text{principal} \cdot \text{rate} \cdot \text{time}$$

Time (t) is in years. When the time is given in months, use a fraction with 12 in the denominator because there are 12 months in a year.

$2800 is deposited at 8% for 3 months. Find the amount of interest.

$$I = p \cdot r \cdot t$$

$$= \underbrace{(2800)(0.08)}\left(\frac{3}{12}\right)$$

$$= \quad (224) \quad \left(\frac{1}{4}\right) = \frac{(224)(1)}{4} = \$56$$

6.8 Finding Compound Amount and Compound Interest

There are three methods for finding the compound amount.

1. Calculate the interest for each compound interest period, then add it back to the principal.

Find the compound amount and interest if $1500 is deposited at 5% interest for 3 years.

1.

	Interest	Compound Amount
Year 1	($1500)(0.05)(1) = $75	
	$1500 + $75 = **$1575**	
Year 2	($1575)(0.05)(1) = $78.75	
	$1575 + $78.75 = **$1653.75**	
Year 3	($1653.75)(0.05) ≈ $82.69	
	$1653.75 + $82.69 = **$1736.44**	

2. Multiply the original deposit by 100% plus the compound interest rate.

2. ($1500)(1.05)(1.05)(1.05) ≈ **$1736.44**

Original deposit   Compound amount

$$100\% + 5\% = 105\% = 1.05$$

3. Use the table on page 459 to find the interest on $1. Then, multiply the table value by the principal.

The compound interest is found with this formula.

$$\begin{matrix} \textbf{compound} \\ \textbf{interest} \end{matrix} = \begin{matrix} \textbf{compound} \\ \textbf{amount} \end{matrix} - \begin{matrix} \textbf{original} \\ \textbf{deposit} \end{matrix}$$

3. Locate 5% across the top of the table and 3 periods at the left. The table value is **1.1576**

$$\text{compound amount} = (\$1500)(1.1576) = \textbf{\$1736.40}^*$$

$$\text{interest} = \$1736.40 - \$1500 = \textbf{\$236.40}$$

---

* The difference in the compound amount results from rounding in the table.

ANSWERS TO TEST YOUR WORD POWER

1. D; *Example:* 50% written as a decimal is 0 50. or 0.5.

2. C; *Example:* 0.25 written as a percent is 0.25 or 25%.

3. D; *Example:* 8% means 8 per 100.

4. B; *Example:* Part = 4, and whole = 25. To find the percent, $\dfrac{4}{25} = \dfrac{x}{100}$; $x = 16$ or 16%.

5. A; *Example:* Percent = 25, and whole = 300. To find the part, $(0.25)(300) = 75$.

6. B; *Example:* Original value = $200, and amount of increase = $40. To find the percent of increase,

   $\dfrac{40}{200} = \dfrac{x}{100}$; $x = 0.20$ or 20% increase.

7. C; *Example:* Principal (p) = $800, rate ($r$) = 5%, and time ($t$) = 1 year. Then, $I = p \cdot r \cdot t$ so $I = (\$800)(0.05)(1) = \$40$.

8. B; *Example:* Principal (p) = $1650, rate ($r$) = 4%, and time ($t$) = $\dfrac{1}{2}$ year. To find the interest (I),

   $$I = (\$1650)(0.04)\left(\dfrac{1}{2}\right) = \$33.$$

Chapter 6 ▷▷▷ Review Exercises

[6.1] *Write each percent as a decimal and each decimal as a percent.*

1. 35%   **0.35**

2. 150%   **1.5**

3. 99.44%   **0.9944**

4. 0.085%   **0.00085**

5. 3.15   **315%**

6. 0.02   **2%**

7. 0.875   **87.5%**

8. 0.002   **0.2%**

[6.2] *Write each percent as a fraction or mixed number in lowest terms and each fraction as a percent.*

9. 15% $\dfrac{3}{20}$

10. 37.5% $\dfrac{3}{8}$

11. 175% $1\dfrac{3}{4}$

12. 0.25% $\dfrac{1}{400}$

13. $\dfrac{3}{4}$ **75%**

14. $\dfrac{5}{8}$ **62.5% or $62\dfrac{1}{2}$%**

15. $3\dfrac{1}{4}$ **325%**

16. $\dfrac{1}{200}$ **0.5%**

Complete this chart.

Fraction	Decimal	Percent
$\dfrac{1}{8}$	**17.** **0.125**	**18.** **12.5%**
19. $\dfrac{1}{4}$	0.25	**20.** **25%**
21. $1\dfrac{4}{5}$	**22.** **1.8**	180%

[6.3] *Find the unknown value in the percent proportion* $\dfrac{part}{whole} = \dfrac{percent}{100}$.

23. part = 25, percent = 10   **whole = 250**

24. whole = 480, percent = 5   **part = 24**

Identify each component and then set up each problem using the percent proportion, $\dfrac{part}{whole} = \dfrac{percent}{100}$.
*Do **not** try to solve for the unknown value.*

25. 35% of 820 mailboxes is 287 mailboxes.

$$\text{Part} \to \dfrac{287}{820} = \dfrac{35}{100} \leftarrow \text{Percent}$$
$$\text{Whole} \to \qquad\qquad \leftarrow \text{Always 100}$$

26. 73 DVDs is what percent of 90 DVDs

$$\dfrac{73}{90} = \dfrac{\text{unknown}}{100}$$

27. Find 14% of 160 Magellan Road Mates.

$$\dfrac{\text{unknown}}{160} = \dfrac{14}{100}$$

28. 418 curtains is 16% of what number of curtains?

$$\dfrac{418}{\text{unknown}} = \dfrac{16}{100}$$

29. A golfer lost three of his eight golf balls. What percent were lost?

$$\dfrac{3}{8} = \dfrac{\text{unknown}}{100}$$

30. Only 88% of the door keys cut will operate properly. If there are 1280 keys cut, find the number of keys that will operate properly.

$$\dfrac{\text{unknown}}{1280} = \dfrac{88}{100}$$

[6.4] *Find the part using the percent proportion or the multiplication shortcut.*

31. 18% of 950 programs

171 programs

32. 60% of 1450 reference books

870 reference books

33. 0.6% of 5200 acres

31.2 acres

34. 0.2% of 1400 kilograms

2.8 kilograms

Find the whole using the percent proportion.

35. 105 crates is 14% of what number of crates?

750 crates

36. 348 test tubes is 15% of what number of test tubes?

2320 test tubes

37. 677.6 miles is 140% of what number of miles?

484 miles

38. 2.5% of what number of cases is 425 cases?

17,000 cases

Find the percent using the percent proportion. Round percent answers to the nearest tenth if necessary.

39. 649 tulip bulbs is what percent of 1180 tulip bulbs?

55%

40. What percent of 1620 dinner rolls is 85 dinner rolls?

5.2% (rounded)

41. What percent of 380 pairs of socks is 36 pairs?

9.5% (rounded)

42. What percent of 650 soup cans is 200 soup cans?

30.8% (rounded)

[6.1–6.4] *Solve each application problem. Round percent answers to the nearest tenth if necessary.*

43. The average cost of 30 seconds of advertising during the Super Bowl three years ago was $2.3 million. If the increase in cost over the last three years has been 17.4%, find the average cost of 30 seconds of advertising during the Super Bowl this year. Round the cost to the nearest tenth of a million. (*Source:* NFL Research; CBS Advertising Sales; www.docsports.com)

$2.7 million (rounded)

44. Scientists tell us that there are 9600 species of birds and that 1000 of these species are in danger of extinction. What percent of the bird species are in danger of extinction?

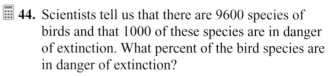

10.4% (rounded)

[6.5] *Use the percent equation to answer each question.*

45. 32% of $454 is what amount?

$145.28

46. 155% of 120 trucks is how many trucks?

186 trucks

47. 0.128 ounce is what percent of 32 ounces?

0.4%

48. 304.5 meters is what percent of 174 meters?

175%

49. 33.6 miles is 28% of what number of miles?

120 miles

50. $92 is 16% of what number?

$575

[6.6] *Find the amount of sales tax or the tax rate and the total cost. Round to the nearest cent if necessary.*

	Amount of Sale	Tax Rate	Amount of Tax	Total Cost
51.	$630	5%	$31.50	$661.50
52.	$780	$7\frac{1}{2}\%$	$58.50	$838.50

Find the commission earned or the rate of commission.

	Sales	Rate of Commission	Commission
53.	$3450	8%	$276
54.	$65,300	$5%	$3265

Find the amount or rate of discount and the sale price. Round to the nearest cent if necessary.

	Original Price	Rate of Discount	Amount of Discount	Sale Price
55.	$112.50	30%	$33.75	$78.75
56.	$252	25%	$63	$189

[6.7] *Find the simple interest due on each loan.*

Principal	Rate	Time in Years	Interest
57. $200	4%	1	$8
58. $1080	5%	$1\frac{1}{4}$	$67.50

Find the simple interest paid on each investment.

Principal	Rate	Time in Months	Interest
59. $400	7%	3	$7
60. $1560	$6\frac{1}{2}\%$	18	$152.10

Find the total amount due on each simple interest loan.

Principal	Rate	Time	Total Amount Due
61. $750	$5\frac{1}{2}\%$	2 years	$832.50
62. $1530	6%	9 months	$1598.85

[6.8] *Find the compound amount and compound interest. Interest is compounded annually. You may use the table on page 459. Round answers to the nearest cent if necessary.*

Principal	Rate	Time in Years	Compound Amount	Compound Interest
63. $4000	3%	10	$5375.60	$1375.60
64. $1870	4%	4	$2187.71 (rounded)	$317.71 (rounded)
65. $3600	8%	3	$4534.92	$934.92
66. $12,500	$5\frac{1}{2}\%$	5	$16,337.50	$3837.50

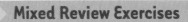 **Mixed Review Exercises**

Find the unknown value in the percent proportion $\dfrac{part}{whole} = \dfrac{percent}{100}$.

67. whole = 80, percent = 15   **part = 12**

68. part = 738, percent = 45   **whole = 1640**

Use the percent proportion or percent equation to answer each question.

69. 12% of 194 meters is how many meters?
23.28 meters

70. 327 cars is what percent of 218 cars?
150%

71. 0.6% of $85 is what amount?
$0.51

72. 99 employees is 5% of what number of employees?
1980 employees

73. 76 chickens is what percent of 190 chickens?
40%

74. 214.484 liters is 43% of what number of liters?
498.8 liters

Write each percent as a decimal and each decimal as a percent.

75. 55% **0.55**

76. 300% **3**

77. 5 **500%**

78. 4.71 **471%**

79. 8.6% **0.086**

80. 0.621 **62.1%**

81. 0.375% **0.00375**

82. 0.0006 **0.06%**

Write each percent as a fraction in lowest terms and each fraction as a percent.

83. $\dfrac{3}{4}$ **75%**

84. 42% $\dfrac{21}{50}$

85. 87.5% $\dfrac{7}{8}$

86. $\dfrac{3}{8}$ **37.5% or $37\frac{1}{2}$ %**

87. $32\frac{1}{2}$% $\dfrac{13}{40}$

88. $\dfrac{3}{5}$ **60%**

89. 0.25% $\dfrac{1}{400}$

90. $3\dfrac{3}{4}$ **375%**

Solve each application problem. Round percent answers to the nearest tenth and money answers to the nearest cent if necessary.

91. Richard Zanotti deposits $20,500 in his credit union savings accounts. If he earns $6\frac{1}{2}$% simple interest for 30 months, how much will be earned?
$3331.25

92. Jim Havey borrows $14,750 at 8% simple interest for 18 months to buy a small Lionel train collection. Find the total amount due.
$16,520

93. A Hotpoint refrigerator has a capacity of 11.5 cubic feet in the refrigerator and 5.5 cubic feet in the freezer. What percent of the total capacity is the capacity of the freezer?   **32.4% (rounded)**

11.5 cubic feet

5.5 cubic feet

94. Tommy Downs invests the $12,500 he inherited from his aunt at 6% interest compounded annually for 4 years.

 (a) Find the compound amount at the end of 4 years. Do not use the table.

 $15,780.96 (rounded)

 (b) Find the amount of interest that he earned.

 $3280.96 (rounded)

95. Tom Dugally, a real estate agent, sold two properties, one for $125,000 and the other for $290,000. He receives a commission of $1\frac{1}{2}$% of total sales. Find the commission that he earned.

$6225

96. Vending machines on campus must include healthy food choices such as fruits, fruit juices, and healthy snacks. Sales of healthy foods in the vending machines increased from 4320 items last month to 5107 items this month. Find the percent of increase.

18.2% (rounded)

97. A Sears Kenmore washer/dryer set priced at $958 is marked down 18%. If the sales tax is 8%, find the cost of the washer/dryer set including the sales tax.

$848.40 (rounded)

98. In a recent insurance company study of boaters who had lost items overboard, 88 boaters or 8% said that they lost their cell phones. Find the total number of boaters in the survey. (*Source:* Progressive Groups of Insurance Companies.)

1100 boaters

99. Jack and Jill Ahearn begin to budget 25% for rent, 20% for food, 7% for education, 5% for clothing, 10% for transportation, 12% for travel and recreation, 4% for miscellaneous, and the remainder for savings. Jack takes home $2850 per month, and Jill takes home $42,300 per year. How much money will the couple save in a year?

$13,005

100. The mileage on a hybrid car dropped from 42.8 miles per gallon in the city to 28.5 miles per gallon on the highway. Find the percent of decrease.

33.4% (rounded)

Write each percent as a decimal and each decimal as a percent.

1. 65%

2. 0.8

3. 1.75

4. 0.875

5. 300%

6. 2%

Write each percent as a fraction in lowest terms.

7. 12.5%

8. 0.25%

Write each fraction or mixed number as a percent.

9. $\dfrac{3}{5}$

10. $\dfrac{5}{8}$

11. $2\dfrac{1}{2}$

Solve each problem.

12. 32 sacks is 4% of what number of sacks?

13. $680 is what percent of $3400?

14. There are still 100,000 households in the United States that do not have electricity. If this is 0.08% of the homes, find the total number of households. (*Source: Time* magazine.)

15. The price of a diamond engagement ring is $3240 plus sales tax of $6\frac{1}{2}$%. Find the total cost of the engagement ring including sales tax.

16. An insurance company pays its salespeople on commission. If a commission of $628 is earned on insurance sales of $7850, find the rate of commission.

1. $\underline{0.65}$

2. $\underline{80\%}$

3. $\underline{175\%}$

4. $\underline{87.5\% \text{ or } 87\frac{1}{2}\%}$

5. $\underline{3.00 \text{ or } 3}$

6. $\underline{0.02}$

7. $\underline{\dfrac{1}{8}}$

8. $\underline{\dfrac{1}{400}}$

9. $\underline{60\%}$

10. $\underline{62.5\% \text{ or } 62\frac{1}{2}\%}$

11. $\underline{250\%}$

12. $\underline{800 \text{ sacks}}$

13. $\underline{20\%}$

14. $\underline{125,000,000 \text{ households}}$

15. $\underline{\$3450.60}$

16. $\underline{8\%}$

17. $\underline{25\%}$

17. Attendance at the homecoming game decreased from 5760 fans last year to 4320 fans this year. Find the percent of decrease.

18. $\dfrac{\text{amount of increase}}{\text{last year's salary}} = \dfrac{p}{100}$

18. A problem includes last year's salary, this year's salary, and asks for the percent of increase. Explain how you would identify the part, the whole, and the percent in the problem. Show the percent proportion that you would use.

A possible answer is: Part is the increase in salary. Whole is last year's salary. Percent of increase is unknown.

19. some possibilities are

$I = (1000)(0.05)\left(\dfrac{9}{12}\right) = \$37.50$

$I = (1000)(0.05)(2.5) = \$125$

19. Write the formula used to find interest. Explain the difference in what to do if the time is expressed in months or in years. Write a problem that involves finding interest for 9 months and another problem that involves finding interest for $2\frac{1}{2}$ years. Use your own numbers for the principal and the rate. Show how to solve your problems.

The interest formula is $I = p \cdot r \cdot t$. If time is in months, it is expressed as a fraction with 12 as the denominator. If time is expressed in years, it is placed over 1 or shown as a decimal number.

Find the amount of discount and the sale price.

	Original Price	Rate of Discount
20.	$96	12%
21.	$280	32.5%

20. $\underline{\$11.52; \$84.48}$

21. $\underline{\$91; \$189}$

Find the simple interest on each loan.

	Principal	Rate	Time
22.	$4200	6%	$1\frac{1}{2}$ years
23.	$6400	9%	4 months

22. $\underline{\$378}$

23. $\underline{\$192}$

24. $\underline{\$5657.75}$

24. A parent borrows $5300 to help her son start college. The loan is for 9 months at 9% interest. Find the total amount due on the loan.

25. (a) $\underline{\$10,668 \text{ (rounded)}}$

(b) $\underline{\$1668}$

25. The River City School PTA Emergency Fund deposited $4000 at 6% interest compounded annually. Two years after the first deposit, they deposit another $5000, also at 6% interest compounded annually. Use the compound interest table on page 459.

(a) What total amount will they have 4 years after their first deposit? Round to the nearest dollar.

(b) What amount of interest will they have earned?

Study Skills

> > > **PREPARING FOR YOUR FINAL EXAM**

Your math final exam is likely to be a **comprehensive exam.** This means that it will cover material from the **entire term.** The end of the term will be less stressful if you **make a plan** for how you will prepare for each of your exams.

First, figure out the **score you need to earn on the final exam** to get the course grade you want. Check your course syllabus for grading policies, or ask your instructor if you are not sure of them. This allows you to set a goal for yourself.

> How many points do you need to earn on your mathematics final exam to get the grade you want? _____
> _____

Second, create a **final exam week plan for your work and personal life.** If you need to make an adjustment in your work schedule, do it in advance, so you aren't scrambling at the last minute. If you have family members to care for, you might want to enlist some help from others so you can spend extra time studying. Try to plan in advance so you don't create additional stress for yourself. You will have to set some priorities, and studying has to be at the top of the list! Although life doesn't stop for finals, some things can be ignored for a short time. You don't want to "burn out" during final exam week; **get enough sleep and healthy food so you can perform your best.**

> What adjustments in your personal life do you need to make for final exam week? _____
> _____

Third, use the following suggestions to guide your studying and reviewing.

▶ **Know exactly which chapters and sections will be on the final exam.**

▶ **Divide up the chapters,** and decide how much you will review each day.

▶ Begin your reviewing **several days** before the exam.

▶ **Use returned quizzes and tests** to review earlier material (if you have them).

▶ **Practice all types of problems,** but emphasize the types that are most difficult for you. Use the **Cumulative Reviews** that are at the end of each chapter in your textbook.

▶ **Rewrite your notes or make mind maps** to create summaries.

▶ **Make study cards for all types of problems.** Be sure to use the same **direction words** (such as *simplify, solve, estimate*) that your exam will use. Carry the cards with you and review them whenever you have a few spare minutes.

OBJECTIVES

1 **Create a final exam week plan.**

2 **Break studying into chunks and study over several days.**

3 **Practice all types of problems.**

Create a Plan

Study and Review

Managing Stress

Of course, a week of final exams produces stress. **Students who develop skills for reducing and managing stress do better on their final exams and are less likely to "bomb" an exam.** You already know the damaging effect of adrenaline on your ability to think clearly. But several days (or weeks) of elevated stress is also harmful to your brain and your body. You will feel better if you make a conscious effort to reduce your stress level. Even if it takes you away from studying for a little while each day, the time will be well spent.

Reducing Physical Stress

Which techniques will you try?

_____

_____

_____

Examples of ways to reduce **physical stress** are listed below. Can you add any of your own ideas to the list?

▶ *Laugh until your eyes water.* Watching your favorite funny movie, exchanging a joke with a friend, or viewing a comedy bit on the Internet are all ways to generate a healthy laugh. Laughing raises the level of *calming* chemicals (endorphins) in your brain.

▶ *Exercise for 20 to 30 minutes.* If you normally exercise regularly, do NOT stop during final exam week! Exercising helps relax muscles, diffuses adrenaline, and raises the level of endorphins in your body. If you don't exercise much, get some gentle exercise, such as a daily walk, to help you relax.

▶ *Practice deep breathing.* Several minutes of deep, smooth breathing will calm you. Close your eyes too.

▶ *Visualize a relaxing scene.* Choose something that you find peaceful and picture it. Imagine what it feels like and sounds like. Try to put yourself in the picture.

▶ If you feel stress in your muscles, such as your shoulders or back, *slowly squeeze the muscles as much as you can, and then release them.* Sometimes we don't realize we are clenching our teeth or holding tension in our shoulders until we consciously work with them. Try to notice what it feels like when they are relaxed and loose. Squeezing and then releasing muscles is also something you can do during an exam if you feel yourself tightening up.

Reducing Mental Stress

Which techniques will you try?

_____

_____

_____

Mental stress reduction is also a powerful tool both before and during an exam. In addition to these suggestions, do you have any of your own techniques?

▶ *Talk positively to yourself.* Tell yourself you will get through it.

▶ *Reward yourself.* Give yourself small breaks, a little treat—something that makes you happy—every day of final exam week.

▶ *Make a list of things to do* and feel the sense of accomplishment when you cross each item off.

▶ When you take time to relax or exercise, *make sure you are relaxing your mind too.* Use your mind for something *completely* different from the kind of thinking you do when you study. Plan your garden, play your favorite music, walk your dog, read a good book.

▶ *Visualize.* Picture yourself completing exams and projects successfully. Picture yourself taking the test calmly and confidently.

Cumulative Review Exercises ▷▷▷ Chapters 1–6

First use front end rounding to round each number and estimate the answer. Then find the exact answer.

1. *Estimate:*   *Exact:*
 80,000   75,078
 − 50,000   − 46,090
 30,000   28,988

2. *Estimate:*   *Exact:*
 8   7.8
 − 4   − 3.5029
 4   4.2971

3. *Estimate:*   *Exact:*
 7000   6538
 × 700   × 708
 4,900,000   4,628,904

4. *Estimate:*   *Exact:*
 70   65.3
 × 9   × 8.7
 630   568.11

5. *Estimate:*   *Exact:*
 1000 902
 $40)\overline{40,000}$ $43)\overline{38,786}$

6. *Estimate:*   *Exact:*
 7 8.45
 $1)\overline{7}$ $0.8)\overline{6.76}$

Use the order of operations to simplify each expression.

7. $6^2 - 3(6)$ 18

8. $\sqrt{49} + 5 \cdot 4 - 8$ 19

9. $9 + 6 \div 3 + 7(4)$ 39

Round each number.

10. 7,583,281 to the nearest hundred-thousand   7,600,000

11. $513.499 to the nearest dollar $513

12. $362.735 to the nearest cent $362.74

Add, subtract, multiply, or divide as indicated. Write answers in lowest terms and as whole or mixed numbers when possible.

13. $5\frac{3}{4}$
 $+ 7\frac{5}{8}$
 $13\frac{3}{8}$

14. $8\frac{3}{8}$
 $- 4\frac{1}{2}$
 $3\frac{7}{8}$

15. $36 \cdot \frac{4}{5}$ $28\frac{4}{5}$

16. $12 \div \frac{3}{4}$ 16

17. The size of a prison cell at Alcatraz Prison in the San Francisco Bay is 5 feet by 9 feet. The average size of a shark cage is 5 feet by $6\frac{1}{2}$ feet. How many more square feet are there in the floor of the prison cell than the shark cage? (*Source:* Discovery Channel, Monster Garage Factoid.)

$12\frac{1}{2}$ ft²

18. To prepare for the state real estate exam, Mia Dawson studied $5\frac{1}{2}$ hours on the first day, $6\frac{1}{4}$ hours on the second day, $3\frac{3}{4}$ hours on the third day, and 7 hours on the fourth day. How many hours did she study altogether?

$22\frac{1}{2}$ hours

Write < or > to make each statement true.

19. $\frac{5}{8}$ < $\frac{2}{3}$

20. $\frac{8}{15}$ < $\frac{11}{20}$

21. $\frac{2}{3}$ > $\frac{7}{12}$

Simplify each expression. Use the order of operations as needed.

22. $\frac{2}{3}\left(\frac{7}{8} - \frac{1}{2}\right)$ $\frac{1}{4}$

23. $\frac{7}{8} \div \left(\frac{3}{4} + \frac{1}{8}\right)$ 1

24. $\left(\frac{5}{6} - \frac{5}{12}\right) - \left(\frac{1}{2}\right)^2 \cdot \frac{2}{3}$ $\frac{1}{4}$

Write each fraction as a decimal. Round to the nearest thousandth if necessary.

25. $\frac{3}{4}$ 0.75

26. $\frac{3}{8}$ 0.375

27. $\frac{7}{12}$ 0.583 (rounded)

28. $\frac{11}{20}$ 0.55

Find the unknown value in each proportion.

29. $\dfrac{1}{5} = \dfrac{x}{30}$ x = 6

30. $\dfrac{224}{32} = \dfrac{28}{x}$ x = 4

31. $\dfrac{8}{x} = \dfrac{72}{144}$ x = 16

32. $\dfrac{x}{120} = \dfrac{7.5}{30}$ x = 30

Write each percent as a decimal. Write each decimal as a percent.

33. 3%   0.03

34. 200%   2.00 or 2

35. 0.87   87%

36. 3.8   380%

Write each percent as a fraction or mixed number in lowest terms. Write each fraction as a percent.

37. 8% $\dfrac{2}{25}$

38. 62.5% $\dfrac{5}{8}$

39. 175% $1\dfrac{3}{4}$

40. $\dfrac{7}{8}$ 87.5% or $87\dfrac{1}{2}$%

41. $4\dfrac{1}{5}$ 420%

Solve.

42. 65% of 3280 DVDs is how many DVDs?   **2132 DVDs**

43. $4\dfrac{1}{2}$% of what number of miles is 76.5 miles?   **1700 miles**

44. 252 hours is what percent of 180 hours?   **140%**

45. Diane McKinney paid $29.90 in sales tax on a $460 purchase. What was the tax rate?   **6.5% or $6\dfrac{1}{2}$%**

46. Find the commission on $12,538 in sales when the commission rate is 8%.   **$1003.04**

47. A box spring and mattress originally priced at $456 is discounted 45%. Find the amount of discount and the sale price.   **$205.20; $250.80**

48. A loan of $22,250 is made at 7% simple interest for 6 months. Find the total amount to be repaid.

$23,028.75

Set up and solve a proportion for each problem.

49. Carol can test the hearing of 11 patients in 4 hours. Find the number of hearing tests that she can do in 12 hours.

33 hearing tests

50. If 12.5 ounces of Roundup weed and grass killer is needed to make 5 gallons of spray, how much Roundup is needed for 102 gallons of spray?

255 ounces

The number of existing single-family homes sold in four regions of the country in the same month of two separate years are shown in the figure below. Use this information to answer Exercises 51–54. Round answers to the nearest tenth of a percent if necessary.

AT HOME

Existing home sales by region.

West 54,000 49,600
Midwest 65,000 66,300
Northeast 32,000 36,000
South 82,000 77,500

= Last year
= This year

51. Find the percent of increase in sales in the northeastern region.

12.5%

52. Find the percent of increase in sales in the midwestern region.

2%

53. What is the percent of decrease in sales in the southern region?

5.5% (rounded)

54. What is the percent of decrease in sales in the western region?

8.1% (rounded)

The world's best-preserved crater made by a meteor crashing into Earth is located in northern Arizona. Notice the buildings at the upper left edge of the crater for size comparison.

7

Measurement

We are constantly measuring things, from very large to very small.

For example, a meteor crashed into Earth in northern Arizona nearly 50,000 years ago. The bowl-shaped crater from that crash is sixty stories deep and 4150 feet across. (See **Section 7.1,** Exercises 63–66.)

On the other hand, computer microchips often measure a tiny 5 mm long by 1 mm wide. (Learn about millimeters in **Section 7.2** and then try Exercise 39.)

A greatly enlarged photo of a computer microchip.

7.1 ▶▶▶ Problem Solving with U.S. Customary Measurements

OBJECTIVES

1 Learn the basic U.S. customary measurement units.

2 Convert among measurement units using multiplication or division.

3 Convert among measurement units using unit fractions.

4 Solve application problems using U.S. customary measurement units.

1 After memorizing the measurement conversions, answer these questions.

(a) 1 c = _____ fl oz

(b) _____ qt = 1 gal

(c) 1 wk = _____ days

(d) _____ ft = 1 yd

(e) 1 ft = _____ in.

(f) _____ oz = 1 lb

(g) 1 ton = _____ lb

(h) _____ min = 1 hr

(i) 1 pt = _____ c

(j) _____ hr = 1 day

(k) 1 min = _____ sec

(l) 1 qt = _____ pt

(m) _____ ft = 1 mi

ANSWERS

1. **(a)** 8 **(b)** 4 **(c)** 7 **(d)** 3 **(e)** 12
(f) 16 **(g)** 2000 **(h)** 60 **(i)** 2 **(j)** 24
(k) 60 **(l)** 2 **(m)** 5280

We measure things all the time: the distance traveled on vacation, the floor area we want to cover with carpet, the amount of milk in a recipe, the weight of the bananas we buy at the store, the number of hours we work, and many more.

In the United States, we still use **U.S. customary measurement units** for many everyday activities. Examples are inches, feet, quarts, ounces, and pounds. However, science, medicine, sports, and manufacturing use the **metric system** (meters, liters, and grams). And, because the rest of the world uses *only* the metric system, U.S. businesses have been changing to the metric system in order to compete internationally.

OBJECTIVE 1 Learn the basic U.S. customary measurement units. Until the switch to the metric system is complete, we still need to know how to use U.S. customary measurement units. The table below lists the relationships you should memorize. The time relationships are used in both the U.S. customary and metric systems.

U.S. Customary Measurement Relationships	
Length	**Weight**
12 inches (in.) = 1 foot (ft)	16 ounces (oz) = 1 pound (lb)
3 feet (ft) = 1 yard (yd)	2000 pounds (lb) = 1 ton
5280 feet (ft) = 1 mile (mi)	
Capacity	**Time**
8 fluid ounces (fl oz) = 1 cup (c)	60 seconds (sec) = 1 minute (min)
2 cups (c) = 1 pint (pt)	60 minutes (min) = 1 hour (hr)
2 pints (pt) = 1 quart (qt)	24 hours (hr) = 1 day
4 quarts (qt) = 1 gallon (gal)	7 days = 1 week (wk)

As you can see, there is no simple way to convert among these various measures. The units evolved over hundreds of years and were based on a variety of "standards." For example, one yard was the distance from the tip of a king's nose to his thumb when his arm was outstretched. An inch was three dried barleycorns laid end to end.

EXAMPLE 1 Knowing U.S. Customary Measurement Units

Memorize the U.S. customary measurement conversions shown above. Then answer these questions.

(a) 24 hr = _____ day Answer: 1 day

(b) 1 yd = _____ ft Answer: 3 ft

◀ *Work Problem* **1** *at the Side.*

OBJECTIVE 2 Convert among measurement units using multiplication or division. You often need to convert from one unit of measure to another. Two methods of converting measurements are shown here. Study each way and use the method you prefer. The first method involves deciding whether to multiply or divide.

> **Converting among Measurement Units**
> 1. *Multiply* when converting from a larger unit to a smaller unit.
> 2. *Divide* when converting from a smaller unit to a larger unit.

EXAMPLE 2 **Converting from One Unit of Measure to Another**

Convert each measurement.

(a) 7 ft to inches

You are converting from a *larger* unit to a *smaller* unit (a *foot* is longer than an *inch*), so multiply.

Because *1 ft = 12 in.*, multiply by 12.

$$7 \text{ ft} = 7 \cdot 12 = 84 \text{ in.}$$

(b) $3\frac{1}{2}$ lb to ounces

You are converting from a *larger* unit to a *smaller* unit (a *pound* is heavier than an *ounce*), so multiply.

Because *1 lb = 16 oz*, multiply by 16.

> Divide 16 and 2 by their common factor of 2.
> $16 \div 2$ is 8.
> $2 \div 2$ is 1.

$$3\frac{1}{2} \text{ lb} = 3\frac{1}{2} \cdot 16 = \frac{7}{2} \cdot \frac{\overset{8}{\cancel{16}}}{1} = \frac{56}{1} = 56 \text{ oz}$$

(c) 20 qt to gallons

You are converting from a *smaller* unit to a *larger* unit (a *quart* is smaller than a *gallon*) so divide.

Because *4 qt = 1 gal*, divide by 4.

$$20 \text{ qt} = \frac{20}{4} = 5 \text{ gal}$$

Divide by 4. ⟍

(d) 45 min to hours

You are converting from a *smaller* unit to a *larger* unit (a *minute* is less than an *hour*), so divide.

Because *60 min = 1 hr*, divide by 60 and write the fraction in lowest terms.

$$45 \text{ min} = \frac{45}{60} = \frac{45 \div 15}{60 \div 15} = \frac{3}{4} \text{ hr} \leftarrow \text{Lowest terms}$$

Divide by 60. ⟍

Work Problem **2** *at the Side.* ▶

OBJECTIVE **3** **Convert among measurement units using unit fractions.** If you have trouble deciding whether to multiply or divide when converting measurements, use *unit fractions* to solve the problem. You'll also use this method in science courses. A **unit fraction** is equivalent to 1. Here is an example.

$$\frac{12 \text{ in.}}{12 \text{ in.}} = \frac{\overset{1}{\cancel{12} \text{ in.}}}{\underset{1}{\cancel{12} \text{ in.}}} = 1$$

Use the table of measurement relationships on the previous page to find that 12 in. is the same as 1 ft. So you can substitute 1 ft for 12 in. in the numerator, or you can substitute 1 ft for 12 in. in the denominator. This makes two useful unit fractions.

$$\frac{1 \text{ ft}}{12 \text{ in.}} = 1 \quad \text{or} \quad \frac{12 \text{ in.}}{1 \text{ ft}} = 1$$

2 Convert each measurement using multiplication or division.

(a) $5\frac{1}{2}$ ft to inches

(b) 64 oz to pounds

(c) 6 yd to feet

(d) 2 tons to pounds

(e) 35 pt to quarts

(f) 20 min to hours

(g) 4 wk to days

ANSWERS

2. **(a)** 66 in. **(b)** 4 lb **(c)** 18 ft
(d) 4000 lb **(e)** $17\frac{1}{2}$ qt **(f)** $\frac{1}{3}$ hr
(g) 28 days

3 First write the unit fraction needed to make each conversion. Then complete the conversion.

(a) 36 in. to feet

$$\left.\begin{array}{c}\text{unit}\\\text{fraction}\end{array}\right\}\dfrac{1\text{ ft}}{12\text{ in.}}$$

(b) 14 ft to inches

$$\left.\begin{array}{c}\text{unit}\\\text{fraction}\end{array}\right\}\dfrac{\text{in.}}{\text{ft}}$$

(c) 60 in. to feet

$$\left.\begin{array}{c}\text{unit}\\\text{fraction}\end{array}\right\}\underline{\hspace{1cm}}$$

(d) 4 yd to feet

$$\left.\begin{array}{c}\text{unit}\\\text{fraction}\end{array}\right\}\underline{\hspace{1cm}}$$

(e) 39 ft to yards

$$\left.\begin{array}{c}\text{unit}\\\text{fraction}\end{array}\right\}\underline{\hspace{1cm}}$$

(f) 2 mi to feet

$$\left.\begin{array}{c}\text{unit}\\\text{fraction}\end{array}\right\}\underline{\hspace{1cm}}$$

To convert from one measurement unit to another, just multiply by the appropriate unit fraction. Remember, a unit fraction is equivalent to 1. Multiplying something by 1 does *not* change its value.

Use these guidelines to choose the correct unit fraction.

> **Choosing a Unit Fraction**
>
> The **numerator** should use the measurement unit you want in the *answer*.
> The **denominator** should use the measurement unit you want to *change*.

EXAMPLE 3 **Using Unit Fractions with Length Measurements**

(a) Convert 60 in. to feet.

Use a unit fraction with feet (the unit for your answer) in the numerator, and inches (the unit being changed) in the denominator. Because *1 ft = 12 in.,* the necessary unit fraction is

$$\dfrac{1\text{ ft}}{12\text{ in.}}\quad\begin{array}{l}\leftarrow\text{ Unit for your answer is feet.}\\\leftarrow\text{ Unit being changed is inches.}\end{array}$$

Next, multiply 60 in. times this unit fraction. Write 60 in. as the fraction $\dfrac{60\text{ in.}}{1}$ and divide out common units and factors wherever possible.

$$60\text{ in.}\cdot\dfrac{1\text{ ft}}{12\text{ in.}}=\dfrac{\overset{5}{\cancel{60}\text{ in.}}}{1}\cdot\dfrac{1\text{ ft}}{\underset{1}{\cancel{12}\text{ in.}}}=\dfrac{5\cdot1\text{ ft}}{1}=5\text{ ft}$$

These units should match.

Divide out inches.

Divide 60 and 12 by 12.

(b) Convert 9 ft to inches.

Select the correct unit fraction to change 9 ft to inches.

$$\dfrac{12\text{ in.}}{1\text{ ft}}\quad\begin{array}{l}\leftarrow\text{ Unit for your answer is inches.}\\\leftarrow\text{ Unit being changed is feet.}\end{array}$$

Multiply 9 ft times the unit fraction.

$$9\text{ ft}\cdot\dfrac{12\text{ in.}}{1\text{ ft}}=\dfrac{9\cancel{\text{ ft}}}{1}\cdot\dfrac{12\text{ in.}}{1\cancel{\text{ ft}}}=\dfrac{9\cdot12\text{ in.}}{1}=108\text{ in.}$$

These units should match.

Divide out feet.

> **CAUTION**
> If no units will divide out, you made a mistake in choosing the unit fraction.

◀ *Work Problem* **3** *at the Side.*

EXAMPLE 4 **Using Unit Fractions with Capacity and Weight Measurements**

(a) Convert 9 pt to quarts.

First select the correct unit fraction.

$$\dfrac{1\text{ qt}}{2\text{ pt}}\quad\begin{array}{l}\leftarrow\text{ Unit for your answer is quarts.}\\\leftarrow\text{ Unit being changed is pints.}\end{array}$$

Continued on Next Page

ANSWERS

3. **(a)** 3 ft **(b)** $\dfrac{12\text{ in.}}{1\text{ ft}}$; 168 in.

(c) $\dfrac{1\text{ ft}}{12\text{ in.}}$; 5 ft **(d)** $\dfrac{3\text{ ft}}{1\text{ yd}}$; 12 ft

(e) $\dfrac{1\text{ yd}}{3\text{ ft}}$; 13 yd **(f)** $\dfrac{5280\text{ ft}}{1\text{ mi}}$; 10,560 ft

Next multiply.

Write as a mixed number.

$$9 \text{ pt} \cdot \frac{1 \text{ qt}}{2 \text{ pt}} = \frac{9 \text{ pt}}{1} \cdot \frac{1 \text{ qt}}{2 \text{ pt}} = \frac{9}{2} \text{ qt} = 4\frac{1}{2} \text{ qt}$$

These units should match.

Divide out pints.

(b) Convert $7\frac{1}{2}$ gal to quarts.

Write as an improper fraction.

$$\frac{7\frac{1}{2} \text{ gal}}{1} \cdot \frac{4 \text{ qt}}{1 \text{ gal}} = \frac{15}{2} \cdot \frac{4}{1} \text{ qt}$$

Divide out gallons.

$$= \frac{15}{\underset{1}{\cancel{2}}} \cdot \frac{\overset{2}{\cancel{4}}}{1} \text{ qt}$$

Divide 4 and 2 by their common factor of 2.
4 ÷ 2 is 2.
2 ÷ 2 is 1.

$$= 30 \text{ qt}$$

(c) Convert 36 oz to pounds.

Notice that **oz** divides out, leaving **lb**, the unit you want for the answer.

$$\frac{\overset{9}{\cancel{36}} \text{ oz}}{1} \cdot \frac{1 \text{ lb}}{\underset{4}{\cancel{16}} \text{ oz}} = \frac{9}{4} \text{ lb} = 2\frac{1}{4} \text{ lb}$$

Note

In Example 4(c) above you get $\frac{9}{4}$ lb. Recall that $\frac{9}{4}$ means $9 \div 4$. If you do $9 \div 4$ on your calculator, you get **2.25** lb. U.S. customary measurements usually use fractions or mixed numbers, like $2\frac{1}{4}$ lb. However, **2.25** lb is also correct and is the way grocery stores often show weights of produce, meat, and cheese.

Work Problem ④ at the Side. ▶

EXAMPLE 5 **Using Several Unit Fractions**

Sometimes you may need to use two or three unit fractions to complete a conversion.

(a) Convert 63 in. to yards.

Use the unit fraction $\dfrac{1 \text{ ft}}{12 \text{ in.}}$ to change inches to feet and the unit fraction $\dfrac{1 \text{ yd}}{3 \text{ ft}}$ to change feet to yards. Notice how all the units divide out except yards, which is the unit you want in the answer.

$$\frac{63 \text{ in.}}{1} \cdot \frac{1 \text{ ft}}{12 \text{ in.}} \cdot \frac{1 \text{ yd}}{3 \text{ ft}} = \frac{63}{36} \text{ yd} = \frac{63 \div 9}{36 \div 9} \text{ yd} = \frac{7}{4} \text{ yd} = 1\frac{3}{4} \text{ yd}$$

Continued on Next Page

④ Convert using unit fractions.

(a) 16 qt to gallons

(b) 3 c to pints

(c) $3\frac{1}{2}$ tons to pounds

(d) $1\frac{3}{4}$ lb to ounces

(e) 4 oz to pounds

ANSWERS

4. (a) 4 gal **(b)** $1\frac{1}{2}$ pt or 1.5 pt **(c)** 7000 lb

(d) 28 oz **(e)** $\frac{1}{4}$ lb or 0.25 lb

5 Convert using two or three unit fractions.

(a) 4 tons to ounces

(b) 3 mi to inches

(c) 36 pt to gallons

(d) 2 wk to minutes

You can also divide out common factors in the numbers.

$$\frac{\overset{7}{\cancel{\overset{21}{63}}}}{1} \cdot \frac{1}{\underset{4}{\cancel{12}}} \cdot \frac{1}{\underset{1}{\cancel{3}}} = \frac{7}{4} = 1\frac{3}{4} \text{ yd}$$

Instead of changing $\frac{7}{4}$ to $1\frac{3}{4}$, you can enter $7 \div 4$ on your calculator to get 1.75 yd. Both answers are correct because 1.75 is equivalent to $1\frac{3}{4}$.

(b) Convert 2 days to seconds.

Use three unit fractions. The first one changes days to hours, the next one changes hours to minutes, and the last one changes minutes to seconds. All the units divide out except seconds, which is the unit you want in your answer.

$$\frac{2 \text{ days}}{1} \cdot \frac{24 \text{ hr}}{1 \text{ day}} \cdot \frac{60 \text{ min}}{1 \text{ hr}} \cdot \frac{60 \text{ sec}}{1 \text{ min}} = 172,800 \text{ sec}$$

Divide out **days**.
Divide out **hr**.
Divide out **min**.

◀ *Work Problem* **5** *at the Side.*

OBJECTIVE 4 Solve application problems using U.S. customary measurement units. To solve application problems, we will use the steps you learned in **Section 1.10.** Those steps are summarized here.

Step 1 **Read** the problem carefully.

Step 2 **Work out a plan.**

Step 3 **Estimate** a reasonable answer.

Step 4 **Solve** the problem.

Step 5 **State the answer.**

Step 6 **Check** your work.

EXAMPLE 6 **Solving U.S. Customary Measurement Applications**

(a) A 36 oz can of coffee is on sale at Jerry's Foods for $7.89. What is the cost per pound, to the nearest cent? (*Source:* Jerry's Foods.)

Step 1 **Read** the problem. The problem asks for the cost per *pound* of coffee.

Step 2 **Work out a plan.** The weight of the coffee is given in *ounces*, but the answer must be cost *per pound*. Convert ounces to pounds. The word *per* indicates division. You need to divide the cost by the number of pounds.

Step 3 **Estimate** a reasonable answer. To estimate, round $7.89 to $8. Then, there are 16 oz in a pound, so 36 oz are a little more than 2 pounds. So, $8 ÷ 2 = $4 per pound as our estimate.

ANSWERS
5. **(a)** 128,000 oz **(b)** 190,080 in.
   (c) $4\frac{1}{2}$ gal or 4.5 gal **(d)** 20,160 min

Continued on Next Page

Step 4   **Solve** the problem. Use a unit fraction to convert 36 oz to pounds.

Notice that **oz** divides out, leaving **lb** for the answer.

$$\frac{\overset{9}{36 \,\cancel{oz}}}{1} \cdot \frac{1 \text{ lb}}{\underset{4}{16 \,\cancel{oz}}} = \frac{9}{4} \text{ lb} = 2.25 \text{ lb}$$

On your calculator, $9 \div 4 = 2.25$

Then divide to find the *cost* per *pound*.

$$\begin{array}{l} \text{Cost} \rightarrow \\ \text{per} \rightarrow \\ \text{pound} \rightarrow \end{array} \frac{\$7.89}{2.25 \text{ lb}} = 3.50\overline{6} \approx 3.51 \quad \text{(rounded)}$$

Step 5   **State the answer.** The coffee costs $3.51 per pound (to the nearest cent).

Step 6   **Check** your work. The exact answer of $3.51 is close to our estimate of $4.

(b) Bilal's favorite cake recipe uses $1\frac{2}{3}$ cups of milk. If he makes six cakes for a bake sale at his son's school, how many quarts of milk will he need?

Step 1   **Read** the problem. The problem asks for the number of *quarts* of milk needed for six cakes.

Step 2   **Work out a plan.** Multiply to find the number of *cups* of milk for six cakes. Then convert *cups* to *quarts* (the unit required in the answer).

Step 3   **Estimate** a reasonable answer. To estimate, round $1\frac{2}{3}$ cups to 2 cups. Then, 2 cups times 6 = 12 cups. There are 4 cups in a quart, so 12 cups ÷ 4 = 3 quarts as our estimate.

Step 4   **Solve** the problem. First multiply. Then use unit fractions to convert.

$$1\frac{2}{3} \cdot 6 = \frac{5}{\underset{1}{\cancel{3}}} \cdot \frac{\overset{2}{\cancel{6}}}{1} = \frac{10}{1} = 10 \text{ cups} \quad \begin{cases} \text{Milk needed for} \\ \text{six cakes} \end{cases}$$

$$\frac{\overset{5}{\cancel{10 \text{ cups}}}}{1} \cdot \frac{1 \text{ pt}}{\underset{1}{2 \,\cancel{\text{cups}}}} \cdot \frac{1 \text{ qt}}{2 \,\cancel{\text{pt}}} = \frac{5}{2} \text{ qt} = 2\frac{1}{2} \text{ qt}$$

Both **cups** and **pt** divide out, leaving **qt**, the unit you want for the answer.

Step 5   **State the answer.** Bilal needs $2\frac{1}{2}$ qt (or 2.5 qt) of milk.

Step 6   **Check** your work. The exact answer of $2\frac{1}{2}$ qt is close to our estimate of 3 qt.

> **Note**
>
> In Step 2 above, we *first multiplied* $1\frac{2}{3}$ cups by 6 to find the number of cups needed, then *converted* 10 cups to $2\frac{1}{2}$ quarts. It would also work to *first convert* $1\frac{2}{3}$ cups to $\frac{5}{12}$ qt, then *multiply* $\frac{5}{12}$ qt by 6 to get $2\frac{1}{2}$ qt.

Work Problem **6** *at the Side.* ▶

6 Solve each application problem using the six problem-solving steps.

(a) Kristin paid $3.29 for 12 oz of extra sharp cheddar cheese. What is the price per pound, to the nearest cent?

(b) A moving company estimates 11,000 lb of furnishings for an average 3-bedroom house. If the company made five such moves last week, how many tons of furnishings did they move? (*Source:* North American Van Lines.)

Math in the Media

GROWING SUNFLOWERS

The front and back of a seed packet for sunflowers are shown at the right. Look at the top of the packet first. (Ignore sales tax in Questions 1 and 2.)

1. There were 42 seeds in the packet. If 40 of the seeds sprouted, what was the cost per sprout, to the nearest cent? **$0.03 (rounded)**

2. If vegetable and flower seeds were on sale at 30% off, what was the cost per sprout, to the nearest cent? **$0.02 (rounded)**

3. What percent of the seeds sprouted, to the nearest whole percent? **95% (rounded)**

4. How many seeds would add up to a weight of 1 gram? **12 seeds**

5. The table at the bottom of the packet uses the symbol (′) for feet, and the symbol (″) for inches.

   (a) How tall will the plants grow, in feet? **8 to 10 ft**

   (b) How tall will they grow in inches? **96 to 120 in.**

   (c) How tall will they grow in yards? **$2\frac{2}{3}$ to $3\frac{1}{3}$ yd**

6. If you plant all 42 seeds in one long row, using the spacing given on the package, how long will your row be in feet? **20.5 ft**

7. How many inches tall should the plants be when you thin them (remove less vigorous plants to give others room to grow)? How tall is that in feet?

   3 in.; $\frac{1}{4}$ ft

8. What is the range in the diameter of the flowers, in inches, and in feet? Diameter is the distance across the circular flower.

   6 in. to 15 in.; $\frac{1}{2}$ ft to $1\frac{1}{4}$ ft

Sunflower, Mammoth Grey Stripe

Tall Plants, Huge Flowers

Net Wt. 3.5 g $1.19

Sunflower, Mammoth Grey Stripe

The stalk of this sunflower will grow to 12′(4 m). Flowers will range from 6″(15cm) to 15″(38cm) in diameter. Sunflowers can thrive in poor soil with little moisture.

Type	Height	Planting Depth	Seed Spacing	Thinning Height	Spacing After Thinning	Days to Germination
Annual	8-10′ 2.4-3 m	1/2″ 13 mm	6″ 15 cm	3″ 8 cm	2′ 61 cm	10-20

Select a sunny or lightly shaded location and plant outdoors, where plants are to remain, after all danger of frost is past. For tallest plants, sow in good soil with moderate moisture.

Stock #1185

7 18964 98119 7

Source: Olds Seed Solutions.

490

Fill in the blanks with the measurement relationships you have memorized. See Example 1.

1. 1 yd = __3__ ft;   __12__ in. = 1 ft

2. 1 ft = __12__ in.;   __5280__ ft = 1 mi

3. __8__ fl oz = 1 c;   1 qt = __2__ pt

4. __4__ qt = 1 gal;   1 pt = __2__ c

5. 1 mi = __5280__ ft;   __3__ ft = 1 yd

6. 1 wk = __7__ days;   __60__ sec = 1 min

7. __2000__ lb = 1 ton;   1 lb = __16__ oz

8. __16__ oz = 1 lb;   1 ton = __2000__ lb

9. 1 min = __60__ sec;   __60__ min = 1 hr

10. 1 day = __24__ hr;   __60__ sec = 1 min

Convert each measurement by multiplying or dividing. See Example 2.

🌐 **11. (a)** 120 sec = __2__ min

  (b) 4 hr = __240__ min

12. (a) 180 min = __3__ hr

  (b) 5 min = __300__ sec

13. (a) 2 qt = $\frac{1}{2}$ or 0.5 gal

  (b) $6\frac{1}{2}$ ft = __78__ in.

14. (a) $4\frac{1}{2}$ gal = __18__ qt

  (b) 12 oz = $\frac{3}{4}$ or 0.75 lb

15. An adult African elephant could weigh 7 to 8 tons. How many pounds could it weigh? (*Source: The Top 10 of Everything.*)

14,000 to 16,000 lb

16. A reticulated python snake is the world's longest snake. It grows to a length of 18 to 33 feet. How many yards long can the snake be? (*Source: The Top 10 of Everything.*)

6 to 11 yd

Convert each measurement in Exercises 17–38 using unit fractions. See Examples 3 and 4.

17. 9 yd = __27__ ft

18. 20,000 lb = __10__ tons

19. 7 lb = __112__ oz

20. 96 oz = __6__ lb

21. 5 qt = __10__ pt

22. 26 pt = __13__ qt

23. 90 min = $\underline{1\frac{1}{2}\text{ or }1.5}$ hr

24. 45 sec = $\underline{\frac{3}{4}\text{ or }0.75}$ min

25. 3 in. = $\underline{\frac{1}{4}\text{ or }0.25}$ ft

26. 30 in. = $\underline{2\frac{1}{2}\text{ or }2.5}$ ft

27. 24 oz = $\underline{1\frac{1}{2}\text{ or }1.5}$ lb

28. 36 oz = $\underline{2\frac{1}{4}\text{ or }2.25}$ lb

29. 5 c = $\underline{2\frac{1}{2}\text{ or }2.5}$ pt

30. 15 qt = $\underline{3\frac{3}{4}\text{ or }3.75}$ gal

Use the information in the bar graph below to answer Exercises 31–32.

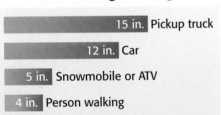

Thickness of Lake Ice Needed for Safe Walking/Driving

15 in.	Pickup truck
12 in.	Car
5 in.	Snowmobile or ATV
4 in.	Person walking

Source: Wisconsin DNR.

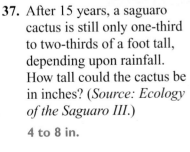

31. If the ice on a lake is $\frac{1}{2}$ ft thick, what will it safely support?

Snowmobile/ATV; person walking

32. How many feet of ice are needed to safely drive a pickup truck on a lake?

$1\frac{1}{4}$ ft

33. $2\frac{1}{2}$ tons = $\underline{5000}$ lb

34. $4\frac{1}{2}$ pt = $\underline{9}$ c

35. $4\frac{1}{4}$ gal = $\underline{17}$ qt

36. $2\frac{1}{4}$ hr = $\underline{135}$ min

37. After 15 years, a saguaro cactus is still only one-third to two-thirds of a foot tall, depending upon rainfall. How tall could the cactus be in inches? (*Source: Ecology of the Saguaro III.*)

4 to 8 in.

38. Yao Ming, an NBA basketball player from China, is $7\frac{1}{2}$ ft tall. What is his height in inches? (*Source:* www.NBA.com)

90 in.

Use two or three unit fractions to make each conversion. See Example 5.

39. 6 yd = __216__ in.

40. 2 tons = __64,000__ oz

41. 112 c = __28__ qt

42. 336 hr = __2__ wk

43. 6 days = __518,400__ sec

44. 5 gal = __80__ c

45. $1\frac{1}{2}$ tons = __48,000__ oz

46. $3\frac{1}{3}$ yd = __120__ in.

47. The statement 8 = 2 is *not* true. But with appropriate measurement units, it *is* true.

$$8 \ quarts = 2 \ gallons$$

Attach measurement units to these numbers to make the statement true.

(a) 1 __pound__ = 16 __ounces__

(b) 10 __quarts/pints__ = 20 __pints/cups__

(c) 120 __minutes/seconds__ = 2 __hours/minutes__

(d) 2 __feet__ = 24 __inches__

(e) 6000 __pounds__ = 3 __tons__

(f) 35 __days__ = 5 __weeks__

48. Explain in your own words why you can add 2 feet + 12 inches to get 3 feet, but you cannot add 2 feet + 12 pounds.

Feet and inches both measure length, so you can add them once you've changed 12 inches into 1 ft. But pounds measure weight and cannot be added to a length measurement.

Convert each measurement. See Example 5.

49. $2\frac{3}{4}$ mi = __174,240__ in.

50. $5\frac{3}{4}$ tons = __184,000__ oz

51. $6\frac{1}{4}$ gal = __800__ fl oz

52. $3\frac{1}{2}$ days = __302,400__ sec

53. 24,000 oz = __0.75 or $\frac{3}{4}$__ ton

54. 57,024 in. = __0.9 or $\frac{9}{10}$__ mi

Solve each application problem. Show your work. See Example 6.

55. Geralyn bought 20 oz of strawberries for $2.29. What was the price per pound for the strawberries, to the nearest cent?

$1.83 (rounded)

56. Zach paid $0.90 for a 0.8 oz candy bar. What was the cost per pound?

$18.00

57. Dan orders supplies for the science labs. Each of the 24 stations in the chemistry lab needs 2 ft of rubber tubing. If rubber tubing sells for $8.75 per yard, how much will it cost to equip all the stations?

$140

58. In 2006, Marquette, Michigan, had 170 inches of snowfall, while Detroit, Michigan, had 15 inches. What was the difference in snowfall between the two cities, in feet? Round to the nearest tenth. (*Source: World Almanac.*)

12.9 ft (rounded)

59. Tropical cockroaches are the fastest land insects. They can run about 5 feet per second. At this rate, how long would it take the cockroach to travel one mile? (*Source: Guinness World Records.*)

Give your answer

(a) in seconds;   1056 sec

(b) in minutes.   17.6 min

60. A snail moves at an average speed of 2 feet every 3 minutes. At that rate, how long would it take the snail to travel one mile? (*Source: Beakman and Jax*.)

Give your answer

(a) in hours;   132 hr

(b) in days. $5\frac{1}{2}$ or 5.5 days

61. At the day care center, each of the 15 toddlers drinks about $\frac{2}{3}$ cup of milk with lunch. The center is open 5 days a week.

(a) How many quarts of milk will the center need for one week of lunches?

$12\frac{1}{2}$ qt

(b) If the center buys milk in gallon containers, how many containers should be ordered for one week?

4 containers, because you can't buy part of a container

62. Bob's Candies in Albany, Georgia, makes 135,000 pounds of candy canes each day. (*Source:* Bob's Candies, Inc.)

(a) How many tons of candy canes are produced during a 5-day workweek?

337.5 tons

(b) The plant operates 24 hours per day. How many tons of candy canes are produced each hour, to the nearest tenth?

2.8 tons (rounded)

Relating Concepts (Exercises 63–66) For Individual or Group Work

On the first page of this chapter, we said that the bowl-shaped crater in northern Arizona made by a meteor crash was sixty stories deep and 4150 ft across. Use this information as you **work Exercises 63–66 in order.** (*Source: The Meteor Crater Story.*)

63. (a) The distance across the crater is what part of a mile, to the nearest tenth?   0.8 mi. (rounded)

(b) The crater is nearly circular. In a circle, the distance around the outside edge is about 3.14 times the distance across the circle. How far is it to walk around the edge of the crater in feet?   13,031 ft

(c) How far is it to walk around the edge of the crater in miles, to the nearest tenth?   2.5 mi (rounded)

64. (a) The crater is 550 ft deep. The depth is how many yards, to the nearest whole number?
183 yd (rounded)

(b) How many inches deep is the crater?   6600 in.

(c) The depth of the crater is what part of a mile, to the nearest tenth?   0.1 mi (rounded)

(d) When we say that the crater is as deep as a sixty-story building, we are assuming that each story is how many feet tall, to the nearest foot?
9 ft (rounded)

65. On one side of the crater there are a few small juniper trees. The trees are 700 years old but only 18 inches to 30 inches tall because of the strong winds and lack of rain.

(a) How tall are the trees in feet?
$1\frac{1}{2}$ to $2\frac{1}{2}$ ft, or 1.5 to 2.5 ft

(b) How many months old are the trees?
8400 months

(c) At this rate of growth, how long would it take a 30-inch tree to reach a height of three feet?
140 more years, for a total of 840 years in all

66. Evidence of two huge meteor crashes has been found on the floor of the Caribbean Sea. One giant circular crater is 90 miles across and the other is 120 miles across.

(a) Using the information about circles in Exercise 63, what is the approximate distance around the edge of the smaller crater, to the nearest mile?

283 mi (rounded)

(b) Around the larger crater?   377 mi (rounded)

7.2 ▶▶▶ The Metric System—Length

Around 1790, a group of French scientists developed the metric system of measurement. It is an organized system based on multiples of 10, like our number system and our money. After you are familiar with metric units, you will see that they are easier to use than the hodgepodge of U.S. customary measurement relationships you used in **Section 7.1.**

> **Note**
> The metric system information in this text is consistent with usage guidelines from the National Institute of Standards and Technology, www.nist.gov/metric.

OBJECTIVE 1 Learn the basic metric units of length. The basic unit of length in the metric system is the **meter** (also spelled *metre*). Use the symbol **m** for meter; do not put a period after it. If you put five of the pages from this textbook side by side, they would measure about 1 meter. Or, look at a yardstick—a meter is just a little longer. A yard is 36 inches long; a meter is about 39 inches long.

book page	book page	book page	book page	book page

|←——— 1 meter (about 39 in.) ———→|
|←——— 1 yard (36 in.) ———→|

In the metric system, you use meters for things like measuring the length of your living room, talking about heights of buildings, or describing track and field athletic events.

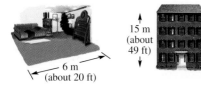

15 m (about 49 ft)

6 m (about 20 ft)

Run the 100 m dash.

Work Problem **1** *at the Side.* ▶

To make longer or shorter length units in the metric system, **prefixes** are written in front of the word *meter*. For example, the prefix *kilo* means **1000**, so a *kilo*meter is **1000** meters. The table below shows how to use the prefixes for length measurements. It is helpful to memorize the prefixes because they are also used with weight and capacity measurements. The blue boxes are the units you will use most often in daily life.

Prefix	kilo-meter	hecto-meter	deka-meter	meter	deci-meter	centi-meter	milli-meter
Meaning	1000 meters	100 meters	10 meters	1 meter	$\frac{1}{10}$ of a meter	$\frac{1}{100}$ of a meter	$\frac{1}{1000}$ of a meter
Symbol	km	hm	dam	m	dm	cm	mm

↑   ↑   ↑   ↑
Length units that are used most often

1 Circle the items that measure about 1 meter.

Length of a pencil

Length of a baseball bat

Height of doorknob from the floor

Height of a house

Basketball player's arm length

Length of a paper clip

ANSWER

1. baseball bat, height of doorknob, basketball player's arm length

2 Write the most reasonable metric unit in each blank. Choose from km, m, cm, and mm.

(a) The woman's height is 168 _____ .

(b) The man's waist is 90 _____ around.

(c) Louise ran the 100 _____ dash in the track meet.

(d) A postage stamp is 22 _____ wide.

(e) Michael paddled his canoe 2 _____ down the river.

(f) The pencil lead is 1 _____ thick.

(g) A stick of gum is 7 _____ long.

(h) The highway speed limit is 90 _____ per hour.

(i) The classroom was 12 _____ long.

(j) A penny is about 18 _____ across.

Here are some comparisons to help you get acquainted with the commonly used length units: km, m, cm, mm.

*Kilo*meters are used instead of miles. A kilometer is **1000** meters. It is about 0.6 mile (a little more than half a mile) or about 5 to 6 city blocks. If you participate in a 10 km run, you'll run about 6 miles.

A meter is divided into 100 smaller pieces called *centi*meters. Each centimeter is $\frac{1}{100}$ of a meter. Centimeters are used instead of inches. A centimeter is a little shorter than $\frac{1}{2}$ inch. The cover of this textbook is about 21 cm wide. A nickel is about 2 cm across. Measure the width and length of your little finger on this centimeter ruler. The width of your little finger is probably about 1 cm, or a little more.

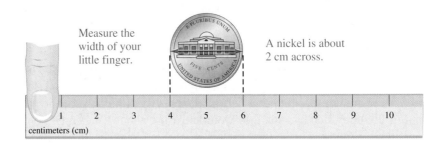

A meter is divided into 1000 smaller pieces called *milli*meters. Each millimeter is $\frac{1}{1000}$ of a meter. It takes 10 mm to equal 1 cm, so it is a very small length. The thickness of a dime is about 1 mm. Measure the width of your pen or pencil and the width of your little finger on this millimeter ruler.

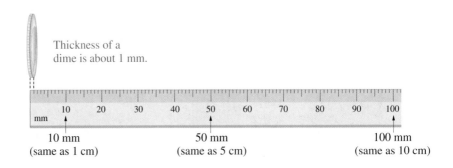

10 mm (same as 1 cm)   50 mm (same as 5 cm)   100 mm (same as 10 cm)

EXAMPLE 1 **Using Metric Length Units**

Write the most reasonable metric unit in each blank. Choose from km, m, cm, and mm.

(a) The distance from home to work is 20 _____ .

20 **km** because kilometers are used instead of miles. 20 km is about 12 miles.

(b) My wedding ring is 4 _____ wide.

4 **mm** because the width of a ring is very small.

(c) The newborn baby is 50 _____ long.

50 **cm**; which is half of a meter; a meter is about 39 inches so half a meter is around 20 inches.

◀ *Work Problem* **2** *at the Side.*

OBJECTIVE 2 **Use unit fractions to convert among units.** You can convert among metric length units using unit fractions. Keep these relationships in mind when setting up the unit fractions.

Metric Length Relationships

1 km = 1000 m so the unit fractions are:	**1 m = 1000 mm** so the unit fractions are:
$\dfrac{1 \text{ km}}{1000 \text{ m}}$ or $\dfrac{1000 \text{ m}}{1 \text{ km}}$	$\dfrac{1 \text{ m}}{1000 \text{ mm}}$ or $\dfrac{1000 \text{ mm}}{1 \text{ m}}$
1 m = 100 cm so the unit fractions are:	**1 cm = 10 mm** so the unit fractions are:
$\dfrac{1 \text{ m}}{100 \text{ cm}}$ or $\dfrac{100 \text{ cm}}{1 \text{ m}}$	$\dfrac{1 \text{ cm}}{10 \text{ mm}}$ or $\dfrac{10 \text{ mm}}{1 \text{ cm}}$

EXAMPLE 2 **Using Unit Fractions to Convert Length Measurements**

Convert each measurement using unit fractions.

(a) 5 km to m

Put the unit for the answer (meters) in the numerator of the unit fraction; put the unit you want to change (km) in the denominator.

Unit fraction equivalent to 1 $\left\{ \dfrac{1000 \text{ m}}{1 \text{ km}} \right.$ ← Unit for answer / ← Unit being changed

Multiply. Divide out common units where possible.

$$5 \text{ km} \cdot \frac{1000 \text{ m}}{1 \text{ km}} = \frac{5 \text{ km}}{1} \cdot \frac{1000 \text{ m}}{1 \text{ km}} = \frac{5 \cdot 1000 \text{ m}}{1} = 5000 \text{ m}$$

These units should match.

Here, **km** divides out leaving **m**, the unit you want for your answer.

5 km = 5000 m

The answer makes sense because a kilometer is much longer than a meter, so 5 km will contain many meters.

(b) 18.6 cm to m

Multiply by a unit fraction that allows you to divide out centimeters.

Unit fraction

$$\frac{18.6 \text{ cm}}{1} \cdot \frac{1 \text{ m}}{100 \text{ cm}} = \frac{18.6}{100} \text{ m} = 0.186 \text{ m}$$

Do **not** write a period here.

18.6 cm = 0.186 m

There are 100 cm in a meter, so 18.6 cm will be a small part of a meter. The answer makes sense.

Work Problem **3** *at the Side.* ▶

---

3 First write the unit fraction needed to make each conversion. Then complete the conversion.

(a) 3.67 m to cm

unit fraction $\left\{ \dfrac{100 \text{ cm}}{1 \text{ m}} \right.$

(b) 92 cm to m

unit fraction $\left\{ \dfrac{\text{m}}{\text{cm}} \right.$

(c) 432.7 cm to m

unit fraction $\left\{ \rule{2cm}{0.4pt} \right.$

(d) 65 mm to cm

unit fraction $\left\{ \rule{2cm}{0.4pt} \right.$

(e) 0.9 m to mm

unit fraction $\left\{ \rule{2cm}{0.4pt} \right.$

(f) 2.5 cm to mm

unit fraction $\left\{ \rule{2cm}{0.4pt} \right.$

ANSWERS

3. **(a)** 367 cm **(b)** $\dfrac{1 \text{ m}}{100 \text{ cm}}$; 0.92 m

(c) $\dfrac{1 \text{ m}}{100 \text{ cm}}$; 4.327 m

(d) $\dfrac{1 \text{ cm}}{10 \text{ mm}}$; 6.5 cm

(e) $\dfrac{1000 \text{ mm}}{1 \text{ m}}$; 900 mm

(f) $\dfrac{10 \text{ mm}}{1 \text{ cm}}$; 25 mm

4 Do each multiplication or division by hand or on a calculator. Compare your answer to the one you get by moving the decimal point.

(a) $(43.5)(10) =$ _____

43.5 gives 435

(b) $43.5 \div 10 =$ _____

43.5 gives _____

(c) $(28)(100) =$ _____

28.00 gives _____

(d) $28 \div 100 =$ _____

28. gives _____

(e) $(0.7)(1000) =$ _____

0.700 gives _____

(f) $0.7 \div 1000 =$ _____

000.7 gives _____

OBJECTIVE 3 Move the decimal point to convert among units.
By now you have probably noticed that conversions among metric units are made by multiplying or dividing by 10, by 100, or by 1000. A quick way to *multiply* by 10 is to move the decimal point one place to the *right*. Move it two places to the right to multiply by 100, three places to multiply by 1000. *Dividing* is done by moving the decimal point to the *left* in the same manner.

◀ *Work Problem* **4** *at the Side.*

An alternate conversion method to unit fractions is moving the decimal point using this **metric conversion line.**

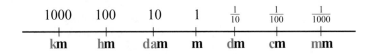

Here are the steps for using the conversion line.

Using the Metric Conversion Line

Step 1 Find the unit you are given on the metric conversion line.

Step 2 Count the number of places to get from the unit you are given to the unit you want in the answer.

Step 3 Move the decimal point the ***same number of places*** and in the ***same direction*** as you did on the conversion line.

EXAMPLE 3 **Using the Metric Conversion Line**

Use the metric conversion line to make the following conversions.

(a) 5.702 km to m

Find **km** on the metric conversion line. To get to **m**, you move *three places* to the *right*. So move the decimal point in 5.702 *three places* to the *right*.

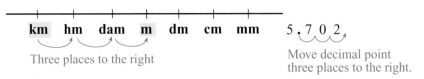

5.702 km = 5702 m

(b) 69.5 cm to m

Find **cm** on the conversion line. To get to **m**, move *two places* to the *left*.

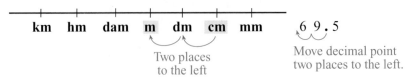

69.5 cm = 0.695 m

—————— **Continued on Next Page**

(c) 8.1 cm to mm

From **cm** to **mm** is *one place* to the *right*.

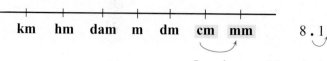

| km | hm | dam | m | dm | cm | mm |

8 . 1

One place   Move decimal point
to the right   one place to the right.

8.1 cm = 81 mm

Work Problem **5** *at the Side.* ▶

EXAMPLE 4 **Practicing Length Conversions**

Convert using the metric conversion line.

(a) 1.28 m to mm

Moving from **m** to **mm** is going *three places to the right.* In order to move the decimal point in 1.28 three places to the right, you must write a 0 as a placeholder.

1.280   Zero is written in as a placeholder.

Move decimal point
three places to the right.

> The answer is a *whole number*, so you do not need to write the decimal point.

1.28 m = 1280 mm

(b) 60 cm to m

From **cm** to **m** is two places to the left. The decimal point in 60 starts at the *far right side* because 60 is a whole number. Then move it two places to the left.

60.
Decimal point starts here.

60.
Move decimal point
two places to the left.

60 cm = 0.60 m, and 0.60 m is equivalent to 0.6 m.

(c) 8 m to km

From **m** to **km** is three places to the left. The decimal point in 8 starts at the far right side. In order to move it three places to the left, you must write two zeros as placeholders.

Two zeros are written
in as placeholders.

008.

8.
Decimal point starts here.

Move decimal point
three places to the left.

8 m = 0.008 km

Work Problem **6** *at the Side.* ▶

---

5 Convert using the metric conversion line.

(a) 12.008 km to m

(b) 561.4 m to km

(c) 20.7 cm to m

(d) 20.7 cm to mm

(e) 4.66 m to cm

(f) 85.6 mm to cm

6 Convert using the metric conversion line.

(a) 9 m to mm

(b) 3 cm to m

(c) 14.6 km to m

(d) 5 mm to cm

(e) 70 m to km

(f) 0.8 m to cm

ANSWERS

5. **(a)** 12,008 m **(b)** 0.5614 km
   (c) 0.207 m **(d)** 207 mm
   (e) 466 cm **(f)** 8.56 cm
6. **(a)** 9000 mm **(b)** 0.03 m
   (c) 14,600 m **(d)** 0.5 cm
   (e) 0.07 km **(f)** 80 cm

Math in the Media

Hair and Nail Growth

Q: How fast do hair and nails grow? Do they grow faster in the summer?

A: Fingernails grow, on average, about one-tenth of a millimeter per day, although there is considerable variation among individuals. Fingernails grow faster than toenails, and nails on the longest fingers appear to grow the fastest.

Fingernails, as well as hair and skin, grow faster in the summer, presumably under the influence of sunlight, which expands blood vessels, bringing more oxygen and nutrients to the area and allowing for faster growth.

The rate the scalp hair grows is 0.3 to 0.4 millimeter per day, or about 6 inches a year.

Source: Minneapolis Star Tribune.

1. **(a)** How much do fingernails grow in one week?
   0.7 mm

   (b) In one month?
   3 mm during a 30-day month

   (c) In one year?
   36.5 mm

2. **(a)** Using metric units, how much does hair grow in one week?
   2.1 to 2.8 mm

   (b) In one month?
   9 to 12 mm

   (c) In one year?
   109.5 to 146 mm

3. When you have finished **Section 7.5,** come back to this article. Is the statement about hair growing 6 inches a year accurate? Explain your answer.   **Answers will vary, but 109.5 mm ≈ 4.3 inches and 146 mm ≈ 5.7 inches.**

4. **(a)** According to *Guinness World Records,* the longest toenails measure 15.2 cm. How many millimeters is that length? How many meters?
   152 mm; 0.152 m

   (b) The longest eyelash is listed as 5.08 cm. How long is the eyelash in millimeters? In meters?
   50.8 mm; 0.0508 m

7.2 ▶▶▶ Exercises

FOR
EXTRA
HELP

MyMathLab

Math XL
PRACTICE

WATCH

DOWNLOAD

READ

REVIEW

Use your knowledge of the meaning of metric prefixes to fill in the blanks.

1. *kilo* means ___1000___ so

    1 km = ___1000___ m

2. *deka* means ___10___ so

    1 dam = ___10___ m

3. *milli* means $\frac{1}{1000}$ or 0.001 ___ so

    1 mm = $\frac{1}{1000}$ or 0.001 ___ m

4. *deci* means $\frac{1}{10}$ or 0.1 ___ so

    1 dm = $\frac{1}{10}$ or 0.1 ___ m

5. *centi* means $\frac{1}{100}$ or 0.01 ___ so

    1 cm = $\frac{1}{100}$ or 0.01 ___ m

6. *hecto* means ___100___ so

    1 hm = ___100___ m

Use this ruler to measure the width of your thumb and hand for Exercises 7–10.

7. The width of your hand in centimeters

Answers will vary; about 8 to 10 cm.

8. The width of your hand in millimeters

10 times the number of cm measured in Exercise 7.

9. The width of your thumb in millimeters

Answers will vary; about 20 to 25 mm.

10. The width of your thumb in centimeters

number of mm measured in Exercise 9 divided by 10

Write the most reasonable metric length unit in each blank. Choose from km, m, cm, and mm. See Example 1.

🌐 **11.** The child was 91 ___cm___ tall.

🌐 **13.** Ming-Na swam in the 200 ___m___ backstroke race.

🌐 **15.** Adriana drove 400 ___km___ on her vacation.

🌐 **17.** An aspirin tablet is 10 ___mm___ across.

19. A paper clip is about 3 ___cm___ long.

21. Dave's truck is 5 ___m___ long.

12. The cardboard was 3 ___mm___ thick.

14. The bookcase is 75 ___cm___ wide.

16. The door is 2 ___m___ high.

18. Lamard jogs 4 ___km___ every morning.

20. My pen is 145 ___mm___ long.

22. Wheelchairs need doorways that are at least 80 ___cm___ wide.

✏ **23.** Describe at least three examples of metric length units that you have come across in your daily life.

Some possible answers are: track and field events, metric auto parts, and lead refills for mechanical pencils.

✏ **24.** Explain one reason the metric system would be easier for a child to learn than the U.S. customary system.

Some possible answers are: conversions can be done using decimals instead of fractions; fewer conversion relationships to memorize.

Convert each measurement. Use unit fractions or the metric conversion line. See Examples 2–4.

25. 7 m to cm
700 cm

26. 18 m to cm
1800 cm

27. 40 mm to m
0.040 m or 0.04 m

28. 6 mm to m
0.006 m

29. 9.4 km to m
9400 m

30. 0.7 km to m
700 m

31. 509 cm to m
5.09 m

32. 30 cm to m
0.3 m

33. 400 mm to cm
40 cm

34. 25 mm to cm
2.5 cm

35. 0.91 m to mm
910 mm

36. 4 m to mm
4000 mm

37. Is 82 cm greater than or less than 1 m? What is the difference in the lengths?
less; 18 cm or 0.18 m

38. Is 1022 m greater than or less than 1 km? What is the difference in the lengths?
greater; 22 m or 0.022 km

39. On the first page of this chapter, we said that computer microchips may be only 5 mm long and 1 mm wide. Using the ruler on the previous page, draw a rectangle that measures 5 mm by 1 mm. Then convert each measurement to centimeters.

A greatly enlarged photo of a computer microchip.

5 mm = 0.5 cm

1 mm = 0.1 cm

40. The world's smallest butterfly has a wingspan of 15 mm. The smallest mouse is 50 mm long. Using the ruler on the previous page, draw a line that is 15 mm long and a line 50 mm long. Then convert each measurement to centimeters. (*Source: Top 10 of Everything.*)

_____ 15 mm = 1.5 cm

_____ 50 mm = 5 cm

41. The Roe River near Great Falls, Montana, is the shortest river in the world, with a north fork that is just under 18 m long. How many kilometers long is the north fork of the river? (*Source: Guinness Book of Amazing Nature.*)
0.018 km

42. There are 60,000 km of blood vessels in the human body. How many meters of blood vessels are in the body? (*Source: Big Book of Knowledge.*)
60,000,000 m

43. The median height for U.S. females who are 20 to 29 years old is about 1.64 m. Convert this height to centimeters and to millimeters. (*Source: U.S. National Center for Health Statistics.*)
164 cm; 1640 mm

44. The median height for 20- to 29-year-old males in the United States is about 177 cm. Convert this height to meters and to millimeters. (*Source: U.S. National Center for Health Statistics.*)
1.77 m; 1770 mm

45. Use two unit fractions to convert 5.6 mm to km.
0.0000056 km

46. Use two unit fractions to convert 16.5 km to mm.
16,500,000 mm

7.3 ▶▶▶ The Metric System—Capacity and Weight (Mass)

We use capacity units to measure liquids, such as the amount of milk in a recipe, the gasoline in our car tank, and the water in an aquarium. (The capacity units in the U.S. customary system are cups, pints, quarts, and gallons.) The basic metric unit for capacity is the **liter** (also spelled *litre*). The capital letter L is the symbol for liter, to avoid confusion with the numeral 1.

OBJECTIVE 1 **Learn the basic metric units of capacity.** The liter is related to metric length in this way: a box that measures 10 cm on every side holds exactly one liter. (The volume of the box is 10 cm • 10 cm • 10 cm = 1000 cubic centimeters. Volume is discussed in **Section 8.7.**) A liter is just a little more than 1 quart.

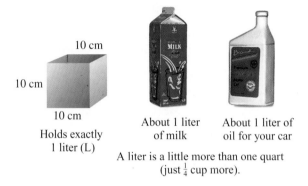

10 cm
10 cm
10 cm
Holds exactly 1 liter (L)

About 1 liter of milk

About 1 liter of oil for your car

A liter is a little more than one quart (just $\frac{1}{4}$ cup more).

In the metric system you use liters for things like buying milk and soda at the store, filling a pail with water, and describing the size of your home aquarium.

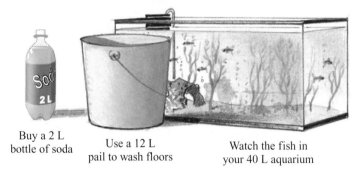

Buy a 2 L bottle of soda

Use a 12 L pail to wash floors

Watch the fish in your 40 L aquarium

Work Problem **1** *at the Side.* ▶

To make larger or smaller capacity units, we use the same **prefixes** as we did with length units. For example, *kilo* means 1000 so a *kilo*meter is 1000 meters. In the same way, a *kilo*liter is 1000 liters.

Prefix	kilo- liter	hecto- liter	deka- liter	liter	deci- liter	centi- liter	milli- liter
Meaning	1000 liters	100 liters	10 liters	1 liter	$\frac{1}{10}$ of a liter	$\frac{1}{100}$ of a liter	$\frac{1}{1000}$ of a liter
Symbol	kL	hL	daL	L	dL	cL	mL

↑   ↑
Capacity units used most often

1 Which things can be measured in liters?

Amount of water in the bathtub

Length of the bathtub

Width of your car

Amount of gasoline you buy for your car

Weight of your car

Height of a pail

Amount of water in a pail

ANSWER

1. water in bathtub, gasoline, water in a pail

2 Write the most reasonable metric unit in each blank. Choose from L and mL.

(a) I bought 8 _____ of soda at the store.

(b) The nurse gave me 10 _____ of cough syrup.

(c) This is a 100 _____ garbage can.

(d) It took 10 _____ of paint to cover the bedroom walls.

(e) My car's gas tank holds 50 _____ .

(f) I added 15 _____ of oil to the pancake mix.

(g) The can of orange soda holds 350 _____ .

(h) My friend gave me a 30 _____ bottle of expensive perfume.

The capacity units you will use most often in daily life are liters (L) and *milli*liters (mL). A tiny box that measures 1 cm on every side holds exactly one milliliter. (In medicine, this small amount is also called 1 cubic centimeter, or 1 cc for short.) It takes 1000 mL to make 1 L. Here are some useful comparisons.

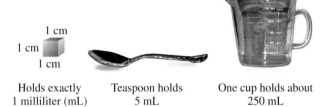

Holds exactly 1 milliliter (mL)	Teaspoon holds 5 mL	One cup holds about 250 mL

EXAMPLE 1 **Using Metric Capacity Units**

Write the most reasonable metric unit in each blank. Choose from L and mL.

(a) The bottle of shampoo held 500 _____ .
500 **mL** because 500 L would be about 500 quarts, which is too much.

(b) I bought a 2 _____ carton of orange juice.
2 **L** because 2 mL would be less than a teaspoon.

◀ *Work Problem* **2** *at the Side.*

OBJECTIVE 2 **Convert among metric capacity units.** Just as with length units, you can convert between milliliters and liters using unit fractions.

Metric Capacity Relationships

1 L = 1000 mL, so the unit fractions are:

$$\frac{1 \text{ L}}{1000 \text{ mL}} \quad \text{or} \quad \frac{1000 \text{ mL}}{1 \text{ L}}$$

Or you can use a metric conversion line to decide how to move the decimal point.

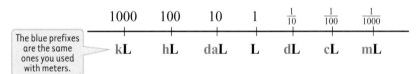

1000	100	10	1	$\frac{1}{10}$	$\frac{1}{100}$	$\frac{1}{1000}$
kL	**hL**	**daL**	**L**	**dL**	**cL**	**mL**

The blue prefixes are the same ones you used with meters.

EXAMPLE 2 **Converting among Metric Capacity Units**

Convert using the metric conversion line or unit fractions.

(a) 2.5 L to mL
Using the metric conversion line:
From **L** to **mL** is *three places* to the *right*.

2.500̮   Write two zeros as placeholders.

2.5 L = 2500 mL

Using unit fractions:

Multiply by a unit fraction that allows you to divide out liters.

$$\frac{2.5 \cancel{\text{ L}}}{1} \cdot \frac{1000 \text{ mL}}{1 \cancel{\text{ L}}} = 2500 \text{ mL}$$

L divides out, leaving **mL** for your answer.

Continued on Next Page

(b) 80 mL to L

Using the metric conversion line:

From **mL** to **L** is *three places* to the *left*.

80.   .080.

↑   ⁓

Decimal point   Move decimal
starts here.   point three places
           to the left.

80 mL = 0.080 L or 0.08 L

Using unit fractions:

Multiply by a unit fraction that allows you to divide out mL.

$$\frac{80 \ \cancel{mL}}{1} \cdot \frac{1 \ L}{1000 \ \cancel{mL}}$$

$$= \frac{80}{1000} \ L = 0.08 \ L$$

> Do **not** write a period here.

Work Problem **3** *at the Side.* ▶

OBJECTIVE 3 Learn the basic metric units of weight (mass).
The **gram** is the basic metric unit for *mass*. Although we often call it "weight," there is a difference. Weight is a measure of the pull of gravity; the farther you are from the center of Earth, the less you weigh. In outer space you become weightless, but your mass, the amount of matter in your body, stays the same regardless of where you are. In science courses, it will be important to distinguish between the weight of an object and its mass. But for everyday purposes, we will use the word *weight*.

The gram is related to metric length in this way: The weight of the water in a box measuring 1 cm on every side is 1 gram. This is a very tiny amount of water (1 mL) and a very small weight. One gram is also the weight of a dollar bill or a single raisin. A nickel weighs 5 grams. A plain, regular-sized hamburger and bun weighs from 175 to 200 grams.

The 1 mL of water
in this tiny box weighs
1 gram.

A nickel weighs
5 grams.

A dollar bill weighs
1 gram.

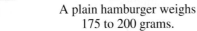

A plain hamburger weighs
175 to 200 grams.

Work Problem **4** *at the Side.* ▶

---

3 Convert.

(a) 9 L to mL

(b) 0.75 L to mL

(c) 500 mL to L

(d) 5 mL to L

(e) 2.07 L to mL

(f) 3275 mL to L

4 Which things would weigh about 1 gram?

A small paper clip

A pair of scissors

One playing card from a deck of cards

A calculator

An average-sized apple

The check you wrote to the cable company

ANSWERS
3. **(a)** 9000 mL **(b)** 750 mL **(c)** 0.5 L
   (d) 0.005 L **(e)** 2070 mL **(f)** 3.275 L
4. paper clip, playing card, check

5 Write the most reasonable metric unit in each blank. Choose from kg, g, and mg.

(a) A thumbtack weighs
800 _____.

(b) A teenager weighs
50 _____.

(c) This large cast-iron frying pan weighs
1 _____.

(d) Jerry's basketball weighed 600 _____.

(e) Tamlyn takes a
500 _____ calcium tablet every morning.

(f) On his diet, Greg can eat
90 _____ of meat for lunch.

(g) One strand of hair weighs
2 _____.

(h) One banana might weigh
150 _____.

ANSWERS

5. **(a)** mg   **(b)** kg   **(c)** kg
   (d) g   **(e)** mg   **(f)** g
   (g) mg   **(h)** g

To make larger or smaller weight units, we use the same **prefixes** as we did with length and capacity units. For example, *kilo* means 1000 so a *kilo*meter is 1000 meters, a *kilo*liter is 1000 liters, and a *kilo*gram is 1000 grams.

Prefix	kilo-gram	hecto-gram	deka-gram	gram	deci-gram	centi-gram	milli-gram
Meaning	1000 grams	100 grams	10 grams	1 gram	$\frac{1}{10}$ of a gram	$\frac{1}{100}$ of a gram	$\frac{1}{1000}$ of a gram
Symbol	kg	hg	dag	g	dg	cg	mg

Weight (mass) units that are used most often

The units you will use most often in daily life are kilograms (kg), grams (g), and milligrams (mg). *Kilo*grams are used instead of pounds. A kilogram is 1000 grams. It is about **2.2** pounds. Two packages of butter plus one stick of butter weigh about 1 kg. An average newborn baby weighs 3 to 4 kg; a college football player might weigh 100 to 130 kg.

1 kilogram is
about **2.2** pounds   100 to 130 kg   3 to 4 kg

Extremely small weights are measured in *milli*grams. It takes 1000 mg to make 1 g. Recall that a dollar bill weighs about 1 g. Imagine cutting it into 1000 pieces; the weight of one tiny piece would be 1 mg. Dosages of medicine and vitamins are given in milligrams. You will also use milligrams in science classes.

Cut a dollar bill into 1000 pieces.
One tiny piece weighs 1 milligram.

EXAMPLE 3 **Using Metric Weight Units**

Write the most reasonable metric unit in each blank.
Choose from kg, g, and mg.

(a) Ramon's suitcase weighed 20 _____.
20 **kg** because kilograms are used instead of pounds.
20 kg is about 44 pounds.

(b) LeTia took a 350 _____ aspirin tablet.
350 **mg** because 350 g would be more than the weight of a hamburger, which is too much.

(c) Jenny mailed a letter that weighed 30 _____.
30 **g** because 30 kg would be much too heavy and 30 mg is less than the weight of a dollar bill.

◀ *Work Problem* **5** *at the Side.*

OBJECTIVE 4 Convert among metric weight (mass) units. As with length and capacity, you can convert among metric weight units by using unit fractions. The unit fractions you need are shown here.

Metric Weight (Mass) Relationships

1 kg = 1000 g so the unit fractions are:

$$\frac{1 \text{ kg}}{1000 \text{ g}} \quad \text{or} \quad \frac{1000 \text{ g}}{1 \text{ kg}}$$

1 g = 1000 mg so the unit fractions are:

$$\frac{1 \text{ g}}{1000 \text{ mg}} \quad \text{or} \quad \frac{1000 \text{ mg}}{1 \text{ g}}$$

Or you can use a metric conversion line to decide how to move the decimal point.

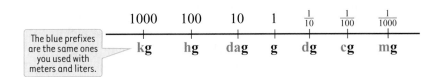

The blue prefixes are the same ones you used with meters and liters.

1000	100	10	1	$\frac{1}{10}$	$\frac{1}{100}$	$\frac{1}{1000}$
kg	**hg**	**dag**	**g**	**dg**	**cg**	**mg**

EXAMPLE 4 Converting among Metric Weight Units

Convert using the metric conversion line or unit fractions.

(a) 7 mg to g

Using the metric conversion line:
From **mg** to **g** is *three places* to the *left*.

7.   007.

Decimal point starts here.   Move decimal point three places to the left.

7 mg = 0.007 g

Using unit fractions:

Multiply by a unit fraction that allows you to divide out mg.

$$\frac{7 \text{ mg}}{1} \cdot \frac{1 \text{ g}}{1000 \text{ mg}} = \frac{7}{1000} \text{ g}$$

$$= 0.007 \text{ g}$$

Three decimal places for thousandths.

(b) 13.72 kg to g

Using the metric conversion line:
From **kg** to **g** is *three places* to the *right*.

13.720   Decimal point moves three places to the right.

13.72 kg = 13,720 g

A comma (not a decimal point)

Using unit fractions:

Multiply by a unit fraction that allows you to divide out kg.

$$\frac{13.72 \text{ kg}}{1} \cdot \frac{1000 \text{ g}}{1 \text{ kg}} = 13,720 \text{ g}$$

A comma (not a decimal point)

---

6 Convert.

(a) 10 kg to g

(b) 45 mg to g

(c) 6.3 kg to g

(d) 0.077 g to mg

(e) 5630 g to kg

(f) 90 g to kg

Work Problem **6** *at the Side.* ▶

ANSWERS

6. (a) 10,000 g **(b)** 0.045 g **(c)** 6300 g
 (d) 77 mg **(e)** 5.63 kg **(f)** 0.09 kg

7 First decide which type of units are needed: length, capacity, or weight. Then write the most appropriate unit in the blank. Choose from km, m, cm, mm, L, mL, kg, g, and mg.

(a) Gail bought a 4 _____ can of paint.

Use _____ units.

(b) The bag of chips weighed 450 _____ .

Use _____ units.

(c) Give the child 5 _____ of cough syrup.

Use _____ units.

(d) The width of the window is 55 _____ .

Use _____ units.

(e) Akbar drives 18 _____ to work.

Use _____ units.

(f) The laptop computer weighs 2 _____ .

Use _____ units.

(g) A credit card is 55 _____ wide.

Use _____ units.

OBJECTIVE 5 Distinguish among basic metric units of length, capacity, and weight (mass). As you encounter things to be measured at home, on the job, or in your classes at school, be careful to use the correct type of measurement unit.

Use *length units* (kilometers, meters, centimeters, millimeters) to measure:

how long	how high	how far away
how wide	how tall	how far around (perimeter)
how deep	distance	

Use *capacity units* (liters, milliliters) to measure liquids (things that can be poured) such as:

water	shampoo	gasoline
milk	perfume	oil
soft drinks	cough syrup	paint

Also use liters and milliliters to describe how much liquid something can hold, such as an eyedropper, measuring cup, pail, or bathtub.

Use *weight units* (kilograms, grams, milligrams) to measure:

the weight of something   how heavy something is

In **Chapter 8** you will use square units (such as square meters) to measure area, and cubic units (such as cubic centimeters) to measure volume.

EXAMPLE 5   Using a Variety of Metric Units

First decide which type of units are needed: length, capacity, or weight. Then write the most appropriate metric unit in the blank. Choose from km, m, cm, mm, L, mL, kg, g, and mg.

(a) The letter needs another stamp because it weighs 40 _____ .

Use _____ units.

The letter weighs 40 **g** because 40 mg is less than the weight of a dollar bill and 40 kg would be about 88 pounds.

Use **weight** units because of the word "weighs."

(b) The swimming pool is 3 _____ deep at the deep end.

Use _____ units.

The pool is 3 **m** deep because 3 cm is only about an inch and 3 km is more than a mile.

Use **length** units because of the word "deep."

(c) This is a 340 _____ can of juice.

Use _____ units.

It is a 340 **mL** can because 340 liters would be more than 340 quarts.

Use **capacity** units because juice is a liquid.

◀ *Work Problem* **7** *at the Side.*

Write the most reasonable metric unit in each blank. Choose from L, mL, kg, g, and mg.
See Examples 1 and 3.

1. The glass held

250 __mL__ of water.

2. Hiromi used 12 __L__ of water to wash the kitchen floor.

3. Dolores can make 10 __L__ of soup in that pot.

4. Jay gave 2 __mL__ of vitamin drops to the baby.

🌐 **5.** Our yellow Labrador dog grew up to weigh 40 __kg__ .

6. A small safety pin weighs 750 __mg__ .

7. Lori caught a small sunfish weighing 150 __g__ .

8. One dime weighs 2 __g__ .

🌐 **9.** Andre donated 500 __mL__ of blood today.

10. Barbara bought the 2 __L__ bottle of cola.

🌐 **11.** The patient received a 250 __mg__ tablet of medication each hour.

12. The 8 people on the elevator weighed a total of 500 __kg__ .

13. The gas can for the lawn mower holds 4 __L__ .

14. Kevin poured 10 __mL__ of vanilla into the mixing bowl.

15. Pam's backpack weighs 5 __kg__ when it is full of books.

16. One grain of salt weighs 2 __mg__ .

Today, medical measurements are usually given in the metric system. Since we convert among metric units of measure by moving the decimal point, it is possible that mistakes can be made. Examine the following dosages and indicate whether they are reasonable or unreasonable. If a dose is unreasonable, indicate whether it is too much or too little.

17. Drink 4.1 L of Kaopectate after each meal.

unreasonable; too much

18. Drop 1 mL of solution into the eye twice a day.

reasonable

19. Soak your feet in 5 kg of Epsom salts per liter of water.

unreasonable; too much

20. Inject 0.5 L of insulin each morning.

unreasonable; too much

21. Take 15 mL of cough syrup every four hours.

reasonable

22. Take 200 mg of vitamin C each day.

reasonable

23. Take 350 mg of aspirin three times a day.

reasonable

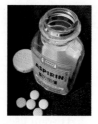

24. Buy a tube of ointment weighing 0.002 g.

unreasonable; too little.

25. Describe at least two examples of metric capacity units and two examples of metric weight units that you have come across in your daily life.

Some capacity examples are 2 L bottles of soda and shampoo bottles marked in mL; weight examples are grams of fat listed on food packages and vitamin doses in milligrams.

26. Explain in your own words how the meter, liter, and gram are related.

A box measuring 1 cm on each side holds 1 mL of water and the water weighs 1 g. Also, a box measuring 10 cm on every side holds exactly 1 L.

27. Describe how you decide which unit fraction to use when converting 6.5 kg to grams.

Unit for your answer (g) is in numerator; unit being changed (kg) is in denominator so it will divide out. The unit fraction is $\dfrac{1000 \text{ g}}{1 \text{ kg}}$.

28. Write an explanation of each step you would use to convert 20 mg to grams using the metric conversion line.

From mg to g is three places to the left on the metric conversion line, so move the decimal point three places left.

$$020. \text{ mg} = 0.02 \text{ g}$$

Convert each measurement. Use unit fractions or the metric conversion line. See Examples 2 and 4.

29. 15 L to mL   **15,000 mL**

30. 6 L to mL   **6000 mL**

31. 3000 mL to L   **3 L**

32. 18,000 mL to L   **18 L**

33. 925 mL to L   **0.925 L**

34. 200 mL to L   **0.2 L**

35. 8 mL to L   **0.008 L**

36. 25 mL to L   **0.025 L**

37. 4.15 L to mL   **4150 mL**

38. 11.7 L to mL   **11,700 mL**

39. 8000 g to kg   **8 kg**

40. 25,000 g to kg   **25 kg**

41. 5.2 kg to g   **5200 g**

42. 12.42 kg to g   **12,420 g**

43. 0.85 g to mg   **850 mg**

44. 0.2 g to mg   **200 mg**

45. 30,000 mg to g   **30 g**

46. 7500 mg to g   **7.5 g**

47. 598 mg to g   **0.598 g**

48. 900 mg to g   **0.9 g**

49. 60 mL to L   **0.06 L**

50. 6.007 kg to g   **6007 g**

51. 3 g to kg   **0.003 kg**

52. 12 mg to g   **0.012 g**

53. 0.99 L to mL   **990 mL**

54. 13,700 mL to L   **13.7 L**

Write the most appropriate metric unit in each blank. Choose from km, m, cm, mm, L, mL, kg, g, and mg. See Example 5.

55. The masking tape is 19 __mm__ wide.

56. The roll has 55 __m__ of tape on it.

57. Buy a 60 __mL__ jar of acrylic paint for art class.

58. One onion weighs 200 __g__ .

59. My waist measurement is 65 __cm__ .

60. Add 2 __L__ of windshield washer fluid to your car.

61. A single postage stamp weighs 90 __mg__ .

62. The hallway is 10 __m__ long.

Solve each application problem. Show your work. (Source for Exercises 63–68: Top 10 of Everything.)

63. Human skin has about 3 million sweat glands, which release an average of 300 mL of sweat per day. How many liters of sweat are released each day?
0.3 L

64. In hot climates, the sweat glands in a person's skin may release up to 3.5 L of sweat in one day. How many milliliters is that?
3500 mL

65. The average weight of a human brain is 1.34 kg. How many grams is that?
1340 g

66. A healthy human heart pumps about 70 mL of blood per beat. How many liters of blood does it pump per beat?
0.07 L

67. On average, we breathe in and out roughly 900 mL of air every 10 seconds. How many liters of air is that?
0.9 L

68. In the Victorian era, people believed that heavier brains meant greater intelligence. They were impressed that Otto von Bismarck's brain weighed 1907 g, which is how many kilograms?
1.907 kg

69. A small adult cat weighs from 3000 g to 4000 g. How many kilograms is that? (*Source: Lyndale Animal Hospital.*)

3 kg to 4 kg

70. If the letter you are mailing weighs 29 g, you must put additional postage on it. How many kilograms does the letter weigh? (*Source: U.S. Postal Service.*)

0.029 kg

71. Is 1005 mg greater than or less than 1 g? What is the difference in the weights?

greater; 5 mg or 0.005 g

72. Is 990 mL greater than or less than 1 L? What is the difference in the amounts?

less; 10 mL or 0.01 L

73. One nickel weighs 5 g. How many nickels are in 1 kg of nickels?

200 nickels

74. The ratio of the total length of all the fish to the amount of water in an aquarium can be 3 cm of fish for every 4 L of water. What is the total length of all fish you can put in a 40 L aquarium? (*Source: Tropical Aquarium Fish.*)

30 cm of fish

Relating Concepts (Exercises 75–78) For Individual or Group Work

*Recall that the prefix **kilo** means 1000, so a **kilo**meter is 1000 meters. You'll learn about other prefixes for numbers greater than 1000 as you **work Exercises 75–78 in order.***

75. (a) The prefix *mega* means one million. Use the symbol M (capitalized) for *mega*. So a megameter (Mm) is how many meters?

1 Mm = _1,000,000_ m

(b) Figure out a unit fraction that you can use to convert megameters to meters. Then use it to convert 3.5 Mm to meters.

$$\frac{3.5 \text{ M\cancel{m}}}{1} \cdot \frac{1,000,000 \text{ m}}{1 \text{ M\cancel{m}}} = 3,500,000 \text{ m}$$

76. (a) The prefix *giga* means one billion. Use the symbol G (capitalized) for *giga*. So a gigameter (Gm) is how many meters?

1 Gm = _1,000,000,000_ m

(b) Figure out a unit fraction you can use to convert meters to gigameters. Then use it to convert 2500 m to gigameters.

$$\frac{2500 \text{ \cancel{m}}}{1} \cdot \frac{1 \text{ Gm}}{1,000,000,000 \text{ \cancel{m}}} = 0.0000025 \text{ Gm}$$

77. (a) The prefix *tera* means one trillion. Use the symbol T (capitalized) for *tera*. So a terameter (Tm) is how many meters?

1 Tm = _1,000,000,000,000_ m.

(b) Think carefully before you fill in the blanks:

1 Tm = _1000_ Gm

1 Tm = _1,000,000_ Mm

78. A computer's memory is measured in *bytes*. A byte can represent a single letter, a digit, or a punctuation mark. The memory for a desktop computer may be measured in megabytes (abbreviated MB) or gigabytes (abbreviated GB). Using the meanings of *mega* and *giga*, it would seem that

1 MB = _1,000,000_ bytes and

1 GB = _1,000,000,000_ bytes.

However, because computers use a base 2 or binary system, 1 MB is actually 2^{20} and 1 GB is 2^{30}. Use your calculator to find the actual values.

2^{20} = _1,048,576_ 2^{30} = _1,073,741,824_

Summary Exercises on U.S. Customary and Metric Units

The most commonly used U.S. customary and metric system units are listed in mixed-up order.
Write each unit in the correct box in the table below.

pound	yard	liter	kilogram	ton	gallon	pint
millimeter	gram	quart	meter	inch	cup	centimeter
milliliter	mile	foot	milligram	kilometer	ounce	fluid ounce

1. U.S. Customary Units			2. Metric System Units		
Length	Weight	Capacity	Length	Weight	Capacity
inch	ounce	fluid ounce	millimeter	milligram	milliliter
foot	pound	cup	centimeter	gram	liter
yard	ton	pint	meter	kilogram	
mile		quart	kilometer		
		gallon			

Fill in the blanks with the measurement relationships that you memorized.

3. (a) 1 __ft__ = 12 in.

(b) 3 ft = 1 __yd__

(c) 1 mi = __5280__ ft

4. (a) 60 sec = 1 __min__

(b) 1 hr = __60__ min

(c) __24__ hr = 1 day

5. (a) 1 cup = __8__ fl oz

(b) 4 qt = 1 __gal__

(c) __2__ pt = 1 qt

6. (a) 16 oz = 1 __lb__

(b) __2000__ lb = 1 ton

(c) 1 lb = __16__ oz

Write the most reasonable metric unit in each blank. Choose from km, m, cm, mm, L, mL, kg, g, and mg.

7. My water bottle holds 450 __mL__ .

8. Michael won the 200 __m__ race today.

9. The child weighed 23 __kg__ .

10. Jifar took a 375 __mg__ aspirin tablet.

11. The red pen is 14 __cm__ long.

12. A quarter is about 25 __mm__ across.

13. Merlene made 12 __L__ of fruit punch for her daughter's birthday party.

14. This cereal has 4 __g__ of protein in each serving.

Convert each measurement using unit fractions or the metric conversion line. Show your work.

15. 45 cm to meters

0.45 m

16. $\frac{3}{4}$ min to seconds

45 sec

17. 0.6 L to milliliters

600 mL

18. 8 g to milligrams

8000 mg

19. 300 mm to centimeters

30 cm

20. 45 in. to feet

$3\frac{3}{4}$ or 3.75 ft

21. 50 mL to liters

0.050 or 0.05 L

22. 18 qt to gallons

$4\frac{1}{2}$ or 4.5 gal

23. 7.28 kg to grams

7280 g

24. $2\frac{1}{4}$ lb to ounces

36 oz

25. 9 g to kilograms

0.009 kg

26. 5 yd to inches

180 in.

Solve Exercises 27–30 using the data in the table at the right on some of the world's tallest people.
(Source: Top 10 of Everything.)

27. What is the height of the tallest person in centimeters? In millimeters?

272 cm; 2720 mm

Robert Wadlow was 2.72 m tall.

28. Find the 4th tallest person's height in meters and in kilometers.

2.64 m; 0.00264 km

WORLD'S TALLEST PEOPLE

Rank	Name/Dates/Country	Height
1st	Robert Wadlow (1918–1940) USA	2.72 m
2nd	John Rogan (1868–1905) USA	268 cm
4th	John Carroll (1932–1969) USA	264 cm
7th	Edouard Beaupré (1881–1904) Canada	2.5 m
10th	Jeng Jinlian (1964–1982) China	248 cm

29. How much taller is the tallest person than the second tallest, in meters? Convert this difference to centimeters and millimeters.

0.04 m; 4 cm; 40 mm

30. What is the difference in height between the 7th tallest and 4th tallest people, in centimeters? Convert the height difference to meters and millimeters.

14 cm; 0.14 m; 140 mm

Solve each application problem. Show your work.

31. Vernice bought a 12 oz bag of chips for $3.49 today. What was the price per pound to the nearest cent?

$4.65 (rounded)

32. In 2006, Miami, Florida received about 64 in. of rain. In 2005, about 59 in. of rain fell. How many feet of rain did Miami get in all during the two years, to the nearest tenth?

10.3 ft (rounded)

7.4 ▶▶▶ Problem Solving with Metric Measurement

OBJECTIVE 1 Solve application problems involving metric measurements. One advantage of the metric system is the ease of comparing measurements in application situations. Just be sure that you are comparing similar units: mg to mg, km to km, and so on.

Once again, we will use the six problem-solving steps that you learned in **Section 1.10.**

OBJECTIVE

1 Solve application problems involving metric measurements.

EXAMPLE 1 Solving a Metric Application

Cheddar cheese is on sale at $8.99 per kilogram. Jake bought 350 g of the cheese. How much did he pay, to the nearest cent?

Step 1 **Read** the problem. The problem asks for the cost of 350 g of cheese.

Step 2 **Work out a plan.** The price is $8.99 per *kilogram*, but the amount Jake bought is given in *grams*. Convert grams to kilograms (the unit in the price). Then multiply the weight by the cost per kilogram.

Step 3 **Estimate** a reasonable answer. Round the cost of 1 kg from $8.99 to $9. There are 1000 g in a kilogram, so 350 g is about $\frac{1}{3}$ of a kilogram. Jake is buying about $\frac{1}{3}$ of a kilogram, so $\frac{1}{3}$ of $9 = $3 as our estimate.

Step 4 **Solve** the problem. Use a unit fraction to convert 350 g to kilograms.

> **g** divides out, leaving **kg** for your answer.

$$\frac{350 \cancel{g}}{1} \cdot \frac{1 \text{ kg}}{1000 \cancel{g}} = \frac{350}{1000} \text{ kg} = 0.35 \text{ kg}$$

Now multiply 0.35 kg times the cost per kilogram.

> Nearest cent is the nearest *hundredth*.

$$\frac{\$8.99}{1 \cancel{kg}} \cdot \frac{0.35 \cancel{kg}}{1} = \$3.1465 \approx \$3.15 \quad \text{(rounded)}$$

Step 5 **State the answer.** Jake paid $3.15, rounded to the nearest cent.

Step 6 **Check** your work. The exact answer of $3.15 is close to our estimate of $3.

Work Problem **1** *at the Side.* ▶

EXAMPLE 2 Solving a Metric Application

Olivia has 2.6 m of lace. How many centimeters of lace can she use to trim each of six hair ornaments? Round to the nearest tenth of a centimeter.

Step 1 **Read** the problem. The problem asks for the number of centimeters of lace for each of six hair ornaments.

Step 2 **Work out a plan.** The given amount of lace is in *meters,* but the answer must be in *centimeters*. Convert meters to centimeters, then divide by 6 (the number of hair ornaments).

Continued on Next Page

1 Solve this problem using the six problem-solving steps.

Satin ribbon is on sale at $0.89 per meter. How much will 75 cm cost, to the nearest cent?

ANSWER

1. $0.67 (rounded)

2 Lucinda's doctor wants her to take 1.2 g of medication each day in three equal doses. How many milligrams should be in each dose? Use the six problem-solving steps.

Step 3 **Estimate** a reasonable answer. To estimate, round 2.6 m of lace to 3 m. Then, 3 m = 300 cm, and 300 cm ÷ 6 = 50 cm as our estimate.

Step 4 **Solve** the problem. On the metric conversion line, moving from **m** to **cm** is two places to the right, so move the decimal point in 2.6 m two places to the right. Then divide by 6.

$$2.60\,\text{m} = 260\text{ cm} \qquad \frac{260\text{ cm}}{6\text{ ornaments}} \approx 43.3\text{ cm per ornament}$$

Step 5 **State the answer.** Olivia can use about 43.3 cm of lace on each ornament.

Step 6 **Check** your work. The exact answer of 43.3 cm is close to our estimate of 50 cm.

◀ Work Problem 2 at the Side.

Note

In Example 1 we used a unit fraction to convert the measurement, and in Example 2 we moved the decimal point. Use whichever method you prefer. Also, there is more than one way to solve an application problem. Another way to solve Example 2 is to divide 2.6 m by 6 to get 0.4333 m of lace for each ornament. Then convert 0.4333 m to 43.3 cm (rounded to the nearest tenth).

3 Andrea has two pieces of fabric. One measures 2 m 35 cm and the other measures 1 m 85 cm. How many meters of fabric does she have in all? Use the six problem-solving steps.

EXAMPLE 3 **Solving a Metric Application**

Rubin measured a board and found that the length was 3 m plus an additional 5 cm. He cut off a piece measuring 1 m 40 cm for a shelf. Find the length in meters of the remaining piece of board.

Step 1 **Read** the problem. Part of a board is cut off. The problem asks what length of board, in meters, is left over. It may help to make a drawing of the board and label the lengths given in the problem.

Step 2 **Work out a plan.** The lengths involve two units, m and cm. Rewrite both lengths in meters (the unit called for in the answer), and then subtract.

Step 3 **Estimate** a reasonable answer. To estimate, 3 m 5 cm can be rounded to 3 m, because 5 cm is less than half of a meter (less than 50 cm). Round 1 m 40 cm down to 1 m. Then, 3 m − 1 m = 2 m as our estimate.

Step 4 **Solve** the problem. Rewrite the lengths in meters. Then subtract.

3 m ⟶	3.00 m	40. ⟶ 1 m ⟶ 1.0 m
plus 5 cm ⟶	+ 0.05 m	plus 40 cm ⟶ + 0.4 m
05.	3.05 m	1.4 m

Subtract to find leftover length.

$$\begin{array}{r} 3.05\text{ m} \leftarrow \text{Board} \\ -\ 1.40\text{ m} \leftarrow \text{Shelf} \\ \hline 1.65\text{ m} \leftarrow \text{Leftover piece} \end{array}$$

Step 5 **State the answer.** The length of the remaining piece is 1.65 m.

Step 6 **Check** your work. The exact answer of 1.65 m is close to our estimate of 2 m.

◀ Work Problem 3 at the Side.

ANSWERS
2. 400 mg per dose
3. 4.2 m

Solve each application problem. Show your work. Round money answers to the nearest cent. See Examples 1–3.

1. Bulk rice at the food co-op is on special at $0.98 per kilogram. Pam scooped some rice into a bag and put it on the scale. How much will she pay for 850 g of rice?

$0.83 (rounded)

2. Lanh is buying a piece of plastic tubing measuring 315 cm for the science lab. The price is $4.75 per meter. How much will Lanh pay?

$14.96 (rounded)

3. A miniature Yorkshire terrier, one of the smallest dogs, may weigh only 500 g. But a St. Bernard, the heaviest dog, could easily weigh 90 kg. What is the difference in the weights of the two dogs, in kilograms? (*Source: Big Book of Knowledge.*)

89.5 kg

4. The world's longest insect is the giant stick insect of Indonesia, measuring 33 cm. The fairy fly, the smallest insect, is just 0.2 mm long. How much longer is the giant stick insect, in millimeters? (*Source: Big Book of Knowledge.*)

329.8 mm

5. An adult human body contains about 5 L of blood. If each beat of the heart pumps 70 mL of blood, how many times must the heart beat to pass all the blood through the heart? Round to the nearest whole number of beats. (*Source: Harper's Index.*)

71 beats (rounded)

6. A floor tile measures 30 cm by 30 cm and weighs 185 g. How many kilograms would a stack of 24 tiles weigh? How much would five stacks of tiles weigh? (*Source: The Tile Shop.*)

4.44 kg; 22.2 kg

7. Each piece of lead for a mechanical pencil has a thickness of 0.5 mm and is 60 mm long. Find the total length in centimeters of the lead in a package with 30 pieces. If the price of the package is $3.29, find the cost per centimeter for the lead. (*Source: Pentel.*)

180 cm; $0.02/cm (rounded)

8. The apartment building caretaker puts 750 mL of chlorine into the swimming pool every day. How many liters should he order to have a one-month (30-day) supply on hand? If chlorine is sold in containers that hold 4 L, how many containers should be ordered for one month? How much chlorine will be left over at the end of the month?

22.5 L; 6 containers, because you cannot buy part of a container; 1.5 L left over.

9. Rosa is building a bookcase. She has one board that is 2 m 8 cm long and another that is 2 m 95 cm long. What is the total length of the two boards in meters?

5.03 m

10. Janet has a piece of fabric that is 10 m 30 cm in length. She wants to make curtains for three windows that are all the same size. What length of fabric is available for each window, to the nearest tenth of a meter?

3.4 m (rounded)

11. In a chemistry lab, each of the 45 students needs 85 mL of acid. How many 1 L bottles of acid need to be ordered?

    4 bottles

12. James needs two 1.3 m pieces and two 85 cm pieces of wood molding to frame a picture. The price is $5.89 per meter plus 7% sales tax. How much will James pay?

    $27.10 (rounded)

Use the bar graph below to answer Exercises 13 and 14.

Caffeine Meter
Average milligrams of caffeine
per 8 oz cup or equivalent

Double espresso 160 mg
Drip coffee 90 mg
Cola 45 mg
25 mg Chocolate bar
5 mg Decaffeinated coffee

Source: Celestial Seasonings.

13. If Agnete usually drinks three 8 oz cups of drip coffee each day, how many grams of caffeine will she consume in one week?

    1.89 g

14. Lorenzo's doctor has suggested that he cut down on caffeine. So Lorenzo switched from drinking four 8 oz cups of cola every day to drinking two 8 oz cups of decaffeinated coffee. How many fewer grams of caffeine is he consuming each week?

    1.19 g

15. During August 2003, Mars moved closer to Earth at a rate of about 10,000 meters per second. How much closer, in kilometers, did Mars get to Earth:

    (a) in one second, **10 km**

    (b) in one minute, **600 km**

    (c) in one hour? **36,000 km**

    (*Source:* NASA.)

16. Some of the newest football stadiums have Field Turf instead of grass. Use the drawing below to find the total thickness in centimeters of the top two layers of Field Turf.

    64 mm fiber grass

    4.5 cm rubber and sand filler

    asphalt base

    10.9 cm

    Source: Sports Facilities Commission.

Relating Concepts (Exercises 17–20) For Individual or Group Work

It is difficult to weigh very light objects, such as a single sheet of paper or a single staple (unless you have an expensive scientific scale). But you can weigh a large number of the items and then divide to find the weight of one item. Before dividing, subtract the weight of the box or wrapper that the items are packaged in to find the net weight. **Work Exercises 17–20 in order**, *to complete the table.*

	Item	Total Weight in Grams	Weight of Packaging	Net Weight	Weight of One Item in Grams	Weight of One Item in Milligrams
17.	Box of 50 envelopes	255 g	40 g	_215 g_	_4.3 g_	_4300 mg_
18.	Box of 1000 staples	350 g	20 g	_330 g_	_0.33 g_	_330 mg_
19.	Ream of paper (500 sheets)	_1550 g_	50 g	_1500 g_	_3 g_	3000 mg
20.	Box of 100 small paper clips	_55 g_	5 g	_50 g_	_0.5 g_	500 mg

7.5 ▶▶▶ Metric–U.S. Customary Conversions and Temperature

OBJECTIVE 1 Use unit fractions to convert between metric and U.S. customary units. Until the United States has switched completely from customary units to the metric system, it will be necessary to make conversions from one system to the other. *Approximate* conversions can be made with the help of the table below, in which the values have been rounded to the nearest hundredth or thousandth. (The only value that is exact, not rounded, is 1 inch = 2.54 cm.)

Metric to U.S. Customary		U.S. Customary to Metric	
1 kilometer	≈ 0.62 mile	1 mile	≈ 1.61 kilometers
1 meter	≈ 1.09 yards	1 yard	≈ 0.91 meter
1 meter	≈ 3.28 feet	1 foot	≈ 0.30 meter
1 centimeter	≈ 0.39 inch	1 inch	= 2.54 centimeters
1 liter	≈ 0.26 gallon	1 gallon	≈ 3.79 liters
1 liter	≈ 1.06 quarts	1 quart	≈ 0.95 liter
1 kilogram	≈ 2.20 pounds	1 pound	≈ 0.45 kilogram
1 gram	≈ 0.035 ounce	1 ounce	≈ 28.35 grams

OBJECTIVES

1 Use unit fractions to convert between metric and U.S. customary units.

2 Learn common temperatures on the Celsius scale.

3 Use formulas to convert between Celsius and Fahrenheit temperatures.

EXAMPLE 1 Converting between Metric and U.S. Customary Length Units

Convert 10 m to yards using unit fractions. Round your answer to the nearest tenth if necessary.

We're changing from a *metric* unit to a *U.S. customary* unit. In the "Metric to U.S. Customary" side of the table above, you see that 1 meter ≈ 1.09 yards. Two unit fractions can be written using that information.

$$\frac{1 \text{ m}}{1.09 \text{ yd}} \quad \text{or} \quad \frac{1.09 \text{ yd}}{1 \text{ m}}$$

Multiply by the unit fraction that allows you to divide out meters (that is, meters is in the denominator).

$$10 \text{ m} \cdot \frac{1.09 \text{ yd}}{1 \text{ m}} = \frac{10 \text{ m}}{1} \cdot \frac{1.09 \text{ yd}}{1 \text{ m}} = \frac{(10)(1.09 \text{ yd})}{1} = 10.9 \text{ yd}$$

These units should match.

Meters (**m**) divide out leaving **yd**, the unit you want for the answer.

10 m ≈ 10.9 yd

Note

In Example 1 above, you could also use the numbers from the "U.S. Customary to Metric" side of the table that involve meters and yards: 1 yard ≈ 0.91 meter.

$$\frac{10 \text{ m}}{1} \cdot \frac{1 \text{ yd}}{0.91 \text{ m}} = \frac{10}{0.91} \text{ yd} ≈ 10.99 \text{ yd}$$

The answer is slightly different because the values in the table are rounded. Also, you have to divide instead of multiply, which is usually more difficult to do without a calculator. We will use the first method in this chapter.

1 Convert using unit fractions. Round your answers to the nearest tenth.

(a) 23 m to yards

(b) 40 cm to inches

(c) 5 mi to kilometers (Look at the "U.S. Customary to Metric" side of the table.)

(d) 12 in. to centimeters

ANSWERS

1. (a) 23 m ≈ 25.1 yd **(b)** 40 cm ≈ 15.6 in.
(c) 5 mi ≈ 8.1 km **(d)** 12 in. ≈ 30.5 cm

Work Problem **1** *at the Side.* ▶

2 Convert. Use the values from the table on the previous page to make unit fractions. Round answers to the nearest tenth.

(a) 17 kg to pounds

(b) 5 L to quarts

(c) 90 g to ounces

(d) 3.5 gal to liters

(e) 145 lb to kilograms

(f) 8 oz to grams

> **EXAMPLE 2** **Converting between Metric and U.S. Customary Weight and Capacity Units**

Convert using unit fractions. Round your answers to the nearest tenth.

(a) 3.5 kg to pounds

Look in the "Metric to U.S. Customary" side of the table on the previous page to see that 1 kilogram $\approx$ 2.20 pounds. Use this information to write a unit fraction that allows you to divide out kilograms.

$$\frac{3.5 \; \cancel{kg}}{1} \cdot \frac{2.20 \; lb}{1 \; \cancel{kg}} = \frac{(3.5)(2.20 \; lb)}{1} = 7.7 \; lb$$

3.5 kg $\approx$ 7.7 lb

> The conversion value is approximate, so use the $\approx$ symbol in your answer.

(b) 18 gal to liters

Look in the "U.S. Customary to Metric" side of the table to see that 1 gallon $\approx$ 3.79 liters. Write a unit fraction that allows you to divide out gallons.

> **gal** divides out, leaving **L** for your answer.

$$\frac{18 \; \cancel{gal}}{1} \cdot \frac{3.79 \; L}{1 \; \cancel{gal}} = \frac{(18)(3.79 \; L)}{1} = 68.22 \; L$$

68.22 rounded to the nearest tenth is 68.2

18 gal $\approx$ 68.2 L

(c) 300 g to ounces

In the "Metric to U.S. Customary" side of the table, 1 gram $\approx$ 0.035 ounce.

$$\frac{300 \; \cancel{g}}{1} \cdot \frac{0.035 \; oz}{1 \; \cancel{g}} = \frac{(300)(0.035 \; oz)}{1} = 10.5 \; oz$$

300 g $\approx$ 10.5 oz

CAUTION

Because the metric and U.S. customary systems were developed independently, almost all comparisons are approximate. Your answers should be written with the "$\approx$" symbol to show they are approximate.

◀ *Work Problem* **2** *at the Side.*

OBJECTIVE **2** **Learn common temperatures on the Celsius scale.** In the metric system, temperature is measured on the **Celsius** scale. On the Celsius scale, water freezes at 0 °C and boils at 100 °C. The small raised circle stands for "degrees" and the capital **C** is for Celsius. Read the temperatures like this:

Water freezes at 0 degrees Celsius (0 °C).

Water boils at 100 degrees Celsius (100 °C).

The U.S. customary temperature system, used only in the United States, is measured on the **Fahrenheit** scale. On this scale:

Water freezes at 32 degrees Fahrenheit (32 °F).

Water boils at 212 degrees Fahrenheit (212 °F).

ANSWERS

2. (a) 17 kg $\approx$ 37.4 lb   **(b)** 5 L $\approx$ 5.3 qt
   (c) 90 g $\approx$ 3.2 oz   **(d)** 3.5 gal $\approx$ 13.3 L
   (e) 145 lb $\approx$ 65.3 kg   **(f)** 8 oz $\approx$ 226.8 g

The thermometer below shows some typical temperatures in both Celsius and Fahrenheit. For example, comfortable room temperature is about 20 °C or 68 °F, and normal body temperature is about 37 °C or 98.6 °F.

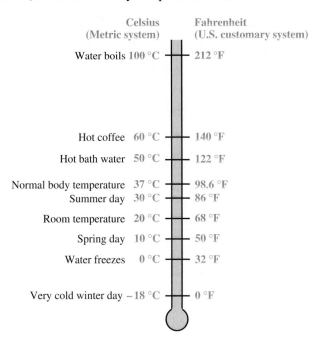

| Celsius (Metric system) | Fahrenheit (U.S. customary system) |

Water boils 100 °C — 212 °F
Hot coffee 60 °C — 140 °F
Hot bath water 50 °C — 122 °F
Normal body temperature 37 °C — 98.6 °F
Summer day 30 °C — 86 °F
Room temperature 20 °C — 68 °F
Spring day 10 °C — 50 °F
Water freezes 0 °C — 32 °F
Very cold winter day −18 °C — 0 °F

Note

The freezing and boiling temperatures are exact. The other temperatures are approximate. Even normal body temperature varies slightly from person to person.

EXAMPLE 3 **Using Celsius Temperatures**

Circle the Celsius temperature that is most reasonable for each situation.

(a) Warm summer day   29 °C   64 °C   90 °C

29 °C is reasonable. 64 °C and 90 °C are too hot; they're both above the temperature of hot bath water (above 122 °F).

(b) Inside a freezer   −10 °C   3 °C   25 °C

−10 °C is the reasonable temperature because it is the only one below the freezing point of water (0 °C). Your frozen foods would start thawing at 3 °C or 25 °C.

Work Problem **3** *at the Side.* ▶

OBJECTIVE **3** **Use formulas to convert between Celsius and Fahrenheit temperatures.** You can use these formulas to convert between Celsius and Fahrenheit temperatures.

Celsius–Fahrenheit Conversion Formulas

Converting from Fahrenheit (F) to Celsius (C)

$$C = \frac{5(F - 32)}{9}$$

Converting from Celsius (C) to Fahrenheit (F)

$$F = \frac{9 \cdot C}{5} + 32$$

3 Circle the Celsius temperature that is *most* reasonable for each situation.

(a) Set the living room thermostat at:
11 °C   21 °C   71 °C

(b) The baby has a fever of:
29 °C   39 °C   49 °C

(c) Wear a sweater outside because it's:
15 °C   25 °C   50 °C

(d) My iced tea is:
−5 °C   5 °C   30 °C

(e) Time to go swimming! It's:
95 °C   65 °C   35 °C

(f) Inside a refrigerator (not the freezer) it's:
−15 °C   0 °C   3 °C

(g) There's a blizzard outside. It's:
12 °C   4 °C   −20 °C

(h) I need hot water to get these clothes clean. It should be:
55 °C   105 °C   200 °C

ANSWERS

3. **(a)** 21 °C **(b)** 39 °C **(c)** 15 °C
(d) 5 °C **(e)** 35 °C **(f)** 3 °C
(g) −20 °C **(h)** 55 °C

As you use these formulas, be sure to follow the order of operations from **Section 1.8.**

4 Convert to Celsius.

(a) 59 °F

> **Order of Operations**
> 1. Do all operations inside **parentheses** or **other grouping symbols.**
> 2. Simplify any expressions with **exponents** and find any **square roots.**
> 3. **Multiply** or **divide,** proceeding from left to right.
> 4. **Add** or **subtract,** proceeding from left to right.

(b) 41 °F

EXAMPLE 4 | **Converting Fahrenheit to Celsius**

Convert 68 °F to Celsius.

Use the formula and follow the order of operations.

(c) 212 °F

$$C = \frac{5(F - 32)}{9}$$ Fahrenheit to Celsius formula.

$$= \frac{5(68 - 32)}{9}$$ Work inside parentheses first; $68 - 32$ is 36.

$$= \frac{5(36)}{9}$$

(d) 98.6 °F

$$= \frac{5(\overset{4}{\cancel{36}})}{\underset{1}{\cancel{9}}}$$ Divide out the common factor. Multiply in the numerator.

$$= 20$$

5 Convert to Fahrenheit.

(a) 100 °C

Thus, 68 °F = 20 °C.

◀ *Work Problem* **4** *at the Side.*

(b) 25 °C

EXAMPLE 5 | **Converting Celsius to Fahrenheit**

Convert 15 °C to Fahrenheit.

Use the formula and follow the order of operations.

(c) 80 °C

$$F = \frac{9 \cdot C}{5} + 32$$ Celsius to Fahrenheit formula.

$$= \frac{9 \cdot 15}{5} + 32$$

$$= \frac{9 \cdot \overset{3}{\cancel{15}}}{\underset{1}{\cancel{5}}} + 32$$ Divide out the common factor. Multiply in the numerator.

(d) 5 °C

$$= 27 + 32$$ Add.

$$= 59$$

Thus, 15 °C = 59 °F.

◀ *Work Problem* **5** *at the Side.*

7.5 ▶▶▶ Exercises

FOR
EXTRA
HELP

MyMathLab

Math XL
PRACTICE

WATCH

DOWNLOAD

READ

REVIEW

Use the table on the first page of this section and unit fractions to make approximate conversions from metric to U.S. customary or U.S. customary to metric. Round your answers to the nearest tenth. See Examples 1 and 2.

1. 20 m to yards
21.8 yd

2. 8 km to miles
5.0 mi

3. 80 m to feet
262.4 ft

4. 85 cm to inches
33.2 in.

5. 16 ft to meters
4.8 m

6. 3.2 yd to meters
2.9 m

7. 150 g to ounces
5.3 oz

8. 2.5 oz to grams
70.9 g

9. 248 lb to kilograms
111.6 kg

10. 7.68 kg to pounds
16.9 lb

11. 28.6 L to quarts
30.3 qt

12. 15.75 L to gallons
4.1 gal

13. For the 2000 Olympics, the 3M Company used 5 g of pure gold to coat Michael Johnson's track shoes. (*Source:* 3M Company.)

(a) How many ounces of gold were used, to the nearest tenth?
about 0.2 oz

(b) Was this enough extra weight to slow him down?   probably not

14. The label on a Van Ness auto feeder for cats and dogs says it holds 1.4 kg of dry food. How many pounds of food does it hold, to the nearest tenth? (*Source:* Van Ness Plastics.)

3.1 pounds (rounded)

15. The heavy-duty wash cycle in a dishwater uses 8.4 gal of water. How many liters does it use, to the nearest tenth? (*Source:* Frigidaire.)

about 31.8 L

16. The rinse-and-hold cycle in a dishwasher uses only 4.5 L of water. How many gallons does it use, to the nearest tenth? (*Source:* Frigidaire.)

about 1.2 gal

17. The smallest pet fish are dwarf gobies, which are half an inch long. How many centimeters long is a dwarf gobie, to the nearest tenth? (*Hint*: Write half an inch in decimal form.) about 1.3 cm

18. The fastest nerve signals in the human body travel 120 meters per second. How many feet per second do the signals travel? (*Source: Big Book of Knowledge.*)

about 393.6 feet per second

Circle the most reasonable Celsius temperature for each situation. See Example 3.

19. A snowy day

12 °C   28 °C   (−8 °C)

20. Brewing coffee

(80 °C)   180 °C   15 °C

21. A high fever

21 °C   (40 °C)   103 °C

22. Swimming pool water

90 °C   78 °C   (25 °C)

23. Oven temperature

(150 °C)   50 °C   30 °C

24. Light jacket weather

0 °C   (10 °C)   −10 °C

25. Would a drop in temperature of 20 Celsius degrees be more or less than a drop of 20 Fahrenheit degrees? Explain your answer.

More. There are 180 degrees between freezing and boiling on the Fahrenheit scale, but only 100 degrees on the Celsius scale, so each Celsius degree is a greater change in temperature.

26. Describe one advantage of switching from the Fahrenheit temperature scale to the Celsius scale. Describe one disadvantage.

Advantage: the rest of the world uses the Celsius scale. Disadvantage: Americans would have to buy new thermometers and get used to a new system.

Use the conversion formulas from this section and the order of operations to convert Fahrenheit temperatures to Celsius and Celsius temperatures to Fahrenheit. Round your answers to the nearest degree if necessary. See Examples 4 and 5.

27. 60 °F

16 °C (rounded)

28. 80 °F

27 °C (rounded)

29. 104 °F

40 °C

30. 36 °F

2 °C (rounded)

🌐 **31.** 8 °C

46 °F (rounded)

32. 18 °C

64 °F (rounded)

33. 35 °C

95 °F

34. 0 °C

32 °F

Solve each application problem. Round your answers to the nearest degree if necessary.

35. The highest temperature ever recorded on Earth was 136 °F at El Azizia, Libya, in 1922. Convert this temperature to Celsius. (*Source: The World Almanac.*)

58 °C (rounded)

36. Hummingbirds have a normal body temperature of 107 °F. But on cold nights they go into a state of torpor where their body temperature drops to 39 °F. What are these temperatures in Celsius? (*Source: Wildbird.*)

42 °C; 4 °C (both rounded)

37. **(a)** Here is the tag on a pair of Sorel boots. In what kind of weather would you wear these boots?

pleasant weather, above freezing but not hot

Comfort
range
24 °C to 4 °C

Source: Sorel.

(b) For what Fahrenheit temperatures are the boots designed?

75 °F to 39 °F (rounded)

(c) What range of metric temperatures do you have in January where you live? Would you be comfortable in these boots?

Answers will vary. In Minnesota, it's 0 °C to −40 °C; in California, 24 °C to 0 °C.

38. Sleeping bags made by Eddie Bauer are sold around the world. Each type of sleeping bag is designed for outdoor camping in certain temperatures.

Junior bag	5 °C or warmer
Removable liner bag	0 °C to 15 °C
Conversion bag	−7 °C to 0 °C

Source: Eddie Bauer.

(a) At what Fahrenheit temperatures should you use the Junior bag?

41 °F or warmer

(b) What Fahrenheit temperatures is the removable liner bag designed for?

32 °F to 59 °F

Relating Concepts (Exercises 39–46) For Individual or Group Work

The article below appeared in American newspapers. However, both Newfoundland (part of Canada) and Ireland use the metric system. Their newspapers would have reported all the measurements in metric units. Complete the conversions to metric, rounding answers to the nearest tenth.

Q.: A recent news brief reported on some men who flew a model airplane from Newfoundland to Ireland. Can you provide some details of the flight?

A.: The model plane is 6 feet long and weighs 11 pounds. Made of balsa wood and mylar, it crossed the Atlantic—the flight path took it 1,888.3 miles—in 38 hours, 23 minutes. It soared at a cruising altitude of 1,000 feet. The plane used a souped-up piston engine and carried less than a gallon of fuel, as mandated by rules of the Federation Aeronautique Internationale, the governing body of model airplane building. When it landed in County Galway, Ireland, it had less than 2 fluid ounces of fuel left. The plane was built by Maynard Hill of Silver Spring, Maryland.

(Source: New York Times.)

39. Length of model plane

about 1.8 m

40. Weight of plane

about 5.0 kg

41. Length of flight path

about 3040.2 km

42. Time of flight

38 hr 23 min

43. Cruising altitude

about 300 m

44. Fuel at the start, in milliliters

less than 3790 mL

45. Fuel left after landing, in milliliters
(*Hint:* First convert 2 fl oz to quarts.)

about 59.4 mL

46. What *percent* of the fuel was left at the end of the flight?

about 1.6%

Chapter 7 Summary

▶ Key Terms

7.1	**U.S. customary measurement units**	The U.S. customary measurement units are used for many daily activities only in the United States. Commonly used units include quarts, pounds, feet, miles, and degrees Fahrenheit.
	metric system	The metric system of measurement is an international system used in manufacturing, science, medicine, sports, and other fields. Commonly used units in this system include meters, liters, grams, and degrees Celsius.
	unit fraction	A unit fraction involves measurement units and is equivalent to 1. Unit fractions are used to convert among different measurements.
7.2	**meter**	The meter is the basic unit of length in the metric system. The symbol **m** is used for meter. One meter is a little longer than a yard.
	prefixes	Attaching a prefix such as *kilo-* or *milli-* to the words meter, liter, or gram gives names of larger or smaller units. For example, the prefix *kilo* means 1000 so a *kilo*meter is 1000 meters.
	metric conversion line	The metric conversion line is a line showing the various metric measurement prefixes and their size relationship to each other.
7.3	**liter**	The liter is the basic unit of capacity in the metric system. The symbol **L** is used for liter. One liter is a little more than one quart.
	gram	The gram is the basic unit of weight (mass) in the metric system. The symbol **g** is used for gram. One gram is the weight of 1 milliliter of water or one dollar bill.
7.5	**Celsius**	The Celsius scale is used to measure temperature in the metric system. Water boils at 100 °C and freezes at 0 °C.
	Fahrenheit	The Fahrenheit scale is used to measure temperature in the U.S. customary system. Water boils at 212 °F and freezes at 32 °F.

▶ New Symbols

Frequently used metric length units

- **km** kilometer
- **m** meter
- **cm** centimeter
- **mm** millimeter

Frequently used metric capacity units

- **L** liter
- **mL** milliliter

Frequently used metric weight (mass) units

- **kg** kilogram
- **g** gram
- **mg** milligram

°C degrees Celsius
(metric temperature unit)

°F degrees Fahrenheit
(U.S. customary temperature unit)

▶ New Formulas

Converting from Celsius to Fahrenheit:

$$F = \frac{9 \cdot C}{5} + 32$$

Converting from Fahrenheit to Celsius:

$$C = \frac{5(F - 32)}{9}$$

▶ Test Your Word Power

See how well you have learned the vocabulary in this chapter. Answers follow the Quick Review.

1. The **metric system**
   A. uses meters, liters, and degrees Fahrenheit
   B. is based on multiples of 10
   C. is used only in the United States
   D. has evolved over centuries.

2. The **U.S. customary measurement units**
   A. are used throughout the world
   B. are based on multiples of 12
   C. include feet, inches, quarts, and pounds
   D. were developed by a group of scientists in 1790.

3. A **unit fraction**
   A. has the unit you want to change in the numerator
   B. has a denominator of 1
   C. must be written in lowest terms
   D. is equivalent to 1.

4. A **gram** is
   A. the weight of 1 mL of water
   B. abbreviated gm
   C. equivalent to 1000 kg
   D. approximately equal to 2.2 pounds.

5. A **meter** is
   A. equivalent to 1000 cm
   B. approximately equal to $\frac{1}{2}$ inch
   C. abbreviated m with no period after it
   D. the basic unit of capacity in the metric system.

6. The **Celsius** temperature scale
   A. shows water freezing at 32°
   B. is used in the U.S. customary system of measurement
   C. shows water boiling at 100°
   D. cannot be converted to the Fahrenheit temperature scale.

▶ Quick Review

Concepts	Examples

7.1 The U.S. Customary Measurement Units

Memorize the basic measurement relationships. Then, to convert units, multiply when changing from a larger unit to a smaller unit; divide when changing from a smaller unit to a larger unit.

Convert each measurement.
(a) 5 ft to inches

$$5 \text{ ft} = 5 \cdot 12 = 60 \text{ in.}$$

(b) 3 lb to ounces

$$3 \text{ lb} = 3 \cdot 16 = 48 \text{ oz}$$

(c) 15 qt to gallons

$$15 \text{ qt} = \frac{15}{4} = 3\frac{3}{4} \text{ gal}$$

7.1 Using Unit Fractions

Another, more useful, conversion method is multiplying by a unit fraction. The unit you want in the answer should be in the numerator. The unit you want to change should be in the denominator.

Convert 32 oz to pounds.

$$32 \text{ oz} \cdot \frac{1 \text{ lb}}{16 \text{ oz}} \Big\} \text{ Unit fraction}$$

These units should match.

$$= \frac{\overset{2}{\cancel{32} \cancel{oz}}}{1} \cdot \frac{1 \text{ lb}}{\underset{1}{\cancel{16} \cancel{oz}}} \quad \begin{array}{l} \text{Divide out ounces.} \\ \text{Divide out common factors.} \end{array}$$

$$= 2 \text{ lb}$$

Concepts	Examples

7.1 **Solving U.S. Customary Measurement Application Problems**

To solve application problems, use the six problem-solving steps.

Step 1 **Read** the problem carefully.

Step 2 **Work out a plan.**

Step 3 **Estimate** a reasonable answer.

Step 4 **Solve** the problem.

Step 5 **State the answer.**
Step 6 **Check** your work.

Use the six steps to solve this problem.
Mr. Green has 10 yd of rope. He is cutting it into eight pieces so his sailing class can practice knot tying. How many feet of rope will each of his eight students get?

Step 1 The problem asks how many feet of rope can be given to each of eight students.

Step 2 Convert 10 yd to feet (the unit required in the answer). Then divide by eight students.

Step 3 There are 3 ft in one yard, so there are 30 ft in 10 yd. Then 30 ft ÷ 8 ≈ 4 ft as our estimate.

Step 4 Use a unit fraction to convert 10 yd to feet, then divide.

$$\frac{10 \text{ yd}}{1} \cdot \frac{3 \text{ ft}}{1 \text{ yd}} = 30 \text{ ft}$$

$$\frac{30 \text{ ft}}{8 \text{ students}} = 3\frac{3}{4} \text{ ft or 3.75 ft per student}$$

Step 5 Each student gets $3\frac{3}{4}$ ft or 3.75 ft of rope.
Step 6 The exact answer of 3.75 ft is close to our estimate of 4 ft.

7.2 **Basic Metric Length Units**

Use approximate comparisons to judge which length units are appropriate:

1 mm is the thickness of a dime.

1 cm is about $\frac{1}{2}$ inch.

1 m is a little more than 1 yard.

1 km is about 0.6 mile.

Write the most reasonable metric unit in each blank. Choose from km, m, cm, and mm.

The room is 6 __m__ long.

A paper clip is 30 __mm__ long.

He drove 20 __km__ to work.

7.2 and 7.3 **Converting within the Metric System**

Using Unit Fractions
One conversion method is to multiply by a unit fraction. Use a fraction with the unit you want in the answer in the numerator and the unit you want to change in the denominator.

Convert.
(a) 9 g to kg

$$\frac{9 \text{ g}}{1} \cdot \frac{1 \text{ kg}}{1000 \text{ g}} = \frac{9}{1000} \text{ kg} = 0.009 \text{ kg}$$

9 g = 0.009 kg

(b) 3.6 m to cm

$$\frac{3.6 \text{ m}}{1} \cdot \frac{100 \text{ cm}}{1 \text{ m}} = 360 \text{ cm}$$

3.6 m = 360 cm

(continued)

Concepts	Example

7.2 and 7.3 Converting within the Metric System
(*continued*)

Using the Metric Conversion Line
Another conversion method is to find the unit you are given on the metric conversion line. Count the number of places to get from the unit you are given to the unit you want. Move the decimal point the same number of places and in the same direction.

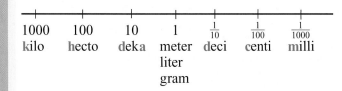

1000	100	10	1	$\frac{1}{10}$	$\frac{1}{100}$	$\frac{1}{1000}$
kilo	hecto	deka	meter liter gram	deci	centi	milli

Convert.
(a) 68.2 kg to g
   From **kg** to **g** is three places to the right.

   6 8.2 0 0   Decimal point is moved three places to the right.

   68.2 kg = 68,200 g

(b) 300 mL to L
   From **mL** to **L** is three places to the left.

   3 0 0.   Decimal point is moved three places to the left.

   300 mL = 0.3 L

(c) 825 cm to m
   From **cm** to **m** is two places to the left.

   8 2 5.   Decimal point is moved two places to the left.

   825 cm = 8.25 m

7.3 Basic Metric Capacity Units

Use approximate comparisons to judge which capacity units are appropriate:

   1 L is a little more than 1 quart.

   1 mL is the amount of water in
       a cube 1 cm on each side.

   5 mL is about one teaspoon.

   250 mL is about one cup.

Write the most reasonable metric unit in each blank. Choose from L and mL.

The pail holds 12 __L____ .

The milk carton from the vending machine holds 250 __mL____ .

7.3 Basic Metric Weight (Mass) Units

Use approximate comparisons to judge which weight units are appropriate:

   1 kg is about 2.2 pounds.

   1 g is the weight of 1 mL of water
       or one dollar bill.

   1 mg is $\frac{1}{1000}$ of a gram; very tiny!

Write the most reasonable metric unit in each blank. Choose from kg, g, and mg.

The wrestler weighed 95 __kg____ .

She took a 500 __mg____ aspirin tablet.

One banana weighs 150 __g____ .

Concepts	Examples

7.4 Solving Metric Measurement Application Problems

Convert units so you are comparing kg to kg, cm to cm, and so on. When a measurement involves two units, such as 6 m 20 cm, write it in terms of the unit called for in the answer (6.2 m or 620 cm).

Use the six problem-solving steps.
Step 1   **Read** the problem carefully.

Step 2   **Work out a plan.**

Step 3   **Estimate** a reasonable answer.

Step 4   **Solve** the problem.

Step 5   **State the answer.**

Step 6   **Check** your work.

Use the six steps to solve this problem.

George cut 1 m 35 cm off of a 3 m board. How long was the leftover piece, in meters?

Step 1   The problem asks for the length of the leftover piece in meters.

Step 2   Convert the cut-off measurement to meters (the unit required in the answer), and then subtract to find the "leftover."

Step 3   To estimate, round 1 m 35 cm to 1 m, because 35 cm is less than half a meter. Then, 3 m − 1 m = 2 m as our estimate.

Step 4   Convert the cut-off measurement to meters, then subtract.

$$
\begin{array}{ll}
\text{1 m} \rightarrow \quad 1.00 \text{ m} & 3.00 \text{ m} \leftarrow \text{Board} \\
\text{plus 35 cm} \rightarrow \underline{+\ 0.35 \text{ m}} & \underline{-\ 1.35 \text{ m}} \leftarrow \text{Cut off} \\
\qquad\qquad\quad 1.35 \text{ m} & 1.65 \text{ m} \leftarrow \text{Left}
\end{array}
$$

Step 5   The leftover piece is 1.65 m long.

Step 6   The exact answer of 1.65 m is close to our estimate of 2 m.

7.5 Converting between Metric and U.S. Customary Units

Write a unit fraction using the values in the table of conversion factors on the first page of **Section 7.5.** Because the values in the table are rounded, your answers will be approximate.

Convert. Round answers to the nearest tenth.

(a) 23 m to yards
From the table, 1 meter ≈ 1.09 yards.

$$\frac{23 \text{ m}}{1} \cdot \frac{1.09 \text{ yd}}{1 \text{ m}} = 25.07 \text{ yd}$$

25.07 rounds to 25.1, so 23 m ≈ 25.1 yd.

(b) 4 oz to grams
From the table, 1 ounce ≈ 28.35 grams.

$$\frac{4 \text{ oz}}{1} \cdot \frac{28.35 \text{ g}}{1 \text{ oz}} = 113.4 \text{ g}$$

4 oz ≈ 113.4 g

Concepts	Examples

7.5 Common Celsius Temperatures

Use approximate and exact comparisons to judge which temperatures are appropriate:

Exact comparisons:
0 °C is the freezing point of water (32 °F).

100 °C is the boiling point of water (212 °F).

Approximate comparisons:
10 °C for a spring day (50 °F)

20 °C for room temperature (68 °F)

30 °C for summer day (86 °F)

37 °C for normal body temperature (98.6 °F)

Circle the Celsius temperature that is most reasonable.

(a) Hot summer day:

(35 °C)   90 °C   110 °C

(b) The first snowy day in winter:

−20 °C   (0 °C)   15 °C

7.5 Converting between Fahrenheit and Celsius Temperatures

Use this formula to convert from Fahrenheit (F) to Celsius (C).

$$C = \frac{5(F - 32)}{9}$$

Convert 176 °F to Celsius.

$$C = \frac{5(176 - 32)}{9} \qquad \text{Replace F with 176.}$$

$$= \frac{5(\overset{16}{\cancel{144}})}{\underset{1}{\cancel{9}}} \qquad \begin{array}{l}\text{Divide out common factors.}\\\text{Then multiply in the numerator.}\end{array}$$

$$= 80$$

$$176 °F = 80 °C$$

Use this formula to convert from Celsius (C) to Fahrenheit (F).

$$F = \frac{9 \cdot C}{5} + 32$$

Convert 80 °C to Fahrenheit.

$$F = \frac{9 \cdot 80}{5} + 32 \qquad \text{Replace C with 80.}$$

$$= \frac{9 \cdot \overset{16}{\cancel{80}}}{\underset{1}{\cancel{5}}} + 32 \qquad \begin{array}{l}\text{Divide out common factors.}\\\text{Then multiply in the}\\\text{numerator.}\end{array}$$

$$= 144 + 32 \qquad \text{Add.}$$

$$= 176$$

$$80 °C = 176 °F$$

ANSWERS TO TEST YOUR WORD POWER

1. B; *Examples:* 10 meters = 1 dekameter; 100 meters = 1 hectometer; 1000 meters = 1 kilometer.

2. C; *Examples:* Feet and inches are used to measure length, quarts to measure capacity, pounds to measure weight.

3. D; *Example:* Because 12 in. = 1 ft, the unit fraction $\dfrac{12 \text{ in.}}{1 \text{ ft}}$ is equivalent to $\dfrac{12 \text{ in.}}{12 \text{ in.}} = 1$.

4. A; *Example:* A small box measuring 1 cm on every edge holds exactly 1 mL of water, and the water weighs 1 g.

5. C; *Example:* A measurement of 16 meters is written 16 m (without a period).

6. C; *Example:* In the metric system, water freezes at 0 °C and boils at 100 °C. The U.S. customary system uses the Fahrenheit temperature scale where water freezes at 32 °F and boils at 212 °F.

Chapter 7   ▶▶▶   Review Exercises

[7.1] *Fill in the blanks with the measurement relationships you have memorized.*

1. 1 lb = __16__ oz

2. __3__ ft = 1 yd

3. 1 ton = __2000__ lb

4. __4__ qt = 1 gal

5. 1 hr = __60__ min

6. 1 c = __8__ fl oz

7. __60__ sec = 1 min

8. __5280__ ft = 1 mi

9. __12__ in. = 1 ft

Convert using unit fractions.

10. 4 ft = __48__ in.

11. 6000 lb = __3__ tons

12. 64 oz = __4__ lb

13. 18 hr = __$\frac{3}{4}$ or 0.75__ day

14. 150 min = __$2\frac{1}{2}$ or 2.5__ hr

15. $1\frac{3}{4}$ lb = __28__ oz

16. $6\frac{1}{2}$ ft = __78__ in.

17. 7 gal = __112__ c

18. 4 days = __345,600__ sec

19. The average depth of the world's oceans is 12,460 ft. (*Source: Handy Ocean Answer Book.*)

   (a) What is the average depth in yards?

   $4153\frac{1}{3}$ **yd (exact) or 4153.3 yd (rounded)**

   ▦ **(b)** What is the average depth in miles, to the nearest tenth? **2.4 mi (rounded)**

20. During the first year of a program to recycle office paper, a company recycled 123,260 pounds of paper. The company received $40 per ton for the paper. How much money did the company make? Use the six problem-solving steps. (*Source: I. C. System.*)

   $2465.20

[7.2] *Write the most reasonable metric length unit in each blank. Choose from km, m, cm, and mm.*

21. My thumb is 20 __mm__ wide.

22. Her waist measurement is 66 __cm__.

23. The two towns are 40 __km__ apart.

24. A basketball court is 30 __m__ long.

25. The height of the picnic bench is 45 __cm__.

Height ↕

26. The eraser on the end of my pencil is 5 __mm__ long.

Convert using unit fractions or the metric conversion line.

27. 5 m to cm

   500 cm

28. 8.5 km to m

   8500 m

29. 85 mm to cm

   8.5 cm

30. 370 cm to m

    3.7 m

31. 70 m to km

    0.07 km

32. 0.93 m to mm

    930 mm

[7.3]   *Write the most reasonable metric unit in each blank. Choose from L, mL, kg, g, and mg.*

33. The eyedropper holds 1 __mL__ .

34. I can heat 3 __L__ of water in this saucepan.

35. Loretta's hammer weighed 650 __g__ .

36. Yongshu's suitcase weighed 20 __kg__ when it was packed.

37. My fish tank holds 80 __L__ of water.

38. I'll buy the 500 __mL__ bottle of mouthwash.

39. Mara took a 200 __mg__ antibiotic pill.

40. This piece of chicken weighs 100 __g__ .

Convert using unit fractions or the metric conversion line.

41. 5000 mL to L

    5 L

42. 8 L to mL

    8000 mL

43. 4.58 g to mg

    4580 mg

44. 0.7 kg to g

    700 g

45. 6 mg to g

    0.006 g

46. 35 mL to L

    0.035 L

[7.4]   *Solve each application problem. Show your work. Use the six problem-solving steps.*

47. Each serving of punch at the wedding reception will be 180 mL. How many liters of punch are needed for 175 servings?

    31.5 L

48. Jason is serving a 10 kg turkey to 28 people. How many grams of meat is he allowing for each person? Round to the nearest whole gram.

    357 g (rounded)

49. Yerald weighed 92 kg. Then he lost 4 kg 750 g. What is his weight now, in kilograms?

    87.25 kg

50. Young-Mi bought 950 g of onions. The price was $1.49 per kilogram. How much did she pay, to the nearest cent?

    $1.42 (rounded)

[7.5]   *Use the table on the first page of **Section 7.5** and unit fractions to make approximate conversions. Round your answers to the nearest tenth if necessary.*

51. 6 m to yards

    6.5 yd (rounded)

52. 30 cm to inches

    11.7 in. (rounded)

53. 108 km to miles

67.0 mi (rounded)

54. 800 mi to kilometers

1288 km

55. 23 qt to liters

21.9 L (rounded)

56. 41.5 L to quarts

44.0 qt (rounded)

*Write the appropriate **metric** temperature in each blank.*

57. Water freezes at _**0 °C**_ .

58. Water boils at _**100 °C**_ .

59. Normal body temperature is about _**37 °C**_ .

60. Comfortable room temperature is about _**20 °C**_ .

*Use the conversion formulas in **Section 7.5** to convert each temperature to Fahrenheit or to Celsius. Round to the nearest degree if necessary.*

61. 77 °F   **25 °C**

62. 92 °F   **33 °C   (rounded)**

63. 6 °C   **43 °F (rounded)**

64. Water coming into a dishwasher should be at least 49 °C to clean the dishes properly. What Fahrenheit temperature is that? (*Source:* Frigidaire.)

120 °F (rounded)

▶▶▶ Mixed Review Exercises

Write the most reasonable metric unit in each blank. Choose from km, m, cm, mm, L, mL, kg, g, and mg.

65. I added 1 _**L**_ of oil to my car.

66. The box of books weighed 15 _**kg**_ .

67. Larry's shoe is 30 _**cm**_ long.

68. Jan used 15 _**mL**_ of shampoo on her hair.

69. My fingernail is 10 _**mm**_ wide.

70. I walked 2 _**km**_ to school.

71. The tiny bird weighed 15 _**g**_ .

72. The new library building is 18 _**m**_ wide.

73. The cookie recipe uses 250 _**mL**_ of milk.

74. Renee's pet mouse weighs 30 _**g**_ .

75. One postage stamp weighs 90 _**mg**_ .

76. I bought 30 _**L**_ of gas for my car.

Convert the following using unit fractions, the metric conversion line, or the temperature conversion formulas.

77. 10.5 cm to millimeters

105 mm

78. 45 min to hours

$\frac{3}{4}$ **hr or 0.75 hr**

79. 90 in. to feet

$7\frac{1}{2}$ **ft or 7.5 ft**

80. 1.3 m to centimeters

130 cm

81. 25 °C to Fahrenheit

77 °F

82. $3\frac{1}{2}$ gal to quarts

14 qt

83. 700 mg to grams

0.7 g

84. 0.81 L to milliliters

810 mL

85. 5 lb to ounces

80 oz

86. 60 kg to grams

60,000 g

87. 1.8 L to milliliters

1800 mL

88. 86 °F to Celsius

30 °C

89. 0.36 m to centimeters

36 cm

90. 55 mL to liters

0.055 L

Solve each application problem. Use the six problem-solving steps in Exercises 91–92.

91. Peggy had a board measuring 2 m 4 cm. She cut off 78 cm. How long is the board now, in meters?

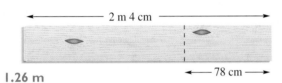

2 m 4 cm

78 cm

1.26 m

92. During the 12-day Minnesota State Fair, one of the biggest in the United States, Sweet Martha's booth sells an average of 3000 pounds of cookies per day. How many tons of cookies are sold in all? (*Source: Minneapolis Star Tribune.*)

18 tons

93. Olivia is sending a recipe to her mother in Mexico. Among other things, the recipe calls for 4 oz of rice and a baking temperature of 350 °F. Convert these measurements to metric, rounding to the nearest gram and nearest degree.

113 g; 177 °C (both rounded)

94. While on vacation in Canada, Jalo became ill and went to a health clinic. They said he weighed 80.9 kg and was 1.83 m tall. Find his weight in pounds and height in feet. Round to the nearest tenth.

178.0 lb; 6.0 ft (both rounded)

The largest two-axle trucks in the world are mining trucks used to haul huge loads of rock and iron ore. Some information about these $2 million trucks is given below. Fill in the blank spaces in the table. Then use the information to answer Exercises 101–102. Round answers to the nearest tenth.

WORLD'S LARGEST TWO-AXLE TRUCK

	Mining Truck	Measurements	
		Metric Units	Customary Units
95.	Length of truck	13.2 m	44 ft
96.	Height of truck	6.7 m	22.0 ft
97.	Height of tire	3.6 m	4 yd
98.	Width of tire tread	102 cm	39.8 in.
99.	Weight of load truck can carry	220,000 kg	484,000 lb
100.	Fuel needed to travel 1 mile	19.0 to 22.7 L	5 to 6 gal

Source: Hull Rust Mahoning Mine.

101. Using U.S. customary measurements:

(a) The truck can carry a load weighing how many tons?

242 tons

(b) How many inches high is each tire?

144 in.

102. Using metric measurements:

(a) What is the width of the tire tread in meters?

1.02 m

(b) How tall is the truck in centimeters?

670 cm

Chapter 7 ▶▶▶ Test

Use the Chapter Test Prep Video CD to see fully worked-out solutions to any of the exercises you want to review.

Convert each measurement.

1. 9 gal = _____ qt

2. 45 ft = _____ yd

3. 135 min = _____ hr

4. 9 in. = _____ ft

5. $3\frac{1}{2}$ lb = _____ oz

6. 5 days = _____ min

Write the most reasonable metric unit in each blank. Choose from km, m, cm, mm, L, mL, kg, g, and mg.

7. My husband weighs 75 _____ .

8. I hiked 5 _____ this morning.

9. She bought 125 _____ of cough syrup.

10. This apple weighs 180 _____ .

11. This page is about 21 _____ wide.

12. My watch band is 10 _____ wide.

13. I bought 10 _____ of soda for the picnic.

14. The bracelet is 16 _____ long.

Convert the following measurements. Show your work.

15. 250 cm to meters

16. 4.6 km to meters

17. 5 mm to centimeters

18. 325 mg to grams

19. 16 L to milliliters

20. 0.4 kg to grams

21. 10.55 m to centimeters

22. 95 mL to liters

1. 36 qt _____

2. 15 yd _____

3. 2.25 hr or $2\frac{1}{4}$ hr _____

4. 0.75 ft or $\frac{3}{4}$ ft _____

5. 56 oz _____

6. 7200 min _____

7. kg _____

8. km _____

9. mL _____

10. g _____

11. cm _____

12. mm _____

13. L _____

14. cm _____

15. 2.5 m _____

16. 4600 m _____

17. 0.5 cm _____

18. 0.325 g _____

19. 16,000 mL _____

20. 400 g _____

21. 1055 cm _____

22. 0.095 L _____

23. 3.2 ft (rounded)

23. The rainiest place in the world is Mount Waialeale in Hawaii, which receives 460 inches of rain each year. What is the average rainfall per month, in feet, to the nearest tenth? (*Source:* National Geographic Society.)

24. (a) 2310 mg

   (b) 500 mg or 0.5 g more

24. A 6-inch Subway Veggie Delite sandwich has 590 mg of sodium. A 6-inch Super Subway Melt sandwich has 2.9 g of sodium. (*Source:* Subway.)

   (a) How much more sodium is in the Super Melt sandwich, in milligrams, than in the Veggie Delite?

   (b) The recommended amount of sodium is less than 2400 mg daily. How much more or less sodium does the Super Melt have than the recommended daily amount?

Pick the metric temperature that is most reasonable in each situation.

25. 95 °C

25. The water is almost boiling.
   210 °C   155 °C   95 °C

26. 0 °C

26. The tomato plants may freeze tonight.
   30 °C   20 °C   0 °C

27. 1.8 m

*Use the table from **Section 7.5** and unit fractions to convert each measurement. Round your answers to the nearest tenth if necessary.*

28. 56.3 kg (rounded)

27. 6 ft to meters

28. 125 lb to kilograms

29. 13 gal

29. 50 L to gallons

30. 8.1 km to miles

30. 5.0 mi (rounded)

31. 23 °C (rounded)

Use the conversion formulas to convert each temperature. Round your answers to the nearest degree if necessary.

31. 74 °F to Celsius

32. 2 °C to Fahrenheit

32. 36 °F (rounded)

6 m ≈ 6.54 yd;
33. $26.03 (rounded)

Solve this application problem. Show your work.

33. Denise is making five matching pillows. She needs 120 cm of braid to trim each pillow. If the braid costs $3.98 per yard, how much will she spend to trim the pillows, to the nearest cent? (First find the number of meters of braid Denise needs.)

34. Possible answers: Use same system as rest of the world; easier system for children to learn; less use of fractional numbers; compete internationally.

34. Describe two benefits the United States would achieve by switching entirely to the metric system.

Cumulative Review Exercises ▷▷▷ Chapters 1–7

First use front end rounding to round each number and estimate the answer. Then find the exact answer.

1. *Estimate:*   *Exact:*

$$\begin{array}{r} 60 \\ \times\ 5 \\ \hline 300 \end{array} \qquad \begin{array}{r} 56.52 \\ \times\ \ 4.7 \\ \hline 265.644 \end{array}$$

2. *Estimate:*   *Exact:*

$$40\overline{)20{,}000} \qquad 37\overline{)19{,}610}$$
$$\ \ \ 500 \qquad\qquad \ \ \ 530$$

3. *Estimate:*   *Exact:*

$$\begin{array}{r} 2 \\ + 4 \\ \hline 6 \end{array} \qquad \begin{array}{r} 1\frac{7}{10} \\ + 3\frac{4}{5} \\ \hline 5\frac{1}{2} \end{array}$$

Simplify. Write answers to fraction problems in lowest terms and as whole or mixed numbers when possible.

4. $3 - 2\frac{5}{16}$

$\frac{11}{16}$

5. $12 \cdot 2\frac{2}{9}$

$26\frac{2}{3}$

6. $0.86 \div 0.066$
Round your answer to the nearest tenth.

13.0 (rounded)

7. $\frac{3}{8} + \frac{5}{6}$ $1\frac{5}{24}$

8. $8 - 0.9207$ **7.0793**

9. $3\frac{3}{4} \div 6$ $\frac{5}{8}$

10. $(2.54)(0.003)$

0.00762

11. $24 - 12 \div 6(8) + (25 - 25)$

8

12. $3^2 + 2^5 \cdot \sqrt{64}$

265

13. Arrange in order from least to greatest.

0.67   0.067   0.6   0.6007

0.067, 0.6, 0.6007, 0.67

 14. Find the best buy on disposable diapers.
Package of 16 for $3.87
Package of 22 for $5.96
Package of 36 for $11.69

16 diapers for $3.87, about $0.242 per diaper

Solve each proportion or percent problem. Round your answers to hundredths if necessary.

15. $\frac{x}{16} = \frac{3}{4}$

$x = 12$

16. $\frac{0.9}{0.75} = \frac{2}{x}$

$x \approx 1.67$ **(rounded)**

17. $4 is what percent of $80?

5%

18. 36 hours is 120% of what number of hours?

30 hours

*Convert the following measurements. Use the table in **Section 7.5** when necessary.*

19. $2\frac{1}{2}$ ft to inches

30 in.

20. 105 sec to minutes

$1\frac{3}{4}$ or **1.75 min**

21. 2.8 m to centimeters

280 cm

 22. 198 km to miles

122.76 mi

Write the most reasonable metric unit in each blank. Choose from km, m, cm, mm, L, mL, kg, g, and mg.

23. Ron bought the tube of toothpaste weighing 100 <u>g</u>.

24. Tia added 125 <u>mL</u> of milk to her cereal.

25. The hallway is 3 <u>m</u> wide.

26. Joe's hammer weighed 1 <u>kg</u>.

Solve each application problem.

27. Calbert works at a Wal-Mart store and used his employee discount to buy a $189.94 digital camera at 10% off. Find the amount of his discount, to the nearest cent, and the sale price. (*Source:* Wal-Mart.)

$18.99 (rounded); $170.95

28. Danielle ordered 45 prints online from her digital camera. Because it was her first order, she got 20 free prints. The rest were $0.12 each plus $6\frac{1}{2}$% sales tax. She also paid $2.95 for shipping. What was the total amount she paid and the cost per print, to the nearest cent?

$6.15 total; $0.14 per print (both rounded)

29. Bags of slivered almonds weigh 4 oz each. They are packed in a carton that weighs 12 oz. How many pounds would a carton containing 48 bags weigh?

$12\frac{3}{4}$ **or 12.75 lb**

30. On the Illinois map, one centimeter represents 12 km. The center of Springfield is 7.8 cm from the center of Bloomington on the map.

(a) What is the actual distance in kilometers?

93.6 km

(b) What is the actual distance in miles, to the nearest tenth?

58.0 mi (rounded)

31. Dimitri took out a $3\frac{1}{2}$ year car loan for $8750 at 9% simple interest. Find the interest and the total amount due on the loan.

$2756.25; $11,506.25

32. On a 35-problem math test, Juana solved 31 problems correctly. What percent of the problems were correct? Round to the nearest tenth of a percent.

88.6% (rounded)

33. Mark bought 650 g of maple sugar candy on his vacation in Montreal. The candy is priced at $14.98 per kilogram. How much did Mark pay, to the nearest cent?

$9.74 (rounded)

34. The Jackson family is making three kinds of holiday cookies that require brown sugar. The recipes call for $2\frac{1}{4}$ cups, $1\frac{1}{2}$ cups, and $\frac{3}{4}$ cup, respectively. They bought two packages of brown sugar, each holding $2\frac{1}{3}$ cups. The amount bought is how much more or less than the amount needed?

$\frac{1}{6}$ **cup more than the amount needed**

35. Akuba is knitting a scarf. Six rows of knitting result in 5 cm of scarf. At that rate, how many rows will she have to knit to make a 100 cm scarf?

120 rows

36. A survey of the 5600 students on our campus found that $\frac{3}{8}$ of the students work 20 hours or more per week. How many students work 20 hours or more?

2100 students

37. A spray-on bandage applies a transparent film over a wound that protects it from bacteria. The cost is $6 for about 40 applications. What is the cost per application? (*Source:* Curad.)

$0.15

38. The average hospital stay is now about 5 days, compared to about 8 days in 1970. What is the percent of decrease in the length of hospital stays? (*Source:* National Center for Health Statistics.)

37.5% decrease

Geometry

An important part of managing our parks and forests is taking an inventory of the trees. The circumference of each tree is measured at chest height. Then, using the circle formulas in **Section 8.6**, the diameter is calculated. This information helps analyze growth patterns and tree age. (See Exercise 31 in **Section 8.6**.)

8.1 ▶▶▶ Basic Geometric Terms

OBJECTIVES

1 Identify and name lines, line segments, and rays.

2 Identify parallel and intersecting lines.

3 Identify and name angles.

4 Classify angles as right, acute, straight, or obtuse.

5 Identify perpendicular lines.

Geometry was developed centuries ago when people needed a way to measure land. The name *geometry* comes from the Greek words *ge,* meaning earth, and *metron,* meaning measure. Today we still use geometry to measure land. It is also important in architecture, construction, navigation, art and design, physics, chemistry, and astronomy. You can use it at home when you buy carpet or wallpaper, hang a picture, or do home repairs. This chapter discusses the basic terms of geometry and the common geometric shapes that are all around us.

Geometry starts with the idea of a point. A **point** is a location in space. It has no length or width. A point is represented by a dot and is named by writing a capital letter next to the dot.

• *P*

Point *P*

OBJECTIVE 1 Identify and name lines, line segments, and rays. A **line** is a straight row of points that goes on forever in both directions. A line is drawn by using arrowheads to show that it never ends. The line is named using the letters of any two points on the line.

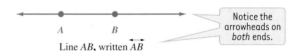

Line *AB*, written $\overleftrightarrow{AB}$

> Notice the arrowheads on *both* ends.

A piece of a line that has two endpoints is called a **line segment.** A line segment is named using its endpoints. The segment with endpoints *P* and *Q* is shown below. It can be named $\overline{PQ}$ or $\overline{QP}$.

Line segment *PQ*, written $\overline{PQ}$

> Notice there are *no* arrowheads.

A **ray** is a part of a line that has only one endpoint and goes on forever in one direction. A ray is named by using the endpoint and some other point on the ray. The endpoint is always mentioned first.

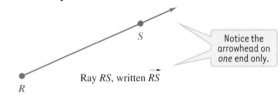

Ray *RS*, written $\overrightarrow{RS}$

> Notice the arrowhead on *one* end only.

---

1 Identify each figure as a line, line segment, or ray, and name it.

(a)

(b)

(c)

(d)

EXAMPLE 1 **Identifying and Naming Lines, Rays, and Line Segments**

Identify each figure below as a line, line segment, or ray, and name it using the appropriate symbol.

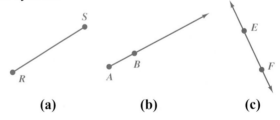

(a)　　　(b)　　　(c)

Figure **(a)** has two endpoints, so it is a *line segment* named $\overline{RS}$ or $\overline{SR}$.

Figure **(b)** starts at point *A* and goes on forever in one direction, so it is a *ray* named $\overrightarrow{AB}$.

Figure **(c)** goes on forever in both directions, so it is a *line* named $\overleftrightarrow{EF}$ or $\overleftrightarrow{FE}$.

◀ *Work Problem* **1** *at the Side.*

ANSWERS

1. (a) line segment named $\overline{EF}$ or $\overline{FE}$
　(b) ray named $\overrightarrow{SR}$
　(c) line named $\overleftrightarrow{WX}$ or $\overleftrightarrow{XW}$
　(d) line segment named $\overline{CD}$ or $\overline{DC}$

OBJECTIVE **2** **Identify parallel and intersecting lines.** A *plane* is an infinitely large flat surface. A floor or a wall is a part of a plane. Lines that are in the *same plane,* but that never intersect (never cross), are called **parallel lines,** while lines that cross are called **intersecting lines.** (Think of an intersection, where two streets cross each other.)

2 Label each pair of lines as appearing to be parallel or as intersecting.

EXAMPLE 2 **Identifying Parallel and Intersecting Lines**

Label each pair of the lines as appearing to be parallel or as intersecting.

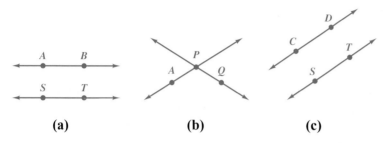

(a) (b) (c)

(a)

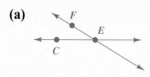

The lines in Figures **(a)** and **(c)** do not intersect; they appear to be *parallel lines.* The lines in Figure **(b)** cross at *P,* so they are *intersecting lines.*

> **CAUTION**
> Appearances may be deceiving! Do not assume that lines are parallel unless it is stated that they are parallel.

Work Problem **2** *at the Side.* ▶

(b)

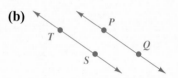

OBJECTIVE **3** **Identify and name angles.** An **angle** is made up of two rays that start at a common endpoint. This common endpoint is called the *vertex.*

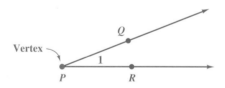

$\overrightarrow{PQ}$ and $\overrightarrow{PR}$ are called the *sides* of the angle. The angle can be named in four different ways, as shown below.

∠1   ∠P   ∠QPR   ∠RPQ

↑   ↑   ↑
Vertex   Vertex in
alone   the middle

(c)

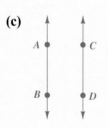

> **Naming an Angle**
> When naming an angle, the vertex is written alone or it is written in the middle of two other points. If two or more angles have the **same vertex,** as in Example 3 on the next page, do **not** use the vertex alone to name an angle.

3 **(a)** Name the highlighted angle in three different ways.

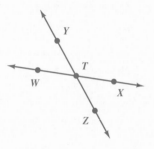

(b) Darken the rays that make up ∠ZTW.

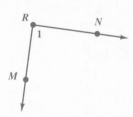

(c) Name this angle in four different ways.

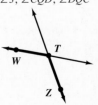

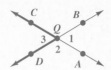
EXAMPLE 3 **Identifying and Naming an Angle**

Name the highlighted angle in three different ways.

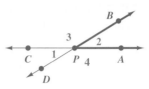

The angle can be named ∠BPA, ∠APB, or ∠2. It cannot be named ∠P, using the vertex alone, because four different angles have P as their vertex.

◀ *Work Problem* **3** *at the Side.*

OBJECTIVE **4** **Classify angles as right, acute, straight, or obtuse.**
Angles can be measured in **degrees.** The symbol for degrees is a small, raised circle °. Think of the minute hand on a clock as a ray of an angle. Suppose it is at 12:00. During one hour of time, the minute hand moves around in a complete circle. It moves 360 *degrees,* or 360°. In half an hour, at 12:30, the minute hand has moved halfway around the circle or 180°. An angle of 180° is called a **straight angle.** When two rays go in opposite directions and form a straight line, then the rays form a straight angle.

Complete circle
360°

Straight angle
(half a circle)
180°

In a quarter of an hour, at 12:15, the minute hand has moved $\frac{1}{4}$ of the way around the circle, or 90°. An angle of 90° is called a **right angle.** Sometimes you hear it called a *square angle*. The minute hands at 12:00 and 12:15 form one corner of a square. So, to show that an angle is a **right angle**, we draw a **small square** at the vertex.

Right angle
($\frac{1}{4}$ of a circle)
90°

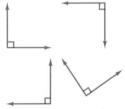

A small square at the vertex identifies right angles.

An angle that measures 1° is shown below. You can see that an angle of 1° is very small.

1° angle

Some other terms used to describe angles are shown below.

Acute angles measure less than 90°.

Examples of acute angles

Obtuse angles measure more than 90° but less than 180°.

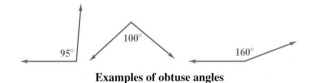

Examples of obtuse angles

Section 10.1 shows you how to use a tool called a *protractor* to measure the number of degrees in an angle.

> **Classifying Angles**
>
> **Acute angles** measure less than 90°.
>
> **Right angles** measure *exactly* 90°.
>
> **Obtuse angles** measure more than 90° but less than 180°.
>
> **Straight angles** measure *exactly* 180°.

> **Note**
>
> Angles can also be measured in radians, which you will learn about in a later math course.

EXAMPLE 4 **Classifying Angles**

Label each angle as acute, right, obtuse, or straight.

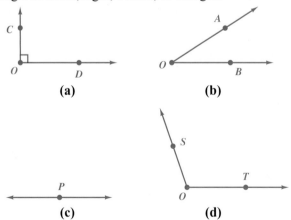

Figure **(a)** shows a *right angle* (exactly 90° and identified by a small square at the vertex).

Figure **(b)** shows an *acute angle* (less than 90°).

Figure **(c)** shows a *straight angle* (exactly 180°).

Figure **(d)** shows an *obtuse angle* (more than 90° but less than 180°).

Work Problem **4** *at the Side.* ▶

---

4 Label each figure as an acute, right, obtuse, or straight angle.

(a)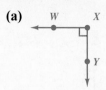

(b)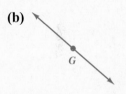

(c)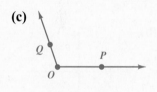

(d)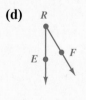

ANSWERS

4. **(a)** right **(b)** straight **(c)** obtuse
   (d) acute

5 Which pair of lines is perpendicular? How can you describe the other pair of lines?

(a)

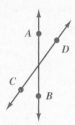

(b)

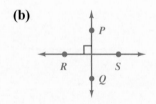

OBJECTIVE 5 Identify perpendicular lines. Two lines are called **perpendicular lines** if they intersect to form a right angle.

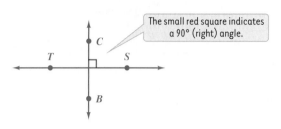

$\overleftrightarrow{CB}$ and $\overleftrightarrow{ST}$ are **perpendicular lines** because they intersect at right angles, as indicated by the small red square in the figure.

Perpendicular lines can be written in the following way: $\overleftrightarrow{CB} \perp \overleftrightarrow{ST}$.

EXAMPLE 5 Identifying Perpendicular Lines

Which pairs of lines are perpendicular?

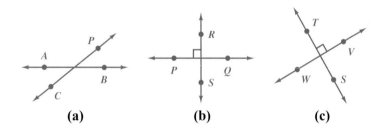

(a)   **(b)**   **(c)**

The lines in Figures **(b)** and **(c)** are *perpendicular* to each other because they intersect at right angles.

The lines in Figure **(a)** are *intersecting lines,* but they are *not* perpendicular because they do *not* form a right angle.

◀ *Work Problem* **5** *at the Side.*

8.1 ▶▶▶ **Exercises**

Identify each figure as a line, line segment, or ray and name it using the appropriate symbol. See Example 1.

1.

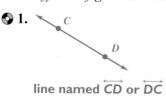

line named $\overleftrightarrow{CD}$ or $\overleftrightarrow{DC}$

2.

ray named $\overrightarrow{AB}$

3.

line segment named $\overline{GF}$ or $\overline{FG}$

4.

line segment named $\overline{EF}$ or $\overline{FE}$

5.

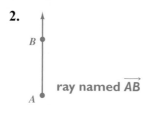

ray named $\overrightarrow{PQ}$

6.

line named $\overleftrightarrow{ST}$ or $\overleftrightarrow{TS}$

Label each pair of lines as appearing to be parallel, as perpendicular, or as intersecting.
See Examples 2 and 5.

7.

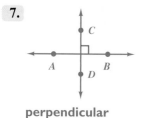

perpendicular

8.

intersecting

9.

parallel

10.

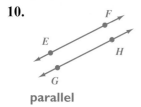

parallel

11.

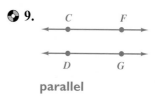

intersecting

12.

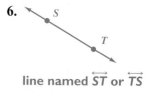

perpendicular

Name each highlighted angle by using the three-letter form of identification. See Example 3.

13.

∠AOS or ∠SOA

14.

∠BOD or ∠DOB

15.

∠CRT or ∠TRC

16.

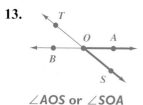

∠CRB or ∠BRC

17.

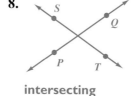

∠AQC or ∠CQA

18.

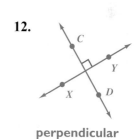

∠FQB or ∠BQF

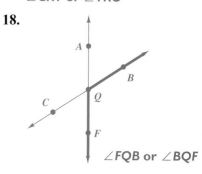
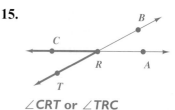
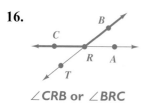
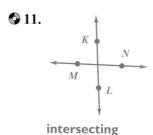
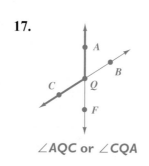
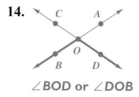

Label each angle as acute, right, obtuse, or straight. For right and straight angles, indicate the number of degrees in the angle. See Example 4.

19.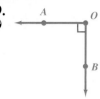

right (90°)

20.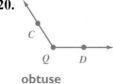

obtuse

🌐 **21.**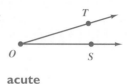

acute

22.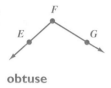

obtuse

🌐 **23.**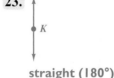

straight (180°)

24.

acute

25. Explain what is happening in each sentence.

   (a) The road was so slippery that my car did a 360.

     The car turned around in a complete circle.

   (b) After the election, the governor's view on taxes took a 180° turn.

     The governor took the opposite view, for example, having once opposed taxes but now supporting them.

26. Find at least four examples of right angles in your home, at work, or on the street. Make a sketch of each example and label the right angle.

   There are many possibilities. Some examples are the corner of a room, a street corner, the corners of a window, the corner of a piece of paper.

Relating Concepts (Exercises 27–32) For Individual or Group Work

*Use the figure below to **work Exercises 27–32 in order.** Decide whether each statement is **true or false.** If it is true, explain why. If it is false, rewrite it to make it a true statement.*

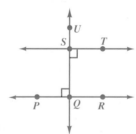

27. $\angle UST$ is 90°.

   True, because $\overleftrightarrow{UQ}$ is perpendicular to $\overleftrightarrow{ST}$.

28. $\overleftrightarrow{SQ}$ and $\overleftrightarrow{PQ}$ are perpendicular.

   True, because they form a 90° angle, as indicated by the small red square.

29. The measure of $\angle USQ$ is less than the measure of $\angle PQR$.

   False; the angles have the same measure (both are 180°).

30. $\overleftrightarrow{ST}$ and $\overrightarrow{PR}$ are intersecting.

   False; $\overleftrightarrow{ST}$ and $\overrightarrow{PR}$ are parallel.

31. $\overleftrightarrow{QU}$ and $\overleftrightarrow{TS}$ are parallel.

   False, $\overleftrightarrow{QU}$ and $\overleftrightarrow{TS}$ are perpendicular.

32. $\angle UST$ and $\angle UQR$ measure the same number of degrees.

   True; both angles are formed by perpendicular lines, so they both measure 90°.

8.2 ▶▶▶ Angles and Their Relationships

OBJECTIVE 1 Identify complementary angles and supplementary angles and find the measure of a complement or supplement of a given angle. Two angles are called **complementary angles** if the sum of their measures is 90°. If two angles are complementary, each angle is the *complement* of the other.

EXAMPLE 1 Identifying Complementary Angles

Identify each pair of complementary angles.

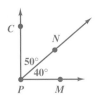

$\angle MPN\,(40°)$ and $\angle NPC\,(50°)$ are complementary angles because

$$40° + 50° = 90°$$

$\angle CAB\,(30°)$ and $\angle FHG\,(60°)$ are complementary angles because

$$30° + 60° = 90°$$

Work Problem **1** *at the Side.* ▶

EXAMPLE 2 Finding the Complement of Angles

Find the complement of each angle.

(a) 30°
Find the complement of 30° by subtracting. $90° - 30° = \mathbf{60°}$ ← Complement

(b) 75°
Find the complement of 75° by subtracting. $90° - 75° = \mathbf{15°}$ ← Complement

Work Problem **2** *at the Side.* ▶

Two angles are called **supplementary angles** if the sum of their measures is 180°. If two angles are supplementary, each angle is the *supplement* of the other.

EXAMPLE 3 Identifying Supplementary Angles

Identify each pair of supplementary angles.

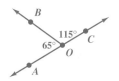 

$\angle BOA$ and $\angle BOC$, because $\quad 65° + 115° = 180°$

$\angle BOA$ and $\angle ERF$, because $\quad 65° + 115° = 180°$

$\angle BOC$ and $\angle MPN$, because $115° + \;\; 65° = 180°$

$\angle MPN$ and $\angle ERF$, because $\quad 65° + 115° = 180°$

Work Problem **3** *at the Side.* ▶

1 Identify each pair of complementary angles.

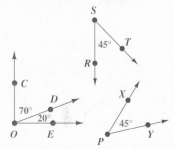

2 Find the complement of each angle.

(a) 35°

(b) 80°

3 Identify each pair of supplementary angles.

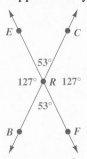

ANSWERS

1. $\angle COD$ and $\angle DOE$; $\angle RST$ and $\angle XPY$

2. (a) 55° **(b)** 10°

3. $\angle CRF$ and $\angle BRF$; $\angle CRE$ and $\angle ERB$; $\angle BRF$ and $\angle BRE$; $\angle CRE$ and $\angle CRF$

4 Find the supplement of each angle.

(a) 175°

(b) 30°

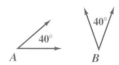

EXAMPLE 4 **Finding the Supplement of Angles**

Find the supplement of each angle.

(a) 70°
Find the supplement of 70° by subtracting. $180° - 70° = \mathbf{110°}$ ←Supplement

(b) 140°
Find the supplement of 140° by subtracting. $180° - 140° = \mathbf{40°}$ ←Supplement

◀ *Work Problem* **4** *at the Side.*

OBJECTIVE **2** **Identify congruent angles and vertical angles and use this knowledge to find the measures of angles.** Two angles are called **congruent angles** if they measure the same number of degrees. If two angles are congruent, this is written as $\angle A \cong \angle B$ and read as, "angle A is congruent to angle B." Here is an example.

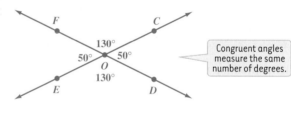

Congruent angles: $\angle A \cong \angle B$

5 Identify the angles that are congruent.

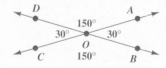

EXAMPLE 5 **Identifying Congruent Angles**

Identify the angles that are congruent.

Congruent angles measure the same number of degrees.

$$\angle FOC \cong \angle EOD \quad \text{and} \quad \angle COD \cong \angle EOF$$

◀ *Work Problem* **5** *at the Side.*

Angles that share a common side and a common vertex are called *adjacent* angles, such as $\angle FOC$ and $\angle COD$ in Example 5 above. Angles that do *not* share a common side are called *nonadjacent* angles. Two nonadjacent angles formed by two intersecting lines are called **vertical angles.**

6 Identify the vertical angles. What is special about vertical angles?

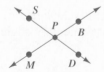

EXAMPLE 6 **Identifying Vertical Angles**

Identify the vertical angles in this figure.

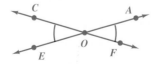

$\angle AOF$ and $\angle COE$ are vertical angles because they do not share a common side and they are formed by two intersecting lines ($\overleftrightarrow{CF}$ and $\overleftrightarrow{EA}$). | $\angle COA$ and $\angle EOF$ are also vertical angles.

◀ *Work Problem* **6** *at the Side.*

ANSWERS

4. **(a)** 5° **(b)** 150°
5. $\angle BOC \cong \angle AOD$; $\angle AOB \cong \angle DOC$
6. $\angle SPB$ and $\angle MPD$; $\angle BPD$ and $\angle SPM$; vertical angles are congruent (they measure the same number of degrees).

Look back at Example 5 on the previous page. Notice that the two *congruent* angles that measure 130° are also *vertical* angles. Also, the two congruent angles that measure 50° are vertical angles. This illustrates the following property.

> **Vertical Angles Are Congruent**
>
> If two angles are *vertical* angles, they are *congruent,* that is, they measure the same number of degrees.

EXAMPLE 7 **Finding the Measures of Vertical Angles**

In the figure below, find the measure of each unlabeled angle. Then write the measure of each angle on the figure.

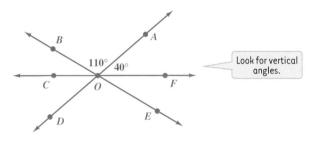

Look for vertical angles.

(a) ∠*COD*

∠*COD* and ∠*AOF* are vertical angles, so they are congruent. This means they measure the same number of degrees.

The measure of ∠*AOF* is 40°   so   the measure of ∠*COD* is 40° also.

(b) ∠*DOE*

∠*DOE* and ∠*BOA* are vertical angles, so they are congruent.

The measure of ∠*BOA* is 110°   so   the measure of ∠*DOE* is 110° also.

(c) ∠*COB*

Look at ∠*COB*, ∠*BOA*, and ∠*AOF*. Notice that $\overrightarrow{OC}$ and $\overrightarrow{OF}$ go in opposite directions. Therefore, ∠*COF* is a straight angle and measures 180°. To find the measure of ∠*COB*, subtract the sum of the other two angles from 180°.

$$180° - (110° + 40°) = 180° - (150°) = 30°$$

The measure of ∠*COB* is 30°.

(d) ∠*EOF*

∠*EOF* and ∠*COB* are vertical angles, so they are congruent. We know from part (c) above that the measure of ∠*COB* is 30° so the measure of ∠*EOF* is 30° also.

The figure with all the angle measures is shown below.

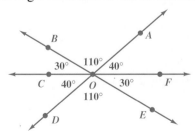

Work Problem **7** *at the Side.* ▶

---

7 In the figure below, find the number of degrees in each unlabeled angle. Then write the angle measures on the figure.

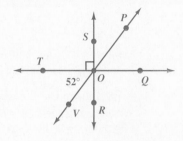

(a) ∠*TOS*

(b) ∠*QOR*

(c) ∠*POQ*

(d) ∠*VOR*

(e) ∠*POS*

ANSWERS

7. **(a)** 90° **(b)** 90° **(c)** 52°
  (d) 38° **(e)** 38°

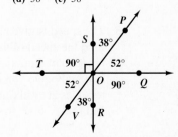

Math in the Media

THE LONDON EYE

The British Airways London Eye was built as a symbol of the turn of the century. It opened on December 31, 1999, and is the largest observation wheel ever designed. It is a unique form of a Ferris wheel with 32 evenly spaced, oval-shaped capsules. Each capsule holds 25 people. On a clear day, passengers can see out for 25 miles.

The London Eye is designed to rotate continuously at walking speed. At one end of the loading platform, passengers can walk into the capsules, and at the other end they can walk right out.

Source: The Bankside Press.

Specifications for the London Eye	
Diameter	135 meters
Speed	0.26 meter per second
Time to revolve	30 minutes

1. **(a)** What is the diameter of the wheel in feet? (*Hint:* Use 1 m ≈ 3.28 ft.) **442.8 ft**

   (b) The circumference of the wheel is the distance around the outside rim. Start with the diameter in feet, from part (a). Then find the circumference by multiplying the diameter times 3.1416 and rounding the answer to the nearest foot. (See **Section 8.6** for more information about circumference.) **1391 ft**

2. Find the distance, to the nearest tenth of a foot, along the rim arc between two adjacent capsules. Recall that there are 32 capsules. **43.5 ft**

3. If the London Eye is operating at full capacity, how many people will be riding in a 90° section of the wheel? in a 45° section? in a 180° section? in a 270° section? in a 360° section? **200 people; 100 people; 400 people; 600 people; 800 people**

4. If the London Eye operates from 8:00 A.M. until 6:00 P.M. at full capacity, how many people per day can ride? **16,000 people**

5. The speed is given as 0.26 meter per second. Find the speed of the London Eye in feet per second. Round the answer to the nearest hundredth. Do you believe that the wheel revolves at "walking speed"? (Try it and see!) **0.85 ft/sec; yes**

8.2 ▶▶▶ Exercises

Identify each pair of complementary angles. See Example 1.

1.

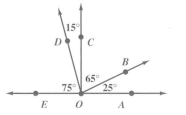

∠EOD and ∠COD; ∠AOB and ∠BOC

2.

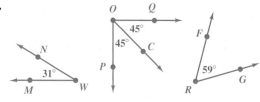

∠COQ and ∠COP; ∠MWN and ∠FRG

Identify each pair of supplementary angles. See Example 3.

3.

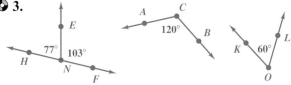

∠HNE and ∠ENF; ∠ACB and ∠KOL

4.

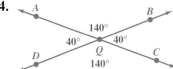

∠AQB and ∠AQD; ∠BQC and ∠CQD;
∠AQB and ∠BQC; ∠CQD and ∠AQD

Find the complement of each angle. See Example 2.

5. 40°   **6.** 35°   **7.** 86°   **8.** 59°
    50°   55°   4°   31°

Find the supplement of each angle. See Example 4.

9. 130°   **10.** 75°   **11.** 90°   **12.** 5°
    50°   105°   90°   175°

In each figure, identify the angles that are congruent. See Examples 5 and 6.

13.

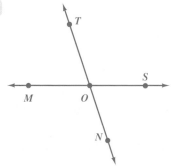

∠SON ≅ ∠TOM; ∠TOS ≅ ∠MON

14.

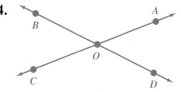

∠AOB ≅ ∠COD; ∠AOD ≅ ∠BOC

Use your knowledge of vertical angles to answer Exercises 15 and 16. See Example 7.

15. In the figure below, ∠*AOH* measures 37° and ∠*COE* measures 63°. Find the measure of each of the other angles.

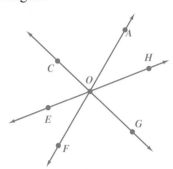

∠**GOH measures 63°;** ∠**EOF measures 37°;**
∠**AOC and** ∠**GOF both measure 80°.**

16. In the figure below, ∠*POU* measures 105° and ∠*UOT* measures 40°. Find the measure of each of the other angles.

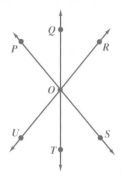

∠**ROS measures 105°;** ∠**QOR measures 40°;**
∠**POQ and** ∠**TOS both measure 35°.**

17. In your own words, write a definition of complementary angles and a definition of supplementary angles. Draw a picture to illustrate each definition.

Two angles are complementary if the sum of their measures is 90°. Two angles are supplementary if the sum of their measures is 180°. Drawings will vary; examples are:

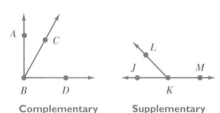

Complementary   Supplementary

18. Make up a test problem in which a student has to use knowledge of vertical angles. Include a drawing with some angles labeled and ask the student to find the measure of each of the remaining angles. Give the correct answer for your problem.

There are many possibilities. One example is:

Find the number of degrees in ∠*AOD* **and** ∠*DOC*.
∠*AOD* **measures 115° and** ∠*DOC* **measures 65°.**

In each figure, $\overrightarrow{BA}$ is parallel to $\overrightarrow{CD}$. Identify two pairs of congruent angles and the number of degrees in each congruent angle.

19.

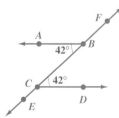

∠*ABF* ≅ ∠*ECD*; **Both are 138°.**
∠*ABC* ≅ ∠*BCD*; **Both are 42°.**

20.

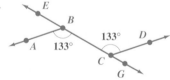

∠*ABC* ≅ ∠*BCD*; **Both are 133°.**
∠*DCG* ≅ ∠*EBA*; **Both are 47°.**

21. Can two obtuse angles be supplementary? Explain why or why not.

No; obtuse angles are >90°, so their sum would be >180°.

22. Can two acute angles be complementary? Explain why or why not.

Yes; acute angles are <90°, so their sum could equal 90°.

8.3 ▶▶▶ Rectangles and Squares

A **rectangle** is a figure with four sides that meet to form 90° angles. Each set of opposite sides is *parallel* and *congruent* (has the same length).

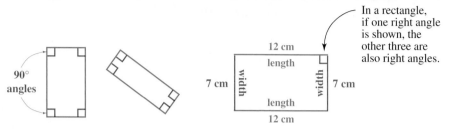

In a rectangle, if one right angle is shown, the other three are also right angles.

Each longer side of a rectangle is called the length (*l*) and each shorter side is called the width (*w*).

Work Problem (**1**) *at the Side.* ▶

OBJECTIVE **1** **Find the perimeter and area of a rectangle.** The distance around the outside edges of a figure is the **perimeter** of the figure. Think of how much fence you would need to put around the sides of a garden plot, or how far you would walk if you walked around the outside edges of your living room. In either case you would add up the lengths of the sides. Look at the rectangle above that has the lengths of the sides labeled. To find its perimeter, you add the lengths of the sides.

Perimeter = **12 cm + 12 cm + 7 cm + 7 cm = 38 cm**

Because the two long sides are both 12 cm, and the two short sides are both 7 cm, you can also use this formula.

> **Finding the Perimeter of a Rectangle**
> Perimeter of a rectangle = length + length + width + width
> $P = \quad (2 \cdot \text{length}) \quad + \quad (2 \cdot \text{width})$
> $P = 2 \cdot l + 2 \cdot w$

EXAMPLE 1 **Finding the Perimeter of a Rectangle**

Find the perimeter of each rectangle.

(a)

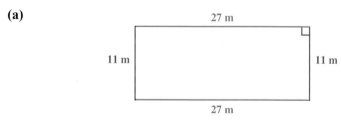

The length of this rectangle is **27 m** and the width is **11 m**.
Use the formula $P = 2 \cdot l + 2 \cdot w$

$P = 2 \cdot \quad l \quad + 2 \cdot \quad w$ Replace *l* with 27 m and *w* with 11 m.

$P = 2 \cdot \textbf{27 m} + 2 \cdot \textbf{11 m}$ Do the multiplications first.

$P = \quad 54\text{ m} \quad + \quad 22\text{ m}$ Add last.

$P = 76\text{ m}$

The perimeter of the rectangle (the distance you would walk around the outside edges of the rectangle) is 76 m.

Continued on Next Page

1 Identify all the rectangles.

(a)

(b)
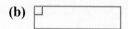

(c)

(d)
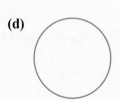

(e)
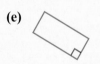

(f)

(g)

ANSWERS

1. **(a)**, **(b)**, and **(e)** are rectangles; **(c)**, **(d)**, **(f)**, and **(g)** are not.

2 Find the perimeter of each rectangle by using the formula or by adding the lengths of the sides.

(a)

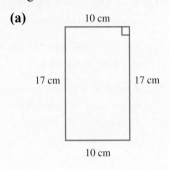

(b)

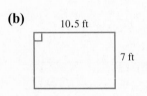

(c) 6 m wide and 11 m long

(d) 0.9 km by 2.8 km

As a check, you can add up the lengths of the four sides.

$$P = 27 \text{ m} + 27 \text{ m} + 11 \text{ m} + 11 \text{ m}$$

$$P = 76 \text{ m} \leftarrow \text{Same result as using the formula}$$

(b) A rectangle 8.9 ft by 12.3 ft
You can use the formula, as shown below.

$$P = 2 \bullet \quad l \quad + 2 \bullet \quad w$$

$$P = \underbrace{2 \bullet 12.3 \text{ ft}} + \underbrace{2 \bullet 8.9 \text{ ft}}$$

$$P = \quad 24.6 \text{ ft} \quad + \quad 17.8 \text{ ft}$$

$$P = 42.4 \text{ ft} \quad \boxed{\text{Be sure to write } ft \text{ in the answer.}}$$

Or, you can add up the lengths of the four sides.

$$P = 12.3 \text{ ft} + 12.3 \text{ ft} + 8.9 \text{ ft} + 8.9 \text{ ft}$$

$$P = 42.4 \text{ ft} \leftarrow \text{Same result as using the formula}$$

Either method will give you the correct result.

◀ *Work Problem* **2** *at the Side.*

The *perimeter* of a rectangle is the distance around the *outside edges*. The **area** of a rectangle is the amount of surface *inside* the rectangle. We measure area by seeing how many squares of a certain size are needed to cover the surface inside the rectangle. Think of covering the floor of a rectangular living room with carpet. Carpet is measured in square yards, that is, square pieces that measure 1 yard along each side. Here is a drawing of a living room floor.

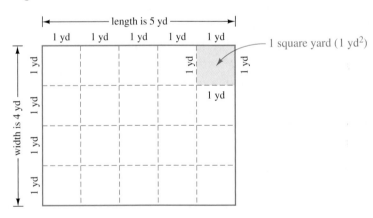

You can see from the drawing that it takes 20 squares to cover the floor. We say that the area of the floor is 20 *square yards*. A shorter way to write square yards is yd².

$$20 \text{ square yards} \quad \text{can be written as} \quad 20 \text{ yd}^2$$

To find the number of squares, you can count them, or you can multiply the number of squares in the length (5) times the number of squares in the width (4) to get 20. The formula is given below.

Finding the Area of a Rectangle

Area of a rectangle = length • width

$$A = l \bullet w$$

Remember to use *square units* when measuring area.

Answers
2. (a) $P = 54$ cm   (b) $P = 35$ ft
(c) $P = 34$ m   (d) $P = 7.4$ km

Squares of many sizes can be used to measure area. For smaller areas, you might use the ones shown below.

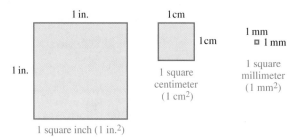

1 in.

1 in.

1 square inch (1 in.²)

1 cm

1 cm

1 square
centimeter
(1 cm²)

1 mm

□ 1 mm

1 square
millimeter
(1 mm²)

Actual-size drawings

Other sizes of squares that are often used to measure area are listed here, but they are too large to draw on this page.

1 square meter (1 m²)	1 square foot (1 ft²)
1 square kilometer (1 km²)	1 square yard (1 yd²)
	1 square mile (1 mi²)

> **CAUTION**
> The raised 2 in 4^2 means that you multiply $4 \cdot 4$ to get 16. The raised 2 in cm² or yd² is a short way to write the word *square*. When you see 5 cm², say "five square centimeters." Do *not* multiply $5 \cdot 5$. The exponent applies to cm, *not* to the number.

EXAMPLE 2   **Finding the Area of a Rectangle**

Find the area of each rectangle.

(a)

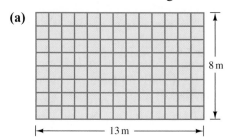

8 m

13 m

The length of this rectangle is 13 m and the width is 8 m. Use the formula $A = l \cdot w$.

$A = \quad l \quad \cdot \quad w$ Replace l with 13 m and w with 8 m.

$A = \textbf{13 m} \cdot \textbf{8 m}$ Multiply.

$A = 104$ square meters

> Write **m²** in the answer.

"Square meters" can be written as m², so the area is 104 m².

(b) A rectangle measuring 7 cm by 21 cm

First make a sketch of the rectangle. The length is 21 cm (the longer measurement) and the width is 7 cm. Then use the formula for the area of a rectangle, $A = l \cdot w$.

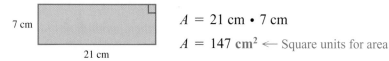

7 cm

21 cm

$A = 21$ cm $\cdot$ 7 cm

$A = \textbf{147 cm}^2 \leftarrow$ Square units for area

The area of the rectangle is 147 cm².

Continued on Next Page

3 Find the area of each rectangle.

(a)

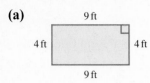

(b) A rectangle is 6 m long and 0.5 m wide. (First make a sketch of the rectangle and label the lengths of the sides.)

(c) A rectangular patio measures 3.5 yd by 2.5 yd. (First make a sketch of the patio and label the lengths of the sides.)

CAUTION

The units for *area* will always be *square* units (cm², m², yd², mi², and so on). The units for *perimeter* will always be *linear* units (cm, m, yd, mi, and so on) *not* square units.

◀ *Work Problem* **3** *at the Side.*

OBJECTIVE **2** **Find the perimeter and area of a square.**
A **square** is a rectangle with all sides the same length. Two squares are shown below. Notice the 90° angles.

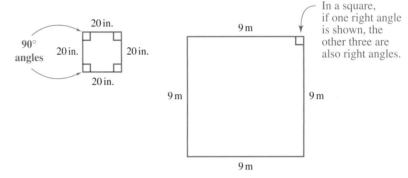

To find the *perimeter* of (distance around) the square on the right, you could add 9 m + 9 m + 9 m + 9 m to get 36 m. A shorter way is to multiply the length of one side times 4, because all four sides are the same length.

> **Finding the Perimeter of a Square**
> Perimeter of a square = side + side + side + side
> or, $P = 4 \cdot$ side
> $P = 4 \cdot s$

As with a rectangle, you can multiply length times width to find the *area* of (surface inside) a square. Because the length and the width are the same in a square, the formula is written as shown below.

> **Finding the Area of a Square**
> Area of a square = side $\cdot$ side
> $A = s \cdot s$
> $A = s^2$
>
> Remember to use *square units* when measuring area.

EXAMPLE 3 **Finding the Perimeter and Area of a Square**

(a) Find the perimeter of the square shown above where each side measures 9 m.

Use the formula. | Or add up the four sides.

$P = 4 \cdot s$ | $P = 9 \text{ m} + 9 \text{ m} + 9 \text{ m} + 9 \text{ m}$

$P = 4 \cdot 9 \text{ m}$ | $P = 36 \text{ m}$

$P = 36 \text{ m}$

Be careful to write **m** in the answer.

Same answer

ANSWERS

3. **(a)** $A = 36 \text{ ft}^2$
   (b) $A = 3 \text{ m}^2$

   (c) $A = 8.75 \text{ yd}^2$

—— **Continued on Next Page**

(b) Find the area of the same square where each side measures 9 m.

$$A = s^2$$

$$A = s \cdot s$$

$$A = 9 \text{ m} \cdot 9 \text{ m}$$

$$A = 81 \text{ m}^2 \leftarrow \text{Square units for area}$$

> **CAUTION**
> Be careful! s^2 means $s \cdot s$. It does *not* mean $2 \cdot s$. In Example 3(b) above, s is 9 m, so s^2 is 9 m $\cdot$ 9 m = 81 m². It is *not* 2 $\cdot$ 9 m = 18 m.

———————————————— *Work Problem* (4) *at the Side.* ▶

OBJECTIVE 3 Find the perimeter and area of a composite figure. As with any other shape, you can find the perimeter of (distance around) an irregular shape by adding up the lengths of the sides. To find the area (surface inside the shape), try to break it up into pieces that are squares or rectangles. Find the area of each piece and then add them together.

> **CAUTION**
> **Perimeter** is the *distance around the outside edges* of a flat shape. It is always measured in *linear units* such as cm, m, yd, and so on.
>
> **Area** is the amount of *surface inside* a flat shape. It is always measured in *square units* such as cm², m², yd², and so on.

EXAMPLE 4 **Finding the Perimeter and Area of a Composite Figure**

The floor of a room has the shape shown below.

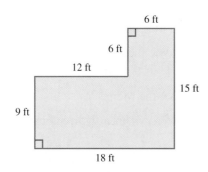

(a) Suppose you want to put a new wallpaper border along the top of all the walls. How much material do you need?

Find the perimeter of the room by adding up the lengths of the sides.

$$P = 9 \text{ ft} + 12 \text{ ft} + 6 \text{ ft} + 6 \text{ ft} + 15 \text{ ft} + 18 \text{ ft}$$

$$P = 66 \text{ ft}$$

You need 66 ft of wallpaper border.

— **Continued on Next Page**

---

4 Find the perimeter and area of each square.

(a)

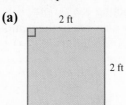

(b) 10.5 cm on each side (Make a sketch of the square.)

(c) 2.1 mi on a side (Make a sketch of the square.)

ANSWERS

4. **(a)** $P = 8$ ft; $A = 4$ ft²
   (b) $P = 42$ cm;
   $A = 110.25$ cm²

   (c) $P = 8.4$ mi;
   $A = 4.41$ mi²

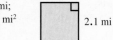

5 Carpet costs $19.95 per square yard. Find the cost of carpeting each room. Round your answers to the nearest cent if necessary.

(a)

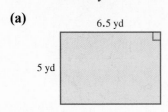

6.5 yd

5 yd

(b)

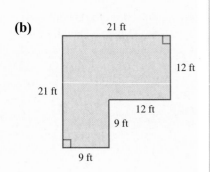

21 ft

12 ft

21 ft

12 ft

9 ft

9 ft

(c) A rectangular classroom is 24 ft long and 18 ft wide. (Make a sketch of the classroom.)

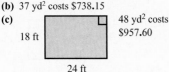
(b) The carpet you like costs $20.50 per square yard. How much will it cost to carpet the room?

First change the measurements from feet to yards, because the carpet is sold in square yards. There are 3 ft in 1 yd, so multiply by the unit fraction that allows you to divide out feet. Let's start with 9 ft.

$$\frac{\overset{3}{\cancel{9}\text{ ft}}}{1} \cdot \frac{1\text{ yd}}{\underset{1}{\cancel{3}\text{ ft}}} = 3\text{ yd}$$

↑ ↑
— Divide out ft.
— Divide 9 and 3 by 3.

Use the same unit fraction to change the other measurements from feet to yards.

$$\frac{\overset{4}{\cancel{12}\text{ ft}}}{1} \cdot \frac{1\text{ yd}}{\underset{1}{\cancel{3}\text{ ft}}} = 4\text{ yd} \qquad \frac{\overset{2}{\cancel{6}\text{ ft}}}{1} \cdot \frac{1\text{ yd}}{\underset{1}{\cancel{3}\text{ ft}}} = 2\text{ yd}$$

$$\frac{\overset{5}{\cancel{15}\text{ ft}}}{1} \cdot \frac{1\text{ yd}}{\underset{1}{\cancel{3}\text{ ft}}} = 5\text{ yd} \qquad \frac{\overset{6}{\cancel{18}\text{ ft}}}{1} \cdot \frac{1\text{ yd}}{\underset{1}{\cancel{3}\text{ ft}}} = 6\text{ yd}$$

Next, break up the room into two pieces. Use just the measurements for the length and width of each piece.

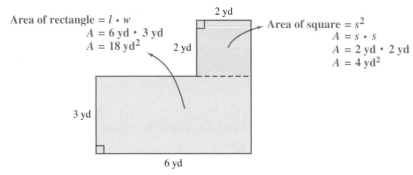

Area of rectangle = $l \cdot w$
$A = 6\text{ yd} \cdot 3\text{ yd}$
$A = 18\text{ yd}^2$

2 yd

2 yd

Area of square = s^2
$A = s \cdot s$
$A = 2\text{ yd} \cdot 2\text{ yd}$
$A = 4\text{ yd}^2$

3 yd

6 yd

Total area = $18\text{ yd}^2 + 4\text{ yd}^2 = 22\text{ yd}^2$

Multiply to find the cost of the carpet.

$$\frac{22\text{ yd}^2}{1} \cdot \frac{\$20.50}{1\text{ yd}^2} = \$451.00$$

It will cost $451.00 to carpet the room.

You could have cut the room into two rectangles as shown below. The total area is the same

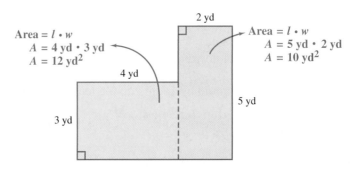

Area = $l \cdot w$
$A = 4\text{ yd} \cdot 3\text{ yd}$
$A = 12\text{ yd}^2$

2 yd

Area = $l \cdot w$
$A = 5\text{ yd} \cdot 2\text{ yd}$
$A = 10\text{ yd}^2$

4 yd

5 yd

3 yd

Total area = $12\text{ yd}^2 + 10\text{ yd}^2 = 22\text{ yd}^2$ ← Same answer as above

◀ *Work Problem* **5** *at the Side.*

8.3 ▶▶▶ Exercises

Find the perimeter and area of each rectangle or square. See Examples 1–3.

1.

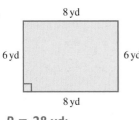

8 yd

6 yd   6 yd

8 yd

P = 28 yd;
A = 48 yd²

2.

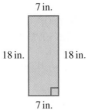

7 in.

18 in.   18 in.

7 in.

P = 50 in.;
A = 126 in.²

3.

0.9 km   0.9 km

0.9 km   0.9 km

P = 3.6 km;
A = 0.81 km²

4.

7.5 m

7.5 m

P = 30 m;
A = 56.25 m²

Draw a sketch of each square or rectangle and label the lengths of the sides. Then find the perimeter and the area. (*Sketches may vary, show your sketches to your instructor.*)

5. 10 ft by 10 ft

P = 40 ft;
A = 100 ft²

6. 8 cm by 17 cm

P = 50 cm;
A = 136 cm²

7. 14 m by 0.5 m

P = 29 m;
A = 7 m²

8. 2.35 km by 8.4 km

P = 21.5 km;
A = 19.74 km²

9. A storage building that is 76.1 ft by 22 ft

P = 196.2 ft;
A = 1674.2 ft²

10. A science lab measuring 12 m by 12 m

P = 48 m;
A = 144 m²

11. A square nature preserve 3 mi wide

P = 12 mi;
A = 9 mi²

12. A square of cardboard 20.3 cm on a side

P = 81.2 cm;
A = 412.09 cm²

Find the perimeter and area of each figure. See Example 4.

13.

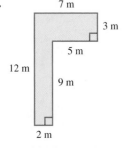

7 m

3 m

5 m

12 m

9 m

2 m

P = 38 m;
A = 39 m²

14.

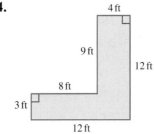

4 ft

9 ft

12 ft

8 ft

3 ft

12 ft

P = 48 ft;
A = 72 ft²

15.

17 m

12 m
4 m
4 m
28 m
4 m
12 m

17 m

P = 98 m;
A = 492 m²

16.

3.5 cm
3 cm
1.5 cm
8 cm
5 cm

5 cm

P = 26 cm;
A = 35.5 cm²

First find the length of the unlabeled side in each figure. Then find the perimeter and area of each figure.

17.

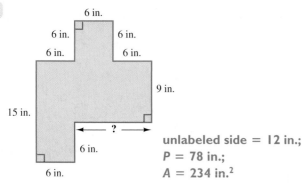

6 in.
6 in.   6 in.
6 in.   6 in.
9 in.
15 in.
?
6 in.
6 in.

unlabeled side = 12 in.;
P = 78 in.;
A = 234 in.²

18.

12 ft
18 ft   20 ft
16 ft   10 ft
40 ft
32 ft
?

unlabeled side = 48 ft;
P = 196 ft;
A = 1472 ft.²

Solve each application problem. In Exercises 19–24, draw a sketch for each problem and label it with the appropriate measurements. (Sketches may vary; show your sketches to your instructor.)

19. Gymnastic floor exercises are performed on a square mat that is 12 meters on a side. Find the perimeter and area of a mat. (*Source:* www.nist.gov)

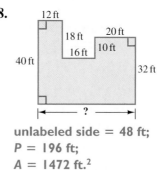

P = 48 m;
A = 144 m²

20. A regulation volleyball court is 18 meters by 9 meters. Find the perimeter and area of a regulation court. (*Source:* www. nist. gov)

P = 54 m;
A = 162 m²

21. The Wang's family room measures 20 ft by 25 ft. They are covering the floor with square tiles that measure 1 ft on a side and cost $0.92 each. How much will they spend on tile?

$460

22. A page in this book measures 27.5 cm from top to bottom and 21 cm from side to side. Find the perimeter and the area of the page.

P = 97 cm;
A = 577.5 cm²

23. Tyra's kitchen is 4.4 m wide and 5.1 m long. She is pasting a decorative border strip that costs $4.99 per meter around the top edge of all the walls. How much will she spend?

$94.81

24. Mr. and Mrs. Gomez are buying carpet for their square-shaped bedroom that is 5 yd wide. The carpet is $23 per square yard and padding and installation is another $6 per square yard. How much will they spend in all?

$725

25. Regulation soccer fields can measure 50 to 100 yards wide and 100 to 130 yards long, depending upon the age and skill level of the players. Find the area of the smallest soccer field and the area of the largest soccer field. What is the difference in the playing room between the smallest and largest fields?

**Area of largest field is 13,000 yd²;
Area of smallest field is 5000 yd²;
difference is 13,000 yd² − 5000 yd² = 8000 yd²**

26. The table below shows information on two tents for camping.

Tents	Coleman Family Dome	Eddie Bauer Dome Tent
Dimensions	$13 \text{ ft} \times 13 \text{ ft}$	$12 \text{ ft} \times 12 \text{ ft}$
Sleeps	8 campers	6 campers
Sale price	$127	$99

Source: target.com

(a) For the Coleman tent, find the perimeter, area, and number of square feet of floor space for each camper. Round to the nearest whole number.

P = 52 ft; A = 169 ft²; 21 ft²/camper (rounded)

(b) Find the same information for the Eddie Bauer tent.

P = 48 ft; A = 144 ft²; 24 ft²/camper

27. A regulation football field is rectangular, 100 yd long (excluding end zones), and has an area of 5300 yd². Find the width of the field. (*Source:* National Football League.)   **53 yd**

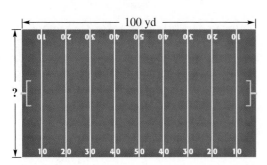

28. There are 14,790 ft² of ice in the rectangular playing area for a major league hockey game (excluding the area behind the goal lines). If the playing area is 85 ft wide, how long is it? (*Source:* National Hockey League.)   **174 ft**

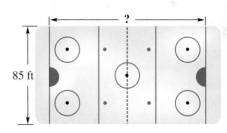

29. A rectangular lot is 124 ft by 172 ft. County rules require that nothing be built on land within 12 ft of any edge of the lot. First, add labels to the sketch of the lot, showing the land that cannot be build on. Then find the area of the land that cannot be built on.

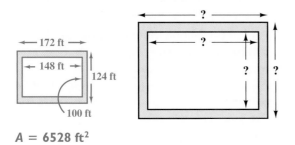

A = 6528 ft²

30. Find the cost of fencing needed for this rectangular field. Fencing along the country roads costs $4.25 per foot. Fencing for the other two sides costs $2.75 per foot.   **$1456**

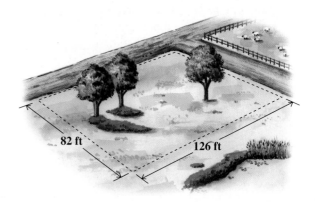

Relating Concepts (Exercises 31–36) For Individual or Group Work

*Use your knowledge of perimeter and area to **work Exercises 31–36 in order.***

31. Suppose you have 12 ft of fencing to make a square or rectangular garden plot. Draw sketches of *all* the possible plots that use exactly 12 ft of fencing and label the lengths of the sides. Use only *whole number* lengths. (*Hint:* There are three possibilities.)

32. (a) Find the area of each plot in Exercise 31.

5 ft by 1 ft has area of 5 ft²; 4 ft by 2 ft has area of 8 ft²; 3 ft by 3 ft has area of 9 ft²

(b) Which plot has the greatest area?

The square plot 3 ft by 3 ft has the greatest area.

33. Repeat Exercise 31 using 16 ft of fencing. Be sure to draw *all* possible plots that have whole number lengths for the sides.

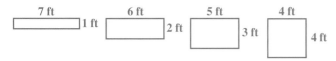

34. (a) Find the area of each plot in Exercise 33.

7 ft²; 12 ft²; 15 ft²; 16 ft²

(b) Compare your results to those from Exercise 32. What do you notice about the plots with the greatest area?

Square plots have the greatest area.

35. (a) Draw a sketch of a rectangular plot 3 ft by 2 ft. Find the perimeter and area.

3 ft
2 ft

$P = 10$ ft;
$A = 6$ ft²

(b) Suppose you *double* the length of the plot and *double* the width. Draw a sketch of the enlarged plot and find the perimeter and area.

6 ft
4 ft

$P = 20$ ft;
$A = 24$ ft²

(c) The *perimeter* of the enlarged plot is how many times greater than the perimeter of the original plot? The *area* of the enlarged plot is how many times greater than the original area?

Perimeter is twice the original; area is four times the original.

36. (a) Refer to part (a) of Exercise 35. Suppose you *triple* the length and width of the original plot. Draw a sketch of the enlarged plot and find the perimeter and area.

9 ft
6 ft

$P = 30$ ft;
$A = 54$ ft²

(b) How many times greater is the *perimeter* of the enlarged plot? How many times greater is the *area* of the enlarged plot?

Perimeter is three times the original; area is nine times the original.

(c) Suppose you make the length and width *four times greater* in the enlarged plot. What would you predict will happen to the perimeter and area, compared to the original plot?

Perimeter will be four times the original; area will be 16 times the original.

8.4 ▶▶▶ Parallelograms and Trapezoids

A **parallelogram** is a four-sided figure with opposite sides parallel, such as the ones shown below. Notice that opposite sides have the same length.

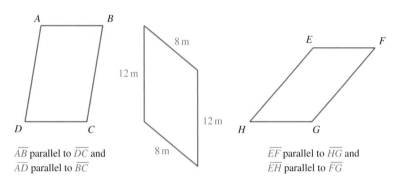

$\overline{AB}$ parallel to $\overline{DC}$ and
$\overline{AD}$ parallel to $\overline{BC}$

$\overline{EF}$ parallel to $\overline{HG}$ and
$\overline{EH}$ parallel to $\overline{FG}$

OBJECTIVE 1 Find the perimeter and area of a parallelogram.
Perimeter is the distance around a flat shape, so the easiest way to find the perimeter of a parallelogram is to add the lengths of the four sides.

EXAMPLE 1 **Finding the Perimeter of a Parallelogram**

Find the perimeter of the middle parallelogram above.

$P = 12\text{ m} + 12\text{ m} + 8\text{ m} + 8\text{ m} = 40\text{ m}$

> You must write **m** in the answer.

Work Problem ① *at the Side.* ▶

To find the area of a parallelogram, first draw a dashed line inside the figure as shown here.

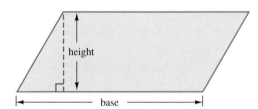

Try this yourself by tracing this parallelogram onto a piece of paper.

The length of the dashed line is the *height* of the parallelogram. It forms a *right angle* with the base. The height is the shortest distance between the base and the opposite side.

Now cut off the triangle created on the left side of the parallelogram above and move it to the right side, as shown below.

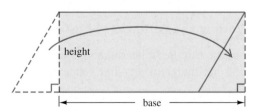

The parallelogram has been made into a rectangle. You can see that the area of the parallelogram and the rectangle are the same.

Equal areas
→ Area of the   rectangle   = length • width
→ Area of the parallelogram = base • height

OBJECTIVES

① **Find the perimeter and area of a parallelogram.**

② **Find the perimeter and area of a trapezoid.**

---

① Find the perimeter of each parallelogram

(a)
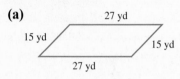
27 yd
15 yd
15 yd
27 yd

(b)
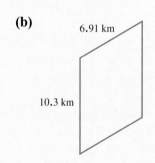
6.91 km
10.3 km

ANSWERS

1. **(a)** $P = 84$ yd   **(b)** $P = 34.42$ km

2 Find the area of each parallelogram.

(a)

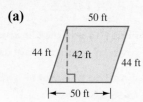

50 ft
44 ft   42 ft   44 ft
|← 50 ft →|

(b)

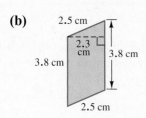

2.5 cm
2.3 cm   3.8 cm
3.8 cm
2.5 cm

(c) A parallelogram with base $12\frac{1}{2}$ mi and height $4\frac{3}{4}$ mi (*Hint:* Write $12\frac{1}{2}$ as 12.5 and $4\frac{3}{4}$ as 4.75.)

3 Find the perimeter of each trapezoid.

(a)

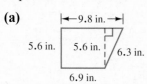

|←9.8 in.→|
5.6 in.   5.6 in.   6.3 in.
6.9 in.

(b)

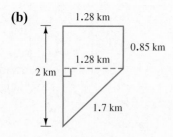

1.28 km
0.85 km
1.28 km
2 km
1.7 km

(c) A trapezoid with sides 39.7 cm, 29.2 cm, 74.9 cm, and 16.4 cm

ANSWERS

2. **(a)** $A = 2100$ ft² **(b)** $A = 8.74$ cm²

   (c) $A = 59\frac{3}{8}$ mi² or 59.375 mi²

3. **(a)** $P = 28.6$ in. **(b)** $P = 5.83$ km

   (c) $P = 160.2$ cm

> **Finding the Area of a Parallelogram**
>
> Area of a parallelogram = base • height
> $$A = b \cdot h$$
> Remember to use *square units* when measuring area.

EXAMPLE 2 **Finding the Area of Parallelograms**

Find the area of each parallelogram.

(a)

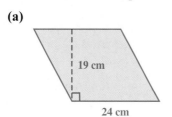

19 cm
24 cm

The base is 24 cm and the height is 19 cm. Use the formula $A = b \cdot h$.

$A = \quad b \quad \cdot \quad h$

$A = 24 \text{ cm} \cdot 19 \text{ cm}$

$A = 456 \text{ cm}^2$

> Careful! Write **cm²** in the answer.

(b)

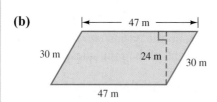

|← 47 m →|
30 m   24 m   30 m
47 m

$A = 47 \text{ m} \cdot 24 \text{ m}$

$A = 1128 \text{ m}^2$

> Write **m²** in the answer.

Notice that you do *not* use the 30 m sides when finding the area. But you would use them when finding the *perimeter* of the parallelogram.

◀ *Work Problem* **2** *at the Side.*

OBJECTIVE **2** **Find the perimeter and area of a trapezoid.**
A **trapezoid** is a four-sided figure with exactly one pair of parallel sides, such as the figures shown below. Unlike parallelograms, opposite sides of a trapezoid might *not* have the same length.

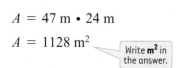

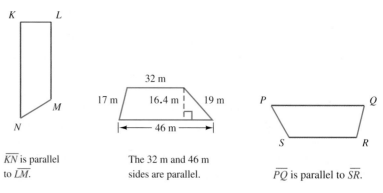

K   L
M
N

$\overline{KN}$ is parallel to $\overline{LM}$.

32 m
17 m   16.4 m   19 m
46 m

The 32 m and 46 m sides are parallel.

P   Q
S   R

$\overline{PQ}$ is parallel to $\overline{SR}$.

EXAMPLE 3 **Finding the Perimeter of a Trapezoid**

Find the perimeter of the middle trapezoid above.
You can find the perimeter of any flat shape by adding the lengths of the sides.

> The height of 16.4 m is **not** one of the sides.

$$P = 17 \text{ m} + 32 \text{ m} + 19 \text{ m} + 46 \text{ m}$$

$$P = 114 \text{ m}$$

Notice that the height (16.4 m) is *not* part of the perimeter, because the height is *not* one of the *outside edges* of the shape.

◀ *Work Problem* **3** *at the Side.*

Use this formula to find the *area* of a trapezoid.

> **Finding the Area of a Trapezoid**
>
> $$\text{Area} = \frac{1}{2} \cdot \text{height} \cdot (\text{short base} + \text{long base})$$
>
> $$A = \frac{1}{2} \cdot h \cdot (b + B)$$
>
> or $A = 0.5 \cdot h \cdot (b + B)$
>
> Remember to use *square units* when measuring area.

EXAMPLE 4 **Finding the Area of a Trapezoid**

Find the area of this trapezoid. The short base and long base are the *parallel* sides.

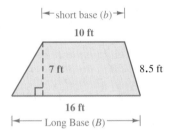

short base (b)
10 ft
7 ft   8.5 ft
16 ft
Long Base (B)

The height (h) is **7 ft**, the short base (b) is **10 ft**, and the long base (B) is **16 ft**. You do *not* need the lengths of the other two sides to find the area.

$$A = \frac{1}{2} \cdot h \cdot (b + B)$$

$$A = \frac{1}{2} \cdot 7 \text{ ft} \cdot (10 \text{ ft} + 16 \text{ ft}) \quad \text{Work inside parentheses first.}$$

$$A = \frac{1}{2} \cdot 7 \text{ ft} \cdot (\overset{13}{\underset{1}{26}} \text{ ft})$$

Be sure to write **ft²** in the answer.

$$A = 91 \text{ ft}^2 \quad \text{Square units for area}$$

You can also find the area by using **0.5**, the decimal equivalent for $\frac{1}{2}$, in the formula.

$$A = 0.5 \cdot h \cdot (b + B)$$
$$A = 0.5 \cdot 7 \cdot (10 + 16)$$
$$A = 0.5 \cdot 7 \cdot \quad 26$$
$$A = 91 \text{ ft}^2 \quad \text{Same answer as above}$$

> **▦ Calculator Tip** Use the parentheses keys on your scientific calculator to work Example 4 above.
>
> 0.5 ⊗ 7 ⊗ ⦅ 10 ⊕ 16 ⦆ ⊜ **91**
>
> What happens if you do *not* use the parentheses keys? What order of operations will the calculator follow then? (Answer: The calculator will multiply 0.5 times 7 times 10, and then add 16, giving an *incorrect* answer of 51.)

Work Problem ④ *at the Side.* ▶

---

4 Find the area of each trapezoid.

(a)

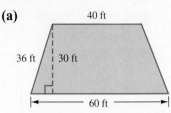

40 ft
36 ft 30 ft
60 ft

(b)

12.1 cm
12 cm
19 cm   14 cm
13 cm

(c) A trapezoid with height 4.7 m, short base 9 m, and long base 10.5 m (First draw a sketch and label the bases and height.)

ANSWERS

4. (a) $A = 1500 \text{ ft}^2$ **(b)** $A = 198 \text{ cm}^2$

   (c) $A = 45.825 \text{ m}^2$

9 m
4.7 m
10.5 m

5 Find the area of each floor.

(a)

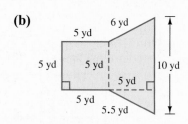

(b)

6 Find the cost of carpeting the floors in Problem 5 above. The cost of carpet is as follows:

(a) Floor (a), $18.50 per square meter.

(b) Floor (b), $28 per square yard.

EXAMPLE 5 **Finding the Area of a Composite Figure**

Find the area of this figure.

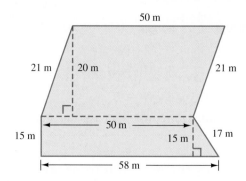

Break the figure into two pieces, a parallelogram and a trapezoid. Find the area of each piece, and then add the areas.

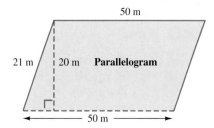

Area of parallelogram

$A = b \cdot h$

$A = 50 \text{ cm} \cdot 20 \text{ cm}$

$A = 1000 \text{ m}^2$

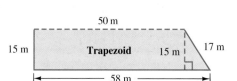

Area of trapezoid

$A = \frac{1}{2} \cdot h \cdot (b + B)$

$A = 0.5 \cdot 15 \text{ m} \cdot (50 \text{ m} + 58 \text{ m})$

$A = 810 \text{ m}^2$

Total area = $1000 \text{ m}^2 + 810 \text{ m}^2 = 1810 \text{ m}^2$

> Write **m²** in the answer.

The area of the figure is 1810 m².

◀ Work Problem **5** at the Side.

EXAMPLE 6 **Applying Knowledge of Area**

Suppose the figure in Example 5 above represents the floor plan of a hotel lobby. What is the cost of labor to install tile on the floor if the labor charge is $35.11 per square meter?

From Example 5, the floor area is 1810 m². To find the labor cost, multiply the number of square meters times the cost of labor per square meter.

$$\text{cost} = \frac{1810 \text{ m}^2}{1} \cdot \frac{\$35.11}{1 \text{ m}^2}$$

$$\text{cost} = \$63,549.10$$

The cost of the labor is $63,549.10.

◀ Work Problem **6** at the Side.

ANSWERS

5. **(a)** $A = 40 \text{ m}^2 + 44 \text{ m}^2 = 84 \text{ m}^2$
   (b) $A = 25 \text{ yd}^2 + 37.5 \text{ yd}^2 = 62.5 \text{ yd}^2$
6. **(a)** $1554 **(b)** $1750

8.4 ▶▶▶ **Exercises**

FOR
EXTRA
HELP   **MyMathLab**   Math XL   🖥   💿   ▬▬   🔄
                         PRACTICE   WATCH   DOWNLOAD   READ   REVIEW

Find the perimeter of each figure. See Examples 1 and 3.

1.
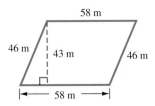
58 m
46 m   43 m   46 m
58 m
P = 208 m

2.
|← 1240 ft →|
1000 ft   930 ft   1000 ft
1240 ft
P = 4480 ft

🌐 **3.**
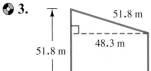
51.8 m
51.8 m   48.3 m
51.8 m
51.8 m
P = 207.2 m

4.
12.6 in.
11.1 in.
14.7 in.   24.8 in.
11.9 in.
P = 64 in.

🌐 **5.**
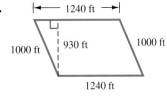
0.8 km
0.4 km   0.95 km
3 km
1.31 km
P = 6.06 km

6.
7.33 cm
2.8 cm
4.3 cm   3 cm
4.17 cm
P = 18.6 cm

Find the area of each figure. See Examples 2 and 4.

7.
31 mm
31 mm
25 mm   31 mm
31 mm
A = 775 mm²

8.
|← 21.4 m →|
20 m   13.2 m   20 m
21.4 m
A = 282.48 m²

🌐 **9.**
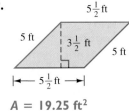
$5\frac{1}{2}$ ft
5 ft $3\frac{1}{2}$ ft   5 ft
|← $5\frac{1}{2}$ ft →|
A = 19.25 ft²

10.
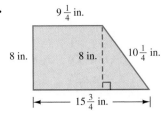
$9\frac{1}{4}$ in.
8 in.   8 in. $10\frac{1}{4}$ in.
|← $15\frac{3}{4}$ in. →|
A = 100 in.²

🌐 **11.**
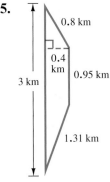
42 cm
61.4 cm   86.2 cm
42 cm
48.8 cm
A = 3099.6 cm²

12.
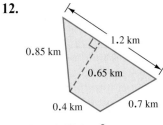
1.2 km
0.85 km
0.65 km
0.4 km   0.7 km
A = 0.52 km²

Solve each application problem. First label the bases and heights on the sketches. See Example 6.

13. The backyard of a new home is shaped like a trapezoid with a height of 45 ft and bases of 80 ft and 110 ft. What is the cost of putting sod on the yard if the landscaper charges $0.33 per square foot for sod?

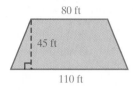

$1410.75

14. A swimming pool is in the shape of a parallelogram with a height of 9.6 m and base of 12.4 m. Find the labor cost to make a custom solar cover for the pool at a cost of $4.92 per square meter.

$585.68 (rounded)

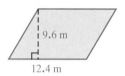

15. A piece of fabric for a quilt design is in the shape of a parallelogram. The base is 5 in. and the height is 3.5 in. What is the total area of the 25 parallelogram pieces needed for the quilt?

A = 437.5 in.²

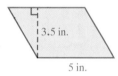

16. An accountant is paying $832 per month to rent an office in an old building. Her office is shaped like a trapezoid, with bases of 32 ft and 20 ft and a height of 20 ft. How much rent is she paying per square foot?

$1.60 per ft²

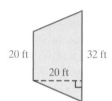

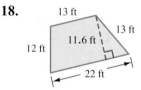 *Find **two** errors in each student's solution below. Write a sentence explaining each error. Then show how to work the problem correctly.*

17.

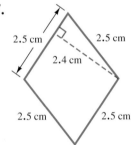

$P = 2.5 \text{ cm} + 2.4 \text{ cm} + 2.5 \text{ cm} + 2.5 \text{ cm} + 2.5 \text{ cm}$
$P = 12.4 \text{ cm}^2$

Height is not part of perimeter; square units are used for area, not perimeter.

$P = 2.5 \text{ cm} + 2.5 \text{ cm} + 2.5 \text{ cm} + 2.5 \text{ cm}$
$P = 10 \text{ cm}$

18.

$A = (0.5)(11.6 \text{ ft}) \cdot (12 \text{ ft} + 13 \text{ ft})$
$A = 145 \text{ ft}$

The bases are the parallel lines, 22 ft and 13 ft; area is measured in square units.

$A = 0.5 \cdot 11.6 \text{ ft} \cdot (22 \text{ ft} + 13 \text{ ft})$
$A = 203 \text{ ft}^2$

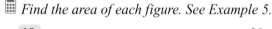 *Find the area of each figure. See Example 5.*

19.

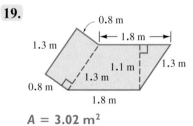

A = 3.02 m²

20.

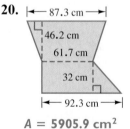

A = 5905.9 cm²

21.

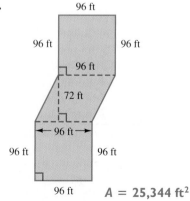

A = 25,344 ft²

8.5 ▷▷▷ Triangles

A **triangle** is a figure with exactly three sides. Some examples are shown below.

 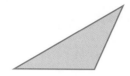 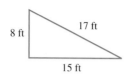

OBJECTIVE 1 Find the perimeter of a triangle. To find the perimeter of a triangle (the distance around the edges), add the lengths of the three sides.

EXAMPLE 1 Finding the Perimeter of a Triangle

Find the perimeter of the triangle above on the right.

$$P = 8 \text{ ft} + 15 \text{ ft} + 17 \text{ ft}$$
$$P = 40 \text{ ft}$$

Careful! Write **ft** in the answer.

Work Problem **1** *at the Side.* ▶

As with parallelograms, you can find the *height* of a triangle by measuring the distance from one vertex of the triangle to the opposite side (the base). The height line must be *perpendicular* to the base; that is, it must form a right angle with the base. Sometimes you have to extend the base in order to draw the height perpendicular to it, as shown below in the figure on the right.

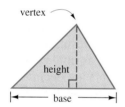 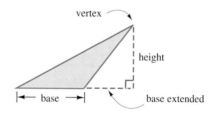 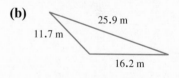

If you cut out two identical triangles and turn one upside down, you can fit them together to form a parallelogram, as shown below.

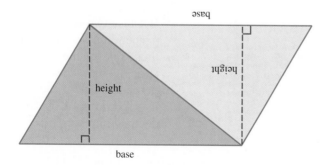

Recall from **Section 8.4** that the area of the parallelogram is *base* times *height*. Because each triangle is *half* of the parallelogram, the area of one triangle is

$$\frac{1}{2} \text{ of base times height}.$$

OBJECTIVES

1. Find the perimeter of a triangle.

2. Find the area of a triangle.

3. Given the measures of two angles in a triangle, find the measure of the third angle.

1 Find the perimeter of each triangle.

(a) 31 mm, 25 mm, 16 mm

(b) 25.9 m, 11.7 m, 16.2 m

(c) A triangle with sides of $6\frac{1}{2}$ yd, $9\frac{3}{4}$ yd, and $11\frac{1}{4}$ yd

ANSWERS

1. **(a)** $P = 72$ mm **(b)** $P = 53.8$ m
   (c) $P = 27\frac{1}{2}$ yd or 27.5 yd

2 Find the area of each triangle.

(a)

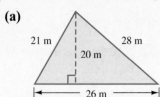

(b)

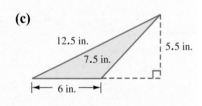

(c)

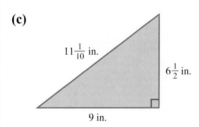

(d)

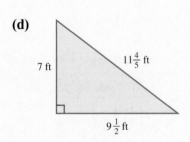

ANSWERS

2. **(a)** $A = 260$ m^2 **(b)** $A = 1.785$ cm^2
   (c) $A = 16.5$ in.^{2}

   (d) $A = 33.25$ ft^2 or $33\frac{1}{4}$ ft^2

OBJECTIVE 2 Find the area of a triangle. Use the following formula to find the *area* of a triangle.

> **Finding the Area of a Triangle**
>
> $$\text{Area of a triangle} = \frac{1}{2} \cdot \text{base} \cdot \text{height}$$
>
> $$A = \frac{1}{2} \cdot b \cdot h$$
>
> or $A = 0.5 \cdot b \cdot h$
>
> Remember to use **square units** when measuring area.

EXAMPLE 2 Finding the Area of Triangles

Find the area of each triangle.

(a)

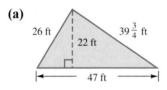

The base is 47 ft and the height is 22 ft. You do *not* need the 26 ft or $39\frac{3}{4}$ ft sides to find the area.

$$A = \frac{1}{2} \cdot b \cdot h$$

$$A = \frac{1}{2} \cdot 47 \text{ ft} \cdot \overset{11}{\cancel{22}} \text{ ft} \quad \begin{array}{l}\text{Divide out}\\ \text{common}\\ \text{factor of 2.}\end{array}$$

> This is *area*, so write **ft²** in the answer.

$$A = 517 \text{ ft}^2 \quad \text{Square units for area}$$

(b)

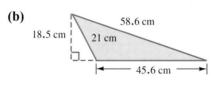

$$A = 0.5 \cdot 45.6 \text{ cm} \cdot 18.5 \text{ cm}$$

$$A = 421.8 \text{ cm}^2$$

The base must be extended to draw the height. However, still use 45.6 cm for b in the formula. Because the measurements are decimal numbers, it is easier to use 0.5 (the decimal equivalent of $\frac{1}{2}$) in the formula.

(c)

Because two sides of the triangle are perpendicular to each other, use those sides as the base and the height. (Recall that the height must be perpendicular to the base.) You can use fractions or use the equivalent decimal numbers.

Using fractions $\quad A = \frac{1}{2} \cdot 9 \text{ in.} \cdot 6\frac{1}{2} \text{ in.} = \frac{1}{2} \cdot \frac{9 \text{ in.}}{1} \cdot \frac{13 \text{ in.}}{2} = \frac{117}{4} \text{ in.}^2 = 29\frac{1}{4} \text{ in.}^2$

Using decimals $\quad A = 0.5 \cdot 9 \text{ in.} \cdot 6.5 \text{ in.} = 29.25 \text{ in.}^2 \leftarrow$ Equivalent

◀ *Work Problem* **2** *at the Side.*

EXAMPLE 3 **Using the Concept of Area**

Find the area of the shaded part in this figure.

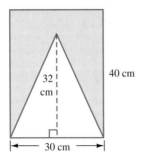

The *entire* figure is a rectangle. Find the area of the rectangle.

$$A = l \cdot w$$

$$A = 30 \text{ cm} \cdot 40 \text{ cm}$$

$$A = 1200 \text{ cm}^2$$

The *un*shaded part is a triangle. Find the area of the triangle.

$$A = \frac{1}{\underset{1}{2}} \cdot \overset{15}{30} \text{ cm} \cdot 32 \text{ cm}$$

$$A = 480 \text{ cm}^2$$

Subtract to find the area of the shaded part.

$$A = \overset{\text{Entire area}}{1200 \text{ cm}^2} - \overset{\text{Unshaded part}}{480 \text{ cm}^2} = \overset{\text{Shaded part}}{720 \text{ cm}^2}$$

> This is *area*, so write **cm²** in the answer.

The area of the shaded part of the figure is 720 cm².

— Work Problem (3) at the Side. ▶

EXAMPLE 4 **Applying the Concept of Area**

The Department of Transportation cuts triangular signs out of rectangular pieces of metal using the measurements shown above in Example 3. If the metal costs \$0.02 per square centimeter, how much does the metal cost for the sign? What is the cost of the metal that is *not* used?

From Example 3 above, the area of the triangle (the sign) is 480 cm². Multiply the area of the sign times the cost per square centimeter.

$$\text{cost of sign} = \frac{480 \text{ cm}^2}{1} \cdot \frac{\$0.02}{1 \text{ cm}^2} = \$9.60$$

The metal that is *not* used is the *shaded* part from Example 3. The unused area is 720 cm².

$$\text{cost of unused mental} = \frac{720 \text{ cm}^2}{1} \cdot \frac{\$0.02}{1 \text{ cm}^2} = \$14.40$$

The cost of the metal for the triangular sign is \$9.60. The unused metal costs \$14.40.

— Work Problem (4) at the Side. ▶

3 Find the area of the shaded part in this figure.

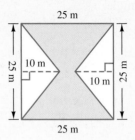

4 Suppose the figure in Problem 3 above is an auditorium floor plan. The shaded part will be covered with carpet costing \$27 per square meter. The rest will be covered with vinyl floor covering costing \$18 per square meter. What is the total cost of covering the floor?

ANSWERS

3. $A = 625 \text{ m}^2 - 125 \text{ m}^2 - 125 \text{ m}^2 = 375 \text{ m}^2$
4. \$10,125 + \$2250 + \$2250 = \$14,625

5 Find the number of degrees in the third angle of each triangle.

(a)
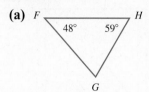

(b)
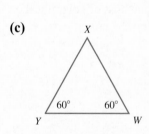

(c)

OBJECTIVE 3 Given the measures of two angles in a triangle, find the measure of the third angle. The *tri* in *triangle* means *three*. So the name tells you that a triangle has three angles. The sum of the measures of the three angles in any triangle is *always* 180°. You can see it by drawing a triangle, cutting off the three angles, and rearranging them to make a straight angle (180°).

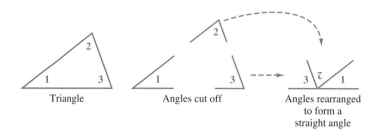

Triangle   Angles cut off   Angles rearranged to form a straight angle

---

Finding the Unknown Angle Measurement in a Triangle

Step 1 Add the number of degrees in the measures of the two given angles.

Step 2 Subtract the sum from 180°.

EXAMPLE 5 **Finding an Angle Measurement in Triangles**

Find the number of degrees in the indicated angle.

(a) Angle R

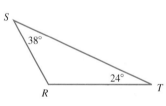

Step 1 Add the two angle measurements you are given.

$$38° + 24° = 62°$$

Step 2 Subtract the sum from 180°.

$$180° - 62° = 118°$$

$\angle R$ measures 118°.

(b) Angle F

$\angle E$ is a right angle, so it measures 90°.

Step 1 $90° + 45° = 135°$

Step 2 $180° - 135° = 45°$

$\angle F$ measures 45°.

> Write a small, raised circle for "degrees."

◄ *Work Problem* **5** *at the Side.*

ANSWERS

1. (a) 73° **(b)** 35° **(c)** 60°

Find the perimeter and area of each triangle. See Examples 1 and 2.

1.
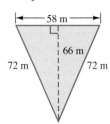
P = 202 m;
A = 1914 m²

2.
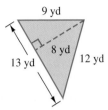
P = 34 yd;
A = 52 yd²

● 3.
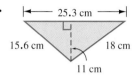
P = 58.9 cm;
A = 139.15 cm²

4.

P = 54.6 in.;
A = 128 in.²

5.
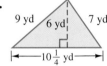
P = 26$\frac{1}{4}$ yd or 26.25 yd

A = 30$\frac{3}{4}$ yd² or 30.75 yd²

6.
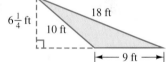
P = 37 ft;
A = 28$\frac{1}{8}$ ft² or 28.125 ft²

7.
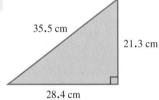
P = 85.2 cm;
A = 302.46 cm²

8.
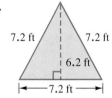
P = 21.6 ft;
A = 22.32 ft²

Find the shaded area in each figure. See Example 3.

9.
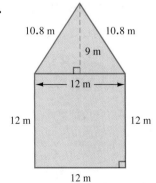
A = 198 m²

10.
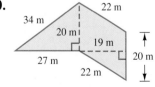
A = 650 m²

11.

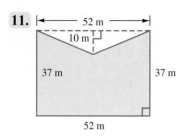

$A = 1664 \text{ m}^2$

12.

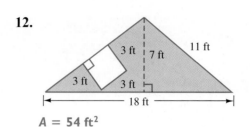

$A = 54 \text{ ft}^2$

Find the number of degrees in the third angle of each triangle. See Example 5.

13.

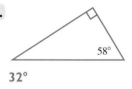

32°
58°

14.

46°
67°
67°

🌐**15.**

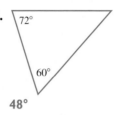

72°
60°
48°

16.

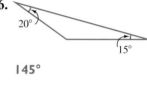

20°
15°
145°

✏ **17.** Can a triangle have two right angles? Explain your answer.

No. Right angles are 90°, so two right angles are 180°, and the sum of all three angles in a triangle equals 180°.

✏ **18.** In your own words, explain where the $\frac{1}{2}$ comes from in the formula for area of a triangle. Draw a sketch to illustrate your explanation.

Two identical triangles form a parallelogram. The area of a parallelogram is base times height, so the area of one of the triangles is $\frac{1}{2}$ • base • height.

Sketches will vary.

Solve each application problem. See Example 4.

19. A triangular tent flap measures $3\frac{1}{2}$ ft along the base and has a height of $4\frac{1}{2}$ ft. How much canvas is needed to make the flap?

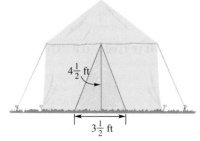

$7\frac{7}{8}$ ft² or 7.875 ft²

20. A wooden sign in the shape of a right triangle has perpendicular sides measuring 1.5 m and 1.2 m. How much surface area does the sign have?

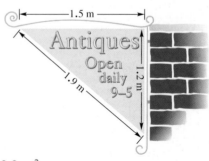

$A = 0.9 \text{ m}^2$

21. A triangular space between three streets has the measurements shown below.

    (a) How much new curbing will be needed to go around the space?   **126.8 m of curb**

    (b) How much sod will be needed to cover the space?   **672 m² of sod**

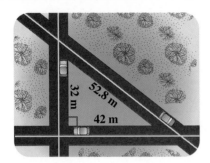

22. Each gable end of a new house has a span of 36 ft and a rise of 9.5 ft. What is the total area of both gable ends of the house?

A = 342 ft²

23. All sides of the house below are congruent and all roof sections are congruent.

    (a) Find the area of one side of the house.

        A = 32 m²

    (b) Find the area of one roof section.   A = 13.41 m²

24. The sketch shows the plan for an office building. The shaded area will be a parking lot. What is the cost of building the parking lot if the contractor charges $35 per square yard for materials and labor?

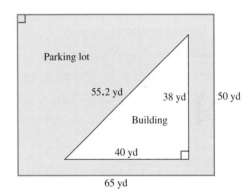

$87,150

25. A city lot with an unusual shape is shown below.

    (a) How much frontage (distance along streets) does the lot have?   **175 yd of frontage**

    (b) What is the area of the lot?   A = 8750 yd²

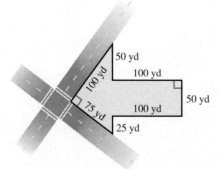

26. A car dealership wants three nylon pennants to hang in its front window. The height of the two smaller pennants is 3.5 ft, and the larger pennant has a height of 6.5 ft. How much nylon fabric is needed for all the pennants?

A = 23.5 ft² of fabric

Math in the Media

SCREEN DISPLAYS

The world's largest plasma screen panel display is now used in airports, sports stadiums, and resorts in the United States, Europe, Japan, and China. The table below shows some data available on-line about the plasma screen. (*Source:* www.eurosell.it)

WORLD'S LARGEST PLASMA SCREEN PANEL DISPLAY

Measurements	Width 2269 mm Height 1277 mm
Weight	Approximately 215 kg
Service life	60,000 hours

Use the data in the table to answer Questions 1–5, rounding to the nearest tenth if necessary.

1. The screen is how many meters wide and how many meters high?

   About 2.3 m wide, 1.3 m high

2. Using your answers from Question 1, convert the width and height to feet, using the fact that 1 m ≈ 3.28 ft.   **Width ≈ 7.5 ft; height ≈ 4.3 ft**

3. Now find the perimeter and area of the screen using your results from Question 2.

   P ≈ 23.6 ft; A ≈ 32.3 ft²

4. Convert the weight to pounds, using the fact that 1 kg ≈ 2.20 lb.   **About 473 lb**

5. (a) If the screen is never turned off, how long is the service life in days?   **2500 days**

   (b) How long is the service life in years? (Use 365 days per year.)   **About 6.8 years**

COMPUTER MONITOR DISPLAYS

Type of Monitor	Width of screen	Height of screen	Viewing area
Dell PC monitor	$12\frac{3}{4}$ in. **12.75 in.**	$9\frac{1}{2}$ in. **9.5 in.**	A ≈ **121 in.²**
ViewSonic LCD monitor	16.1 in.	10.0 in.	A = **161 in.²**

Source: Companies' Web sites.

6. The table above shows the measurements of several typical, rectangular, computer monitor screens. Find the viewing area of each screen, to the nearest whole number, and write it in the table. For the Dell PC monitor, first change the measurements to decimal numbers before finding the viewing area.

7. Draw a sketch of the Dell monitor. Label the lengths of the sides. Label the right angles. Which sides are parallel?

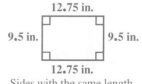

12.75 in.

9.5 in.   9.5 in.

12.75 in.

Sides with the same length are also parallel

Panaso c

8.6 ▶▶▶ Circles

OBJECTIVE 1 Find the radius and diameter of a circle. Suppose you start with one dot on a piece of paper. Then you draw many dots that are each 2 cm away from the first dot. If you draw enough dots (points) you'll end up with a *circle.* Each point on the circle is exactly 2 cm away from the *center* of the circle. The 2 cm distance is called the *radius, r,* of the circle. The distance across the circle (passing through the center) is called the *diameter, d,* of the circle.

OBJECTIVES

1 Find the radius and diameter of a circle.

2 Find the circumference of a circle.

3 Find the area of a circle.

4 Become familiar with Latin and Greek prefixes used in math terminology.

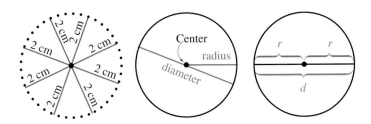

Circle, Radius, and Diameter

A **circle** is a two-dimensional (flat) figure with all points the same distance from a fixed center point.

The **radius** (r) is the distance from the center of the circle to any point on the circle.

The **diameter** (d) is the distance across the circle passing through the center.

Using the circle above on the right as a model, you can see some relationships between the radius and diameter.

Finding the Diameter and Radius of a Circle

$$\text{diameter} = 2 \cdot \text{radius}$$
$$d = 2 \cdot r$$

and $\text{radius} = \dfrac{\text{diameter}}{2}$ or $r = \dfrac{d}{2}$

EXAMPLE 1 **Finding the Diameter and Radius of a Circle**

Find the unknown length of the diameter or radius in each circle.

(a)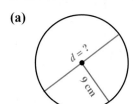

Because the radius is 9 cm, the diameter is twice as long.

$$d = 2 \cdot r$$
$$d = 2 \cdot 9 \text{ cm}$$
$$d = 18 \text{ cm}$$

Continued on Next Page

1 Find the unknown length of the diameter or radius in each circle.

(a)

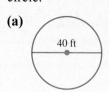

40 ft

(b)

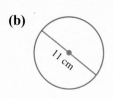

11 cm

(c)

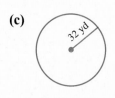

32 yd

(d)

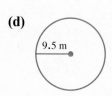

9.5 m

(b)

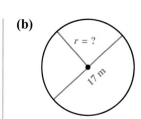

r = ?

17 m

The radius is half the diameter.

$$r = \frac{d}{2}$$

$$r = \frac{17 \text{ m}}{2}$$

$$r = 8.5 \text{ m} \quad \text{or} \quad 8\frac{1}{2} \text{ m}$$

> $8.5 = 8\frac{1}{2}$ because $8\frac{5}{10}$ simplifies to $8\frac{1}{2}$.

◀ Work Problem **1** at the Side.

OBJECTIVE 2 Find the circumference of a circle. The perimeter of a circle is called its **circumference**. Circumference is the distance around the edge of a circle.

The diameter of the can in the drawing is about 10.6 cm, and the circumference of the can is about 33.3 cm. Dividing the circumference of the circle by the diameter gives an interesting result.

$$\frac{\text{circumference}}{\text{diameter}} = \frac{33.3}{10.6} \approx 3.14 \qquad \text{Rounded to the nearest hundredth}$$

Dividing the circumference of *any* circle by its diameter *always* gives an answer close to 3.14. This means that going around the edge of any circle is a little more than 3 times as far as going straight across the circle.

This ratio of circumference to diameter is called π (the Greek letter **pi**, pronounced PIE). There is no decimal that is exactly equal to π, but here is the *approximate* value.

$$\pi \approx 3.14159265359$$

> **Rounding the Value of Pi (π)**
> We usually round π to 3.14. Therefore, calculations involving π will give approximate answers and should be written using the $\approx$ symbol.

Use the following formulas to find the *circumference* of a circle.

> **Finding the Circumference (Distance around a Circle)**
> $$\text{Circumference} = \pi \cdot \text{diameter}$$
> $$C = \pi \cdot d$$
> or, because $d = 2 \cdot r$, then $C = \pi \cdot 2 \cdot r$ usually written $C = 2 \cdot \pi \cdot r$
>
> Remember to use linear units such as ft, yd, m, and cm when measuring circumference (**not** square units).

ANSWERS

1. (a) $r = 20$ ft (b) $r = 5.5$ cm
   (c) $d = 64$ yd (d) $d = 19$ m

EXAMPLE 2   **Finding the Circumference of Circles**

Find the circumference of each circle. Use 3.14 as the approximate value for π. Round answers to the nearest tenth.

(a)

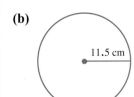

38 m

The *diameter* is 38 m, so use the formula with *d* in it.

$$C = \pi \cdot d$$

$$C \approx 3.14 \cdot 38 \text{ m}$$

> Be sure to write **m** in the answer.

$$C \approx 119.3 \text{ m} \quad \text{Rounded}$$

(b)

11.5 cm

In this example, the *radius* is labeled, so it is easier to use the formula with *r* in it.

$$C = 2 \cdot \pi \cdot r$$

$$C \approx 2 \cdot 3.14 \cdot 11.5 \text{ cm}$$

> Write **cm** in the answer.

$$C \approx 72.2 \text{ cm} \quad \text{Rounded}$$

▦ **Calculator Tip**   Many *scientific* calculators have a ⓟ key. Try pressing it. With a 10-digit display, you'll see the value of π to the nearest billionth.

3.141592654

But this is still an approximate value, although it is more precise than rounding π to 3.14. Try finding the circumference in Example 2(a) above using the ⓟ key.

ⓟ ⓧ 38 ⓔ   Answer is 119.3805208; rounds to 119.4

When you used 3.14 as the approximate value of π, the result rounded to 119.3, so the answers are slightly different. In this book, we will use 3.14 instead of the ⓟ key. Our measurements of radius and diameter are given as whole numbers or with tenths, so it is acceptable to round π to hundredths. And, some students may be using standard calculators without a ⓟ key or doing the calculations by hand.

Work Problem **2** *at the Side.* ▶

OBJECTIVE **3** **Find the area of a circle.**   To find the formula for the area of a circle, start by cutting two circles into many pie-shaped pieces.

Circumference (distance around) is $2 \cdot \pi \cdot r$.

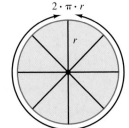

Unfold the circles, much as you might "unfold" a peeled orange, and put them together as shown here.

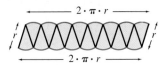

(*continued*)

---

2 Find the circumference of each circle. Use 3.14 as the approximate value for π. Round answers to the nearest tenth.

(a)

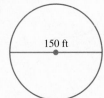

150 ft

(b)

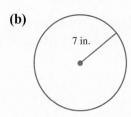

7 in.

(c)   diameter 0.9 km

(d)   radius 4.6 m

3 Find the area of each circle. Use 3.14 for π. Round your answers to the nearest tenth.

(a)

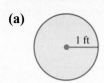
1 ft

(b)

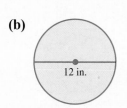
12 in.

(*Hint:* The diameter is 12 in. so $r =$ _____ in.)

(c)

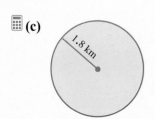
1.8 km

(d)

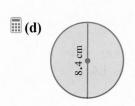
8.4 cm

The figure is approximately a parallelogram with height r (the radius of the original circle) and base $2 \cdot \pi \cdot r$ (the circumference of the original circle). The area of the "parallelogram" is base times height.

$$\text{Area} = \quad b \quad \cdot h$$
$$\text{Area} = 2 \cdot \pi \cdot r \cdot r$$
$$\text{Area} = 2 \cdot \pi \cdot r^2 \leftarrow \text{Recall that } r \cdot r \text{ is } r^2$$

Because the "parallelogram" was formed from *two* circles, the area of *one* circle is half as much.

$$\frac{1}{2} \cdot 2 \cdot \pi \cdot r^2 = 1 \cdot \pi \cdot r^2 \quad \text{or simply} \quad \pi \cdot r^2$$

Finding the Area of a Circle

$$\text{Area of a circle} = \pi \cdot \text{radius} \cdot \text{radius}$$
$$A = \pi \cdot r^2$$

Remember to use **square units** when measuring area.

EXAMPLE 3 **Finding the Area of Circles**

Find the area of each circle. Use 3.14 for π. Round your answers to the nearest tenth.

(a) A circle with a radius of 8.2 cm
Use the formula $A = \pi \cdot r^2$, which means $\pi \cdot r \cdot r$.

$$A = \pi \cdot r \cdot r$$
$$A \approx 3.14 \cdot 8.2 \text{ cm} \cdot 8.2 \text{ cm}$$

This is *area*, so write **cm²** in the answer.

$$A \approx 211.1 \text{ cm}^2 \quad \text{Rounded}$$

(b)

10 ft

To use the area formula, you need to know the *radius* (r). In this circle, the *diameter* is 10 ft. First find the radius.

$$r = \frac{d}{2}$$

You *cannot* use the *diameter* measurement in the area formula, so find the *radius*.

$$r = \frac{10 \text{ ft}}{2} = 5 \text{ ft}$$

Now find the area.

$$A \approx 3.14 \cdot 5 \text{ ft} \cdot 5 \text{ ft}$$
$$A \approx 78.5 \text{ ft}^2$$

Write square units (**ft²**) in your answer.

CAUTION
When finding *circumference*, you can start with either the radius or the diameter. When finding *area*, you must use the *radius*. If you are given the diameter, divide it by 2 to find the radius. Then find the area.

ANSWERS
3. (a) $A \approx 3.1$ ft²
   (b) $r = 6$ in.; $A \approx 113.0$ in.²
   (c) $A \approx 10.2$ km²   **(d)** $A \approx 55.4$ cm²

◀ *Work Problem* **3** *at the Side.*

⊞ **Calculator Tip**   You can use your calculator to find the area of the circle in Example 3(a) on the previous page. The first method works on both scientific and standard calculators:

3.14 ⊗ 8.2 ⊗ 8.2 ⊜   Answer is 211.1336

You round the answer to 211.1 (nearest tenth).

On a *scientific* calculator you can also use the ⊗ x^2 key, which automatically squares the number you enter (that is, multiplies the number times itself):

3.14 ⊗ 8.2 x^2 67.24 ⊜   Answer is 211.1336

Appears automatically;
8.2 × 8.2 is 67.24

In the next example, we find the area of a *semicircle,* which is half the area of a circle.

┌─ **EXAMPLE 4**   **Finding the Area of a Semicircle**

Find the area of the semicircle. Use 3.14 for π. Round your answer to the nearest tenth.

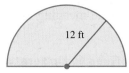

First, find the area of a whole circle with a radius of 12 ft.

$$A = \pi \cdot r \cdot r$$

$$A \approx 3.14 \cdot 12 \text{ ft} \cdot 12 \text{ ft}$$

$$A \approx 452.16 \text{ ft}^2 \leftarrow \text{Do not round yet.}$$

Divide the area of the whole circle by 2 to find the area of the semicircle.

$$\frac{452.16 \text{ ft}^2}{2} = 226.08 \text{ ft}^2$$

The *last* step is rounding 226.08 to the nearest tenth.

Write **ft²** in the answer.

Area of semicircle ≈ 226.1 ft²   Rounded

───────────────── Work Problem **4** at the Side. ▶

┌─ **EXAMPLE 5**   **Applying the Concept of Circumference**

A circular rug is 8 ft in diameter. The cost of fringe for the edge is $2.25 per foot. What will it cost to add fringe to the rug? Use 3.14 for π.

$$\text{Circumference} = \pi \cdot d$$

$$C \approx 3.14 \cdot 8 \text{ ft}$$

$$C \approx 25.12 \text{ ft}$$

$$\text{cost} = \text{cost per foot} \cdot \text{circumference}$$

$$\text{cost} = \frac{\$2.25}{1 \text{ ft}} \cdot \frac{25.12 \text{ ft}}{1}$$

$$\text{cost} = \$56.52$$

The cost of adding fringe to the rug is $56.25.

───────────────── Work Problem **5** at the Side. ▶

4 Find the area of each semicircle. Use 3.14 for π. Round your answers to the nearest tenth.

(a)

24 m

(b)

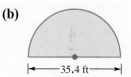

├◄─── 35.4 ft ───►┤

(c)

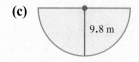

9.8 m

5 Find the cost of binding around the edge of a circular rug that is 3 m in diameter. The binder charges $4.50 per meter. Use 3.14 for π.

ANSWERS

4. **(a)** $A \approx 904.3 \text{ m}^2$ **(b)** $A \approx 491.9 \text{ ft}^2$
   (c) $A \approx 150.8 \text{ m}^2$
5. $42.39

6 Find the cost of covering the underside of the rug in margin problem 5 with a nonslip rubber backing. The rubber backing costs $3.89 per square meter.

EXAMPLE 6 Applying the Concept of Area

Find the cost of covering the rug in Example 5 (on the previous page) with a plastic cover. The material for the cover costs $1.50 per square foot. Use 3.14 for π.

First find the radius. $\quad r = \dfrac{d}{2} = \dfrac{8 \text{ ft}}{2} = 4 \text{ ft}$

Then find the area. $\quad A = \pi \cdot r^2$

$$A \approx 3.14 \cdot 4 \text{ ft} \cdot 4 \text{ ft}$$

$$A \approx 50.24 \text{ ft}^2$$

$$\text{cost} = \frac{\$1.50}{1 \text{ ft}^2} \cdot \frac{50.24 \text{ ft}^2}{1} = \$75.36$$

The cost of the plastic cover is $75.36.

◄ Work Problem **6** at the Side.

OBJECTIVE 4 Become familiar with Latin and Greek prefixes used in math terminology. Many English words are built from Latin or Greek root words and prefixes. Knowing the meaning of the more common ones can help you figure out the meaning of terms in many subject areas, including mathematics.

EXAMPLE 7 Using Prefixes to Understand Math Terms

7 (a) Here are some more prefixes you have seen in this textbook. List at least one math term and one nonmathematical word that use each prefix.

dia- (through):

fract- (break):

par- (beside):

per- (divide):

peri- (around);

rad- (ray):

rect- (right):

sub- (below):

(a) Listed below are some Latin and Greek root words and prefixes with their meanings in parentheses. You've already seen math terms in this textbook that use these prefixes. List at least one math term and one nonmathematical word that use each prefix or root word.

cent- (100): *cent*imeter; *cent*ury

circum- (around): *circum*ference; *circum*vent

de- (down): *de*nominator; *de*cline

dec- (10): *dec*imal; *Dec*ember (originally the 10th month in the old calendar)

There are many answers. These are some of the possibilities.

(b) Suppose you have trouble remembering which part of a fraction is the denominator. How could your knowledge of prefixes help in this situation?

The *de-* prefix in *de*nominator means "down" so the denominator is the number *down* below the fraction bar.

◄ Work Problem **7** at the Side.

(b) How could you use your knowledge of prefixes to remember the difference between perimeter and area?

Note

Here are some additional prefixes and root words and their meanings that you will see in the rest of **Chapter 8,** in **Chapter 9,** and in other math classes. An example of a math term and a nonmathematical word are shown for each one.

equ- (equal): *equ*ation; *equi*nox

hemi- (half): *hemi*sphere; *hemi*trope

lateral (side): quadri*lateral*; bi*lateral*

re- (back or again): *re*ciprocal; *re*duce

 **8.6** ▸▸▸ **Exercises**

Find the unknown length in each circle. See Example 1.

1.

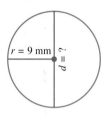

$r = 9$ mm $d = ?$

$d = 18$ mm

2.

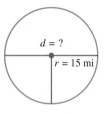

$d = ?$
$r = 15$ mi

$d = 30$ mi

3.

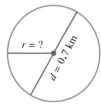

$r = ?$ $d = 0.7$ km

$r = 0.35$ km

4.

$d = 6.1$ cm $r = ?$

$r = 3.05$ cm

Find the circumference and area of each circle. Use 3.14 as the approximate value for π. Round your answers to the nearest tenth. See Examples 2 and 3.

5.

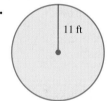

11 ft

$C \approx 69.1$ ft;
$A \approx 379.9$ ft^2

6.

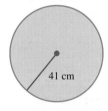

41 cm

$C \approx 257.5$ cm;
$A \approx 5278.3$ cm^2

7.

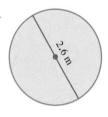

2.6 m

$C \approx 8.2$ m;
$A \approx 5.3$ m^2

8.

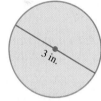

3 in.

$C \approx 9.4$ in.;
$A \approx 7.1$ in.^{2}

Find the circumference and area of circles having the following diameters. Use 3.14 as the approximate value of π. Round your answers to the nearest tenth. See Examples 2 and 3.

9. $d = 15$ cm

$C \approx 47.1$ cm;
$A \approx 176.6$ cm^2

10. $d = 39$ ft

$C \approx 122.5$ ft;
$A \approx 1194.0$ ft^2

11. $d = 7\frac{1}{2}$ ft

$C \approx 23.6$ ft;
$A \approx 44.2$ ft^2

12. $d = 4\frac{1}{2}$ yd

$C \approx 14.1$ yd;
$A \approx 15.9$ yd^2

13. $d = 8.65$ km

$C \approx 27.2$ km;
$A \approx 58.7$ km^2

14. $d = 19.5$ mm

$C \approx 61.2$ mm
$A \approx 298.5$ mm^2

Find each shaded area. Note that Exercises 15–18 all contain semicircles. Use 3.14 as the approximate value of π. Round your answers to the nearest tenth if necessary. See Example 4.

15.

7 in.

$A \approx 76.9$ in.^{2}

16.

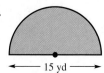

15 yd

$A \approx 88.3$ yd^2

17.

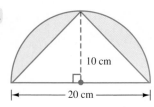

10 cm

20 cm

$A \approx 57 \text{ cm}^2$

18.

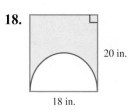

20 in.

18 in.

$A \approx 232.8 \text{ in.}^2$

19. How would you explain π to a friend who is not in your math class? Write an explanation. Then make up a test question that requires the use of π, and show how to solve it.

π is the ratio of circumference of a circle to its diameter. If you divide the circumference of any circle by its diameter, the answer is always a little more than 3. The approximate value is 3.14, which we call π (pi). Your test question could involve finding the circumference or the area of a circle.

20. Explain how circumference and perimeter are alike. How are they different? Make up two problems, one involving perimeter, the other circumference. Show how to solve your problems.

Circumference and perimeter are both the distance around a flat shape and are both measured in linear units like ft, yd, cm, or m. However, circumference applies *only* to circles. Perimeter can apply to many shapes such as squares, rectangles, triangles, and so on. Your circumference problem should use the formula $C = 2\pi r$ or $C = \pi d$. To find perimeter, add up the lengths of the sides of the shape.

Solve each application problem. Use 3.14 *as the approximate value of* π. *Round your answers to the nearest tenth. See Examples 5 and 6.*

21. An irrigation system moves around a center point to water a circular area for crops. If the irrigation system is 50 yd long, how large is the watered area?

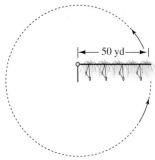

50 yd

$A \approx 7850 \text{ yd}^2$

22. If you swing a ball held at the end of a string 20 cm long, how far will the ball travel on each turn?

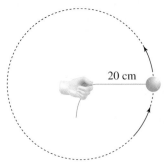

20 cm

$C \approx 125.6 \text{ cm}$

23. A Michelin Cross Terrain SUV tire has an overall diameter of 29.10 inches. How far will a point on the tire tread move in one complete turn? (*Source:* Michelin.) *Bonus question:* How many revolutions does the tire make per mile? Round to the nearest whole number.

$C \approx 91.4 \text{ in.}$;
Bonus, 693 revolutions/mile (rounded)

24. In August 2005, hurricane Katrina slammed into coastal Mississippi and flooded New Orleans. The diameter of the circular storm was 210 miles. In July 2005, hurricane Dennis, with a diameter of only 80 miles, hit the Florida Panhandle. What was the area of each storm, to the nearest hundred square miles? (*Source*: Associated Press.)

Katrina: $A \approx 34{,}600 \text{ mi}^2$;
Dennis: $A \approx 5000 \text{ mi}^2$

🖩 *For Exercises 25–30, first draw a circle and label the radius or diameter. Then solve the problem. Use 3.14 as the approximate value of π and round answers to the nearest tenth, money answers to the nearest cent.*

25. A radio station can be heard 150 miles in all directions during evening hours. How many square miles are in the station's broadcast area?

$A \approx 70{,}650 \text{ mi}^2$

26. An earthquake was felt by people 900 km away in all directions from the epicenter (the source of the earthquake). How much area was affected by the quake?

$A \approx 2{,}543{,}400 \text{ km}^2$

27. The diameter of Diana Hestwood's wristwatch is 1 in. and the radius of the clock face on her kitchen wall is 3 in. Find the circumference and the area of the watch face and the clock face.

$C \approx 3.1 \text{ in.}; A \approx 0.8 \text{ in.}^2$ $C \approx 18.8 \text{ in.}; A \approx 28.3 \text{ in.}^2$

28. The diameter of the largest known ball of twine is 12 ft 9 in. The sign posted near the ball says it has a circumference of 40 ft. Is the sign correct? *Hint:* First change 9 in. to feet and add it to 12 ft. (*Source: Guinness World Records.*)

$C \approx 40.0 \text{ ft}$; the sign is correct.

29. Blaine Fenstad wants to buy a pair of two-way radios. Some models have a range of 2 miles under ideal conditions. More expensive models have a range of 5 miles. What is the difference in the area covered by the 2-mile and 5-mile models? (*Source:* Best Buy.)

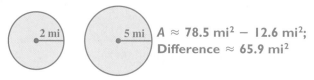

$A \approx 78.5 \text{ mi}^2 - 12.6 \text{ mi}^2$; Difference $\approx 65.9 \text{ mi}^2$

30. The National Audubon Society holds an end-of-year bird count. Volunteers count all the birds they see in a circular area during a 24-hour period. Each circle has a diameter of 15 miles. About 1700 circular areas are counted across the United States each December. What is the total area covered by the count?

Total area $\approx 300{,}262.5 \text{ mi}^2$

31. On the first page of this chapter, you read about a forester measuring the circumferences of trees.

(a) If the circumference of one tree is 144 cm, what is the diameter?

$d \approx 45.9 \text{ cm}$

📝 (b) Explain how you solved part (a).

Divide the circumference by π (3.14).

32. In Atlanta, Interstate 285 circles the city and is known as the "perimeter." If the circumference of the circle made by the highway is 62.8 miles, find:

(a) the diameter of the circle.

$d \approx 20 \text{ mi}$

(b) the area inside the circle.

$A \approx 314 \text{ mi}^2$

(*Source: Greater Atlanta Newcomer's Guide.*)

33. Find the cost of sod, at $0.49 per square foot, for this playing field that has a semicircle on each end.

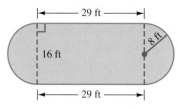

$325.83 (rounded)

34. Find the area of this skating rink.

$A \approx 2971.9 \text{ ft}^2$

Use the information about prefixes in Example 7 to answer Exercises 35 and 36.

35. Explain how you could use the information about prefixes to remember the difference between radius, diameter, and circumference.

The prefix *rad* tells you that radius is a ray from the center of the circle. The prefix *dia* means the diameter goes through the circle, and the prefix *circum* means the circumference is the distance around.

36. Explain how you could use the information about prefixes to avoid confusion between parallel and perpendicular lines.

The prefix *par* means beside, so parallel lines are beside each other. (Perpendicular lines cross so that they form a right angle.)

Relating Concepts (Exercises 37–42) For Individual or Group Work

*Use the table below to **work Exercises 37–42 in order.***

Find the best buy for each type of pizza. The best buy is the lowest cost per square inch of pizza. All the pizzas are circular in shape, and the measurement given on the menu board is the diameter of the pizza in inches. Use 3.14 as the approximate value of π. Round the area to the nearest tenth. Round cost per square inch to the nearest thousandth.

Pizza Menu	Small 7½"	Medium 13"	Large 16"
Cheese only	$2.80	$ 6.50	$ 9.30
"The Works"	$3.70	$ 8.95	$14.30
Deep-dish combo	$4.35	$10.95	$15.65

37. Find the area of a small pizza.

$A \approx 44.2 \text{ in.}^2$

38. Find the area of a medium pizza.

$A \approx 132.7 \text{ in.}^2$

39. Find the area of a large pizza.

$A \approx 201.0 \text{ in.}^2$

40. What is the cost per square inch for each size of cheese pizza? Which size is the best buy?

small: $0.063 (rounded);
medium: $0.049 (rounded);
large: $0.046 (rounded)   **Best Buy**

41. What is the cost per square inch for each size of "The Works" pizza? Which size is the best buy?

small: $0.084 (rounded);
medium: $0.067 (rounded)   **Best Buy**;
large: $0.071 (rounded)

42. You have a coupon for 95¢ off any small pizza. What is the cost per square inch for each size of deep-dish combo pizza? Which size is the best buy?

small: $0.077 (rounded)   **Best Buy**;
medium: $0.083 (rounded);
large: $0.078 (rounded)

Summary Exercises on Perimeter, Circumference, and Area

1. Draw a sketch of each of these shapes: **(a)** square, **(b)** rectangle, **(c)** parallelogram. On each sketch, indicate 90° angles and show which sides are the same length.

(a)   **(b)**   **(c)**
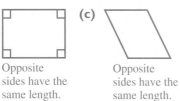

All sides have the same length.   Opposite sides have the same length.   Opposite sides have the same length.

2. (a) Draw a sketch of a circle and show the radius.

(b) Draw another circle and show the diameter.

(c) Describe the relationship between the radius and diameter of a circle.

The diameter is twice the radius, or the radius is half the diameter.

3. Describe how you can find the perimeter of any flat shape with straight sides.

Add up the lengths of all the sides to find the perimeter.

4. In your own words, describe the difference between finding the perimeter of a shape and finding the area of the shape.

Perimeter is the total distance around the outside edges of a shape. Area is the number of square units needed to cover the space inside the shape.

5. Match each shape to its corresponding area formula.

Shapes

parallelogram __f__

square __e__

trapezoid __d__

circle __b__

rectangle __c__

triangle __a__

Area Formulas

(a) $A = \dfrac{1}{2} \cdot b \cdot h$

(b) $A = \pi \cdot r^2$

(c) $A = l \cdot w$

(d) $A = \dfrac{1}{2} \cdot h \cdot (b + B)$

(e) $A = s^2$

(f) $A = b \cdot h$

6. (a) If you know the *radius* of a circle, which formula do you use to find its circumference?

$C = 2 \cdot \pi \cdot r$

(b) If you know the *diameter* of a circle, which formula do you use to find its circumference?

$C = \pi \cdot d$

(c) If you know the *diameter* of a circle, what must you do *before* using the formula for finding the area?

Divide the diameter by 2 to find the radius.

7. Find the perimeter and area of each triangle, to the nearest tenth.

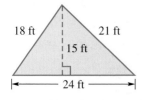

18 ft   21 ft   15 ft   24 ft

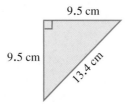

9.5 cm   9.5 cm   13.4 cm

$P = 63$ ft;
$A = 180$ ft²

$P = 32.4$ cm;
$A \approx 45.1$ cm²
(rounded)

8. Some answers from a student's test paper are listed below. The *number* part of each answer is correct, but the *units* are not. Rewrite each answer with the correct units.

(a) $A = 12$ cm

$A = 12$ cm²

(b) $P = 6\dfrac{1}{2}$ ft²

$P = 6\dfrac{1}{2}$ ft

(c) $C \approx 28.5$ m²

$C \approx 28.5$ m

(d) $A = 307$ in.

$A = 307$ in.²

Name each figure and find its perimeter and area. Round answers to the nearest tenth.

9.

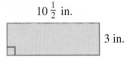

10 ½ in.

3 in.

Rectangle;
P = 27 in.;
A = 31 ½ in.² or 31.5 in.²

10.

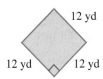

12 yd

12 yd   12 yd

Square;
P = 48 yd;
A = 144 yd²

11.

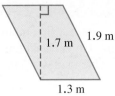

1.9 m

1.7 m

1.3 m

Parallelogram;
P = 6.4 m;
A ≈ 2.2 m² (rounded)

12.

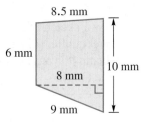

8.5 mm

6 mm

8 mm

10 mm

9 mm

Trapezoid;
P = 33.5 mm;
A = 64 mm²

*For each circle, find **(a)** the diameter or radius, **(b)** the circumference, and **(c)** the area.*
Use 3.14 as the approximate value of π. Round answers to the nearest tenth.

13.

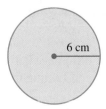

6 cm

d = 12 cm;
C ≈ 37.7 cm (rounded);
A ≈ 113.0 cm² (rounded)

14.

30 mi

r = 15 mi;
C ≈ 94.2 mi;
A ≈ 706.5 mi²

15.

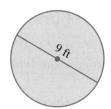

9 ft

r = 4.5 ft;
C ≈ 28.3 ft (rounded);
A ≈ 63.6 ft.² (rounded)

Find the area of each shaded region. Note that Exercise 18 contains semicircles. Use 3.14
as the approximate value of π. Round answers to the nearest tenth.

16.

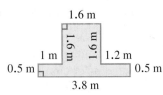

1.6 m

1.6 m   1.6 m

1 m   1.2 m

0.5 m   0.5 m

3.8 m

A ≈ 4.5 m² (rounded)

17.

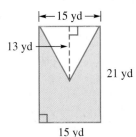

|← 15 yd →|

13 yd

21 yd

15 yd

A = 217.5 yd²

18.

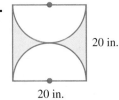

20 in.

20 in.

A ≈ 86 in.²

Solve each application problem. Use 3.14 as the approximate value of π. Round answers
to the nearest tenth.

19. The Mormons traveled west to Utah by covered wagon in 1847. They tied a rag to a wagon wheel to keep track of the distance they traveled. The radius of the wheel was 2.33 ft. How far did the rag travel each time the wheel made a complete revolution? (*Source: Trail of Hope.*) *Bonus question:* How many wheel revolutions equaled one mile?

C ≈ 14.6 ft;
Bonus: About
361.6 revolutions;
the Mormons counted
360 revolutions, which is
the answer you get using
2 ⅓ ft instead of 2.33 ft as
the radius.

20. The rectangular front door for a new log home is 0.9 m wide and 2 m high. How much will it cost for weather strip material to go around all edges of the door, if weather strip costs $0.77 per meter?

$4.47 (rounded)

2 m

|← 0.9 m →|

8.7 ▷▷▷ Volume

OBJECTIVE 1 Find the volume of a rectangular solid. A shoe box and a cereal box are examples of three-dimensional (or solid) figures. The three dimensions are length, width, and height. (A rectangle or square is a two-dimensional figure. The two dimensions are length and width.)

If you want to know how much a shoe box will hold, you find its *volume*. We measure volume by seeing how many cubes of a certain size will fill the space inside the box. Three sizes of *cubic units* are shown below. Notice that all the edges of a cube have the same length and all the sides meet at right angles.

OBJECTIVES

Find the volume of a

1 **rectangular solid;**

2 **sphere;**

3 **cylinder;**

4 **cone and pyramid.**

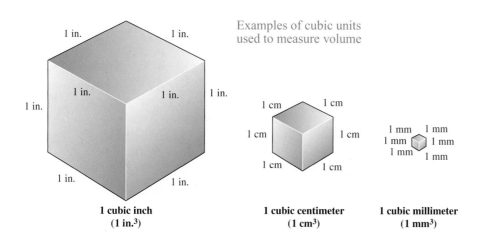

Examples of cubic units used to measure volume

1 cubic inch
(1 in.³)

1 cubic centimeter
(1 cm³)

1 cubic millimeter
(1 mm³)

Some other sizes of cubes that are used to measure volume are 1 cubic foot (1 ft³), 1 cubic yard (1 yd³), and 1 cubic meter (1 m³).

CAUTION
The raised 3 in 4^3 means that you multiply 4 • 4 • 4 to get 64. The raised 3 in cm³ or ft³ is a short way to write the word *cubic*. When you see 5 cm³, say "five *cubic* centimeters." Do *not* multiply 5 • 5 • 5. The exponent applies to cm, *not* to the number.

Volume
Volume is a measure of the space inside a solid shape. The volume of a solid is how many cubic units it takes to fill the solid.

Use the formula below to find the *volume* of *rectangular solids* (box-like shapes).

Finding the Volume of Rectangular Solids
    Volume of a rectangular solid = length • width • height
$$V = l \cdot w \cdot h$$
Remember to use **cubic units** when measuring volume.

---

1 Find the volume of each box. Round your answers to the nearest tenth if necessary.

(a)

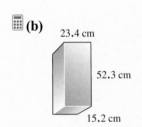

3 m

8 m

3 m

(b)

23.4 cm

52.3 cm

15.2 cm

(c) Length $6\frac{1}{4}$ ft,

width $3\frac{1}{2}$ ft, height 2 ft

ANSWERS

1. **(a)** $V = 72 \text{ m}^3$ **(b)** $V \approx 18{,}602.1 \text{ cm}^3$

(c) $V = 43\frac{3}{4} \text{ ft}^3$ or $V \approx 43.8 \text{ ft}^3$

EXAMPLE 1 **Finding the Volume of Rectangular Solids**

Find the volume of each box.

(a)

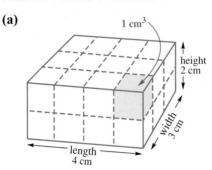

1 cm³

height 2 cm

width 3 cm

length 4 cm

Each cube that fits in the box is 1 cubic centimeter (1 cm³). To find the volume, you can count the number of cubes.

Bottom layer has 12 cubes.

Top layer has 12 cubes.

total of 24 cubes (24 cm³)

Or, you can use the formula for rectangular solids.

$$V = l \cdot w \cdot h$$

$$V = 4 \text{ cm} \cdot 3 \text{ cm} \cdot 2 \text{ cm}$$

$$V = 24 \text{ cm}^3 \quad \text{Cubic units for volume}$$

(b)

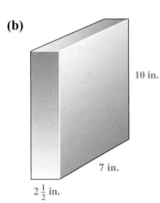

10 in.

7 in.

$2\frac{1}{2}$ in.

Use the formula $V = l \cdot w \cdot h$.

$$V = 7 \text{ in.} \cdot 2\frac{1}{2} \text{ in.} \cdot 10 \text{ in.}$$

$$V = \frac{7 \text{ in.}}{1} \cdot \frac{\overset{5}{\cancel{5}} \text{ in.}}{\underset{1}{\cancel{2}}} \cdot \frac{\overset{5}{\cancel{10}} \text{ in.}}{1} = 175 \text{ in.}^3$$

If you like, use 2.5 instead of $2\frac{1}{2}$; they are equivalent.

$$V = 7 \text{ in.} \cdot 2.5 \text{ in.} \cdot 10 \text{ in.} = 175 \text{ in.}^3$$

Write **in.**³ for volume.

◀ *Work Problem* **1** *at the Side.*

OBJECTIVE **2** **Find the volume of a sphere.** A *sphere* is shown below. Examples of spheres include baseballs, oranges, and Earth. (They aren't perfect spheres, but they're close.)

r

As with circles, the *radius* of a sphere is the distance from the center to the edge of the sphere. Use the following formula to find the *volume* of a *sphere*.

Finding the Volume of a Sphere

$$\text{Volume of a sphere} = \frac{4}{3} \cdot \pi \cdot r \cdot r \cdot r$$

$$V = \frac{4}{3} \cdot \pi \cdot r^3 \quad \text{or} \quad \frac{4 \cdot \pi \cdot r^3}{3}$$

Remember to use **cubic units** when measuring volume.

EXAMPLE 2 **Finding the Volume of Spheres**

Find the volume of each sphere with the help of a calculator. Use 3.14 as the approximate value of π. Round your answers to the nearest tenth.

(a)

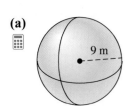

9 m

$$V = \frac{4}{3} \cdot \pi \cdot r^3$$

$$V \approx \frac{4 \cdot 3.14 \cdot 9 \text{ m} \cdot 9 \text{ m} \cdot 9 \text{ m}}{3}$$

$V \approx 3052.08 \text{ m}^3$ Now round to tenths.

$V \approx 3052.1 \text{ m}^3$ Cubic units for volume

(b)

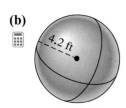

4.2 ft

$$V \approx \frac{4 \cdot 3.14 \cdot 4.2 \text{ ft} \cdot 4.2 \text{ ft} \cdot 4.2 \text{ ft}}{3}$$

$V \approx 310.18176 \text{ ft}^3$ Now round to tenths.

$V \approx 310.2 \text{ ft}^3$ Cubic units for volume

Calculator Tip   You can find the volume of the sphere in Example 2(b) above on your calculator. The first method works on both scientific and standard calculators:

4 ⓧ 3.14 ⓧ 4.2 ⓧ 4.2 ⓧ 4.2 ⓓ 3 ⓔ   Answer is 310.18176

Round the answer to 310.2 ft^3.

On a *scientific* calculator you can use the ⓨˣ key to calculate r^3 (to multiply the radius times itself three times).

4 ⓧ 3.14 ⓧ 4.2 ⓨˣ 3 ⓓ 3 ⓔ   Answer is 310.18176
          ⎵
          r^3

Recall that we are using 3.14 as the approximate value for π instead of using the ⓟ key.

You can also use the ⓨˣ key with other exponents. For example:

To find 2^5, press 2 ⓨˣ 5 ⓔ   Answer is 32

To find 6^4, press 6 ⓨˣ 4 ⓔ   Answer is 1296

——————— Work Problem **2** at the Side. ▶

Half a sphere is called a *hemisphere*. The volume of a hemisphere is *half* the volume of a sphere. Use the following formula to find the *volume* of a hemisphere.

Finding the Volume of a Hemisphere

Volume of a hemisphere $= \dfrac{1}{\underset{1}{\cancel{2}}} \cdot \dfrac{\overset{2}{\cancel{4}}}{3} \cdot \pi \cdot r^3$

$$V = \frac{2}{3} \cdot \pi \cdot r^3 \quad \text{or} \quad \frac{2 \cdot \pi \cdot r^3}{3}$$

Remember to use **cubic units** when measuring volume.

---

2 Find the volume of each sphere. Use 3.14 for π. Round your answers to the nearest tenth.

(a)

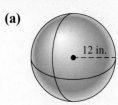

12 in.

(b)

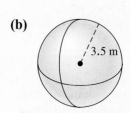

3.5 m

(c) Sphere with a radius of 2.7 cm

3 Find the volume of each hemisphere. Use 3.14 for π. Round your answers to the nearest tenth.

(a)

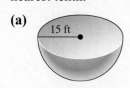

15 ft

(b)

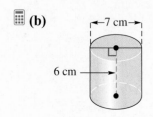

6 cm

4 Find the volume of each cylinder. Use 3.14 for π. Round your answers to the nearest tenth.

(a)

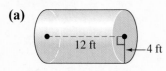

12 ft

4 ft

(b)

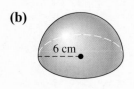

7 cm

6 cm

EXAMPLE 3 Finding the Volume of a Hemisphere

Find the volume of the hemisphere with the help of a calculator. Use 3.14 for π. Round your answer to the nearest tenth.

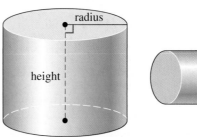

7 m

$$V = \frac{2 \cdot \pi \cdot r^3}{3}$$

$$V \approx \frac{2 \cdot 3.14 \cdot 7\text{ m} \cdot 7\text{ m} \cdot 7\text{ m}}{3}$$

Write **m³** for volume.

$$V \approx 718.0 \text{ m}^3 \qquad \text{Rounded to nearest tenth}$$

◀ *Work Problem* **3** *at the Side.*

OBJECTIVE 3 Find the volume of a cylinder. Several *cylinders* are shown below.

The height must be perpendicular to the circular top and bottom of the cylinder.

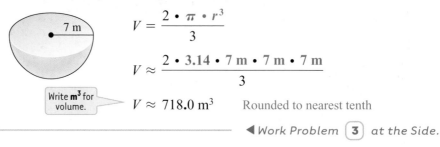

radius

height

These are called *right circular cylinders* because the top and bottom are circles, and the side makes a right angle with the top and bottom. Examples of cylinders are a soup can, a home water heater, and a piece of pipe.

Use the formula below to find the *volume* of a *cylinder*. Notice that the first part of the formula, $\pi \cdot r^2$, is the area of the circular base.

Finding the Volume of a Cylinder

Volume of a cylinder $= \pi \cdot r \cdot r \cdot h$

$$V = \pi \cdot r^2 \cdot h$$

Remember to use **cubic units** when measuring volume.

EXAMPLE 4 Finding the Volume of Cylinders

Find the volume of each cylinder. Use 3.14 as the approximate value of π. Round your answers to the nearest tenth if necessary.

(a)

20 m

9 m

The diameter is 20 m so the radius is $\frac{20\text{ m}}{2} = 10$ m. The height is 9 m. Use the formula to find the volume.

$$V = \pi \cdot \overbrace{r^2} \cdot h$$

$$V \approx 3.14 \cdot \overbrace{10\text{ m} \cdot 10\text{ m}} \cdot 9\text{ m}$$

$$V \approx 2826 \text{ m}^3 \qquad \text{Cubic units for volume}$$

(b)

6.2 cm

38.4 cm

$$V \approx 3.14 \cdot 6.2\text{ cm} \cdot 6.2\text{ cm} \cdot 38.4\text{ cm}$$

$$V \approx 4634.94144 \qquad \text{Now round to tenths.}$$

$$V \approx 4634.9 \text{ cm}^3 \qquad \text{Cubic units for volume}$$

◀ *Work Problem* **4** *at the Side.*

ANSWERS

3. (a) $V \approx 7065 \text{ ft}^3$ **(b)** $V \approx 452.2 \text{ cm}^3$
4. (a) $V \approx 602.9 \text{ ft}^3$ **(b)** $V \approx 230.8 \text{ cm}^3$

OBJECTIVE **4** **Find the volume of a cone and a pyramid.** A cone and a pyramid are shown below. Notice that the height line is perpendicular to the base in both solids.

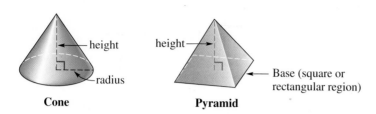

Cone

Pyramid

Use the formula below to find the *volume* of a cone.

Finding the Volume of a Cone

$$\text{Volume of a cone} = \frac{1}{3} \cdot B \cdot h$$

$$\text{or} \quad V = \frac{B \cdot h}{3}$$

where B is the area of the circular base of the cone and h is the height of the cone.

Remember to use *cubic units* when measuring volume.

EXAMPLE 5 **Finding the Volume of a Cone**

Find the volume of the cone. Use 3.14 for π. Round your answer to the nearest tenth.

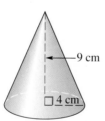

9 cm

4 cm

First find the value of B in the formula, which is the *area of the circular base*. Recall that the formula for the area of a circle is πr^2.

$$B = \pi \cdot r \cdot r$$

$$B \approx 3.14 \cdot 4 \text{ cm} \cdot 4 \text{ cm}$$

$$B \approx 50.24 \text{ cm}^2 \leftarrow \text{Do not round to tenths yet.}$$

Now find the volume. The height is 9 cm.

$$V = \frac{B \cdot h}{3}$$

$$V \approx \frac{50.24 \text{ cm}^2 \cdot 9 \text{ cm}}{3}$$

$$V \approx 150.72 \text{ cm}^3 \qquad \text{Now round to tenths.}$$

Write **cm³** in the answer. $\quad V \approx 150.7 \text{ cm}^3 \qquad$ Cubic units for volume

Work Problem **5** *at the Side.* ▶

5 Find the volume of a cone with base radius 2 ft and height 11 ft. Use 3.14 for π. Round your answer to the nearest tenth.

ANSWER

5. $V \approx 46.1 \text{ ft}^3$

6 Find the volume of a pyramid with a square base 10 m by 10 m. The height of the pyramid is 8 m. Round your answer to the nearest tenth.

Use the same formula to find the *volume* of a *pyramid* as you did to find the *volume* of a *cone*.

Finding the Volume of a Pyramid

$$\text{Volume of a pyramid} = \frac{1}{3} \cdot B \cdot h$$

$$\text{or} \quad V = \frac{B \cdot h}{3}$$

where B is the area of the square or rectangular base of the pyramid and h is the height of the pyramid.

Remember to use **cubic units** when measuring volume.

Note

In this book, we will work only with pyramids that have a base with four sides (square or rectangle). In later math courses you may work with pyramids that have a base with three sides (triangle), five sides (pentagon), six sides (hexagon), and so on.

EXAMPLE 6 **Finding the Volume of a Pyramid**

Find the volume of this pyramid with a rectangular base. Round your answer to the nearest tenth.

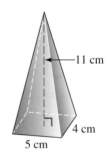

First find the value of B in the formula, which is the *area of the rectangular base*. Recall that the area of a rectangle is found by multiplying length times width.

$$B = 5 \text{ cm} \cdot 4 \text{ cm}$$

$$B = 20 \text{ cm}^2 \quad \boxed{\text{Find the } \textit{area of} \text{ the } \textit{base} \text{ first.}}$$

Now find the volume.

$$V = \frac{B \cdot h}{3}$$

$$V = \frac{20 \text{ cm}^2 \cdot 11 \text{ cm}}{3} \quad \boxed{\text{Now find the } \textit{volume} \text{ of the pyramid.}}$$

$$V \approx 73.3 \text{ cm}^3 \qquad \text{Rounded to nearest tenth}$$

◀ *Work Problem* **6** *at the Side.*

ANSWER

6. $V \approx 266.7 \text{ m}^3$

Name each solid and find its volume. Use 3.14 *as the approximate value of* π. *Round your
answers to the nearest tenth if necessary. See Examples 1–6.*

1.

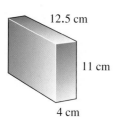

12.5 cm

11 cm

4 cm

Rectangular solid;
V = 550 cm³

2.

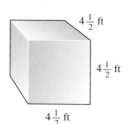

$4\frac{1}{2}$ ft

$4\frac{1}{2}$ ft

$4\frac{1}{2}$ ft

Rectangular solid or cube;
V = $91\frac{1}{8}$ ft³ or 91.1 ft³ (rounded)

3.

22 m

Sphere;
V ≈ 44,579.6 m³

4.

1.53 m

Sphere; V ≈ 15.0 m³

5.

12 in.

Hemisphere; V ≈ 3617.3 in³

6.

7.4 in.

Hemisphere; V ≈ 848.3 in.³

7.

5 ft

6 ft

Cylinder; V ≈ 471 ft³

8.

12 in.

21 in.

Cylinder; V ≈ 9495.4 in.³

9.

16 m

5 m

Cone; V ≈ 418.7 m³

10.

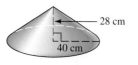

28 cm

40 cm

Cone; V ≈ 46,890.7 cm³

11.

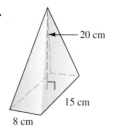

20 cm

15 cm

8 cm

Pyramid; V = 800 cm³

12.

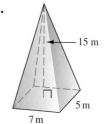

15 m

5 m

7 m

Pyramid; V = 175 m³

Solve each application problem. Use 3.14 *as the approximate value of* π. *Round your final
answers to the nearest tenth if necessary.*

13. A pencil box measures 3 in. by 8 in. by $\frac{3}{4}$ in. high.
Find the volume of the box. (*Source:* Faber Castell.)

  V = 18 in.³

14. A train is being loaded with shipping crates. Each
one is 6 m long, 3.4 m wide, and 2 m high. How
much space will each crate take? **V = 40.8 m³**

15. An oil candle globe made of hand-blown glass has a diameter of 16.8 cm. What is the volume of the globe?

V ≈ **2481.5 cm³**

16. A metal sphere used as part of a fountain has a diameter of $6\frac{1}{2}$ ft. Find its volume.

V ≈ **143.7 ft³**

17. One of the ancient stone pyramids in Egypt has a square base that measures 145 m on each side. The height is 93 m. What is the volume of the pyramid? (*Source: The Columbia Encyclopedia.*)

V = **651,775 m³**

18. A cylindrical woven basket made by a Northwest Coast tribe is 8 cm high and has a diameter of 11 cm. What is the volume of the basket?

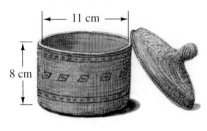

V ≈ **759.9 cm³**

19. A city sewer pipe has a diameter of 5 ft and a length of 200 ft. Find the volume of the pipe.

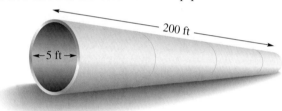

V ≈ **3925 ft³**

20. An ice cream cone has a diameter of 2 in. and a height of 4 in. Find its volume.

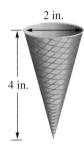

V ≈ **4.2 in.³**

21. Explain the *two* errors made by a student in finding the volume of a cylinder with a diameter of 7 cm and a height of 5 cm. Find the correct answer.

$$V \approx 3.14 \cdot 7 \cdot 7 \cdot 5$$

$$V \approx 769.3 \text{ cm}^2$$

Student used diameter of 7 cm; should use radius of 3.5 cm in formula. Units for volume are cm³, not cm². Correct answer is *V* ≈ 192.3 cm³.

22. Compare the steps in finding the volume of a cylinder and a cone. How are they similar? Suppose you know the volume of a cylinder. How can you find the volume of a cone with the same radius and height by doing just a one-step calculation?

Both involve finding the area of a circular base and multiplying by the height. To find the volume of the cone, divide the volume of the cylinder by 3.

23. Find the volume.

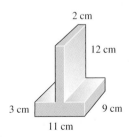

V = **513 cm³**

24. Find the volume. (*Hint:* Notice the square hole that goes through the center of the shape.)

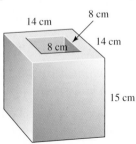

V = **1980 cm³**

8.8 ▷▷▷ Pythagorean Theorem

In **Section 8.3** you used this formula for the area of a square, $A = s^2$. The blue square below has an area of 25 cm² because 5 cm • 5 cm = 25 cm².

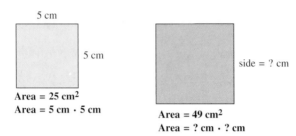

5 cm

5 cm

Area = 25 cm²
Area = 5 cm · 5 cm

side = ? cm

Area = 49 cm²
Area = ? cm · ? cm

OBJECTIVES

1 Find square roots using the square root key on a calculator.

2 Find the unknown length in a right triangle.

3 Solve application problems involving right triangles.

The red square above has an area of 49 cm². To find the length of a side, ask yourself, "What number can be multiplied by itself to give 49?" Because 7 • 7 = 49, the length of each side is 7 cm.

Also, because 7 • 7 = 49, we say that 7 is the *square root* of 49, or $\sqrt{49} = 7$. Also, $\sqrt{81} = 9$, because 9 • 9 = 81. (See **Section 1.8** for further review.)

Work Problem **1** *at the Side.* ▶

A number that has a whole number as its square root is called a *perfect square*. For example, 9 is a perfect square because $\sqrt{9} = 3$, and 3 is a whole number.

The first few perfect squares are listed below.

1 Find each square root.

(a) $\sqrt{36}$

(b) $\sqrt{25}$

(c) $\sqrt{9}$

The First Twelve Perfect Squares

$\sqrt{1} = 1$	$\sqrt{16} = 4$	$\sqrt{49} = 7$	$\sqrt{100} = 10$
$\sqrt{4} = 2$	$\sqrt{25} = 5$	$\sqrt{64} = 8$	$\sqrt{121} = 11$
$\sqrt{9} = 3$	$\sqrt{36} = 6$	$\sqrt{81} = 9$	$\sqrt{144} = 12$

OBJECTIVE **1** **Find square roots using the square root key on a calculator.** If a number is *not* a perfect square, then you can find its *approximate* square root by using a calculator with a square root key.

(c) $\sqrt{9}$

(d) $\sqrt{100}$

🖩 **Calculator Tip** To find a square root, use the $\boxed{\sqrt{}}$ key on a standard calculator or the $\boxed{\sqrt{x}}$ key on a scientific calculator. In either case, you do *not* need to use the $\boxed{=}$ key. Try these. Jot down your answers.

To find $\sqrt{16}$ press: 16 $\boxed{\sqrt{x}}$ Answer is 4

To find $\sqrt{7}$ press: 7 $\boxed{\sqrt{x}}$ Answer is 2.645751311

For $\sqrt{7}$, your calculator shows 2.645751311, which is an *approximate* answer. (Some calculators show more or fewer digits.) We will be rounding to the nearest thousandth, so $\sqrt{7} \approx 2.646$. To check, multiply 2.646 times 2.646. Do you get 7 as the result? No, you get 7.001316, which is very close to 7. The difference is due to rounding.

(e) $\sqrt{121}$

ANSWERS

1. **(a)** 6 **(b)** 5 **(c)** 3 **(d)** 10 **(e)** 11

◀ *Work Problem* (**2**) *at the Side.*

2 Use a calculator with a square root key to find each square root. Round to the nearest thousandth if necessary.

(a) $\sqrt{11}$

(b) $\sqrt{40}$

(c) $\sqrt{56}$

(d) $\sqrt{196}$

(e) $\sqrt{147}$

EXAMPLE 1 **Finding the Square Root of Numbers**

Use a calculator to find each square root. Round answers to the nearest thousandth.

> Your calculator may show more or fewer digits.

(a) $\sqrt{35}$ Calculator shows 5.916079783; round to 5.916

(b) $\sqrt{124}$ Calculator shows 11.13552873; round to 11.136

(c) $\sqrt{200}$ Calculator shows 14.14213562; round to 14.142

OBJECTIVE **2** **Find the unknown length in a right triangle.** One place you will use square roots is when working with the *Pythagorean Theorem.* This theorem applies only to *right* triangles (triangles with a 90° angle). The longest side of a right triangle is called the **hypotenuse.** It is opposite the right angle. The other two sides are called *legs.* The legs form the right angle.

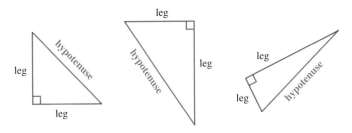

Examples of right triangles

> **Pythagorean Theorem**
>
> $$(\text{hypotenuse})^2 = (\text{leg})^2 + (\text{leg})^2$$
>
> In other words, square the length of each side. After you have squared all the sides, the sum of the squares of the two legs will equal the square of the hypotenuse. An example is shown below.
>
>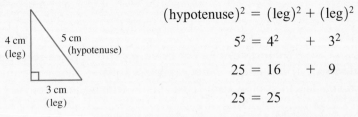
>
> $$(\text{hypotenuse})^2 = (\text{leg})^2 + (\text{leg})^2$$
> $$5^2 = 4^2 + 3^2$$
> $$25 = 16 + 9$$
> $$25 = 25$$

The theorem is named after Pythagoras, a Greek mathematician who lived about 2500 years ago. He and his followers may have used floor tiles to prove the theorem, as shown below.

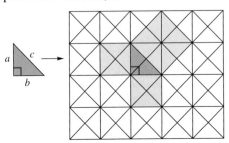

The green right triangle in the center of the floor tiles has sides a, b, and c. The pink square drawn on side a contains four triangular tiles. The pink square on side b contains four tiles. The blue square on side c contains eight tiles. The number of tiles in the square on side c equals the sum of the number of tiles in the squares on sides a and b, that is, 8 tiles = 4 tiles + 4 tiles. As a result, you often see the Pythagorean Theorem written as $c^2 = a^2 + b^2$.

If you know the lengths of any two sides in a right triangle, you can use the Pythagorean Theorem to find the length of the third side.

> **Formulas Based on the Pythagorean Theorem**
> To find the hypotenuse, use this formula:
> $$\text{hypotenuse} = \sqrt{(\text{leg})^2 + (\text{leg})^2}$$
> To find a leg, use this formula:
> $$\text{leg} = \sqrt{(\text{hypotenuse})^2 - (\text{leg})^2}$$

> **CAUTION**
> *Remember:* A small square drawn in one angle of a triangle indicates a right angle. You can use the Pythagorean Theorem *only* on triangles that have a right angle.

EXAMPLE 2   **Finding the Unknown Length in Right Triangles**

Find the unknown length in each right triangle. Round answers to the nearest tenth if necessary.

(a)

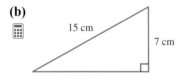

The unknown length is the side opposite the right angle, which is the hypotenuse. Use the formula for finding the hypotenuse.

$$\text{hypotenuse} = \sqrt{(\text{leg})^2 + (\text{leg})^2} \quad \text{Find the hypotenuse.}$$
$$\text{hypotenuse} = \sqrt{(3)^2 + (4)^2} \quad \text{Legs are 3 and 4}$$
$$= \sqrt{9 + 16} \quad 3 \cdot 3 \text{ is } 9 \quad \text{and} \quad 4 \cdot 4 \text{ is } 16$$
$$= \sqrt{25}$$
$$= 5 \qquad \boxed{\text{Write ft in the answer (not ft}^2\text{).}}$$

The hypotenuse is 5 ft long.

(b)

We *do* know the length of the hypotenuse (15 cm), so it is the length of one of the legs that is unknown. Use the formula for finding a leg.

$$\text{leg} = \sqrt{(\text{hypotenuse})^2 - (\text{leg})^2} \quad \text{Find a leg.}$$
$$\text{leg} = \sqrt{(15)^2 - (7)^2} \quad \text{Hypotenuse is 15, one leg is 7}$$
$$= \sqrt{225 - 49} \quad 15 \cdot 15 \text{ is } 225 \quad \text{and} \quad 7 \cdot 7 \text{ is } 49$$
$$= \sqrt{176} \quad \text{Use calculator to find } \sqrt{176}$$
$$\approx 13.3 \quad \text{Round } 13.26649916 \text{ to } 13.3$$

The length of the leg is approximately 13.3 cm.

Work Problem **3** *at the Side.* ▶

---

3 Find the unknown length in each right triangle. Round your answers to the nearest tenth if necessary.

(a)

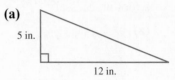

(b)

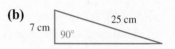

(c)

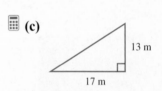

(d)

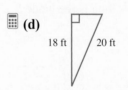

(e)

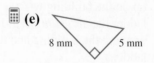

ANSWERS

3. **(a)** $\sqrt{169} = 13$ in.   **(b)** $\sqrt{576} = 24$ cm
   (c) $\sqrt{458} \approx 21.4$ m   **(d)** $\sqrt{76} \approx 8.7$ ft
   (e) $\sqrt{89} \approx 9.4$ mm

4 These problems show ladders leaning against buildings. Find the unknown lengths. Round to the nearest tenth of a foot if necessary.

(a)

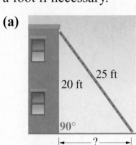

How far away from the building is the bottom of the ladder?

(b)

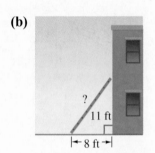

How long is the ladder?

(c) A 17 ft ladder is leaning against a building. The bottom of the ladder is 10 ft from the building. How high up on the building will the ladder reach? (*Hint:* Start by drawing the building and the ladder.)

ANSWERS
4. (a) leg = $\sqrt{225}$ = 15 ft
   (b) hypotenuse = $\sqrt{185}$ ≈ 13.6 ft
   (c) leg = $\sqrt{189}$ ≈ 13.7 ft

OBJECTIVE 3 **Solve application problems involving right triangles.** The next example shows an application of the Pythagorean Theorem.

EXAMPLE 3 **Using the Pythagorean Theorem**

A television antenna is on the roof of a house, as shown below. Find the length of the support wire. Round your answer to the nearest tenth of a meter if necessary.

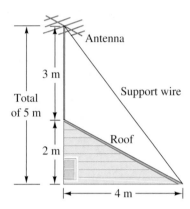

A right triangle is formed. The total length of the leg on the left is 3 m + 2 m = 5 m.

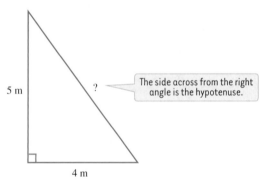

The side across from the right angle is the hypotenuse.

Notice that the support wire is opposite the right angle, so it is the hypotenuse of the right triangle.

$$\text{hypotenuse} = \sqrt{(\text{leg})^2 + (\text{leg})^2} \quad \text{Find the hypotenuse.}$$

$$\text{hypotenuse} = \sqrt{(5)^2 + (4)^2} \quad \text{Legs are 5 and 4}$$

$$= \sqrt{25 + 16} \quad 5^2 \text{ is 25 and } 4^2 \text{ is 16}$$

$$= \sqrt{41} \quad \text{Use a calculator to find } \sqrt{41}$$

$$\approx 6.4 \quad \text{Round 6.403124237 to 6.4}$$

The length of the support wire is approximately 6.4 m. Write **m** in the answer (**not** m²).

CAUTION
You use the Pythagorean Theorem to find the *length* of one side, *not* the area of the triangle. Your answer will be in linear units, such as ft, yd, cm, m, and so on (*not* ft², yd², cm², m²).

◄ *Work Problem* **4** *at the Side.*

8.8 ▶▶▶ **Exercises**

FOR EXTRA HELP

MyMathLab

Math XL
PRACTICE

WATCH

DOWNLOAD

READ

REVIEW

Find each square root. Starting with Exercise 5, use the square root key on a calculator. Round your answers to the nearest thousandth if necessary. See Example 1.

1. $\sqrt{16}$ 4

2. $\sqrt{4}$ 2

3. $\sqrt{64}$ 8

4. $\sqrt{81}$ 9

5. $\sqrt{11} \approx$ 3.317

6. $\sqrt{23} \approx$ 4.796

7. $\sqrt{5} \approx$ 2.236

8. $\sqrt{2} \approx$ 1.414

9. $\sqrt{73} \approx$ 8.544

10. $\sqrt{80} \approx$ 8.944

11. $\sqrt{101} \approx$ 10.050

12. $\sqrt{125} \approx$ 11.180

13. $\sqrt{190} \approx$ 13.784

14. $\sqrt{160} \approx$ 12.649

15. $\sqrt{1000} \approx$ 31.623

16. $\sqrt{2000} \approx$ 44.721

17. You know that $\sqrt{25} = 5$ and $\sqrt{36} = 6$. Using just that information (no calculator), describe how you could *estimate* $\sqrt{30}$. How would you estimate $\sqrt{26}$ or $\sqrt{35}$? Now check your estimates using a calculator.

30 is about halfway between 25 and 36, so $\sqrt{30}$ should be about halfway between 5 and 6, or about 5.5. Using a calculator, $\sqrt{30} \approx 5.477$. Similarly, $\sqrt{26}$ should be a little more than $\sqrt{25}$; by calculator $\sqrt{26} \approx 5.099$. And $\sqrt{35}$ should be a little less than $\sqrt{36}$; by calculator $\sqrt{35} \approx 5.916$.

18. Explain the relationship between *squaring* a number and finding the *square root* of a number. Include two examples to illustrate your explanation.

Squaring a number is multiplying the number times itself. Finding the square root is the opposite operation and "undoes" squaring. Examples will vary; one possibility is $7^2 = 7 \cdot 7 = 49$, so $\sqrt{49} = 7$.

Find the unknown length in each right triangle. Use a calculator to find square roots. Round your answers to the nearest tenth if necessary. See Example 2.

19.

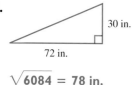

15 ft   90°   36 ft

$\sqrt{1521} = 39$ ft

20.

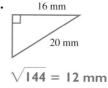

9 cm   12 cm

$\sqrt{225} = 15$ cm

21.

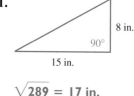

8 in.   90°   15 in.

$\sqrt{289} = 17$ in.

22.

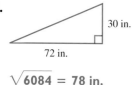

30 in.   72 in.

$\sqrt{6084} = 78$ in.

23.

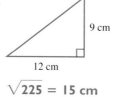

16 mm   20 mm

$\sqrt{144} = 12$ mm

24.

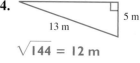

13 m   5 m

$\sqrt{144} = 12$ m

25.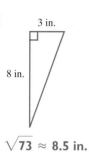

3 in.

8 in.

$\sqrt{73} \approx 8.5$ in.

26.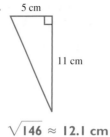

5 cm

11 cm

$\sqrt{146} \approx 12.1$ cm

27.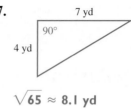

7 yd

90°

4 yd

$\sqrt{65} \approx 8.1$ yd

28.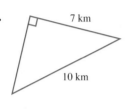

7 km

10 km

$\sqrt{51} \approx 7.1$ km

29.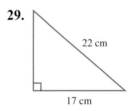

22 cm

17 cm

$\sqrt{195} \approx 14.0$ cm

30.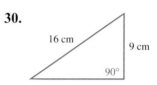

16 cm

9 cm

90°

$\sqrt{175} \approx 13.2$ cm

31.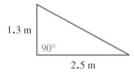

1.3 m

90°

2.5 m

$\sqrt{7.94} \approx 2.8$ m

32.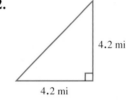

4.2 mi

4.2 mi

$\sqrt{35.28} \approx 5.9$ mi

33.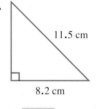

11.5 cm

8.2 cm

$\sqrt{65.01} \approx 8.1$ cm

34.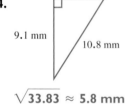

9.1 mm

10.8 mm

$\sqrt{33.83} \approx 5.8$ mm

35.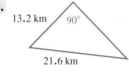

13.2 km   90°

21.6 km

$\sqrt{292.32} \approx 17.1$ km

36.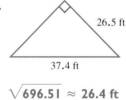

26.5 ft

37.4 ft

$\sqrt{696.51} \approx 26.4$ ft

▦ *Solve each application problem. Round your answers to the nearest tenth if necessary. See Example 3.*

37. Find the length of this loading ramp.

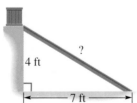

4 ft

?

7 ft

hypotenuse = $\sqrt{65} \approx 8.1$ ft

38. Find the unknown length in this roof plan.

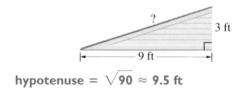

?

3 ft

9 ft

hypotenuse = $\sqrt{90} \approx 9.5$ ft

39. How high is the airplane above the ground?

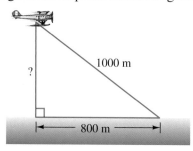

leg = $\sqrt{360{,}000}$ = 600 m

40. Find the height of this farm silo.

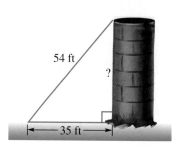

leg = $\sqrt{1691}$ ≈ 41.1 ft

41. How long is the diagonal brace on this rectangular gate?

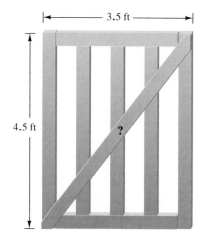

hypotenuse = $\sqrt{32.5}$ ≈ 5.7 ft

42. Find the height of this rectangular television screen. (*Source:* Sears.)

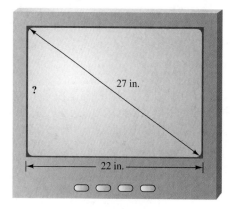

leg = $\sqrt{245}$ ≈ 15.7 in.

43. To reach his ladylove, a knight placed a 12 ft ladder against the castle wall. If the base of the ladder is 3 ft from the building, how high on the castle will the top of the ladder reach? Draw a sketch of the castle and ladder and solve the problem.

leg = $\sqrt{135}$ ≈ 11.6 ft

44. William drove his car 15 miles north, then made a right turn and drove 7 miles east. How far is he, in a straight line, from his starting point? Draw a sketch to illustrate the problem and solve it.

hypotenuse = $\sqrt{274}$ ≈ 16.6 mi

45. Explain the *two* errors made by a student in solving this problem. Also find the correct answer. Round to the nearest tenth.

$$? = \sqrt{(9)^2 + (7)^2}$$
$$= \sqrt{18 + 14}$$
$$= \sqrt{32} \approx 5.657 \text{ in.}$$

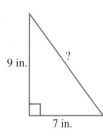

9 in.

?

7 in.

The student did not square the numbers correctly: 9^2 is 81 and 7^2 is 49. Also, the final answer is rounded to thousandths instead of tenths. Correct answer is $\sqrt{130} \approx 11.4$ in.

46. Explain the *two* errors made by a student in solving this problem. Also find the correct answer. Round to the nearest tenth.

$$? = \sqrt{(13)^2 + (20)^2}$$
$$= \sqrt{169 + 400}$$
$$= \sqrt{569} \approx 23.9 \text{ m}^2$$

?

13 m

20 m

The student used the formula for finding the hypotenuse, but the unknown side is a leg; so $? = \sqrt{(20)^2 - (13)^2}$. Also, the final answer should be m, not m^2. Correct answer is $\sqrt{231} \approx 15.2$ m.

Relating Concepts (Exercises 47–50) For Individual or Group Work

Use your knowledge of the Pythagorean Theorem to **work Exercises 47–50 in order.** *Round answers to the nearest tenth.*

47. A major league baseball diamond is a square shape measuring 90 ft on each side. If the catcher throws a ball from home plate to second base, how far is he throwing the ball? (*Source:* American League of Professional Baseball Clubs.)

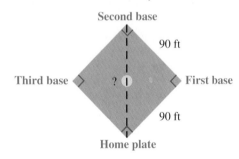

Second base

90 ft

Third base   ?   First base

90 ft

Home plate

$\sqrt{16200} \approx 127.3$ ft

48. A softball diamond is only 60 ft on each side. (*Source:* Amateur Softball Association.)

(a) Draw a sketch of the softball diamond and label the bases and the lengths of the sides.

Second base

60 ft

Third base   First base

60 ft

Home plate

(b) How far is it to throw a ball from home plate to second base?

$\sqrt{7200} \approx 84.9$ ft

49. Look back at your answer to Exercise 47. Explain how you can tell the distance from third base to first base without doing any further calculations.

The distance from third to first is the same as the distance from home to second because the baseball diamond is a square.

50. Show how you could set up a proportion to answer Exercise 48 instead of using the Pythagorean Theorem. (You'll need your answer from Exercise 47.)

One possibility is

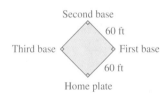

major league $\left\{ \dfrac{90 \text{ ft}}{127.3 \text{ ft}} = \dfrac{60 \text{ ft}}{x} \right\}$ softball

$x \approx 84.9$ ft

8.9 ▶▶▶ Similar Triangles

Two triangles with the same *shape* (but not necessarily the same size) are called **similar triangles.** Three pairs of similar triangles are shown below.

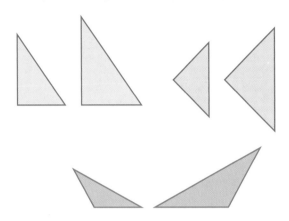

OBJECTIVE 1 Identify corresponding parts in similar triangles.
The two triangles shown below are different sizes but have the same shape, so they are *similar triangles.* Angles A and P measure the same number of degrees and are called *corresponding angles.* Angles B and Q are corresponding angles, as are angles C and R. The triangles have the same shape because the corresponding angles have the same measure.

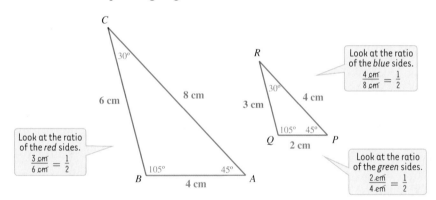

$\overline{PR}$ and $\overline{AC}$ are called *corresponding sides* because they are *opposite* corresponding angles. Also, $\overline{QR}$ and $\overline{BC}$ are corresponding sides, as are $\overline{PQ}$ and $\overline{AB}$. Although corresponding angles measure the same number of degrees, corresponding sides do *not* need to be the same length. In the triangles here, each side in the smaller triangle is *half* the length of the corresponding side in the larger triangle.

Work Problem ① *at the Side.* ▶

OBJECTIVE 2 Find the unknown lengths of sides in similar triangles.   Similar triangles are useful because of the following definition.

Similar Triangles

If two triangles are **similar,** then

1. Corresponding angles have the same measure, and
2. The ratios of the lengths of corresponding sides are equal.

---

1 Identify corresponding angles and sides in these similar triangles.

(a)

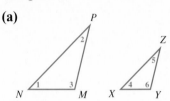

Angles:

1 and _____

2 and _____

3 and _____

Sides:

$\overline{PN}$ and _____

$\overline{PM}$ and _____

$\overline{NM}$ and _____

(b)

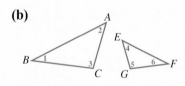

Angles:

1 and _____

2 and _____

3 and _____

Sides:

$\overline{AB}$ and _____

$\overline{BC}$ and _____

$\overline{AC}$ and _____

ANSWERS

1. (a) 4; 5; 6; $\overline{ZX}$; $\overline{ZY}$; $\overline{XY}$
   (b) 6; 4; 5; $\overline{EF}$; $\overline{FG}$; $\overline{EG}$

2 Find the length of $\overline{EF}$ in Example 1 by setting up and solving a proportion. Let x represent the unknown length.

EXAMPLE 1 **Finding the Unknown Lengths of Sides in Similar Triangles**

Find the length of y in the smaller triangle. Assume the triangles are similar.

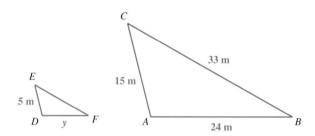

The length you want to find in the smaller triangle is $\overline{DF}$, and it corresponds to $\overline{AB}$ in the larger triangle. Then, notice that $\overline{ED}$ in the smaller triangle corresponds to $\overline{CA}$ in the larger triangle, and you know both of their lengths. Because the *ratios* of the lengths of corresponding sides are equal, you can set up a proportion. (Recall that a proportion states that two ratios are equal.)

$$\text{Corresponding sides} \left\{ \begin{array}{l} DF \to \\ AB \to \end{array} \frac{y}{24} = \frac{5}{15} \begin{array}{l} \leftarrow ED \\ \leftarrow CA \end{array} \right\} \text{Corresponding sides}$$

$$\frac{y}{24} = \frac{1}{3} \qquad \text{Write } \tfrac{5}{15} \text{ in lowest terms as } \tfrac{1}{3}.$$

Find the cross products.

$$\frac{y}{24} \overset{\times}{=} \frac{1}{3} \qquad \begin{array}{l} 24 \cdot 1 = 24 \\[6pt] y \cdot 3 \end{array}$$

$$y \cdot 3 = 24 \qquad \text{Show that the cross products are equivalent.}$$

$$\frac{y \cdot \overset{1}{\cancel{3}}}{\cancel{3}_{1}} = \frac{24}{3} \qquad \text{Divide both sides by 3.}$$

$$y = 8$$

> Write **m** in the answer.

$\overline{DF}$ has a length of 8 m.

◀ *Work Problem* **2** *at the Side.*

EXAMPLE 2 **Finding an Unknown Length and the Perimeter**

Find the perimeter of the smaller triangle. Assume the triangles are similar.

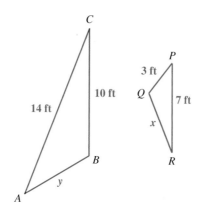

3 **(a)** Find the perimeter of triangle ABC in Example 2 at the left.

First find x, the length of $\overline{QR}$, then add the lengths of all three sides to find the perimeter.

The smaller triangle is turned "upside down" compared to the larger triangle, so be careful when identifying corresponding sides. $\overline{PR}$ is the longest side in the smaller triangle, and $\overline{AC}$ is the longest side in the larger triangle. So $\overline{PR}$ and $\overline{AC}$ are corresponding sides and you know both of their lengths. $\overline{QR}$, the length you want to find in the smaller triangle, corresponds to $\overline{BC}$ in the larger triangle. The ratios of the lengths of corresponding sides are equal, so you can set up a proportion.

$$\begin{array}{l} QR \rightarrow \\ BC \rightarrow \end{array} \frac{x}{10} = \frac{7}{14} \begin{array}{l} \leftarrow PR \\ \leftarrow AC \end{array}$$

$$\frac{x}{10} = \frac{1}{2} \qquad \text{Write } \tfrac{7}{14} \text{ in lowest terms as } \tfrac{1}{2}.$$

Find the cross products.

$$\frac{x}{10} \times \frac{1}{2} \qquad \begin{array}{l} 10 \cdot 1 = 10 \\[4pt] x \cdot 2 \end{array}$$

$$x \cdot 2 = 10 \qquad \begin{array}{l}\text{Show that the cross products} \\ \text{are equivalent.}\end{array}$$

$$\frac{x \cdot \overset{1}{\cancel{2}}}{\underset{1}{\cancel{2}}} = \frac{10}{2} \qquad \text{Divide both sides by 2.}$$

$$x = 5$$

(b) Find the perimeter of each triangle. Assume the triangles are similar.

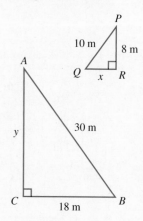

> Write **ft** for the length.

$\overline{QR}$ has a length of 5 ft.

Now add the lengths of all three sides to find the perimeter of the smaller triangle.

$$\text{Perimeter} = 5 \text{ ft} + 3 \text{ ft} + 7 \text{ ft} = 15 \text{ ft}$$

Work Problem **3** *at the Side.* ▶

4 Find the height of each flagpole.

(a)

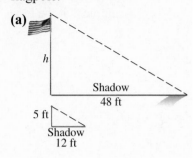

OBJECTIVE **3** **Solve application problems involving similar triangles.** The next example shows an application of similar triangles.

EXAMPLE 3 **Using Similar Triangles in an Application**

A flagpole casts a shadow 99 m long at the same time that a pole 10 m tall casts a shadow 18 m long. Find the height of the flagpole.

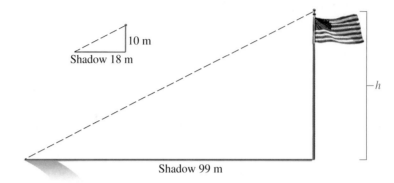

The triangles shown are similar, so write a proportion to find h.

Height in larger triangle $\rightarrow$ $\dfrac{h}{10} = \dfrac{99}{18}$ $\leftarrow$ Shadow in larger triangle
Height in smaller triangle $\rightarrow$ $\phantom{\dfrac{h}{10}}$ $\leftarrow$ Shadow in smaller triangle

Find the cross products and show that they are equivalent.

$$h \cdot 18 = 10 \cdot 99$$

$$h \cdot 18 = 990$$

$$\frac{h \cdot \overset{1}{\cancel{18}}}{\underset{1}{\cancel{18}}} = \frac{990}{18} \qquad \text{Divide both sides by 18.}$$

$$h = 55$$

The flagpole is 55 m high.

(b)

> **Note**
>
> There are several other correct ways to set up the proportion in Example 3 above. One way is to simply flip the ratios on *both* sides of the equal sign.
>
> $$\frac{10}{h} = \frac{18}{99}$$
>
> But there is another option, shown below.
>
> Height in larger triangle $\rightarrow$ $\dfrac{h}{99} = \dfrac{10}{18}$ $\leftarrow$ Height in smaller triangle
> Shadow in larger triangle $\rightarrow$ $\phantom{\dfrac{h}{99}}$ $\leftarrow$ Shadow in smaller triangle
>
> Notice that both ratios compare *height* to *shadow* in the same order. The ratio on the left describes the larger triangle, and the ratio on the right describes the smaller triangle.

◀ *Work Problem* **4** *at the Side.*

ANSWERS

4. **(a)** $h = 20$ ft **(b)** $h = 18$ m

8.9 ▶▶▶ Exercises

FOR
EXTRA
HELP

MyMathLab

Math XL
PRACTICE

WATCH

DOWNLOAD

READ

REVIEW

Which pairs of triangles appear to be similar? Write similar *or* not similar *for each pair.*

1.

similar

2.

similar

3.

not similar

4.

not similar

5.

similar

6.

not similar

Name the corresponding angles and the corresponding sides in each pair of similar triangles. See Margin Problem 1.

🌐 7.

∠1 and ∠4;
∠2 and ∠5;
∠3 and ∠6;
$\overline{AB}$ and $\overline{PQ}$;
$\overline{BC}$ and $\overline{QR}$;
$\overline{AC}$ and $\overline{PR}$

8.

∠1 and ∠5;
∠2 and ∠4;
∠3 and ∠6;
$\overline{SR}$ and $\overline{YX}$;
$\overline{ST}$ and $\overline{YZ}$;
$\overline{RT}$ and $\overline{XZ}$

9.

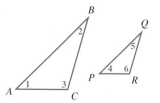

∠1 and ∠6;
∠2 and ∠5;
∠3 and ∠4;
$\overline{MP}$ and $\overline{QS}$;
$\overline{MN}$ and $\overline{QR}$;
$\overline{NP}$ and $\overline{RS}$

10.

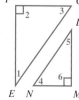

∠1 and ∠5;
∠2 and ∠6;
∠3 and ∠4;
$\overline{FG}$ and $\overline{MN}$;
$\overline{FE}$ and $\overline{ML}$;
$\overline{EG}$ and $\overline{LN}$

Write the ratio for each pair of corresponding sides in the similar triangles shown below.
Write the ratios as fractions in lowest terms. See Examples 1 and 2.

11. $\dfrac{AB}{PQ}$; $\dfrac{AC}{PR}$; $\dfrac{BC}{QR}$

$\dfrac{3}{2}$; $\dfrac{3}{2}$; $\dfrac{3}{2}$

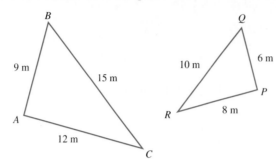

12. $\dfrac{AB}{PQ}$; $\dfrac{AC}{PR}$; $\dfrac{BC}{QR}$

$\dfrac{2}{3}$; $\dfrac{2}{3}$; $\dfrac{2}{3}$

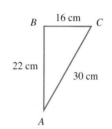
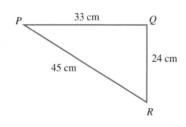

Find the unknown lengths in each pair of similar triangles. See Example 1.

13.

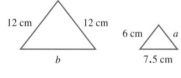

$a = 6$ cm;
$b = 15$ cm

14.

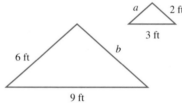

$a = 2$ ft;
$b = 6$ ft

15.

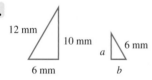

$a = 5$ mm;
$b = 3$ mm

16.

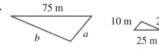

$a = 30$ m;
$b = 60$ m

Find the perimeter of each triangle. Assume the triangles are similar. See Example 2.

17.

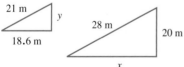

$x = 24.8$ m; $P = 72.8$ m;
$y = 15$ m; $P = 54.6$ m

18.

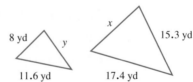

$x = 12$ yd; $P = 44.7$ yd;
$y = 10.2$ yd; $P = 29.8$ yd

19. Triangles *CDE* and *FGH* are similar. Find the perimeter and area of triangle *FGH*. *Note:* The heights of similar triangles have the same ratio as corresponding sides. Round to the nearest tenth when necessary.

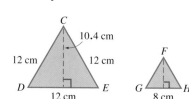

P = 8 cm + 8 cm + 8 cm = 24 cm;
A ≈ 0.5 • 8 cm • 6.9 cm ≈ 27.6 cm²

20. Triangles *JKL* and *MNO* are similar. Find the perimeter and area of triangle *MNO*.

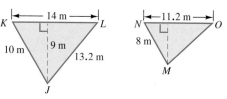

P = 11.2 m + 8 m + 10.56 m = 29.76 m;
A = 0.5 • 11.2 m • 7.2 m = 40.32 m²

Solve each application problem. See Example 3.

🌐 21. The height of the house shown here can be found by comparing its shadow to the shadow cast by a 3 ft stick. Find the height of the house by writing a proportion and solving it. *h* = 24 ft

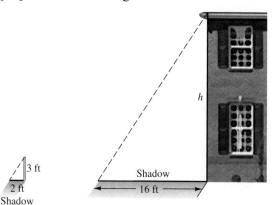

22. A fire lookout tower provides an excellent view of the surrounding countryside. The height of the tower can be found by lining up the top of the tower with the top of a 2-meter stick. Use similar triangles to find the height of the tower. *h* = 32 m

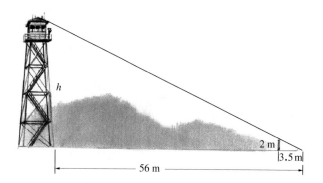

23. Refer to the building in Exercise 21. Later in the day, the same building had a shadow 6 ft long. How long would the stick's shadow be at that time?

$\frac{3}{4}$ ft or 0.75 ft

24. Refer to the lookout tower in Exercise 22.

 (a) How far away from the tower was the 2 m stick?

 56 m − 3.5 m = 52.5 m from the tower

 (b) To use a 1-meter stick and have it line up with the same endpoint, how far from the tower would it have to be?

 56 m − 1.75 m = 54.25 m from the tower

✏ 25. Look up the word *similar* in a dictionary. What is the nonmathematical definition of this word? Describe two examples of similar objects at home, school, or work.

One dictionary definition is "resembling, but not identical." Examples of similar objects are sets of different-size pots or measuring cups; small- and large-size cans of beans; child's tennis shoe and adult's tennis shoe.

✏ 26. *Congruent* objects have the *same shape* and the *same size.* Sketch a pair of congruent triangles. Describe two examples of congruent objects at home, school, or work.

Congruent triangles

Examples of congruent objects include two matching chairs, two contact lenses, and two pieces of paper from the same notebook.

Find the unknown length in Exercises 27–30. Round your answers to the nearest tenth. Note: When a line is drawn parallel to one side of a triangle, the smaller triangle that is formed will be similar to the original triangle. In Exercises 27–28, the red segments are parallel.

27.

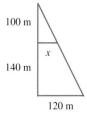

100 m

x

140 m

120 m

Hint: Redraw the two triangles and label the sides.

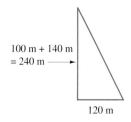

100 m

x

100 m + 140 m = 240 m →

120 m

$x = 50$ m

28.

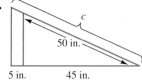

c

50 in.

5 in.   45 in.

Hint: Redraw the two triangles and label the sides.

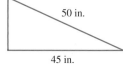

c

50 in.

5 in. + 45 in. = 50 in.

45 in.

$c \approx 55.6$ in.

29. Use similar triangles and a proportion to find the length of the lake shown here. (*Hint:* The side 100 m long in the smaller triangle corresponds to side of 100 m + 120 m = 220 m in the larger triangle.)

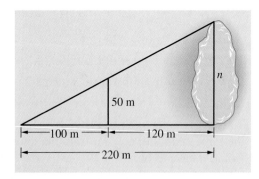

n

50 m

|← 100 m →|← 120 m →|

|← 220 m →|

$n = 110$ m

30. To find the height of the tree, find y and then add $5\frac{1}{2}$ ft for the distance from the ground to the person's eye level.

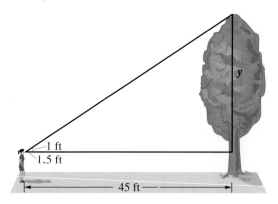

y

1 ft

1.5 ft

|← 45 ft →|

$y = 30$ ft;

tree height $= 30$ ft $+ 5\frac{1}{2}$ ft $= 35\frac{1}{2}$ ft or 35.5 ft

Chapter 8 ▷▷▷ Summary

▶ Key Terms

8.1

point	A point is a location in space. *Example:* Point *P* at the right.	• *P*
line	A line is a straight row of points that goes on forever in both directions. *Example:* Line *AB*, written $\overleftrightarrow{AB}$, at the right.	
line segment	A line segment is a piece of a line with two endpoints. *Example:* Line segment *PQ,* written $\overline{PQ}$, at the right.	
ray	A ray is a part of a line that has one endpoint and extends forever in one direction. *Example:* Ray *RS,* written $\overrightarrow{RS}$, at the right.	
angle	An angle is made up of two rays that have a common endpoint called the vertex. *Example:* Angle 1 at the right.	
degrees	Degrees are used to measure angles; a complete circle is 360 degrees, written 360°.	
right angle	A right angle is an angle that measures *exactly* 90°. *Example:* Angle *AOB* at the right.	
acute angle	An acute angle is an angle that measures less than 90°. *Example:* Angle *E* at the right.	
obtuse angle	An obtuse angle is an angle that measures more than 90° but less than 180°. *Example:* Angle *F* at the right.	
straight angle	A straight angle is an angle that measures *exactly* 180°; its sides form a straight line. *Example:* Angle *G* at the right.	
intersecting lines	Intersecting lines cross. *Example:* $\overleftrightarrow{RQ}$ intersects $\overleftrightarrow{AB}$ at point *P* at the right.	
perpendicular lines	Perpendicular lines are two lines that intersect to form a right angle. *Example:* $\overleftrightarrow{PQ}$ is perpendicular to $\overleftrightarrow{RS}$ at the right.	
parallel lines	Parallel lines are two lines in the same plane that never intersect (never cross). *Example:* $\overleftrightarrow{AB}$ is parallel to $\overleftrightarrow{ST}$ at the right.	

8.2

complementary angles	Complementary angles are two angles whose measures add up to 90°.
supplementary angles	Supplementary angles are two angles whose measures add up to 180°.
congruent angles	Congruent angles are angles that measure the same number of degrees.
vertical angles	Vertical angles are two nonadjacent congruent angles formed by two intersecting lines. *Example:* ∠*COA* and ∠*EOF* are vertical angles at the right.

8.3– 8.5

perimeter	Perimeter is the distance around the outside edges of a flat shape. It is measured in linear units such as ft, yd, cm, m, km, and so on.

8.3– 8.6

area	Area is the surface inside a two-dimensional (flat) shape. It is measured by determining the number of squares of a certain size needed to cover the surface inside the shape. Some of the commonly used units for measuring area are square inches (in.^{2}), square feet (ft^2), square yards (yd^2), square centimeters (cm^2), and square meters (m^2).

8.3	**rectangle**	A rectangle is a four-sided figure with all sides meeting at 90° angles. The opposite sides are the same length. *Example:* The rectangle measuring 12 cm by 7 cm at the right.	

square	A square is a rectangle with all four sides the same length. *Example:* The square with a side measurement of 20 inches at the right.	

8.4	**parallelogram**	A parallelogram is a four-sided figure with both pairs of opposite sides parallel and equal in length. *Example:* See the parallelogram at the right with sides measuring 8 m and 12 m.	

trapezoid	A trapezoid is a four-sided figure with exactly one pair of parallel sides. *Example:* Trapezoid *PQRS* at the right; $\overline{PQ}$ is parallel to $\overline{SR}$.	

8.5	**triangle**	A triangle is a figure with exactly three sides. *Example:* Triangle *ABC* at the right.	

8.6	**circle**	A circle is a figure with all points the same distance from a fixed center point. *Example:* See figure at the right.	

radius	Radius is the distance from the center of a circle to any point on the circle. *Example:* See the red radius in the circle at the right.	

diameter	Diameter is the distance across a circle, passing through the center. *Example:* See the blue diameter in the circle at the right.

circumference	Circumference is the distance around a circle.

π (pi)	π is the ratio of the circumference to the diameter of any circle. It is approximately equal to 3.14.

8.7	**volume**	Volume is a measure of the space inside a three-dimensional (solid) shape. Volume is measured in cubic units such as in.^{3}, ft^3, yd^3, mm^3, cm^3, and so on.

8.8	**hypotenuse**	The hypotenuse is the side of a right triangle opposite the 90° angle; it is the longest side. *Example:* See the red side in the triangle at the right.	

8.9	**similar triangles**	Similar triangles are triangles with the same shape but not necessarily the same size; corresponding angles measure the same number of degrees, and the *ratios* of the lengths of corresponding sides are equal.

▶ New Symbols

$\overleftrightarrow{AB}$	line *AB*	**right angle:** (90° angle)		**Square units:** (for measuring area)	in.^{2} mm^2	ft^2 cm^2	yd^2 m^2	mi^2 km^2
$\overline{EF}$	line segment *EF*							
$\overrightarrow{RS}$	ray *RS*	$\perp$	is perpendicular to	**Cubic units:** (for measuring volume)	in.^{3} mm^3	ft^3 cm^3	yd^3 m^3	
$\angle MRN$	angle *MRN*	$\cong$	is congruent to					
1°	one degree	π	Greek letter pi; ratio of circumference to diameter of any circle					

▶ New Formulas

Perimeter of a rectangle: $P = 2 \cdot l + 2 \cdot w$

Area of a rectangle: $A = l \cdot w$

Perimeter of a square: $P = 4 \cdot s$

Area of a square: $A = s^2$

Area of a parallelogram: $A = b \cdot h$

Area of a trapezoid: $A = \dfrac{1}{2} \cdot h \cdot (b + B)$

or $A = 0.5 \cdot h \cdot (b + B)$

Area of a triangle: $A = \dfrac{1}{2} \cdot b \cdot h$

or $A = 0.5 \cdot b \cdot h$

Diameter of a circle: $d = 2 \cdot r$

Radius of a circle: $r = \dfrac{d}{2}$

Circumference of a circle: $C = \pi \cdot d$

or $C = 2 \cdot \pi \cdot r$

Area of a circle: $A = \pi \cdot r^2$

Area of semicircle: $A = \dfrac{\pi \cdot r^2}{2}$

Volume of rectangular solid: $V = l \cdot w \cdot h$

Volume of a sphere: $V = \dfrac{4}{3} \cdot \pi \cdot r^3$

or $V = \dfrac{4 \cdot \pi \cdot r^3}{3}$

Volume of a hemisphere: $V = \dfrac{2}{3} \cdot \pi \cdot r^3$

or $V = \dfrac{2 \cdot \pi \cdot r^3}{3}$

Volume of a cylinder: $V = \pi \cdot r^2 \cdot h$

Volume of a cone: $V = \dfrac{1}{3} \cdot B \cdot h$ or $V = \dfrac{B \cdot h}{3}$

Volume of a pyramid: $V = \dfrac{1}{3} \cdot B \cdot h$ or $V = \dfrac{B \cdot h}{3}$

Right triangle: $\text{hypotenuse} = \sqrt{(\text{leg})^2 + (\text{leg})^2}$

$\text{leg} = \sqrt{(\text{hypotenuse})^2 - (\text{leg})^2}$

▶ Test Your Word Power

See how well you have learned the vocabulary in this chapter. Answers follow the Quick Review.

1. Two angles that are **complementary**
   A. have measures that add up to 180°
   B. are always congruent
   C. form a straight angle
   D. have measures that add up to 90°.

2. The **perimeter** of a flat shape is
   A. measured in square units
   B. the distance around the outside edges
   C. the number of squares needed to cover the space inside the shape
   D. measured in cubic units.

3. An **obtuse angle**
   A. is formed by perpendicular lines
   B. is congruent to a right angle
   C. measures more than 90° but less than 180°
   D. measures less than 90°.

4. The **hypotenuse** is
   A. the long base in a trapezoid
   B. the height line in a parallelogram
   C. the longest side in a right triangle
   D. the distance across a circle, passing through the center.

5. π is the ratio of
   A. the diameter to the radius of a circle
   B. the circumference to the diameter of a circle
   C. the circumference to the radius of a circle
   D. the diameter to the circumference of a circle.

6. **Perpendicular lines**
   A. intersect to form a right angle
   B. intersect to form an acute angle
   C. never intersect
   D. have a common endpoint called the vertex.

7. In a pair of **similar triangles,**
   A. corresponding sides have the same length
   B. all the angles have the same measure
   C. the perimeters are equal
   D. the ratios of the lengths of corresponding sides are equal.

8. The **area of a rectangle** is found by
   A. using the formula $P = 2 \cdot l + 2 \cdot w$
   B. multiplying length times width
   C. adding the lengths of the sides
   D. using the formula $V = l \cdot w \cdot h$.

▶ Quick Review

Concepts	Examples

8.1 Lines

A *line* is a straight row of points that goes on forever in both directions. If a piece of a line has one endpoint, it is a *ray*. If it has two endpoints, it is a *line segment*.

Identify each of the following as a line, line segment, or ray and name it using the appropriate symbol.

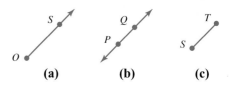

(a)   (b)   (c)

Figure **(a)** shows a ray named $\overrightarrow{OS}$.

Figure **(b)** shows a line named $\overleftrightarrow{PQ}$ or $\overleftrightarrow{QP}$.

Figure **(c)** shows a line segment named $\overline{ST}$ or $\overline{TS}$.

If two lines intersect at right angles, they are *perpendicular.*

If two lines in the same plane never intersect, they are *parallel.*

Label each pair of lines as appearing to be parallel or as perpendicular.

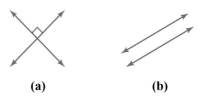

(a)   (b)

Figure **(a)** shows perpendicular lines (they intersect at 90°).

Figure **(b)** shows lines that appear to be parallel (they never intersect).

8.2 Angles

If the sum of the measures of two angles is 90°, they are *complementary.*

If the sum of the measures of two angles is 180°, they are *supplementary.*

Find the complement and supplement of a 35° angle.

$$90° - 35° = 55° \text{ (the complement)}$$
$$180° - 35° = 145° \text{ (the supplement)}$$

If two angles measure the same number of degrees, the angles are *congruent*. The symbol for congruent is ≅.

Two nonadjacent angles formed by two intersecting lines are called *vertical angles*. Vertical angles are congruent.

Identify the vertical angles in this figure. Which angles are congruent?

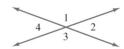

∠1 and ∠3 are vertical angles.

∠2 and ∠4 are vertical angles.

Vertical angles are congruent, so ∠1 ≅ ∠3 and ∠2 ≅ ∠4.

Concepts	Examples

8.3 Rectangles and Squares

Use this formula to find the perimeter of a *rectangle*.

$$P = 2 \cdot l + 2 \cdot w$$

Use this formula to find the area of a rectangle.

$$A = l \cdot w$$

Area is measured in **square units**.

Use these formulas to find the perimeter and area of a *square*.

$$P = 4 \cdot s$$
$$A = s^2$$

Area is measured in **square units**.

Find the perimeter and area of this rectangle.

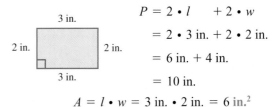

$$P = 2 \cdot l \quad + 2 \cdot w$$
$$= 2 \cdot 3 \text{ in.} + 2 \cdot 2 \text{ in.}$$
$$= 6 \text{ in.} + 4 \text{ in.}$$
$$= 10 \text{ in.}$$
$$A = l \cdot w = 3 \text{ in.} \cdot 2 \text{ in.} = 6 \text{ in.}^2$$

Find the perimeter and area of this square.

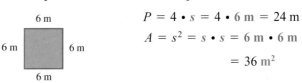

$$P = 4 \cdot s = 4 \cdot 6 \text{ m} = 24 \text{ m}$$
$$A = s^2 = s \cdot s = 6 \text{ m} \cdot 6 \text{ m}$$
$$= 36 \text{ m}^2$$

8.4 Parallelograms

Use these formulas to find the perimeter and area of a *parallelogram*.

$$P = \text{sum of the lengths of the sides}$$
$$A = b \cdot h$$

Area is measured in **square units**.

Find the perimeter and area of this parallelogram.

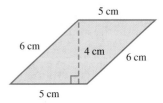

$$P = 5 \text{ cm} + 6 \text{ cm} + 5 \text{ cm} + 6 \text{ cm} = 22 \text{ cm}$$
$$A = 5 \text{ cm} \cdot 4 \text{ cm} = 20 \text{ cm}^2$$

8.4 Trapezoids

Use these formulas to find the perimeter and area of a *trapezoid*.

$$P = \text{sum of the lengths of the sides}$$
$$A = \frac{1}{2} \cdot h \cdot (b + B)$$

where *b* is the short base and *B* is the long base.

Area is measured in **square units**.

Find the perimeter and area of this trapezoid.

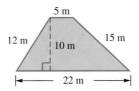

$$P = 5 \text{ m} + 15 \text{ m} + 22 \text{ m} + 12 \text{ m} = 54 \text{ m}$$
$$A = \frac{1}{2} \cdot \overset{5}{\cancel{10}} \text{ m} \cdot (5 \text{ m} + 22 \text{ m})$$
$$= 5 \text{ m} \cdot (27 \text{ m}) = 135 \text{ m}^2$$

8.5 Triangles

Use these formulas to find the perimeter and area of a *triangle*.

$$P = \text{sum of the lengths of the sides}$$
$$A = \frac{1}{2} \cdot b \cdot h$$
$$\text{or} \quad A = 0.5 \cdot b \cdot h$$

Area is measured in **square units**.

Find the perimeter and area of this triangle.

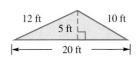

$$P = 12 \text{ ft} + 10 \text{ ft} + 20 \text{ ft} = 42 \text{ ft}$$
$$A = \frac{1}{2} \cdot b \cdot h$$
$$A = \frac{1}{2} \cdot \overset{10}{\cancel{20}} \text{ ft} \cdot 5 \text{ ft} = 50 \text{ ft}^2$$
$$\text{or} \quad A = 0.5 \cdot 20 \text{ ft} \cdot 5 \text{ ft} = 50 \text{ ft}^2$$

Concepts	Examples

8.6 Circles

Use this formula to find the *diameter* of a circle when you are given the radius.

$$d = 2 \cdot r$$

Find the diameter of a circle if the radius is 7 yd.

$$d = 2 \cdot r = 2 \cdot 7 \text{ yd} = 14 \text{ yd}$$

Use this formula to find the *radius* of a circle when you are given the diameter.

$$r = \frac{d}{2}$$

Find the radius of a circle if the diameter is 5 cm.

$$r = \frac{d}{2} = \frac{5 \text{ cm}}{2} = 2.5 \text{ cm}$$

Use these formulas to find the *circumference* of a circle.

When you know the radius, use $C = 2 \cdot \pi \cdot r$

When you know the diameter, use $C = \pi \cdot d$

Use 3.14 as the approximate value for π.

Find the circumference of a circle with a radius of 3 cm.

$$C = 2 \cdot \pi \cdot r$$
$$C \approx 2 \cdot 3.14 \cdot 3 \text{ cm} \approx 18.8 \text{ cm} \leftarrow \text{Rounded}$$

Use this formula to find the *area* of a circle.

$$A = \pi \cdot r^2$$

Area is measured in **square units**.

Find the area of this circle.

$$A = \pi \cdot r^2$$
$$A \approx 3.14 \cdot 3 \text{ cm} \cdot 3 \text{ cm}$$
$$A \approx 28.3 \text{ cm}^2 \leftarrow \text{Rounded;}$$
square units for area

8.7 Volume of a Rectangular Solid

Use this formula to find the volume of *rectangular solids* (box-like solids).

$$V = l \cdot w \cdot h$$

Volume is measured in **cubic units**.

Find the volume of this box.

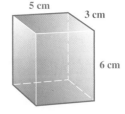

$$V = l \cdot w \cdot h$$
$$V = 5 \text{ cm} \cdot 3 \text{ cm} \cdot 6 \text{ cm}$$
$$V = 90 \text{ cm}^3 \quad \text{Cubic units for volume}$$

8.7 Volume of a Sphere and Hemisphere

Use this formula to find the volume of a *sphere* (a ball-shaped solid).

$$V = \frac{4}{3} \cdot \pi \cdot r^3$$

or $V = \frac{4 \cdot \pi \cdot r^3}{3}$

where r is the radius of the sphere.

Volume is measured in **cubic units**.

Find the volume of a sphere with a radius of 5 m.

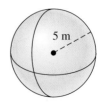

$$V = \frac{4 \cdot \pi \cdot (\text{radius})^3}{3}$$
$$V \approx \frac{4 \cdot 3.14 \cdot 5 \text{ m} \cdot 5 \text{ m} \cdot 5 \text{ m}}{3}$$
$$V \approx 523.3 \text{ m}^3 \leftarrow \text{Rounded}$$

(continued)

Concepts	Examples

8.7 Volume of a Sphere and Hemisphere (continued)

Use this formula to find the volume of a *hemisphere* (half of a sphere).

$$V = \frac{2}{3} \cdot \pi \cdot r^3$$

or $\quad V = \frac{2 \cdot \pi \cdot r^3}{3}$

where *r* is the radius of the hemisphere.
Volume is measured in **cubic units**.

Find the volume of a hemisphere with a radius of 20 cm.

$$V = \frac{2 \cdot \pi \cdot (\text{radius})^3}{3}$$

$$V \approx \frac{2 \cdot 3.14 \cdot 20 \text{ cm} \cdot 20 \text{ cm} \cdot 20 \text{ cm}}{3}$$

$V \approx 16{,}746.7 \text{ cm}^3 \;\leftarrow$ Rounded

8.7 Volume of a Cylinder

Use this formula to find the volume of a *cylinder*.

$$V = \pi \cdot r^2 \cdot h$$

where *r* is the radius of the circular base and *h* is the height of the cylinder.
Volume is measured in **cubic units**.

Find the volume of this cylinder.

First, find the radius. $\quad r = \dfrac{8 \text{ m}}{2} = 4 \text{ m}$

$$V = \pi \cdot r^2 \cdot h$$
$$V \approx 3.14 \cdot 4 \text{ m} \cdot 4 \text{ m} \cdot 10 \text{ m}$$
$$V \approx 502.4 \text{ m}^3$$

8 m

10 m

8.7 Volume of a Cone

Use this formula to find the volume of a *cone*.

$$V = \frac{1}{3} \cdot B \cdot h$$

or $\quad V = \dfrac{B \cdot h}{3}$

where *B* is the area of the circular base and *h* is the height of the cone.

Volume is measured in **cubic units**.

Find the volume of this cone.
Area of circular *Base* $\approx 3.14 \cdot 4$ in. $\cdot$ 4 in.

$$B \approx 50.24 \text{ in.}^2$$

$$V \approx \frac{B \cdot h}{3}$$

$$V \approx \frac{50.24 \text{ in.}^2 \cdot 9 \text{ in.}}{3}$$

$V \approx 150.7 \text{ in.}^3 \;\leftarrow$ Rounded

9 in.

4 in.

8.7 Volume of a Pyramid

Use this formula to find the volume of a *pyramid*.

$$V = \frac{1}{3} \cdot B \cdot h$$

or $\quad V = \dfrac{B \cdot h}{3}$

where *B* is the area of the square or rectangular base and *h* is the height of the pyramid.

Volume is measured in **cubic units**.

Find the volume of this pyramid.
Area of square *Base* = 2 cm $\cdot$ 2 cm

$$B = 4 \text{ cm}^2$$

$$V = \frac{B \cdot h}{3}$$

$$V = \frac{4 \text{ cm}^2 \cdot 6 \text{ cm}}{3}$$

$$V = 8 \text{ cm}^3$$

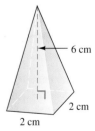

6 cm

2 cm

2 cm

8.8 Finding the Square Root of a Number

Use the square root key on a calculator, 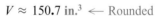 or .
Round to the nearest thousandth if necessary.

$\sqrt{64} = 8 \qquad$ A perfect square

$\sqrt{43} \approx 6.557 \qquad$ 6.557438524 is rounded to nearest thousandth.

Concepts	Examples

8.8 Finding the Unknown Length in a Right Triangle

To find the *hypotenuse,* use this formula.

$$\text{hypotenuse} = \sqrt{(\text{leg})^2 + (\text{leg})^2}$$

The hypotenuse is the side opposite the right angle; it is the longest side in a right triangle.

Find the unknown length in this right triangle. Round to the nearest tenth.

$$\begin{aligned} \text{hypotenuse} &= \sqrt{(6)^2 + (5)^2} \\ &= \sqrt{36 + 25} \\ &= \sqrt{61} \approx 7.8 \text{ m} \end{aligned}$$

To find a *leg,* use this formula.

$$\text{leg} = \sqrt{(\text{hypotenuse})^2 - (\text{leg})^2}$$

The legs are the sides that form the right angle.

Find the unknown length in this right triangle. Round to the nearest tenth.

$$\begin{aligned} \text{leg} &= \sqrt{(25)^2 - (16)^2} \\ &= \sqrt{625 - 256} \\ &= \sqrt{369} \approx 19.2 \text{ cm} \end{aligned}$$

8.9 Finding the Unknown Lengths in Similar Triangles

Use the fact that in similar triangles, the *ratios* of the lengths of corresponding sides are equal. Write a proportion. Then find the cross products and show that they are equivalent. Finish solving for the unknown length.

Find the unknown lengths in this pair of similar triangles.

$$\frac{x}{8} = \frac{5}{10}$$

$$x \cdot 10 = 8 \cdot 5$$

$$\frac{x \cdot \overset{1}{\cancel{10}}}{\cancel{10}_{1}} = \frac{40}{10}$$

$$x = 4 \text{ m}$$

$$\frac{y}{12} = \frac{5}{10}$$

$$y \cdot 10 = 12 \cdot 5$$

$$\frac{y \cdot \overset{1}{\cancel{10}}}{\cancel{10}_{1}} = \frac{60}{10}$$

$$y = 6 \text{ m}$$

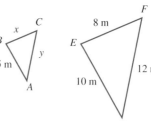

ANSWERS TO TEST YOUR WORD POWER

1. D; *Example:* If $\angle 1$ measures 35° and $\angle 2$ measures 55°, the angles are complementary because 35° + 55° = 90°.
2. B; *Example:* If a square measures 5 ft on each side, then the perimeter is 5 ft + 5 ft + 5 ft + 5 ft = 20 ft.
3. C; *Examples:* Angles that measure 91°, 120°, and 175° are all obtuse angles.
4. C; *Example:* In triangle *ABC* at the right, side *AC* is the hypotenuse; sides *AB* and *BC* are the legs.
5. B; *Example:* The ratio of a circumference of 12.57 cm to a diameter of 4 cm is $\dfrac{12.57}{4} \approx 3.14$ (rounded).
6. A; *Example:* $\overleftrightarrow{EF}$ is perpendicular to $\overleftrightarrow{GH}$, at the right.
7. D; *Example:* Triangle *ABC* is similar to triangle *DEF,* so the ratios of corresponding sides are equal.

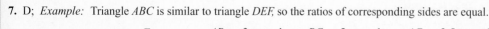

$$\frac{AB}{DE} = \frac{3 \text{ m}}{6 \text{ m}} = \frac{1}{2} \qquad \frac{BC}{EF} = \frac{2 \text{ m}}{4 \text{ m}} = \frac{1}{2} \qquad \frac{AC}{DF} = \frac{3.5 \text{ m}}{7 \text{ m}} = \frac{1}{2}$$

8. B; *Example:* In a rectangle with a length of 8 in. and a width of 5 in., Area = 8 in. • 5 in. = 40 in.²

Chapter 8 ▶▶▶ Review Exercises

[8.1] Identify each figure as a line, line segment, or ray, and name it using the appropriate symbol.

1.

line segment named $\overline{AB}$ or $\overline{BA}$

2.

line named $\overleftrightarrow{CD}$ or $\overleftrightarrow{DC}$

3.

ray named $\overrightarrow{OP}$

Label each pair of lines as appearing to be parallel, as perpendicular, or as intersecting.

4.

parallel

5.

perpendicular

6.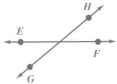

intersecting

Label each angle as acute, right, obtuse, or straight. For right and straight angles, indicate the number of degrees in the angle.

7.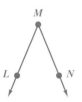

acute

8.

obtuse

9.

straight; 180°

10.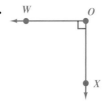

right; 90°

[8.2] In each figure you are given the measures of two of the angles. Find the measure of each of the other angles.

11.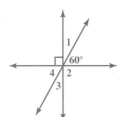

∠1 and ∠3 measure 30°;

∠2 measures 90°;

∠4 measures 60°

12.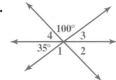

∠1 measures 100°;

∠2 and ∠4 measure 45°;

∠3 measures 35°

Name the pairs of supplementary angles in each figure.

13.

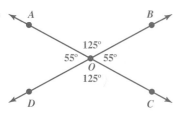

∠AOB and ∠BOC; ∠BOC and ∠COD;

∠COD and ∠DOA; ∠DOA and ∠AOB

14.

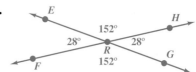

∠ERH and ∠HRG; ∠HRG and ∠GRF;

∠FRG and ∠FRE; ∠FRE and ∠ERH

Find the complement or supplement of each angle.

15. Find the complement of:

    (a) 80°   10°

    (b) 45°   45°

    (c) 7°   83°

16. Find the supplement of:

    (a) 155°   25°

    (b) 90°   90°

    (c) 33°   147°

[8.3] *Find the perimeter of each rectangle or square.*

17.

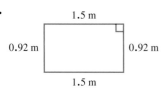

P = 4.84 m

18.

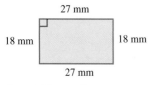

P = 128 in.

19. A square-shaped pillow measures 38 cm along each side. How much lace is needed to trim all the edges?

152 cm

20. A rectangular garden plot is $8\frac{1}{2}$ ft wide and 12 ft long. How much fencing is needed to surround the garden?

41 ft

Find the area of each rectangle or square. Round your answers to the nearest tenth when necessary.

21.

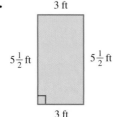

A = 486 mm²

22.

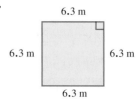

$$A = 16.5 \text{ ft}^2 \text{ or } 16\frac{1}{2}\text{ ft}^2$$

23.

6.3 m

6.3 m   6.3 m

6.3 m

A ≈ 39.7 m²

[8.4] *Find the perimeter and area of each parallelogram or trapezoid. Round your answers to the nearest tenth when necessary.*

24.

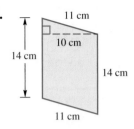

P = 50 cm;
A = 140 cm²

25.

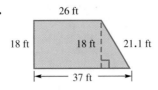

P = 102.1 ft;
A = 567 ft²

26.

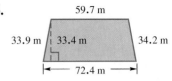

P = 200.2 m;
A ≈ 2206.1 m²

[8.5] *Find the perimeter and area of each triangle.*

27.

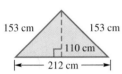

P = 518 cm;
A = 11,660 cm²

28.

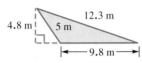

P = 27.1 m;
A = 23.52 m²

29.

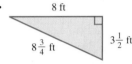

$P = 20\frac{1}{4}$ ft or 20.25 ft;
A = 14 ft²

Find the number of degrees in the third angle of each triangle.

30.

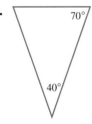

70°

31.

24°

[8.6] *Find the unknown length.*

32. The radius of a circular irrigation field is 68.9 m. What is the diameter of the field?

d = 137.8 m

33. The diameter of a juice can is 3 in. What is the radius of the can?

$r = 1\frac{1}{2}$ in. or 1.5 in.

Find the circumference and area of each circle. Use 3.14 as the approximate value for π. Round your answers to the nearest tenth.

34.

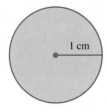

C ≈ 6.3 cm;
A ≈ 3.1 cm²

35.

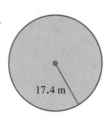

C ≈ 109.3 m;
A ≈ 950.7 m²

36.

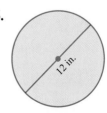

C ≈ 37.7 in.;
A ≈ 113.0 in.²

[8.3–8.6] *Find each shaded area. Note that Exercises 44 and 45 contain semicircles. Use* 3.14 *as the approximate value for* π. *Round your answers to the nearest tenth when necessary.*

37.

A ≈ 20.3 m²

38.

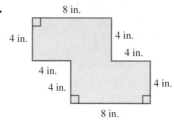

A = 64 in.²

39.

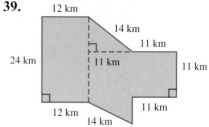

A = 673 km²

40.

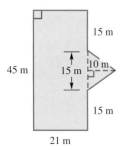

A = 1020 m²

41.

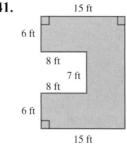

A = 229 ft²

42.

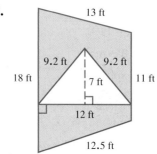

A = 132 ft²

43.

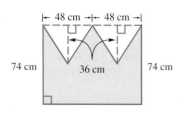

A = 5376 cm²

44.

A ≈ 498.9 ft²

45.

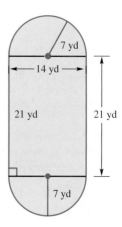

A ≈ 447.9 yd²

[8.7] *Name each solid and find its volume. Use 3.14 as the approximate value for π.*
Round your answers to the nearest tenth when necessary.

46.

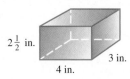

Rectangular solid;
V = 30 in.³

47.

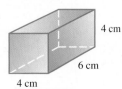

Rectangular solid;
V = 96 cm³

48.

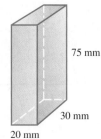

Rectangular solid;
V = 45,000 mm³

49.

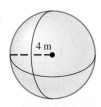

Sphere;
V ≈ 267.9 m³

50.

Hemisphere;
V ≈ 452.2 ft³

51.

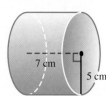

Cylinder;
V ≈ 549.5 cm³

52.

Cylinder;
V ≈ 1808.6 m³

53.

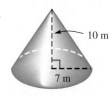

Cone;
V ≈ 512.9 m³

54.

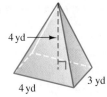

Pyramid;
V = 16 yd³

[8.8] *Find each square root. Round your answers to the nearest thousandth when necessary.*

55. $\sqrt{49}$

7

56. $\sqrt{8}$

2.828 (rounded)

57. $\sqrt{3000}$

54.772 (rounded)

58. $\sqrt{144}$

12

59. $\sqrt{58}$

7.616 (rounded)

60. $\sqrt{625}$

25

61. $\sqrt{105}$

10.247 (rounded)

62. $\sqrt{80}$

8.944 (rounded)

Find the unknown length in each right triangle. Use a calculator to find square roots.
Round your answers to the nearest tenth when necessary.

63.

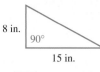

$\sqrt{289} = 17$ in.

64.

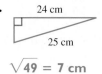

$\sqrt{49} = 7$ cm

65.

$\sqrt{104} \approx 10.2$ cm

66.

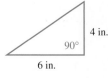

$\sqrt{52} \approx 7.2$ in.

67.

$\sqrt{6.53} \approx 2.6$ m

68.

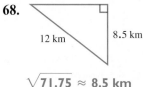

$\sqrt{71.75} \approx 8.5$ km

[8.9] *Find the unknown lengths in each pair of similar triangles. Then find the perimeter of the larger triangle in each pair.*

69.

y = 30 ft;
x = 34 ft;
P = 104 ft

70.

y = 7.5 m;
x = 9 m;
P = 22.5 m

71.

x = 12 mm;
y = 7.5 mm;
P = 38 mm

▶▶▶ **Mixed Review Exercises**

Name each figure and find its perimeter (or circumference) and area. Use 3.14 as the approximate value for π. Round your answers to the nearest tenth when necessary.

72.

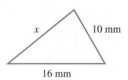

Square; P = 18 in.;

$A \approx 20.3$ in.^{2} or $A = 20\frac{1}{4}$ in.^{2}

73.

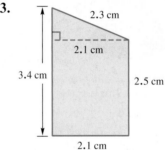

Trapezoid; P = 10.3 cm;
$A \approx 6.2$ cm^2

▦ **74.**

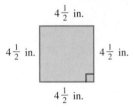

Circle; $C \approx 40.8$ m;
$A \approx 132.7$ m^2

75.

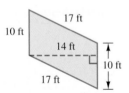

Parallelogram; P = 54 ft;
$A = 140$ ft^2

76.

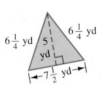

Triangle; P = 20 yd;

$A = 18\frac{3}{4}$ yd^2 or 18.8 yd^2 (rounded)

77.

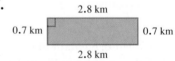

Rectangle; P = 7 km;
$A \approx 2.0$ km^2

78.

▦

Circle; $C \approx 53.4$ m;
$A \approx 226.9$ m^2

79.

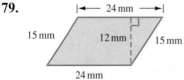

Parallelogram; P = 78 mm;
$A = 288$ mm^2

80.

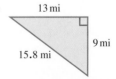

Triangle; P = 37.8 mi;
$A = 58.5$ mi^2

Label each figure. Choose from these labels: line segment, ray, parallel lines, perpendicular lines, intersecting lines, acute angle, right angle, straight angle, obtuse angle. Indicate the number of degrees in the right angle and the straight angle.

81.

parallel lines

82.

line segment

83.

acute angle

84.

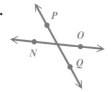

intersecting lines

85.

right angle; 90°

86.

ray

87.

straight angle; 180°

88.

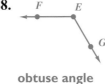

obtuse angle

89.

perpendicular lines

90. What is the complement of an angle measuring 9°?

81°

91. What is the supplement of an angle measuring 42°?

138°

Find the perimeter and area of each figure. In Exercise 92, assume that all angles are 90°.

92.

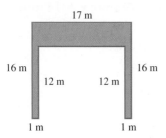

P = 90 m; *A* = 92 m²

93.

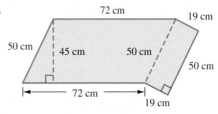

P = 282 cm; *A* = 4190 cm²

Find the volume of each solid. Use 3.14 as the approximate value for π.
Round your answers to the nearest tenth when necessary.

94.

8 ft — 2 ft

$V \approx 100.5 \text{ ft}^3$

95.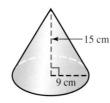

$1\frac{1}{2}$ in.

$1\frac{1}{2}$ in.

$1\frac{1}{2}$ in.

$V \approx 3.4 \text{ in.}^3 \text{ or } V = 3\frac{3}{8} \text{ in.}^3$

96.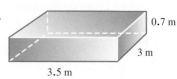

0.7 m

3 m

3.5 m

$V \approx 7.4 \text{ m}^3$

97.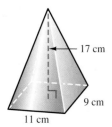

17 cm

9 cm

11 cm

$V = 561 \text{ cm}^3$

98.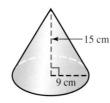

15 cm

9 cm

$V \approx 1271.7 \text{ cm}^3$

99.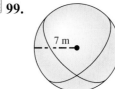

7 m

$V \approx 1436.0 \text{ m}^3$

Find the unknown angle or side measurement. Round your answers to the nearest tenth when necessary.

100.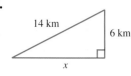

14 km

6 km

x

$x \approx 12.6 \text{ km}$

101.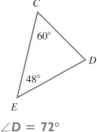

C

60°

D

48°

E

$\angle D = 72°$

102. similar triangles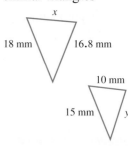

x

18 mm

16.8 mm

10 mm

15 mm

y

$x = 12 \text{ mm}; y = 14 \text{ mm}$

103. Explain how you could use the information about prefixes from **Section 8.6**
to solve a problem that asks, "How many decades are in two centuries?"

The prefix *dec-* in *decade* means 10 and the prefix *cent-* in *century* means 100, so divide 200 (two centuries) by 10. The answer is 20 decades.

Chapter 8 ▶▶▶ Test

Test Prep VIDEO CD Use the Chapter Test Prep Video CD to see fully worked-out solutions to any of the exercises you want to review.

Choose the figure that matches each label. For right and straight angles, indicate the number of degrees in the angle.

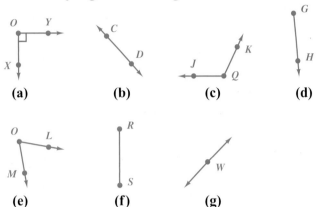

(a)   (b)   (c)   (d)

(e)   (f)   (g)

1. Acute angle is figure _____.

2. Right angle is figure _____ and its measure is _____.

3. Ray is figure _____.

4. Straight angle is figure _____ and its measure is _____.

🖉 5. Write a definition of parallel lines and a definition of perpendicular lines. Make a sketch to illustrate each definition.

6. Find the complement of an 81° angle.

7. Find the supplement of a 20° angle.

8. Find the measure of each unlabeled angle in the figure at the right.

Name each figure and find its perimeter and area.

9.

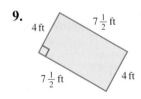
4 ft $7\frac{1}{2}$ ft $7\frac{1}{2}$ ft   4 ft

10.

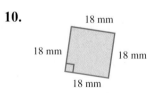
18 mm   18 mm   18 mm   18 mm

11.

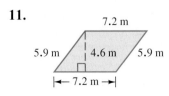
7.2 m   5.9 m   4.6 m   5.9 m   7.2 m

12.

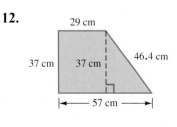
29 cm   37 cm   37 cm   46.4 cm   57 cm

Answers column:

1. (e) _____

2. (a); 90° _____

3. (d) _____

4. (g); 180° _____

5. **Parallel lines are lines in the same plane that never intersect. Perpendicular lines intersect to form a right angle.**

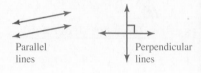

Parallel lines   Perpendicular lines

6. **9°** _____

7. **160°** _____

8. **∠1 measures 50°; ∠3 measures 90°; ∠2 and ∠4 measure 40°** _____

9. **Rectangle; P = 23 ft; A = 30 ft²** _____

10. **Square; P = 72 mm; A = 324 mm²** _____

11. **Parallelogram; P = 26.2 m; A = 33.12 m²** _____

12. **Trapezoid; P = 169.4 cm; A = 1591 cm²** _____

13. $P = 32.45$ m;
$A = 48$ m^2

14. $P = 37.8$ yd or $37\frac{4}{5}$ yd;
$A = 58.5$ yd^2

15. $55°$

16. $r = 12.5$ in. or $12\frac{1}{2}$ in.

17. $C \approx 5.7$ km (rounded)

18. $A \approx 206.0$ cm^2 (rounded)

19. $A \approx 39.3$ m^2 (rounded)

20. Rectangular solid;
$V = 6480$ m^3

21. Sphere;
$V \approx 33.5$ ft^3 (rounded)

22. Cylinder; $V \approx 5086.8$ ft^3

23. $\sqrt{85} \approx 9.2$ cm (rounded)

24. $y = 12$ cm; $z = 6$ cm

25. Linear units like cm are used to measure perimeter, radius, diameter, and circumference. Area is measured in square units like cm^2 (squares that measure 1 cm on each side). Volume is measured in cubic units like cm^3.

Find the perimeter and area of each triangle.

13.

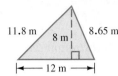

14.

15. A triangle has angles that measure 90° and 35°. What does the third angle measure?

In Problems 16–22, use 3.14 as the approximate value for π. Round your answers to the nearest tenth when necessary.

16. Find the radius.

17. Find the circumference.

📟 *Find the area of each figure.*

18.

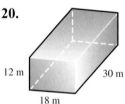

19.

20.

Name each solid and find its volume.

20.

📟 **21.**

📟 **22.**

Find the unknown lengths. Round your answers to the nearest tenth when necessary.

📟 **23.**

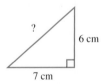

24. Similar triangles

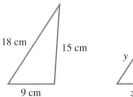

✍ **25.** Explain the difference between cm, cm^2, and cm^3. In what types of geometry problems might you use each of these units?

Cumulative Review Exercises ▶▶▶ Chapters 1–8

Simplify. Write answers to fraction problems in lowest terms and as whole or mixed numbers when possible.

1. $3\dfrac{3}{5} \div 8$

$\dfrac{9}{20}$

2. $1 - 0.0868$

0.9132

3. $(0.006)(0.013)$

0.000078

4. $0.7 \div 0.036$ Round answer to the nearest hundredth.

19.44 (rounded)

5. $6\dfrac{1}{6} - 1\dfrac{3}{4}$

$4\dfrac{5}{12}$

6. $16 - (10 - 2) \div 2(3) + 5$

9

7. Write 0.0208 in words.

two hundred eight ten-thousandths

8. Arrange in order from least to greatest.

2.55   2.505   2.055   2.5005

2.055; 2.5005; 2.505; 2.55

Solve each proportion or percent problem. Round your answer to hundredths if necessary.

9. $\dfrac{5}{13} = \dfrac{x}{91}$

x = 35

10. $\dfrac{4.5}{x} = \dfrac{6.7}{3}$

x ≈ 2.01 (rounded)

11. 72 patients is what percent of 45 patients?

160%

12. $18 is 3% of what number of dollars?

$600

Convert each measurement.

13. $2\dfrac{1}{4}$ hours to minutes

135 min

14. 40 oz to pounds

$2\dfrac{1}{2}$ or 2.5 lb

15. 8 cm to meters

0.08 m

16. 1.8 L to milliliters

1800 mL

Write the most reasonable metric unit in each blank. Choose from km, m, cm, mm, L, mL, kg, g, and mg.

17. Her wristwatch strap is 15 <u>mm</u> wide.

18. Jon added 2 <u>L</u> of oil to his car.

19. The child weighs 15 <u>kg</u>.

20. The bookcase is 90 <u>cm</u> high.

Name each figure and find its perimeter (or circumference) and area. Use 3.14 *as the approximate value of* π.

21.

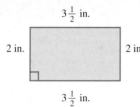

$3\frac{1}{2}$ in.

2 in.   2 in.

$3\frac{1}{2}$ in.

Rectangle;
P = 11 in.;
A = 7 in.²

22.

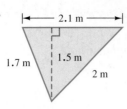

|← 2.1 m →|

1.7 m   1.5 m

2 m

Triangle;
P = 5.8 m;
A = 1.575 m²

23.

5 ft

Circle;
C ≈ 31.4 ft;
A ≈ 78.5 ft²

24.

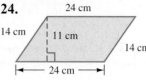

24 cm

14 cm   11 cm

14 cm

|← 24 cm →|

Parallelogram;
P = 76 cm;
A = 264 cm²

---

25. Find the unknown length to the nearest tenth.

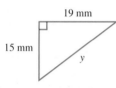

19 mm

15 mm

y

y ≈ 24.2 mm

26. Mei Ling must earn 90 credits to receive an associate of arts degree. She has 53 credits. What percent of the necessary credits does she have? Round to the nearest whole percent.

59% (rounded)

27. A Folger's coffee can has a diameter of 13 cm and a height of 17 cm. Find the volume of the can. Use 3.14 for π and round your answer to the nearest tenth. (*Source:* Folger's.)

V ≈ 2255.3 cm³

28. Which bag of chips is the best buy: Brand T is $15\frac{1}{2}$ oz for $2.99, Brand F is 14 oz for $2.49, and Brand H is 18 oz for $3.89. You have a coupon for 40¢ off Brand H and another for 30¢ off Brand T.

Brand T at 15.5 oz for $2.99 − $0.30 coupon, about $0.174/oz

29. Steven bought a $4\frac{1}{2}$ yd length of canvas material to repair the tents used by the scout troop. He used $1\frac{2}{3}$ yd on one tent and $1\frac{3}{4}$ yd on another. What length of material is left?

$1\frac{1}{12}$ **yd**

30. The cooks at a homeless shelter used 25 lb of meat, 35 lb of potatoes, and 15 lb of carrots to make stew for 140 people. At that rate, how much of each item is needed for stew to feed 200 people? Round to the nearest tenth.

35.7 lb meat (rounded); 50 lb potatoes;
21.4 lb carrots (rounded)

31. Graciela needs 85 cm of yarn to make a tassel for one corner of a pillow. How many meters of yarn does she need to put a tassel on each corner of a square-shaped pillow?

3.4 m

32. Swimsuits are on sale in August at 65% off the regular price. How much will Lanece pay for a suit that has a regular price of $64?

$22.40

9

Basic Algebra

The weather—we all talk about it and often complain about it. When reporting temperatures, we need both *positive* and *negative* numbers. If you live in Chicago, the highest temperature ever recorded is 104 °F and the lowest is −27 °F, a difference of 131 degrees! In Atlanta, the record high and low are 105 °F and −8 °F. And in Barrow, Alaska, the range of extreme temperatures is 79 °F to −56 °F, a difference of 135 degrees. To make things even more uncomfortable, humidity can make hot temperatures feel hotter and wind can make cold temperatures feel colder. Learn more about *windchill* as you work Exercises 69–72 in **Section 9.2.** (*Source:* National Climatic Data Center.)

9.1 ▶▶▶ Signed Numbers

All the numbers you have studied so far in this book have been either 0 or greater than 0. Numbers *greater* than 0 are called *positive numbers*. For example, you have worked with these positive numbers.

Salary of $45,000
Temperature of 98.6 °F
Length of $3\frac{1}{2}$ feet

OBJECTIVE 1 Write negative numbers. Not all numbers are positive. For example, "15 degrees below 0" or "a loss of $500" is expressed with a number *less* than 0. Numbers less than 0 are called **negative numbers**. Zero is neither positive nor negative.

> **Writing a Negative Number**
> To write a negative number, put a ***negative sign,*** − , in front of it.

For example, "15 degrees below 0" is written with a negative sign, as −15°. And "a loss of $500" is written −$500.

◀ *Work Problem* **1** *at the Side.*

1 Write each number.

(a) The temperature at the North Pole is 70 degrees below 0.

(b) Your checking account is overdrawn by 15 dollars.

(c) A submarine dived to 284 ft below sea level.

OBJECTIVE 2 Graph signed numbers on a number line. In **Section 3.5** you graphed positive numbers on a number line. Negative numbers can also be shown on a number line. Zero separates the positive numbers from the negative numbers on the number line. The number −5 is read "negative five."

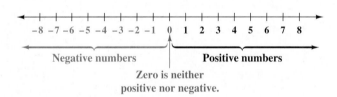

> **Note**
> For every *positive* number on a number line, there is a corresponding *negative* number on the *opposite* side of 0.

When you work with both positive and negative numbers (and 0), we say you are working with **signed numbers.**

2 Write *positive, negative,* or *neither* for each number.

(a) −8

(b) $-\frac{3}{4}$

(c) 1

(d) 0

> **Writing a Positive Number**
> A positive number can be written in two ways.
> 1. Use a "+" sign. For example, +2 is "positive two."
> 2. Do not write any sign. For example, when you write 3, it is assumed to be "positive three."

ANSWERS
1. **(a)** −70° **(b)** −$15 **(c)** −284 ft
2. **(a)** negative **(b)** negative
   (c) positive **(d)** neither

◀ *Work Problem* **2** *at the Side.*

The next example shows you how to graph signed numbers on a number line.

EXAMPLE 1 **Graphing Signed Numbers**

Graph these numbers on the number line.

(a) -4 **(b)** 3 **(c)** -1 **(d)** 0 **(e)** $1\frac{1}{4}$

Place a dot at the correct location for each number.

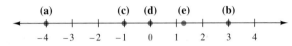

Work Problem ③ *at the Side.* ▶

OBJECTIVE **3** **Use the** $<$ **and** $>$ **symbols.** When you look at the number line below, 3 is to the *left* of 5.

3 is to the *left* of 5.
3 is *less than* 5.

Recall the following symbols for comparing two numbers.

$<$ means "**is less than.**"

$>$ means "**is greater than.**"

Use these symbols to write "3 is less than 5" as follows.

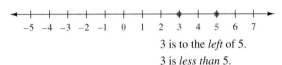

This example suggests the following.

> The ***lesser*** of two numbers is the one farther to the ***left*** on a number line. The ***greater*** of two numbers is the one farther to the ***right*** on a number line.

EXAMPLE 2 **Using the Symbols** $<$ **and** $>$

Use this number line to compare each pair of numbers. Then write $>$ or $<$ to make each statement true.

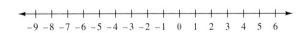

(a) Compare 2 and 6. $2 < 6$ (read "2 is less than 6") because 2 is to the *left* of 6 on the number line.

(b) Compare -9 and -4. $-9 < -4$ because -9 is to the *left* of -4.

> Be careful when comparing two *negative* numbers.

(c) Compare 2 and -1. $2 > -1$ because 2 is to the *right* of -1.

(d) Compare -4 and 0. $-4 < 0$ because -4 is to the *left* of 0.

—— **Continued on Next Page**

③ Graph each set of numbers.

(a) $-1, 1, -3, 3$

(b) $-2, 4, 0, -1, -4$

ANSWERS

3. **(a)**

   (b)

4 Write $<$ or $>$ in each blank to make a true statement.

(a) 4 _____ 0

(b) -1 _____ 0

(c) -3 _____ -1

(d) -8 _____ -9

(e) 0 _____ -3

5 Simplify each absolute value expression.

(a) $|5|$

(b) $|-5|$

(c) $|-17|$

(d) $-|-9|$

(e) $-|2|$

> **Note**
> When using $>$ and $<$, the *smaller* pointed end of the symbol points to the *smaller* (lesser) number.

◀ *Work Problem* **4** *at the Side.*

OBJECTIVE **4** **Find absolute value.** In order to graph a number on the number line, you need to answer two questions:

1. Which *direction* is it from 0? It can be in a *positive* direction or a *negative* direction. You can tell the direction by looking for a positive sign or a negative sign (or no sign, which is positive).

2. How *far* is it from 0? The *distance* from 0 is the **absolute value** of a number.

Absolute value is indicated by two vertical bars. For example, $|6|$ is read "the **absolute value** of 6."

> **Note**
> Absolute value is never negative because it is the *distance* from 0. A distance is never negative.

EXAMPLE 3 **Finding Absolute Values**

Simplify each absolute value expression.

(a) $|8|$ The *distance* from 0 to 8 is 8, so $|8| = 8$.

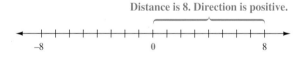

Distance is 8. Direction is positive.

(b) $|-8|$ The *distance* from 0 to -8 is also 8, so $|-8| = 8$.

Distance is 8. Direction is negative.

(c) $|0| = 0$

(d) $-|-3|$ First, $|-3|$ is 3. But there is also a negative sign *outside* the absolute value bars. So, -3 is the simplified expression. *[Watch for a negative sign* outside *the absolute value bars.]*

> **CAUTION**
> A negative sign *outside* the absolute value bars is *not* affected by the absolute value bars. Therefore, your final answer is negative, as in Example 3(d) above.

ANSWERS
4. (a) $>$ (b) $<$ (c) $<$ (d) $>$ (e) $>$
5. (a) 5 (b) 5 (c) 17 (d) -9 (e) -2

◀ *Work Problem* **5** *at the Side.*

OBJECTIVE 5 Find the opposite of a number. Two numbers that are the *same distance* from 0 on a number line but on *opposite sides* of 0 are called **opposites** of each other. As this number line shows, -3 and 3 are opposites of each other.

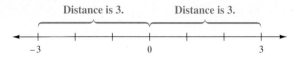

Distance is 3.   Distance is 3.

To indicate the opposite of a number, write a negative sign in front of the number.

EXAMPLE 4 Finding Opposites

Find the opposite of each number.

Number	Opposite
5	$-(5) = -5$ — Write a negative sign
9	$-(9) = -9$
$\frac{4}{5}$	$-\left(\frac{4}{5}\right) = -\frac{4}{5}$
0	$-(0) = 0$ — No negative sign

The opposite of 0 is 0. Zero is neither positive nor negative.

Work Problem **6** *at the Side.* ▶

Some numbers have two negative signs. An example is shown below.

$$-(-3)$$

The negative sign in front of (-3) means the *opposite* of -3. The opposite of -3 is 3.

$$-(-3) = 3$$

Use the following rule to find the opposite of a negative number.

Double Negative Rule

The opposite of a negative number is positive.

For example, $-(-3) = 3$ and $-(-10) = 10$.

EXAMPLE 5 Finding Opposites

Find the opposite of each number.

Number	Opposite	
-2	$-(-2) = 2$	By the double negative rule
-9	$-(-9) = 9$	
$-\frac{1}{2}$	$-\left(-\frac{1}{2}\right) = \frac{1}{2}$	Remember, the *opposite* of any *negative* number is *positive*.

Work Problem **7** *at the Side.* ▶

---

6 Find the opposite of each number.

(a) 4

(b) 10

(c) 49

(d) $\frac{2}{5}$

(e) 0

7 Find the opposite of each number.

(a) -4

(b) -10

(c) -25

(d) -1.9

(e) -0.85

(f) $-\frac{3}{4}$

ANSWERS

6. **(a)** -4 **(b)** -10 **(c)** -49
   (d) $-\frac{2}{5}$ **(e)** 0

7. **(a)** 4 **(b)** 10 **(c)** 25 **(d)** 1.9
   (e) 0.85 **(f)** $\frac{3}{4}$

Math in the Media

GOLF SCOREBOARDS

At the 2007 PGA Championship golf tournament in Atlanta, Georgia, 72 strokes was the "par" score for each round of play. A negative score indicates that the player had less than 72 strokes for that round, and a positive score indicates that the player had more than 72 strokes. The scoreboard below shows the scores for several of the players.

Player	Round 1	Round 2	Round 3	Round 4
Tiger Woods	−6	−7	−6	−4
Zach Johnson	+1	−4	−10	−2
Sergio Garcia	−2	−6	−6	0

Source: www.pgatour.com

1. Identify the round and the player who had each score. There may be more than one correct answer.

    (a) Seven strokes less than par.   **Round 2, Woods**

    (b) One stroke more than par.   **Round 1, Johnson**

    (c) Six strokes less than par.   **Round 1, Woods; Round 2, Garcia; Round 3, Woods and Garcia**

2. What score did Garcia have in Round 4? Is the score positive or negative? What does that score tell you?   **0; neither positive nor negative; Garcia had exactly 72 strokes (par) for Round 4.**

3. Graph Zach Johnson's scores on the number line below. Label the round for each score that you graph.

    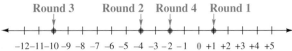

4. **(a)** List Johnson's scores from least to greatest.
       −10, −4, −2, +1

    (b) List Woods' scores from least to greatest.
       −7, −6, −6, −4

    (c) List Garcia's scores from least to greatest.
       −6, −6, −2, 0

5. To win at golf, you need to get the least number of strokes. Which round was the best round for **(a)** Woods; **(b)** Johnson; **(c)** Garcia.   **(a) Round 2;   (b) Round 3;   (c) Rounds 2 and 3 tied for best.**

6. Bonus question: Find the total score for each player on the scoreboard. (You'll learn more about adding signed numbers in **Section 9.2**.)   **Woods, −23; Johnson, −15; Garcia, −14**

Tiger Woods won the 2007 PGA Championship.

9.1 ▶▶▶ Exercises

Write a signed number for each situation.

1. Water freezes at 32 degrees above 0 on the Fahrenheit temperature scale.

  +32 or 32 degrees

2. She made a profit of $920.

  +$920 or $920

3. The price of the stock fell $12.

  −$12

4. His checking account is overdrawn by $30.79.

  −$30.79

5. Keith lost $6\frac{1}{2}$ lb while he was sick with the flu.

  $-6\frac{1}{2}$ lb

6. The river is 20 ft above flood stage.

  +20 ft or 20 ft

7. Mount McKinley, the tallest mountain in the United States, rises 20,320 ft above sea level. (*Source: World Almanac.*) **+20,320 ft or 20,320 ft**

8. The bottom of Lake Baykal in central Asia is 5315 ft below the surface of the water. (*Source: World Almanac.*) **−5315 ft**

Depth below surface

Graph each set of numbers on the number line. See Example 1.

9. $4, -1, 2, 0, -5$

10. $-2, 1, -3, 5, 0$

11. $-\frac{1}{2}, -3, \frac{7}{4}, -4\frac{1}{2}, 3\frac{1}{4}$

12. $-4, -\frac{3}{4}, 1, -1\frac{1}{4}, \frac{5}{2}$

13. $3, 4.5, -1.5, 2.2, -0.5$

14. $3.25, -1, -4.5, 1.25, 2$

Write < or > in each blank to make a true statement. See Example 2.

15. 9 __<__ 14

16. 6 __<__ 11

17. 0 __>__ −2

18. 0 __<__ 2

19. −6 __<__ 3

20. −9 __<__ 9

21. 1 __>__ −1

22. −1 __<__ 0

23. −11 __<__ −2

24. −5 __<__ −1

25. −72 __>__ −75

26. −50 __>__ −60

Relating Concepts (Exercises 27–30) For Individual or Group Work

A special X-ray test (DEXA) is a quick way to screen patients for osteoporosis (brittle bone disease). Use the information on the Patient Report Form to **work Exercises 27–30 in order.**

Patient Report Form

Your T score measures your bone density compared to that of a young, healthy woman when peak bone mass is achieved.

T Score	Interpretation
Above 0	Above normal
0 to –1.0	Normal
Below –1	You may be at risk; further discussion with your health provider is recommended.

Source: Health Partners, Inc.

27. Here are the T scores for four patients. Draw a number line and graph the four scores.

Patient A: −1.2

Patient B: 0.6

Patient C: −0.5

Patient D: 0

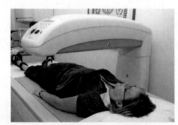

28. List the patients' scores in order from lowest to highest.

−1.2, −0.5, 0, 0.6

29. What is the interpretation of each patient's score?

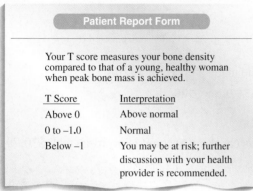

Taking a DEXA test of a patient's spine.

A: may be at risk; B: above normal; C: normal; D: normal

30. (a) Explain what could happen if patient A did not understand the importance of a negative sign.

The patient would think the interpretation was "above normal" and wouldn't get treatment.

(b) For which patient does the sign of the score make no difference? Explain your answer.

Patient D's score of 0; because 0 is neither positive nor negative.

Simplify each absolute value expression. See Example 3.

31. $|3|$ 3

32. $|9|$ 9

33. $|-10|$ 10

34. $|-2|$ 2

35. $|0|$ 0

36. $\left|-\dfrac{1}{2}\right|$ $\dfrac{1}{2}$

37. $-|-18|$ −18

38. $-|-5|$ −5

39. $-|32|$ −32

40. $-|20|$ −20

Find the opposite of each number. See Examples 4 and 5.

41. 7   −7

42. 1   −1

43. −14   14

44. −5   5

45. $\dfrac{2}{3}$ $-\dfrac{2}{3}$

46. 0   0

47. −8.3   8.3

48. 0.2   −0.2

49. $-\dfrac{1}{6}$ $\dfrac{1}{6}$

50. $-\dfrac{3}{10}$ $\dfrac{3}{10}$

Write true *or* false *for each statement.*

51. $|-5| > 0$

true

52. $|-12| > |-15|$

false

53. $0 < -(-6)$

true

54. $-9 < -(-9)$

true

55. $-|-4| < -|-7|$

false

56. $-|-0| > 0$

false

9.2 ▶▶▶ Adding and Subtracting Signed Numbers

You can show a *positive* number on a number line by drawing an arrow pointing to the *right*. On the number lines below, both arrows represent positive 4 units.

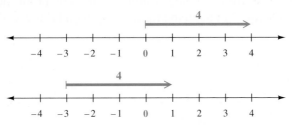

Draw arrows pointing to the *left* to show *negative* numbers. Both of the arrows below represent −3 units.

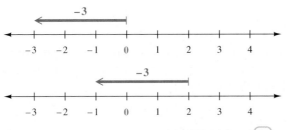

Work Problem **1** *at the Side.* ▶

OBJECTIVE **1** **Add signed numbers by using a number line.**   You can use a number line to add signed numbers. For example, this number line shows how to add 2 and 3.

Add 2 and 3 by starting at 0 and drawing an arrow 2 units to the right. From the end of this arrow, draw another arrow 3 units to the right. This second arrow ends at 5, showing that the sum of 2 + 3 is 5.

$$2 + 3 = 5$$

EXAMPLE 1 **Adding Signed Numbers Using a Number Line**

Add using a number line.

(a) 4 + (−1)

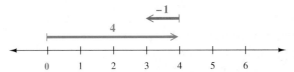

Start at 0 and draw an arrow 4 units to the right. From the end of this arrow, draw an arrow 1 unit to the *left*. (Remember to go to the left for a negative number.) This second arrow ends at 3, so

$$4 + (−1) = 3$$

CAUTION
Always start at 0 when adding on the number line.

Continued on Next Page

OBJECTIVES

1 Add signed numbers by using a number line.

2 Add signed numbers without using a number line.

3 Find the additive inverse of a number.

4 Subtract signed numbers.

5 Add or subtract a series of signed numbers.

1 Complete each arrow so it represents the indicated number of units.

(a)
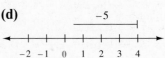

(b)

(c)

(d)

ANSWERS

1. (a)
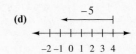

(b)

(c)

(d)

2 Draw arrows to find each sum.

(a) $3 + (-2)$

(b) $-4 + 1$

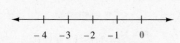

(c) $-3 + 7$

(d) $-1 + (-4)$

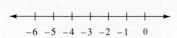

ANSWERS

2. (a) $3 + (-2) = 1$

(b) $-4 + 1 = -3$

(c) $-3 + 7 = 4$

(d) $-1 + (-4) = -5$

(b) $-6 + 2$

Always start at 0.

Draw an arrow from 0 going 6 units to the left. From the end of this arrow, draw an arrow 2 units to the right. This second arrow ends at -4, so

$$-6 + 2 = -4$$

(c) $-3 + (-5)$

As the arrows along the number line show,

$$-3 + (-5) = -8$$

> **Note**
>
> When a negative number *follows* an operation sign, we usually write parentheses around the negative number to prevent confusion. You can see this in Example 1(c) above.
>
> $$-3 + (-5) = -8$$
>
> Operation sign ⎯ Parentheses around the negative number that follows

◀ *Work Problem* **2** *at the Side.*

OBJECTIVE **2** **Add signed numbers without using a number line.**
After working with number lines for a while, you will see ways to add signed numbers without drawing arrows. You already know how to add two positive numbers (from **Chapter 1**). Here are the steps for adding two negative numbers.

> **Adding Two Negative Numbers**
>
> **Step 1** Add the absolute values of the numbers.
>
> **Step 2** Write a negative sign in front of the sum.

EXAMPLE 2 **Adding Two Negative Numbers**

Add (without using number lines).

(a) $-4 + (-12)$

The absolute value of -4 is **4.**

The absolute value of -12 is **12.**

Add the absolute values.

$$4 + 12 = 16$$

Write a negative sign in front of the sum.

Write a negative sign in front of 16

$$-4 + (-12) = -16$$

(b) $-5 + (-25) = -30$ ← Sum of absolute values, with a negative sign written in front of 30

(c) $-11 + (-46) = -57$

Continued on Next Page

(d) $-\dfrac{3}{4} + \left(-\dfrac{1}{2}\right)$

The absolute value of $-\frac{3}{4}$ is $\frac{3}{4}$, and the absolute value of $-\frac{1}{2}$ is $\frac{1}{2}$. Add the absolute values. Check that the sum is in lowest terms.

The common denominator for 4 and 2 is 4. $\qquad \dfrac{3}{4} + \dfrac{1}{2} = \dfrac{3}{4} + \dfrac{2}{4} = \dfrac{5}{4}$ ← Lowest terms

Write a negative sign in front of the sum.

$$-\dfrac{3}{4} + \left(-\dfrac{1}{2}\right) = -\dfrac{5}{4} \quad \text{Write a negative sign.}$$

> **Note**
>
> In algebra, we always write fractions in lowest terms, but usually do *not* change improper fractions to mixed numbers, because improper fractions are easier to work with. In Example 2(d) above, we checked that $-\frac{5}{4}$ was in lowest terms but did *not* rewrite it as $-1\frac{1}{4}$.

——————— *Work Problem* **3** *at the Side.* ▶

Use the following steps to add two numbers with *different* signs.

> **Adding Two Numbers with Different Signs**
>
> **Step 1** **Subtract** the lesser absolute value from the greater absolute value.
>
> **Step 2** Write the sign of the number with the **greater** absolute value in front of the answer.

EXAMPLE 3 **Adding Two Numbers with Different Signs**

Find each sum.

(a) $8 + (-3)$

First find this sum with a number line.

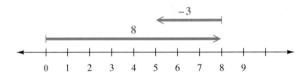

Because the second arrow ends at 5,

$$8 + (-3) = 5$$

Now find the sum by using the above rule. First, find the absolute value of each number.

$$|8| = 8 \quad \text{and} \quad |-3| = 3$$

Subtract the lesser absolute value from the greater absolute value.

$$8 - 3 = 5$$

The *positive* number 8 has the greater absolute value, so the answer is *positive*.

$$8 + (-3) = 5 \leftarrow \text{Positive answer}$$

Continued on Next Page

3 Add.

(a) $-4 + (-4)$

(b) $-3 + (-20)$

(c) $-31 + (-5)$

(d) $-10 + (-8)$

(e) $-\dfrac{9}{10} + \left(-\dfrac{3}{5}\right)$

ANSWERS

3. **(a)** -8 **(b)** -23 **(c)** -36
   (d) -18 **(e)** $-\dfrac{3}{2}$

4 Add.

(a) $10 + (-2)$

(b) $-7 + 8$

(c) $-11 + 11$

(d) $23 + (-32)$

(e) $-\dfrac{7}{8} + \dfrac{1}{4}$

(b) $4 + (-12)$

First, find the absolute values.

$$|4| = 4 \quad \text{and} \quad |-12| = 12$$

Subtract the lesser absolute value from the greater absolute value.

$$12 - 4 = 8$$

The *negative* number -12 has the greater absolute value, so the answer is *negative*.

$$4 + (-12) = -8$$

Write a negative sign in front of the answer because -12 has the greater absolute value.

(c) $-21 + 15 = -6$

Write a negative sign in front of the answer because -21 has the greater absolute value.

(d) $13 + (-9) = 4 \leftarrow$ Positive answer because the positive number 13 has the greater absolute value

(e) $-\dfrac{1}{2} + \dfrac{2}{3}$

The absolute value of $-\frac{1}{2}$ is $\frac{1}{2}$, and the absolute value of $\frac{2}{3}$ is $\frac{2}{3}$. Subtract the lesser absolute value from the greater absolute value.

The common denominator for 3 and 2 is 6.

$$\frac{2}{3} - \frac{1}{2} = \frac{4}{6} - \frac{3}{6} = \frac{1}{6}$$

Because the *positive* number $\frac{2}{3}$ has the greater absolute value, the answer is *positive*.

$$-\frac{1}{2} + \frac{2}{3} = \frac{1}{6} \leftarrow \text{Positive answer}$$

◀ *Work Problem* **4** *at the Side.*

OBJECTIVE **3** **Find the additive inverse of a number.** Recall that the opposite of 9 is -9, and the opposite of -4 is $-(-4)$, or 4. Adding opposites gives the results shown below.

$$9 + (-9) = 0 \quad \text{and} \quad -4 + 4 = 0$$

The sum of a number and its opposite is always 0. For this reason, opposites are also called *additive inverses* of each other.

> **Additive Inverse**
>
> The opposite of a number is called its **additive inverse.**
> The sum of a number and its opposite is 0.

EXAMPLE 4   **Finding Additive Inverses**

This table shows you several numbers and the additive inverse of each.

Number	Additive Inverse	Sum of Number and Inverse
6	-6	$6 + (-6) = 0$
-8	$-(-8)$ or 8	$(-8) + 8 = 0$
4	-4	$4 + (-4) = 0$
-3	$-(-3)$ or 3	$-3 + 3 = 0$
$\dfrac{5}{8}$	$-\dfrac{5}{8}$	$\dfrac{5}{8} + \left(-\dfrac{5}{8}\right) = 0$
0	0	$0 + 0 = 0$

Work Problem **5** *at the Side.* ▶

OBJECTIVE **4** **Subtract signed numbers.**   You may have noticed that negative numbers are often written with parentheses, like (-8). This is especially helpful when subtracting because the $-$ sign is used both to indicate a **negative number** and to indicate **subtraction**.

Example	How to Say It
(-5)	**negative five**
8	positive eight
$-3 - 2$	**negative three minus** positive two
$6 - (-4)$	positive six **minus negative four**
$-7 - (-1)$	**negative seven minus negative one**
$8 - 3$	positive eight **minus** positive three

CAUTION
Be sure you understand when the "$-$" sign means "subtract," and when it means "negative number."

Work Problem **6** *at the Side.* ▶

When working with signed numbers, it is helpful to write a subtraction problem as an addition problem. For example, you know that $6 - 4 = 2$. But you get the same result by *adding* 6 and the *opposite* of 4, that is, $6 + (-4)$.

$$6 - 4 = 2$$
$$6 + (-4) = 2$$
Same result

This suggests the following definition of subtraction.

Defining Subtraction
The difference of two numbers, a and b, is
$$a - b = a + (-b)$$
In other words, subtract two numbers by adding the first number and the opposite (additive inverse) of the second number.

---

5 Give the additive inverse of each number. Then find the sum of the number and its inverse.

(a) 12

(b) -9

(c) 3.5

(d) $-\dfrac{7}{10}$

(e) 0

6 Write each example in words.

(a) $-7 - 2$

(b) -10

(c) $3 - (-5)$

(d) 4

(e) $-8 - (-6)$

(f) $2 - 9$

ANSWERS
5. **(a)** $-12; 12 + (-12) = 0$
   (b) $9; -9 + 9 = 0$
   (c) $-3.5; 3.5 + (-3.5) = 0$
   (d) $\dfrac{7}{10}; -\dfrac{7}{10} + \dfrac{7}{10} = 0$
   (e) $0; 0 + 0 = 0$
6. **(a)** negative seven minus positive two
   (b) negative ten
   (c) positive three minus negative five
   (d) positive four
   (e) negative eight minus negative six
   (f) positive two minus positive nine

7 Subtract.

(a) −6 − 5

(b) 3 − (−10)

(c) −8 − (−2)

(d) 4 − 9

(e) −7 − (−15)

(f) 100 − 300

Subtracting Signed Numbers

To subtract two signed numbers, add the opposite of the second number to the first number. Use these steps.

Step 1 Change the subtraction sign to addition.

Step 2 Change the sign of the second number to its opposite.

Step 3 Proceed as in addition.

Note

The pattern in a subtraction problem is shown below.

$$\begin{array}{ccccccc} \text{1st} & & \text{2nd} & & \text{1st} & & \text{opposite of} \\ \text{number} & - & \text{number} & = & \text{number} & + & \text{2nd number} \end{array}$$

EXAMPLE 5 **Subtracting Signed Numbers**

Subtract.

(a) 8 − 11

The first number, 8, stays the same. Change the subtraction sign to addition. Change the sign of the second number to its opposite.

$$8 - 11$$
$$\downarrow \quad \downarrow$$
$$8 + (-11)$$

Positive 8 stays the same. Positive 11 is changed to its opposite, −11

Subtraction is changed to addition.

Now add.

$$8 + (-11) = -3$$
$$\text{So} \quad 8 - 11 = -3 \quad \text{also.}$$

(b) $-9 - 15$
$$\downarrow \quad \downarrow$$
$$-9 + (-15) = -24$$

Subtraction is changed to addition.
Positive 15 is changed to its opposite, (−15).

(c) $-5 - (-7)$
$$\downarrow \quad \downarrow$$
$$-5 + (+7) = 2$$

Subtraction is changed to addition.
Negative 7 is changed to its opposite, (+7).

(d) $10 - (-6)$
$$\downarrow \quad \downarrow$$
$$10 + (+6) = 16$$

Subtraction is changed to addition.
Negative 6 is changed to its opposite, (+6).

(e) $-25 - (-20)$
$$\downarrow \quad \downarrow$$
$$-25 + (+20) = -5$$

Subtraction is changed to addition.
Negative 20 is changed to its opposite, (+20).

◀ Work Problem **7** at the Side.

EXAMPLE 6	**Subtracting Signed Decimal and Fraction Numbers**

Subtract.

(a) $7.6 - (-8.3)$ Subtraction is changed to addition.
Negative 8.3 is changed to its opposite, $(+8.3)$.

$7.6 + (+8.3) = 15.9$

Once the subtraction is changed to addition, notice that both numbers are positive. You can add them as you did in **Chapter 4.**

Line up decimal points.

$$\begin{array}{r} 7.6 \\ +\ 8.3 \\ \hline 15.9 \end{array}$$

(b) $-0.7 -\ 16.5$ Subtraction is changed to addition.
Positive 16.5 is changed to its opposite, (-16.5).

$-0.7 + (-16.5) = -17.2$

To add two *negative* numbers, first add the absolute values.

$|-0.7|$ is 0.7
$|-16.5|$ is 16.5

$$\begin{array}{r} 16.5 \\ +\ 0.7 \\ \hline 17.2 \end{array}$$

Then write a *negative* sign in front of 17.2
$$-0.7 + -16.5 = -17.2$$

(c) $\dfrac{5}{8} - \dfrac{3}{4}$ Subtraction is changed to addition.
Positive $\frac{3}{4}$ is changed to its opposite, $\left(-\frac{3}{4}\right)$.

$$\dfrac{5}{8} + \left(-\dfrac{3}{4}\right) = \dfrac{5}{8} + \left(-\dfrac{6}{8}\right) = -\dfrac{1}{8}$$

Once the subtraction is changed to addition, notice that the numbers have *different* signs. So, we subtract the lesser absolute value from the greater absolute value.

$\left|\dfrac{5}{8}\right|$ is $\dfrac{5}{8}$ and $\left|-\dfrac{6}{8}\right|$ is $\dfrac{6}{8}$ Then $\dfrac{6}{8} - \dfrac{5}{8} = \dfrac{1}{8}$

Lesser absolute value   Greater absolute value

The *negative* number $-\frac{6}{8}$ has the greater absolute value, so the answer is *negative*.

(d) $-\dfrac{1}{4} - \left(-\dfrac{5}{6}\right)$ Subtraction is changed to addition.
Negative $\frac{5}{6}$ is changed to its opposite, $\left(+\frac{5}{6}\right)$.

$$-\dfrac{3}{12} + \left(+\dfrac{5}{6}\right) = -\dfrac{3}{12} + \left(+\dfrac{10}{12}\right) = \dfrac{7}{12}$$

Once the subtraction is changed to addition, notice that the numbers have *different* signs. So we subtract the lesser absolute value from the greater absolute value.

$\left|-\dfrac{3}{12}\right|$ is $\dfrac{3}{12}$ and $\left|+\dfrac{10}{12}\right|$ is $\dfrac{10}{12}$ Then $\dfrac{10}{12} - \dfrac{3}{12}$ is $\dfrac{7}{12}$

The *positive* number $\frac{10}{12}$ has the greater absolute value so the answer is *positive*.

Work Problem **8** *at the Side.* ▶

8 Subtract.

(a) $-0.8 - 0.5$

(b) $4 - 12.03$

(c) $-1.65 - (-20.4)$

(d) $\dfrac{1}{2} - \left(-\dfrac{2}{3}\right)$

(e) $\dfrac{1}{10} - \dfrac{4}{5}$

(f) $-\dfrac{7}{9} - \left(-\dfrac{1}{6}\right)$

ANSWERS

8. (a) -1.3 **(b)** -8.03 **(c)** 18.75
(d) $\dfrac{7}{6}$ **(e)** $-\dfrac{7}{10}$ **(f)** $-\dfrac{11}{18}$

9 Perform the addition and subtraction from left to right.

(a) $6 - 7 + (-3)$

(b) $-2 + (-3) - (-5)$

(c) $-3 - (-9) - (-5)$

(d) $8 - (-2) + (-6)$

(e) $-1 - 2 + 3 - 4$

(f) $7 - 6 - 5 + (-4)$

(g) $-6 + (-15) - (-19)$ $+ (-25)$

(h) -19.2
-6.7
15.8
17.1
-5.4

OBJECTIVE 5 Add or subtract a series of signed numbers. If a problem involves both addition and subtraction, use the order of operations, from **Section 1.8.**

EXAMPLE 7 **Combining Addition and Subtraction of Signed Numbers**

According to the last step in the order of operations, perform addition and subtraction from left to right.

(a) $\underbrace{-6 + (-11)}\ - \quad 5$

$\quad\quad -17 \quad - \quad 5$ Change subtraction to addition; change positive 5 to its opposite, (-5).

$\quad\quad \underbrace{-17 \quad + \quad (-5)}$

$\quad\quad\quad\quad -22$

(b) $4 - (-3) + (-9)$ Change subtraction to addition; change -3 to its opposite, $(+3)$.

$\quad 4 + (+3) + (-9)$

$\quad \underbrace{7 \quad\quad + (-9)}$

$\quad\quad\quad -2$

(c) $\underbrace{14 + (-9)}\ - (-8) + 10$

$\quad\quad 5 \quad - (-8) + 10$ Change subtraction to addition; change -8 to its opposite, $(+8)$.

$\quad\quad \underbrace{5 \quad + (+8)}\ + 10$

$\quad\quad\quad \underbrace{13 \quad\quad + 10}$

$\quad\quad\quad\quad 23$

(d) $\quad -6.3$
$\quad -14.9$
$\quad 8.5$
$\quad -7.4$
$\quad \underline{5.2}$

Start at the top.

$\begin{array}{llll} -6.3 \rbrace \\ -14.9 \end{array} \rightarrow \begin{array}{l} -21.2 \rbrace \\ 8.5 \end{array} \rightarrow \begin{array}{l} -12.7 \rbrace \\ -7.4 \end{array} \rightarrow \begin{array}{l} -20.1 \\ \underline{5.2} \\ -14.9 \end{array}$

◀ *Work Problem* **9** *at the Side.*

Add by using the number line. See Example 1.

1. $-2 + 5$ **3**

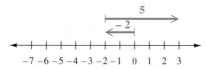

2. $-3 + 4$ **1**

3. $-5 + (-2)$ **−7**

4. $-2 + (-2)$ **−4**

5. $3 + (-4)$ **−1**

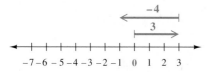

6. $5 + (-1)$ **4**

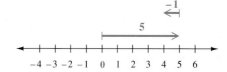

Add. See Examples 2 and 3.

7. $-8 + 5$ **−3**   **8.** $-3 + 2$ **−1**   **9.** $-1 + 8$ **7**   **10.** $-4 + 10$ **6**

🌐 **11.** $-2 + (-5)$ **−7**   **12.** $-7 + (-3)$ **−10**   **13.** $6 + (-5)$ **1**   **14.** $11 + (-3)$ **8**

15. $4 + (-12)$ **−8**   **16.** $9 + (-10)$ **−1**   **17.** $-10 + (-10)$ **−20**   **18.** $-5 + (-20)$ **−25**

Write and solve an addition problem for each situation.

19. The football team gained 13 yd on the first play and lost 17 yd on the second play.

$13 + (-17) = -4$ yd

20. At penguin breeding grounds on Antarctic islands, winter temperatures routinely drop to $-15\,°C$; but at the interior of the continent, temperatures may easily drop another $60°$ below that.

$-15 + (-60) = -75\,°C$

21. Nicole's checking account was overdrawn by $52.50. She deposited $50 in the account.

$-\$52.50 + \$50 = -\$2.50$

22. $48.40 was stolen from Jay's car. He got $30 of it back.

$-\$48.40 + \$30 = -\$18.40$

23. Use the score sheet to find each player's point total after three rounds in a card game.

	Jeff	Terry
Round 1	Lost 20 pts	Won 42 pts
Round 2	Won 75 pts	Lost 15 pts
Round 3	Lost 55 pts	Won 20 pts

Jeff: $-20 + 75 + (-55) = 0$ pts
Terry: $42 + (-15) + 20 = 47$ pts

24. Use the information in the table on flood water depths to find the new flood level for each river.

	Red River	Mississippi
Monday	Rose 8 ft	Rose 4 ft
Tuesday	Fell 3 ft	Rose 7 ft
Wednesday	Fell 5 ft	Fell 13 ft

Red River: $8 + (-3) + (-5) = 0$ ft
Mississippi: $4 + 7 + (-13) = -2$ ft

Add.

25. $7.8 + (-14.6)$ -6.8 **26.** $4.9 + (-8.1)$ -3.2 **27.** $-\dfrac{1}{2} + \dfrac{3}{4}$ $\dfrac{1}{4}$ **28.** $-\dfrac{2}{3} + \dfrac{5}{6}$ $\dfrac{1}{6}$

29. $-\dfrac{7}{10} + \dfrac{2}{5}$ $-\dfrac{3}{10}$ **30.** $-\dfrac{3}{4} + \dfrac{3}{8}$ $-\dfrac{3}{8}$ **31.** $-\dfrac{7}{3} + \left(-\dfrac{5}{9}\right)$ $-\dfrac{26}{9}$ **32.** $-\dfrac{8}{5} + \left(-\dfrac{3}{10}\right)$ $-\dfrac{19}{10}$

Give the additive inverse of each number. See Example 4.

33. 3 -3 **34.** 4 -4 **35.** -9 9 **36.** -14 14

37. $\dfrac{1}{2}$ $-\dfrac{1}{2}$ **38.** $\dfrac{7}{8}$ $-\dfrac{7}{8}$ **39.** -6.2 6.2 **40.** -0.5 0.5

Subtract by changing subtraction to addition. See Examples 5 and 6.

41. $19 - 5$ 14 **42.** $24 - 11$ 13 **43.** $10 - 12$ -2 **44.** $1 - 8$ -7

45. $7 - 19$ -12 **46.** $2 - 17$ -15 ⊕ **47.** $-15 - 10$ -25 **48.** $-10 - 4$ -14

49. $-9 - 14$ -23 **50.** $-3 - 11$ -14 **51.** $-3 - (-8)$ 5 **52.** $-1 - (-4)$ 3

53. $6 - (-14)$ 20 **54.** $8 - (-1)$ 9 **55.** $1 - (-10)$ 11 **56.** $6 - (-1)$ 7

57. $-30 - 30$ -60 **58.** $-25 - 25$ -50 **59.** $-16 - (-16)$ 0 **60.** $-20 - (-20)$ 0

61. $-\dfrac{7}{10} - \dfrac{4}{5}$ $-\dfrac{3}{2}$ **62.** $-\dfrac{8}{15} - \dfrac{3}{10}$ $-\dfrac{5}{6}$ **63.** $\dfrac{1}{2} - \dfrac{9}{10}$ $-\dfrac{2}{5}$

64. $\dfrac{2}{3} - \dfrac{11}{12}$ $-\dfrac{1}{4}$ **65.** $-8.3 - (-9)$ 0.7 **66.** $-2 - (-3.9)$ 1.9

67. Explain and correct the mistakes in these subtraction problems.

(a) $-6 - 6$
$\downarrow \quad \downarrow$
$-6 + 6 = 0$

(b) $-9 - 5$
$\downarrow \quad \downarrow$
$9 + (-5) = 4$

(a) Change 6 to its opposite, −6.
Correct answer is $-6 + (-6) = -12$.

(b) Do not change 9 to its opposite.
Correct answer is $-9 + (-5) = -14$.

68. Explain the purpose of the "−" sign in each of these examples.

(a) $6 - 9$ **(b)** (-9) **(c)** $-(-2)$

(a) 6 *minus* 9
(b) *negative* 9
(c) the *opposite* of *negative* 2

The windchill table below shows how wind increases a person's heat loss in cold weather. For example, suppose the temperature is 15 °F (along the top of the table) and the wind speed is 20 mph (along the left side of the table). This column and row intersect at −2 °F, the "windchill" temperature. The actual temperature is 15 °F but the wind makes it feel like −2 °F. Use the table to find the windchill temperature under each set of conditions in Exercises 69–72. Then write and solve a subtraction problem to calculate actual temperature minus windchill temperature.

WINDCHILL
Temperature (degrees Fahrenheit)

Calm	40	35	30	25	20	15	10	5	0	−5	−10	−15	−20	−25	−30
5	36	31	25	19	13	7	1	−5	−11	−16	−22	−28	−34	−40	−46
10	34	27	21	15	9	3	−4	−10	−16	−22	−28	−35	−41	−47	−53
15	32	25	19	13	6	0	−7	−13	−19	−26	−32	−39	−45	−51	−58
20	30	24	17	11	4	−2	−9	−15	−22	−29	−35	−42	−48	−55	−61
25	29	23	16	9	3	−4	−11	−17	−24	−31	−37	−44	−51	−58	−64
30	28	22	15	8	1	−5	−12	−19	−26	−33	−39	−46	−53	−60	−67
35	28	21	14	7	0	−7	−14	−21	−27	−34	−41	−48	−55	−62	−69
40	27	20	13	6	−1	−8	−15	−22	−29	−36	−43	−50	−57	−64	−71

Wind Speed (miles per hour)

Shaded area: Frostbite occurs in 15 minutes or less.

Source: National Weather Service

69. (a) 30 °F; 10 mph wind

21 °F; 30 − 21 = 9 degrees difference

(b) 15 °F; 15 mph wind

0 °F; 15 − 0 = 15 degrees difference

70. (a) 40 °F; 20 mph wind

30 °F; 40 − 30 = 10 degrees difference

(b) 20 °F; 35 mph wind

0 °F; 20 − 0 = 20 degrees difference

71. (a) 5 °F; 25 mph wind

−17 °F; 5 − (−17) = 22 degrees difference

(b) −10 °F; 35 mph wind

−41 °F; −10 − (−41) = 31 degrees difference

72. (a) 10 °F; 15 mph wind

−7 °F; 10 − (−7) = 17 degrees difference

(b) −5 °F; 30 mph wind

−33 °F; −5 − (−33) = 28 degrees difference

Follow the order of operations to work each problem. See Example 7.

73. $-2 + (-11) - (-3)$
−10

74. $-5 - (-2) + (-6)$
−9

75. $4 - (-13) + (-5)$
12

76. $6 - (-1) + (-10)$
−3

77. $-12 - (-3) - (-2)$
−7

78. $-1 - (-7) - (-4)$
10

79. $4 - (-4) - 3$
5

80. $5 - (-2) - 8$
−1

81. $\dfrac{1}{2} - \dfrac{2}{3} + \left(-\dfrac{5}{6}\right)$
−1

82. $\dfrac{2}{5} - \dfrac{7}{10} + \left(-\dfrac{3}{2}\right)$
$-\dfrac{9}{5}$

83. $-5.7 - (-9.4) - 8.1$
−4.4

84. $-6.5 - (-11.2) - 1.4$
3.3

85. $-2 + (-11) + |-2|$ -11

86. $|-7 + 2| + (-2) + 4$ 7
(*Hint:* Work inside the absolute value bars first.)

87. $-3 - (-2 + 4) + (-5)$ -10
(*Hint:* Work inside parentheses first.)

88. $5 - 8 - (6 - 7) + 1$ -1

Find the balance in each checking account.

89. LaVerle had $37 in her checking account. She wrote a $689 check for tuition, deposited a $908 paycheck, and got $60 in cash at an ATM machine.

$196

90. Yvonne had $478 in her checking account. She deposited a $212 tax refund and wrote an $89 check for electricity and a $605 check for rent.

$-$4

91. Rod's account was overdrawn by $89.62. He wrote checks for $110.70 and $99.68 before depositing three $100 bills into his account.

$0

92. Dwayne's checking account was overdrawn by $23.77, so the bank charged a $16 fee. He deposited his $583.29 paycheck before withdrawing $50 in cash.

$493.52

Relating Concepts (Exercises 93–96) For Individual or Group Work

✒ *Use your knowledge of addition and subtraction of signed numbers to **work Exercises 93–96 in order.***

93. Work each pair of examples.

  (a) $-5 + 3$ = $\underline{-2}$ $3 + (-5) =$ $\underline{-2}$

  (b) $-2 + (-6) =$ $\underline{-8}$ $-6 + (-2) =$ $\underline{-8}$

  (c) $17 + (-7) =$ $\underline{10}$ $-7 + 17$ = $\underline{10}$

Explain why the answers to each pair are the same.

Addition is commutative, so changing the order of the addends does *not* change the sum.

94. Work each pair of examples

  (a) $-3 - 5$ = $\underline{-8}$ $5 - (-3) =$ $\underline{8}$

  (b) $-4 - (-6) =$ $\underline{2}$ $-6 - (-4) =$ $\underline{-2}$

  (c) $3 - 10$ = $\underline{-7}$ $10 - 3$ = $\underline{7}$

Explain what happens when you try to apply the commutative property to subtraction.

Changing the order of the numbers in subtraction *does change* the result.

95. Look back at your answers in Exercises 94.

  (a) Describe how the two answers for each pair are similar, and how they are different.

The answers have the same absolute value but opposite signs.

  (b) Write a rule that explains what happens when you switch the order of the numbers in a subtraction problem.

When the order of the numbers in a subtraction problem is switched, change the sign on the answer to its opposite.

96. Work each set of exercises to see how 0 functions in addition and subtraction.

  (a) Describe the pattern in these addition answers.

    $-18 + 0$ = $\underline{-18}$ $20 + 0 =$ $\underline{20}$

    $0 + (-5) =$ $\underline{-5}$ $0 + 4 =$ $\underline{4}$

Adding 0 to any number leaves the number unchanged.

  (b) Describe the pattern in these subtraction answers.

    $2 - 0 =$ $\underline{2}$ $0 - 10$ = $\underline{-10}$

    $-3 - 0 =$ $\underline{-3}$ $0 - (-7) =$ $\underline{7}$

Subtracting 0 from a number leaves the number unchanged, but subtracting a *number from 0* changes the sign of the number.

9.3 ▶▶▶ Multiplying and Dividing Signed Numbers

In mathematics, the rules or patterns must be consistent. We can use this idea to see how to multiply two numbers with *different* signs. Look for a pattern in this list of products.

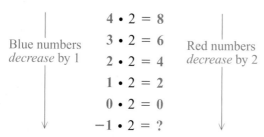

Blue numbers *decrease* by 1

$$4 \cdot 2 = 8$$
$$3 \cdot 2 = 6$$
$$2 \cdot 2 = 4$$
$$1 \cdot 2 = 2$$
$$0 \cdot 2 = 0$$
$$-1 \cdot 2 = \,?$$

Red numbers *decrease* by 2

As the numbers in **blue** decrease by 1, the numbers in **red** decrease by 2. To continue the red pattern, replace **?** with a number 2 *less than* 0, which is -2. We get this result.

$$-1 \cdot 2 = -2$$

OBJECTIVE 1 Multiply or divide two numbers with opposite signs.   The pattern above suggests a rule for multiplying two numbers with different signs.

> **Multiplying Two Numbers with Different Signs**
> The product of two numbers with *different* signs is *negative.*

EXAMPLE 1 **Multiplying Numbers with Different Signs**

Multiply.

(a) $-8 \cdot 4 = -32$ Factors have *different* signs, so the product is *negative.*
    Negative   Positive

(b) $6(-3) = -18$ Factors have *different* signs, so the product is *negative.*
    Negative   Positive

(c) $(-5)(11) = -55$

(d) $12(-7) = -84$

Work Problem **1** *at the Side.* ▶

For two numbers with the *same* sign, look at this pattern.

Blue numbers *decrease* by 1

$$4 \cdot (-2) = -8$$
$$3 \cdot (-2) = -6$$
$$2 \cdot (-2) = -4$$
$$1 \cdot (-2) = -2$$
$$0 \cdot (-2) = 0$$
$$-1 \cdot (-2) = \,?$$

Red numbers *increase* by 2

This time, as the numbers in **blue** *decrease* by 1, the products *increase* by 2. To continue the **red** pattern, replace **?** with a number 2 *greater than* 0, which is **positive 2**. Therefore, we get this result.

$$-1 \cdot (-2) = 2$$

OBJECTIVE 2 Multiply or divide two numbers with the same sign.   In the pattern above, a negative number times a negative number gave a positive result.

1 Multiply.

(a) $5 \cdot (-4)$

(b) $(-9)(15)$

(c) $12(-1)$

(d) $-6(6)$

(e) $\left(-\dfrac{7}{8}\right)\left(\dfrac{4}{3}\right)$

ANSWERS

1. (a) -20 **(b)** -135 **(c)** -12 **(d)** -36
    (e) $-\dfrac{7}{6}$

2 Multiply.

(a) $(-5)(-5)$

(b) $(-14)(-1)$

(c) $-7(-8)$

(d) $3(12)$

(e) $\left(-\dfrac{2}{3}\right)\left(-\dfrac{6}{5}\right)$

3 Divide.

(a) $\dfrac{-20}{4}$

(b) $\dfrac{-50}{-5}$

(c) $\dfrac{44}{2}$

(d) $\dfrac{6}{-6}$

(e) $\dfrac{-15}{-1}$

(f) $\dfrac{-\frac{3}{5}}{\frac{9}{10}}$

(g) $\dfrac{-35}{0}$

Multiplying Two Numbers with the Same Sign

The product of two numbers with the *same* sign is *positive.*

EXAMPLE 2 **Multiplying Two Numbers with the Same Sign**

Multiply.

(a) $(-9)(-2)$ The factors have the same sign (both are negative).

$(-9)(-2) = 18 \leftarrow$ The product is positive.

(b) $-7(-4) = 28$ (c) $(-6)(-2) = 12$

(d) $(-10)(-5) = 50$ (e) $7(5) = 35$

> When signs *match*, the product is *positive.*

◀ *Work Problem* **2** *at the Side.*

You can use the same sign rules for dividing signed numbers as you use for multiplying signed numbers.

Dividing Signed Numbers

When two nonzero numbers with *different* signs are divided, the result is *negative.*

When two nonzero numbers with the *same* sign are divided, the result is *positive.*

Division involving 0 works the same as it did for whole numbers (see **Section 1.5**). Division by 0 cannot be done; we say it is *undefined.* But 0 divided by any other number is 0.

EXAMPLE 3 **Dividing Signed Numbers**

Divide.

(a) $\dfrac{-15}{5}$ ← Numbers have *different* signs, so the quotient is negative.

$\dfrac{-15}{5} = -3$

(b) $\dfrac{-8}{-4}$ ← Numbers have the *same* sign (both negative), so the quotient is positive.

$\dfrac{-8}{-4} = 2$

(c) $\dfrac{-75}{-25} = 3$

(d) $\dfrac{-6}{0}$ is undefined.

> Division by 0 cannot be done. Write "undefined."

(e) $\dfrac{0}{-5} = 0$

> 0 can be divided by a non-zero number.

(f) $\dfrac{90}{-9} = -10$

(g) $\dfrac{-\frac{2}{3}}{-\frac{5}{9}} = \left(-\dfrac{2}{3}\right)\left(-\dfrac{9}{5}\right)$ Use the reciprocal of $-\frac{5}{9}$, which is $-\frac{9}{5}$

$= \left(-\dfrac{2}{\cancel{3}_{1}}\right)\left(-\dfrac{\cancel{9}^{3}}{5}\right)$ Divide out the common factor. Then multiply.

$= \dfrac{6}{5}$ Both numbers were negative, so the quotient is *positive.*

> When signs *match*, the quotient is *positive.*

◀ *Work Problem* **3** *at the Side.*

ANSWERS

2. (a) 25 (b) 14 (c) 56 (d) 36 (e) $\dfrac{4}{5}$

3. (a) -5 (b) 10 (c) 22 (d) -1 (e) 15

  (f) $-\dfrac{2}{3}$ (g) undefined

9.3 ▷▷▷ Exercises

Multiply. See Examples 1 and 2.

1. $-5 \cdot 7$
-35

2. $-10 \cdot 2$
-20

3. $(-5)(9)$
-45

4. $(-9)(4)$
-36

5. $3(-6)$
-18

6. $8(-6)$
-48

7. $10(-5)$
-50

8. $5(-11)$
-55

9. $(-1)(40)$
-40

10. $(75)(-1)$
-75

11. $-8(-4)$
32

12. $-3(-9)$
27

13. $11(7)$
77

14. $4(25)$
100

15. $-19(-7)$
133

16. $-21(-3)$
63

17. $-13(-1)$
13

18. $-1(-31)$
31

19. $(0)(-25)$
0

20. $(-50)(0)$
0

21. $-\dfrac{1}{2}(-8)$
4

22. $\dfrac{1}{3}(-15)$
-5

23. $-10\left(\dfrac{2}{5}\right)$
-4

24. $-25\left(-\dfrac{7}{10}\right)$
$\dfrac{35}{2}$

25. $\left(\dfrac{3}{5}\right)\left(-\dfrac{1}{6}\right)$
$-\dfrac{1}{10}$

26. $\left(-\dfrac{7}{9}\right)\left(-\dfrac{3}{4}\right)$
$\dfrac{7}{12}$

27. $-\dfrac{7}{5}\left(-\dfrac{10}{3}\right)$
$\dfrac{14}{3}$

28. $-\dfrac{9}{10}\left(\dfrac{5}{4}\right)$
$-\dfrac{9}{8}$

29. $-\dfrac{7}{15} \cdot \dfrac{25}{14}$
$-\dfrac{5}{6}$

30. $-\dfrac{5}{9} \cdot \dfrac{18}{25}$
$-\dfrac{2}{5}$

31. $-\dfrac{5}{2}\left(-\dfrac{7}{10}\right)$
$\dfrac{7}{4}$

32. $-\dfrac{8}{5}\left(-\dfrac{15}{16}\right)$
$\dfrac{3}{2}$

33. $9(-4.7)$

-42.3

34. $15(-6.3)$

-94.5

35. $(-0.5)(-12)$

6

36. $(-3.15)(-5)$

15.75

37. $(-6.2)(5.1)$

-31.62

38. $(-4.3)(9.7)$

-41.71

39. $-1.25(-3.6)$

4.5

40. $6.33(0.2)$

1.266

41. $(-8.23)(-1)$

8.23

42. $(-1)(-0.69)$

0.69

43. $0(-58.6)$

0

44. $-91.3(0)$

0

Divide. See Example 3.

45. $\dfrac{-14}{7}$

-2

46. $\dfrac{-8}{2}$

-4

47. $\dfrac{30}{-6}$

-5

48. $\dfrac{21}{-7}$

-3

49. $\dfrac{-28}{0}$

undefined

50. $\dfrac{-40}{0}$

undefined

51. $\dfrac{14}{-1}$

-14

52. $\dfrac{25}{-1}$

-25

53. $\dfrac{-20}{-2}$

10

54. $\dfrac{-80}{-4}$

20

55. $\dfrac{-48}{-12}$

4

56. $\dfrac{-30}{-15}$

2

57. $\dfrac{-18}{18}$

-1

58. $\dfrac{50}{-50}$

-1

59. $\dfrac{-573}{-3}$

191

60. $\dfrac{-580}{-5}$

116

61. $\dfrac{0}{-9}$

0

62. $\dfrac{0}{-4}$

0

63. $\dfrac{-30}{-30}$

1

64. $\dfrac{-25}{-25}$

1

65. $\dfrac{-\dfrac{5}{7}}{-\dfrac{15}{14}}$

$\dfrac{2}{3}$

66. $\dfrac{-\dfrac{3}{4}}{-\dfrac{9}{16}}$

$\dfrac{4}{3}$

67. $-\dfrac{2}{3} \div (-2)$

$\dfrac{1}{3}$

68. $-\dfrac{3}{4} \div (-9)$

$\dfrac{1}{12}$

69. $5 \div \left(-\dfrac{5}{8}\right)$

-8

70. $7 \div \left(-\dfrac{14}{15}\right)$

$-\dfrac{15}{2}$

71. $-\dfrac{7}{5} \div \dfrac{3}{10}$

$-\dfrac{14}{3}$

72. $-\dfrac{4}{9} \div \dfrac{8}{3}$

$-\dfrac{1}{6}$

73. $\dfrac{-18.92}{-4}$

4.73

74. $\dfrac{-22.75}{-7}$

3.25

75. $\dfrac{-7.05}{1.5}$

−4.7

76. $\dfrac{-17.02}{7.4}$

−2.3

77. $\dfrac{45.58}{-8.6}$

−5.3

78. $\dfrac{6.27}{-0.3}$

−20.9

Following the order of operations, work from left to right in each exercise.

79. $(-4)(-6)\dfrac{1}{2}$

12

80. $(-9)(-3)\dfrac{2}{3}$

18

81. $(-0.6)(-0.2)(-3)$

−0.36

82. $(-4)(-1.2)(-0.7)$

−3.36

83. $\left(-\dfrac{1}{2}\right)\left(\dfrac{2}{5}\right)\left(\dfrac{7}{8}\right)$

$-\dfrac{7}{40}$

84. $\left(\dfrac{3}{4}\right)\left(-\dfrac{5}{6}\right)\left(\dfrac{2}{3}\right)$

$-\dfrac{5}{12}$

85. $-36 \div (-2) \div (-3) \div (-3) \div (-1)$

−2

86. $-48 \div (-8) \cdot (-4) \div (-4) \div (-3)$

−2

87. $|-8| \div (-4) \cdot |-5|$

−10

88. $-6 \cdot |-3| \div |9| \cdot (-2)$

4

Solve each application problem. Be sure to indicate whether the answer is a positive or negative number.

89. A new computer-software store had losses of $9950 during each month of its first year. What was the total loss for the year?

−$119,400

90. A college's enrollment dropped by 3245 students over the last 11 years. What was the average drop in enrollment each year?

−295 students

91. The greatest ocean depth is 36,198 ft below sea level. If an unmanned research sub dives to that depth in 18 equal steps, how far does it dive in each step?

−2011 ft

92. Pat ate a dozen crackers as a snack. Each cracker had 17 calories. How many calories did Pat eat? (*Source:* Nabisco, Inc.)

204 calories

93. Tuition at the state university is $182 per credit for undergraduates. How much tuition will Wei Chen pay for 13 credits?

$2366

94. A long-distance phone company estimates that it is losing 95 customers each week. How many customers will it lose in a year?

−4940 customers

95. There is a 3-degree drop in temperature for every thousand feet that an airplane climbs into the sky. If the temperature on the ground is 50 degrees, what will be the temperature when the plane reaches an altitude of 24,000 ft? (*Source:* Lands' End.)

−22 degrees

96. An unmanned submarine descends to 150 ft below the surface of the ocean. Then it continues to go deeper, taking a water sample every 25 ft. What is its depth when it takes the 15th sample?

−525 ft

Relating Concepts (Exercises 97–98) For Individual or Group Work

Use your knowledge of multiplying and dividing signed numbers and look for patterns as you **work Exercises 97 and 98 in order.**

97. Write three numerical examples for each situation.

(a) A positive number multiplied by −1

Examples will vary. Some possibilities:
$6(-1) = -6; 2(-1) = -2; 15(-1) = -15$

(b) A negative number multiplied by −1

Examples will vary. Some possibilities:
$-6(-1) = 6; -2(-1) = 2; -15(-1) = 15$

Now write a rule that explains what happens when you multiply a signed number by −1.

The result of multiplying any nonzero number times −1 is the number with the opposite sign.

98. Write three numerical examples for each situation.

(a) A negative number divided by −1

Examples will vary. Some possibilities:
$\frac{-6}{-1} = 6; \frac{-2}{-1} = 2; \frac{-15}{-1} = 15$

(b) A positive number divided by −1

Examples will vary. Some possibilities:
$\frac{6}{-1} = -6; \frac{2}{-1} = -2; \frac{15}{-1} = -15$

(c) A negative number divided by itself

Examples will vary. Some possibilities:
$\frac{-6}{-6} = 1; \frac{-2}{-2} = 1; \frac{-15}{-15} = 1$

Now write a rule that explains what happens when you divide a signed number by −1. Write another rule for a negative number divided by itself.

When dividing by −1, the sign of the number changes to its opposite. A negative number divided by itself gives a result of 1.

9.4 ▶▶▶ Order of Operations

In the last two sections, you worked examples that mixed either addition and subtraction or multiplication and division. In those situations, you worked from left to right. Here are two more examples as a review.

Work additions and subtractions from left to right.

$$\underbrace{-8 - (-6)}_{-2} + (-11)$$
$$\underbrace{-2 \quad + (-11)}_{-13}$$

Work multiplications and divisions from left to right.

$$\underbrace{(-15) \div (-3)}_{\quad \searrow 5} (6) \qquad \text{Divide.}$$
$$\underbrace{5 (6)}_{30} \qquad \text{Multiply.}$$

Work Problem (**1**) *at the Side.* ▶

OBJECTIVE 1 Use the order of operations. Before working examples that include a mix of operations and/or parentheses, let's review the order of operations from **Section 1.8.**

> **Order of Operations**
> 1. Do all operations inside *parentheses* or *other grouping symbols.*
> 2. Simplify any expressions with *exponents* and find any *square roots.*
> 3. Do the remaining *multiplications and divisions* as they occur from left to right.
> 4. Do the remaining *additions and subtractions* as they occur from left to right.

EXAMPLE 1 **Using the Order of Operations**

Use the order of operations to simplify this expression.

$$4 - 10 \div 2 + 7 \qquad \text{Check for parentheses: none.}$$

Check for exponents and square roots: none.

Move from left to right, checking for multiplying and dividing.

Yes, here is dividing. Use the number on each side of the ÷ sign.

$$4 - \mathbf{10 \div 2} + 7$$
$$\downarrow\downarrow \quad \searrow \quad \downarrow\downarrow$$
$$4 - \quad 5 \quad + 7$$

$10 \div 2$ is 5. Bring down the other numbers and signs you haven't used.

Move from left to right, checking for adding and subtracting.

Yes, here is subtracting.

$$4 - \quad 5 \quad + 7$$
$$\downarrow \quad \searrow \quad \downarrow$$
$$4 + (-5) \quad + 7$$

Change subtraction to addition; change 5 to its opposite, (-5).

Add $4 + (-5)$ to get -1

$$\underbrace{4 + (-5)}_{-1} + 7$$
$$\underbrace{-1 \quad + 7}_{6}$$

Add $-1 + 7$ to get 6

Work Problem (**2**) *at the Side.* ▶

OBJECTIVES

1 Use the order of operations.

2 Use the order of operations with exponents.

3 Use the order of operations with fraction bars.

1 Simplify.

(a) $-9 + (-15) + (-3)$

(b) $-8 - (-2) + (-6)$

(c) $-2 - (-7) - (-4)$

(d) $3(-4) \div (-6)$

(e) $-18 \div 9(-4)$

2 Use the order of operations to simplify each expression.

(a) $10 + 8 \div 2$ Divide first!

(b) $4 - 6(-2)$ Multiply first!

(c) $-3 + (-5) \cdot 2 - 1$

(d) $-6 \div 2 + 3(-2)$

(e) $7 - 6(2) \div (-3)$

ANSWERS
1. **(a)** -27 **(b)** -12 **(c)** 9 **(d)** 2 **(e)** 8
2. **(a)** 14 **(b)** 16 **(c)** -14 **(d)** -9
   (e) 11

3 Simplify.

(a) $2 + 40 \div (-5 + 3)$

(b) $5 - 3(2 + 4)$

(c) $(-24 \div 2) + (15 - 3)$

(d) $-3(2 - 8) - 5(4 - 3)$

(e) $3(3) - 10(3) \div 5$

(f) $6 - (2 + 7) \div (-4 + 1)$

EXAMPLE 2 **Parentheses and the Order of Operations**

Use the order of operations to simplify each expression.

(a) $-8(7 - 5) - 9$ Work inside parentheses first.
Bring down the other numbers and signs you haven't used yet.

$-8(2) \quad - 9$ Check for exponents and square roots: none.

Move from left to right, looking for multiplying and dividing. Here is multiplying.

$-8(2) \quad - 9$

$-16 \quad - 9$ Move from left to right, doing any adding and subtracting. Change subtraction to addition.

$-16 \quad + (-9)$ Add $-16 + (-9)$ to get -25

-25

> $3 + 2$ is **not** done first.

(b) $3 + 2\underline{(6 - 8)}\,(15 \div 3)$ Work inside first set of parentheses; change $6 - 8$ to $6 + (-8)$ to get (-2).

$3 + 2(-2)\underline{(15 \div 3)}$ Work inside second set of parentheses; $15 \div 3$ is 5

$3 + \underline{2(-2)}\quad (5)$ Multiply and divide from left to right. First multiply $2(-2)$ to get -4

$3 + \underline{(-4)\quad(5)}$ Then multiply $-4 \cdot 5$ to get -20

$3 + \quad (-20)$ Add last; $3 + (-20)$ gives -17

-17

◀ *Work Problem* **3** *at the Side.*

OBJECTIVE 2 **Use the order of operations with exponents.**
Recall from **Section 1.8** that 2^3 means 2 is used as a factor 3 times.

$$2^3 = 2 \cdot 2 \cdot 2 = 8$$

The small raised 3 in 2^3 is called an *exponent.* Exponents are also used with signed numbers. Here are three examples.

$$(-3)^2 = (-3) \cdot (-3) = 9$$

$$(-4)^3 = \underline{(-4) \cdot (-4)} \cdot (-4)$$ Be careful! Multiply two numbers at a time. Watch the signs.
$$= \quad 16 \quad \cdot (-4)$$
$$= \quad -64 \longleftarrow \text{Negative product}$$

$$\left(-\frac{1}{2}\right)^4 = \underline{\left(-\frac{1}{2}\right) \cdot \left(-\frac{1}{2}\right)} \cdot \left(-\frac{1}{2}\right) \cdot \left(-\frac{1}{2}\right)$$
$$= \underline{\frac{1}{4} \cdot \left(-\frac{1}{2}\right)} \cdot \left(-\frac{1}{2}\right)$$
$$= \underline{-\frac{1}{8} \cdot \left(-\frac{1}{2}\right)}$$
$$= \frac{1}{16}$$

Be very careful with exponents and signed numbers. For example,

$$(-3)^2 = (-3) \cdot (-3) = 9 \longleftarrow \text{Positive 9}$$

But the expression -3^2, with *no parentheses,* is different.

$$-3^2 = -(3 \cdot 3) = -9 \longleftarrow \text{Negative 9}$$

CAUTION
$(-3)^2$ is *not* the same as -3^2.

$$(-3)^2 = (-3) \cdot (-3) = 9 \quad but \quad -3^2 = -(3 \cdot 3) = -9$$

You will need this information as you take more algebra classes.

EXAMPLE 3 **Exponents and the Order of Operations**

Simplify.

(a) $\underbrace{4^2 - (-3)^2}$ There are parentheses around (-3), but no work
   can be done inside these parentheses.

   Apply the exponents: $4^2 = 4 \cdot 4 = 16$ and
   $(-3)^2 = (-3) \cdot (-3) = 9$

$\underbrace{16 - 9}$ Subtract: $16 - 9 = 7$

7

(b) $(-5)^2 - \underbrace{(4 - 6)}^2(-3)$ Work inside parentheses first.

$\underbrace{(-5)^2} - \underbrace{(-2)^2}(-3)$ Apply the exponents next.

$25 - \underbrace{4(-3)}$ Multiply.

$25 - (-12)$ Change subtraction to addition.

$\underbrace{25 + (+12)}$ Add.

37

(c) $\left(\dfrac{2}{3} - \dfrac{1}{6}\right)^2 \div \left(-\dfrac{3}{8}\right)$ Inside parentheses: $\frac{2}{3} - \frac{1}{6} = \frac{4}{6} - \frac{1}{6} = \frac{3}{6} = \frac{1}{2}$

$\left(\dfrac{1}{2}\right)^2 \div \left(-\dfrac{3}{8}\right)$ Apply the exponent: $(\frac{1}{2})^2 = \frac{1}{2} \cdot \frac{1}{2} = \frac{1}{4}$

$\dfrac{1}{4} \div \left(-\dfrac{3}{8}\right)$ Divide by using the reciprocal of the divisor:
   $-\frac{3}{8}$ becomes $-\frac{8}{3}$

$\underbrace{\dfrac{1}{4} \cdot \left(-\dfrac{8}{3}\right)}$ Divide out common factors, then

   multiply: $\frac{1}{4} \cdot -\frac{\overset{2}{8}}{3} = -\frac{2}{3}$

$-\dfrac{2}{3}$

Work Problem **4** *at the Side.* ▶

Note
Parentheses can be used in several different ways.

To indicate multiplication: $(4)(-3) = -12$

To separate a negative number $8 - (-2) = 10$
from a minus sign:

To indicate which operation $35 + \underbrace{(6 - 2)}$
to do first:
    $\underbrace{35 + 4}$

39

---

4 Simplify.

(a) $2^3 - 3^2$

(b) $4^2 - 3^2(5 - 2)$

(c) $-18 \div (-3)(2^3)$

(d) $(-3)^3 + (3 - 8)^2$

(e) $\dfrac{3}{8} + \left(-\dfrac{1}{2}\right)^2 \div \dfrac{1}{4}$

ANSWERS
4. **(a)** -1 **(b)** -11 **(c)** 48 **(d)** -2
   (e) $\dfrac{11}{8}$

5 Simplify.

(a) $\dfrac{-3(2^3)}{-10 - 6 + 8}$

A fraction bar indicates division, as in $\frac{-6}{2}$, which means $-6 \div 2$. In an expression such as

$$\frac{-5 + 3^2}{16 - 7(2)}$$

the fraction bar also acts as a grouping symbol, like parentheses. It tells us to do the work in the numerator, and then the work in the denominator. The last step is to divide the results.

$$\frac{-5 + 3^2}{16 - 7(2)} \longrightarrow \frac{-5 + 9}{16 - 14} \longrightarrow \frac{4}{2} \longrightarrow \text{Now divide.} \quad 4 \div 2 = 2$$

The final simplified result is 2.

(b) $\dfrac{(-10)(-5)}{-6 \div 3(5)}$

EXAMPLE 4 **Fraction Bars and the Order of Operations**

Simplify.

$$\frac{-8 + 5(4 - 6)}{4 - 4^2 \div 8}$$

First do the work in the numerator.

$$-8 + 5\underbrace{(4 - 6)}_{} \qquad \text{Work inside parentheses.}$$

$$-8 + \underbrace{5(-2)}_{} \qquad \text{Multiply.}$$

$$\underbrace{-8 + (-10)}_{} \qquad \text{Add.}$$

$$\text{Numerator} \longrightarrow -18$$

> When *adding* two negative numbers, the sum is also negative.

(c) $\dfrac{6 + 18 \div (-2)}{(1 - 10) \div 3}$

Now do the work in the denominator.

$$4 - \underbrace{4^2}_{} \div 8 \qquad \text{No parentheses; apply the exponent.}$$

$$4 - \underbrace{16 \div 8}_{} \qquad \text{Divide.}$$

$$\underbrace{4 - \quad 2}_{} \qquad \text{Subtract.}$$

$$\text{Denominator} \longrightarrow 2$$

(d) $\dfrac{6^2 - 3^2(4)}{5 + (3 - 7)^2}$

The last step is the division.

$$\begin{array}{c} \text{Numerator} \longrightarrow \\ \text{Denominator} \longrightarrow \end{array} \frac{-18}{2} = -9$$

> This is *division*. Signs do *not* match, so the quotient is *negative*.

◀ **Work Problem** **5** *at the Side.*

9.4 ▶▶▶ Exercises

Simplify. See Examples 1–3.

1. $6 + 3(-4)$
   -6

2. $10 - 30 \div 2$
   -5

3. $-1 + 15 + 2(-7)$
   0

4. $9 + (-5) + 2(-2)$
   0

5. $6^2 + 4^2$
   52

6. $3^2 + 8^2$
   73

7. $10 - 7^2$
   -39

8. $5 - 5^2$
   -20

9. $(-2)^5 + 2$
   -30

10. $(-2)^4 - 7$
   9

11. $4^2 + 3^2 + (-8)$
   17

12. $5^2 + 2^2 + (-12)$
   17

13. $2 - (-5) + 3^2$
   16

14. $6 - (-9) + 2^3$
   23

15. $(-4)^2 + (-3)^2 + 5$
   30

16. $(-5)^2 + (-6)^2 + 12$
   73

17. $3 + 5(6 - 2)$
   23

18. $4 + 3(8 - 3)$
   19

19. $-7 + 6(8 - 14)$
   -43

20. $-3 + 5(9 - 12)$
   -18

21. $-6 + (-5)(9 - 14)$
   19

22. $-5 + (-3)(6 - 7)$
   -2

23. $(-5)(7 - 13) \div (-10)$
   -3

24. $(-4)(9 - 17) \div (-8)$
   -4

25. $9 \div (-3)^2 + (-1)$
0

26. $-48 \div (-4)^2 + 3$
0

27. $2 - (-5)(-3)^2$
47

28. $1 - (-10)(-2)^3$
-79

29. $(-2)(-7) + 3(9)$
41

30. $4(-2) + (-3)(-5)$
7

31. $30 \div (-5) - 36 \div (-9)$
-2

32. $8 \div (-4) - 42 \div (-7)$
4

33. $2(5) - 3(4) + 5(3)$
13

34. $9(3) - 6(4) + 3(7)$
24

35. $4(3^2) + 7(3 + 9) - (-6)$
126

36. $5(4^2) - 6(1 + 4) - (-3)$
53

Simplify. See Example 4.

37. $\dfrac{-1 + 5^2 - (-3)}{-6 - 9 + 12}$

$\dfrac{27}{-3} = -9$

38. $\dfrac{-6 + 3^2 - (-7)}{7 - 9 - 3}$

$\dfrac{10}{-5} = -2$

39. $\dfrac{-2(4^2) - 4(6 - 2)}{-4(8 - 13) \div (-5)}$

$\dfrac{-48}{-4} = 12$

40. $\dfrac{3(3^2) - 5(9 - 2)}{8(6 - 9) \div (-3)}$

$\dfrac{-8}{8} = -1$

41. $\dfrac{2^3(-2 - 5) + 4(-1)}{4 + 5(-6 \cdot 2) + (5 \cdot 11)}$

$\dfrac{-60}{-1} = 60$

42. $\dfrac{3^3 + 4(-1 - 2) - 25}{-4 + 4(3 \cdot 5) + (-6 \cdot 9)}$

$\dfrac{-10}{2} = -5$

Simplify each expression.

43. $(-4)^2 (7 - 9)^2 \div 2^3$

8

44. $(-5)^2 (9 - 17)^2 \div (-10)^2$

16

45. $(-0.3)^2 + (-0.5)^2 + 0.9$

1.24

46. $(0.2)^3 - (-0.4)^2 + 3.02$

2.868

47. $(-0.75)(3.6 - 5)^2$

-1.47

48. $(-0.3)(4 - 6.8)^2$

-2.352

49. $(0.5)^2 (-8) - (0.31)$

-2.31

50. $(0.3)^3 (-5) - (-2.8)$

2.665

51. $\dfrac{2}{3} \div \left(-\dfrac{5}{6}\right) - \dfrac{1}{2}$

$-\dfrac{13}{10}$

52. $\dfrac{5}{8} \div \left(-\dfrac{10}{3}\right) - \dfrac{3}{4}$

$-\dfrac{15}{16}$

53. $\left(-\dfrac{1}{2}\right)^2 - \left(\dfrac{3}{4} - \dfrac{7}{4}\right)$

$\dfrac{5}{4}$

54. $\left(-\dfrac{2}{3}\right)^2 - \left(\dfrac{1}{6} - \dfrac{11}{6}\right)$

$\dfrac{19}{9}$

55. $\dfrac{3}{5}\left(-\dfrac{7}{6}\right) - \left(\dfrac{1}{6} - \dfrac{5}{3}\right)$

$\dfrac{4}{5}$

56. $\dfrac{2}{7}\left(-\dfrac{14}{5}\right) - \left(\dfrac{4}{3} - \dfrac{13}{9}\right)$

$-\dfrac{31}{45}$

57. $5^2 (9 - 11)(-3)(-2)^3$

-1200

58. $4^2 (13 - 17)(-2)(-3)^2$

1152

59. $1.6(-0.8) \div (-0.32) \div 2^2$

1

60. $6.5(-4.8) \div (-0.3) \div (-2)^3$

-13

✏ **61.** Simplify.

$(-2)^2 = $ ___4___ $(-2)^6 = $ ___64___

$(-2)^3 = $ ___-8___ $(-2)^7 = $ ___-128___

$(-2)^4 = $ ___16___ $(-2)^8 = $ ___256___

$(-2)^5 = $ ___-32___ $(-2)^9 = $ ___-512___

(a) Describe the pattern you see in the sign of the answers.

When a negative number is raised to an even power, the answer is positive; when raised to an odd power, the answer is negative.

(b) What would be the sign of $(-2)^{10}$? of $(-2)^{15}$? of $(-2)^{24}$?

positive; negative; positive

✏ **62.** Explain the difference between -5^2 and $(-5)^2$.

-5^2 is the opposite of 5^2 or $-(5 \cdot 5) = -25$. The parentheses in $(-5)^2$ mean that -5 is multiplied times itself. So $(-5)^2 = (-5) \cdot (-5) = 25$. One answer is negative and the other is positive.

Simplify.

63. $\dfrac{-9 + 18 \div (-3)(-6)}{5 - 4(12) \div 3(2)}$

$\dfrac{27}{-27} = -1$

64. $\dfrac{-20 - 15(-4) - (-40)}{4 + 27 \div 3(-2) - 6}$

$\dfrac{80}{-20} = -4$

65. $-7\left(6 - \dfrac{5}{8} \cdot 24 + 3 \cdot \dfrac{8}{3}\right)$

7

66. $(-0.3)^2(-5)(3) + (6 \div 2)(0.4)$

-0.15

67. $|-12| \div 4 + 2(3^2) \div 6$

6

68. $6 - (2 - 12) + 4^2 \div (-2)\left(\dfrac{5}{2}\right) + (2)^2$

0

Summary Exercises on Operations with Signed Numbers

Simplify each expression.

1. $2 - 8$ **−6**

2. $(-16)(0)$ **0**

3. $-14 - (-7)$ **−7**

4. $\dfrac{-42}{6}$ **−7**

5. $-9(-7)$ **63**

6. $\dfrac{-12}{12}$ **−1**

7. $(1)(-56)$ **−56**

8. $1 + (-23)$ **−22**

9. $5 - (-7)$ **12**

10. $-\dfrac{8}{3} \div \left(-\dfrac{4}{9}\right)$ **6**

11. $-18 + 5$ **−13**

12. $\dfrac{0}{-10}$ **0**

13. $-40 - 40$ **−80**

14. $-17 + 0$ **−17**

15. $8(-6)$ **−48**

16. $-\dfrac{1}{10} - \dfrac{9}{10}$ **−1**

17. $\left(-\dfrac{5}{6}\right)\left(\dfrac{1}{5}\right)$ **$-\dfrac{1}{6}$**

18. $\dfrac{30}{0}$ **undefined**

19. $0 - 14.6$ **−14.6**

20. $\dfrac{1.8}{-3}$ **−0.6**

21. $-4(-6)(2)$
48

22. $-2 + (-12) + (-5)$
−19

23. $-60 \div 10 \div (-3)$
2

24. $-8 - 4 - 8$
−20

25. $64(0) \div (-8)$
0

26. $2 - (-5) + 3^2$
16

27. $-9 + 8 + (-2)$
−3

28. $(-6)(-2)(-3)$
−36

29. $8 + 6 + (-8)$
6

30. $-72 \div (-9) \div (-4)$
−2

31. $-7 + 28 + (-56) + 3$
−32

32. $9 - 6 - 3 - 5$
−5

33. $-6(-8) \div (-5 - 7)$
−4

34. $-1(9732)(-1)(-1)$
−9732

35. $-80 \div 4(-5)$
100

36. $-10 - 4 + 0 + 18$
4

37. $-7 \cdot |7| \cdot |-7|$
−343

38. $5 - |-3| + 3$
5

39. $-2(-3)(7) \div (-7)$
−6

40. $-3 - (-2 + 4) - 5$
−10

41. $0 - |-7 + 2|$
−5

42. $(-4)^2(7 - 9)^2 \div 2^3$
8

43. $12 \div 4 + 2(-2)^2 \div (-4)$
1

44. $\dfrac{-9 + 24 \div (-4)(-6)}{32 - 4(12) \div 3(2)}$

$\dfrac{27}{0}$ **is undefined.**

45. $\dfrac{5 - |2 - 4 \cdot 4| + (-5)^2 \div 5^2}{-9 \div 3(2 - 2) - (-8)}$

$\dfrac{-8}{8} = -1$

Solve each application problem. Be sure to indicate whether the final answer is a positive or negative number.

46. When Ashwini discovered that her checking account was overdrawn by $238, she quickly transferred $450 from her savings to her checking account. What is the balance in her checking account?

+$212 or $212

47. A discount store found that 174 items were lost to shoplifting last month. The total loss was $4176. What was the average loss on each item?

−$24

48. In a laboratory experiment, a mixture started at a temperature of −102 degrees. First the temperature was raised 37 degrees and then raised 52 degrees. What was the final temperature?

−13 degrees

49. A plane ascended an average of 730 ft each minute during a 37-minute takeoff. How far did the plane ascend during the takeoff?

+27,010 ft or 27,010 ft

50. The Tigers gained a total of 148 yd during the first half of the football game. During the second half they lost 191 yd. How many yards did they gain or lose during the entire game?

−43 yd

51. A scuba diver was photographing fish at 65 ft below the surface of the ocean. She swam up 24 ft and then swam down 49 ft. What was her final depth?

−90 ft

Elena Sanchez opened a shop that does alterations and designs custom clothing. Use the table of her income and expenses to answer Exercises 52–55.

Month	Income	Expenses	Profit or Loss
January	$2400	−$3100	−$700;
February	$1900	−$2000	−$100;
March	$2500	−$1800	$700
April	$2300	−$1400	$900
May	$1600	−$1600	$0
June	$1900	−$1200	$700

52. Complete the table by finding Elena's profit or loss for each month.

January: −$700; February: −$100; March: $700; April: $900; May: $0; June: $700

53. Which month had the greatest loss? Which month had the greatest profit?

January; April

54. What was Elena's average monthly income?

$2100

55. What was the average monthly amount of expenses?

−$1850

56. Explain in your own words how to add two numbers with different signs. Include two examples in your explanation, one that has a positive answer and one that has a negative answer.

Subtract the lesser absolute value from the greater absolute value. The answer has the same sign as the addend with the greater absolute value. Examples: −6 + 2 = −4 and 6 + (−2) = 4

57. Explain what is different and what is similar between multiplying and dividing signed numbers.

Similar: If the signs match, the result is positive. If the signs are different, the result is negative. Different: Multiplication is commutative, division is not. You can multiply by 0, but dividing by 0 cannot be done.

9.5 ▶▶▶   Evaluating Expressions and Formulas

In formulas, you have seen that numbers can be represented by letters. For example, you used this formula for finding simple interest in **Section 6.7.**

$$I = p \cdot r \cdot t$$

In this formula, p (principal) represents the amount of money borrowed, r is the rate of interest, and t is the time in years. In algebra, we often write multiplication without the multiplication dots. If there is no operation sign written between two letters, or between a letter and a number, you assume it is multiplication.

OBJECTIVES

1 Define variable and expression.

2 Find the value of an expression when values of the variables are given.

Showing Multiplication in Algebra

If there is no operation sign, it is understood to be multiplication. Here are some examples.

$I = p \cdot r \cdot t$	is written	$I = prt$
$2 \cdot r$	is written	$2r$
$3 \cdot x + 4 \cdot y$	is written	$3x + 4y$

OBJECTIVE 1 Define variable and expression. Letters (such as the I, p, r, and t used above) that represent numbers are called **variables.** A combination of operations on letters and numbers is an **expression.** Three examples of expressions are shown here.

$$9 + p \qquad 8r \qquad 7k - 2m$$

OBJECTIVE 2 Find the value of an expression when values of the variables are given. The value of an expression changes depending on the value of each variable. To find the value of an expression, replace the variables with their values. It is helpful to write each value inside parentheses when multiplication is involved.

---

1 Find the value of $5x - 3y$, if:

(a) $x = 5, \quad y = 2$

(b) $x = -3, \quad y = 4$

(c) $x = 0, \quad y = 6$

EXAMPLE 1 Finding the Value of an Expression

Find the value of $5x - 3y$, if $x = 2$ and $y = 7$.

Replace x with **2**. Replace y with **7**.

Using the order of operations, do all multiplications first. Then subtract.

$$5\,x \;-\; 3\,y$$
$$5\,(2) \;-\; 3\,(7)$$
$$\underbrace{10 \;-\; 21}$$
$$-11$$

> $10 - 21$ becomes $10 + (-21)$

Work Problem **1** *at the Side.* ▶

EXAMPLE 2 Finding the Value of an Expression

What is the value of $\dfrac{6k + 2r}{5s}$, if $k = -2$, $r = 5$, and $s = -1$?

Continued on Next Page

ANSWERS

1. (a) 19 **(b)** -27 **(c)** -18

2 Find the value of $\dfrac{3k + r}{2s}$ if:

(a) $k = 1$, $r = 1$, $s = 2$

(b) $k = 8$, $r = -2$, $s = -4$

(c) $k = -3$, $r = 1$, $s = -2$

3 Find the value of each expression.

(a) $-x - 6y$, if $x = -1$ and $y = -4$

(b) $-4a - b$, if $a = 4$ and $b = -2$

(c) $-w - 3x - y$, if $w = 10$, $x = -5$, and $y = -6$

4 Using the given values, evaluate each formula.

(a) $A = \dfrac{1}{2}bh$, when $b = 6$ yd, $h = 12$ yd

(b) $P = 2l + 2w$, when $l = 10$ cm, $w = 8$ cm

(c) $C = 2\pi r$, when $\pi \approx 3.14$, $r = 6$ ft

ANSWERS

2. (a) $\dfrac{4}{4} = 1$ (b) $\dfrac{22}{-8} = -\dfrac{11}{4}$ (c) $\dfrac{-8}{-4} = 2$
3. (a) 25 (b) −14 (c) 11
4. (a) $A = 36$ yd^2 (b) $P = 36$ cm
   (c) $C \approx 37.68$ ft

Replace k with -2, r with 5, and s with -1.

$$\frac{6k + 2r}{5s} = \frac{6(-2) + 2(5)}{5(-1)}$$ Do all the multiplications first.

$$= \frac{-12 + 10}{-5}$$ Add in the numerator.

$$= \frac{-2}{-5}$$ Dividing two numbers with the same sign gives a positive answer.

$$= \frac{2}{5}$$

◀ *Work Problem* **2** *at the Side.*

The next example shows how to evaluate an expression with negative signs and/or subtraction, when the value of the variable is also negative.

EXAMPLE 3 **Evaluating an Expression with Negative Signs and Negative Values**

Find the value of $-c - 5b$ when $c = -2$ and $b = -3$.

Replace c with -2.
Replace b with -3.

$-(-2)$ is the opposite of (-2), which is $(+2)$.

$$-(-2) - 5(-3)$$ Multiply $5 \cdot (-3)$.

$$+2 \quad - (-15)$$ Change subtraction to addition; change (-15) to its opposite, $(+15)$.

$$2 \quad + (+15)$$

$$17$$

CAUTION
Watch the signs carefully when there are negative signs in the expression and the value of a variable is also negative, as in Example 3 above.

◀ *Work Problem* **3** *at the Side.*

EXAMPLE 4 **Evaluating a Formula**

The formula you used in **Chapter 8** for the area of a triangle can now be written without the multiplication dots.

$$A = \frac{1}{2} \cdot b \cdot h \quad \text{is written} \quad A = \frac{1}{2}bh$$

In this formula, b is the length of the base of the triangle and h is the height of the triangle. What is the area if $b = 9$ cm and $h = 24$ cm?

$$A = \frac{1}{2} \quad b \quad h$$

$$A = \frac{1}{2}(9 \text{ cm})(24 \text{ cm})$$ Replace b with 9 cm and h with 24 cm.

$$A = \frac{1}{2}(9 \text{ cm})(\overset{12}{24} \text{ cm})$$ Divide out any common factors.

$$A = 108 \text{ cm}^2$$ Remember, *area* is measured in *square* units; cm^2 means *square centimeters*.

The area of the triangle is 108 cm^2.

◀ *Work Problem* **4** *at the Side.*

Find the value of the expression $2r + 4s$ for each set of values of r and s. See Example 1.

1. $r = 2, \quad s = 6$
28

2. $r = 6, \quad s = 1$
16

3. $r = 1, \quad s = -3$
−10

4. $r = 7, \quad s = -2$
6

5. $r = -4, \quad s = 4$
8

6. $r = -3, \quad s = 5$
14

7. $r = -1, \quad s = -7$
−30

8. $r = -3, \quad s = -5$
−26

9. $r = 0, \quad s = -2$
−8

10. $r = -7, \quad s = 0$
−14

Use the given values of the variables to find the value of each expression. See Examples 1 and 2.

11. $8x - y$
$x = 1, \quad y = 8$
0

12. $a - 5b$
$a = 10, \quad b = 2$
0

13. $6k + 2s$
$k = 1, \quad s = -2$
2

14. $7p + 7q$
$p = -4, \quad q = 1$
−21

15. $\dfrac{-m + 5n}{2s + 2}$
$m = 4, \quad n = -8, \quad s = 0$
−22

16. $\dfrac{2y - z}{x - 2}$
$y = 0, \quad z = 5, \quad x = 1$
5

17. $-m - 3n$
$m = \dfrac{1}{2}, \quad n = \dfrac{3}{8}$
$-\dfrac{13}{8}$

18. $7k - 3r$
$k = \dfrac{2}{3}, \quad r = \dfrac{1}{3}$
$\dfrac{11}{3}$

Be careful when an expression has a negative sign and the value of the variable is also negative. Use the given values to find the value of each expression. See Example 3.

19. $-c - 5b$
$c = -8, \quad b = -4$
28

20. $-c - 5b$
$c = -1, \quad b = -2$
11

21. $-4x - y$
$x = 5, \quad y = -15$
-5

22. $-4x - y$
$x = 3, \quad y = -8$
-4

23. $-k - m - 8n$
$k = 6, \quad m = -9, \quad n = 0$
3

24. $-k - m - 8n$
$k = 0, \quad m = -7, \quad n = -1$
15

25. $\dfrac{-3s - t - 4}{-s + 6 + t}$
$s = -1, \quad t = -13$
$\dfrac{12}{-6} = -2$

26. $\dfrac{-3s - t - 4}{-s - 20 - t}$
$s = -3, \quad t = -6$
$\dfrac{11}{-11} = -1$

Using the given values, evaluate each formula. See Example 4.

27. $P = 4s; \quad s = 7.5$
$P = 30$

28. $P = 4s; \quad s = 0.8$
$P = 3.2$

29. $P = 2l + 2w; \quad l = 9, \quad w = 5$
$P = 28$

30. $P = 2l + 2w; \quad l = 12, \quad w = 2$
$P = 28$

31. $A = \pi r^2$; $\pi \approx 3.14$, $r = 5$

$A \approx 78.5$

32. $A = \pi r^2$; $\pi \approx 3.14$, $r = 10$

$A \approx 314$

33. $A = \frac{1}{2}bh$; $b = 15$, $h = 3$

$A = \frac{45}{2}$ or $22\frac{1}{2}$

34. $A = \frac{1}{2}bh$; $b = 5$, $h = 11$

$A = \frac{55}{2}$ or $27\frac{1}{2}$

● 35. $V = \frac{1}{3}Bh$; $B = 30$, $h = 60$

$V = 600$

36. $V = \frac{1}{3}Bh$; $B = 105$, $h = 5$

$V = 175$

37. $d = rt$; $r = 53$, $t = 6$

$d = 318$

38. $d = rt$; $r = 180$, $t = 5$

$d = 900$

39. $C = 2\pi r$; $\pi \approx 3.14$, $r = 4$

$C \approx 25.12$

40. $C = 2\pi r$; $\pi \approx 3.14$, $r = 18$

$C \approx 113.04$

Solve each application problem.

41. The expression for finding the perimeter of a triangle with sides of equal length is $3s$, where s is the length of one side. Evaluate the expression when

(a) the length of one side is 11 in.

P = 33 in.

(b) the length of one side is 3 ft.

P = 9 ft

42. The expression for finding the perimeter of a pentagon with sides of equal length is $5s$, where s is the length of one side. Evaluate the expression when

(a) the length of one side is 25 cm

P = 125 cm

(b) the length of one side is 8 in.

P = 40 in.

43. The expression for figuring a student's average test score is $\frac{p}{t}$, where p is the total points earned on all the tests and t is the number of tests. Evaluate the expression when

(a) 332 points were earned on 4 tests

average score = 83

(b) there were 7 tests and 637 points were earned.

average score = 91

44. The expression for deciding how many buses are needed for a group trip is $\frac{p}{b}$, where p is the total number of people and b is the number of people that one bus will hold. Evaluate the expression when

(a) 176 people are going on a trip and one bus holds 44 people

buses needed = 4

(b) a bus holds 36 people and 72 people are going on a trip.

buses needed = 2

45. Find and correct the error made by the student who solved this example:

Find the value of $-x - 4y$ if $x = -3$ and $y = -1$.

$$-x - 4y$$
$$-3 - 4(-1)$$
$$-3 - (-4)$$
$$-3 + (+4)$$
$$1$$

After stating the error, rework the problem and write a sentence next to each step, explaining what is being done in that step.

The error was made when replacing x with -3; should be $-(-3)$, not -3.

$-x - 4y$	Replace x with (-3) and y with (-1).
$-(-3) - 4(-1)$	Opposite of (-3) is $+3$.
$(+3) - (-4)$	Change subtraction to addition.
$3 + (+4)$	Add.
7	

46. Go back to **Chapter 8** and find each of the formulas listed below. Pick values for the variables and then find the value of A or V. Then pick different values for the variables and again find the value of A or V.

(a) Area of a trapezoid: pick values for h, B, and b.

Area of trapezoid $= \dfrac{1}{2}h(b + B)$

Selected values will vary. Here is one possibility: If h is 12, b is 6, and B is 10, then

$$A = \frac{1}{2}(12)(6 + 10)$$

$$A = \frac{1}{2}\left(\frac{\overset{6}{\cancel{12}}}{1}\right)(16)$$

$$A = 96$$

(b) Volume of a rectangular solid: pick values for l, w, and h.

Volume of rectangular solid $= lwh$

Selected values will vary. One possibility is: If $l = 5$, $w = 3$, and $h = 4$, then

$$V = (5)(3)(4)$$

$$V = 60$$

Relating Concepts (Exercises 47–52) For Individual or Group Work

*Work Exercises 47–52 in order. Use the given formula and values of the variables to find the value of the remaining variable. If you studied **Chapters 7 and 8,** write a sentence telling when you would use each formula.*

47. $F = \dfrac{9C}{5} + 32;\quad C = -40$

F = −40; convert a Celsius temperature to Fahrenheit

48. $C = \dfrac{5(F - 32)}{9};\quad F = -4$

C = −20; convert a Fahrenheit temperature to Celsius

49. $V = \dfrac{4\pi r^3}{3};\quad \pi \approx 3.14,\quad r = 3$

V ≈ 113.04; find the volume of a sphere

50. $c^2 = a^2 + b^2;\quad a = 3,\quad b = 4$

c = 5; Pythagorean Theorem for finding the hypotenuse or a leg in a right triangle.

51. $A = \dfrac{1}{2}h(b + B);\quad h = 7,\quad b = 4,\quad B = 12$

A = 56; find the area of a trapezoid

52. $V = \dfrac{\pi r^2 h}{3};\quad \pi \approx 3.14,\quad r = 6,\quad h = 10$

V ≈ 376.8; find the volume of a cone

9.6 ▶▶▶ Solving Equations

An **equation** is a statement that says two expressions are equal. Examples of equations are shown here.

$$x + 1 = 9 \qquad 20 = 5k \qquad 6r - 1 = 17$$

The **equal sign** in an equation divides the equation into two parts, the *left side* and the *right side*. In $6r - 1 = 17$, the left side is $6r - 1$, and the right side is 17. The equal sign tells us that the two sides are equivalent.

$$6r - 1 = 17$$
Left side = Right side

You solve an equation by finding all numbers that can be substituted for the variable to make the equation true. These numbers are called **solutions** of the equation.

OBJECTIVE 1 Determine whether a number is a solution of an equation. To decide whether a number is a solution of an equation, substitute the number in the equation to see whether the result is true.

> **EXAMPLE 1** **Determining Whether a Number Is a Solution of an Equation**
>
> Is 7 a solution of either one of these equations?
>
> **(a)** $12 = x + 5$
> Replace x with 7.
>
> $$12 = x + 5 \qquad \text{Replace } x \text{ with 7} \quad \boxed{\text{7 is the solution, } \textbf{not } 12.}$$
> $$12 = \underbrace{7 + 5}$$
> $$12 = \quad 12 \qquad \text{True}$$
>
> Because the statement is true, 7 is a solution of the equation $12 = x + 5$.
>
> **(b)** $2y + 1 = 16$
> Replace y with 7.
>
> $$2y + 1 = 16$$
> $$2(7) + 1 = 16$$
> $$\underbrace{14 + 1} = 16$$
> $$15 \quad = 16 \qquad \text{False}$$
>
> The *false* statement shows that 7 is *not* a solution of $2y + 1 = 16$.

Work Problem **1** *at the Side.* ▶

OBJECTIVE 2 Solve equations using the addition property of equations. If the equation $a = b$ is true, and if a number c is added to both a and b, the new equation is also true. This rule, called the **addition property of equations,** means that you can add the *same* number to *both* sides of an equation and still have a true equation.

> **Addition Property of Equations**
> If $a = b$, then $a + c = b + c$.
>
> In other words, you may add the *same* number to *both* sides of an equation.

OBJECTIVES

1 Determine whether a number is a solution of an equation.

2 Solve equations using the addition property of equations.

3 Solve equations using the multiplication property of equations.

1 Decide whether the given number is a solution of the equation.

(a) $p + 1 = 8$; 7

(b) $30 = 5r$; 6

(c) $3k - 2 = 4$; 3

(d) $23 = 4y + 3$; 5

ANSWERS

1. **(a)** solution **(b)** solution
   (c) not a solution **(d)** solution

You can use the addition property to solve equations. The idea is to get the variable (the letter) by itself on one side of the equal sign and a number by itself on the other side.

> ### EXAMPLE 2 Solving Equations Using the Addition Property
>
> Solve each equation.
>
> **(a)** $k - 4 = 6$
>
>   To get k by itself on the left side, add 4 to the left side, because $k - 4 + 4$ gives $k + 0$. You must then add 4 to the right side also.
>
> $$k - 4 = 6 \quad\quad \leftarrow \text{Original equation}$$
> $$k \underbrace{- 4 + 4}_{} = 6 + 4 \quad\quad \text{Add 4 to both sides.}$$
> $$\underbrace{k + 0}_{} = 10 \quad\quad \text{On the left side, } -4 + 4 \text{ is } 0$$
> $$k = 10 \qu\quad\quad \text{On the left side, } k + 0 \text{ is } k.$$
>
> The solution is 10. To check the solution, replace k with 10 in the original equation.
>
> $$k - 4 = 6 \leftarrow \text{Original equation}$$
> $$10 - 4 = 6 \quad\quad \text{Replace } k \text{ with } 10$$
> $$6 = 6 \quad\quad \text{True, so 10 is the correct solution.}$$
>
> > 10 is the solution, **not** 6.
>
> This result is true, so **10 is the solution.**
>
> > ### CAUTION
> > When checking the solution in Example 2(a) above, we ended up with $6 = 6$. Notice that 6 is **not** the solution. The solution is 10, the number used to replace k in the original equation.
>
> **(b)** $2 = z + 8$
>
>   To get z by itself on the right side, add -8 to both sides.
>
> $$2 = z + 8 \quad\quad \leftarrow \text{Original equation}$$
> $$2 + (-8) = z + \underbrace{8 + (-8)}_{} \quad\quad \text{Add } (-8) \text{ to both sides.}$$
> $$-6 = z + \quad 0$$
> $$-6 = z \quad\quad \leftarrow \text{The solution is } -6$$
>
> > ### Note
> > Notice that we *added* -8 to both sides of the equation to get z by itself. We can accomplish the same thing by *subtracting* 8 from both sides. Recall from **Section 9.2** that subtraction is defined in terms of addition. On the left side of the equation above, $2 - 8$ gives the same result as $2 + (-8)$.
>
> Check the solution by replacing z with -6 in the original equation.
>
> $$2 = z + 8 \quad\quad \leftarrow \text{Original equation}$$
> $$2 = -6 + 8 \quad\quad \text{Replace } z \text{ with } -6$$
> $$2 = 2 \quad\quad \text{True, so } -6 \text{ is the correct solution.}$$
>
> > -6 is the solution, **not** 2.
>
> The result is true, so **−6 is the solution** (*not* 2).
>
> ── **Continued on Next Page**

Here is a summary of the rules you can use to solve equations using the addition property. In these rules, x is the variable and a and b represent numbers.

> **Solving an Equation Using the Addition Property**
>
> Solve $x - a = b$ by adding a to *both* sides.
>
> Solve $x + a = b$ by subtracting a from *both* sides.

Work Problem 2 *at the Side.* ▶

OBJECTIVE 3 **Solve equations using the multiplication property of equations.** As long as you do the *same* thing to *both* sides of an equation, it will still be a true equation. So far you have added or subtracted on both sides. Now we will multiply or divide on both sides.

> **Multiplication Property of Equations**
>
> If $a = b$ and c does not equal 0, then we can say the following:
>
> $$a \cdot c = b \cdot c \quad \text{and} \quad \frac{a}{c} = \frac{b}{c}$$
>
> In other words, you may multiply or divide *both* sides of an equation by the *same* number. (The only exception is you *cannot* divide by 0.)

EXAMPLE 3 **Solving Equations Using the Multiplication Property**

Solve each equation.

(a) $9p = 63$

You want to get the variable, p, by itself on the left side. The expression $9p$ means $9 \cdot p$. To undo the multiplication and get p by itself, *divide* both sides by 9.

$$9p = 63$$

On the left side, divide out 9.

$$\frac{\overset{1}{\cancel{9}} \cdot p}{\underset{1}{\cancel{9}}} = \frac{63}{9} \qquad \text{Divide both sides by 9}$$

$$p = 7 \quad \leftarrow \text{The solution is 7}$$

Check

$$9p = 63 \quad \leftarrow \text{Original equation}$$

$$9 \cdot 7 = 63 \qquad \text{Replace } p \text{ with 7} \qquad \boxed{\text{7 is the solution, \textbf{not} 63.}}$$

$$63 = 63 \qquad \text{True}$$

The result is true, so **7** is the solution (***not*** 63).

——— **Continued on Next Page**

2 Solve each equation. Check each solution.

(a) $n - 5 = 8$

(b) $5 = r - 10$

(c) $3 = z + 1$

(d) $k + 9 = 0$

(e) $-2 = y + 9$

(f) $x - 2 = -6$

ANSWERS

2. **(a)** $n = 13$ **(b)** $r = 15$ **(c)** $z = 2$
   (d) $k = -9$ **(e)** $y = -11$ **(f)** $x = -4$

3 Solve each equation. Check each solution.

(a) $2y = 14$

(b) $42 = 7p$

(c) $-8a = 32$

(d) $-3r = -15$

(e) $-60 = -6k$

(f) $10x = 0$

(b) $-4r = 24$

Divide *both* sides by -4 to get r by itself on the left side.

$$\frac{\overset{1}{-\cancel{4}} \cdot r}{\underset{1}{-\cancel{4}}} = \frac{24}{-4} \qquad \text{Divide both sides by } -4$$

$$r = -6 \quad \leftarrow \text{The solution is } -6$$

Check

$$-4r = 24 \quad \leftarrow \text{Original equation}$$

$$-4(-6) = 24 \qquad \text{Replace } r \text{ with } -6 \quad \boxed{-6 \text{ is the solution, \textbf{not} } 24.}$$

$$24 = 24 \qquad \text{True}$$

The result is true, so **-6** is the solution (**not** 24).

(c) $-55 = -11m$

Divide *both* sides by -11 to get m by itself on the right side.

$$\frac{-55}{-11} = \frac{\overset{1}{-\cancel{11}} \cdot m}{\underset{1}{-\cancel{11}}}$$

$$5 = m$$

Check

$$-55 = -11m \quad \leftarrow \text{Original equation}$$

$$-55 = -11(5) \qquad \text{Replace } m \text{ with } 5 \quad \boxed{5 \text{ is the solution, \textbf{not} } -55.}$$

$$-55 = -55 \qquad \text{True}$$

The result is true, so **5** is the solution (**not** -55).

◀ Work Problem **3** at the Side.

EXAMPLE 4 Solving Equations Using the Multiplication Property

Solve each equation.

(a) $\dfrac{x}{2} = 9$

Replace $\dfrac{x}{2}$ with $\dfrac{1}{2}x$, because dividing x by 2 is the same as multiplying x by $\dfrac{1}{2}$. Then, to get x by itself, multiply both sides by the reciprocal of $\dfrac{1}{2}$, which is $\dfrac{2}{1}$. (Recall from **Section 2.7** that when two numbers are reciprocals, their product is 1.)

$$\frac{1}{2}x = 9$$

$$\frac{\overset{1}{\cancel{2}}}{1} \cdot \frac{1}{\underset{1}{\cancel{2}}}x = 2 \cdot 9 \qquad \text{Multiply both sides by } \tfrac{2}{1} \text{ (which equals 2).}$$

$$1x = 18$$

$$x = 18 \quad \leftarrow \text{The solution is 18}$$

Continued on Next Page

Check $\dfrac{x}{2} = 9$ ← Original equation

$\dfrac{18}{2} = 9$ Replace x with 18   ⟨ 18 is the solution, **not** 9. ⟩

$9 = 9$ True, so 18 is the correct solution.

18 is the correct solution (***not*** 9).

(b) $-\dfrac{2}{3}r = 4$

Multiply both sides by the reciprocal of $-\frac{2}{3}$, which is $-\frac{3}{2}$.

$$-\dfrac{2}{3}r = 4$$

$$-\dfrac{\overset{1}{\cancel{3}}}{\underset{1}{\cancel{2}}} \cdot \left(-\dfrac{\overset{1}{\cancel{2}}}{\underset{1}{\cancel{3}}}r\right) = -\dfrac{3}{\underset{1}{\cancel{2}}} \cdot \dfrac{\overset{2}{\cancel{4}}}{1}$$ Multiply both sides by $-\frac{3}{2}$

$$r = -6$$ ← The solution is -6

Check by replacing r with -6 in the original equation. Write -6 as $\frac{-6}{1}$.

$$-\dfrac{2}{3}r = 4 \leftarrow \text{Original equation}$$

$$-\dfrac{2}{\underset{1}{\cancel{3}}} \cdot \dfrac{\overset{-2}{\cancel{-6}}}{1} = 4$$ Replace r with -6 ⟨ -6 is the solution, **not** 4. ⟩

$$4 = 4$$ True, so -6 is the correct solution.

-6 is the correct solution (***not*** 4).

Here is a summary of the rules for using the multiplication property. In these rules, x is the variable and a, b, and c represent numbers. (Note that a and b are not equal to 0.)

Solving Equations Using the Multiplication Property

Solve the equation $ax = b$ by dividing *both* sides by a.

Solve the equation $\dfrac{a}{b}x = c$ by multiplying *both* sides by $\dfrac{b}{a}$.

Work Problem **4** *at the Side.* ▶

4 Solve each equation. Check each solution.

(a) $\dfrac{a}{4} = 2$

(b) $\dfrac{y}{7} = -3$

(c) $-8 = \dfrac{k}{6}$

(d) $8 = -\dfrac{4}{5}z$

(e) $-\dfrac{5}{8}p = -10$

ANSWERS

4. **(a)** $a = 8$ **(b)** $y = -21$ **(c)** $k = -48$
   (d) $z = -10$ **(e)** $p = 16$

Math in the Media

EXPRESSIONS

Algebraic expressions are often useful in the real world. For example, traffic engineers use special expressions to decide how long to have the green, yellow, and red lights showing on a traffic signal.

To find the number of seconds for the green light, the engineers use this expression.

$$2.1n + 3.7$$

The variable n stands for the average number of vehicles traveling in each lane of the street during one complete light cycle (green to yellow to red, and back to green again).

1. How many seconds should the green light be on if a street averages 5 vehicles in each lane per light cycle? Round to the nearest whole number.

   14 seconds (rounded)

2. How many seconds for the green light, to the nearest whole number, if the average is 10 vehicles per light cycle? If the average is 15 vehicles per light cycle?

   25 seconds (rounded); 35 seconds (rounded)

3. As the average number of vehicles increases, what is happening to the number of seconds for the green light?

   The number of seconds is also increasing.

4. Based on the answers to Questions 1 and 2, make a guess on the number of seconds for the green light when the average is 20 vehicles per light cycle. Then calculate the number of seconds using the expression.

   Guesses will vary; calculated answer is 46 seconds (rounded).

To decide the number of seconds that a yellow light should be on, the traffic engineers use this expression.

$$\frac{5v}{100} + 1$$

where v is the speed limit in miles per hour (mph).

5. How many seconds should the yellow light be on if the speed limit is 20 mph? 40 mph? 60 mph?

   2 sec; 3 sec; 4 sec

6. Based on the answers you just calculated, how could you estimate the time for a yellow light if the speed limit is 30 mph? 50 mph? **Calculate the average time. 2.5 sec; 3.5 sec**

7. Use the given expression to find the number of seconds that the yellow light should be on if the speed limit is 30 mph and 50 mph. Did you get the same result as in Question 6? **yes**

Source: Applying Mathematics: A Consumer/Career Approach.

9.6 ▶▶▶ Exercises

FOR EXTRA HELP **MyMathLab**   Math XL PRACTICE   WATCH   DOWNLOAD READ   REVIEW

Determine whether the given number is a solution of the equation. See Example 1.

1. $x + 7 = 11$; 4
yes

2. $k - 2 = 7$; 9
yes

3. $4y = 28$; 7
yes

4. $5p = 30$; 6
yes

5. $2z - 1 = -15$; -8
no

6. $6r - 3 = -14$; -2
no

Solve each equation by using the addition property. Check each solution. See Example 2.

7. $p + 5 = 9$
$p = 4$

8. $a + 3 = 12$
$a = 9$

9. $k + 15 = 0$
$k = -15$

10. $y + 6 = 0$
$y = -6$

11. $z - 5 = 3$
$z = 8$

12. $x - 9 = 4$
$x = 13$

13. $8 = r - 2$
$r = 10$

14. $3 = b - 5$
$b = 8$

15. $-5 = n + 3$
$n = -8$

16. $-1 = a + 8$
$a = -9$

17. $7 = r + 13$
$r = -6$

18. $12 = z + 7$
$z = 5$

19. $-4 + k = 14$
$k = 18$

20. $-9 + y = 7$
$y = 16$

21. $-12 + x = -1$
$x = 11$

22. $-3 + m = -9$
$m = -6$

23. $-5 = -2 + r$
$r = -3$

24. $-1 = -10 + y$
$y = 9$

25. $d + \dfrac{2}{3} = 3$

$d = \dfrac{7}{3}$ or $2\dfrac{1}{3}$

26. $x + \dfrac{1}{2} = 4$

$x = \dfrac{7}{2}$ or $3\dfrac{1}{2}$

27. $z - \dfrac{7}{8} = 10$

$z = \dfrac{87}{8}$ or $10\dfrac{7}{8}$

28. $m - \dfrac{3}{4} = 6$

$m = \dfrac{27}{4}$ or $6\dfrac{3}{4}$

29. $\dfrac{1}{2} = k - 2$

$k = \dfrac{5}{2}$ or $2\dfrac{1}{2}$

30. $\dfrac{3}{5} = t - 1$

$t = \dfrac{8}{5}$ or $1\dfrac{3}{5}$

31. $m - \dfrac{7}{5} = \dfrac{11}{4}$

$m = \dfrac{83}{20}$ or $4\dfrac{3}{20}$

32. $z - \dfrac{7}{3} = \dfrac{32}{9}$

$z = \dfrac{53}{9}$ or $5\dfrac{8}{9}$

33. $x - 0.8 = 5.07$

$x = 5.87$

34. $a - 3.82 = 7.9$

$a = 11.72$

35. $3.25 = 4.76 + r$

$r = -1.51$

36. $8.9 = 10.5 + b$

$b = -1.6$

Solve each equation. Check each solution. See Example 3.

37. $6z = 12$

$z = 2$

38. $8k = 24$

$k = 3$

39. $48 = 12r$

$r = 4$

40. $99 = 11m$

$m = 9$

41. $3y = 0$

$y = 0$

42. $5a = 0$

$a = 0$

43. $-6k = 36$

$k = -6$

44. $-7y = 70$

$y = -10$

45. $-36 = -4p$

$p = 9$

46. $-54 = -9r$

$r = 6$

47. $-1.2m = 8.4$

$m = -7$

48. $-5.4z = 27$

$z = -5$

49. $-8.4p = -9.24$

$p = 1.1$

50. $-3.2y = -16.64$

$y = 5.2$

Solve each equation. Check each solution. See Example 4.

51. $\dfrac{k}{2} = 17$

$k = 34$

52. $\dfrac{y}{3} = 5$

$y = 15$

53. $11 = \dfrac{a}{6}$

$a = 66$

54. $5 = \dfrac{m}{8}$

$m = 40$

55. $\dfrac{r}{3} = -12$

$r = -36$

56. $\dfrac{z}{9} = -3$

$z = -27$

57. $-\dfrac{2}{5}p = 8$

$p = -20$

58. $-\dfrac{5}{6}k = 15$

$k = -18$

59. $-\dfrac{3}{4}m = -3$

$m = 4$

60. $-\dfrac{9}{10}b = -18$

$b = 20$

61. $6 = \dfrac{3}{8}x$

$x = 16$

62. $4 = \dfrac{2}{3}a$

$a = 6$

63. $\dfrac{y}{2.6} = 0.5$

$y = 1.3$

64. $\dfrac{k}{0.7} = 3.2$

$k = 2.24$

65. $\dfrac{z}{-3.8} = 1.3$

$z = -4.94$

66. $\dfrac{m}{-5.2} = 2.1$

$m = -10.92$

🖉 **67.** Explain the addition property of equations. Then show an example of an equation where you would use the addition property to solve it. Make the equation so it has -3 as the solution.

You may add or subtract the same number on both sides of an equation. Many different equations could have -3 as the solution. One possibility is:

$$x + 5 = 2$$
$$x + 5 - 5 = 2 - 5$$
$$x = -3$$

🖉 **68.** Explain the multiplication property of equations. Then show an example of an equation where you would use the multiplication property to solve it. Make the equation so it has $+6$ as the solution.

You may multiply or divide both sides of an equation by the same number (except you cannot divide by 0). Many different equations could have $+6$ as the solution. One possibility is:

$$8r = 48$$
$$\frac{\cancel{8} \cdot r}{\cancel{8}} = \frac{48}{8}$$
$$r = 6$$

Solve each equation.

69. $x - 17 = 5 - 3$

$x = 19$

70. $y + 4 = 10 - 9$

$y = -3$

71. $3 = x + 9 - 15$

$x = 9$

72. $-1 = y + 7 - 9$

$y = 1$

73. $\dfrac{7}{2}x = \dfrac{4}{3}$

$x = \dfrac{8}{21}$

74. $\dfrac{3}{4}x = \dfrac{5}{3}$

$x = \dfrac{20}{9}$

75. $\dfrac{1}{2} - \dfrac{3}{4} = \dfrac{a}{5}$

$a = -\dfrac{5}{4}$

76. $\dfrac{2}{3} - \dfrac{8}{9} = \dfrac{c}{6}$

$c = -\dfrac{4}{3}$

77. $m - 2 + 18 = |-3 - 4| + 5$

$m = -4$

78. $10 - |0 - 8| = n + 1 - 4$

$n = 5$

9.7 ▶▶▶ Solving Equations with Several Steps

OBJECTIVE 1 Solve equations with several steps. You cannot solve the equation $5m + 1 = 16$ by just adding the same number to both sides, nor by just dividing both sides by the same number. Instead, you use a combination of operations. Here are the steps.

> **Solving an Equation Using the Addition and Multiplication Properties**
>
> **Step 1** Add or subtract the *same* amount on *both* sides of the equation so that the variable term ends up by itself on one side.
>
> **Step 2** Multiply or divide *both* sides by the *same* number to find the solution.
>
> **Step 3** Check the solution in the original equation.

EXAMPLE 1   Solving an Equation with Several Steps

Solve $5m + 1 = 16$.

Step 1 Subtract 1 from *both* sides so that $5m$ will be by itself on the left side.

$$5m + 1 - 1 = 16 - 1$$
$$5m = 15$$

Step 2 Divide *both* sides by 5.

$$\frac{\overset{1}{\cancel{5}} \cdot m}{\underset{1}{\cancel{5}}} = \frac{15}{5}$$

$$m = 3 \leftarrow \text{The solution is 3}$$

Step 3 Check the solution.

$$5m + 1 = 16 \leftarrow \text{Original equation}$$
$$5(3) + 1 = 16 \qquad \text{Replace } m \text{ with 3} \qquad \boxed{\text{3 is the solution, not 16.}}$$
$$15 + 1 = 16$$
$$16 = 16 \qquad \text{True, so 3 is the correct solution.}$$

3 is the correct solution (**not** 16).

───────── *Work Problem* 1 *at the Side.* ▶

OBJECTIVE 2 Use the distributive property. We can use the order of operations to simplify these two expressions.

$$2(6 + 8) \qquad \text{and} \qquad 2 \cdot 6 + 2 \cdot 8$$
$$2(14) \qquad\qquad\qquad 12 + 16$$
$$28 \qquad\qquad\qquad\qquad 28$$

Because both answers are the same, the two expressions are equivalent.

$$2(6 + 8) = 2 \cdot 6 + 2 \cdot 8$$

This is an example of the **distributive property.**

> **Distributive Property**
>
> $$a(b + c) = ab + ac$$

1 Solve each equation. Check each solution.

(a) $2r + 7 = 13$

(b) $20 = 6y - 4$

(c) $7m + 9 = 9$

(d) $-2 = 4p + 10$

(e) $-10z - 9 = 11$

ANSWERS

1. **(a)** $r = 3$ **(b)** $y = 4$ **(c)** $m = 0$
   (d) $p = -3$ **(e)** $z = -2$

2 Use the distributive property.

(a) $3(2 + 6)$

(b) $8(k - 3)$

(c) $-6(r + 5)$

(d) $-9(s - 8)$

3 Combine like terms.

(a) $5y + 11y$

(b) $10a - 28a$

(c) $3x + 3x - 9x$

(d) $k + k$

(e) $6b - b - 7b$

EXAMPLE 2 **Using the Distributive Property**

Simplify each expression by using the distributive property.

(a) $9(4 + 2) = 9 \cdot 4 + 9 \cdot 2 = 36 + 18 = 54$

Multiplying is done before adding.

The 9 outside the parentheses is *distributed* over the 4 and the 2 inside the parentheses. That means that *every* number inside the parentheses is multiplied by 9.

(b) $-3(k + 9) = -3 \cdot k + (-3) \cdot 9 = -3k + (-27) = -3k - 27$

(c) $6(y - 5) = 6 \cdot y - 6 \cdot 5 = 6y - 30$

(d) $-2(x - 3) = -2 \cdot x - (-2) \cdot 3 = -2x - (-6) = -2x + 6$

> **CAUTION**
> Notice how the final step in Example 2(b) above uses the definition of subtraction "in reverse."
>
> $$-3k + (-27)$$
> Change addition to subtraction; change (-27) to its opposite, 27
> $$-3k - 27$$

◀ *Work Problem* **2** *at the Side.*

OBJECTIVE **3** **Combine like terms.** A single letter or number, or the product of a variable and a number, makes up a *term*. Here are six examples of terms.

$$3y \qquad 5 \qquad -9 \qquad 8r \qquad 10r^2 \qquad x$$

Terms with exactly the same variable and the same exponent are called **like terms.**

$5x$	and	$3x$	like terms
$5x$	and	$3m$	*not* like terms; variables are different
$5x^2$	and	$5x^3$	*not* like terms; exponents are different
$5x^4$	and	$3x^4$	like terms

The distributive property can be used to simplify a sum of like terms such as $6r + 3r$.

$$6r + 3r = (6 + 3)r = 9r$$

This process of *adding* the like terms is called *combining like terms.*

EXAMPLE 3 **Combining Like Terms**

Use the distributive property to combine like terms.

(a) $5k + 11k = (5 + 11)k = 16k$

(b) $10m - 14m + 2m = (10 - 14 + 2)m = -2m$

(c) $-5x + x$ can be written $-5x + 1x = (-5 + 1)x = -4x$

x is equivalent to $1 \cdot x$ or $1x$.

◀ *Work Problem* **3** *at the Side.*

OBJECTIVE **4** **Solve more difficult equations.** The next examples show you how to solve more difficult equations using the addition, multiplication, and distributive properties.

EXAMPLE 4 **Solving Equations**

Solve each equation. Check each solution.

(a) $6r + 3r = 36$

You can combine $6r$ and $3r$ because they are *like terms*. $6r + 3r$ is $9r$, so the equation becomes

> $6r + 3r$ is $(6 + 3)r$ or $9r$.

$$9r = 36$$

$$\frac{\overset{1}{\cancel{9}} \cdot r}{\underset{1}{\cancel{9}}} = \frac{36}{9} \qquad \text{Divide both sides by 9}$$

$$r = 4 \longleftarrow \text{The solution is 4}$$

Check

$$6r + 3r = 36 \longleftarrow \text{Original equation}$$

$$6(4) + 3(4) = 36 \qquad \text{Replace } r \text{ with 4} \quad \boxed{\text{4 is the solution, } \textbf{not } 36.}$$

$$24 + 12 = 36$$

$$36 = 36 \qquad \text{True, so 4 is the correct solution.}$$

4 is the correct solution (**not** 36).

(b) $2k - 2 = 5k - 11$

One way to get the variable term by itself on one side is to subtract $5k$ from both sides.

$$2k - 2 - 5k = 5k - 11 - 5k$$

> $2k - 5k$ is $(2 - 5)k$ or $-3k$

$$2k - 5k - 2 = 5k - 5k - 11$$

$$-3k - 2 = -11$$

Next, add 2 to both sides.

$$-3k - 2 + 2 = -11 + 2$$

$$-3k = -9$$

Finally, divide both sides by -3.

$$\frac{\overset{1}{\cancel{-3}} \cdot k}{\underset{1}{\cancel{-3}}} = \frac{-9}{-3} \qquad \boxed{\text{In division, when signs } \textit{match}, \text{ the result is } \textit{positive}.}$$

$$k = 3 \longleftarrow \text{The solution is 3}$$

Check

$$2k - 2 = 5k - 11 \longleftarrow \text{Original equation}$$

$$2(3) - 2 = 5(3) - 11 \qquad \text{Replace } k \text{ with 3}$$

$$6 - 2 = 15 - 11$$

$$4 = 4 \qquad \text{True, so 3 is the correct solution.}$$

3 is the correct solution (**not** 4).

Work Problem **4** *at the Side.* ▶

---

4 Solve each equation. Check each solution.

(a) $3y - 1 = 2y + 7$

(b) $5a + 7 = 3a - 9$

(c) $3p - 2 = p - 6$

Now that you know about the distributive property and combining like terms, here is a summary of all the steps you can use to solve an equation.

5 Solve each equation. Check each solution.

(a) $-12 = 4(y - 1)$

> **Solving an Equation**
>
> *Step 1*   If possible, use the *distributive property* to remove parentheses.
>
> *Step 2*   *Combine any like terms* on the left side of the equation. Combine any like terms on the right side of the equation.
>
> *Step 3*   *Add or subtract* the same amount on *both* sides of the equation so that the variable term ends up by itself on one side.
>
> *Step 4*   *Multiply or divide* *both* sides by the same number to find the solution.
>
> *Step 5*   *Check* your solution by going back to the *original equation*. Replace the variable with your solution. Follow the order of operations to complete the calculations. If the two sides of the equation are equal, your solution is correct.

(b) $5(m + 4) = 20$

EXAMPLE 5   **Solving Equations Using the Distributive Property**

Solve $-6 = 3(y - 2)$

Step 1   Use the distributive property on the right side of the equation.

$$3(y - 2) \quad \text{becomes} \quad 3 \cdot y - 3 \cdot 2 \quad \text{or} \quad 3y - 6$$

Now the equation looks like this.

$$-6 = 3y - 6$$

Step 2   Combine like terms. Check the left side of the equation. There are no like terms. Check the right side. No like terms there either, so go on to Step 3.

Step 3   Add 6 to both sides to get the variable term by itself on the right side.

$$-6 + 6 = 3y - 6 + 6$$
$$0 = 3y$$

(c) $6(t - 2) = 18$

Step 4   Divide both sides by 3.

> You can divide 0 by a non-zero number. The result is 0.

$$\frac{0}{3} = \frac{\overset{1}{\cancel{3}} \cdot y}{\underset{1}{\cancel{3}}}$$

$$0 = y \leftarrow \text{The solution is 0}$$

Step 5   Check. Go back to the original equation.

$$-6 = 3(y - 2) \leftarrow \text{Original equation}$$
$$-6 = 3(0 - 2) \qquad \text{Replace } y \text{ with 0}$$
$$-6 = 3(-2)$$

> 0 is the solution, **not** -6.

$$-6 = -6 \qquad \text{True, so 0 is the correct solution.}$$

0 is the correct solution (*not* -6).

◀ *Work Problem* **5** *at the Side.*

9.7 ▶▶▶ Exercises

Solve each equation. Check each solution. See Example 1.

🌐 **1.** $7p + 5 = 12$

$p = 1$

2. $6k + 3 = 15$

$k = 2$

3. $2 = 8y - 6$

$y = 1$

4. $10 = 11p - 12$

$p = 2$

5. $-3m + 1 = 1$

$m = 0$

6. $-4k + 5 = 5$

$k = 0$

7. $28 = -9a + 10$

$a = -2$

8. $5 = -10p + 25$

$p = 2$

9. $-5x - 4 = 16$

$x = -4$

10. $-12a - 3 = 21$

$a = -2$

11. $-\dfrac{1}{2}z + 2 = -1$

$z = 6$

12. $-\dfrac{5}{8}r + 4 = -6$

$r = 16$

13. $-0.7 = 5b - 5.2$

$b = 0.9$

14. $0.25 = -3c + 0.85$

$c = 0.2$

Use the distributive property to simplify. See Example 2.

🌐 **15.** $6(x + 4)$

$6x + 24$

16. $8(k + 5)$

$8k + 40$

17. $7(p - 8)$

$7p - 56$

18. $9(t - 4)$

$9t - 36$

19. $-3(m + 6)$

$-3m - 18$

20. $-5(a + 2)$

$-5a - 10$

🌐 **21.** $-2(y - 3)$

$-2y + 6$

22. $-4(r - 7)$

$-4r + 28$

23. $-8(c + 8)$

$-8c - 64$

24. $-6(n + 5)$

$-6n - 30$

25. $-10(w - 9)$

$-10w + 90$

26. $-11(x - 11)$

$-11x + 121$

Combine like terms. See Example 3.

27. $11r + 6r$

  17r

28. $2m + 5m$

  7m

29. $8z - 7z$

  1z or z

30. $10x - 2x$

  8x

31. $y - 3y$

  −2y

32. $-10a + a$

  −9a

33. $-4t + t - 4t$

  −7t

34. $3y - y - 4y$

  −2y

35. $7p - 9p + 2p$

  0p = 0

36. $-6c - c + 7c$

  0c = 0

37. $\dfrac{5}{2}b - \dfrac{11}{2}b$

  −3b

38. $\dfrac{3}{8}d - \dfrac{9}{8}d$

  $-\dfrac{3}{4}d$

Solve each equation. Check each solution. See Example 4.

39. $4k + 6k = 50$

  k = 5

40. $3a + 2a = 15$

  a = 3

41. $54 = 10m - m$

  m = 6

42. $28 = x + 6x$

  x = 4

43. $2b - 6b = 24$

  b = −6

44. $3r - 9r = 18$

  r = −3

45. $-12 = 6y - 18y$

  y = 1

46. $-5 = 10z - 15z$

  z = 1

47. $6p - 2 = 4p + 6$

  p = 4

48. $5y - 5 = 2y + 10$

  y = 5

49. $9 + 7z = 9z + 13$

  z = −2

50. $8 + 4a = 2a + 2$

  a = −3

51. $-2y + 6 = 6y - 10$

$y = 2$

52. $5x - 4 = -3x + 4$

$x = 1$

53. $b + 3.05 = 2$

$b = -1.05$

54. $t + 0.8 = -1.7$

$t = -2.5$

55. $2.5r + 9 = -1$

$r = -4$

56. $0.5x - 6 = 2$

$x = 16$

Solve each equation by using the distributive property. Check each solution. See Example 5.

57. $-10 = 2(y + 4)$

$y = -9$

58. $-3 = 3(x + 6)$

$x = -7$

59. $-4(t + 2) = 12$

$t = -5$

60. $-5(k + 3) = 25$

$k = -8$

61. $6(x - 5) = -30$

$x = 0$

62. $7(r - 5) = -35$

$r = 0$

63. Solve $-2t - 10 = 3t + 5$. Show each step you take to solve it. Next to each step, write a sentence that explains what you did in that step. Be sure to tell when you used the addition property of equations and when you used the multiplication property of equations.

$-2t - 10 = 3t + 5$

$-2t - 3t - 10 = 3t - 3t + 5$ **Subtract 3t from both sides (addition property).**

$-5t - 10 = 5$

$-5t - 10 + 10 = 5 + 10$ **Add 10 to both sides (addition property).**

$-5t = 15$

$\dfrac{-5 \cdot t}{-5} = \dfrac{15}{-5}$ **Divide both sides by -5 (multiplication property).**

$t = -3$

64. Here is one student's solution to an equation.

$3(2x + 5) = -7$

$6x + 5 = -7$

$6x + 5 - 5 = -7 - 5$

$6x = -12$

$x = -2$

Show how to check the solution. If the solution does not check, find and correct the error.

Check: $3(2(-2) + 5) = -7$

$3(-4 + 5) = -7$

$3(1) = -7$

$3 = -7$ **False**

The first step on the left side should have been $3(2x + 5)$ is $6x + 15$. The correct solution is $x = -\dfrac{11}{3}$.

Solve each equation.

65. $30 - 40 = -2x + 7x - 4x$

$x = -10$

66. $-6 - 5 + 14 = -50a + 51a$

$a = 3$

67. $0 = -2(y - 2)$

$y = 2$

68. $0 = -9(b - 1)$

$b = 1$

69. $\dfrac{y}{2} - 2 = \dfrac{y}{4} + 3$

$y = 20$

70. $\dfrac{z}{3} + 1 = \dfrac{z}{2} - 3$

$z = 24$

71. $-3(w - 2) = |0 - 13| + 4w$

$w = -1$

72. $-5b - |47 - 7| = -8(b + 8)$

$b = -8$

73. $2(a + 0.3) = 1.2(a - 4)$

$a = -6.75$

74. $-0.5(c - 4) = -3(c - 2.5)$

$c = 2.2$

9.8 ▶▶▶ Using Equations to Solve Application Problems

It is rare for an application problem to be presented as an equation. Usually, the problem is given in words. You need to *translate* these words into an equation that you can solve.

OBJECTIVE 1 Translate word phrases into expressions with variables. Examples 1 and 2 show you how to translate word phrases into algebraic expressions.

EXAMPLE 1 Translating Word Phrases into Expressions with Variables

Write each word phrase in symbols, using x as the variable.

Words	Algebraic Expression
A number **plus** 2	$x + 2$ or $2 + x$
The **sum** of 8 and a number	$8 + x$ or $x + 8$
5 **more than** a number	$x + 5$ or $5 + x$
-35 **added to** a number	$-35 + x$ or $x + (-35)$
A number **increased by** 6	$x + 6$ or $6 + x$
9 **less than** a number	$x - 9$
A number **subtracted from** 3	$3 - x$
A number **decreased by** 4	$x - 4$
10 **minus** a number	$10 - x$

CAUTION
Recall that addition can be done in any order, so $x + 2$ gives the same result as $2 + x$. This is ***not*** true in subtraction, so be careful. $10 - x$ does ***not*** give the same result as $x - 10$.

Work Problem **1** *at the Side.* ▶

EXAMPLE 2 Translating Word Phrases into Expressions with Variables

Write each word phrase in symbols, using x as the variable.

Words	Algebraic Expression
8 **times** a number	$8x$
The **product** of 12 and a number	$12x$
Double a number (meaning "2 times")	$2x$
Twice a number (meaning "2 times")	$2x$
The **quotient** of 6 and a number	$\dfrac{6}{x}$
A number **divided by** 10	$\dfrac{x}{10}$
One-third of a number	$\dfrac{1}{3}x$ or $\dfrac{x}{3}$
The result **is**	$=$

Work Problem **2** *at the Side.* ▶

OBJECTIVES

1 Translate word phrases into expressions with variables.

2 Translate sentences into equations.

3 Solve application problems.

---

1 Write each word phrase in symbols, using x as the variable.

(a) 15 less than a number

(b) 12 more than a number

(c) A number increased by 13

(d) A number minus 8

(e) -10 plus a number

(f) A number subtracted from 6

2 Write each word phrase in symbols, using x as the variable.

(a) Double a number

(b) The product of -8 and a number

(c) The quotient of 15 and a number

(d) One-half of a number

ANSWERS

1. **(a)** $x - 15$ **(b)** $x + 12$ or $12 + x$
   (c) $x + 13$ or $13 + x$ **(d)** $x - 8$
   (e) $-10 + x$ or $x + (-10)$ **(f)** $6 - x$
2. **(a)** $2x$ **(b)** $-8x$ **(c)** $\dfrac{15}{x}$ **(d)** $\dfrac{1}{2}x$ or $\dfrac{x}{2}$

3 Translate each sentence into an equation and solve it. Check your solution by going back to the words in the original problem.

(a) If 3 times a number is added to 4, the result is 19. Find the number.

(b) If −6 times a number is added to 5, the result is −13. Find the number.

(c) If twice a number is subtracted from 65, the result is −21.

OBJECTIVE **2** **Translate sentences into equations.** The next example shows you how to translate a sentence into an equation that you can solve.

EXAMPLE 3 **Translating a Sentence into an Equation**

If 5 times a number is added to 11, the result is 26. Find the number.

Let *x* represent the unknown number.
Use the information in the problem to write an equation.

> **Note**
> The phrase "the result is" translates to "= ."

Next, solve the equation.

$$5x + 11 - 11 = 26 - 11 \qquad \text{Subtract 11 from both sides.}$$
$$5x = 15$$
$$\frac{\overset{1}{\cancel{5}}x}{\underset{1}{\cancel{5}}} = \frac{15}{5} \qquad \text{Divide both sides by 5}$$
$$x = 3 \longleftarrow \text{The solution is 3}$$

The unknown number is 3.
To check the solution, go back to the words of the *original* problem.

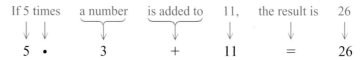

Does 5 • 3 + 11 really equal 26? Yes, 15 + 11 = 26. So 3 is the correct solution because it "works" when you put it back into the original problem.

◄ *Work Problem* **3** *at the Side.*

OBJECTIVE **3** **Solve application problems.** In **Section 1.10,** you learned how to solve application problems using six steps. Now that you know how to use variables and solve equations, we can include those tools in the problem-solving process. Here is a revised list of the six steps we will use.

Solving an Application Problem Using Algebra

Step 1 **Read** the problem carefully to see what it is about.

Step 2 **Assign a variable** by identifying the unknown(s). Write down what the variable represents. If there are several unknowns, let the variable represent the one you know the least about.

Step 3 **Write an equation** using the variable expression(s).

Step 4 **Solve** the equation.

Step 5 **State the answer.**

Step 6 **Check** the solution in the words of the original problem.

Notice that Steps 1, 5, and 6 are nearly identical to what you have been using. Steps 2, 3, and 4 introduce the use of variables and equations.

ANSWERS

3. **(a)** $3x + 4 = 19$
     $x = 5$
   (b) $-6x + 5 = -13$
     $x = 3$
   (c) $65 - 2x = -21$
     $x = 43$

EXAMPLE 4 **Solving an Application Problem with One Unknown**

Jennifer put some money aside in an envelope for small household expenses. Yesterday she took out $30 for groceries. Today a friend paid back a $42 loan, and Jennifer put the cash in the envelope. Now she has $51 in the envelope. How much was in the envelope at the start?

Step 1 **Read.** The problem asks for the amount of money in the envelope at the start.

Step 2 **Assign a variable.** There is only *one* unknown: the amount of money in the envelope at the start. Let *x* represent the money at the start.

Step 3 **Write an equation.**

Money at the start	Took out $30	Put in $42	Ended up with $51
x	$- \$30$	$+ \$42$	$= \$51$

Step 4 **Solve.**

> On the left side, $-30 + 42$ is $+12$.

$$x \underbrace{- 30 + 42} = 51$$

$$x \quad + 12 \quad = 51$$

$$x + 12 - 12 = 51 - 12 \qquad \text{Subtract 12 from both sides.}$$

$$x = 39 \longleftarrow \text{The solution is 39}$$

Step 5 **State the answer.** Jennifer had $39 in the envelope at the start.

> Be sure to write a $ in the answer.

Step 6 **Check.** Use the words of the *original* problem.

	Took out	Put in	Now has
Had			
$\$39$	$- \$30$	$+ \$42$	$= \$51$
	$\$9$	$+ \$42$	$= \$51$
		$\$51$	$= \$51$

So $39 is the correct solution because it "works" in the original problem.

Work Problem **4** *at the Side.* ▶

4 A college bookstore ordered some red pens. The store sold a dozen of the pens last week and two dozen this week. There are 48 red pens left. How many were originally ordered?

Show your work for each of the six problem-solving steps.

5 Susan donated $10 more than twice what LuAnn donated. If Susan donated $22, how much did LuAnn donate?

Show your work for each of the six problem-solving steps.

EXAMPLE 5 **Solving an Application Problem with One Unknown**

Michael has 5 less than three times as many lab experiments completed as David. If Michael has completed 13 experiments, how many lab experiments has David completed?

Step 1 **Read.** The problem asks for the number of lab experiments completed by David.

Step 2 **Assign a variable.** There is only *one* unknown: David's number of experiments.
Let x represent David's number of experiments.

Step 3 **Write an equation.**

The number Michael did	is	5 less than 3 times David's number.
13	=	$3x - 5$

> Be careful when subtracting! $5 - 3x$ does **not** mean the same thing as $3x - 5$.

Step 4 **Solve.**

$$13 = 3x - 5$$
$$13 + 5 = 3x - 5 + 5 \qquad \text{Add 5 to both sides.}$$
$$18 = 3x$$
$$\frac{18}{3} = \frac{\overset{1}{\cancel{3}}x}{\underset{1}{\cancel{3}}} \qquad \text{Divide both sides by 3}$$
$$6 = x \longleftarrow \text{The solution is 6}$$

Step 5 **State the answer.** David completed 6 lab experiments.

Step 6 **Check.** Use the words of the *original problem*.

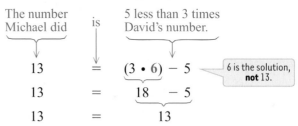

The number Michael did	is	5 less than 3 times David's number.
13	=	$(3 \cdot 6) - 5$
13	=	$18 - 5$
13	=	13

> 6 is the solution, **not** 13.

So 6 is the correct solution because it "works" in the original problem.

> **Note**
>
> In **Step 3** above, you may also write the equation as $3x - 5 = 13$, with the two sides switched. The solution will be the same.

◀ *Work Problem* **5** *at the Side.*

EXAMPLE 6 **Solving an Application Problem with Two Unknowns**

During the day, Sheila drove 72 km more than Russell. The total distance traveled by them both was 232 km. Find the distance traveled by each person.

Step 1 **Read.** The problem asks how far Sheila drove and how far Russell drove.

Step 2 **Assign a variable.** There are *two* unknowns: Sheila's distance and Russell's distance.

Let x be the distance traveled by Russell, because you know less about his distance than Sheila's distance.

Since Sheila drove 72 km *more* than Russell, the distance she traveled is $x + 72$ km, that is, Russell's distance (x) plus 72 km.

Step 3 **Write an equation.**

Distance for Russell	plus	distance for Sheila	is	total distance.
x	$+$	$x + 72$	$=$	232

Step 4 **Solve.**

$$x + x + 72 = 232$$ On the left side, $x + x$ is $1x + 1x$, which is $2x$.

$$2x + 72 = 232$$

$$2x + 72 - 72 = 232 - 72$$ Subtract 72 from both sides.

$$2x = 160$$

$$\frac{\overset{1}{\cancel{2}}x}{\underset{1}{\cancel{2}}} = \frac{160}{2}$$ Divide both sides by 2

$$x = 80 \longleftarrow \text{The solution is 80}$$

Step 5 **State the answer.**

Russell's distance is x, so Russell traveled 80 km.

Sheila's distance is $x + 72$,

so Sheila traveled 80 km + 72 km = 152 km.

> Be sure to write **km** in the answers.

Step 6 **Check.** Use the words of the *original* problem.

"Sheila drove 72 km more than Russell."

Sheila's 152 km is 72 km more than Russell's 80 km, so that checks.

"The total distance traveled by them both was 232 km."

Sheila's 152 km + Russell's 80 km = 232 km, so that checks.

Work Problem ⑥ *at the Side.* ▶

---

⑥ Solve each application using the six problem-solving steps.

(a) In a day of work, Keonda made $12 more than her daughter. Together they made $182. Find the amount made by each person. (*Hint:* Which amount do you know the least about, Keonda's or her daughter's? Let x be that amount.)

(b) A rope is 21 yd long. Marcos cut it into two pieces, so that one piece is 3 yd longer than the other. Find the length of each piece.

7 Make a drawing and use the six steps to solve this problem. The length of Ann's rectangular garden plot is 3 yd more than the width. She used 22 yd of fencing around the edge. Find the length and the width of the garden.

EXAMPLE 7 **Solving a Geometry Application Problem**

The length of a rectangle is 2 cm more than the width. The perimeter is 68 cm. Find the length and width.

Step 1 **Read.** The problem asks for the length and the width of a rectangle.

Step 2 **Assign a variable.** There are *two* unknowns, length and width. You know the least about the width, so let x represent the **width.**

Since the length is 2 cm *more* than the width, the **length** is $x + 2$. A drawing of the rectangle will help you see these relationships.

width is x
length is $x + 2$
perimeter is 68

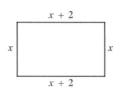

Step 3 **Write an equation.**

Use the formula for perimeter of a rectangle, $P = 2l + 2w$.

$$P = 2 \cdot l \quad + 2 \cdot w$$

Step 4 **Solve.**

$$68 = \underbrace{2(x + 2)}_{} + \underbrace{2 \cdot x}_{} \quad \text{Use the distributive property.}$$

$$68 = 2x + 4 \quad + \quad 2x \quad \text{Combine like terms.}$$

$$68 = 4x + 4$$

$$68 - 4 = 4x + 4 - 4 \quad \text{Subtract 4 from both sides.}$$

$$64 = 4x$$

$$\frac{64}{4} = \frac{\overset{1}{\cancel{4}} \cdot x}{\underset{1}{\cancel{4}}} \quad \text{Divide both sides by 4}$$

$$16 = x \longleftarrow \text{The solution is 16}$$

Step 5 **State the answer.**

The width is x, so the width is 16 cm.

The length is $x + 2$, so the length is 16 + 2 or 18 cm.

Be sure to write **cm** in the answers.

The width is 16 cm and the length is 18 cm.

Step 6 **Check.** Use the words of the original problem. It says the length is 2 cm more than the width. 18 cm is 2 cm more than 16 cm, so that part checks.

The original problem also says the perimeter is 68 cm. Use 18 cm and 16 cm to find the perimeter.

$$P = \underbrace{2 \cdot 18 \text{ cm}}_{} + \underbrace{2 \cdot 16 \text{ cm}}_{}$$

$$P = \quad 36 \text{ cm} \quad + \quad 32 \text{ cm} \quad = 68 \text{ cm} \leftarrow \text{Checks}$$

◀ *Work Problem* **7** *at the Side.*

ANSWER

7.

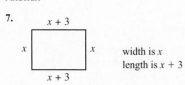

width is x
length is $x + 3$

$22 = 2(x + 3) + 2 \cdot x$
width is 4 yd
length is 7 yd
Check: 7 yd is 3 more than 4 yd.

$$P = 2 \cdot 7 \text{ yd} + 2 \cdot 4 \text{ yd}$$
$$P = 14 \text{ yd} + 8 \text{ yd}$$
$$P = 22 \text{ yd} \leftarrow \text{Checks}$$

9.8 ▶▶▶ **Exercises**

FOR EXTRA HELP

 PRACTICE   WATCH   DOWNLOAD   READ REVIEW

Write each word phrase in symbols, using x as the variable. See Examples 1 and 2.

1. 14 plus a number
$14 + x$ or $x + 14$

2. The sum of a number and -8
$x + (-8)$ or $-8 + x$

3. -5 added to a number
$-5 + x$ or $x + (-5)$

4. 16 more than a number
$x + 16$ or $16 + x$

5. 20 minus a number
$20 - x$

6. A number decreased by 25
$x - 25$

7. 9 less than a number
$x - 9$

8. A number subtracted from -7
$-7 - x$

9. Subtract 4 from a number
$x - 4$

10. 3 fewer than a number
$x - 3$

11. Six times a number
$6x$

12. The product of -3 and a number
$-3x$

13. Double a number
$2x$

14. Half a number
$\dfrac{x}{2}$ or $\dfrac{1}{2}x$

15. A number divided by 2
$\dfrac{x}{2}$

16. 4 divided by a number
$\dfrac{4}{x}$

17. Twice a number added to 8
$8 + 2x$ or $2x + 8$

18. Five times a number plus 5
$5x + 5$ or $5 + 5x$

19. 10 fewer than seven times a number
$7x - 10$

20. 12 less than six times a number
$6x - 12$

21. The sum of twice a number and the number
$2x + x$ or $x + 2x$

22. Triple a number subtracted from the number
$x - 3x$

23. In your own words, write a definition for each of these words: (a) variable; (b) expression; (c) equation. Give three examples to illustrate each definition.

    (a) **A variable is a letter that represents an unknown quantity. Examples:** x, w, p

    (b) **An expression is a combination of operations on variables and numbers. Examples:** $6x, w - 5, 2p + 3x$

    (c) **An equation has an = sign and shows that two expressions are equal. Examples:** $2y = 14$; $x + 5 = 2x$; $8p - 10 = 54$

24. "You can use any letter to represent the unknown in an application problem." Is this statement true or false? Explain your answer.

    True. Choose any letter you like, although it may help to choose a letter that reminds you of what it represents, such as w for an unknown width.

Translate each sentence into an equation and solve it. See Example 3.

25. If four times a number is decreased by 2, the result is 26. Find the number.

    $4n - 2 = 26$;
    $n = 7$

26. The sum of 8 and five times a number is 53. Find the number.

    $8 + 5n = 53$;
    $n = 9$

27. If twice a number is added to the number, the result is -15. What is the number?

    $2n + n = -15$;
    $n = -5$

28. If a number is subtracted from three times the number, the result is -8. What is the number?

    $3n - n = -8$;
    $n = -4$

29. If the product of some number and 5 is increased by 12, the result is seven times the number. Find the number.

    $5n + 12 = 7n$;
    $n = 6$

30. If eight times a number is subtracted from eleven times the number, the result is -9. Find the number.

    $11n - 8n = -9$;
    $n = -3$

31. When three times a number is subtracted from 30, the result is 2 plus the number. What is the number?

    $30 - 3n = 2 + n$;
    $n = 7$

32. When twice a number is decreased by 8, the result is the number increased by 7. Find the number.

    $2n - 8 = n + 7$;
    $n = 15$

33. If half a number is added to twice the number, the answer is 50. Find the number.

    $\frac{1}{2}n + 2n = 50$;
    $n = 20$

34. If one-third of a number is added to three times the number, the result is 30. Find the number.

    $\frac{1}{3}n + 3n = 30$;
    $n = 9$

Solve each application problem by using the six problem-solving steps you learned in this section. See Examples 4, 5, and 6.

35. Ricardo gained 15 pounds over the winter. He went on a diet and lost 28 pounds. Then he regained 5 pounds and weighed 177 pounds. How much did he weigh originally?

Let x be Ricardo's original weight.
$x + 15 - 28 + 5 = 177$
He weighed 185 pounds originally.

36. Mr. Chee deposited $80 into his checking account. Then, after writing a $43 check for gas and a $190 check for his child's day care, the balance in his account was $47. How much was in his account before he made the deposit?

Let x be the original amount in the account.
$x + 80 - 43 - 190 = 47$
He had $200 in his account.

37. While shopping for clothes, Consuelo spent $3 less than twice what Brenda spent. Consuelo spent $81. How much did Brenda spend?

Let x be the amount Brenda spent.
$81 = 2x - 3$
Brenda spent $42.

38. Dennis weighs 184 lb. His weight is 2 lb less than six times his child's weight. How much does his child weigh?

Let x be child's weight.
$184 = 6x - 2$
Child weighs 31 lb.

39. My sister is 9 years younger than I am. The sum of our ages is 51. Find our ages.

Let x be my age; $x - 9$ is my sister's age.
$x + x - 9 = 51$
I am 30; my sister is 21.

40. Ed and Marge were candidates for city council. Marge won, with 93 more votes than Ed. The total number of votes cast in the election was 587. Find the number of votes received by each candidate.

Let x be Ed's votes; $x + 93$ is Marge's votes.
$x + x + 93 = 587$
Ed had 247 votes; Marge had 340 votes.

41. Last year, Lien earned $1500 more than her husband. Together they earned $37,500. How much did each of them earn?

Let x be husband's earnings;
$x + 1500$ **is Lien's earnings.**
$x + x + 1500 = 37,500$
Husband earned $18,000; Lien earned $19,500.

42. A $149,000 estate is to be divided between two charities so that one charity receives $18,000 less than the other. How much will each charity receive?

Let x be one charity's amount;
$x - 18,000$ **is the other charity's amount.**
$x + x - 18,000 = 149,000$
One charity receives $83,500 and the other receives $65,500.

43. Jason paid five times as much for his computer as he did for his printer. He paid a total of $1320 for both items. What did each item cost?

Let x be printer's cost; $5x$ is computer's cost.
$x + 5x = 1320.$
The printer cost $220; the computer cost $1100.

44. The attendance at the Saturday night baseball game was three times the attendance at Sunday's game. In all, 56,000 fans attended the games. How many fans were at each game?

Let x be Sunday's attendance;
$3x$ **is Saturday's attendance.**
$x + 3x = 56,000$
14,000 fans on Sunday; 42,000 fans on Saturday

45. A board is 78 cm long. Rosa cut the board into two pieces, with one piece 10 cm longer than the other. Find the length of both pieces. (*Hint:* Make a sketch of the board.)

longer piece | shorter piece

x + 10   x

Let *x* be length of shorter piece;
x + 10 is length of longer piece.
x + *x* + 10 = 78
Shorter piece is 34 cm; longer piece is 44 cm.

46. A rope is 50 ft long. Juan cut it into two pieces so that one piece is 8 ft longer than the other. Find the length of each piece (*Hint:* Make a sketch of the rope.)

longer piece | shorter piece

x + 8   x

Let *x* be length of shorter piece;
x + 8 is length of longer piece.
x + *x* + 8 = 50
Shorter piece is 21 ft; longer piece is 29 ft.

47. A wire is cut into two pieces, with one piece 7 ft shorter than the other. The wire was 31 ft long before it was cut. How long was each piece?

Let *x* be one piece, *x* − 7 the other piece.
x + *x* − 7 = 31
One piece is 19 ft; the other piece is 12 ft.

48. A 90 cm pipe is cut into two pieces so that one piece is 6 cm shorter than the other. Find the length of each piece.

Let *x* be length of one piece;
x − 6 is length of the other piece.
x + *x* − 6 = 90
One piece is 48 cm; the other piece is 42 cm.

In Exercises 49–54, use the formula for the perimeter of a rectangle, P = 2l + 2w.
Make a drawing to help you solve each problem, and use the six problem-solving steps.
See Example 7.

49. The perimeter of a rectangle is 48 yd. The width is 5 yd. Find the length.

48 = 2(5) + 2x
Length is 19 yd.

x
5 [ ] 5
x

50. The length of a rectangle is 27 cm, and the perimeter is 74 cm. Find the width of the rectangle.

74 = 2(27) + 2w
Width is 10 cm.

27
w [ ] w
27

51. A rectangular dog pen is twice as long as it is wide. The perimeter of the pen is 36 ft. Find the length and the width of the pen.

36 = 2(2w) + 2w
Length is 12 ft;
width is 6 ft.

2w
w [ ] w
2w

52. A new city park has a rectangular shape. The length is triple the width. It will take 240 yd of fencing to go around the park. Find the length and width of the park.

240 = 2(3w) + 2w
Length is 90 yd;
width is 30 yd.

3w
w [ ] w
3w

53. The length of a rectangular jewelry box is 3 inches more than twice the width. The perimeter is 36 inches. Find the length and the width.

36 = 2(2x + 3) + 2x
Length is 13 in.;
width is 5 in.

2x + 3
x [ ] x
2x + 3

54. The perimeter of a rectangular house is 122 ft. The width is 5 ft less than the length. Find the length and the width.

122 = 2(x − 5) + 2x
Length is 33 ft;
width is 28 ft.

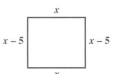

x
x − 5 [ ] x − 5
x

Chapter 9 ▶▶▶ Summary

▶ Key Terms

9.1	**negative numbers**	Negative numbers are numbers that are less than 0.
	signed numbers	Signed numbers include positive numbers, negative numbers, and 0.
	absolute value	Absolute value is the distance of a number from 0 on a number line. Absolute value is never negative.
	opposite of a number	The opposite of a number is a number the same distance from 0 on a number line, but on the opposite side of 0.
9.2	**additive inverse**	The additive inverse is the opposite of a number. The sum of a number and its additive inverse is always 0.
9.5	**variables**	Variables are letters that represent numbers.
	expression	An expression is a combination of operations on variables and numbers.
9.6	**equation**	An equation is a statement that says two expressions are equal. An equation contains an equal sign.
	solution	The solution of an equation is a number that can replace the variable so that the equation is true.
	addition property of equations	The addition property of equations states that the same number can be added or subtracted on both sides of an equation.
	multiplication property of equations	The multiplication property of equations states that both sides of an equation can be multiplied or divided by the same number, except division by 0 cannot be done.
9.7	**distributive property**	If a, b, and c are three numbers, the distributive property says that $a(b + c) = ab + ac$.
	like terms	Like terms are terms with exactly the same variables and the same exponents.

▶ New Symbols

-6 negative sign; read as "negative 6"; indicates the number is *less* than 0.

$|4|$ and $|-2|$ absolute value of 4; absolute value of -2

$a, b, c, d,$ etc.   variables that represent numbers

▶ Test Your Word Power

See how well you have learned the vocabulary in this chapter. Answers follow the Quick Review.

1. An **expression**
   - A. has an equal sign
   - B. contains like terms with the same variables and exponents
   - C. follows the order of operations
   - D. is a combination of operations on variables and numbers.

2. An **equation**
   - A. always has more than one solution
   - B. can be graphed on a number line
   - C. is evaluated by adding the same number to both sides
   - D. contains an equal sign.

3. The **absolute value** of a number is
   - A. its distance from 0
   - B. never positive
   - C. used when multiplying
   - D. less than 0.

4. **Like terms**
   - A. are always positive
   - B. are the same distance from 0 but in opposite directions
   - C. have matching variables and exponents
   - D. can be multiplied but not added.

5. The **opposite** of a number is
   - A. never negative
   - B. called the additive inverse
   - C. called the absolute value
   - D. at the same point on the number line.

6. A **variable** is
   - A. a solution of an equation
   - B. a negative number
   - C. a letter representing a number
   - D. one of several like terms.

▶ Quick Review

Concepts	Examples

9.1 Graphing Signed Numbers

Place a dot at the correct location on the number line. Positive numbers are to the right of 0; negative numbers are to the left of 0.

Graph $-2, 1, 0,$ and $2\frac{1}{2}$.

9.1 Using the $<$ and $>$ Symbols

Place the symbols $<$ (less than) or $>$ (greater than) between two numbers to make the statement true. The smaller pointed end of the symbol points to the smaller number.

Use the symbol $<$ or $>$ to make each statement true.

$2 \underline{\quad > \quad} 1$

$-3 \underline{\quad > \quad} -5$

$-6 \underline{\quad < \quad} 2$

9.1 Finding the Absolute Value of a Number

Determine the distance from 0 to the given number on the number line. Because the absolute value is a distance, it is never negative.

Simplify each absolute value expression.

(a) $|8|$ **(b)** $|-7|$ **(c)** $-|-5|$

$|8| = 8$ $|-7| = 7$ $\rightarrow -5$

9.1 Finding the Opposite of a Number

Find the number that is the same distance from 0 as the given number, but on the opposite side of 0 on the number line.

Find the opposite of each number.

(a) -6 **(b)** $+9$

$-(-6) = 6$ $-(+9) = -9$

9.2 Adding Two Signed Numbers

Adding Two Positive Numbers
Add the absolute values. The sum is positive.

Add.

(a) $8 + 6 = 14$

(b) $-8 + (-6)$

Adding Two Negative Numbers
Add the absolute values and write a negative sign in front of the sum.

Find absolute values. $|-8| = 8$ $|-6| = 6$

Add absolute values: $8 + 6 = 14$.
Write a negative sign in front of the sum.

$$-8 + (-6) = -14$$

Adding Two Numbers with Different Signs
Subtract the lesser absolute value from the greater absolute value. Write the sign of the number with the greater absolute value in front of the answer.

(c) $5 + (-7)$

Find absolute values. $|5| = 5$ $|-7| = 7$

Subtract the lesser absolute value from the greater:
$7 - 5 = 2$.

The number with the greater absolute value is -7. Its sign is negative, so write a negative sign in front of the answer.

$$5 + (-7) = -2$$

9.2 Subtracting Two Signed Numbers

To subtract two numbers, add the first number to the opposite of the second number. Follow these steps.
Step 1   Change the subtraction sign to addition.
Step 2   Change the sign of the second number to its opposite.
Step 3   Proceed as in addition.

Subtract.

(a) $-6 - \quad 5$
          ↓   ↓
$-6 + (-5) = -11$

(b) $5 - (-8)$
          ↓   ↓
$5 + (+8) = 13$

Concepts	Examples

9.3 Multiplying Signed Numbers

The product of two numbers with the *same* sign is *positive*.
The product of two numbers with *different* signs is *negative*.

Multiply.
(a) $7 \cdot 3 = 21$ **(b)** $(-3) \cdot 4 = -12$
(c) $5(-8) = -40$ **(d)** $(-9)(-6) = 54$

9.3 Dividing Signed Numbers

Use the same sign rules as for multiplying signed numbers.
When two numbers have the *same* sign, the quotient is *positive*.
When two numbers have *different* signs, the quotient is *negative*.

Divide
(a) $\dfrac{8}{4} = 2$ **(b)** $\dfrac{-20}{5} = -4$

(c) $\dfrac{50}{-5} = -10$ **(d)** $\dfrac{-12}{-6} = 2$

9.4 Using the Order of Operations to Simplify Numerical Expressions

Use the order of operations to simplify numerical expressions.
1. Do all operations inside **parentheses** or **other grouping symbols.**
2. Simplify any expressions with **exponents** and find any **square roots.**
3. Do the remaining **multiplications and divisions** as they occur from left to right.
4. Do the remaining **additions and subtractions** as they occur from left to right.

Simplify.
(a) $-4 + 6 \div (-2)$ **(b)** $3^2 + 3(8 \div 2)$
$\quad\;\; -4 + \;\; (-3)$ $\quad\;\; 3^2 + 3 \;\; (4)$
$\qquad\quad -7$ $\qquad 9 + 3(4)$
$\qquad\qquad\qquad\qquad\qquad 9 + 12$
$\qquad\qquad\qquad\qquad\qquad\quad 21$

9.5 Evaluating Expressions

Replace each variable in the expression with the specified number.
Use the order of operations to simplify the expression.
Remember that when no operation sign is written between a number and a letter, you assume it is multiplication. For example, $7x$ means $7 \cdot x$.

What is the value of $6p - 5s$, if $p = -3$ and $s = -4$?
$$6\,p \; - 5\,s$$
$$6(-3) - 5(-4)$$
$$-18 - (-20)$$
$$2$$

9.6 Determining Whether a Number Is a Solution of an Equation

Substitute the number for the variable in the equation.
If the resulting statement is *true,* the number is a solution of the equation.
If the resulting statement is *false,* the number is *not* a solution of the equation.

Is 4 a solution of the equation $3x - 5 = 7$?

Replace x with 4.
$$3(4) - 5 = 7$$
$$12 - 5 = 7$$
$$7 = 7 \qquad \text{True}$$

The result is true, so 4 is a solution of $3x - 5 = 7$.

9.6 Using the Addition Property of Equations to Solve an Equation

Add or subtract the same number on both sides of the equation so that you get the variable by itself on one side.

Solve each equation.
(a) $\qquad x - 6 = 9$
$\qquad x - 6 + 6 = 9 + 6 \qquad$ Add 6 to both sides.
$\qquad\qquad x + 0 = 15$
$\qquad\qquad\qquad x = 15 \qquad \leftarrow$ The solution is 15

(b) $\qquad -7 = x + 9$
$\qquad -7 - 9 = x + 9 - 9 \qquad$ Subtract 9 from both sides.
$\qquad\quad -16 = x + 0$
$\qquad\quad -16 = x \qquad \leftarrow$ The solution is -16

Concepts	Examples

9.6 Using the Multiplication Property of Equations to Solve an Equation

Multiply or divide both sides of the original equation by the same number so that you get the variable by itself on one side. (Do not divide by 0.)

Solve each equation.

(a) $-54 = 6x$

$$\frac{-54}{6} = \frac{\overset{1}{\cancel{6}} \cdot x}{\cancel{6}}$$

$$-9 = x$$

(b) $\frac{1}{3}x = 8$

$$\frac{\cancel{3}}{1} \cdot \frac{1}{\cancel{3}}x = 3 \cdot 8$$

$$1x = 24$$

$$x = 24$$

9.7 Solving Equations with Several Steps

Use the following steps.

Step 1 Add or subtract the same amount on both sides of the equation so that the variable ends up by itself on one side.

Step 2 Multiply or divide both sides by the same number to find the solution.

Step 3 Check the solution by going back to the *original* equation.

Solve: $2p - 3 = 9$.

$2p - 3 + 3 = 9 + 3$ Add 3 to both sides.

$$2p = 12$$

$$\frac{\overset{1}{\cancel{2}} \cdot p}{\cancel{2}} = \frac{12}{2}$$ Divide both sides by 2

$$p = 6$$ ← The solution is 6

Check $2p - 3 = 9$ ← Original equation

$2(6) - 3 = 9$ Replace p with 6

$$12 - 3 = 9$$

$$9 = 9$$ True, so 6 is the solution.

6 is the solution (**not** 9).

9.7 Using the Distributive Property

To simplify expressions, use the distributive property.

$$a(b + c) = ab + ac$$

Simplify: $-2(x + 4)$

$$= -2 \cdot x + (-2) \cdot 4$$

$$= -2x + (-8) = -2x - 8$$

9.7 Combining Like Terms

To combine like terms, add or subtract the number parts of the terms. The variable part stays the same.

Combine like terms.
(a) $6p + 7p = (6 + 7)p = 13p$
(b) $8m - 11m = (8 - 11)m = -3m$

9.8 Translating Word Phrases into Expressions with Variables

Use x (or any other letter) as a variable and symbolize the operations described by the words.

Write each word phrase in symbols, using x as the variable.
(a) Two more than a number $x + 2$ or $2 + x$
(b) 8 subtracted from a number $x - 8$
(c) The product of a number and 15 $15x$
(d) A number divided by 9 $\dfrac{x}{9}$

ANSWERS TO TEST YOUR WORD POWER

1. D; *Examples:* $3x - 5$; $2y^2$; $4a + b$.
2. D; *Examples:* $2x + 6 = -4$; $-8 + y = 7y - 14$.
3. A; *Example:* $|-3| = 3$ and $|3| = 3$ because the distance from 0 for both -3 and 3 is 3.
4. C; *Examples:* $2y$ and $-7y$ are like terms; $4x^2$ and x^2 are like terms; $3y^2$ and $3y^3$ are not like terms.
5. B; *Example:* The opposite of 5 is -5; it is the additive inverse because $5 + (-5) = 0$.
6. C; *Examples:* x, y, and d are variables.

Chapter 9 ▶▶▶ Review Exercises

[9.1] *Graph each set of numbers on the number line.*

1. $2, -3, 4, 1, 0, -5$

2. $-2, 5, -4, -1, 3, -6$

3. $-1\dfrac{1}{4}, -\dfrac{5}{8}, -3\dfrac{3}{4}, 2\dfrac{1}{8}, \dfrac{3}{2}, -2\dfrac{1}{8}$

4. $0, -\dfrac{3}{4}, \dfrac{5}{4}, -4\dfrac{1}{2}, \dfrac{7}{8}, -7\dfrac{2}{3}$

Write $<$ or $>$ in each blank to make a true statement.

5. $0 \underline{\quad > \quad} -2$

6. $-5 \underline{\quad < \quad} 0$

7. $-1 \underline{\quad > \quad} -4$

8. $-9 \underline{\quad < \quad} -6$

Simplify each absolute value expression.

9. $|8|$ **8**

10. $|-19|$ **19**

11. $-|-7|$ **−7**

12. $-|15|$ **−15**

[9.2] *Add.*

13. $-4 + 6$ **2**

14. $-10 + 3$ **−7**

15. $-11 + (-8)$ **−19**

16. $-9 + (-24)$ **−33**

17. $12 + (-11)$ **1**

18. $1 + (-20)$ **−19**

19. $\dfrac{9}{10} + \left(-\dfrac{3}{5}\right)$ $\dfrac{3}{10}$

20. $-\dfrac{7}{8} + \dfrac{1}{2}$ $-\dfrac{3}{8}$

21. $-6.7 + 1.5$ **−5.2**

22. $-0.8 + (-0.7)$ **−1.5**

Give the additive inverse (opposite) of each number

23. 6
  −6

24. −14
  14

25. $-\dfrac{5}{8}$

  $\dfrac{5}{8}$

26. 3.75
  −3.75

Subtract.

27. 4 − 10
  −6

28. 7 − 15
  −8

29. −6 − 1
  −7

30. −12 − 5
  −17

31. 8 − (−3)
  11

32. 2 − (−9)
  11

33. −1 − (−14)
  13

34. −10 − (−4)
  −6

35. −40 − 40
  −80

36. −15 − (−15)
  0

37. $\dfrac{1}{3} - \dfrac{5}{6}$

  $-\dfrac{1}{2}$

38. 2.8 − (−6.2)
  9

[9.3] *Multiply or divide.*

39. −4 (6)
  −24

40. 5 • (−4)
  −20

41. −3 (−5)
  15

42. −8 (−8)
  64

43. $\dfrac{80}{-10}$

  −8

44. $\dfrac{-9}{3}$

  −3

45. $\dfrac{-25}{-5}$

  5

46. $\dfrac{-120}{-6}$

  20

47. (−37) (0)
  0

48. (−1) (81)
  −81

49. $\dfrac{0}{-10}$

  0

50. $\dfrac{-20}{0}$

  undefined

51. $\left(\dfrac{2}{3}\right)\left(-\dfrac{6}{7}\right)$

  $-\dfrac{4}{7}$

52. $-\dfrac{4}{5} \div \left(-\dfrac{2}{15}\right)$

  6

53. (−0.5) (−2.8)
  1.4

54. $\dfrac{-5.28}{0.8}$

  −6.6

[9.4] *Simplify each expression.*

55. $2 - 11(-5)$

57

56. $(-4)(-8) - 9$

23

57. $48 \div (-2)^3 - (-5)$

-1

58. $-36 \div (-3)^2 - (-2)$

-2

59. $5(4) - 7(6) + 3(-4)$

-34

60. $2(8) - 4(9) + 2(-6)$

-32

61. $3^3(-4) - 2(5 - 9)$

-100

62. $6(-4)^2 - 3(7 - 14)$

117

63. $\dfrac{3 - (5^2 - 4^2)}{14 + 24 \div (-3)}$

-1

64. $(-0.8)^2(0.2) - (-1.2)$

1.328

65. $\left(-\dfrac{1}{3}\right)^2 + \dfrac{1}{4}\left(-\dfrac{4}{9}\right)$

0

66. $\dfrac{12 \div (2 - 5) + 12(-1)}{2^3 - (-4)^2}$

2

[9.5] *Find the value of each expression using the given values of the variables.*

67. $3k + 5m$

$k = 4, \quad m = 3$

27

68. $3k + 5m$

$k = -6, \quad m = 2$

-8

69. $2p - q$

$p = -5, \quad q = -10$

0

70. $2p - q$

$p = 6, \quad q = -7$

19

71. $\dfrac{5a - 7y}{2 + m}$

$a = 1, \quad y = 4, \quad m = -3$

23

72. $\dfrac{5a - 7y}{2 + m}$

$a = 2, \quad y = -2, \quad m = -26$

-1

Using the given values, evaluate each formula.

73. $P = a + b + c; \quad a = 9, \quad b = 12, \quad c = 14$

$P = 35$

74. $A = \dfrac{1}{2}bh; \quad b = 6, \quad h = 9$

$A = 27$

[9.6–9.7] *Solve each equation. Check each solution.*

75. $y + 3 = 0$
$y = -3$

76. $a - 8 = 8$
$a = 16$

77. $-5 = z - 6$
$z = 1$

78. $-8 = -9 + r$
$r = 1$

79. $-\dfrac{3}{4} + x = -2$

$x = -\dfrac{5}{4}$ or $-1\dfrac{1}{4}$

80. $12.92 + k = 4.87$
$k = -8.05$

81. $-8r = 56$
$r = -7$

82. $3p = 24$
$p = 8$

83. $\dfrac{z}{4} = 5$

$z = 20$

84. $\dfrac{a}{5} = -11$

$a = -55$

85. $20 = 3y - 7$
$y = 9$

86. $-5 = 2b + 3$
$b = -4$

Use the distributive property to simplify each expression.

87. $6(r - 5)$
$6r - 30$

88. $11(p + 7)$
$11p + 77$

89. $-9(z - 3)$
$-9z + 27$

90. $-8(x + 4)$
$-8x - 32$

Combine like terms.

91. $3r + 8r$
$11r$

92. $10z - 15z$
$-5z$

93. $3p - 12p + p$
$-8p$

94. $-6x - x + 9x$
$2x$

Solve each equation. Check each solution.

95. $-4z + 2z = 18$
$z = -9$

96. $-35 = 9k - 2k$
$k = -5$

97. $4y - 3 = 7y + 12$
$y = -5$

98. $b + 6 = 3b - 8$
$b = 7$

99. $-14 = 2(a - 3)$
$a = -4$

100. $42 = 7(t + 6)$
$t = 0$

[9.8] *Write each word phrase in symbols, using x to represent the variable.*

101. 18 plus a number

$18 + x$ or $x + 18$

102. Half a number

$\frac{1}{2}x$ or $\frac{x}{2}$

103. -5 times a number

$-5x$

104. A number subtracted from 20

$20 - x$

Translate each sentence into an equation and solve it.

105. The sum of four times a number and 6 is -14. Find the number.

$4x + 6 = -14;$
$x = -5$

106. If a number is subtracted from five times the number, the result is 100. Find the number.

$5x - x = 100;$
$x = 25$

Solve each application problem using the six problem-steps.

107. A \$30,000 scholarship is being divided between two students so that one student receives three times as much as the other. How much will each student receive?

Let x be one student's scholarship, 3x is the other student's scholarship.

$x + 3x = 30,000$

One student receives \$7,500; the other student receives \$22,500.

108. The perimeter of a rectangular game board is 48 in. The length is 4 in. more than the width. Find the length and width of the game board. (Draw a sketch of the board.)

$48 = 2(w + 4) + 2w$

Width is 10 in.; length is 14 in.

> > > **Mixed Review Exercises**

Add, subtract, multiply, or divide as indicated.

109. $-6 - (-9)$

3

110. $(-8)(-5)$

40

111. $-12 + 11$

-1

112. $\dfrac{-70}{10}$

-7

113. $-4(4)$

-16

114. $5 - 14$

-9

115. $\dfrac{-42}{-7}$

6

116. $16 + (-11)$

5

117. $-10 - 10$

-20

118. $\dfrac{-5}{0}$

undefined

119. $-\dfrac{2}{3} + \dfrac{1}{9}$

$-\dfrac{5}{9}$

120. $0.7(-0.5)$

-0.35

121. $|-6| + 2 - 3(-8) - 5^2$

7

122. $9 \div |-3| + 6(-5) + 2^3$

-19

Solve each equation. Show your work.

123. $-45 = -5y$

$y = 9$

124. $b - 8 = -12$

$b = -4$

125. $6z - 3 = 3z + 9$

$z = 4$

126. $-5 = r + 5$

$r = -10$

127. $-3x = 33$

$x = -11$

128. $2z - 7z = -15$

$z = 3$

129. $3(k - 6) = 6 - 12$

$k = 4$

130. $6(t + 3) = -2 + 20$

$t = 0$

131. $-10 = \dfrac{a}{5} - 2$

$a = -40$

132. $4 + 8p = 4p + 16$

$p = 3$

Solve each application problem using the six problem-solving steps.

133. The recommended daily intake of iron for an adult female is 4 mg less than twice the recommended amount for a newborn infant. The amount for an adult female is 18 mg. How much should the infant receive? (*Source:* Food and Nutrition Board.)

Let x be the daily amount for an infant.

$2x - 4 = 18$

An infant should receive 11 mg of iron.

134. In the U.S. Congress, the number of Representatives is 65 less than five times the number of Senators. There are a total of 535 members of Congress. Find the number of Senators and the number of Representatives. (*Source: World Almanac.*)

Let x be the number of Senators;
5x − 65 is the number of Representatives.

$x + 5x - 65 = 535$

100 Senators; 435 Representatives

135. A cheetah's sprinting speed is 25 miles per hour (mph) faster than a zebra can run. The sum of their running speeds is 111 mph. How fast can each animal run? (*Source: Grolier Multimedia Encyclopedia.*)

Let x be the zebra's speed;
x + 25 is the cheetah's speed.

$x + x + 25 = 111$

Zebra's speed is 43 mph; cheetah's speed is 68 mph.

136. A rectangular park has a perimeter of 1320 yd. The length of the park is five times the width. Find the length and width of the park.

Let w be the width; 5w is the length.

$1320 = 2(5w) + 2(w)$

Width is 110 yd; length is 550 yd.

Work each problem.

1. Graph -4, -1, $1\frac{1}{2}$, 3, and 0 on the number line at the right.

2. Write $<$ or $>$ in each blank to make a true statement.

   -3 _____ 0 -4 _____ -8

3. Find each absolute value: $|-7|$ and $|15|$

Add, subtract, multiply, or divide.

4. $-8 + 7$

5. $-11 + (-2)$

6. $6.7 + (-1.4)$

7. $8 - 15$

8. $4 - (-12)$

9. $-\frac{1}{2} - \left(-\frac{3}{4}\right)$

10. $8(-4)$

11. $-7(-12)$

12. $(-16)(0)$

13. $\frac{-100}{4}$

14. $\frac{-24}{-3}$

15. $-\frac{1}{4} \div \frac{5}{12}$

Simplify each expression.

16. $-5 + 3(-2) - (-12)$

17. $2 - (6 - 8) - (-5)^2$

Find the value of $8k - 3m$, given each set of values.

18. $k = -4$, $m = 2$

19. $k = 3$, $m = -1$

20. In Exercises 18 and 19, you were evaluating an expression. Explain the difference between evaluating an expression and solving an equation.

   When evaluating, you are given specific values to replace each variable. When solving an equation, you are not given the value of the variable. You must find a value that "works"; that is, when your solution is substituted for the variable, the two sides of the equation are equal.

Answers:

1.
   $-5\ -4\ -3\ -2\ -1\ \ 0\ \ 1\ \ 2\ \ 3\ \ 4\ \ 5$

2. $<, >$

3. $7, 15$

4. -1

5. -13

6. 5.3

7. -7

8. 16

9. $\frac{1}{4}$

10. -32

11. 84

12. 0

13. -25

14. 8

15. $-\frac{3}{5}$

16. 1

17. -21

18. -38

19. 27

20. See left.

21. 110

21. The formula for the area of a triangle is $A = \frac{1}{2}bh$. Find A, if $b = 20$ and $h = 11$.

Solve each equation. Show your work.

22. $x = 5$

22. $x - 9 = -4$

23. $30 = -1 + r$

23. $r = 31$

24. $t = 5$

24. $3t - 8t = -25$

25. $\frac{p}{5} = -3$

25. $p = -15$

26. $a = -3$

26. $-15 = 3(a - 2)$

27. $3m - 5 = 7m - 13$

27. $m = 2$

Solve each application problem using the six problem-solving steps.

28. Let x be shorter piece; $x + 4$ is longer piece.
$x + x + 4 = 118$

Shorter piece is 57 cm; longer piece is 61 cm.

28. A board is 118 cm long. Karin cut it into two pieces, with one piece 4 cm longer than the other. Find the length of both pieces.

29. Width is 42 ft; length is 168 ft.

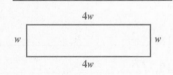

$420 = 2(4w) + 2w$

29. The perimeter of a rectangular building is 420 ft. The length is four times as long as the width. Find the length and the width. Make a drawing to help solve this problem.

30. Let x be Marcella's time; $x - 3$ is Tim's time.
$x + x - 3 = 19$

Marcella spent 11 hr.
Tim spent 8 hr.

30. Marcella and her husband, Tim, spent a total of 19 hours redecorating their living room. Tim spent 3 hours less time than Marcella. How long did each person work on the room?

Cumulative Review Exercises ▶▶▶ Chapters 1–9

First use front end rounding to round each number and estimate an answer. Then find the exact answer. Write answers to fraction problems in lowest terms and as whole or mixed numbers when possible.

1. *Estimate:*

$$\overset{500}{80\overline{)40{,}000}}$$

Exact:

$$\overset{503}{78\overline{)39{,}234}}$$

2. *Exact:*

$$5\frac{5}{6} \cdot \frac{9}{10} = 5\frac{1}{4}$$

Estimate:

$$\underline{6} \cdot \underline{1} = \underline{6}$$

3. *Exact:*

$$4\frac{1}{6} \div 1\frac{2}{3} = 2\frac{1}{2}$$

Estimate:

$$\underline{4} \div \underline{2} = \underline{2}$$

Simplify.

4. $17 - 8.094$

8.906

5. $4.06 \div 0.072$
Round to nearest tenth.

56.4 (rounded)

6. $\dfrac{30}{-6}$

−5

7. $-3 - (-7)$

4

8. $3.2 + (-4.5)$

−1.3

9. $\dfrac{1}{4} - \dfrac{3}{4}$

$-\dfrac{1}{2}$

10. $45 \div \sqrt{25} - 2(3) + (10 \div 5)$

5

11. $-6 - (4 - 5) + (-3)^2$

4

Solve each proportion or percent problem.

12. $\dfrac{x}{12} = \dfrac{1.5}{45}$

$x = 0.4$

13. 90 cars is 180% of what number of cars?

50 cars

14. $5.80 is what percent of $145?

4%

Convert these measurements.

15. $3\frac{1}{2}$ gal to quarts

14 qt

16. 72 hr to days

3 days

17. 3.75 kg to grams

3750 g

18. 40 cm to meters

0.4 m

Name each figure and find its perimeter (or circumference) and the area. Use 3.14 as the approximate value of π.

19.

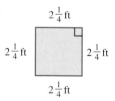

$2\frac{1}{4}$ ft

Square;
P = 9 ft:
$A = 5\frac{1}{16}$ or 5.0625 ft²

20.

9 mm

Circle;
C ≈ 28.26 mm;
A ≈ 63.585 mm²

21.

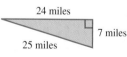

24 miles

7 miles

25 miles

Right triangle;
P = 56 mi;
A = 84 mi²

22.

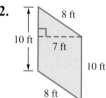

8 ft

10 ft

7 ft

10 ft

8 ft

Parallelogram;
P = 36 ft;
A = 70 ft²

Solve each equation. Show your work.

23. $-2t - 6t = 40$

$t = -5$

24. $3x + 5 = 5x - 11$

$x = 8$

25. $6(p + 3) = -6$

$p = -4$

26. Write an equation and solve it. If 40 is added to four times a number, the result is zero. Find the number.

$4x + 40 = 0$;
$x = -10$

27. Use the six problem-solving steps. $1000 in prize money is being split between Reggie and Donald. Donald should get $300 more than Reggie. How much will each man receive?

Let _m_ be Reggie's money; _m_ + 300 is Donald's money.
_m_ + _m_ + 300 = 1,000;
$350 for Reggie; $650 for Donald

28. Make a drawing to help solve this problem. The length of a photograph is 5 cm more than the width. The perimeter of the photograph is 82 cm. Find the length and the width.

$82 = 2(w + 5) + 2w$;
length is 23 cm; width is 18 cm.

29. Find the unknown length in this figure. Round your answer to the nearest tenth if necessary.

_y_ ≈ 13.2 yd (rounded)

Solve each application problem. Round money answers to the nearest cent.

30. Portia bought two CDs at $14.98 each. The sales tax rate is $6\frac{1}{2}\%$. Find the total amount charged to Portia's credit card.

$31.91 (rounded)

31. Rich spent 25 minutes reading 14 pages in his sociology textbook. At that rate, how long will it take him to read 30 pages? Round to the nearest whole number of minutes.

54 min (rounded)

32. A packing crate measures 2.4 m long, 1.2 m wide, and 1.2 m high. A trucking company wants crates that hold 4 m³. The crate's volume is how much more or less than 4 m³?

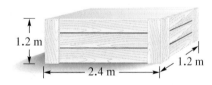

0.544 m³ less

33. Jackie drove her car 364 miles on 14.5 gallons of gas. Maya used 16.3 gallons to drive 406 miles. Naomi drove 300 miles on 11.9 gallons. Which car had the highest number of miles per gallon? How many miles per gallon did that car get, rounded to the nearest tenth?

Naomi's car; 25.2 miles per gallon (rounded)

34. The local food pantry received 2480 lb of food this month. Their goal was 2000 lb. What percent of their goal was received?

124%

35. Brian's spaghetti sauce recipe calls for $3\frac{1}{3}$ cups of tomato sauce. He wants to make $2\frac{1}{2}$ times the usual amount. How much tomato sauce does he need?

$8\frac{1}{3}$ **cups**

Statistics

10

The old saying "A picture is worth a thousand words" was never more true than when applied to the understanding of data. As an information society, we are constantly being bombarded with facts and numbers. The ability to understand and interpret the many types of information and the many ways it is presented has become essential. For example, if you own or manage a business, you need to understand and interpret the sales history of the business and the performance of employees to make wise managerial decisions. (See **Section 10.2**, Exercises 25–32 and 35–42, and **Section 10.3**, Exercises 29–36.)

10.1 ▸▸▸ Circle Graphs

OBJECTIVES

1 **Read and understand a circle graph.**

2 **Use a circle graph.**

3 **Draw a circle graph.**

1 Use the circle graph to answer each question.

(a) The greatest number of hours is spent in which activity?

(b) How many more hours are spent working than studying?

(c) Find the total number of hours spent studying, working, and attending classes.

2 Use the circle graph to find each ratio. Write the ratios as fractions in lowest terms.

(a) Hours spent driving to whole day

(b) Hours spent sleeping to whole day

(c) Hours spent attending class and studying to whole day

(d) Hours spent driving and working to whole day

ANSWERS

1. (a) sleeping (b) 2 hr (c) 13 hr
2. (a) $\frac{1}{12}$ (b) $\frac{7}{24}$ (c) $\frac{7}{24}$ (d) $\frac{1}{3}$

The word *statistics* originally came from words that mean *state numbers*. State numbers refer to numerical information, or *data,* gathered by the government such as the number of births, deaths, or marriages in a population. Today, the word *statistics* has a much broader application; data from the fields of economics, social science, and business can all be organized and studied under the branch of mathematics called *statistics.*

OBJECTIVE 1 Read and understand a circle graph. It can be hard to understand a large collection of data. The graphs described in this section help you make sense of such data. For example, a **circle graph** shows how a total amount is divided into parts. The circle graph below shows you how 24 hours in the life of a college student are divided among different activities.

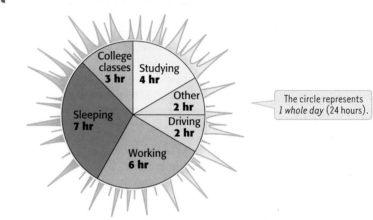

THE DAY OF A COLLEGE STUDENT

College classes **3 hr** | Studying **4 hr** | Other **2 hr** | Driving **2 hr** | Working **6 hr** | Sleeping **7 hr**

The circle represents *1 whole day* (24 hours).

◀ *Work Problem* 1 *at the Side.*

OBJECTIVE 2 Use a circle graph. The above circle graph uses pie-shaped pieces called *sectors* to show the amount of time spent on each activity (the total must be one day, which is 24 hours); the circle graph can therefore be used to compare the time spent on any one activity to the total number of hours in the day.

EXAMPLE 1 **Using a Circle Graph**

Find the ratio of hours spent in college classes to the total number of hours in the day. Write the ratio as a fraction in lowest terms. (See **Section 5.1.**)

The circle graph shows that 3 of the 24 hours in a day are spent in class. The ratio of class time to the hours in a day is shown below.

$$\frac{3 \text{ hours (college classes)}}{24 \text{ hours (whole day)}} = \frac{3 \text{ hours}}{24 \text{ hours}} = \frac{3 \div 3}{24 \div 3} = \frac{1}{8} \longleftarrow \text{Lowest terms}$$

◀ *Work Problem* 2 *at the Side.*

The circle graph above can also be used to find the ratio of the time spent on one activity to the time spent on any other activity.

EXAMPLE 2   **Finding a Ratio from a Circle Graph**

Use the circle graph on a student's day to find the ratio of study time to class time. Write the ratio as a fraction in lowest terms.

The circle graph shows 4 hours spent studying and 3 hours spent in class. The ratio of study time to class time is shown below.

$$\frac{4 \text{ hours (study)}}{3 \text{ hours (class)}} = \frac{4 \text{ hours}}{3 \text{ hours}} = \frac{4}{3}$$

The common units (hours) divide out.

Work Problem 3 *at the Side.* ▶

A circle graph often shows data as percents. For example, suppose that the yearly vending machine snack food sales in the United States were $36 billion. The circle graph below shows how sales were divided among various types of snack foods. The entire circle represents the total $36 billion in sales. Each sector represents the sales of one snack item as a percent of the total sales (the total must be 100%).

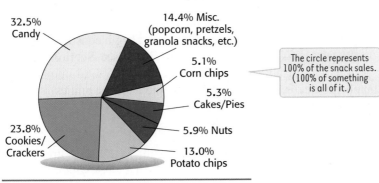

YEARLY U.S. SNACK MARKET SALES
($36 BILLION)

32.5% Candy

14.4% Misc. (popcorn, pretzels, granola snacks, etc.)

5.1% Corn chips

The circle represents 100% of the snack sales. (100% of something is all of it.)

5.3% Cakes/Pies

23.8% Cookies/Crackers

5.9% Nuts

13.0% Potato chips

Source: Natural Choice—USA.

EXAMPLE 3   **Calculating an Amount Using a Circle Graph**

Use the circle graph above on vending machine snack sales to find the amount spent on candy for the year.

Recall the percent equation.

part = percent • whole

The total sales are $36 billion, so the whole is $36 billion. The percent is 32.5% or, as a decimal, 0.325. Find the part.

part = percent • whole

32.5% = 0.325 → $x = (0.325)(36 \text{ billion})$

$x = 11.7 \text{ billion}$

The amount spent on candy was $11.7 billion or $11,700,000,000.

Work Problem 4 *at the Side.* ▶

3 Use the circle graph on a student's day to find the following ratios. Write the ratios as fractions in lowest terms.

(a) Hours spent in class to hours spent studying

(b) Hours spent working to hours spent sleeping

(c) Hours spent driving to hours spent working

(d) Hours spent in class to hours spent for "Other"

4 Use the circle graph on vending machine snack sales to find the following.

(a) The amount spent on corn chips

(b) The amount spent on miscellaneous (popcorn, pretzels, granola snacks, etc.)

(c) The amount spent on cakes/pies

(d) The amount spent on cookies/crackers

ANSWERS

3. (a) $\frac{3}{4}$ (b) $\frac{6}{7}$ (c) $\frac{1}{3}$ (d) $\frac{3}{2}$

4. (a) $1.836 billion or $1,836,000,000
   (b) $5.184 billion or $5,184,000,000
   (c) $1.908 billion or $1,908,000,000
   (d) $8.568 billion or $8,568,000,000

OBJECTIVE 3 Draw a circle graph. Last year, Goodwill Industries donors helped fund programs that let nearly 1 million people take their first financial steps toward new and better jobs and financial independence. The following table shows those who were served.

A HELPING HAND

Group Served	Percent of Total
People with disabilities	25%
Welfare recipients	15%
Working poor	10%
Ex-offenders	10%
At-risk youth	5%
Unemployed	35%
Total	100%

Source: Goodwill Industries.

You can show these percents visually by using a circle graph. The entire circle will represent all of the groups served (all 100%).

EXAMPLE 4 **Drawing a Circle Graph**

Using the data in the table, find the number of degrees in the sector that would represent the "people with disabilities" and begin constructing a circle graph.

Recall that a complete circle has 360° (see **Section 8.1**). Because the "people with disabilities" make up 25% of the total number of people, the number of degrees needed for the "people with disabilities" sector of the circle graph is 25% of 360°.

$$(360°)(25\%) = (360°)(0.25) = 90°$$

Use a tool called a **protractor** to make a circle graph. First, using a straightedge, draw a line from the center of a circle to the left edge. Place the hole in the protractor over the center of the circle, making sure that 0 on the protractor lines up with the line that was drawn. Find 90° and make a mark as shown in the illustration. Then remove the protractor and use the straightedge to draw a line from the center of the circle to the 90° mark at the edge of the circle. This sector is 90° and represents "people with disabilities." Label the sector with the group name and percent.

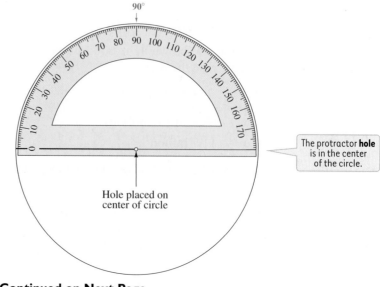

90°

Hole placed on center of circle

The protractor **hole** is in the center of the circle.

Continued on Next Page

To draw the "Welfare recipients" sector, begin by finding the number of degrees in the sector. From the table, you see that "welfare recipients" represent 15% of the total.

$$(360°)(15\%) = (360°)(0.15) = 54°$$

Again, place the hole of the protractor over the center of the circle, but this time align 0 on the second line that was drawn. Make a mark at 54° and draw a line as before. This sector is 54° and represents those who are "Welfare recipients."

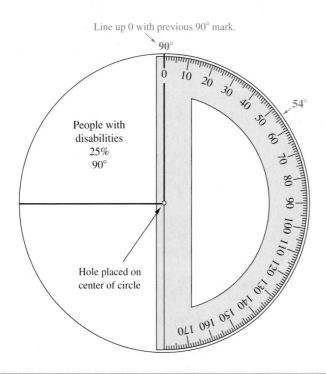

Line up 0 with previous 90° mark.

Hole placed on center of circle

People with disabilities 25% 90°

CAUTION
You must be certain that the hole in the protractor is placed over the exact center of the circle each time you measure the size of a sector.

Work Problem **5** *at the Side.* ▶

Use this circle for Problem 5 at the side.

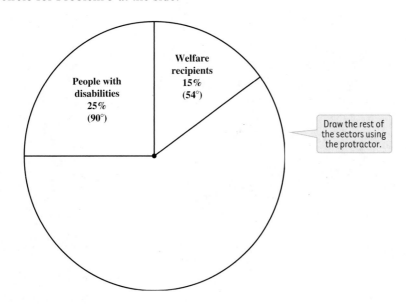

Welfare recipients 15% (54°)

People with disabilities 25% (90°)

Draw the rest of the sectors using the protractor.

5 Using the information in the table on the groups served, find the number of degrees needed for each sector. Complete the circle graph at the bottom left. Label each sector with the group name and percent.

(a) Working poor

(b) Ex-offenders

(c) At-risk youth

(d) Unemployed

ANSWERS

5. (a) 36° **(b)** 36° **(c)** 18° **(d)** 126°

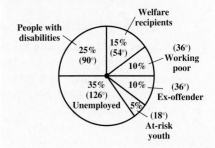

Welfare recipients

People with disabilities 25% (90°) 15% (54°) (36°) Working poor

35% (126°) Unemployed 10% 10% (36°) Ex-offender

5% (18°) At-risk youth

Math in the Media

The *Antiques Road Show*, a popular international television program, is shown on the Public Broadcasting System (PBS) in the United States. People bring antique items and collectables to the show and receive free appraisals of value from a professional antiques appraiser. Popular collectables appraised on the show have been antique clocks, some of which have dated back to the 1800s and have ranged in value from $1000 to $100,000. (*Source:* www.antiquesroadshow.com)

Antique clocks usually chime either on the hour or on both the hour and the half hour. The mechanisms that control the chimes are a set of gears called the count plate and hammer wheel and a lever called the count hook.

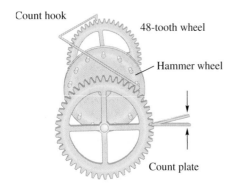

Count hook
48-tooth wheel
Hammer wheel
Count plate

Hour-Striking Clocks

1. If a clock chimes only on the hour, what will be the total number of chimes in a 12-hour period? (*Hint:* For example, it will chime 6 times at 6:00.)
$1 + 2 + \cdots + 12 = 78$ **chimes**

2. If the count plate has one gear tooth for each chime, what fractional part of the count plate is one tooth? $\dfrac{1}{78}$

3. How many degrees correspond to one gear tooth?
4.62° (rounded) or $4\dfrac{8}{13}$**° (exact)**

4. The mechanism designed to move the count plate has two wheels. The hammer wheel has 13 pins, which is combined with a wheel with 48 teeth and a pinion that fits 8 teeth. That gives a 6:1 ratio. What is significant about the combination of 13 and 6 that causes the count plate to move one gear tooth?
$13 \times 6 = 78$**, so the gear advances** $\dfrac{1}{78}$
of the circle.

Hour- and Half-Hour-Striking Clocks

5. If a clock chimes on the hour and also one time on each half hour, what will be the total number of chimes in a 12-hour period?
$78 + 12 = 90$ **chimes**

6. If the count plate has one tooth for each chime, what fractional part of the count plate makes one tooth? $\dfrac{1}{90}$

7. How many degrees correspond to one gear tooth? **4°**

8. If the count plate has a diameter of 2 in., what is the circumference of the count plate, rounded to the nearest hundredth? What is the width of each gear tooth, rounded to the nearest hundredth? (*Note:* Use $\pi \approx 3.14$.)
$C \approx 6.28$ **in.; width of gear tooth** ≈ 0.07 **in.**

10.1 ▶▶▶ Exercises

This circle graph shows the number of pets owned in the United States. Use this circle graph to answer Exercises 1–6. Write ratios as fractions in lowest terms. See Examples 1 and 2.

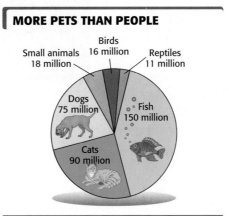

MORE PETS THAN PEOPLE

Birds 16 million
Small animals 18 million
Reptiles 11 million
Dogs 75 million
Fish 150 million
Cats 90 million

Source: American Pet Product Manufacturers Association

1. Find the number of pets owned in the United States.

360 million or 360,000,000 pets

2. Which type of pet is owned by the greatest number of people? How many of these pets were owned?

fish; 150 million or 150,000,000

3. Find the ratio of the number of cats owned to the total number of pets.

$$\frac{90}{360} = \frac{1}{4}$$

4. Find the ratio of the number of small animals owned to the total number of pets.

$$\frac{18}{360} = \frac{1}{20}$$

5. Find the ratio of the number of cats owned to the number of dogs.

$$\frac{90}{75} = \frac{6}{5}$$

6. Find the ratio of the number of fish owned to the number of cats.

$$\frac{150}{90} = \frac{5}{3}$$

This circle graph, adapted from USA Today, *shows the number of people in a survey who gave various reasons for eating dinner at restaurants. Use this circle graph to answer Exercises 7–14. See Examples 1 and 2.*

ON THE TOWN

When asked in a survey why they ate dinner in restaurants, a group of people gave these reasons.

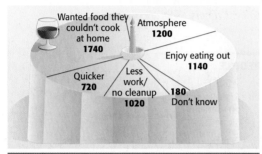

Wanted food they couldn't cook at home **1740**
Atmosphere **1200**
Enjoy eating out **1140**
Quicker **720**
Less work/no cleanup **1020**
180 Don't know

Source: Market Facts for Tyson Foods.

7. Which reason was given by the least number of people? How many gave that reason?

Don't know; 180 people

8. Which reason was given by the second-highest number of people? How many gave that reason?

Atmosphere; 1200 people

Answer Exercises 9–14 by writing a ratio as a fraction in lowest terms.

9. Those who said dining out is "Quicker" to total people in the survey

$$\frac{720}{6000} = \frac{3}{25}$$

10. Those who said "Enjoy eating out" to the total people in the survey

$$\frac{1140}{6000} = \frac{19}{100}$$

11. Those who said "Less work/no cleanup" to those who said "Atmosphere"

$$\frac{1020}{1200} = \frac{17}{20}$$

12. Those who said "Don't know" to those who said "Quicker"

$$\frac{180}{720} = \frac{1}{4}$$

13. Those who said "Wanted food they couldn't cook at home" to those who said "Less work/no cleanup"

$$\frac{1740}{1020} = \frac{29}{17}$$

14. Those who said "Atmosphere" to those who said "Enjoy eating out"

$$\frac{1200}{1140} = \frac{20}{19}$$

This circle graph shows the favorite hot dog toppings in the United States. Each topping is expressed as a percent of the 3200 people in the survey. Use the graph to find the number of people in the survey who favored each of the toppings in Exercises 15–20. See Example 3.

15. Onions

160 people

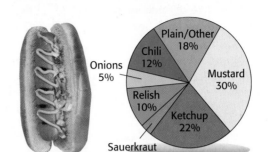

FAVORITE HOT DOG TOPPINGS

Source: National Hot Dog and Sausage Council.

16. Ketchup

704 people

17. Sauerkraut

96 people

18. Relish

320 people

19. Mustard

960 people

20. Chili

384 people

The National Academy of Sciences recommends that adults ages 19–50 get 1000 mg of calcium daily, or about three 8-ounce glasses of milk. The circle graph, adapted from USA Today, *shows the daily consumption of milk products by adults. If 5540 adults were surveyed in this study, find the number of people giving each response in Exercises 21–26. Round to the nearest whole number. See Example 3.*

21. Consume none or very few milk products

1219 people (rounded)

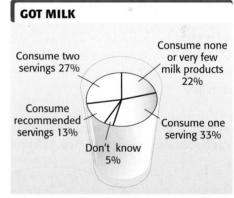

GOT MILK

Source: Market Facts for Milk Mustache Mobile 100-City Cruise for Calcium.

22. Consume recommended servings

720 people (rounded)

23. Consume one serving

1828 people (rounded)

24. Consume two servings

1496 people (rounded)

25. Don't know

277 people

26. Consume less than the recommended servings (Do not include those who don't know.)

4543 people (rounded)

27. Describe the procedure for determining how large each sector must be to represent each of the items in a circle graph.

   First find the percent of the total that is to be represented by each item. Next, multiply the percent by 360° to find the size of each sector. Finally, use a protractor to draw each sector.

28. A protractor is the tool used to draw a circle graph. Give a brief explanation of what the protractor does and how you would use it to measure and draw each sector in the circle graph.

   A protractor is used to measure the number of degrees in a sector. First, you must draw a line from the center of the circle to the left edge. Next, place the hole of the protractor over the center of the circle, making sure that the 0 on the protractor is on the line. Finally, make a mark at the desired number of degrees. This gives you the size of the sector.

During one month the Orangevale Parks and Recreation District spent $5460 for the activities shown in the following chart. Find all numbers missing from the chart.

Item	Dollar Amount	Percent of Total	Degrees of a Circle
29. Adult sports	$1365	25%	90°
30. Children's sports	$1092	20%	72°
31. Day camp	$546	10%	36°
32. Senior fitness	$546	10%	36°
33. Annual egg hunt	$819	15%	54°
34. Arts and crafts	$273	5%	18°
35. Mommy and baby exercise	$819	15%	54°

36. Draw a circle graph by using the information from Exercises 29–35. Label each sector in your graph. See Example 4.

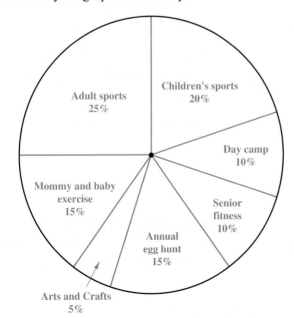

37. White Water Rafting Company divides its annual sales into five categories as follows.

Category	Annual Sales
Adventure classes	$12,500
Grocery and provision sales	$40,000
Equipment rentals	$60,000
Rafting tours	$50,000
Equipment sales	$37,500

(a) Find the total sales for the year.

$200,000

(b) Find the number of degrees in a circle graph for each item.

22.5°; 72°; 108°; 90°; 67.5°

(c) Make a circle graph showing the percent for each category. Label each sector in your graph.

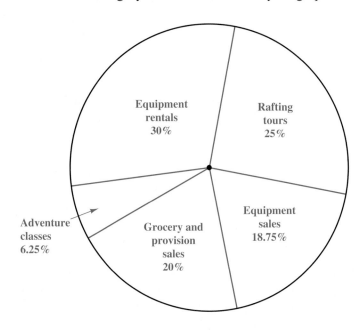

38. Online retail sales in the United States are growing rapidly. The top online retail sales categories are 50% for travel; 20% for apparel, accessories, and footwear; 15% for computer hardware and software; 10% for autos and auto parts; and 5% for home furnishings. (*Source*: Forester Research.)

(a) Find the number of degrees in a circle graph for each online retail sales category.

180°; 72°; 54°; 36°; 18°

(b) Draw a circle graph showing the percent for each category. Label each sector in your graph.

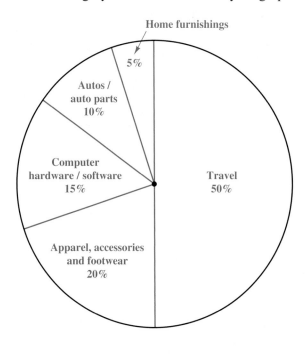

39. The Pathfinder Research Group asked 4488 Americans how they fall asleep, and the results are shown in the figure on the right.

(a) Use this information to complete the chart and draw a circle graph. Round to the nearest whole percent and to the nearest degree. Label each sector in your graph.

Sleeping Position	Number of Americans	Percent of Total	Number of Degrees
Side	2464	55% (rounded)	198°
Not sure	220	5% (rounded)	18°
Stomach	536	12% (rounded)	43° (rounded)
Varies	520	12% (rounded)	43° (rounded)
Back	748	17% (rounded)	61° (rounded)

(b) Add up the percents. Is the total 100%? Explain why or why not.

No. The total is 101% due to rounding.

(c) Add up the degrees. Is the total 360°? Explain why or why not.

No. The total is 363° due to rounding.

SET TO SLEEP

Number of Americans surveyed who fall asleep on their:

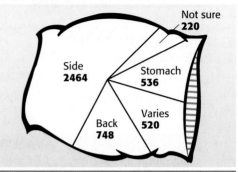

Source: Pathfinder Research Group.

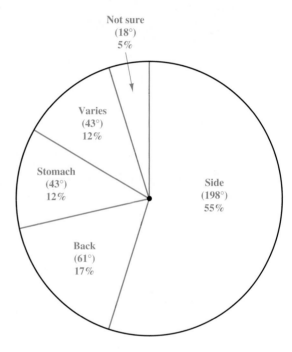

10.2 ▶▶▶ Bar Graphs and Line Graphs

OBJECTIVE 1 Read and understand a bar graph. Bar graphs are useful when showing comparisons. For example, the bar graph below compares the total number of members in all the Fitness Center locations during each of five years.

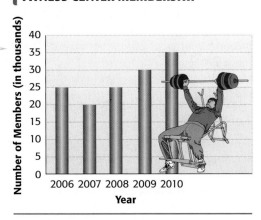

FITNESS CENTER MEMBERSHIP

Notice that the label says "in thousands."

EXAMPLE 1   Using a Bar Graph

How many members did the Fitness Center have in 2008?

The bar for 2008 rises to 25. Notice the label along the left side of the graph that says "Number of Members (in thousands)." The phrase *in thousands* means you have to multiply 25 by 1000 to get 25,000. So, there were 25,000 (not 25) members in the Fitness Center locations in 2008.

Work Problem **1** *at the Side.* ▶

OBJECTIVE 2 Read and understand a double-bar graph. A **double-bar graph** can be used to compare two sets of data. The graph below shows the number of DSL (digital subscriber line) installations each quarter for two different years.

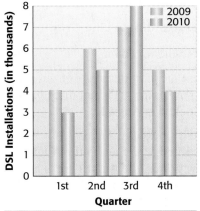

DSL INSTALLATIONS
(FOR HIGH-SPEED INTERNET ACCESS)

2009
2010

The color key shows which bars represent each year.

1 Use the bar graph at the left to find the number of members in the Fitness Centers in each of these years.

(a) 2006

(b) 2007

(c) 2009

(d) 2010

2 Use the double-bar graph to find the number of DSL installations in 2009 and 2010 for each quarter.

(a) 1st quarter

(b) 3rd quarter

(c) 4th quarter

(d) Find the greatest number of installations. Identify the quarter and the year in which they occurred.

3 Use the line graph at the right to find the number of trout stocked in each month.

(a) June

(b) May

(c) April

(d) July

EXAMPLE 2 Reading a Double-Bar Graph

Use the double-bar graph on the previous page to find the following.

(a) The number of DSL installations in the second quarter of 2009
There are two bars for the second quarter. The color code in the upper right-hand corner of the graph tells you that the **red bars** represent 2009. So the **red bar** on the *left* is for the 2nd quarter of 2009. It rises to 6. Multiply 6 by 1000 because the label on the left side of the graph says *in thousands*. So there were 6000 DSL installations for the second quarter in 2009.

(b) The number of DSL installations in the second quarter of 2010
The **green bar** for the second quarter rises to 5 and 5 times 1000 is 5000. So, in the second quarter of 2010, there were 5000 DSL installations.

> **CAUTION**
> Use a ruler or straightedge to line up the top of the bar with the number on the left side of the graph.

─── ◀ *Work Problem* **2** *at the Side.*

OBJECTIVE 3 Read and understand a line graph. A line graph is often useful for showing a trend. The line graph below shows the number of trout stocked along the Feather River over a 5-month period. Each dot indicates the number of trout stocked during the month directly below that dot.

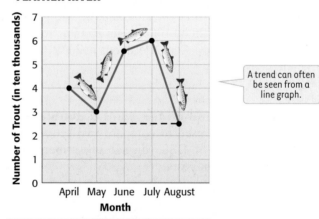

TROUT STOCKED IN THE FEATHER RIVER

A trend can often be seen from a line graph.

EXAMPLE 3 **Understanding a Line Graph**

Use the line graph to find the following.

(a) In which month were the least number of trout stocked?
The lowest point on the graph is the dot directly over August, so the least number of trout were stocked in August.

(b) How many trout were stocked in August?
Use a ruler or straightedge to line up the August dot with the numbers along the left edge of the graph. The August dot is halfway between the 2 and the 3. Notice that the label on the left side says *in ten thousands*. So August is halfway between (2 • 10,000) and (3 • 10,000). It is halfway between 20,000 and 30,000. That means 25,000 trout were stocked in August.

─── ◀ *Work Problem* **3** *at the Side.*

ANSWERS

2. **(a)** 4000; 3000 installations
  (b) 7000; 8000 installations
  (c) 5000; 4000 installations
  (d) 8000 installations; 3rd quarter of 2010
3. **(a)** 55,000 trout **(b)** 30,000 trout
  (c) 40,000 trout **(d)** 60,000 trout

OBJECTIVE 4 Read and understand a comparison line graph.
Two sets of data can also be compared by drawing two line graphs together as a **comparison line graph.** For example, the line graph below compares the number of plasma high-definition televisions (Plasma HDTVs) and the number of liquid crystal diode high-definition televisions (LCD HDTVs) sold during each of 5 years.

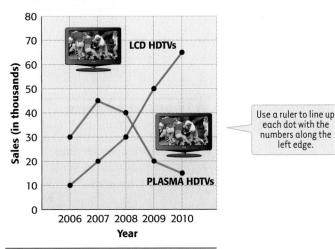

4 Use the comparison line graph at the left to find the following.

(a) The number of plasma HDTVs sold in 2006, 2008, 2009, and 2010

(b) The number of LCD HDTVs sold in 2006, 2007, 2008, and 2009

EXAMPLE 4 **Interpreting a Comparison Line Graph**

Use the comparison line graph above to find the following.

(a) The number of plasma HDTVs sold in 2007
    Find the dot on the blue line above 2007. Use a ruler or straightedge to line up the dot with the numbers along the left edge. The dot is halfway between 40 and 50, which is 45. Then, 45 times 1000 is 45,000 plasma HDTVs sold in 2007.

(b) The number of LCD HDTVs sold in 2010
    The red line on the graph shows that 65,000 LCD HDTVs were sold in 2010.

> **Note**
> Both the double-bar graph and the comparison line graph are used to compare two or more sets of data.

Work Problem **4** *at the Side.* ▶

(c) The first full year in which the number of LCD HDTVs sold was greater than the number of plasma HDTVs sold

Math in the Media

SNACKS: HOW LOW CAN THEY GO?

Packaged-food giants from Quaker to Kraft had such great sales success with their 100-calorie snack packs that they are now going even lower. The bar graph shows the number of individual snack products introduced each year in the United States with fewer than 100 calories. The table shows some of the new snack-pack products that have been introduced to the market.

NEW SNACKS UNDER 100 CALORIES

Manufacturer	Product Name	Number of Calories
Quaker	Mini Delight Bar	90
Kellogg	Special K Bliss Bar	70
ConAgra	Hunts Fat Free Pudding	80
Kraft	Jet-Puffed Marshmallows	90
General Mills	Yoplait Fiber One Yogurt	80
Hershey's	Hershey Sticks in four flavors	60
Del Monte Pet Products	Pup-Peroni (for dogs)	50

Source: USA Today.

UNDER 100 CALORIES
Number of individual products* introduced in the USA at fewer than 100 calories per serving:

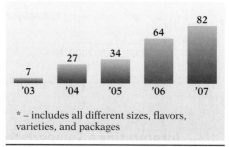

* – includes all different sizes, flavors, varieties, and packages

Source: Datamonitor's Production Online.

1. What is the title of the bar graph? Is the number of new snack products having fewer than 100 calories increasing or decreasing?   **Under 100 Calories; increasing**

2. The greatest number of new products introduced was in which year? How many new products were introduced in that year?   **2007; 82 new products**

3. What were the two new snack products with the highest number of calories? How many calories do they have?   **Mini Delight Bar and Jet-Puffed Marshmallows: 90 calories**

4. Which product has the fewest number of calories? This product will most likely be eaten by _____.   **Pup-Peroni; dogs**

5. Sometimes people "jump to conclusions" without enough information. Which of these conclusions are reasonable, based on the information given in the table?

    (a) The snacks in the table are healthy snacks.
    unreasonable

    (b) Some snacks with fewer than 100 calories could be healthy.   **reasonable**

    (c) A person can have a snack without consuming a large number of calories   **reasonable**

    (d) Low-calorie snacks don't taste good.   **unreasonable**

    (e) Of the snacks included in the table, dogs will eat only Pup-Peronis.   **unreasonable**

6. Conduct a survey of your class members. Make a list of their favorite snacks that are healthy while also being low in calories.   **Answers will vary.**

734

10.2 ▶▶▶ Exercises

FOR EXTRA HELP

MyMathLab

Math XL
PRACTICE

WATCH

DOWNLOAD

READ

REVIEW

The American Farm Bureau Federation reports that the average adult in the United States will work 40 days (rounded to the nearest day) to earn enough to pay the annual household food bill. This was found by multiplying the average percent of household income spent on food by 365 (the number of days in a year). This bar graph shows the percent of income spent in various countries of the world. Use this graph to answer Exercises 1–6. See Example 1.

TAKING A BITE OUT OF HOUSEHOLD INCOME

The average American adult will work 40 days each year to earn enough to pay the household food bill. Percent of household income spent on food in:

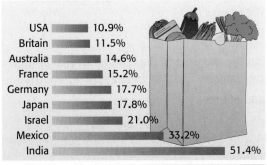

USA	10.9%
Britain	11.5%
Australia	14.6%
France	15.2%
Germany	17.7%
Japan	17.8%
Israel	21.0%
Mexico	33.2%
India	51.4%

Source: American Farm Bureau Federation.

1. In which country is the highest percent of income spent on food? What percent is this?

   India; 51.4%

2. In which country is the lowest percent of income spent on food? What percent is this?

   USA; 10.9%

3. List all countries in the graph in which less than 15% of household income is spent, on average, for food.

   USA, Britain, and Australia

4. List all countries in the graph in which more than 20% of household income is spent, on average, for food.

   Israel, Mexico, and India

5. How many days each year will the average adult have to work to earn enough to pay for food in Mexico? Round to the nearest day.

   121 days (rounded)

6. How many days each year will the average adult have to work to earn enough to pay for food in Israel? Round to the nearest day.

   77 days (rounded)

This double-bar graph shows the number of outdoor plants shipped by Capital Growers during the first six months of 2009 and 2010. Use this graph to answer Exercises 7–12. See Example 2.

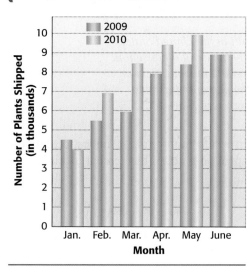

OUTDOOR PLANTS SHIPPED

7. In which month in 2010 were the greatest number of plants shipped? What was the total number of plants shipped in that month?

   May; 10,000 plants

8. How many plants were shipped in January of 2009?

   4500 plants

9. How many more plants were shipped in February of 2010 than in February of 2009?

   1500 plants

10. How many fewer plants were shipped in March of 2009 than in March of 2010?

    2500 plants

11. Find the increase in the number of plants shipped from February 2009 to April 2010.

    4000 plants

12. Find the increase in the number of plants shipped from January 2010 to June 2010.

    5000 plants

This double-bar graph shows sales of super unleaded and supreme unleaded gasoline at a service station for each of 5 years. Use this graph to answer Exercises 13–18. See Example 2.

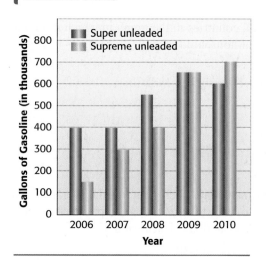

GASOLINE SALES

13. How many gallons of supreme unleaded gasoline were sold in 2006?

    150,000 gal

14. How many gallons of super unleaded gasoline were sold in 2009?

    650,000 gal

15. In which year did the greatest difference in sales between super unleaded and supreme unleaded gasoline occur? Find the difference.

    2006; 250,000 gal

16. In which year did the sales of supreme unleaded gasoline surpass the sales of super unleaded gasoline?

    2010

17. Find the increase in supreme unleaded gasoline sales from 2006 to 2010.

    550,000 gal

18. Find the increase in super unleaded gasoline sales from 2006 to 2010.

    200,000 gal

This line graph shows how the personal computer (PC) has evolved over the quarter century of its existence. What began as a technician's dream is now a common tool of business and home life. Use this line graph to answer Exercises 19–24. See Example 3.

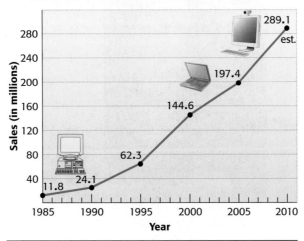

A QUARTER CENTURY OF COMPUTING
WORLDWIDE PC SALES FROM 1985 TO 2010

Source: Gartner Dataquest and ARS Technic.

19. Find the number of PCs shipped in 1990.

24.1 million or 24,100,000 PCs

20. What was the number of PCs shipped in 1995?

62.3 million or 62,300,000 PCs

21. Find the increase in the number of PCs shipped in 2010 from the number shipped in 2000.

144.5 million or 144,500,000 PCs

22. How many more PCs were shipped in 2000 than in 1990?

120.5 million or 120,500,000 PCs

23. Give two possible explanations for the increase in the number of PCs shipped.

Answers will vary. Some possibilities are: Greater demand as a result of lower price; more uses and applications for the owner, improved technology; multiple computers in each location; more laptop computers sold.

24. Give two possible conditions that could result in a decrease in PC shipments in the future.

Answers will vary. Some possibilities are: Fewer people will want a computer because they already own one, new technology will replace the computer with something better.

This comparison line graph shows the number of Apple iPods sold by two different chain stores during each of 5 years. Use this graph to find the annual number of Apple iPods sold each year in Exercises 25–30. See Example 4.

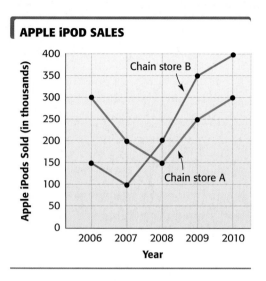

APPLE iPOD SALES

Chain store B

Chain store A

25. Store A in 2010

300,000 iPods

26. Store A in 2009

250,000 iPods

27. Store A in 2008

150,000 iPods

28. Store B in 2010

400,000 iPods

29. Store B in 2009

350,000 iPods

30. Store B in 2008

200,000 iPods

31. Looking at the comparison line graph above, which store would you like to own? Explain why. Based on the graph, what amount of sales would you predict for your store in 2011?

Probably Store B with greater sales. Predicted sales might be 450,000 iPods to 500,000 iPods in 2011.

32. In the comparison line graph above, Store B used to have lower sales than Store A. What might have happened to cause this change? Give two possible explanations.

Some possible answers are that Store B may have started to: do more advertising; keep longer store hours; give better training to their staff; employ more help; give better service than Store A.

33. Explain in your own words why a bar graph or a line graph (not a double-bar graph or comparison line graph) can be used to show only one set of data.

A single bar or a single line must be used for each set of data. To show multiple sets of data, multiple sets of bars or lines must be used.

34. The double-bar graph and the comparison line graph are both useful for comparing two sets of data. Explain how this works and give your own example.

You would use a set of bars or a set of lines for each set of data. Examples will vary.

This comparison line graph shows the sales and profits of Tacos-to-Go for each of 4 years. Use the graph to answer Exercises 35–42. See Example 4.

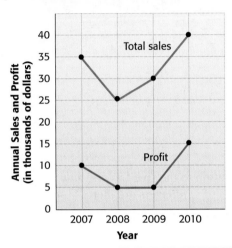

TACOS-TO-GO ANNUAL SALES

35. Total sales in 2010 $40,000

36. Total sales in 2009 $30,000

37. Total sales in 2008 $25,000

38. Profit in 2010 $15,000

39. Profit in 2009 $5000

40. Profit in 2008 $5000

41. Give two possible explanations for the decrease in sales from 2007 to 2008 and two possible explanations for the increase in sales from 2008 to 2010.

Answers will vary. Some possibilities are that the decrease in sales may have resulted from poor service or greater competition. The increase in sales may have been a result of more advertising or better service.

42. Based on the graph, what conclusion can you make about the relationship between sales and profits?

As sales increase or decrease, so do profits.

Relating Concepts (Exercises 43—48) For Individual or Group Work

This newspaper clipping, adapted from USA Today, *shows some statistics regarding*
LifeSavers. Use this information to **work Exercises 43–48 in order.**

ROLL WITH IT

Sweet-toothed fans recently voted to change three
of the original flavors. The new five-flavor roll
includes cherry, watermelon, pineapple, raspberry,
and blackberry. The orange, lemon, and lime
flavors have been replaced.

LifeSavers by the numbers:

Year first flavor invented: **1912 (Pep-O-Mint)**

Year five-flavor roll invented: **1935**

Total number of flavors today: **25**

Number of candies per roll: **14**

LifeSavers produced daily: **3 million**

Pounds of sugar used per day: **250,000**

Number of miniature rolls given out at
Halloween: **88 million**

Source: *USA Today* and Kraft Foods.

43. The first LifeSaver flavor was Pep-O-Mint.
How long after the Pep-O-Mint LifeSaver was
invented was the five-flavor roll invented?

23 years

44. In addition to the five flavors, how many more
flavors of LifeSavers are there?

20 additional flavors

45. Find the number of rolls of LifeSavers
produced daily. Round to the nearest
whole number.

214,286 rolls each day (rounded)

46. Use your answer from Exercise 45 to find the
amount of sugar in one roll of LifeSavers. Round
to the nearest hundredth of a pound.

1.17 lb (rounded)

47. Check your work in Exercise 46. Is the answer
reasonable? Explain why or why not.

**The weight of one roll of LifeSavers is much less
than 1.17 lb. The answer is "correct" using the
information given, but some of the data given
must not be accurate. Perhaps 3 million rolls
of lifesavers are produced daily.**

48. Name three possible causes of errors in statistics.

**Answers will vary. Possible answers are:
Misprints or typographical errors; careless
reporting of data; math errors.**

Summary Exercises on Graphs

The circle graph shows the number of people in a recent survey who preferred each of the five top ice cream flavors. Use the graph to answer Exercises 1–6.

ICE CREAM – FAVORITE FLAVORS

All other flavors 640 people

Vanilla 660 people

Strawberry 80 people

Neopolitan 100 people

Nut caramel 140 people

Chocolate 380 people

Source: The NPD Group, National Eating Trends Group

1. Find the total number of people in the favorite ice cream flavor survey.

2000 people

2. Which ice cream flavor is preferred by the greatest number of people in the survey? How many people picked this flavor?

Vanilla; 660 people

3. Find the ratio of the number of people preferring chocolate to the total number of people in the survey.

$$\frac{380}{2000} = \frac{19}{100}$$

4. Find the ratio of the number of people preferring neopolitan to the total number of people in the survey.

$$\frac{100}{2000} = \frac{1}{20}$$

5. Find the ratio of the number of people preferring all other flavors to those preferring strawberry.

$$\frac{640}{80} = \frac{8}{1}$$

6. Find the ratio of the number of people preferring vanilla to those preferring nut caramel.

$$\frac{660}{140} = \frac{33}{7}$$

The bar graph shows the projected growth in world population. Use the graph to answer Exercises 7–11.

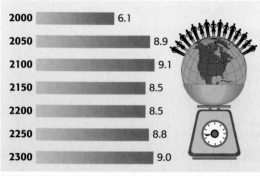

WORLD POPULATION GROWTH

The United Nations projects the world's population to grow as shown (in billions).

2000	6.1
2050	8.9
2100	9.1
2150	8.5
2200	8.5
2250	8.8
2300	9.0

DATA: United Nations.

7. When is the world population expected to reach its peak and what is the estimated population?

2100; 9.1 billion or 9,100,000,000 people

8. Find the increase in world population from 2000 to 2300

2.9 billion or 2,900,000,000 people

9. Find the decrease in population from 2100 to 2200.

0.6 billion or 600,000,000 people

10. Find the percent of increase in world population from 2000 to 2050 to the nearest tenth of a percent.

45.9% (rounded)

11. In 2250, assume that 24% of the world's population live in China and 4.5% of the world's population live in the United States. Find the population of each country.

2.112 billion or 2,112,000,000 in China;

0.396 billion or 396,000,000 in the United States.

The line graph shows the historical trends in marital status in the United States. Use the graph to answer Exercises 12–16.

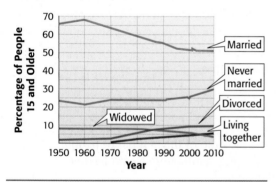

TRENDS IN MARRIAGE

The percentage of people who live together is rising while the marriage rate is in decline.

DATA: U.S. Census Bureau.

12. The percent of people who were married was greatest in what year? What percent were married?

1960; about 68%

13. What percent of the population was divorced in 1950? Has the percent of those divorced more than doubled since 1950?

about $2\frac{1}{2}$%; yes

14. The percent of those who are married has decreased over the years. The percent of what other group has decreased?

widowed

15. What percent of the people were living together in 2000?

about 4%

16. The line graph shows that 0% of the people were living together in 1970. How would you explain this?

Answers will vary. Some possibilities are: no one lived together in 1970; this category was very small; no one kept track of this group; no one would admit it.

10.3 ▶▶▶ Frequency Distributions and Histograms

The owner of Towne Insurance Agency has kept track of her personal phone sales call activity over the past 50 weeks. The number of sales calls made for each of the weeks is given below. Read down the columns, beginning with the left column, for successive weeks of the year.

75	65	40	50	45	30	30	35	45	25
75	70	60	55	30	25	44	30	35	30
75	70	50	30	50	20	30	30	20	25
60	62	45	45	48	40	35	25	20	25
75	45	50	40	35	40	40	30	27	40

OBJECTIVES

1 Understand a frequency distribution.

2 Arrange data in class intervals.

3 Read and understand a histogram.

OBJECTIVE 1 Understand a frequency distribution. A long list of numbers can be confusing. You can make the data easier to read by putting it in a special type of table called a **frequency distribution.**

EXAMPLE 1 Preparing a Frequency Distribution

Using the data above, construct a table that shows each possible number of sales calls. Then go through the original data and place a *tally* mark (I) in the tally column next to each corresponding value. Total the tally marks and place the totals in the third column. The result is a frequency distribution table.

Number of Sales Calls	Tally	Frequency	Number of Sales Calls	Tally	Frequency
20	III	3	48	I	1
25	IHH	5	50	IIII	4
27	I	1	55	I	1
30	IHH IIII	9	60	II	2
35	IIII	4	62	I	1
40	IHH I	6	65	I	1
44	I	1	70	II	2
45	IHH	5	75	IIII	4

Work Problem ① *at the Side.* ▶

OBJECTIVE 2 Arrange data in class intervals. The frequency distribution given in Example 1 above contains a great deal of information— perhaps too much to digest. It can be simplified by combining the number of sales calls into groups, forming the class intervals shown below in the left column of the table.

GROUPED DATA

Class Intervals (Number of Sales Calls)	Class Frequency (Number of Weeks)
20–29	9
30–39	13
40–49	13
50–59	5
60–69	4
70–79	6

In the table above, look at how many weeks had 20 to 29 sales calls: $3 + 5 + 1 = 9$ weeks.

1 Use the frequency distribution table at the left to find the following.

(a) The least number of sales calls made in a week

(b) The most common number of sales calls made in a week

(c) The number of weeks in which 35 calls were made

(d) The number of weeks in which 45 calls were made

ANSWERS

1. **(a)** 20 calls   **(b)** 30 calls
   (c) 4 weeks   **(d)** 5 weeks

2 Use the grouped data for the insurance agency on the previous page to answer each question.

(a) During how many weeks were fewer than 50 calls made?

(b) During how many weeks were 50 or more calls made?

> **Note**
> The number of class intervals in the left column of the grouped data table is arbitrary. Grouped data usually has between 5 and 15 class intervals.

EXAMPLE 2 **Analyzing a Frequency Distribution**

Use the grouped data for the insurance agency (on the preceding page) to answer the following questions.

(a) During how many weeks were fewer than 30 calls made?
  The first interval in the grouped data table (20–29) is the number of weeks during which fewer than 30 calls were made. Therefore, the owner made fewer than 30 calls during 9 weeks out of the 50 weeks shown.

(b) During how many weeks were 40 or more calls made?
  The last four intervals in the grouped data table are the number of weeks during which 40 or more calls were made.

$$13 + 5 + 4 + 6 = \ 28 \text{ weeks}$$

◀ *Work Problem* **2** *at the Side.*

OBJECTIVE 3 **Read and understand a histogram.** The results in the grouped data table have been used to draw the special bar graph below, which is called a **histogram**. In a histogram, the width of each bar represents a range of numbers (*class interval*). The height of each bar in a histogram gives the *class frequency,* that is, the number of occurrences in each class interval.

3 Use the histogram at the right to answer each question.

(a) During how many weeks were fewer than 60 calls made?

(b) During how many weeks were 60 or more calls made?

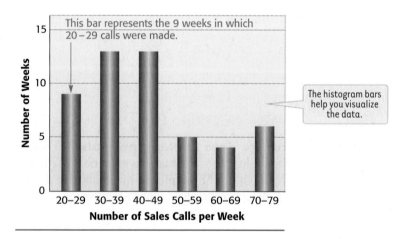

SALES CALL DATA FOR THE PAST 50 WEEKS (GROUPED DATA)

This bar represents the 9 weeks in which 20–29 calls were made.

The histogram bars help you visualize the data.

Number of Weeks (vertical axis)
Number of Sales Calls per Week (horizontal axis)

EXAMPLE 3 **Using a Histogram**

Use the histogram to find the number of weeks in which fewer than 40 calls were made.
  Because 20–29 calls were made during 9 of the weeks and 30–39 calls were made during 13 of the weeks, the number of weeks in which fewer than 40 calls were made is $9 + 13 = 22$ weeks.

◀ *Work Problem* **3** *at the Side.*

ANSWERS

2. (a) 35 weeks **(b)** 15 weeks
3. (a) 40 weeks **(b)** 10 weeks

FOR
EXTRA
HELP **MyMathLab** Math XL
PRACTICE WATCH DOWNLOAD READ REVIEW

The Quilters Club of America recorded the ages of its members and used the results to construct this histogram. Use the histogram to answer Exercises 1–6. See Example 3.

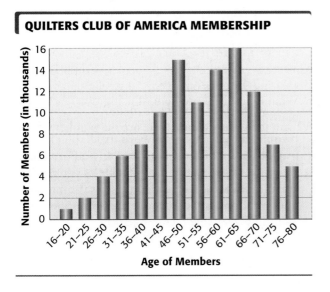

QUILTERS CLUB OF AMERICA MEMBERSHIP

Number of Members (in thousands)

Age of Members

1. The greatest number of members are in which age group? How many members are in that group?

61–65 years;
16,000 members

2. The least number of members are in which age group? How many members are in that group?

16–20 years; 1000 members

3. Find the number of members 35 years of age and under.

13,000 members

4. Find the number of members ages 61 to 80.

40,000 members

5. How many members are 41 to 60 years of age?

50,000 members

6. How many members are 46 to 55 years of age?

26,000 members

This histogram shows the annual earnings for the part-time employees of Wally World Amusement Park. Use this histogram to answer Exercises 7–12. See Example 3.

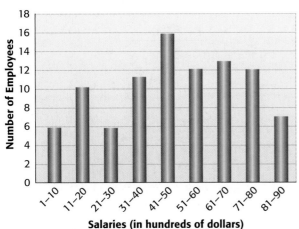

**ANNUAL EARNINGS AT WALLY WORLD
(PART-TIME EMPLOYEES)**

Number of Employees

Salaries (in hundreds of dollars)

7. The greatest number of employees are in which earnings group? How many are in that group?

$4100 to $5000; 16 employees

8. The fewest number of employees are in which earnings groups? How many are in each group?

$100 to $1000, 6 employees; $2100 to $3000, 6 employees

9. Find the number of employees who earn $3100 to $4000.

11 employees

10. Find the number of employees who earn $1100 to $2000.

10 employees

11. How many employees earn $5000 or less?

49 employees

12. How many employees earn $6100 or more?

32 employees

13. Describe class interval and class frequency. How are they used when preparing a histogram?

Class intervals are the result of combining data into groupings. Class frequency is the number of data items that fit in each class interval. These are used to group data and to have multiple responses (frequency) in a class interval—this makes the data easier to interpret.

14. What might be a problem of using two few or too many class intervals?

If too few class intervals were used, the class frequencies would be high and any differences in the data might not be observable. If too many class intervals were used, interpretation might become impossible because class frequencies would be very low or nonexistent.

This list shows the number of new accounts opened annually by the employees of the Schools Credit Union. Use it to complete the table. See Example 1.

186	191	144	198	147	158	174
193	142	155	174	162	151	178
145	151	199	182	147	195	146

	Class Intervals (Number of New Accounts)	Tally	Class Frequency (Number of Employees)
15.	140–149	⊮℟ I	6
16.	150–159	IIII	4
17.	160–169	I	1
18.	170–179	III	3
19.	180–189	II	2
20.	190–199	⊮℟	5

A college professor asked her 30 students how many hours they worked each week. Use her list of student responses to complete the following table. See Examples 1–3.

14	8	12	28	33	14
6	34	17	20	13	20
25	33	32	4	7	14
0	6	10	8	35	31
25	4	32	18	0	24

	Class Intervals (Number of Hours Worked)	Tally	Class Frequency (Number of Students)
21.	0–5	IIII	4
22.	6–10	IIII I	6
23.	11–15	IIII	5
24.	16–20	IIII	4
25.	21–25	III	3
26.	26–30	I	1
27.	31–35	IIII II	7

28. Construct a histogram by using the data in Exercises 21–27.

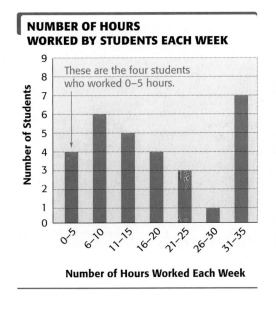

NUMBER OF HOURS WORKED BY STUDENTS EACH WEEK

These are the four students who worked 0–5 hours.

Number of Students

Number of Hours Worked Each Week

Century 21 All Professional Realty has 80 salespeople spread over its five offices. The number of homes sold by each of these salespeople during the past year is shown below. Use these numbers to complete the following table. See Example 1.

9	33	14	8	17	10	25	11	4	16	3	9	5	7	14	18
15	24	19	30	16	31	21	20	30	2	6	6	27	17	3	32
3	8	5	11	15	26	7	18	29	10	7	3	12	9	25	15
11	6	10	4	2	35	10	25	5	19	34	2	4	14	11	28
8	13	25	15	23	26	12	4	22	12	21	12	22	10	18	21

	Class Intervals (New Homes Sold)	**Tally**	**Class Frequency (Number of Salespeople)**
29.	1–5	𝍸𝍸 IIII	14
30.	6–10	𝍸𝍸𝍸 II	17
31.	11–15	𝍸𝍸𝍸 I	16
32.	16–20	𝍸𝍸	10
33.	21–25	𝍸𝍸 I	11
34.	26–30	𝍸 II	7
35.	31–35	𝍸	5

36. Make a histogram showing the results from Exercises 29–35.

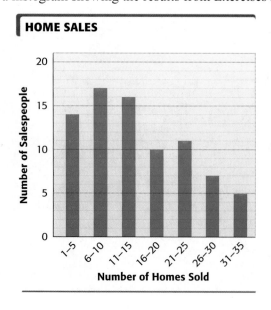

10.4 ▶▶▶ Mean, Median, and Mode

Businesses, governments, laboratories, colleges, and others working with lists of numbers are often faced with the problem of analyzing great amounts of raw data. The measures discussed in this section are helpful for such analyses.

OBJECTIVE 1 Find the mean of a list of numbers. When analyzing data, one of the first things to look for is a *measure of central tendency*— a single number that we can use to represent the entire list of numbers. One such measure is the *average* or **mean.** The mean can be found with the following formula.

Finding the Mean (Average)

$$\text{mean} = \frac{\text{sum of all values}}{\text{number of values}}$$

EXAMPLE 1 Finding the Mean

David had test scores of 84, 90, 95, 98, and 88. Find the mean (average) of his scores.

Use the formula for finding the mean. Add up all the test scores and then divide by the number of tests.

$$\text{mean} = \frac{84 + 90 + 95 + 98 + 88}{5} \quad \text{Add up all the test scores.}$$

$$\text{mean} = \frac{455}{5} \quad \textit{Divide.} \quad \text{Divide by the number of tests.}$$

$$\text{mean} = 91$$

David has a mean score of 91.

Work Problem ① at the Side. ▶

EXAMPLE 2 Applying the Average or Mean

Milk sales at a local 7-Eleven for each of the days last week were

$$\$86, \$103, \$118, \$117, \$126, \$158, \text{ and } \$149$$

Find the mean milk sales (rounded to the nearest cent) as shown below.

$$\text{mean} = \frac{\$86 + \$103 + \$118 + \$117 + \$126 + \$158 + \$149}{7}$$

$$\text{mean} = \frac{\$857}{7}$$

$$\text{mean} \approx \$122.43$$

The mean daily sales amount for milk was $122.43.

Work Problem ② at the Side. ▶

OBJECTIVES

1 Find the mean of a list of numbers.

2 Find a weighted mean.

3 Find the median.

4 Find the mode.

---

1 Tanya has test scores of 96, 98, 84, 88, 82, and 92. Find her mean (average) score.

2 Find the mean for each list of numbers. Round to the nearest cent if necessary.

(a) Monthly gasoline expenses of $50.28, $85.16, $110.50, $78, $120.70, $58.64, $73.80, $86.24, $67.85, $96.56, $138.65, $48.90

(b) The attendance at eight major league home games this season was: 48,076; 58,595; 37,874; 46,289; 29,235; 59,311; 45,675; 39,721

ANSWERS

1. 90

2. **(a)** $\dfrac{\$1015.28}{12} \approx \$84.61$ (rounded)

   (b) $\dfrac{364{,}776}{8} = 45{,}597$ people

3 The numbers below show the amount that Pat Dunn spent for lunch and the number of days that he spent that amount. Find the weighted mean.

Value	Frequency
$ 2	4
$ 4	6
$ 6	5
$ 8	6
$10	12
$12	5
$14	8
$16	4

OBJECTIVE 2 Find a weighted mean. Some items in a list might appear more than once. In this case, we find a **weighted mean,** in which each value is "weighted" by multiplying it by the number of times it occurs.

EXAMPLE 3 Understanding the Weighted Mean

The following table shows the family size and the number of families (frequency) who were given groceries in one morning at the Twin Lakes Food Bank. Find the weighted mean.

Family Size	Frequency	
1	7	← 7 families had 1 person
2	12	
3	8	
4	8	
5	5	
6	9	← 9 families had 6 people
7	1	
8	6	

The same number of people were in more than one family: for example, there were 2 people in 12 of the families and there were 4 people in 8 of the families. There were 7 people in just 1 of the families. To find the mean, multiply the family size by its frequency. Then add the products. Next, add the numbers in the frequency column to find the total number of families.

Size	Frequency	Product
1	7	$(1 \cdot 7) = 7$
2	12	$(2 \cdot 12) = 24$
3	8	$(3 \cdot 8) = 24$
4	8	$(4 \cdot 8) = 32$
5	5	$(5 \cdot 5) = 25$
6	9	$(6 \cdot 9) = 54$
7	1	$(7 \cdot 1) = 7$
8	6	$(8 \cdot 6) = 48$
Totals	56	221

Finally, divide the totals. Round to the nearest hundredth.

$$\text{mean} = \frac{221}{56} \approx 3.95 \text{ (rounded)}$$

The mean family size of those using the food bank was 3.95 people.

◀ *Work Problem* **3** *at the Side.*

ANSWER

3. $\text{mean} = \dfrac{\$466}{50} = \$9.32$

A common use of the weighted mean is to find a student's *grade point average* (GPA), as shown by the next example.

EXAMPLE 4 **Applying the Weighted Mean**

Find the grade point average for a student earning the following grades. Assume A = 4, B = 3, C = 2, D = 1, and F = 0. The number of credits determines how many times the grade is counted (the frequency).

Course	Credits	Grade	Credits · Grade
Mathematics	4	A (= 4)	$4 \cdot 4 = 16$
Speech	3	C (= 2)	$3 \cdot 2 = 6$
English	3	B (= 3)	$3 \cdot 3 = 9$
Computer science	2	A (= 4)	$2 \cdot 4 = 8$
Art history	2	D (= 1)	$2 \cdot 1 = 2$
Totals	14		41

It is common to round grade point averages to the nearest hundredth. So the grade point average for this student is rounded to 2.93.

$$GPA = \frac{41}{14} \approx 2.93$$

Work Problem **4** *at the Side.* ▶

OBJECTIVE 3 Find the median.   Because it can be affected by extremely high or low numbers, the mean is often a poor indicator of central tendency for a list of numbers. In cases like this, another measure of central tendency, called the *median,* can be used. The **median** divides a group of numbers in half; half the numbers lie above the median, and half lie below the median.

Find the median by listing the numbers *in order* from *smallest* to *largest.* If the list contains an *odd* number of items, the median is the *middle number.*

EXAMPLE 5 **Finding the Median for an Odd Number of Items**

Find the median for the following list of prices for women's T-shirts.

$$\$9, \$23, \$15, \$8, \$18, \$12, \$24$$

First arrange the numbers in numerical order from smallest to largest.

Smallest → 8, 9, 12, 15, 18, 23, 24 ← Largest

Next, find the middle number in the list.

8, 9, 12, **15**, 18, 23, 24

Three are below. Middle number. Three are above.

Remember to list the numbers from smallest to largest **before** finding the median.

The median price is $15.

Work Problem **5** *at the Side.* ▶

---

4 Find the grade point average (GPA) for Greg Barnes, who earned the following grades last semester. Round to the nearest hundredth.

Course	Credits	Grade
Mathematics	3	A (= 4)
P.E.	1	C (= 2)
English	3	C (= 2)
Keyboarding	2	B (= 3)
Biology	4	B (= 3)

5 Find the median for the following weights of bagged groceries.

14 lb, 18 lb, 10 lb, 17 lb, 15 lb, 19 lb, 20 lb

ANSWERS

4. GPA $\approx$ 2.92
5. 17 lb (the middle number when the numbers are arranged from smallest to largest)

6 Find the median for the following list of measurements.

125 m, 87 m, 96 m, 108 m, 136 m, 74 m

If a list contains an *even* number of items, there is no single middle number. In this case, the median is defined as the mean (average) of the *middle two* numbers.

EXAMPLE 6   **Finding the Median for an Even Number of Items**

Find the median for the following list of ages.

74, 7, 15, 13, 25, 28, 47, 59, 32, 68

First arrange the numbers in numerical order from least to greatest. Then find the middle two numbers.

Smallest → 7, 13, 15, 25, 28, 32, 47, 59, 68, 74 ← Largest

Middle two numbers

The median age is the mean of the two middle numbers.

$$\text{median} = \frac{28 + 32}{2} = \frac{60}{2} = 30 \text{ years}$$

◄ *Work Problem* **6** *at the Side.*

7 Find the mode for each list of numbers.

(a) Ages of summer work applicants (in years): 19, 18, 22, 20, 18

(b) Total points on a screening exam: 312, 219, 782, 312, 219, 426

(c) Monthly commissions of sales people: $1706, $1289, $1653, $1892, $1301, $1782

OBJECTIVE **4** **Find the mode.**   The last important statistical measure is the **mode,** the number that occurs *most often* in a list of numbers. For example, if the test scores for 10 students were

74, 81, 39, 74, 82, 80, 100, 92, 74, and 85,

then the mode is 74. Three students earned a score of 74, so 74 appears more times on the list than any other score. (It is not necessary to place the numbers in numerical order when looking for the mode.)

A list can have two modes; such a list is sometimes called **bimodal.** If no number occurs more frequently than any other number in a list, the list has *no mode.*

EXAMPLE 7   **Finding the Mode**

Find the mode for each list of numbers.

The **mode** is the value that occurs most often.

(a) 51, 32, 49, 73, 49, 90

The number **49** occurs more often than any other number; therefore, 49 is the mode.

(b) 482, 485, 483, 485, 487, 487, 489

Because both **485** and **487** occur twice, each is a mode. This list is *bimodal.*

(c) $10,708; $11,519; $10,972; $12,546; $13,905; $12,182

No number occurs more than once. This list has *no mode.*

◄ *Work Problem* **7** *at the Side.*

Measures of Central Tendency

The **mean** is the sum of all the values divided by the number of values. It is the mathematical average.

The **median** is the middle number (or the average of the middle two numbers) in a group of values that are listed from least to greatest. It divides a group of numbers in half.

The **mode** is the value that occurs most often in a group of values.

ANSWERS

6. $\frac{96 + 108}{2} = \frac{204}{2} = 102$ m

7. (a) 18 yr
   (b) bimodal, 219 points and 312 points (this list has two modes)
   (c) no mode (no number occurs more than once)

10.4 ▶▶▶ Exercises

*Find the mean for each list of numbers. Round answers to the nearest tenth if necessary.
See Example 1.*

1. Shopping center ages (in years) of 6, 22, 15, 2, 8, 13

   11 yr

2. Minutes of cell phone use each day: 53, 77, 38, 29, 46, 48, 52

   49 minutes

3. Inches of rain per month of 3.1, 1.5, 2.8, 0.8, 4.1

   2.5 in. of rain (rounded)

4. Algebra quiz scores of 32, 26, 30, 19, 51, 46, 38, 39

   35.1 (rounded)

5. Annual salaries of $38,500; $39,720; $42,183; $21,982; $43,250

   $37,127

6. Numbers of students enrolled in Community Colleges: 27,500; 18,250; 17,357; 14,298; 33,110

   22,103 people

Solve each application problem. See Example 2.

7. The Sunrise Pharmacy filled prescriptions that sold at the following amounts: $18.38, $168.75, $28.63, $72.85, $39.60, $183.74, $15.82, $33.18, $87.45, $98.72, and $50.70. Find the average price (mean) of the prescriptions sold.

   $72.53

8. In one evening, a waitress collected the following checks from her dinner customers: $30.10, $42.80, $91.60, $51.20, $88.30, $21.90, $43.70, $51.20. Find the average (mean) dinner check amount.

   $52.60

Find the weighted mean. Round answers to the nearest tenth. See Example 3.

9.

Customers Each Hour	Frequency
8	2
11	12
15	5
26	1

12.5 customers (rounded)

10.

Deliveries Each Week	Frequency
4	1
8	3
16	5
20	1

12.8 deliveries

11.

Fish per Boat	Frequency
12	4
13	2
15	5
19	3
22	1
23	5

17.2 fish (rounded)

12.

Patients per Clinic	Frequency
25	1
26	2
29	5
30	4
32	3
33	5

30.2 patients (rounded)

Solve each application problem. See Example 4.

13. The table below shows the face value (policy amount) of life insurance policies sold and the number of policies sold for each amount by the New World Life Company during one week. Find the weighted mean amount for the policies sold.

Policy Amount	Number of Policies Sold
$ 10,000	6
$ 20,000	24
$ 25,000	12
$ 30,000	8
$ 50,000	5
$100,000	3
$250,000	2

$35,500

14. A national health survey provided the information for this table. It shows how often each of the adults in the survey engages in a vigorous, leisure time activity. Find the weighted mean to determine the weekly hours of vigorous activity for the adults surveyed, to the nearest tenth.

Hours per week	Number of Adults
0	224
1	86
2	62
3	45
4	25
5	18
6	24
7	16

1.6 hours per week (rounded)

Find the median for each list of numbers. See Examples 5 and 6.

15. Number of books loaned: 125, 100, 150, 135, 114

125 books

16. Number of hits for a World Wide Web site (in thousands): 140, 85, 122, 114, 98

114,000 hits

17. Calories in fast-food menu items: 501, 412, 521, 515, 298, 621, 346, 528

508 calories

18. Number of cars in the parking lot each day: 520, 523, 513, 1283, 338, 509, 290, 420, 320, 980

511 cars

Find the mode or modes for each list of numbers. See Example 7.

19. Porosity of soil samples: 21%, 18%, 21%, 28%, 22%, 21%, 25%

21%

20. Low daily temperatures (in degrees Fahrenheit): 21, 32, 46, 32, 49, 32, 49

32 degrees Fahrenheit

21. Ages of residents (in years) at Leisure Village: 74, 68, 68, 68, 75, 75, 74, 74, 70

68 and 74 yr (bimodal)

22. Number of pages read: 86, 84, 79, 75, 88, 66, 72, 85, 71

no mode

23. When is the median a better measure of central tendency than the mean to describe a set of data? Make up a list of numbers to illustrate your explanation. Calculate both the mean and the median.

The median is a better measure of central tendency when the list contains one or more extreme values. Find the mean and the median of the following home values: $182,000; $164,000; $191,000; $115,000; $982,000.
mean home value = $326,800;
median home value = $182,000

24. Suppose you own a hat shop and can order a certain hat in only one size. You look at last year's sales to decide on the size to order. Should you find the mean, median, or mode for these sales? Explain your answer.

The size to order is the mode. The mode is the size most worn by customers, and it would be wise to order most hats in this size.

Find the grade point average for students earning the following grades. Assume A = 4, B = 3, C = 2, D = 1, *and* F = 0. *Round answers to the nearest hundredth.*

25.

Credits	Grade
4	B
2	C
2	A
1	C
3	D

2.42 (rounded)

26.

Credits	Grade
1	C
3	A
4	B
3	C
2	A

3.08 (rounded)

27.

Credits	Grade
4	B
2	A
5	C
1	F
3	B

2.60

28.

Credits	Grade
3	A
3	B
4	B
3	C
3	C

2.81 (rounded)

29.

Credits	Grade
2	A
3	C
4	A
1	C
4	B

3.14 (rounded)

30.

Credits	Grade
3	A
2	A
5	B
4	A
1	A

3.67 (rounded)

Relating Concepts (Exercises 31—40) For Individual or Group Work

*Gluco Industries manufactures and sells glucose monitors and other diabetes-related products. The number of sales calls made over an 8-week period by two sales representatives, Scott Samuels and Rob Stricker, is shown below. Use this information to **work Exercises 31–40 in order.***

	Number of Calls	
Week	Samuels	Stricker
1	39	21
2	15	22
3	40	20
4	22	23
5	13	19
6	22	24
7	17	25
8	8	22

31. Find the total number of sales calls made by each of the sales representatives.

176 sales calls

32. Find the mean number of sales calls made by Samuels and by Stricker.

mean for Samuels: 22;

mean for Stricker: 22

33. Find the median number of sales calls made by Samuels and by Stricker.

median for Samuels: 19.5;

median for Stricker: 22

34. Find the mode for the number of sales calls for each of the sales representatives.

mode for Samuels: 22;

mode for Stricker: 22

35. How do the mean, median, and mode for the two sales representatives compare?

The mean and mode are identical for both sales representatives and the medians are close.

36. Describe how the pattern in the number of weekly sales calls made by Samuels differs from the pattern in Stricker's weekly sales calls.

The number of weekly sales calls made by Samuels varies greatly from week to week while the number of weekly sales calls made by Stricker remains fairly constant.

To show the variation *or* spread *of the number of sales calls made by each of the sales representatives requires some **measure of the dispersion,** or spread of the numbers around the mean. A common measure of the dispersion is the **range**. The range is the* difference between the largest value and the smallest value in the set of numbers.

37. Find the range for the number of sales calls made by Samuels.

range = 40 − 8 = 32

38. Find the range for the number of sales calls made by Stricker.

range = 25 − 19 = 6

39. Is the performance data for these two sales representatives sufficient to accurately determine which is the best? What else might you want to know?

No, not with any certainty. There probably are additional questions that need to be answered, such as the dollar amount of sales, number of repeat customers, and so on.

40. List three possible explanations for the wide variation (range) in the number of sales calls for Samuels.

Answers will vary. Some possible answers are: He works hard one week, then takes it easy the next week; the characteristics of the sales territories vary greatly; illness or personal problems may be affecting performance.

Chapter 10 ▷▷▷ Summary

▶ Key Terms

10.1	**circle graph**	A circle graph shows how a total amount is divided into parts or sectors. It is based on percents of 360°.
	protractor	A protractor is a device (usually in the shape of a half-circle) used to measure the number of degrees in angles or parts of a circle.
10.2	**bar graph**	A bar graph uses bars of various heights or lengths to show quantity or frequency.
	double-bar graph	A double-bar graph compares two sets of data by showing two sets of bars.
	line graph	A line graph uses dots connected by lines to show trends.
	comparison line graph	A comparison line graph shows how two sets of data relate to each other by showing a line graph for each set of data.
10.3	**frequency distribution**	A frequency distribution is a table that includes a column showing each possible number in the data collected. The original data is then entered in another column using a tally mark for each corresponding value. The tally marks are counted and the totals are placed in a third column.
	histogram	A histogram is a bar graph in which the width of each bar represents a range of numbers (class interval) and the height represents the quantity or frequency of items that fall within the interval.
10.4	**mean**	The mean is the sum of all the values divided by the number of values. It is often called the *average*.
	weighted mean	The weighted mean is a mean calculated so that each value is multiplied by its frequency.
	median	The median is the middle number in a group of values that are listed from smallest to largest. It divides a group of values in half. If there are an even number of values, the median is the mean (average) of the two middle values.
	mode	The mode is the value that occurs most often in a group of values.
	bimodal	A list of numbers is bimodal when it has two modes. The two values occur the same number of times.
	dispersion	The dispersion is the variation or spread of the numbers around the mean.
	range	The range is a common measure of the dispersion of numbers. It is the difference between the largest value and the smallest value in the set of numbers.

▶ New Formula

Mean or average: $\quad \text{mean} = \dfrac{\text{sum of all values}}{\text{number of values}}$

▶ Test Your Word Power

See how well you have learned the vocabulary in this chapter. Answers follow the Quick Review.

1. A **circle graph**
   A. uses bars of various heights to show quantity or frequency
   B. shows how a total amount is divided into parts or sectors
   C. uses dots connected by lines to show trends
   D. uses bars of various widths to represent a range of numbers.

2. A **bar graph**
   A. uses bars of various heights to show quantity or frequency
   B. shows how a total amount is divided into parts or sectors
   C. uses dots connected by lines to show trends
   D. uses bars of various widths to represent a range of numbers.

3. A **histogram** is a graph in which
   A. tally marks are used to record original data
   B. two sets of data are compared using two sets of bars
   C. dots are connected by lines to show trends
   D. the width of each bar represents a range of numbers and the height represents the frequency of items within that range.

4. A **protractor** is a device used to
   A. construct a histogram
   B. calculate measures of central tendency
   C. measure the number of degrees in angles or parts of a circle
   D. compare two sets of data.

5. The **mean** is
   A. calculated so that each value is multiplied by its frequency
   B. the sum of all values divided by the number of values
   C. the middle number in a group of values that are listed from smallest to largest
   D. the value that occurs most often in a group of values.

6. The **mode** is
   A. calculated so that each value is multiplied by its frequency
   B. the sum of all values divided by the number of values
   C. the middle number in a group of values that are listed from smallest to largest
   D. the value that occurs most often in a group of values.

▶ Quick Review

Concepts	Examples

10.1 **Constructing a Circle Graph**

Step 1 Determine the percent of the total for each item.
Step 2 Find the number of degrees out of 360° that each percent represents.
Step 3 Use a protractor to measure the number of degrees for each item in the circle.

Construct a circle graph from the following table, which lists the costs of options (billing) on a new luxury sports car.

Item	Amount
Leather interior	$1600
Wheels/tires	$2400
Sun roof	$1200
Sport package	$2800
Total	**$8000**

Item	Amount	Percent of Total	Sector Size
Leather interior	$1600	$\dfrac{\$1600}{\$8000} = \dfrac{1}{5} = 20\%$ so $360° \cdot 20\%$ $= 360 \cdot 0.20$	$= 72°$
Wheels/tires	$2400	$\dfrac{\$2400}{\$8000} = \dfrac{3}{10} = 30\%$ so $360° \cdot 30\%$ $= 360 \cdot 0.30$	$= 108°$
Sun roof	$1200	$\dfrac{\$1200}{\$8000} = \dfrac{3}{20} = 15\%$ so $360° \cdot 15\%$ $= 360 \cdot 0.15$	$= 54°$
Sport package	$2800	$\dfrac{\$2800}{\$8000} = \dfrac{7}{20} = 35\%$ so $360° \cdot 35\%$ $= 360 \cdot 0.35$	$= 126°$

See completed circle graph on the next page.

(*continued*)

Concepts	Examples

10.1 Constructing a Circle Graph (*continued*)

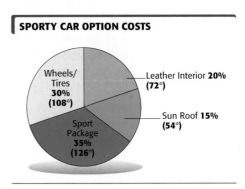

SPORTY CAR OPTION COSTS

Wheels/Tires **30%** (**108°**)

Leather Interior **20%** (**72°**)

Sun Roof **15%** (**54°**)

Sport Package **35%** (**126°**)

10.2 Reading a Bar Graph

The height of the bar is used to show the quantity or frequency (number) in a specific category. Use a ruler or straightedge to line up the top of each bar with the numbers on the left side of the graph.

Use the bar graph below to determine the number of students who earned each letter grade.

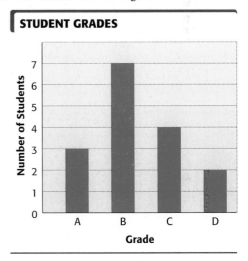

STUDENT GRADES

Number of Students / Grade (A, B, C, D)

Grade of A: 3 students; B: 7 students; C: 4 students; D: 2 students.

10.2 Reading a Line Graph

A dot is used to show the number or quantity in a specific class. The dots are connected with lines. This kind of graph is used to show a trend.

The line graph below shows the annual sales for the Fabric Supply Center for each of 4 years.

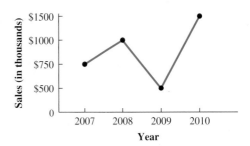

Sales (in thousands) / Year (2007, 2008, 2009, 2010)

Find the sales in 2009.
The dot above 2009 lines up with $500 on the left edge.
Then $500 • 1000 = $500,000 in sales.

10.3 **Preparing a Frequency Distribution and a Histogram from Raw Data**

Step 1 Construct a table listing each value, and the number of times this value occurs.
Step 2 Divide the data into groups, categories, or classes.
Step 3 Draw bars representing these groups to make a histogram.

Draw a histogram for these student quiz scores.

12	15	15	14
13	20	10	12
11	9	10	12
17	20	16	17
14	18	19	13

Quiz Score	Tally	Frequency	
9	I	1	1st
10	II	2	class
11	I	1	interval
12	III	3	2nd
13	II	2	class
14	II	2	interval
15	II	2	3rd
16	I	1	class
17	II	2	interval
18	I	1	4th
19	I	1	class
20	II	2	interval

Class Interval (Quiz Scores)	Frequency (Number of Students)
9–11	4
12–14	7
15–17	5
18–20	4

STUDENT QUIZ SCORES

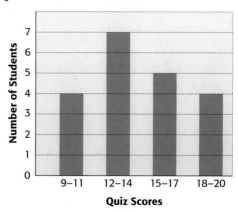

10.4 **Finding the Mean (Average) of a Set of Numbers**

Step 1 Add all values to obtain a total.
Step 2 Divide the total by the number of values.

$$\text{mean (average)} = \frac{\text{sum of all values}}{\text{number of values}}$$

The test scores for Keith Zagorin in his algebra course were as follows:

80	92	92	94
76	88	84	93

Find Keith's mean (average) test score to the nearest tenth.

$$\text{mean} = \frac{80 + 92 + 92 + 94 + 76 + 88 + 84 + 93}{8}$$

$$= \frac{699}{8} \approx 87.4$$

Keith's mean test score is approximately 87.4.

10.4 **Finding the Weighted Mean**

Step 1 Multiply frequency by value.
Step 2 Add all the products from Step 1.
Step 3 Divide the sum in Step 2 by the total number of pieces of data.

This table shows the distribution of the number of school-age children in a survey of 30 families.

Number of School-Age Children	Frequency (Number of Families)
0	12
1	6
2	7
3	3
4	2
Total of 30 families	

Find the mean number of school-age children per family. Round to the nearest hundredth.

Value	Frequency	Product
0	12	$(0 \cdot 12) = 0$
1	6	$(1 \cdot 6) = 6$
2	7	$(2 \cdot 7) = 14$
3	3	$(3 \cdot 3) = 9$
4	2	$(4 \cdot 2) = 8$
Totals	30	37

$$\text{mean} = \frac{37}{30} \approx 1.23$$

The mean number of school-age children per family is approximately 1.23.

Concepts	Examples

10.4 **Finding the Median of a Set of Numbers**

Step 1 Arrange the data from least to greatest.
Step 2 Select the middle value, or, if there is an even number of values, find the average of the two middle values.

Find the median for Keith Zagorin's test scores from the previous page.

The data arranged from smallest to largest is as follows:

76   80   84   88   92   92   93   94

Middle values

The middle two values are 88 and 92. The average of these two values is

$$\frac{88 + 92}{2} = 90$$

Keith's median test score is 90.

10.4 **Finding the Mode of a Set of Values**

Find the value that appears most often in the list of values. If no value appears more than once, there is no mode. If two different values appear the same number of times, the list is bimodal.

Find the mode for Keith's test scores shown above.

The most frequently occurring score is 92 (it occurs twice). Therefore, the mode is 92.

ANSWERS TO TEST YOUR WORD POWER

1. (B) *Example:*

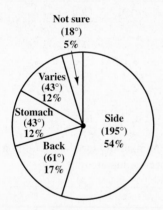

2. (A) *Example:*

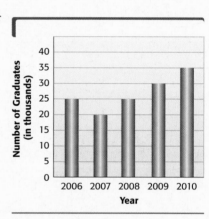

3. (D) *Example:*

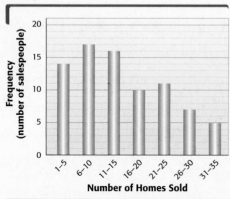

4. (C) *Example:*

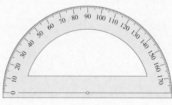

5. (B) *Example:* The mean of the values $5, $9, $7, $5, $2, and $8, is

$$\frac{\$5 + \$9 + \$7 + \$5 + \$2 + \$8}{6} = \frac{\$36}{6} = \$6.$$

6. (D) *Example:* The mode of the values $5, $9, $7, $5, $2, and $8 is $5 because $5 appears twice in the list.

Chapter 10 ▶▶▶ Review Exercises

[10.1] **1.** The number of girls participating in high school sports in the United States exceeds 3 million. The circle graph shows the number of high school girls' teams in the most popular sports. What women's sport has the greatest number of teams? How many are there?

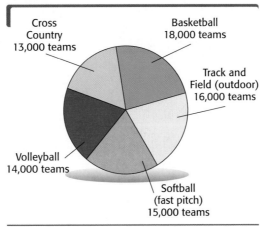

Cross Country 13,000 teams

Basketball 18,000 teams

Track and Field (outdoor) 16,000 teams

Softball (fast pitch) 15,000 teams

Volleyball 14,000 teams

Source: National Federation of State High School Associations

Basketball; 18,000 teams

Using the circle graph in Exercise 1, find each ratio. Write the ratios as fractions in lowest terms.

2. Number of track and field teams to the total number of teams.

$$\frac{16,000}{76,000} = \frac{4}{19}$$

3. Number of softball teams to the total number of teams.

$$\frac{15,000}{76,000} = \frac{15}{76}$$

4. Number of volleyball teams to the total number of teams.

$$\frac{14,000}{76,000} = \frac{7}{38}$$

5. Number of basketball teams to the number of track and field teams.

$$\frac{18,000}{16,000} = \frac{9}{8}$$

6. Number of track and field teams to the number of volleyball teams.

$$\frac{16,000}{14,000} = \frac{8}{7}$$

[10.2] *This bar graph shows the most frequently offered "work perks" and the percent of the responding companies offering them. The survey was conducted on-line and included 4800 companies ranging in size from 2 to 5000 employees. Use this graph to find the number of companies offering each work perk listed in Exercises 7–10 and to answer Exercises 11 and 12. (Source: Work Perks Survey, Ceridian Employer Services.)*

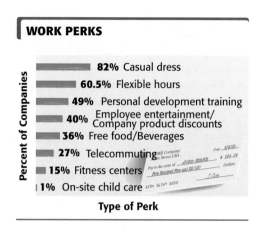

WORK PERKS

- **82%** Casual dress
- **60.5%** Flexible hours
- **49%** Personal development training
- **40%** Employee entertainment/Company product discounts
- **36%** Free food/Beverages
- **27%** Telecommuting
- **15%** Fitness centers
- **1%** On-site child care

Type of Perk (horizontal axis)

Percent of Companies (vertical axis)

7. Casual dress

3936 companies

8. Free food/Beverages

1728 companies

9. Fitness centers

720 companies

10. Flexible hours

2904 companies

11. Which two work perks do companies offer least often? Give one possible explanation why these work perks are not offered.

On-site child care and Fitness centers Answers will vary. Perhaps employers feel that they are not needed or would not be used. Or it may be that they would be too expensive for the benefit derived.

12. Which two work perks do companies offer most often? Give one possible explanation why these work perks are so popular.

Flexible hours and casual dress Answers will vary. Perhaps employees request them and appreciate them. Or, it may be that neither of them cost the employer anything to offer.

This double-bar graph shows the number of acre-feet of water in Lake Natoma for each of the first six months of 2009 and 2010. Use this graph to answer Exercises 13–18.

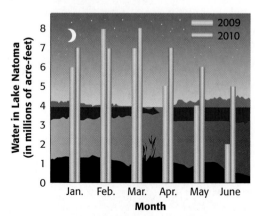

WATER IN LAKE NATOMA

Water in Lake Natoma (in millions of acre-feet)

Month: Jan. Feb. Mar. Apr. May June

Legend: 2009, 2010

13. During which month in 2010 was the greatest amount of water in the lake? How much was there?

March; 8,000,000 acre-feet

14. During which month in 2009 was the least amount of water in the lake? How much was there?

June; 2,000,000 acre-feet

15. How many acre-feet of water were in the lake in June of 2010?

5,000,000 acre-feet

16. How many acre-feet of water were in the lake in May of 2009?

4,000,000 acre-feet

17. Find the decrease in the amount of water in the lake from March 2009 to June 2009.

5,000,000 acre-feet

18. Find the decrease in the amount of water in the lake from April 2010 to June 2010.

2,000,000 acre-feet

This comparison line graph shows the annual floor-covering sales of two different home improvement centers during each of 5 years. Use this graph to find the amount of annual floor-covering sales in each year shown in Exercises 19–22 and to answer Exercises 23 and 24.

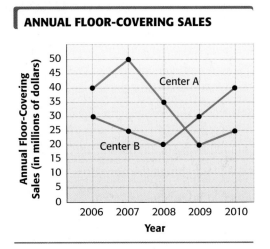

19. Center A in 2007

$50,000,000

20. Center A in 2009

$20,000,000

21. Center B in 2008

$20,000,000

22. Center B in 2010

$40,000,000

✎ **23.** What trend do you see in center A's sales from 2007 to 2010? Why might this have happened?

The floor-covering sales decreased for 2 years and then moved up slightly. Answers will vary. Perhaps there is less new construction, remodeling, and home improvement in the area near center A, or better product selection and service have reversed the decline in sales.

✎ **24.** What trend do you see in center B's sales starting in 2008? Why might this have happened?

The floor-covering sales are increasing. Answers will vary. Perhaps new construction and home remodeling have increased in the area near center B, or greater advertising has attracted more customers.

[10.4] *Find the mean for each list of numbers. Round answers to the nearest tenth if necessary.*

25. Digital cameras sold: 18, 12, 15, 24, 9, 42, 54, 87, 21, 3

28.5 digital cameras

26. Number of harassment complaints filed: 31, 9, 8, 22, 46, 51, 48, 42, 53, 42

35.2 complaints

Find the weighted mean for each list. Round to the nearest tenth if necessary.

27.

Dollar Value	Frequency
$42	3
$47	7
$53	2
$55	3
$59	5

$51.05

28.

Total Points	Frequency
243	1
247	3
251	5
255	7
263	4
271	2
279	2

257.3 points (rounded)

Find the median for each list of numbers.

29. The number of accident forms filed: 43, 37, 13, 68, 54, 75, 28, 35, 39

39 forms

30. Commissions of $576, $578, $542, $151, $559, $565, $525, $590

$562

Find the mode or modes for each list of numbers.

31. Running shoes priced at $79, $56, $110, $79, $72, $86, $79

$79

32. Boat launchings: 18, 25, 63, 32, 28, 37, 32, 26, 18

18 and 32 launchings (bimodal)

▶▶▶ **Mixed Review Exercises**

In her senior year of college Ally Romao had expenses of $17,920. This amount was spent as shown below. Find all the missing numbers in Exercises 33–37.

Item	Dollar Amount	Percent of Total	Degrees of Circle
33. Books and supplies	$1792	10%	36°
34. Rent	$6272	35%	126°
35. Food	$3584	20%	72°
36. Tuition/fees	$4480	25%	90°
37. Miscellaneous	$1792	10%	36°

38. Draw a circle graph using the information in Exercises 33–37.

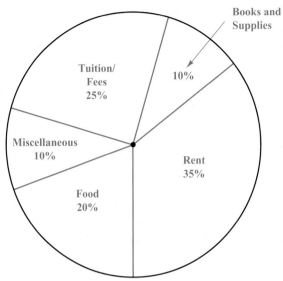

Find the mean for each list of numbers. Round answers to the nearest tenth if necessary.

39. Number of volunteers for the project: 48, 72, 52, 148, 180

100 volunteers

40. Number of flu vaccinations in a day: 122, 135, 146, 159, 128, 147, 168, 139, 158

144.7 vaccinations (rounded)

Find the mode or modes for each list of numbers.

41. Job applicants meeting the qualifications: 48, 43, 46, 47, 48, 48, 43

48 applicants

42. Number of two-bedroom apartments in each building: 26, 31, 31, 37, 43, 51, 31, 43, 43

31 and 43 two-bedroom apartments (bimodal)

Find the median for each list of numbers.

43. Hours worked: 4.7, 3.2, 2.9, 5.3, 7.1, 8.2, 9.4, 1.0

5.0 hr

44. Number of e-mails each day: 35, 51, 9, 2, 17, 12, 46, 23, 3, 19, 39, 27

21 e-mails

Here are the scores of 40 students on a computer science exam. Complete the table.

78	89	36	59	78	99	92	86
73	78	85	57	99	95	82	76
63	93	53	76	92	79	72	62
74	81	77	76	59	84	76	94
58	37	76	54	80	30	45	38

	Class Intervals (Scores)	Tally	Class Frequency (Number of Students)
45.	30–39	IIII	4
46.	40–49	I	1
47.	50–59	ⅲ I	6
48.	60–69	II	2
49.	70–79	ⅲ ⅲ III	13
50.	80–89	ⅲ II	7
51.	90–99	ⅲ II	7

52. Construct a histogram by using the data in Exercises 45–51.

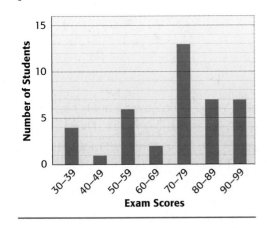

COMPUTER SCIENCE EXAM SCORES

Find each weighted mean. Round answers to the nearest tenth if necessary.

53.   **64.5 (rounded)**

Test Score	Frequency
46	4
54	10
62	8
70	12
78	10

54.   **118.8 units (rounded)**

Units Sold	Frequency
104	6
112	14
115	21
119	13
123	22
127	6
132	9

The circle graph shows the sources of electricity generated in the United States. If the total cost of all electricity generated in one year was $298 billion, find the dollar amount spent on electricity generated by each of these sources. Round to the nearest tenth of a billion.

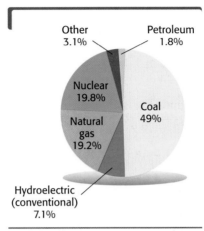

Source: Official Energy Statistics from the U.S. Government.

1. Nuclear

    1. $59.0 billion (rounded)

2. Hydroelectric (conventional)

    2. $21.2 billion (rounded)

3. Petroleum

    3. $5.4 billion (rounded)

4. Natural gas

    4. $57.2 billion (rounded)

5. Coal

    5. $146.0 billion (rounded)

6. Other

    6. $9.2 billion (rounded)

During a one-year period, Big 5 Sporting Goods had the following sales in each department. Find all numbers missing from the chart.

Item	Dollar Amount	Percent of Total	Degrees of a Circle
7. Team sports	$432,000	30%	108°
8. Golf	$144,000	10%	36°
9. Hunting and fishing	$288,000	20%	72°
10. Athletic shoes	$504,000	35%	126°
11. Water sports	$72,000	5%	18°

7. 108°

8. 36°

9. 72°

10. 126°

11. 5%

12.

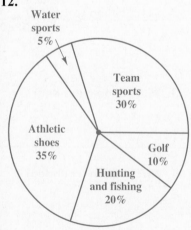

12. Draw a circle graph using the information in Problems 7–11. Label each sector of the graph.

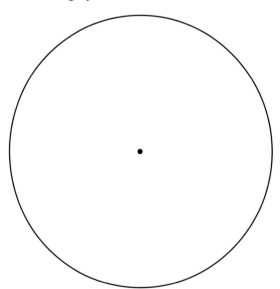

Here are the profits for each of the past 20 weeks from Alan's Snack Bar vending machines. Complete the table.

$142 $137 $125 $132 $147 $129 $151 $172 $175 $129
$159 $148 $173 $160 $152 $174 $169 $163 $149 $173

Profit	Number of Weeks	
13. $120–129	_____	**13.** <u>3</u>
14. $130–139	_____	**14.** <u>2</u>
15. $140–149	_____	**15.** <u>4</u>
16. $150–159	_____	**16.** <u>3</u>
17. $160–169	_____	**17.** <u>3</u>
18. $170–179	_____	**18.** <u>5</u>

19. Use the information in Problems 13–18 to draw a histogram.

19. <u>See histogram at left.</u>

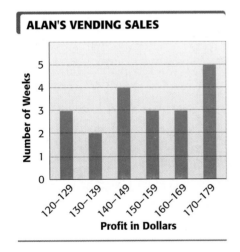

ALAN'S VENDING SALES

Find the mean for each list of numbers. Round answers to the nearest tenth if necessary.

20. Number of miles run each week while training: 52, 61, 68, 69, 73, 75, 79, 84, 91, 98

20. <u>75 mi</u>

21. Weight in pounds for the largest bass caught in the lake: 11, 14, 12, 14, 20, 16, 17, 18

21. <u>15.3 lb (rounded)</u>

22. Airplane speeds in miles per hour: 458, 432, 496, 491, 500, 508, 512, 396, 492, 504

22. <u>478.9 mph</u>

Credits	Grade		
3	A	$3 \cdot 4 =$	12
2	C	$2 \cdot 2 =$	4
4	B	$4 \cdot 3 =$	12
$\overline{9}$			$\overline{28}$

23. $28 \div 9 \approx 3.11$

23. Explain why a weighted mean is used to find a student's grade point average. Calculate your own grade point average for last semester or quarter. If you are a new student, make up a grade point average problem of your own and solve it. Round to the nearest hundredth.

The weighted mean is used because different classes are worth different numbers of credits. GPA problems will vary; one possibility is shown at the left.

8, 17, 23, 32, 64

24. ⎿— Median

24. Explain in your own words the procedure for finding the median when there are an odd number of values in a list. Make up a problem with a list of five numbers and solve for the median.

Arrange the values in order, from smallest to largest. When there is an odd number of values in a list, the median is the middle value. Students' problems will vary; one possibility is shown at the left.

Find the weighted mean for the following. Round answers to the nearest whole number if necessary.

25. $26 (rounded)

25.

Cost	Frequency
$12	5
$20	6
$22	8
$28	4
$38	6
$48	2

26. 174 points (rounded)

26.

Points Scored	Frequency
150	15
160	17
170	21
180	28
190	19
200	7

Find the median for each list of numbers.

27. 31.5 degrees

27. Lowest daily temperatures in degrees Fahrenheit:
32, 41, 28, 28, 37, 35, 16, 31

28. 10.0 meters

28. The length of steel beams in meters:
7.6, 11.4, 6.2, 12.5, 31.7, 22.8, 9.1, 10.0, 9.5

Find the mode or modes for each list of numbers.

29. 52 milliliters

29. Blood sample amounts in milliliters:
72, 46, 52, 37, 28, 18, 52, 61

30. 103 and 104 degrees (bimodal)

30. Hot tub temperatures in degrees Fahrenheit of
96, 104, 103, 104, 103, 104, 91, 74, 103

Cumulative Review Exercises ▶▶▶ Chapters 1–10

First use front end rounding to round each number and estimate the answer. Then find the exact answer.

1. Estimate: $\begin{array}{r} 900 \\ -\ 60 \\ \hline 840 \end{array}$ Exact: $\begin{array}{r} 875.62 \\ -\ 63.757 \\ \hline 811.863 \end{array}$

2. Estimate: $\begin{array}{r} 7000 \\ \times\ 600 \\ \hline 4{,}200{,}000 \end{array}$ Exact: $\begin{array}{r} 7064 \\ \times\ 635 \\ \hline 4{,}485{,}640 \end{array}$

3. Estimate: $\begin{array}{r} 15 \\ 4{\overline{)60}} \end{array}$ Exact: $\begin{array}{r} 14.72 \\ 4.25{\overline{)62.56}} \end{array}$

Simplify. Write answers in lowest terms and as whole or mixed numbers when possible.

4. $4\frac{3}{5} + 5\frac{2}{3}$ $10\frac{4}{15}$

5. $6\frac{2}{3} - 4\frac{3}{4}$ $1\frac{11}{12}$

6. $\left(9\frac{3}{5}\right)\left(4\frac{5}{8}\right)$ $44\frac{2}{5}$

7. $22\left(\frac{2}{5}\right)$ $8\frac{4}{5}$

8. $3\frac{1}{3} \div 8\frac{3}{4}$ $\frac{8}{21}$

9. $\frac{2}{3}\left(\frac{7}{8} - \frac{3}{4}\right)$ $\frac{1}{12}$

10. $4 + 10 \div 2 + 7(2)$ 23

11. $\sqrt{81} - 4(2) + 9$ 10

12. $2^2 \cdot 3^3$ 108

Write in order, from least to greatest.

13. 0.218, 0.22, 0.199, 0.207, 0.2215
   0.199, 0.207, 0.218, 0.22, 0.2215

14. $0.6319, \frac{5}{8}, 0.608, \frac{13}{20}, 0.58$
   $0.58, 0.608, \frac{5}{8}, 0.6319, \frac{13}{20}$

Write each ratio in lowest terms. Be sure to make all necessary conversions.

15. $5\frac{1}{2}$ in. to 44 in. $\frac{1}{8}$

16. 3 hr to 45 min $\frac{4}{1}$

Find the unknown number in each proportion.

17. $\frac{1}{5} = \frac{x}{30}$ x = 6

18. $\frac{15}{x} = \frac{390}{156}$ x = 6

19. $\frac{200}{135} = \frac{24}{x}$ x = 16.2

20. $\frac{x}{208} = \frac{6.5}{26}$ x = 52

Solve each percent problem.

21. Find 5.4% of 6000 homes.
   324 homes

22. $8\frac{1}{2}$% of what number of people is 238 people?
   2800 people

23. What percent of $555 is $1443?
   260%

Convert each measurement using unit fractions or the metric conversion line.

24. 28 qt = __7__ gal

25. 400 mm to cm
   40 cm

26. 230 g to kg
   0.23 kg

27. __12,000__ lb = 6 tons

Name each shape and find its area. Use 3.14 as the approximate value of π. Round answers to the nearest tenth.

28.
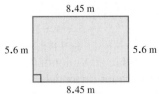
Rectangle; A ≈ 47.3 m² (rounded)

29.

Triangle; A ≈ 9.6 ft² (rounded)

📱 **30.**

Circle: A ≈ 132.7 cm² (rounded)

In Exercises 31 and 32, name each solid and find its volume. Use 3.14 as the approximate value for π.
In Exercise 33, find the unknown length. Round answers to the nearest tenth if necessary.

31.

Cylinder;
$V \approx 549.8 \text{ cm}^3$
(rounded)

32.

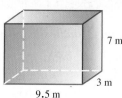

Rectangular solid;
$V = 199.5 \text{ m}^3$

33.

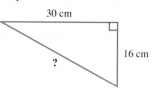

34 cm

Add, subtract, multiply, or divide as indicated.

34. $-12 + (-10)$
-22

35. $-5.7 - (-12.6)$
6.9

36. $(-14.6)(-5.7)$
83.22

37. $\dfrac{-34.04}{14.8}$
-2.3

Solve each equation. Show your work.

38. $3x - 5 = 16$
$x = 7$

39. $-12 = 3(x + 2)$
$x = -6$

40. $3.4x + 6 = 1.4x - 8$
$x = -7$

Find the mean, the median, and the mode for each list of numbers. Round to the nearest tenth if necessary.

41. Cable hookups per installer: 16, 37, 27, 31, 19, 25, 15, 38, 43, 19

mean = 27 hookups; median = 26 hookups;
mode = 19 hookups

42. Number of acres plowed each hour: 10.3, 4.3, 1.65, 2.85, 5.3, 5.7, 2.3, 4.35, 2.85

mean = 4.4 acres; median = 4.3 acres;
mode = 2.85 acres

Solve each application problem.

43. In a study of 2082 workers it was found that only 874 of them used all of their paid time-off. What percent used all of their time-off? Round to the nearest tenth of a percent. (*Source:* Hudson Time-Off Survey.)

42.0%

44. The average increase in residential winter heating bills will be 9.8%. If the increase amounts to $87.20, find (a) the average heating bill before the increase, and (b) the average heating bill after the increase. (*Source:* Energy Information Administration.)

(a) $889.80 (b) $977

45. Breathe Right™ nasal strips are sold in four package sizes: 12 nasal strips $6.50; 24 nasal strips $7.50; 30 nasal strips $8.95; 38 nasal strips $9.95. You have a $2-off coupon for the 12-strip size and a $1-off coupon for the 30-strip size. Which choice is the best buy?

38 nasal strips
for $9.95

46. The sketch below shows the plans for a lobby in a large commercial complex. What is the cost of carpeting the lobby, excluding the atrium, if the contractor charges $43.50 per square yard? Use 3.14 for π.

$59,225.25

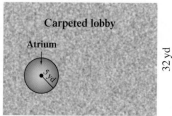

47. The Spa Service Center services 140 spas. If each spa needs 125 mL of muriatic acid, how many liters of acid are needed for all the spas?

17.5 L

48. Linda Vincent has $11,400 in an 8-month CD (certificate of deposit) account earning $4\frac{3}{4}\%$ interest. Find the total amount in her account at maturity.

$11,761

Appendix: Inductive and Deductive Reasoning

Inductive and Deductive Reasoning

Appendix

OBJECTIVE 1 Use inductive reasoning to analyze patterns. In many scientific experiments, conclusions are drawn from specific outcomes. After many repetitions and similar outcomes, the findings are generalized into statements that appear to be true. When general conclusions are drawn from specific observations, we are using a type of reasoning called **inductive reasoning.** The next several examples illustrate this type of reasoning.

OBJECTIVES

1 **Use inductive reasoning to analyze patterns.**

2 **Use deductive reasoning to analyze arguments.**

3 **Use deductive reasoning to solve problems.**

EXAMPLE 1 Using Inductive Reasoning

Find the next number in the sequence 3, 7, 11, 15, ...

To discover a pattern, calculate the difference between each pair of successive numbers.

$$7 - 3 = 4$$
$$11 - 7 = 4$$
$$15 - 11 = 4$$

Notice that the difference is always 4. Each number is 4 greater than the previous one. Thus, the next number in the pattern is $15 + 4$, or 19.

Work Problem **1** *at the Side.* ▶

1 Find the next number in the sequence 2, 8, 14, 20, ... Describe the pattern.

EXAMPLE 2 Using Inductive Reasoning

Find the next number in this sequence.

$$7, 11, 8, 12, 9, 13, ...$$

The pattern in this example involves addition and subtraction.

$$7 + 4 = 11$$
$$11 - 3 = 8$$
$$8 + 4 = 12$$
$$12 - 3 = 9$$
$$9 + 4 = 13$$

To get the second number, we add 4 to the first number. To get the third number, we subtract 3 from the second number. To obtain subsequent numbers, we continue the pattern. The next number is $13 - 3 = 10$.

Work Problem **2** *at the Side.* ▶

2 Find the next number in the sequence 6, 11, 7, 12, 8, 13, ... Describe the pattern.

ANSWERS

1. 26; add 6 each time.
2. 9; add 5, subtract 4.

A–1

3 Find the next number in the sequence 2, 6, 18, 54, ... Describe the pattern.

EXAMPLE 3   **Using Inductive Reasoning**

Find the next number in the sequence 1, 2, 4, 8, 16, ...

Each number after the first is obtained by multiplying the previous number by 2. So the next number would be $16 \cdot 2 = 32$.

◀ *Work Problem* **3** *at the Side.*

EXAMPLE 4   **Using Inductive Reasoning**

(a) Find the next geometric shape in this sequence.

The figures alternate between a blue circle and a red triangle. Also, the number of dots increases by 1 in each subsequent figure. Thus, the next figure should be a blue circle with five dots inside it.

(b) Find the next geometric shape in this sequence.

The first two shapes consist of vertical lines with horizontal lines at the bottom extending first *left* and then *right*. The third shape is a vertical line with a horizontal line at the top extending to the *left*. Therefore, the next shape should be a vertical line with a horizontal line at the top extending to the *right*.

◀ *Work Problem* **4** *at the Side.*

---

4 Find the next shape in this sequence.

OBJECTIVE **2** **Use deductive reasoning to analyze arguments.**
In the previous discussion, specific cases were used to find patterns and predict the next event. There is another type of reasoning called **deductive reasoning,** which moves from general cases to specific conclusions.

EXAMPLE 5   **Using Deductive Reasoning**

Does the conclusion follow from the premises in this argument?

$$\begin{array}{ll} \text{All Buicks are automobiles.} & \leftarrow \text{Premise} \\ \underline{\text{All automobiles have horns.}} & \leftarrow \text{Premise} \\ \therefore \quad \text{All Buicks have horns.} & \leftarrow \text{Conclusion} \end{array}$$

In this example, the first two statements are called *premises* and the third statement (below the line) is called a *conclusion*. The symbol $\therefore$ is a mathematical symbol meaning "**therefore.**" The entire set of statements is called an *argument*.

Continued on Next Page

ANSWERS

3. 162; multiply by 3.

4.

The focus of deductive reasoning is to determine whether the conclusion follows (is valid) from the premises. A set of circles called **Euler circles** is used to analyze the argument.

In Example 5, the statement "All Buicks are automobiles" can be represented by two circles, one for Buicks and one for automobiles. Note that the circle representing Buicks is totally inside the circle representing automobiles because the first premise states that *all* Buicks are automobiles.

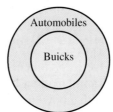

Now, a circle is added to represent the second statement, vehicles with horns. This circle must completely surround the circle representing automobiles because the second premise states that *all* automobiles have horns.

To analyze the conclusion, notice that the circle representing Buicks is *completely* inside the circle representing vehicles with horns. Therefore, it must follow that all Buicks have horns. ***The conclusion is valid.***

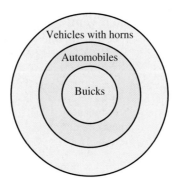

Work Problem **5** *at the Side.* ▶

5 Does the conclusion follow from the premises in the following argument?

All cars have four wheels.
All Fords are cars.

∴ All Fords have four wheels.

6 Does each conclusion follow from the premises?

(a) All animals are wild.
All cats are animals.

∴ All cats are wild.

EXAMPLE 6 **Using Deductive Reasoning**

Does the conclusion follow from the premises in this argument?

All tables are round.

All glasses are round.

∴   All glasses are tables.

Use Euler circles. Draw a circle representing tables *inside* a circle representing round objects, because the first premise states that *all* tables are round.

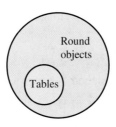

The second statement requires that a circle representing glasses must now be drawn inside the circle representing round objects, but not necessarily inside the circle representing tables. Therefore, the conclusion does *not* follow from the premises. This means that ***the conclusion is invalid.***

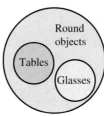

(b) All students use math.
All adults use math.

∴ All adults are students.

Work Problem **6** *at the Side.* ▶

7 In a college class of 100 students, 35 take both math and history, 50 take history, and 40 take math. How many take neither math nor history? Draw a Venn diagram.

8 A Chevy, BMW, Cadillac, and Ford are parked side by side. The known facts are:

(a) The Ford is on the right end.

(b) The BMW is next to the Cadillac.

(c) The Chevy is between the Ford and the Cadillac.

Which car is parked on the left end?

OBJECTIVE **3** **Use deductive reasoning to solve problems.**
Another type of deductive reasoning problem occurs when a set of facts is given in a problem and a conclusion must be drawn using these facts.

EXAMPLE 7 **Using Deductive Reasoning**

There were 25 students enrolled in a ceramics class. During the class, 10 of the students made a bowl and 8 students made a birdbath. Three students made both a bowl and a birdbath. How many students did not make either a bowl or a birdbath?

This type of problem is best solved by organizing the data using a drawing called a **Venn diagram.** Two overlapping circles are drawn, with each circle representing one item made by the students, as shown below.

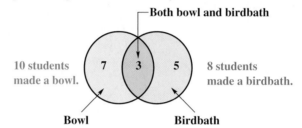

In the region where the circles overlap, write the number of students who made *both* items, namely, 3. In the remaining portion of the birdbath circle, write the number 5, which when added to 3 will give the total number of students who made a birdbath, namely, 8. In a similar manner, write 7 in the remaining portion of the bowl circle, since $7 + 3 = 10$, the total number of students who made a bowl. The total of all three numbers written in the circles is 15. Since there were 25 students in the class, this means $25 - 15$ or 10 students did not make either a birdbath or a bowl.

◀ *Work Problem* **7** *at the Side.*

EXAMPLE 8 **Using Deductive Reasoning**

Four cars in a race finish first, second, third, and fourth. The following facts are known.

(a) Car A beat Car C.

(b) Car D finished between Cars C and B.

(c) Car C beat Car B.

In which order did the cars finish?

To solve this type of problem, it is helpful to use a line diagram.

1. *Write A before C,* because Car A beat Car C (fact **a**).

$$A \quad C$$

2. *Write B after C,* because Car C beat Car B (fact **c**).

$$A \quad C \quad B$$

3. *Write D between C and B,* because Car D finished between Car C and Car B (fact **b**).

The correct order of finish is shown below.

$$A \quad C \quad D \quad B$$

◀ *Work Problem* **8** *at the Side.*

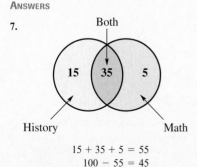

Find the next number in each sequence. Describe the pattern in each sequence. See Examples 1–3.

1. 2, 9, 16, 23, 30, . . .
37; add 7.

2. 5, 8, 11, 14, 17, . . .
20; add 3.

3. 0, 10, 8, 18, 16, . . .
26; add 10, subtract 2.

4. 3, 9, 7, 13, 11, . . .
17; add 6, subtract 2.

5. 1, 2, 4, 8, . . .
16; multiply by 2.

6. 1, 4, 16, 64, . . .
256; multiply by 4.

7. 1, 3, 9, 27, 81, . . .
243; multiply by 3.

8. 3, 6, 12, 24, 48, . . .
96; multiply by 2.

9. 1, 4, 9, 16, 25, . . .
36; add 3, add 5, add 7, etc.; or 1^2, 2^2, 3^2, etc.

10. 6, 7, 9, 12, 16, . . .
21; add 1, add 2, add 3, and so on.

Find the next shape in each sequence. See Example 4.

11.

12.

13.

14.

For each argument, draw Euler circles and then state whether or not the conclusion follows from the premises. See Examples 5 and 6.

15. All animals are wild.
All lions are animals.
∴ All lions are wild.

Conclusion follows.

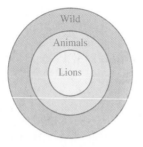

16. All students are hard workers.
All business majors are students.
∴ All business majors are hard workers.

Conclusion follows.

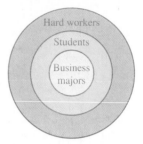

17. All teachers are serious.
All mathematicians are serious.
∴ All mathematicians are teachers.

Conclusion does not follow.

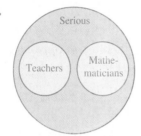

18. All boys ride bikes.
All Americans ride bikes.
∴ All Americans are boys.

Conclusion does not follow.

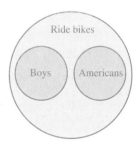

Solve each application problem. See Examples 7 and 8.

19. In a given 30-day period, a husband watched television 20 days and his wife watched television 25 days. If they watched television together 18 days, how many days did neither watch television? Draw a Venn diagram.

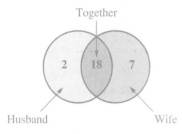

$2 + 18 + 7 = 27$
$30 - 27 = 3$

Neither watched TV on 3 days.

20. In a class of 40 students, 21 students take both calculus and physics. If 30 students take calculus and 25 students take physics, how many do not take either calculus or physics? Draw a Venn diagram.

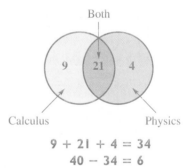

$9 + 21 + 4 = 34$
$40 - 34 = 6$

6 students take neither one.

21. Tom, Dick, Mary, and Joan all work for the same company. One is a secretary, one is a computer operator, one is a receptionist, and one is a mail clerk.

   (a) Tom and Joan eat dinner with the computer operator.

   (b) Dick and Mary carpool with the secretary.

   (c) Mary works on the same floor as the computer operator and the mail clerk.

   Who is the computer operator?

   Dick is the computer operator.

22. Four cars—a Ford, a Buick, a Mercedes, and an Audi—are parked in a garage in four spaces.

   (a) The Ford is in the last space.

   (b) The Buick and Mercedes are next to each other.

   (c) The Audi is next to the Ford but not next to the Buick.

   Which car is in the first space?

   Buick

In this section we provide the answers that we think most students will obtain when they work the exercises using the methods explained in the text. If your answer does not look exactly like the one given here, it is not necessarily wrong. In many cases there are equivalent forms of the answer that are correct. For example, if the answer section shows $\frac{3}{4}$ and your answer is 0.75 you have obtained the right answer but written it in a different (yet equivalent) form. Unless the directions specify otherwise, 0.75 is just as valid an answer as $\frac{3}{4}$.

In general, if your answer does not agree with the one given in the text, see whether it can be transformed into the other form. If it can, then it is the correct answer. If you still have doubts, talk with your instructor.

Chapter 1 Whole Numbers

Section 1.1 (pages 7–8)

1. 3; 6 **3.** 1; 0 **5.** 8; 2 **7.** 3; 561; 435 **9.** 60; 000; 502; 109
11. Evidence suggests that this is true. It is common to count using fingers. **13.** twenty-three thousand, one hundred fifteen
15. three hundred forty-six thousand, nine **17.** twenty-five million, seven hundred fifty-six thousand, six hundred sixty-five
19. 63,163 **21.** 10,000,223 **23.** 3,200,000 **25.** 50,051,507
27. 35,079,448 **29.** 800,000,621,020,215 **31.** public transportation; six million, sixty-nine thousand, five hundred eighty-nine **33.** seven million, eight hundred ninety-four thousand, nine hundred eleven

Section 1.2 (pages 17–20)

1. 97 **3.** 89 **5.** 889 **7.** 889 **9.** 7785 **11.** 1589 **13.** 7676
15. 78,446 **17.** 8928 **19.** 59,224 **21.** 150 **23.** 155
25. 121 **27.** 145 **29.** 102 **31.** 1651 **33.** 1154 **35.** 413
37. 1771 **39.** 1410 **41.** 6391 **43.** 11,624 **45.** 17,611
47. 15,954 **49.** 10,648 **51.** 15,594 **53.** 11,557 **55.** 12,078
57. 4250 **59.** 12,268 **61.** correct **63.** incorrect; should be 769
65. correct **67.** incorrect; should be 11,577 **69.** correct
71. Changing the order in which numbers are added does not change the sum. You can add from bottom to top when checking addition. **73.** 33 miles **75.** 38 miles **77.** $16,342
79. 699 people **81.** 20,157 students **83.** 970 ft **85.** 72 ft
87. 9421 **88.** 1249 **89.** 77,762 **90.** 22,267 **91.** 9,994,433
92. 3,334,499 **93.** Write the largest digits on the left, using the smaller digits as you move right. **94.** Write the smallest digits on the left, using the larger digits as you move right.

Section 1.3 (pages 27–30)

1. 16 **3.** 33 **5.** 17 **7.** 213 **9.** 101 **11.** 7111 **13.** 3412
15. 2111 **17.** 13,160 **19.** 41,110 **21.** correct **23.** incorrect; should be 62 **25.** incorrect; should be 121 **27.** correct
29. incorrect; should be 7222 **31.** 38 **33.** 45 **35.** 19
37. 281 **39.** 519 **41.** 7059 **43.** 7589 **45.** 8859 **47.** 3
49. 23 **51.** 19 **53.** 2833 **55.** 7775 **57.** 503 **59.** 156
61. 2184 **63.** 5687 **65.** 19,038 **67.** 31,556 **69.** 6584
71. correct **73.** correct **75.** correct **77.** correct **79.** Possible answers are 1. $3 + 2 = 5$ could be changed to $5 - 2 = 3$ or $5 - 3 = 2$
2. $6 - 4 = 2$ could be changed to $2 + 4 = 6$ or $4 + 2 = 6$.
81. 47 calories **83.** 367 ft **85.** 467 passengers **87.** 3270 jobs eliminated **89.** 9539 flags **91.** $263 **93.** 758 more people visited on Tuesday **95.** $57,500 **97.** 330 fewer calories; 27 fewer fat grams **99.** 380 calories; 10 fat grams

Section 1.4 (pages 37–40)

1. 24 **3.** 48 **5.** 0 **7.** 24 **9.** 40 **11.** 0 **13.** Factors may be multiplied in any order to get the same answer. They are the same; you may add or multiply numbers in any order. **15.** 210 **17.** 238
19. 3210 **21.** 1872 **23.** 8612 **25.** 10,084 **27.** 20,488
29. 258,447 **31.** 280 **33.** 480 **35.** 2220 **37.** 3600 **39.** 3750
41. 65,400 **43.** 270,000 **45.** 86,000,000 **47.** 48,500 **49.** 720,000
51. 1,940,000 **53.** 476 **55.** 2400 **57.** 3735 **59.** 2378 **61.** 6164
63. 15,792 **65.** 21,665 **67.** 15,730 **69.** 82,320 **71.** 183,996
73. 2,468,928 **75.** 66,005 **77.** 86,028 **79.** 19,422,180
81. 2,278,410 **83.** To multiply by 10, 100, or 1000, just add one, two, or three zeros, respectively, to the number you are multiplying and that's your answer. **85.** 3000 balls **87.** 540 eggs **89.** 24,090 gallons
91. $600 **93.** $1560 **95.** $112,888 **97.** 50,568 **99.** 38,250 trees
101. 7,623,663 people **103.** $7390 **105. (a)** 452 **(b)** 452
106. commutative **107. (a)** 281 **(b)** 281 **108.** associative
109. (a) 15,840 **(b)** 15,840 **110.** commutative **111. (a)** 6552
(b) 6552 **112.** associative **113.** No. Some examples are 1. $7 - 5 = 2$, but $5 - 7$ does not equal 2   2. $12 - 6 = 6$, but $6 - 12$ does not equal 6
3. $(8 - 2) - 5 = 1$, but $8 - (2 - 5)$ does not equal 1.   **114.** No. Some examples are 1. $10 \div 2 = 5$, but $2 \div 10$ does not equal 5
2. $(16 \div 8) \div 2 = 1$, but $16 \div (8 \div 2)$ does not equal 1.

Section 1.5 (pages 51–54)

1. $4)\overline{24} \quad \frac{24}{4} = 6$ **3.** $9)\overline{45} \quad 45 \div 9 = 5$ **5.** $16 \div 2 = 8 \quad \frac{16}{2} = 8$
7. 1 **9.** 7 **11.** undefined **13.** 24 **15.** 0 **17.** undefined **19.** 15
21. 8 **23.** 25 **25.** 18 **27.** 304 **29.** 627 R1 **31.** 1522 R5 **33.** 309
35. 3005 **37.** 5006 **39.** 811 R1 **41.** 2589 R2 **43.** 7324 R2
45. 3157 R2 **47.** 2630 **49.** 12,458 R3 **51.** 10,253 R5 **53.** 18,377 R6
55. correct **57.** incorrect; should be 1908 R1 **59.** incorrect; should be 670 R2 **61.** incorrect; should be 3568 R1 **63.** correct
65. correct **67.** incorrect; should be 9628 R3 **69.** correct
71. Multiply the quotient by the divisor and add any remainder. The result should be the dividend. **73.** 328 tables **75.** 9600 each hour **77.** $48,500 **79.** 135 acres **81.** $1,137,500 **83.** $5429
85. ✓ ✓ ✓ **87.** ✓ X X X **89.** X X ✓ X
91. X ✓ X X **93.** ✓ ✓ X X **95.** X X X X

Section 1.6 (pages 61–63)

1. 53 **3.** 250 **5.** 120 R7 **7.** 1105 R5 **9.** 7134 R12 **11.** 900 R100
13. 73 R5 **15.** 476 R15 **17.** 2407 R1 **19.** 1146 R15 **21.** 3331 R82
23. 850 **25.** incorrect; should be 101 R14 **27.** incorrect; should be 658
29. incorrect; should be 62 **31.** When dividing by 10, 100, or 1000, drop the same number of zeros from the dividend as there are in the divisor to get the quotient. One example is $2500 \div 100 = 25$. **33.** 50 miles
35. 56 floor clocks **37.** $355 **39.** 43,200 rings **41.** $39 per week
43. $0 **44.** 0 **45.** undefined **46.** impossible; if you have 6 cookies, it is not possible to divide them among 0 people. **47. (a)** 14 **(b)** 17
(c) 38 **48.** Yes. Some examples are $18 \cdot 1 = 18$; $26 \cdot 1 = 26$;
$43 \cdot 1 = 43$. **49. (a)** 3200 **(b)** 320 **(c)** 32 **50.** Drop the same number of zeros that appear in the divisor. The result is the quotient. With the divisor 10, drop one 0; with 100, drop two zeros; with 1000, drop three zeros.

Summary Exercises on Whole Numbers (pages 65–66)

1. 3; 4 **2.** 6; 0 **3.** 1; 6 **4.** eighty-six thousand, two **5.** four hundred twenty-five million, two hundred eight thousand, seven hundred thirty-three
6. 97 **7.** 905 **8.** 21 **9.** 409 **10.** 17,573 **11.** 82,164 **12.** 677

13. 37,674 **14.** 35,889 **15.** 560 **16.** 5600 **17.** 350,000
18. 252,000 **19.** 6,617,418,351 **20.** 24,657 **21.** 1 **22.** 0
23. undefined **24.** 15 **25.** 56 **26.** 0 **27.** 96 **28.** 304
29. 2750 R2 **30.** 761 R3 **31.** 3380 **32.** 220,545 **33.** 2016
34. 1476 **35.** 78 **36.** 210 **37.** 18,038,816 **38.** 506 R28 **39.** 52
40. 1208 R3 **41.** 573 R3 **42.** 41 **43.** 208,530 **44.** 1,101,744

Section 1.7 (pages 73–76)

1. 620 **3.** 860 **5.** 6800 **7.** 86,800 **9.** 28,500 **11.** 6000
13. 16,000 **15.** 78,000 **17.** 8,000,000,000 **19.** 10,000 **21.** 600,000
23. 5,000,000 **25.** 4480; 4500; 4000 **27.** 3370; 3400; 3000
29. 6050; 6000; 6000 **31.** 5340; 5300; 5000 **33.** 19,540; 19,500;
20,000 **35.** 26,290; 26,300; 26,000 **37.** 93,710; 93,700; 94,000
39. 1. Locate the place to be rounded and underline it. 2. Look only at
the next digit to the right. If this digit is 5 or more, increase the underlined
digit by 1. 3. Change all digits to the right of the underlined place to
zeros. **41.** 30 60 50 80 220; 219 **43.** 80 40 40; 35 **45.** 70 30 2100;
2278 **47.** 900 700 400 800 2800; 2828 **49.** 900 400 500; 435
51. 800 400 320,000; 282,000 **53.** 8000 60 700 4000 12,760; 12,605
55. 700 500 200; 158 **57.** 900 30 27,000; 27,231 **59.** Perhaps the
best explanation is that 3492 is closer to 3500 than 3400, but 3492 is
closer to 3000 than to 4000. **61.** 80 million people; 300 million people
63. 349,000; 350,000 **65.** 1,670,000 pounds; 1,700,000 pounds;
2,000,000 pounds **67.** $25,765,500,000; $25,800,000,000;
$26,000,000,000 **69.** 71,500 **70.** 72,499 **71.** 7500 **72.** 8499
73. 3930; 11,240; 15,970; 17,920; 534,880; 2,788,000 **74.** 4000; 10,000;
20,000; 20,000; 500,000; 3,000,000 **75. (a)** When using front end
rounding, all digits are 0 except the first digit. These numbers are easier
to work with when estimating answers. **(b)** When using front end
rounding to estimate an answer, the estimated answer can vary greatly
from the exact answer.

Section 1.8 (pages 79–82)

1. 2; 3; 9 **3.** 2; 5; 25 **5.** 2; 8; 64 **7.** 2; 15; 225 **9.** 4 **11.** 8
13. 10 **15.** 12 **17.** 36; 36 **19.** 400; 400 **21.** 1225; 1225
23. 625; 625 **25.** 10,000; 10,000 **27.** A perfect square is the square
of a whole number. The number 25 is the square of 5 because $5 \cdot 5 = 25$.
The number 50 is not a perfect square. There is no whole number that can
be squared to get 50. **29.** 12 **31.** 15 **33.** 4 **35.** 20 **37.** 45
39. 41 **41.** 118 **43.** 22 **45.** 30 **47.** 102 **49.** 9 **51.** 63
53. 33 **55.** 70 **57.** 7 **59.** 17 **61.** 55 **63.** 108 **65.** 26
67. 26 **69.** 27 **71.** 16 **73.** 16 **75.** 21 **77.** 7 **79.** 20
81. 14 **83.** 25 **85.** 16 **87.** 23 **89.** 233

Section 1.9 (pages 87–90)

1. 4500 stores **3.** Dollar General; about 5750 stores **5.** 1250 fewer
stores **7.** 9 people **9. (a)** Saw ad **(b)** 25 people **11.** 9 people
13. 2011; 7000 trees **15.** 4500 trees **17.** Possible answers are
1. shortage of trees to plant 2. lack of qualified workers 3. poor economy
4. less demand for planting. **19.** $(7 - 2) \cdot 3 - 6$ **20.** $(4 + 2) \cdot (5 + 1)$

21. $36 \div (3 \cdot 3) \cdot 4$ **22.** $56 \div (2 \cdot 2 \cdot 2) + \dfrac{0}{6}$

23. (a) $7920 + 1320 + 2640 + (5280 - 1320 - 1320) + 2640 + 1320 + 7920 + 5280$ **(b)** $31,680 \times 3 = 95,040$ ft **(c)** 18 miles

Section 1.10 (pages 95–98)

1. *Estimate:* $600 + 900 + 1000 + 800 + 2000 = 5300$ sandwiches;
Exact: 5208 sandwiches **3.** *Estimate:* $70 - 60 = 10$ more recalls;
Exact: 13 more recalls **5.** *Estimate:* $200 \times 20 = 4000$ kits;
Exact: 5664 kits **7.** *Estimate:* $3000 \div 700 \approx 4$ toys; *Exact:* 4 toys
9. *Estimate:* $8000 - 4000 = 4000$ people; *Exact:* 4174 people
11. *Estimate:* $30 \times 5 = 150$; *Exact:* $170
13. *Estimate:* $50,000 - 40,000 = 10,000$; *Exact:* $14,100
15. (a) *Estimate:* $50,000 + 40,000 = 90,000$ White;
$40,000 + 50,000 = 90,000$ Easterly; *Exact:* $86,950 White;
$93,370, Easterly; Mr. and Mrs. Easterly **(b)** *Estimate:* $90,000 -
$90,000 = $0; *Exact:* $6420 **17.** *Estimate:* $2000 - $700 -
$300 - $400 - $400 - $200 = $200; *Exact:* $350

19. *Estimate:* $40,000 \times 100 = 4,000,000$ square feet; *Exact:* 6,011,280
square feet **21.** *Estimate:* $400 + $1000 + $200 + $400 + $200 +
$200 = $2400; *Exact:* $2680 **23.** *Estimate:* $400 + $200 + $200 +
$400 = $1200; $1200 - $1000 = $200; *Exact:* $140
25. *Estimate:* $($1000 \times 6) + ($900 \times 20) = $24,000; *Exact:* $20,961
27. Possible answers are Addition: more; total; gain of Subtraction:
less; loss of; decreased by Multiplication: twice; of; product Division:
divided by; goes into; per Equals: is; are **29.** Estimating the answer
can help you avoid careless mistakes like decimal or calculation errors.
Examples of reasonable answers in daily life might be a $35 bag of
groceries, $50 to fill the gas tank, or $45 for a phone bill. **31.** $20,009
33. 2477 pounds **35.** $378 **37.** $375 **39.** 20 seats

Chapter 1 Review Exercises (pages 103–110)

1. 6; 573 **2.** 36; 215 **3.** 105; 724 **4.** 1; 768; 710; 618 **5.** seven
hundred twenty-eight **6.** fifteen thousand, three hundred ten **7.** three
hundred nineteen thousand, two hundred fifteen **8.** sixty-two million, five
hundred thousand, five **9.** 10,008 **10.** 200,000,455 **11.** 110 **12.** 121
13. 5464 **14.** 15,657 **15.** 10,986 **16.** 9845 **17.** 40,602 **18.** 49,855
19. 36 **20.** 27 **21.** 189 **22.** 184 **23.** 6849 **24.** 4327 **25.** 224
26. 25,866 **27.** 49 **28.** 0 **29.** 32 **30.** 64 **31.** 45 **32.** 42 **33.** 56
34. 81 **35.** 40 **36.** 45 **37.** 48 **38.** 8 **39.** 0 **40.** 42 **41.** 48
42. 0 **43.** 84 **44.** 368 **45.** 522 **46.** 98 **47.** 5000 **48.** 2992
49. 5396 **50.** 45,815 **51.** 14,912 **52.** 20,160 **53.** 465,525
54. 174,984 **55.** 875 **56.** 2368 **57.** 1176 **58.** 5100 **59.** 15,576
60. 30,184 **61.** 887,169 **62.** 500,856 **63.** $360 **64.** $1064
65. $20,352 **66.** $684 **67.** 14,000 **68.** 23,800 **69.** 206,800
70. 318,500 **71.** 128,000,000 **72.** 90,300,000 **73.** 5 **74.** 7 **75.** 6
76. 2 **77.** 6 **78.** 4 **79.** 7 **80.** 0 **81.** undefined **82.** 0 **83.** 8
84. 9 **85.** 82 **86.** 98 **87.** 4422 **88.** 352 **89.** 150 R4 **90.** 124 R25
91. 820 **92.** 15,200 **93.** 21,000 **94.** 70,000 **95.** 3490; 3500; 3000
96. 20,070; 20,100; 20,000 **97.** 98,200; 98,200; 98,000 **98.** 352,120;
352,100; 352,000 **99.** 4 **100.** 7 **101.** 12 **102.** 14 **103.** 3; 7; 343
104. 6; 3; 729 **105.** 3; 5; 125 **106.** 5; 4; 1024 **107.** 34 **108.** 26
109. 9 **110.** 4 **111.** 9 **112.** 6 **113.** 8 parents **114.** 5 parents
115. Keeping bedroom clean; 25 parents **116.** Hanging up wet bath towels;
3 parents **117.** *Estimate:* 40 million $\times 400 = 16,000$ million or
16,000,000,000 checks; *Exact:* 14,600 million or 14,600,000,000 checks
118. *Estimate:* $1000 \times 60 = 60,000$ revolutions; *Exact:* 84,000 revolutions **119.** *Estimate:* $100 \times 20 = 2000$ forks; *Exact:* 2160 forks
120. *Estimate:* $6000 \times 30 = 180,000$ brackets; *Exact:* 180,000 brackets
121. *Estimate:* $2000 \times 10 = 20,000$ hr; *Exact:* 24,000 hr
122. *Estimate:* $80 \times 5 = 400$ mi; *Exact:* 400 mi **123.** *Estimate:*
$(30 \times $2000) + (30 \times $900) = $87,000; *Exact:* $74,052
124. *Estimate:* $(60 \times $20) + (20 \times $7) = $1340; *Exact:* $1139
125. *Estimate:* $600 - $100 = $500; *Exact:* $513
126. *Estimate:* $30,000 \div 1000 = 30$ hr; *Exact:* 33 hr **127.** *Estimate:*
$9000 \div 200 = 45$ pounds; *Exact:* 50 pounds **128.** *Estimate:*
$2000 - $500 - $400 = $1100; *Exact:* $1019 **129.** *Estimate:*
$30,000 \div 600 = 50$ acres; *Exact:* 52 acres **130.** *Estimate:*
$6000 \div 200 = 30$ homes; *Exact:* 32 homes **131.** 332 **132.** 448
133. 253 **134.** 588 **135.** 1041 **136.** 1661 **137.** 32,062
138. 24,947 **139.** 3 **140.** 7 **141.** 93,635 **142.** 83,178
143. undefined **144.** 7 **145.** 6900 **146.** 2310 **147.** 1,079,040
148. 130,212 **149.** 108 **150.** 207 **151.** three hundred seventy-six
thousand, eight hundred fifty-three **152.** four hundred eight thousand,
six hundred ten **153.** 8700 **154.** 401,000 **155.** 8 **156.** 9
157. $5544 **158.** $31,080 **159.** $2288 **160.** $15,782 **161.** 468 cards
162. 3600 textbooks **163.** $280 **164.** $114,635 **165.** $1905
166. $12,420 **167.** 869 ft **168.** 162 ft **169. (a)** 8393 ft
(b) greater than a mile by 3113 ft **170.** 4 football fields

Chapter 1 Test (pages 111–112)

1. nine thousand, two hundred five **2.** twenty-five thousand,
sixty-five **3.** 426,005 **4.** 8530 **5.** 112,630 **6.** 1045 **7.** 6206

8. 168 **9.** 171,000 **10.** 1615 **11.** 4,450,743 **12.** 7047 **13.** undefined
14. 458 R5 **15.** 160 **16.** 6350 **17.** 77,000 **18.** 41 **19.** 28
20. *Estimate:* $500 + $500 + $500 + $400 − $800 = $1100;
Exact: $1140 **21.** *Estimate:* 90,000 ÷ 400 = 225 acres; *Exact:* 231 acres
22. *Estimate:* $2000 − $500 − $200 − $200 = $1100; *Exact:* $948
23. *Estimate:* $(50 \times 60 \times 4) + (40 \times 60 \times 3) = 19{,}200$ chicks; *Exact:*
18,000 chicks **24.** 1. Locate the place to which you are rounding and
underline it. 2. Look only at the next digit to the right. If this digit is a
4 or less, do not change the underlined digit. If the digit is 5 or more,
increase the underlined digit by 1. 3. Change all digits to the right of
the underlined place to zeros. Each person's rounding example will vary.
25. 1. Read the problem carefully. 2. Work out a plan. 3. Estimate a
reasonable answer. 4. Solve the problem. 5. State the answer.
6. Check your work.

Chapter 2 Multiplying and Dividing Fractions

Section 2.1 (pages 117–118)

1. $\dfrac{3}{4}; \dfrac{1}{4}$ **3.** $\dfrac{1}{3}; \dfrac{2}{3}$ **5.** $\dfrac{7}{5}; \dfrac{3}{5}$ **7.** $\dfrac{5}{6}; \dfrac{4}{6}; \dfrac{2}{6}$ **9.** $\dfrac{8}{25}$ **11.** $\dfrac{303}{520}$ **13.** 4; 5

15. 9; 8 **17.** Proper $\dfrac{1}{3}, \dfrac{5}{8}, \dfrac{7}{16}$ Improper $\dfrac{8}{5}, \dfrac{6}{6}, \dfrac{12}{2}$

19. Proper $\dfrac{3}{4}, \dfrac{9}{11}, \dfrac{7}{15}$ Improper $\dfrac{3}{2}, \dfrac{5}{5}, \dfrac{19}{18}$

21. One possibility is

$\dfrac{3}{4}$ ← Numerator
   ← Denominator

The denominator shows the number of equal parts in the whole and the
numerator shows how many of the parts are being considered.

23. 3; 8 **25.** 5; 24

Section 2.2 (pages 123–126)

1. $\dfrac{5}{4}$ **3.** $\dfrac{23}{5}$ **5.** $\dfrac{13}{2}$ **7.** $\dfrac{33}{4}$ **9.** $\dfrac{18}{11}$ **11.** $\dfrac{19}{3}$ **13.** $\dfrac{81}{8}$ **15.** $\dfrac{43}{4}$

17. $\dfrac{27}{8}$ **19.** $\dfrac{43}{5}$ **21.** $\dfrac{54}{11}$ **23.** $\dfrac{131}{4}$ **25.** $\dfrac{221}{12}$ **27.** $\dfrac{269}{15}$ **29.** $\dfrac{187}{24}$

31. $1\dfrac{1}{3}$ **33.** $2\dfrac{1}{4}$ **35.** 8 **37.** $7\dfrac{3}{5}$ **39.** $4\dfrac{7}{8}$ **41.** 9 **43.** $15\dfrac{3}{4}$

45. $5\dfrac{2}{9}$ **47.** $8\dfrac{1}{8}$ **49.** $16\dfrac{4}{5}$ **51.** 28 **53.** $26\dfrac{1}{7}$

55. Multiply the denominator by the whole number and add the numerator. The result becomes the new numerator, which is placed over the original denominator.

$$2\dfrac{1}{2} \qquad 2 \times 2 + 1 = 5 \qquad \dfrac{5}{2}$$

57. $\dfrac{501}{2}$ **59.** $\dfrac{1000}{3}$ **61.** $\dfrac{4179}{8}$ **63.** $154\dfrac{1}{4}$ **65.** 171

67. $122\dfrac{13}{32}$ **69.** $\dfrac{2}{3}, \dfrac{4}{5}, \dfrac{3}{4}, \dfrac{7}{10}$ **70.** (a) numerator; denominator

(b)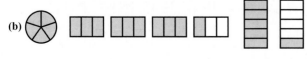

(c) less **71.** $\dfrac{5}{5}, \dfrac{10}{3}, \dfrac{6}{5}$ **72.** (a) numerator; denominator

(b)

(c) greater **73.** $\dfrac{5}{3} = 1\dfrac{2}{3}; \dfrac{7}{7} = 1; \dfrac{11}{6} = 1\dfrac{5}{6}$

74. (a) improper; greater than or equal to

(b)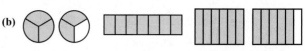

(c) Divide the numerator by the denominator. Use the quotient as the
whole number and place the remainder over the denominator.

Section 2.3 (pages 131–132)

1. 1, 2, 4, 8 **3.** 1, 3, 5, 15 **5.** 1, 2, 3, 4, 6, 8, 12, 16, 24, 48
7. 1, 2, 3, 4, 6, 9, 12, 18, 36 **9.** 1, 2, 4, 5, 8, 10, 20, 40
11. 1, 2, 4, 8, 16, 32, 64 **13.** composite **15.** prime **17.** composite
19. prime **21.** prime **23.** composite **25.** composite **27.** composite
29. 2^3 **31.** $2^2 \cdot 5$ **33.** $2^2 \cdot 3^2$ **35.** 5^2 **37.** $2^2 \cdot 17$ **39.** $2^3 \cdot 3^2$
41. $2^2 \cdot 11$ **43.** $2^2 \cdot 5^2$ **45.** 5^3 **47.** $2^2 \cdot 3^2 \cdot 5$ **49.** $2^6 \cdot 5$
51. $2^3 \cdot 3^2 \cdot 5$ **53.** A composite number has a factor(s) other than itself
or 1. Examples include 4, 6, 8, 9, 10. A prime number is a whole number
that has exactly two *different* factors, itself and 1. Examples include 2, 3,
5, 7, 11. The numbers 0 and 1 are neither prime nor composite.
55. All the possible factors of 24 are 1, 2, 3, 4, 6, 8, 12, and 24. This list
includes both prime numbers and composite numbers. The prime factors
of 24 include only prime numbers. The prime factorization of 24 is
$2 \cdot 2 \cdot 2 \cdot 3 = 2^3 \cdot 3$ **57.** $2 \cdot 5^2 \cdot 7$ **59.** $2^6 \cdot 3 \cdot 5$
61. $2^3 \cdot 3 \cdot 5 \cdot 13$ **63.** $2^2 \cdot 3^2 \cdot 5 \cdot 7$
65. 2, 3, 5, 7, 11, 13, 17, 19, 23, 29, 31, 37, 41, 43, 47
66. A prime number is a whole number that is evenly divisible by itself
and 1 only. **67.** No. Every other even number is divisible by 2 in addition to being divisible by itself and 1. **68.** No. A multiple of a prime
number can never be prime because it will always be divisible by the
prime number. **69.** $2 \cdot 2 \cdot 3 \cdot 5 \cdot 5 \cdot 7$ **70.** $2^2 \cdot 3 \cdot 5^2 \cdot 7$

Section 2.4 (pages 137–138)

1. ✓ ✓ ✓ **3.** ✓ ✓ ✗ ✗ **5.** ✓ ✗ ✓ ✓ **7.** ✓ ✓ ✗ ✗ **9.** $\dfrac{3}{4}$

11. $\dfrac{1}{4}$ **13.** $\dfrac{3}{5}$ **15.** $\dfrac{6}{7}$ **17.** $\dfrac{7}{8}$ **19.** $\dfrac{6}{7}$ **21.** $\dfrac{4}{7}$ **23.** $\dfrac{1}{50}$ **25.** $\dfrac{8}{11}$

27. $\dfrac{5}{9}$ **29.** $\dfrac{\cancel{2} \cdot \cancel{3} \cdot 3}{\cancel{2} \cdot 2 \cdot 2 \cdot \cancel{3}} = \dfrac{3}{4}$ **31.** $\dfrac{\cancel{5} \cdot 7}{2 \cdot 2 \cdot 2 \cdot \cancel{5}} = \dfrac{7}{8}$

33. $\dfrac{\cancel{2} \cdot \cancel{3} \cdot \cancel{3} \cdot \cancel{5}}{\cancel{2} \cdot 2 \cdot \cancel{3} \cdot \cancel{3} \cdot \cancel{5}} = \dfrac{1}{2}$ **35.** $\dfrac{\cancel{2} \cdot \cancel{2} \cdot \cancel{3} \cdot 3}{\cancel{2} \cdot \cancel{2} \cdot \cancel{3}} = 3$

37. $\dfrac{2 \cdot 2 \cdot 2 \cdot \cancel{3} \cdot \cancel{3}}{\cancel{3} \cdot \cancel{3} \cdot 5 \cdot 5} = \dfrac{8}{25}$ **39.** $\dfrac{1}{2} = \dfrac{1}{2}$; equivalent

41. $\dfrac{5}{12} \neq \dfrac{2}{5}$; not equivalent **43.** $\dfrac{5}{8} \neq \dfrac{35}{52}$; not equivalent

45. $\dfrac{7}{8} = \dfrac{7}{8}$; equivalent **47.** $8 \neq 9$; not equivalent

49. $\dfrac{5}{6} = \dfrac{5}{6}$; equivalent **51.** A fraction is in lowest terms when the
numerator and the denominator have no common factors other than 1.
Some examples are $\dfrac{1}{2}, \dfrac{3}{8}$, and $\dfrac{2}{3}$. **53.** $\dfrac{5}{8}$ **55.** $\dfrac{2}{1} = 2$

Summary Exercises on Fraction Basics (pages 141–142)

1. $\dfrac{5}{6}; \dfrac{1}{6}$ **2.** $\dfrac{1}{3}; \dfrac{2}{3}$ **3.** $\dfrac{5}{8}; \dfrac{3}{8}$ **4.** 3; 4 **5.** 8; 5 **6.** Proper $\dfrac{3}{5}, \dfrac{4}{25}, \dfrac{1}{32}$

Improper $\dfrac{8}{2}, \dfrac{16}{7}, \dfrac{8}{8}$ **7.** $\dfrac{7}{22}$ **8.** $\dfrac{8}{22}$ **9.** $\dfrac{5}{22}$ **10.** $\dfrac{15}{22}$ **11.** $2\dfrac{1}{2}$ **12.** $1\dfrac{3}{8}$

13. $1\dfrac{2}{7}$ **14.** $2\dfrac{2}{3}$ **15.** 8 **16.** 17 **17.** $7\dfrac{1}{5}$ **18.** $4\dfrac{7}{10}$ **19.** $\dfrac{10}{3}$

20. $\dfrac{43}{8}$ **21.** $\dfrac{34}{5}$ **22.** $\dfrac{53}{5}$ **23.** $\dfrac{51}{4}$ **24.** $\dfrac{62}{13}$ **25.** $\dfrac{71}{6}$ **26.** $\dfrac{189}{8}$

27. $2 \cdot 5$ **28.** $5 \cdot 11$ **29.** $2^2 \cdot 3^2$ **30.** 3^4 **31.** $2^3 \cdot 5 \cdot 7$
32. $2^3 \cdot 3^2 \cdot 5$ **33.** $\frac{1}{3}$ **34.** $\frac{1}{2}$ **35.** $\frac{1}{4}$ **36.** $\frac{3}{5}$ **37.** $\frac{3}{4}$ **38.** $\frac{5}{6}$
39. $\frac{7}{8}$ **40.** $\frac{1}{50}$ **41.** $\frac{1}{8}$ **42.** $\frac{5}{9}$ **43.** $\frac{4}{7}$ **44.** $\frac{5}{9}$

45. $\dfrac{\cancel{2} \cdot \cancel{2} \cdot 2 \cdot \cancel{3}}{\cancel{2} \cdot \cancel{2} \cdot 3 \cdot \cancel{3}} = \dfrac{2}{3}$ **46.** $\dfrac{\cancel{2} \cdot \cancel{2} \cdot \cancel{2} \cdot \cancel{2} \cdot 2 \cdot \cancel{5}}{\cancel{2} \cdot \cancel{2} \cdot \cancel{2} \cdot \cancel{2} \cdot 2 \cdot \cancel{5}} = \dfrac{1}{2}$

47. $\dfrac{\cancel{2} \cdot \cancel{3} \cdot 3 \cdot \cancel{7}}{\cancel{2} \cdot \cancel{3} \cdot \cancel{7}} = 3$ **48.** $\dfrac{\cancel{2} \cdot \cancel{2} \cdot \cancel{2} \cdot \cancel{2} \cdot 2 \cdot 3}{\cancel{2} \cdot \cancel{2} \cdot \cancel{2} \cdot \cancel{2} \cdot 7} = \dfrac{6}{7}$

Section 2.5 (pages 149–152)

1. $\frac{1}{4}$ **3.** $\frac{2}{35}$ **5.** $\frac{3}{4}$ **7.** $\frac{1}{4}$ **9.** $\frac{5}{12}$ **11.** $\frac{9}{32}$ **13.** $\frac{2}{5}$ **15.** $\frac{13}{32}$ **17.** $\frac{21}{128}$
19. 4 **21.** 40 **23.** 24 **25.** $13\frac{1}{2}$ **27.** $31\frac{1}{2}$ **29.** 80 **31.** $94\frac{2}{3}$
33. 400 **35.** 810 **37.** $\frac{1}{4}$ mi² **39.** 9 m² **41.** $\frac{3}{10}$ in.²

43. Multiply the numerators and multiply the denominators. An example is
$\frac{3}{4} \cdot \frac{1}{2} = \frac{3 \cdot 1}{4 \cdot 2} = \frac{3}{8}$ **45.** $1\frac{1}{2}$ yd² **47.** $3\frac{1}{2}$ mi² **49.** They are both the
same size: $\frac{3}{64}$ mi² **51.** 60,000 stores **52.** 64,615 stores
53. *Estimate:* $\frac{2}{3} \cdot 9000 = 6000$; *Exact:* 6267 stores (rounded)
54. *Estimate:* $\frac{2}{5} \cdot 8000 = 3200$; *Exact:* 3013 stores (rounded)
55. $\frac{2}{3} \cdot 9300 = 6200$ stores
56. $\frac{2}{5} \cdot 7500 = 3000$ stores

Section 2.6 (pages 157–160)

1. $\frac{1}{2}$ yd² **3.** $\frac{8}{9}$ ft² **5.** $\frac{3}{10}$ yd² **7.** \$2568 **9.** \$36
11. 910 women **13.** Don't know; 50 people
15. 100 much less + 250 somewhat less = 350 people
17. Because everyone is included and fractions are given for *all* groups,
the sum of the fractions must be *1* or *all* of the people.
19. \$58,000 **21.** \$11,600 **23.** \$3625

25. The correct solution is $\dfrac{9}{10} \times \dfrac{20}{21} = \dfrac{\cancel{9}^{3}}{\cancel{10}} \times \dfrac{\cancel{20}^{2}}{\cancel{21}_{7}} = \dfrac{6}{7}$

27. \$750 **29.** 2 ft **31.** 9000 votes **33.** $\frac{1}{32}$ of the estate

Section 2.7 (pages 167–170)

1. $\frac{8}{3}$ **3.** $\frac{6}{5}$ **5.** $\frac{5}{8}$ **7.** $\frac{1}{4}$ **9.** $\frac{2}{3}$ **11.** $2\frac{5}{8}$ **13.** $\frac{9}{20}$ **15.** 4 **17.** 6
19. $\frac{13}{16}$ **21.** $\frac{4}{5}$ **23.** 18 **25.** 24 **27.** $\frac{1}{14}$ **29.** $\frac{2}{9}$ quart **31.** 15 times
33. 88 dispensers **35.** 60 trips **37.** You can divide two fractions by
using the reciprocal of the second fraction (divisor) and then multiplying
39. 12 pounds **41.** 208 homes **43.** 124 visits **45.** 2432 towels
47. double, twice, times, product, twice as much **48.** goes into, divide,
per, quotient, divided by **49.** reciprocal **50.** $\frac{4}{3}$; $\frac{8}{7}$; $\frac{1}{5}$; $\frac{19}{12}$
51. (a) Multiply the length of one side by 3, 4, 5, or 6.

(b) $\dfrac{15}{16} \times 4 = \dfrac{15}{\cancel{16}_{4}} \times \dfrac{\cancel{4}^{1}}{1} = \dfrac{15}{4} = 3\frac{3}{4}$ in.

52. $\frac{225}{256}$ in.²; Multiply the length by the width.

Section 2.8 (pages 177–180)

1. *Exact:* $7\frac{7}{8}$; *Estimate:* $5 \cdot 2 = 10$ **3.** *Exact:* $4\frac{1}{2}$;
Estimate: $2 \cdot 3 = 6$ **5.** *Exact:* 4; *Estimate:* $3 \cdot 1 = 3$
7. *Exact:* 50; *Estimate:* $8 \cdot 6 = 48$ **9.** *Exact:* $49\frac{1}{2}$;
Estimate: $5 \cdot 2 \cdot 5 = 50$ **11.** *Exact:* 12;
Estimate: $3 \cdot 2 \cdot 3 = 18$ **13.** *Exact:* $\frac{1}{3}$; *Estimate:* $1 \div 4 = \frac{1}{4}$
15. *Exact:* $\frac{5}{6}$; *Estimate:* $3 \div 3 = 1$ **17.** *Exact:* $3\frac{3}{5}$;
Estimate: $9 \div 3 = 3$ **19.** *Exact:* $\frac{5}{12}$; *Estimate:* $1 \div 2 = \frac{1}{2}$
21. *Exact:* $\frac{3}{10}$; *Estimate:* $2 \div 6 = \frac{1}{3}$ **23.** *Exact:* $\frac{17}{18}$; *Estimate:*
$6 \div 6 = 1$ **25. (a)** *Estimate:* $1 \cdot 3 = 3$ cups; *Exact:* $1\frac{7}{8}$ cups of
applesauce **(b)** *Estimate:* $1 \cdot 3 = 3$ teaspoons;
Exact: $1\frac{1}{4}$ teaspoons of salt **(c)** *Estimate:* $2 \cdot 3 = 6$ cups;
Exact: $4\frac{3}{8}$ cups of flour **27. (a)** *Estimate:* $1 \div 2 = \frac{1}{2}$ teaspoon;
Exact: $\frac{1}{4}$ teaspoon vanilla extract **(b)** *Estimate:* $1 \div 2 = \frac{1}{2}$ cup;
Exact: $\frac{3}{8}$ cup applesauce **(c)** *Estimate:* $2 \div 2 = 1$ cup;
Exact: $\frac{7}{8}$ cup of flour **29.** *Estimate:* $1316 \div 12 \approx 110$ units;
Exact: 112 units **31.** *Estimate:* $135 \cdot 20 = 2700$ in.;
Exact: $2632\frac{1}{2}$ in. **33.** The answer should include *Step 1* Change mixed
numbers to improper fractions. *Step 2* Multiply the fractions. *Step 3* Write
the answer in lowest terms, changing to mixed or whole numbers where
possible. **35.** *Estimate:* $51{,}460 \div 20 = 2573$ tires; *Exact:* 2480 tires
37. *Estimate:* $10 \div 1 = 10$ spacers; *Exact:* 13 spacers **39.** *Estimate:*
(a) $18 \cdot 4 = 72$ in.; *Exact:* 71 in. **(b)** No, because 6 ft = 72 in.
41. *Estimate:* $7 \cdot 26 = 182$ gallons; *Exact:* $172\frac{1}{8}$ gallons

Chapter 2 Review Exercises (pages 187–190)

1. $\frac{1}{3}$ **2.** $\frac{5}{8}$ **3.** $\frac{2}{4}$ **4.** Proper $\frac{1}{8}, \frac{3}{4}, \frac{2}{3}$; Improper $\frac{4}{3}, \frac{5}{5}$
5. Proper $\frac{15}{16}, \frac{1}{8}$; Improper $\frac{6}{5}, \frac{16}{13}, \frac{5}{3}$ **6.** $\frac{19}{4}$ **7.** $\frac{59}{6}$ **8.** $3\frac{3}{8}$
9. $12\frac{3}{5}$ **10.** 1, 2, 3, 6 **11.** 1, 2, 3, 4, 6, 8, 12, 24 **12.** 1, 5, 11, 55
13. 1, 2, 3, 5, 6, 9, 10, 15, 18, 30, 45, 90 **14.** 3^3 **15.** $2 \cdot 3 \cdot 5^2$
16. $2^2 \cdot 3 \cdot 5 \cdot 7$ **17.** 25 **18.** 288 **19.** 1728 **20.** 2048
21. $\frac{24}{24} = 1$ **22.** $\frac{18}{24} = \frac{3}{4}$ **23.** $\frac{14}{24} = \frac{7}{12}$ **24.** $\frac{10}{24} = \frac{5}{12}$

25. $\dfrac{\cancel{5} \cdot 5}{2 \cdot 2 \cdot 3 \cdot \cancel{5}}; \dfrac{5}{12}$ **26.** $\dfrac{\cancel{2} \cdot \cancel{2} \cdot \cancel{2} \cdot \cancel{2} \cdot \cancel{2} \cdot 2 \cdot 2 \cdot \cancel{3}}{\cancel{2} \cdot \cancel{2} \cdot \cancel{2} \cdot \cancel{2} \cdot \cancel{2} \cdot \cancel{3}}; 4$

27. equivalent **28.** not equivalent **29.** equivalent
30. $\frac{3}{5}$ **31.** $\frac{3}{16}$ **32.** $\frac{1}{7}$ **33.** $\frac{4}{21}$ **34.** 15 **35.** 625 **36.** $1\frac{1}{3}$
37. $\frac{5}{3} = 1\frac{2}{3}$ **38.** $\frac{5}{2} = 2\frac{1}{2}$ **39.** 2 **40.** 8 **41.** 24 **42.** $\frac{5}{24}$
43. $\frac{2}{15}$ **44.** $\frac{4}{13}$ **45.** $\frac{9}{40}$ ft² **46.** $\frac{7}{12}$ in.² **47.** 7857 ft²
48. $5\frac{1}{2}$ ft² **49.** *Exact:* $6\frac{7}{8}$; *Estimate:* $6 \cdot 1 = 6$

50. *Exact:* $21\frac{3}{8}$; *Estimate:* $2 \cdot 7 \cdot 1 = 14$

51. *Exact:* $5\frac{1}{6}$; *Estimate:* $16 \div 3 = 5\frac{1}{3}$

52. *Exact:* $\frac{3}{4}$; *Estimate:* $5 \div 6 = \frac{5}{6}$

53. 512 bins   **54.** $\frac{2}{15}$ of the estate

55. *Estimate:* $158 \div 4 \approx 40$ pull cords; *Exact:* 36 pull cords

56. *Estimate:* $8 \cdot 2 \cdot 50 = 800$ pounds; *Exact:* $833\frac{1}{3}$ pounds

57. 25 pounds   **58.** \$186   **59.** $\frac{3}{32}$ of the budget

60. $\frac{4}{25}$ ton   **61.** $\frac{3}{8}$ **62.** $\frac{2}{5}$ **63.** $31\frac{1}{4}$ **64.** $28\frac{1}{8}$ **65.** $\frac{1}{10}$ **66.** $\frac{5}{32}$

67. 30   **68.** $2\frac{3}{5}$ **69.** $1\frac{3}{5}$ **70.** $38\frac{1}{4}$ **71.** $\frac{17}{3}$ **72.** $\frac{307}{8}$

73. $\dfrac{\cancel{2} \cdot \cancel{2} \cdot 2}{\cancel{2} \cdot \cancel{2} \cdot 3} = \dfrac{2}{3}$ **74.** $\dfrac{\cancel{2} \cdot 2 \cdot \cancel{3} \cdot 3 \cdot 3}{\cancel{2} \cdot \cancel{3} \cdot 5 \cdot 7} = \dfrac{18}{35}$ **75.** $\frac{5}{6}$ **76.** $\frac{2}{3}$ **77.** $\frac{2}{5}$

78. $\frac{1}{3}$ **79.** *Estimate:* $4 \cdot 44 = 176$ ounces; *Exact:* $157\frac{1}{2}$ ounces

80. *Estimate:* $7 \cdot 26 = 182$ quarts; *Exact:* $184\frac{7}{8}$ quarts

81. $1\frac{17}{32}$ in.²   **82.** $\frac{5}{16}$ yd²

Chapter 2 Test (pages 193–194)

1. $\frac{5}{6}$ **2.** $\frac{3}{8}$ **3.** $\frac{2}{3}, \frac{6}{7}, \frac{1}{4}, \frac{5}{8}$ **4.** $\frac{27}{8}$ **5.** $30\frac{3}{4}$ **6.** 1, 2, 3, 6, 9, 18

7. $3^2 \cdot 5$ **8.** $2^4 \cdot 3^2$ **9.** $2^2 \cdot 5^3$ **10.** $\frac{3}{4}$ **11.** $\frac{5}{6}$

12. $\dfrac{56}{84} = \dfrac{\cancel{2} \cdot \cancel{2} \cdot 2 \cdot \cancel{7}}{\cancel{2} \cdot \cancel{2} \cdot 3 \cdot \cancel{7}} = \dfrac{2}{3}$

13. Answers will vary. One possibility is: Multiply fractions by multiplying the numerators and multiplying the denominators. Divide two fractions by using the reciprocal of the second fraction (divisor) and then multiplying.

14. $\frac{1}{3}$ **15.** 36   **16.** $\frac{5}{12}$ yd²   **17.** 5475 seedlings   **18.** $\frac{9}{10}$ **19.** $15\frac{3}{4}$

20. 24 pieces   **21.** *Estimate:* $4 \cdot 4 = 16$;   *Exact:* $14\frac{7}{16}$

22. *Estimate:* $2 \cdot 4 = 8$;   *Exact:* $7\frac{17}{18}$ **23.** *Estimate:* $10 \div 2 = 5$;

Exact: $4\frac{4}{15}$ **24.** *Estimate:* $9 \div 2 = 4\frac{1}{2}$;   *Exact:* $5\frac{1}{10}$

25. *Estimate:* $3 \cdot 12 = 36$;   *Exact:* $30\frac{5}{8}$ grams

Cumulative Review Exercises: Chapters 1–2 (pages 197–198)

1. hundreds 7; tens 8   **2.** millions 8; ten thousands 2   **3.** 177
4. 149,199   **5.** 4452   **6.** 3,221,821   **7.** 747   **8.** 72   **9.** 2,168,232
10. 450,400   **11.** 9   **12.** 7581   **13.** 8471 R2   **14.** 22 R26
15. 6580; 6600; 7000   **16.** 76,270; 76,300; 76,000   **17.** 8   **18.** 5
19. 540 gallons   **20.** 60 decibels   **21.** 47,000 hairs   **22.** 188 hours
23. \$699   **24.** 21 pieces   **25.** proper   **26.** improper   **27.** proper
28. $\frac{27}{8}$ **29.** $\frac{32}{5}$ **30.** 2   **31.** $12\frac{7}{8}$ **32.** $2^3 \cdot 3^2$ **33.** $2 \cdot 3^2 \cdot 7$
34. $2 \cdot 5^2 \cdot 7$ **35.** 64   **36.** 288   **37.** 640   **38.** $\frac{7}{8}$ **39.** $\frac{2}{3}$ **40.** $\frac{5}{9}$
41. $\frac{3}{8}$ **42.** 12   **43.** 25   **44.** $\frac{24}{25}$ **45.** $\frac{7}{12}$ **46.** $2\frac{2}{5}$

Chapter 3 Adding and Subtracting Fractions

Section 3.1 (pages 203–204)

1. $\frac{5}{8}$ **3.** $\frac{5}{6}$ **5.** $\frac{1}{2}$ **7.** $1\frac{1}{5}$ **9.** $\frac{1}{3}$ **11.** $\frac{13}{20}$ **13.** $\frac{11}{15}$ **15.** $1\frac{1}{2}$

17. $\frac{11}{27}$ **19.** $\frac{3}{8}$ **21.** $\frac{6}{11}$ **23.** $\frac{3}{5}$ **25.** $1\frac{1}{7}$ **27.** $\frac{1}{5}$ **29.** $1\frac{1}{6}$ **31.** $1\frac{1}{10}$

33. Three steps to add like fractions are: 1. Add the numerators of the fractions to find the numerator of the sum (the answer).   2. Use the denominator of the fractions as the denominator of the sum.   3. Write the answer in lowest terms.   **35.** $\frac{7}{9}$ **37.** $\frac{5}{7}$ **39.** $\frac{1}{2}$ acre

Section 3.2 (pages 213–216)

1. 6   **3.** 15   **5.** 36   **7.** 14   **9.** 30   **11.** 100   **13.** 20   **15.** 60

17. 36   **19.** 120   **21.** 180   **23.** 720   **25.** $\frac{16}{24}$ **27.** $\frac{18}{24}$ **29.** $\frac{20}{24}$

31. 3   **33.** 12   **35.** 28   **37.** 12   **39.** 32   **41.** 72   **43.** 84   **45.** 96
47. 27   **49.** It probably depends on how large the numbers are. If the numbers are small, the method using multiples of the largest number seems best. If the numbers are larger, or there are more than two numbers, then the factorization method will be better.   **51.** 3600   **53.** 10,584
55. like; unlike   **56.** numerators; denominator; lowest   **57.** least; smallest
58. 40 is the least common multiple.   **59.** 70   **60.** 450   **61.** 240 is a common multiple but twice as large as the least common multiple; 120 is the LCM.   **62.** The least common multiple can be no smaller than the largest number in a group and the number 1760 is a multiple of 55.

Section 3.3 (pages 221–224)

1. $\frac{7}{8}$ **3.** $\frac{8}{9}$ **5.** $\frac{3}{4}$ **7.** $\frac{39}{40}$ **9.** $\frac{23}{36}$ **11.** $\frac{14}{15}$ **13.** $\frac{29}{36}$ **15.** $\frac{17}{20}$

17. $\frac{23}{30}$ **19.** $\frac{3}{8}$ **21.** $\frac{23}{48}$ **23.** $\frac{1}{2}$ **25.** $\frac{1}{2}$ **27.** $\frac{7}{15}$ **29.** $\frac{1}{6}$ **31.** $\frac{19}{45}$

33. $\frac{3}{40}$ **35.** $\frac{17}{48}$ **37.** $\frac{1}{4}$ in.   **39.** $\frac{17}{40}$ for reserved seating   **41.** $\frac{31}{40}$ in.

43. $\frac{1}{24}$ of the tank   **45.** You cannot add or subtract until all the fractional pieces are the same size. For example, halves are larger than fourths, so you cannot add $\frac{1}{2} + \frac{1}{4}$ until you rewrite $\frac{1}{2}$ as $\frac{2}{4}$.   **47.** $\frac{7}{24}$

49. work and travel; 8 hr; $\frac{1}{2}$ **51.** $\frac{3}{16}$ in.

Section 3.4 (pages 231–238)

1. *Estimate:* $6 + 3 = 9$; *Exact:* $8\frac{5}{6}$

3. *Estimate:* $7 + 4 = 11$; *Exact:* $11\frac{1}{2}$

5. *Estimate:* $1 + 4 = 5$; *Exact:* $4\frac{5}{24}$

7. *Estimate:* $25 + 19 = 44$; *Exact:* $43\frac{2}{3}$

9. *Estimate:* $34 + 19 = 53$; *Exact:* $52\frac{1}{10}$

11. *Estimate:* $23 + 15 = 38$; *Exact:* $38\frac{5}{28}$

13. *Estimate:* $13 + 19 + 15 = 47$; *Exact:* $45\frac{5}{6}$

15. *Estimate:* $15 - 12 = 3$; *Exact:* $2\frac{5}{8}$

17. *Estimate:* $13 - 1 = 12$; *Exact:* $11\frac{7}{15}$

19. *Estimate:* $28 - 6 = 22$; *Exact:* $22\frac{7}{30}$

21. *Estimate:* $17 - 7 = 10$; *Exact:* $10\frac{3}{8}$

23. *Estimate:* $19 - 6 = 13$; *Exact:* $12\frac{19}{20}$

25. *Estimate:* $20 - 12 = 8$; *Exact:* $7\frac{11}{12}$

27. $9\frac{1}{8}$ **29.** $11\frac{1}{2}$ **31.** $3\frac{5}{6}$ **33.** $6\frac{11}{12}$ **35.** $8\frac{1}{8}$ **37.** $\frac{5}{6}$ **39.** $2\frac{7}{8}$

41. $2\frac{7}{12}$ **43.** $5\frac{9}{20}$ **45.** $3\frac{16}{21}$ **47.** Find the least common denominator. Change the fraction parts so that they have the same denominator. Add the fraction parts. Add the whole number parts. Write the answer as a mixed number.

49. *Estimate:* $26 - 15 = 11$ ft; *Exact:* $11\frac{1}{4}$ ft

51. *Estimate:* $23 - 19 = 4$ in.; *Exact:* $4\frac{1}{8}$ in.

53. *Estimate:* $23 + 22 + 21 = 66$ in.; *Exact:* 66 in.

55. *Estimate:* $3 - 1 = 2$ in.; *Exact:* $2\frac{3}{16}$ in.

57. *Estimate:* $16 + 19 + 24 + 31 = 90$ ft; *Exact:* $87\frac{8}{4} = 89$ ft

59. *Estimate:* $24 + 35 + 24 + 35 = 118$ in.; *Exact:* $116\frac{1}{2}$ in.

61. *Estimate:* $100 - 10 - 14 - 9 - 19 - 12 - 10 - 14 =$ 12 gallons; *Exact:* $12\frac{5}{8}$ gallons

63. *Estimate:* $527 - 108 - 151 - 139 = 129$ ft; *Exact:* 130 ft

65. *Estimate:* $59 + 24 + 17 + 29 + 58 = 187$ tons; *Exact:* $186\frac{13}{24}$ tons

67. $4\frac{11}{16}$ in. **69.** $21\frac{3}{8}$ in. **71. (a)** 30 **(b)** 28 **(c)** 25 **(d)** 264

72. least common denominator **73. (a)** $\frac{23}{24}$ **(b)** $\frac{8}{15}$ **(c)** $\frac{43}{48}$ **(d)** $\frac{4}{21}$

74. fraction parts **75.** improper; large

76. (a) $4\frac{5}{8} + 3\frac{6}{8} = 7\frac{11}{8} = 8\frac{3}{8}; \frac{37}{8} + \frac{30}{8} = \frac{67}{8} = 8\frac{3}{8}$

(b) $11\frac{56}{40} - 8\frac{35}{40} = 3\frac{21}{40}; \frac{496}{40} - \frac{355}{40} = \frac{141}{40} = 3\frac{21}{40}$

Section 3.5 (pages 243–246)

1.–12.

$\frac{1}{4}$ $\frac{1}{2}$ $\frac{7}{8}$ $\frac{5}{4}$ $\frac{3}{2}$ $1\frac{7}{8}$ $2\frac{1}{6}$ $\frac{7}{3}$ $\frac{11}{4}$ $3\frac{1}{4}$ $\frac{7}{2}$ $3\frac{4}{5}$

0 | 1 | 2 | 3 | 4

2. 1. **10.** 4. **3.** 12. **7.** 5. **6.** **11.** 9. **8.**

13. $>$ **15.** $<$ **17.** $>$ **19.** $<$ **21.** $>$ **23.** $>$ **25.** $\frac{1}{9}$ **27.** $\frac{25}{64}$

29. $\frac{9}{16}$ **31.** $\frac{64}{125}$ **33.** $\frac{81}{16} = 5\frac{1}{16}$ **35.** $\frac{81}{256}$

37. A number line is a horizontal line with a range of equally spaced whole numbers placed on it. The lowest number is on the left and the greatest number is on the right. It can be used to compare the size or value of numbers.

0 $\frac{1}{2}$ 1 $1\frac{1}{2}$ 2 $2\frac{1}{2}$ 3

39. 4 **41.** 10 **43.** 1 **45.** $\frac{3}{16}$ **47.** $\frac{4}{9}$ **49.** $\frac{1}{3}$ **51.** $\frac{1}{2}$ **53.** $\frac{3}{8}$

55. $\frac{1}{4}$ **57.** $1\frac{1}{2}$ **59.** $\frac{1}{12}$ **61.** 3 **63.** $\frac{5}{16}$ **65.** $\frac{1}{4}$ **67.** $\frac{1}{32}$

69. $\frac{5}{30}$ in Atlanta is greater. **71.** $<$; $>$ **72. (a)** like; numerators; numerator **(b)** Answers will vary. **73.** parentheses; exponents; square; multiply; divide; add; subtract **74.** $\frac{2}{45}$

75.–80.

| 77. | 75. | | 78. | 79. | 76. | 80. |
| $\frac{27}{125}$ | $\frac{4}{9}$ | | $1\frac{9}{16}$ | 2 | $2\frac{1}{4}$ | $2\frac{57}{64}$ |

0 | 1 | 2 | 3

Summary Exercises on Fractions (pages 247–248)

1. proper **2.** improper **3.** improper **4.** proper **5.** $\frac{5}{6}$ **6.** $\frac{7}{8}$

7. $\frac{3}{7}$ **8.** $\frac{23}{47}$ **9.** $\frac{1}{2}$ **10.** $\frac{3}{8}$ **11.** 35 **12.** $\frac{5}{6}$ **13.** $1\frac{1}{6}$ **14.** 56

15. $1\frac{13}{24}$ **16.** $1\frac{13}{16}$ **17.** $2\frac{1}{12}$ **18.** $\frac{1}{12}$ **19.** $\frac{11}{24}$ **20.** $\frac{2}{15}$

21. *Exact:* $7\frac{7}{8}$; *Estimate:* $4 \cdot 2 = 8$

22. *Exact:* $17\frac{15}{32}$; *Estimate:* $5 \cdot 3 = 15$

23. *Exact:* $107\frac{2}{3}$; *Estimate:* $8 \cdot 6 \cdot 2 = 96$

24. *Exact:* $1\frac{1}{6}$; *Estimate:* $4 \div 4 = 1$

25. *Exact:* $3\frac{7}{16}$; *Estimate:* $7 \div 2 = 3\frac{1}{2}$

26. *Exact:* $6\frac{1}{6}$; *Estimate:* $5 \div 1 = 5$

27. *Estimate:* $6 + 4 = 10$; *Exact:* $9\frac{11}{12}$

28. *Estimate:* $18 + 10 = 28$; *Exact:* $28\frac{1}{6}$

29. *Estimate:* $15 + 11 = 26$; *Exact:* $25\frac{4}{15}$

30. *Estimate:* $9 - 4 = 5$; *Exact:* $4\frac{19}{20}$

31. *Estimate:* $14 - 7 = 7$; *Exact:* $6\frac{5}{8}$

32. *Estimate:* $32 - 23 = 9$; *Exact:* $9\frac{1}{4}$

33. $\frac{1}{12}$ **34.** $\frac{9}{10}$ **35.** $\frac{5}{18}$ **36.** 40 **37.** 72 **38.** 84

39. 35 **40.** 12 **41.** 55 **42.** $<$ **43.** $>$ **44.** $>$

Chapter 3 Review Exercises (pages 253–258)

1. $\frac{6}{7}$ **2.** $\frac{7}{9}$ **3.** $\frac{3}{4}$ **4.** $\frac{1}{8}$ **5.** $\frac{4}{5}$ **6.** $\frac{1}{6}$ **7.** $\frac{13}{31}$ **8.** $\frac{1}{3}$

9. $\frac{11}{12}$ of his total income **10.** $\frac{1}{4}$ Web page less **11.** 10 **12.** 12

13. 60 **14.** 24 **15.** 120 **16.** 180 **17.** 8 **18.** 21 **19.** 10

20. 45 **21.** 32 **22.** 20 **23.** $\frac{5}{6}$ **24.** $\frac{7}{8}$ **25.** $\frac{5}{8}$ **26.** $\frac{5}{12}$ **27.** $\frac{13}{24}$

28. $\frac{17}{36}$ **29.** $\frac{9}{10}$ of the students **30.** $\frac{23}{24}$ of her business

31. *Estimate:* $19 + 14 = 33$; *Exact:* $32\frac{3}{8}$

32. *Estimate:* $23 + 15 = 38$; *Exact:* $38\frac{1}{9}$

33. *Estimate:* $13 + 9 + 10 = 32$; *Exact:* $31\frac{43}{80}$

34. *Estimate:* $32 - 15 = 17$; *Exact:* $17\frac{1}{12}$

35. *Estimate:* $34 - 16 = 18$; *Exact:* $18\frac{1}{3}$

36. *Estimate:* $215 - 136 = 79$; *Exact:* $79\frac{7}{16}$

37. $9\frac{1}{10}$ **38.** $10\frac{5}{12}$ **39.** $3\frac{1}{4}$ **40.** $1\frac{2}{3}$ **41.** $5\frac{1}{2}$ **42.** $2\frac{19}{24}$

43. *Estimate:* $19 - 6 - 7 = 6$ miles; *Exact:* $5\frac{19}{24}$ miles

44. *Estimate:* $29 + 25 = 54$ tons; *Exact:* $53\frac{5}{12}$ tons

45. *Estimate:* $9 + 10 + 7 = 26$ pounds; *Exact:* $25\frac{1}{8}$ pounds

46. *Estimate:* $9 - 2 - 3 = 4$ acres; *Exact:* $4\frac{1}{16}$ acres

47.–50.

51. $<$ **52.** $<$ **53.** $>$ **54.** $>$ **55.** $<$ **56.** $>$ **57.** $<$

58. $>$ **59.** $\frac{1}{4}$ **60.** $\frac{4}{9}$ **61.** $\frac{27}{1000}$ **62.** $\frac{81}{4096}$ **63.** $\frac{1}{2}$ **64.** $6\frac{3}{4}$

65. $\frac{1}{16}$ **66.** 1 **67.** $\frac{3}{16}$ **68.** $1\frac{25}{64}$ **69.** $\frac{3}{4}$ **70.** $\frac{2}{5}$ **71.** $\frac{19}{32}$

72. $\frac{11}{16}$ **73.** $2\frac{1}{6}$ **74.** $26\frac{1}{4}$ **75.** $5\frac{3}{8}$ **76.** $11\frac{43}{80}$ **77.** $15\frac{5}{12}$

78. $\frac{8}{11}$ **79.** $\frac{1}{250}$ **80.** $\frac{1}{2}$ **81.** $\frac{2}{9}$ **82.** $\frac{11}{27}$ **83.** $>$ **84.** $<$ **85.** $<$

86. $>$ **87.** 36 **88.** 120 **89.** 126 **90.** 18 **91.** 108 **92.** 60

93. *Estimate:* $93 - 14 - 22 = 57$ ft; *Exact:* $56\frac{7}{8}$ ft

94. *Estimate:* $4 \cdot 50 = 200$ pounds of sugar; $200 - 69 - 77 - 33 = 21$ pounds; *Exact:* $21\frac{5}{8}$ pounds

Chapter 3 Test (pages 259–260)

1. $\frac{3}{4}$ **2.** $\frac{1}{2}$ **3.** $\frac{2}{5}$ **4.** $\frac{1}{6}$ **5.** 12 **6.** 30 **7.** 108 **8.** $\frac{5}{8}$

9. $\frac{23}{36}$ **10.** $\frac{5}{24}$ **11.** $\frac{1}{40}$

12. *Estimate:* $8 + 5 = 13$; *Exact:* $12\frac{1}{2}$

13. *Estimate:* $16 - 12 = 4$; *Exact:* $4\frac{11}{15}$

14. *Estimate:* $19 + 9 + 12 = 40$; *Exact:* $40\frac{29}{60}$

15. *Estimate:* $24 - 18 = 6$; *Exact:* $5\frac{5}{8}$

16. Answers will vary. Probably addition and subtraction of fractions is more difficult because you have to find the least common denominator and then change the fractions to the same denominator.   **17.** Answers will vary. Round mixed numbers to the nearest whole number. Then add or subtract to estimate the answer. The estimate may vary from the exact answer but it lets you know if your answer is reasonable.

18. *Estimate:* $10 + 85 + 37 + 8 = 140$ pounds; *Exact:* $140\frac{1}{24}$ pounds

19. *Estimate:* $148 - 69 - 37 - 6 = 36$ gal; *Exact:* $35\frac{7}{8}$ gal

20. $>$ **21.** $>$ **22.** 2 **23.** $\frac{13}{48}$ **24.** $1\frac{3}{4}$ **25.** $1\frac{1}{3}$

Cumulative Review Exercises: Chapters 1–3 (pages 263–264)

1. $5, 3, 9, 2$ **2.** $59,800$; $59,800$; $60,000$
3. *Estimate:* $20,000 - 10,000 = 10,000$;
Exact: $14,389$ **4.** *Estimate:* $100,000 \div 40 = 2500$; *Exact:* 3211
5. $1,255,609$ **6.** $2,801,695$ **7.** 160 **8.** $369,408$ **9.** 135 **10.** 2693 R2
11. *Estimate:* $20 + 9 + 5 + 20 + 9 + 5 = 68$ ft; *Exact:* 64 ft
12. *Estimate:* $20 \cdot 10 = 200$ ft^2; *Exact:* 252 ft^2

13. *Estimate:* $2 \cdot 3 = 6$ yd^2; *Exact:* $4\frac{2}{3}$ yd^2

14. *Estimate:* $70 \cdot 130 = 9100$ in.; *Exact:* $9132\frac{1}{2}$ in.

15. 144 **16.** 9 **17.** 44 **18.** $\frac{4}{45}$

19. $\frac{9}{10}$ **20.** $1\frac{1}{16}$ **21.** $\frac{1}{2}$ **22.** $36\frac{3}{4}$ **23.** $2\frac{3}{16}$ **24.** $13\frac{1}{2}$
25. *Estimate:* $3 + 5 = 8$; *Exact:* $7\frac{7}{8}$

26. *Estimate:* $22 + 4 = 26$; *Exact:* $26\frac{7}{24}$

27. *Estimate:* $5 - 2 = 3$; *Exact:* $2\frac{5}{8}$

28.–31.

32. $<$ **33.** $>$ **34.** $<$

Chapter 4 Decimals

Section 4.1 (pages 271–274)

1. $7; 0; 4$ **3.** $5; 1; 8$ **5.** $4; 7; 0$ **7.** $1; 6; 3$ **9.** $1; 8; 9$ **11.** $6; 2; 1$

13. 410.25 **15.** 6.5432 **17.** 5406.045 **19.** $\frac{7}{10}$ **21.** $13\frac{2}{5}$ **23.** $\frac{1}{4}$

25. $\frac{33}{50}$ **27.** $10\frac{17}{100}$ **29.** $\frac{3}{50}$ **31.** $\frac{41}{200}$ **33.** $5\frac{1}{500}$ **35.** $\frac{343}{500}$

37. five tenths   **39.** seventy-eight hundredths   **41.** one hundred five thousandths   **43.** twelve and four hundredths   **45.** one and seventy-five thousandths   **47.** 6.7 **49.** 0.32 **51.** 420.008 **53.** 0.0703 **55.** 75.030
57. Anne should not say "and" because that denotes a decimal point.

59. ten thousandths inch; $\frac{10}{1000} = \frac{1}{100}$ inch   **61.** 12 pounds   **63.** 3-C

65. 4-A   **67.** One and six hundred two thousandths centimeters
69. millionths, ten-millionths, hundred-millionths, billionths; these match the words on the left side of the chart with "ths" attached.   **70.** The first place to the left of the decimal point is ones, so the first place to the right could be one*ths*, like tens and ten*ths*. But anything that is 1 or more is to the *left* of the decimal point.   **71.** Seventy-two million four hundred thirty-six thousand nine hundred fifty-five hundred-millionths   **72.** six hundred seventy-eight thousand five hundred fifty-four billionths   **73.** eight thousand six and five hundred thousand one millionths   **74.** twenty thousand sixty and five hundred five millionths   **75.** 0.0302040 **76.** $9,876,543,210.100200300$

Section 4.2 (pages 281–282)

1. 16.9 **3.** 0.956 **5.** 0.80 **7.** 3.661 **9.** 794.0 **11.** 0.0980
13. 49 **15.** 9.09 **17.** 82.0002 **19.** $0.82 **21.** $1.22 **23.** $0.50
25. $48,650 **27.** $310 **29.** $849 **31.** $500 **33.** $1.00 **35.** $1000
37. (a) 322 miles per hour **(b)** 107 miles per hour **39. (a)** 186.0
miles per hour **(b)** 763.0 miles per hour **41.** Rounds to $0 (zero dollars)
because $0.499 is closer to $0 than to $1. **42.** Round amounts less than
$1.00 to the nearest cent instead of the nearest dollar. **43.** Rounds to
$0.00 (zero cents) because $0.0015 is closer to $0.00 than to $0.01.
44. Both round to $0.60. Rounding to nearest thousandth (tenth of a
cent) would allow you to identify $0.597 as less than $0.601.

Section 4.3 (pages 287–290)

1. 17.48 **3.** 23.013 **5.** 7.763 **7.** 77.006 **9.** 20.104
11. 0.109 **13.** 330.86895 **15. (a)** 24.75 in. **(b)** 3.95 in.
17. (a) 62.27 in. **(b)** 0.39 in. **19.** 6 should be written 6.00; sum is
46.22. **21.** *Estimate:* $20 − 7 = $13; *Exact:* $13.16 **23.** *Estimate:*
400 + 1 + 20 = 421; *Exact:* 414.645 **25.** *Estimate:* 9 − 4 = 5; *Exact:* 4.849 **27.** *Estimate:* 60 + 500 + 6 = 566; *Exact:* 608.4363 **29.**
0.275 **31.** 6.507 **33.** 1.81 **35.** 6056.7202 **37.** *Estimate:*
30 − 20 = 10 million people; *Exact:* 12.1 million people **39.** *Estimate:*
200 + 100 + 90 + 50 + 30 + 20 + 20 = 510 million people; *Exact:*
527.0 million people **41.** *Estimate:* 2 + 2 + 2 = 6 meters; *Exact:*
6.1 meters, which is less than the rhino by 0.3 meter **43.** *Estimate:*
11 − 10 = 1 ounce; *Exact:* 0.65 ounce **45.** *Estimate:*
20 + 6 + 20 + 6 = 52 in.; *Exact:* 52.1 in. **47.** *Estimate:*
$5 − $5 = $0; *Exact:* $0.30
49. *Estimate:* $19 + 2 + 2 + 10 + 2 = $35; *Exact:* $35.25
51. $1939.36 **53.** $3.97 **55.** $598.22 **57.** $b = 1.39$ centimeters
59. $q = 23.843$ ft

Section 4.4 (pages 293–296)

1. 0.1344 **3.** 159.10 **5.** 15.5844 **7.** $34,500.20 **9.** 43.2
11. 0.432 **13.** 0.0432 **15.** 0.00432 **17.** 0.0000312 **19.** 0.000025
21. 59.6; 32; 4.76; 803.5; 7226; 9. Multiplying by 10, decimal point
moves one place to the right; by 100, two places to the right; by 1000,
three places to the right. **22.** 5.96; 0.32; 0.0476; 8.035; 6.5; 52.3.
Multiplying by 0.1, decimal point moves one place to the left; by 0.01,
two places to the left; by 0.001, three places to the left. **23.** *Estimate:*
40 × 5 = 200; *Exact:* 190.08 **25.** *Estimate:* 40 × 40 = 1600; *Exact:*
1558.2 **27.** *Estimate:* 7 × 5 = 35; *Exact:* 30.038 **29.** *Estimate:*
3 × 7 = 21; *Exact:* 19.24165 **31.** unreasonable; $289.00
33. reasonable **35.** unreasonable; $4.19 **37.** unreasonable; 9.5 pounds
39. $945.87 (rounded) **41.** $2.45 (rounded) **43.** $90.02 (rounded)
45. $20,265 **47. (a)** Area before 1929 ≈ 23.2 in.² ; Area today ≈ 16.0 in.²
(b) 7.2 in.² **49. (a)** 0.43 in. **(b)** 4.3 in. **51.** $984.04; $2207.80
53. $76.50 **55.** $4.09 (rounded) **57.** $129.25 **59. (a)** $70.05
(b) $25.80

Summary Exercises on Decimals (pages 297–298)

1. $\frac{4}{5}$ **2.** $6\frac{1}{250}$ **3.** $\frac{7}{20}$ **4.** ninety-four and five tenths **5.** two and
three ten-thousandths **6.** seven hundred six thousandths **7.** 0.05
8. 0.0309 **9.** 10.7 **10.** 6.19 **11.** 1.0 **12.** 0.420 **13.** $0.89
14. $3.00 **15.** $100 **16.** 0.945 **17.** 49.6199 **18.** 1.845
19. $93.50 **20.** 0.00488 **21.** 2.15 **22.** 10.955 **23.** 18.4009
24. 4.3043 **25.** 0.87 **26.** $110.84 **27.** 82.84 (rounded)
28. $P = 3.1$ in.; $A \approx 0.5$ in.² **29.** $P = 3.25$ in.; $A \approx 0.66$ in.²
30. (a) Squash is heaviest; strawberry is lightest **(b)** 961.5 pounds
31. 92.253 pounds **32. (a)** 40.5 pounds; **(b)** 6.0 or 6 pounds
33. 3.6875 pounds **34. (a)** 4; 7; 16; 2; 69; 962; 1; 8 (all pounds)
(b) 1069 pounds **35.** $63.28

Section 4.5 (pages 305–308)

1. 3.9 **3.** 0.47 **5.** 400.2 **7.** 36 **9.** 0v06 **11.** 6000 **13.** 25.3
15. 516.67 (rounded) **17.** 26.756 (rounded) **19.** 10,082.647
(rounded) **21.** 0.377; 0.91; 0.0886; 3.019; 40.65; 662.57. **(a)** Dividing
by 10, decimal point moves one place to the left; by 100, two places to the
left; by 1000, three places to the left. **(b)** The decimal point moved to the
right when *multiplying* by 10 or 100 or 1000. Here it moves to the *left*
when *dividing* by those numbers. **22.** 402; 71; 3.39; 157.7; 460; 8730.
(a) Dividing by 0.1, decimal point moves one place to the right; by 0.01,
two places to the right; by 0.001, three places to the right. **(b)** The decimal
point moved to the *left* when *multiplying* by 0.1 or 0.01 or 0.001. Here the
decimal point moves to the *right* when *dividing* by those numbers.
23. unreasonable; 40 ÷ 8 = 5; Correct answer is 4.725. **25.** reasonable;
50 ÷ 50 = 1 **27.** unreasonable; 300 ÷ 5 = 60; Correct answer is 60.2.
29. unreasonable; 9 ÷ 1 = 9; Correct answer is 7.44. **31.** $4.00
(rounded) **33.** $67.08 **35.** $0.30 **37.** $11.92 per hour
39. 21.2 miles per gallon (rounded) **41.** 7.37 meters (rounded)
43. 0.08 meter **45.** 22.49 meters **47.** 14.25 **49.** 73.4 **51.** 1.205
53. 0.334 **55.** $0.03 per can (rounded) **57. (a)** 1,583,333 pieces
(rounded) **(b)** 26,389 pieces (rounded) **(c)** 440 pieces (rounded).
59. 100,000 box tops **61.** 2632 box tops (rounded)

Section 4.6 (pages 313–316)

1. 0.5 **3.** 0.75 **5.** 0.3 **7.** 0.9 **9.** 0.6 **11.** 0.875 **13.** 2.25
15. 14.7 **17.** 3.625 **19.** 6.333 (rounded) **21.** 0.833 (rounded)
23. 1.889 (rounded) **25. (a)** A proper fraction like $\frac{5}{9}$ is less than 1, so it
cannot be equivalent to a decimal number that is greater than 1.
(b) $\frac{5}{9}$ means $5 \div 9$ or $9\overline{)5}$ so correct answer is 0.556 (rounded). This
makes sense because both the fraction and decimal are less than 1.
26. (a) $2.035 = 2\frac{35}{1000} = 2\frac{7}{200}$, not $2\frac{7}{20}$. **(b)** Adding the whole
number part gives 2 + 0.35, which is 2.35, not 2.035. To check,
$2.35 = 2\frac{35}{100} = 2\frac{7}{20}$. **27.** Just add the whole number part to 0.375.
So $1\frac{3}{8} = 1.375$; $3\frac{3}{8} = 3.375$; $295\frac{3}{8} = 295.375$. **28.** It works only
when the fraction part has a one-digit numerator and a denominator of 10,
a two-digit numerator and a denominator of 100, and so on.
29. $\frac{2}{5}$ **31.** $\frac{5}{8}$ **33.** $\frac{7}{20}$ **35.** 0.35 **37.** $\frac{1}{25}$ **39.** $\frac{3}{20}$ **41.** 0.2
43. $\frac{9}{100}$ **45.** shorter; 0.72 inch **47.** too much; 0.005 gram
49. 0.9991 cm, 1.0007 cm **51.** more; 0.05 inch
53. 0.5399, 0.54, 0.5455 **55.** 5.0079, 5.79, 5.8, 5.804
57. 0.6009, 0.609, 0.628, 0.62812 **59.** 2.8902, 3.88, 4.876, 5.8751
61. 0.006, 0.043, $\frac{1}{20}$, 0.051 **63.** 0.37, $\frac{3}{8}$, $\frac{2}{5}$, 0.4001 **65.** red box
67. green box **69.** 1.4 in. (rounded) **71.** 0.3 in. (rounded)
73. 0.4 in. (rounded)

Chapter 4 Review Exercises (pages 321–324)

1. 0; 5 **2.** 0; 6 **3.** 8; 9 **4.** 5; 9 **5.** 7; 6 **6.** $\frac{1}{2}$ **7.** $\frac{3}{4}$ **8.** $4\frac{1}{20}$
9. $\frac{7}{8}$ **10.** $\frac{27}{1000}$ **11.** $27\frac{4}{5}$ **12.** eight tenths **13.** four hundred and
twenty-nine hundredths **14.** twelve and seven thousandths **15.** three
hundred six ten-thousandths **16.** 8.3 **17.** 0.205 **18.** 70.0066
19. 0.30 **20.** 275.6 **21.** 72.79 **22.** 0.160 **23.** 0.091 **24.** 1.0
25. $15.83 **26.** $0.70 **27.** $17,625.79 **28.** $350 **29.** $130
30. $100 **31.** $29 **32.** *Estimate:* 6 + 400 + 20 = 426; *Exact:* 444.86

33. *Estimate:* 80 + 1 + 100 + 1 + 30 = 212; *Exact:* 233**.**515
34. Estimate: 300 − 20 = 280; Exact: 290**.**7
35. Estimate: 9 − 8 = 1; *Exact:* 1**.**2684
36. *Estimate:* 90 million − 20 million = 70 million; *Exact:* 73**.**9 million
more cats **37.** *Estimate:* $300 − $200 − $40 = $60; Exact: $45**.**80
38. *Estimate:* $2 + $5 + $20 = $27; $30 − $27 = $3;
Exact: $4**.**14 **39.** *Estimate:* 2 + 4 + 5 = 11 kilometers;
Exact: 11**.**55 kilometers **40.** *Estimate:* 6 × 4 = 24; *Exact:* 22**.**7106
41. *Estimate:* 40 × 3 = 120; *Exact:* 141**.**57 **42.** 0**.**0112 **43.** 0**.**000355
44. reasonable; 700 ÷ 10 = 70 **45.** unreasonable; 30 ÷ 3 = 10;
Correct answer is 9**.**5. **46.** 14**.**467 (rounded) **47.** 1200 **48.** 0**.**4
49. $708 (rounded) **50.** $2**.**99 (rounded) **51.** 133 shares (rounded)
52. $3**.**47 (rounded) **53.** 29**.**215 **54.** 10**.**15 **55.** 3**.**8 **56.** 0**.**64
57. 1**.**875 **58.** 0**.**111 (rounded) **59.** 3**.**6008, 3**.**68, 3**.**806
60. 0**.**209, 0**.**2102, 0**.**215, 0**.**22 **61.** $\frac{1}{8}, \frac{3}{20}$, 0**.**159, 0**.**17 **62.** 404**.**865
63. 254**.**8 **64.** 3583**.**261 (rounded) **65.** 29**.**0898 **66.** 0**.**03066
67. 9**.**4 **68.** 175**.**675 **69.** 9**.**04 **70.** 19**.**50 **71.** 8**.**19
72. 0**.**928 **73.** 35 **74.** 0**.**259 **75.** 0**.**3 **76.** $3**.**00 (rounded)
77. $2**.**17 (rounded) **78.** $35**.**96 **79.** $199**.**71 **80.** $78**.**50
81. **(a)** baked potato with skin **(b)** cup of orange juice
(c) 0**.**59 milligram **82.** **(a)** 2**.**06 milligrams
(b) more, by 0**.**06 milligram

Chapter 4 Test (pages 325–326)

1. $18\frac{2}{5}$ **2.** $\frac{3}{40}$ **3.** sixty and seven thousandths **4.** two hundred eight
ten-thousandths **5.** 725**.**6 **6.** 0**.**630 **7.** $1**.**49 **8.** $7860 **9.** *Estimate:*
8 + 80 + 40 + 128; *Exact:* 129**.**2028 **10.** *Estimate:* 80 − 4 = 76;
Exact: 75**.**498 **11.** *Estimate:* 6(1) = 6; *Exact:* 6**.**948 **12.** *Estimate:*
20 ÷ 5 = 4; *Exact:* 4**.**175 **13.** 839**.**762 **14.** 669**.**004 **15.** 0**.**0000483
16. 480 **17.** 2**.**625 **18.** 0**.**44, $\frac{9}{20}$, 0**.**4506, 0**.**451 **19.** 35**.**49
20. $446**.**87 **21.** pintails, wigeons, gadwalls **22.** $5**.**35 (rounded)
23. 2**.**8 degrees **24.** $1**.**79 per foot (rounded) **25.** Answer varies.

Cumulative Review Exercises:
Chapters 1–4 (pages 329–330)

1. 500,000 **2.** 602**.**49 **3.** $710 **4.** $0**.**05 **5.** 9**.**671 **6.** $1\frac{4}{9}$
7. $1\frac{2}{5}$ **8.** 4914 **9.** 93**.**603 **10.** 404 R3 **11.** $1\frac{17}{24}$ **12.** 233,728
13. 0**.**03264 **14.** 8 **15.** 45 **16.** $\frac{4}{31}$ **17.** $\frac{2}{3}$ **18.** 0**.**51 (rounded)
19. 4 **20.** 14 **21.** forty and thirty-five thousandths
22. 0**.**0306 **23.** 7**.**005, 7**.**5, 7**.**5005, 7**.**505 **24.** 0**.**8, 0**.**8015, $\frac{21}{25}, \frac{7}{8}$
25. *Estimate:* $40 − $18 − $1 = $21; *Exact:* $21**.**17
26. *Estimate:* 50 − 47 = 3 in.; *Exact:* $3\frac{3}{8}$ in.
27. *Estimate:* $10 × 20 = $200; *Exact:* $191**.**90 (rounded)
28. *Estimate:* (8 × 20) + (10 × 30) = 160 + 300 = 460 students;
Exact: 488 students **29.** *Estimate:* 2 + 4 = 6 yards; *Exact:* $6\frac{5}{24}$ yards
30. *Estimate:* $30 + $200 − $40 − $20 = $170; *Exact:* $191**.**50
31. *Estimate:* $6 ÷ 3 = $2 per pound; *Exact:* $2**.**29 per pound (rounded)
32. *Estimate:* $80,000 ÷ 100 = $800; Exact $729 (rounded)
33. size M (medium) **34.** XXS $1\frac{1}{2}$ in.; XS $\frac{3}{4}$ in.; S $\frac{3}{4}$ in.; M $\frac{7}{8}$ in.; L $\frac{3}{4}$ in.

35. 21**.**125 − 20**.**25 = 0**.**875 in.; 7 ÷ 8 = 0**.**875 in.
or $\frac{875}{1000} = \frac{875 \div 125}{1000 \div 125} = \frac{7}{8}$ in. **36.** Answers will vary. Many people
prefer using decimals because you do not need to find a common
denominator or rewrite answers in lowest terms.

Chapter 5 Ratio and Proportion

Section 5.1 (pages 337–340)

1. $\frac{8}{9}$ **3.** $\frac{2}{1}$ **5.** $\frac{1}{3}$ **7.** $\frac{8}{5}$ **9.** $\frac{3}{8}$ **11.** $\frac{9}{7}$ **13.** $\frac{6}{1}$ **15.** $\frac{5}{6}$ **17.** $\frac{8}{5}$
19. $\frac{1}{12}$ **21.** $\frac{5}{16}$ **23.** $\frac{4}{1}$ **25.** $\frac{1}{2}$ **27.** $\frac{36}{1}$ **29.** Answers will vary. One
possibility is stocking cards of various types in the same ratios as those in
the table. **31.** $\frac{6}{5}; \frac{36}{17}$ **33.** *White Christmas* to *It's Now or Never; White
Christmas* to *I Will Always Love You; Candle in the Wind* to *I Want to Hold
Your Hand.* **35. (a)** $\frac{6}{1}$ **(b)** $\frac{5}{2}$ **(c)** $\frac{7}{16}$ **37.** $\frac{7}{5}$ **39.** $\frac{6}{1}$ **41.** $\frac{38}{17}$ **43.** $\frac{1}{4}$
45. $\frac{34}{35}$ **47.** $\frac{1}{1}$; as long as the sides all have the same length, any measure-
ment you choose will maintain the ratio. **48.** Answers will vary. Some
possibilities are $\frac{4}{5} = \frac{8}{10} = \frac{12}{15} = \frac{16}{20} = \frac{20}{25} = \frac{24}{30} = \frac{28}{35}$. **49.** It is not
possible. Amelia would have to be older than her mother to have a ratio of
5 to 3. **50.** Answers will vary, but a ratio of 3 to 1 means your income is
3 times your friend's income.

Section 5.2 (pages 345–348)

1. $\frac{5 \text{ cups}}{3 \text{ people}}$ **3.** $\frac{3 \text{ feet}}{7 \text{ seconds}}$ **5.** $\frac{18 \text{ miles}}{1 \text{ gallon}}$
7. $12 per hour or $12/hour **9.** 1**.**25 pounds/person **11.** $103**.**30/day
13. 325**.**9; 21**.**0 (rounded) **15.** 338**.**6; 20**.**9 (rounded) **17.** 4 ounces for
$3**.**65, about $0**.**913/ounce **19.** 14 ounces for $2**.**89, about $0**.**206/ounce
21. 18 ounces for $1**.**79, about $0**.**099/ounce **23.** Answers will vary. For
example, you might choose Brand B because you like more chicken, so the
cost per chicken chunk may actually be the same or less than Brand A.
25. 1**.**75 pounds/week **27.** $12**.**26/hour **29. (a)** Radiant, $0**.**44;
IDT, $0**.**25; Access: $0**.**235 **(b)** Radiant, $0**.**088/min; IDT, $0**.**05/min;
Access, $0**.**047/min; Access America is the best buy. **31.** For a 15-minute
call: Radiant, $0**.**036/min; IDT, $0**.**031/min; Access, $0**.**047/min; IDT is the
best buy. For a 25-minute call: Radiant, $0**.**026/min; IDT, $0**.**028/min;
Access, $0**.**047/min; Radiant is the best buy. **33.** one battery for $1**.**79;
like getting 3 batteries so $1**.**79 ÷ 3 ≈ $0**.**597 per battery **35.** Brand P
with the 50¢ coupon is the best buy. ($3**.**39 − $0**.**50 = $2**.**89;
$2**.**89 ÷ 16**.**5 ounces ≈ $0**.**175 per ounce) **37.** $2**.**92 (rounded) per
month for Verizon, T-Mobile, and Nextel; $3 per month for Sprint.
38. Average number of weekdays per month is about 21**.**7;
Verizon ≈ 18 min/weekday; T-Mobile ≈ 28 min/weekday; Nextel
and Sprint ≈ 23 min/weekday. **39.** Round to hundredths (nearest
cent) to see that T-Mobile is best buy at $0**.**07 per "anytime minute";
Verizon ≈ $0**.**16; Sprint ≈ $0**.**12; Nextel ≈ $0**.**10.
40. Verizon ≈ $1**.**65 per "anytime minute"; T-Mobile ≈ $1**.**57;
Nextel ≈ $1**.**63; Sprint = $1**.**48.

Section 5.3 (pages 353–354)

1. $\frac{\$9}{12 \text{ cans}} = \frac{\$18}{24 \text{ cans}}$ **3.** $\frac{200 \text{ adults}}{450 \text{ children}} = \frac{4 \text{ adults}}{9 \text{ children}}$ **5.** $\frac{120}{150} = \frac{8}{10}$
7. $\frac{3}{5} = \frac{3}{5}$; true **9.** $\frac{5}{8} = \frac{5}{8}$; true **11.** $\frac{3}{4} \neq \frac{2}{3}$; false **13.** $\frac{14}{5} = \frac{14}{5}$; true

15. $\dfrac{16}{9} = \dfrac{16}{9}$; true   **17.** $\dfrac{7}{6} \neq \dfrac{9}{8}$; false   **19.** $54 = 54$; True

21. $336 \neq 320$; False   **23.** $2880 \neq 2970$; False   **25.** $28 = 28$; True

27. $44.8 \neq 45$; False   **29.** $66 = 66$; True   **31.** $68\dfrac{1}{4} = 68\dfrac{1}{4}$; True

33. $5\dfrac{2}{5} \neq 5\dfrac{1}{3}$; False   **35.** $2\dfrac{1}{80} \neq 2\dfrac{7}{100}$ or $2.0125 \neq 2.07$; False

37. $\dfrac{17 \text{ hits}}{50 \text{ at bats}} = \dfrac{153 \text{ hits}}{450 \text{ at bats}}$ $\begin{array}{l} 50 \cdot 153 = 7650 \\ 17 \cdot 450 = 7650 \end{array}$ Cross products are *equal* so the proportion is *true;* they hit equally well.

Section 5.4 (pages 359–360)

1. $x = 4$ **3.** $x = 2$ **5.** $x = 88$ **7.** $x = 91$ **9.** $x = 5$ **11.** $x = 10$
13. $x \approx 24.44$ (rounded)   **15.** $x = 50.4$ **17.** $x \approx 17.64$ (rounded)
19. $x = 1$ **21.** $x = 3\dfrac{1}{2}$ **23.** $x = 0.2$ or $x = \dfrac{1}{5}$
25. $x = 0.005$ or $x = \dfrac{1}{200}$
27. Find cross products: $20 \neq 30,$ so the proportion is false.
$\dfrac{6\frac{2}{3}}{4} = \dfrac{5}{3}$ or $\dfrac{10}{6} = \dfrac{5}{3}$ or $\dfrac{10}{4} = \dfrac{7.5}{3}$ or $\dfrac{10}{4} = \dfrac{5}{2}$
28. Find cross products: $192 \neq 180,$ so the proportion is false.
$\dfrac{6.4}{8} = \dfrac{24}{30}$ or $\dfrac{6}{7.5} = \dfrac{24}{30}$ or $\dfrac{6}{8} = \dfrac{22.5}{30}$ or $\dfrac{6}{8} = \dfrac{24}{32}$

Summary Exercises on Ratios, Rates, and Proportions (pages 361–362)

1. $\dfrac{6}{5}$ **2.** $\dfrac{2}{7}$ **3.** $\dfrac{2}{1}$ **4.** $\dfrac{13}{11}$ **5.** Comparing the violin to piano, guitar, organ, clarinet, and drums gives ratios of $\dfrac{1}{11}, \dfrac{1}{10}, \dfrac{1}{3}, \dfrac{1}{2},$ and $\dfrac{2}{3}$ respectively.

6. (a) guitar to clarinet **(b)** organ to drums, or clarinet to violin
7. 2.1 points/min; 0.5 min/point   **8.** 1.6 points/min; 0.6 min/point
9. \$16.32/hour; \$24.48/hour for overtime   **10.** \$0.50/channel;
\$0.39/channel; \$0.32/channel   **11.** 12 ounces for \$7.24, about \$0.60/ounce
12. Brand P with the \$2 coupon is the best buy at \$0.57 per pound.
13. $\dfrac{4}{3} = \dfrac{4}{3}$ or $924 = 924$; true   **14.** $2.0125 \neq 2.07$; false
15. $68\dfrac{1}{4} = 68\dfrac{1}{4}$; true   **16.** $x = 28$ **17.** $x = 3.2$ **18.** $x = 182$
19. $x \approx 3.64$ (rounded)   **20.** $x \approx 0.93$ (rounded)   **21.** $x = 1.56$
22. $x \approx 0.05$ (rounded)   **23.** $x = 1$ **24.** $x = \dfrac{3}{4}$

Section 5.5 (pages 367–370)

1. 22.5 hours   **3.** \$7.20   **5.** 42 pounds   **7.** \$403.68   **9.** 10 ounces
(rounded)   **11.** 5 quarts   **13.** 14 ft, 10 ft   **15.** 14 ft, 8 ft
17. 96 pieces chicken; 33.6 pounds lasagna; 10.8 pounds deli meats; $5\dfrac{3}{5}$
pounds cheese; 7.2 dozen (about 86) buns; 14.4 pounds of salad
19. 2065 students (reasonable); about 4214 students with incorrect setup
(only 2950 students in the group)   **21.** about 83 people (reasonable);
about 750 people with incorrect setup (only 250 people attended)
23. 110,838,000 households (reasonable); about 115,408,163 households
with incorrect setup (only 113,100,000 U.S. households).   **25.** 625 stocks
27. 4.06 meters (rounded)   **29.** 311 calories (rounded)   **31.** 10.53 meters
(rounded)   **33.** You cannot solve this problem using a proportion because
the ratio of age to weight is not constant. As Jim's age increases, his weight
may decrease, stay the same, or increase.   **35.** 5050 students use cream
37. 120 calories and 12 grams of fiber   **39.** $1\dfrac{3}{4}$ cups water, 3 Tbsp
margarine, $\dfrac{3}{4}$ cup milk, 2 cups flakes   **40.** $5\dfrac{1}{4}$ cups water, 9 Tbsp margarine,

$2\dfrac{1}{4}$ cups milk, 6 cups flakes   **41.** $\dfrac{7}{8}$ cup water, $1\dfrac{1}{2}$ Tbsp margarine,
$\dfrac{3}{8}$ cup milk, 1 cup flakes   **42.** $2\dfrac{5}{8}$ cups water, $4\dfrac{1}{2}$ Tbsp margarine,
$1\dfrac{1}{8}$ cups milk, 3 cups flakes

Chapter 5 Review Exercises (pages 377–380)

1. $\dfrac{3}{4}$ **2.** $\dfrac{4}{1}$ **3.** great white shark to whale shark; whale shark to
blue whale   **4.** $\dfrac{2}{1}$ **5.** $\dfrac{2}{3}$ **6.** $\dfrac{5}{2}$ **7.** $\dfrac{1}{6}$ **8.** $\dfrac{3}{1}$ **9.** $\dfrac{3}{8}$ **10.** $\dfrac{4}{3}$
11. $\dfrac{1}{9}$ **12.** $\dfrac{10}{7}$ **13.** $\dfrac{7}{5}$ **14.** $\dfrac{5}{6}$ **15.** $\dfrac{\$11}{1 \text{ dozen}}$ **16.** $\dfrac{12 \text{ children}}{5 \text{ families}}$
17. 0.2 page/minute or $\dfrac{1}{5}$ page/minute; 5 minutes/page
18. \$8/hour; 0.125 hour/dollar or $\dfrac{1}{8}$ hour/dollar
19. 8 ounces for \$4.98, about \$0.623/ounce   **20.** 17.6 pounds for
\$18.69 $-$ \$1 coupon, about \$1.005/pound
21. $\dfrac{3}{5} = \dfrac{3}{5}$ or $90 = 90$; true   **22.** $\dfrac{1}{8} \neq \dfrac{1}{4}$ or $432 \neq 216$; false
23. $\dfrac{47}{10} \neq \dfrac{49}{10}$ or $980 \neq 940$; false   **24.** $\dfrac{16}{9} = \dfrac{16}{9}$ or $3456 = 3456$;
true   **25.** $4.8 = 4.8$; true   **26.** $14 = 14$; true   **27.** $x = 1575$
28. $x = 20$ **29.** $x = 400$ **30.** $x = 12.5$ **31.** $x \approx 14.67$ (rounded)
32. $x \approx 8.17$ (rounded)   **33.** $x = 50.4$ **34.** $x \approx 0.57$ (rounded)
35. $x \approx 2.47$ (rounded)   **36.** 27 cats   **37.** 46 hits   **38.** \$15.63 (rounded)
39. 3299 students (rounded)   **40.** 68 feet   **41.** $27\dfrac{1}{2}$ hours or 27.5 hours
42. 511 calories (rounded)   **43.** 14.7 milligrams   **44.** $x = 105$
45. $x = 0$ **46.** $x = 128$ **47.** $x \approx 23.08$ (rounded)   **48.** $x = 6.5$
49. $x \approx 117.36$ (rounded)   **50.** $1440 \neq 1485$; False
51. $10.8 \neq 10.864$; False   **52.** $2\dfrac{1}{3} = 2\dfrac{1}{3}$; True   **53.** $\dfrac{8}{5}$
54. $\dfrac{33}{80}$ **55.** $\dfrac{15}{4}$ **56.** $\dfrac{4}{1}$ **57.** $\dfrac{4}{5}$ **58.** $\dfrac{37}{7}$ **59.** $\dfrac{3}{8}$ **60.** $\dfrac{1}{12}$
61. $\dfrac{45}{13}$ **62.** 24,900 fans (rounded)   **63.** $\dfrac{8}{3}$ **64.** 75 ft for \$1.99 $-$ \$0.50
coupon, about \$0.020/ft   **65.** 21 ft long; 15 ft wide   **66.** 7.5 hours or
$7\dfrac{1}{2}$ hours   **67.** $\dfrac{1}{2}$ teaspoon or 0.5 teaspoon   **68.** 21 points (rounded)
69. Set up the proportion to compare teaspoons to pounds on both sides.
$\dfrac{1.5 \text{ teaspoons}}{24 \text{ pounds}} = \dfrac{x \text{ teaspoons}}{8 \text{ pounds}}$
Show that cross products are equal. $(24)(x) = (1.5)(8)$
Divide both sides by 24. $\dfrac{\overset{1}{\cancel{24}}(x)}{\underset{1}{\cancel{24}}} = \dfrac{12}{24}$
so $x = \dfrac{1}{2}$ teaspoon or 0.5 teaspoon
70. (a) 1400 milligrams   **(b)** 100 milligrams

Chapter 5 Test (pages 381–382)

1. $\dfrac{4}{5}$ **2.** $\dfrac{20 \text{ miles}}{1 \text{ gallon}}$ **3.** $\dfrac{\$1}{5 \text{ minutes}}$ **4.** $\dfrac{9}{2}$ **5.** $\dfrac{15}{4}$
6. Best buy is 8 in. sub, about \$0.724/inch
7. 16 ounces for \$1.89 $-$ \$0.50 coupon, about \$0.087/ounce
8. You earned less this year. An example is: $\begin{array}{l} \text{Last year} \longrightarrow \$30,000 \\ \text{This year} \longrightarrow \$20,000 \end{array} = \dfrac{3}{2}$
9. $\dfrac{3}{7} \neq \dfrac{2}{5}$ or $252 \neq 270$; false   **10.** $5.88 = 5.88$; true   **11.** $x = 25$

12. $x \approx 2.67$ (rounded) **13.** $x = 325$ **14.** $x = 10\frac{1}{2}$ **15.** 24 orders
16. 3.6 ounces **17.** 87 students (rounded) **18.** No, 4875 cannot be correct because there are only 650 students in the whole school.
19. 23.8 grams (rounded) **20.** 60 ft

Cumulative Review Exercises: Chapters 1–5 (pages 383–384)

1. 9900 **2.** 617.1 **3.** $100 **4.** $3.06 **5.** 29.34 **6.** 610 R27
7. 0.0076 **8.** 2312 **9.** 68.381 **10.** 55.6 **11.** 39 **12.** 18 **13.** 64
14. 0.95 **15.** one hundred five ten-thousandths **16.** 60.071
17. $\frac{1}{5}$ **18.** $\frac{4}{1}$ **19.** $1\frac{1}{8}$, 1.25, $1\frac{3}{8}$, 1.5, $1\frac{9}{16}$ (all inches)
20. $\frac{1}{16}$ inch or 0.063 inch (rounded) **21.** $x = 21$ **22.** $x \approx 17.14$ (rounded) **23.** $x \approx 0.98$ (rounded) **24.** 250 pounds
25. 26.7 centimeters (rounded) **26.** $7\frac{3}{20}$ miles **27.** 34 servings for $3.85 − $0.50 coupon, about $0.099/serving **28.** 140 residents
29. $1\frac{1}{4}$ teaspoons **30.** $0.028, $0.04, $0.043, $0.05, $0.052, $0.106, $0.12
31. 233 minutes (rounded) **32.** 83 minutes (rounded) per $10 card, so buy 3 cards to cover 240 minutes (4 hours). **33.** $\frac{250}{200} = \frac{5}{4}$

Chapter 6 Percent

Section 6.1 (pages 391–396)

1. 0.12 **3.** 0.70 or 0.7 **5.** 0.25 **7.** 1.40 or 1.4 **9.** 0.055
11. 1.00 or 1 **13.** 0.005 **15.** 0.0035 **17.** 60% **19.** 58%
21. 1% **23.** 12.5% **25.** 37.5% **27.** 200% **29.** 370%
31. 3.12% **33.** 416.2% **35.** 0.28% **37.** Answers will vary. Some possibilities are: No common denominators are needed with percents. The denominator is always 100 with percent, which makes comparisons easier to understand. **39.** 0.38 **41.** 0.47 **43.** 3.5% **45.** 200%
47. 0.5% **49.** 1.536 **51.** 20 children **53.** 420 employees
55. 270 chairs **57.** $377.50 **59.** 820 commuters **61.** 26 plants
63. (a) Since 100% means 100 parts out of 100 parts, 100% is all of the number. **(b)** Answers will vary. For example, 100% of $72 is $72.
65. (a) Since 200% is two times a number, find 200% of the number by multiplying the number by 2 (double it). **(b)** Answers will vary. For example, 200% of $20 is 2 • $20 = $40. **67. (a)** Since 10% means 10 parts out of 100 parts or $\frac{1}{10}$, the shortcut for finding 10% of a number is to move the decimal point in the number one place to the left. **(b)** Answers will vary. For example, 10% of $90 is $9. **69.** 12%; 0.12 **71. (a)** witch costume **(b)** 4.5%; 0.045 **73.** 26%; 0.26
75. (a) Marine Corps **(b)** 5%; 0.05 **77.** 16.8%; 0.168 **(a)** Germany; Bulgaria **(b)** 16.5%; 0.165 **81.** 95% shaded; 5% unshaded
83. 30% shaded; 70% unshaded **85.** 55% shaded; 45% unshaded
87. 64% shaded; 36% unshaded

Section 6.2 (pages 403–408)

1. $\frac{1}{4}$ **3.** $\frac{3}{4}$ **5.** $\frac{17}{20}$ **7.** $\frac{5}{8}$ **9.** $\frac{1}{16}$ **11.** $\frac{1}{6}$ **13.** $\frac{1}{15}$ **15.** $\frac{1}{200}$
17. $1\frac{4}{5}$ **19.** $3\frac{3}{4}$ **21.** 50% **23.** 80% **25.** 70% **27.** 37%
29. 62.5% **31.** 87.5% **33.** 48% **35.** 46% **37.** 35% **39.** 83.3% (rounded) **41.** 55.6% (rounded) **43.** 14.3% (rounded) **45.** $\frac{1}{2}$; 50%
47. $\frac{7}{8}$; 0.875 **49.** $\frac{4}{5}$; 80% **51.** 0.167 (rounded); 16.7% (rounded)

53. $\frac{7}{10}$; 70% **55.** $\frac{1}{8}$; 0.125 **57.** 0.667 (rounded); 66.7% (rounded)
59. 0.06; 6% **61.** 0.08; 8% **63.** 0.005; 0.5% **65.** $2\frac{1}{2}$; 250%
67. 3.25; 325% **69.** There are many possible answers. Examples 2 and 3 show the steps that students should include in their answers.
71. $\frac{9}{50}$; 0.18; 18% **73.** $\frac{13}{100}$; 0.13; 13% **75.** $\frac{1}{5}$; 0.2; 20% **77.** $\frac{1}{5}$; 0.20; 20% **79.** $\frac{1}{10}$; 0.1; 10% **81.** $\frac{1}{4}$; 0.25; 25% **83.** $\frac{1}{20}$; 0.05; 5%
85. 100; 100 **86. (a)** 765 workers **(b)** 96 letters **(c)** 21 videos
87. 50; 100; half or $\frac{1}{2}$ **88.** 10; 100; 1; left **89.** 1; 100; 2; left
90. (a) 525 homes **(b)** 37 printers **(c)** $0.08
91. Find 10% of $160, then add $\frac{1}{2}$ of 10%.

$$10\% + 5\% = 15\%$$
$$\downarrow \quad\quad \downarrow \quad\quad \downarrow$$
$$\$16 + \$8 = \$24$$

92. Find 100% of $160, then add 50% of 160.
$$100\% + 50\% = 150\%$$
$$\downarrow \quad\quad \downarrow \quad\quad \downarrow$$
$$\$160 + \$80 = \$240$$

93. From 100% of $450, subtract 10% of $450.
$$100\% - 10\% = 90\%$$
$$\downarrow \quad\quad \downarrow \quad\quad \downarrow$$
$$\$450 - \$45 = \$405$$

94. To 100% of $800, add 100% of $800, and then add 10% of $800.
$$100\% + \$100\% + 10\% = 210\%$$
$$\downarrow \quad\quad \downarrow \quad\quad \downarrow \quad\quad \downarrow$$
$$\$800 + \$800 + \$80 = \$1680$$

Section 6.3 (pages 413–416)

1. whole = 50 **3.** whole = 150 **5.** whole = 70 **7.** percent = 25%
9. percent = 150% **11.** percent = 33.3% (rounded) **13.** part = 26
15. part = 21.6 **17.** percent = 26.5% (rounded) **19.** 115
21. 0.4% **23.** 2.5% **25.** 0.3%
27. Part → $\frac{60}{\text{unknown}} = \frac{10}{100}$ ← Percent, Whole → unknown ← Always 100
29. $\frac{600}{800} = \frac{75}{100}$ **31.** $\frac{\text{unknown}}{970} = \frac{25}{100}$ **33.** $\frac{12}{\text{unknown}} = \frac{20}{100}$
35. $\frac{34}{68} = \frac{50}{100}$ **37.** $\frac{177}{296} = \frac{\text{unknown}}{100}$ **39.** $\frac{54.34}{\text{unknown}} = \frac{3.25}{100}$
41. $\frac{\text{unknown}}{487} = \frac{0.68}{100}$
43. Percent—the ratio of the part to the whole. It appears with the word *percent* or "%" after it. Whole—the entire quantity. Often appears after the word *of*. Part—the part being compared with the whole.
45. $\frac{535}{1082} = \frac{\text{unknown}}{100}$ **47.** $\frac{86}{142} = \frac{\text{unknown}}{100}$ **49.** $\frac{\text{unknown}}{610} = \frac{23}{100}$
51. $\frac{680}{2000} = \frac{\text{unknown}}{100}$ **53.** $\frac{\text{unknown}}{480} = \frac{55}{100}$ **55.** $\frac{\text{unknown}}{822} = \frac{49.5}{100}$
57. $\frac{504}{\text{unknown}} = \frac{16.8}{100}$ **59.** $\frac{\text{unknown}}{680} = \frac{45}{100}$

Section 6.4 (pages 425–430)

1. 42 test tubes **3.** 1836 military personnel **5.** 4.8 ft **7.** 315 files
9. 819 trucks **11.** $3.28 **13.** 1530 tables **15.** 182 cell phones
17. $21.60 **19.** 320 e-mails **21.** 160 hay bales **23.** 550 students
25. 330 mountain bikes **27.** 2800 **29.** 50% **31.** 52% **33.** 8%

35. 1.5% **37.** 18.6% (rounded) **39.** 9.2% **41.** 150% of $30 cannot be less than $30 because 150% is greater than 1 (100%). The answer must be greater than $30. 25% of $16 cannot be greater than $16 because 25% is less than 1 (100%). The answer must be less than $16. **43.** $52.80 **45.** 44 females (rounded) **47.** 220 students **49. (a)** 28.6% female lawyers **(b)** 313,099 female lawyers (rounded) **51.** 33% **53.** 2074 children **55.** 2% **57.** $3200 monthly earnings; $38,400 yearly earnings **59.** Breyers **61.** 76 people (rounded) **63.** Hardee's **65.** $18.06 billion **67.** $645 **69.** 2156 products **71.** whole; 100 **72.** whole **73.** 108 calories **74.** 300 grams **75.** 67 grams **76.** Yes, since they would eat 28% × 4 servings = 112% of the daily value. **77.** 17 packages (rounded). It may be possible but would result in a diet that is high in total fat, saturated fat, sodium, and total carbohydrates.

Section 6.5 (pages 435–438)

1. 270 donors **3.** 1350 bath towels **5.** 83.2 quarts **7.** 3500 airbags **9.** 1029.2 meters **11.** $4.16 **13.** 160 patients **15.** 325 salads **17.** 680 circuits **19.** 1080 people **21.** 300 gallons **23.** 50% **25.** 76% **27.** 125% **29.** 1.5% **31.** 250% **33.** You must first change the fraction in the percent to a decimal, then divide the percent by 100 to change it to a decimal.

$$\underbrace{2\tfrac{1}{2}\% = 2.5\%}_{\text{Write } 2\tfrac{1}{2} \text{ as } 2.5.} = 0.025 \leftarrow 2.5\% \text{ as a decimal}$$

35. 3.78 million or 3,780,000 office workers **37.** 91.5 million homes (rounded) **39. (a)** 17,640 have a refrigerator **(b)** 360 do not have a refrigerator **41. (a)** 4% **(b)** 96% **43.** 36.9% (rounded) **45.** 2106 people **47.** 2916 people **49.** $1817.2 million (rounded) **51.** 478,175 Mustangs (rounded) **53.** $510,390 **55.** $439.61

Summary Exercises on Percent (pages 439–440)

1. 0.0625 **2.** 3.80 or 3.8 **3.** 37.5% **4.** 0.6% **5.** $\frac{7}{8}$ **6.** $1\frac{3}{5}$ **7.** 62.5% **8.** 0.8% **9.** $\frac{13}{50}$; 0.26; 26% **10.** $\frac{17}{100}$; 0.17; 17% **11.** $\frac{3}{50}$; 0.06; 6% **12.** $\frac{9}{100}$; 0.09; 9% **13.** $46.80 **14.** 500 rolls **15.** 55% **16.** 28 screening exams **17.** 363 acres **18.** 2.7% (rounded) **19.** 168 people **20.** $429.92 **21.** 1280 people **22.** 1749 owners **23.** 40% **24.** 3.7% (rounded) **25.** 5,700,500 subscribers **26.** 3419 ski-lift tickets

Section 6.6 (pages 447–450)

1. $0.24; $6.24 **3.** 3%; $437.75 **5.** 5%; $298.20 **7.** $672.60 (rounded); $12,901.60 (rounded) **9.** $22.40 **11.** 20% **13.** $185.51 (rounded) **15.** $6615 **17.** $20 (rounded); $179.99 (rounded) **19.** 30%; $126 **21.** $4.38 (rounded); $13.12 **23.** $8.76; $49.64 **25.** On the basis of commission alone you would choose Company A. Other considerations might be: reputation of the company; expense allowances; other fringe benefits; travel; promotion and training, to name a few. **27.** $203.93 (rounded) **29.** $90.62 (rounded) **31.** 5% **33.** 35.9% (rounded) **35.** 22.5% (rounded) **37.** $134 **39.** $568.80 **41.** 4% **43.** $85.80; $304.20 **45.** 5.6% (rounded) **47.** $18.43 (rounded) **49.** $4276.97 (rounded) **51.** $14,032.12 (rounded) **53.** percent; whole **54.** rate (percent) of tax; cost of item **55.** $146.25 (rounded) **56.** Yes, the same; (12.4% • $123) + (6.5% • $123) = $23.25 (rounded) and (12.4% + 6.5%) • ($123) = $23.25 **57.** $1358.12 **58.** No. In Exercise 57 the excise tax of $13.40 cannot be added to the sales tax rate of $7\frac{3}{4}$% because the excise tax is an amount, not a percent.

Section 6.7 (pages 453–456)

1. $6 **3.** $105 **5.** $28.80 **7.** $488.75 **9.** $2227.50 **11.** $10 **13.** $49.20 **15.** $42.30 **17.** $16.84 (rounded) **19.** $647.06 (rounded) **21.** $210 **23.** $773.30 **25.** $2043 **27.** $3412.50 **29.** $17,797.81 (rounded) **31.** The answer should include: amount of principal—This is the amount of money borrowed or loaned. Interest rate—This is the percent used to calculate the interest. Time of loan—The length of time that money is loaned or borrowed is an important factor in determining interest. **33.** $41.25 **35.** $26,250 **37.** $1240 **39.** $277.50 **41.** $159.50 **43.** $12,254.69 (rounded) **45.** $2165.63 (rounded)

Section 6.8 (pages 461–467)

1. $540.80 **3.** $1966.91 (rounded) **5.** $4587.79 (rounded) **7.** $1102.50 **9.** $1873.52 (rounded) **11.** $2027.46 (rounded) **13.** $17,360.41 (rounded) **15.** $1216.70; $216.70 **17.** $14,326.40; $6326.40 **19.** $10,976.01 (rounded); $2547.84 (rounded) **21.** Interest paid on past interest as well as on the principal. Many people describe compound interest as "interest on interest." **23.** $13,467.61 **25. (a)** $128,402 **(b)** $52,402 **27. (a)** $87,786 **(b)** $17,786 **29.** principal; rate; time p; r; t **30.** principal; interest **31.** principal; interest **32.** principal; interest **33. (a)** $254.48 **(b)** No; it has more than doubled (about five times). **(c)** Examples will vary. **34. (a)** the compound interest account; $1091.20 more **(b)** Greater amounts of interest are earned with compound interest than simple interest because interest is earned on both principal and past interest.

Chapter 6 Review Exercises (pages 471–474)

1. 0.35 **2.** 1.5 **3.** 0.9944 **4.** 0.00085 **5.** 315% **6.** 2% **7.** 87.5% **8.** 0.2% **9.** $\frac{3}{20}$ **10.** $\frac{3}{8}$ **11.** $1\frac{3}{4}$ **12.** $\frac{1}{400}$ **13.** 75% **14.** 62.5% or $62\frac{1}{2}$% **15.** 325% **16.** 0.5% **17.** 0.125 **18.** 12.5% **19.** $\frac{1}{4}$ **20.** 25% **21.** $1\frac{4}{5}$ **22.** 1.8 **23.** whole = 250 **24.** part = 24 **25.** $\dfrac{\text{Part} \rightarrow 287}{\text{Whole} \rightarrow 820} = \dfrac{35}{100} \dfrac{\leftarrow \text{Percent}}{\leftarrow \text{Always 100}}$ **26.** $\dfrac{73}{90} = \dfrac{\text{unknown}}{100}$ **27.** $\dfrac{\text{unknown}}{160} = \dfrac{14}{100}$ **28.** $\dfrac{418}{\text{unknown}} = \dfrac{16}{100}$ **29.** $\dfrac{3}{8} = \dfrac{\text{unknown}}{100}$ **30.** $\dfrac{\text{unknown}}{1280} = \dfrac{88}{100}$ **31.** 171 programs

32. 870 reference books **33.** 31.2 acres **34.** 2.8 kilograms

35. 750 crates **36.** 2320 test tubes **37.** 484 miles **38.** 17,000 cases **39.** 55% **40.** 5.2% (rounded) **41.** 9.5% (rounded) **42.** 30.8% (rounded) **43.** $2.7 million (rounded) **44.** 10.4% (rounded) **45.** $145.28 **46.** 186 trucks **47.** 0.4% **48.** 175% **49.** 120 miles **50.** $575 **51.** $31.50; $661.50 **52.** $7\frac{1}{2}$%; $838.50 **53.** $276 **54.** 5% **55.** $33.75; $78.75 **56.** 25%; $189 **57.** $8 **58.** $67.50 **59.** $7 **60.** $152.10 **61.** $832.50 **62.** $1598.85 **63.** $5375.60; $1375.60 **64.** $2187.71 (rounded); $317.71 (rounded) **65.** $4534.92; $934.92 **66.** $16,337.50; $3837.50 **67.** part = 12 **68.** whole = 1640 **69.** 23.28 meters **70.** 150% **71.** $0.51 **72.** 1980 employees **73.** 40% **74.** 498.8 liters **75.** 0.55 **76.** 3 **77.** 500% **78.** 471% **79.** 0.086 **80.** 62.1% **81.** 0.00375 **82.** 0.06% **83.** 75% **84.** $\frac{21}{50}$ **85.** $\frac{7}{8}$ **86.** 37.5% or $37\frac{1}{2}$% **87.** $\frac{13}{40}$ **88.** 60% **89.** $\frac{1}{400}$ **90.** 375%

91. $3331.25 **92.** $16,520 **93.** 32.4% (rounded)

94. (a) $15,780.96 (rounded) **(b)** $3280.96 (rounded) **95.** $6225 **96.** 18.2% (rounded) **97.** $848.40 (rounded) **98.** 1100 boaters **99.** $13,005 **100.** 33.4% (rounded)

Chapter 6 Test (pages 477–478)

1. 0.65 **2.** 80% **3.** 175% **4.** 87.5% or $87\frac{1}{2}$% **5.** 3.00 or 3

6. 0.02 **7.** $\frac{1}{8}$ **8.** $\frac{1}{400}$ **9.** 60% **10.** 62.5% or $62\frac{1}{2}$%

11. 250% **12.** 800 sacks **13.** 20% **14.** 125,000,000 households

15. $3450.60 **16.** 8% **17.** 25%

18. $\dfrac{\text{amount of increase}}{\text{last year's salary}} = \dfrac{p}{100}$

A possible answer is: Part is the increase in salary. Whole is last year's salary. Percent of increase is unknown.

19. $I = p \cdot r \cdot t$

Some possibilities are

$$I = (1000)(0.05)\left(\frac{9}{12}\right) = \$37.50$$

$$I = (1000)(0.05)(2.5) = \$125$$

The interest formula is $I = p \cdot r \cdot t$. If time is in months, it is expressed as a fraction with 12 as the denominator. If time is expressed in years, it is placed over 1 or shown as a decimal number. **20.** $11.52; $84.48
21. $91; $189 **22.** $378 **23.** $192 **24.** $5657.75
25. **(a)** $10,668 (rounded) **(b)** $1668

Cumulative Review Exercises: Chapters 1–6 (pages 481–482)

1. *Estimate:* $80,000 - 50,000 = 30,000$; *Exact:* 28,988

2. *Estimate:* $8 - 4 = 4$; *Exact:* 4.2971
3. *Estimate:* $7000 \times 700 = 4,900,000$; *Exact:* 4,628,904
4. *Estimate:* $70 \times 9 = 630$; *Exact:* 568.11
5. *Estimate:* $40,000 \div 40 = 1000$; *Exact:* 902
6. *Estimate:* $7 \div 1 = 7$; *Exact:* 8.45
7. 18 **8.** 19 **9.** 39 **10.** 7,600,000 **11.** $513 **12.** $362.74
13. $13\frac{3}{8}$ **14.** $3\frac{7}{8}$ **15.** $28\frac{4}{5}$ **16.** 16 **17.** $12\frac{1}{2}$ ft² **18.** $22\frac{1}{2}$ hours
19. < **20.** < **21.** > **22.** $\frac{1}{4}$ **23.** 1 **24.** $\frac{1}{4}$ **25.** 0.75 **26.** 0.375

27. 0.583 (rounded) **28.** 0.55 **29.** $x = 6$ **30.** $x = 4$ **31.** $x = 16$
32. $x = 30$ **33.** 0.03 **34.** 2.00 or 2 **35.** 87% **36.** 380% **37.** $\frac{2}{25}$
38. $\frac{5}{8}$ **39.** $1\frac{3}{4}$ **40.** 87.5% or $87\frac{1}{2}$% **41.** 420% **42.** 2132 DVDs

43. 1700 miles **44.** 140% **45.** 6.5% or $6\frac{1}{2}$% **46.** $1003.04
47. $205.20; $250.80 **48.** $23,028.75 **49.** 33 hearing tests
50. 255 ounces **51.** 12.5% **52.** 2% **53.** 5.5% (rounded)
54. 8.1% (rounded)

Chapter 7

Section 7.1 (pages 491–494)

1. 3; 12 **3.** 8; 2 **5.** 5280; 3 **7.** 2000; 16 **9.** 60; 60 **11.** **(a)** 2;
(b) 240 **13.** **(a)** $\frac{1}{2}$ or 0.5 **(b)** 78 **15.** 14,000 to 16,000 lb

17. 27 **19.** 112 **21.** 10 **23.** $1\frac{1}{2}$ or 1.5 **25.** $\frac{1}{4}$ or 0.25

27. $1\frac{1}{2}$ or 1.5 **29.** $2\frac{1}{2}$ or 2.5 **31.** snowmobile/ATV; person
walking **33.** 5000 **35.** 17 **37.** 4 to 8 in. **39.** 216 **41.** 28
43. 518,400 **45.** 48,000 **47.** **(a)** pound/ounces **(b)** quarts/pints or
pints/cups **(c)** minutes/hours or seconds/minutes **(d)** feet/inches
(e) pounds/tons **(f)** days/weeks **49.** 174,240 **51.** 800

53. 0.75 or $\frac{3}{4}$ **55.** $1.83 (rounded) **57.** $140 **59.** **(a)** 1056 sec;

(b) 17.6 min **61.** **(a)** $12\frac{1}{2}$ qt **(b)** 4 containers, because you can't buy
part of a container **63.** **(a)** 0.8 mi (rounded) **(b)** 13,031 ft **(c)** 2.5 mi
(rounded) **64.** **(a)** 183 yd (rounded) **(b)** 6600 in. **(c)** 0.1 mi (rounded)

(d) 9 ft (rounded) **65.** **(a)** $1\frac{1}{2}$ to $2\frac{1}{2}$ ft, or 1.5 to 2.5 ft **(b)** 8400 months

(c) 140 more years, for a total of 840 years in all **66.** **(a)** 283 mi (rounded)
(b) 377 mi (rounded)

Section 7.2 (pages 501–502)

1. 1000; 1000 **3.** $\frac{1}{1000}$ or 0.001; $\frac{1}{1000}$ or 0.001 **5.** $\frac{1}{100}$ or 0.01;

$\frac{1}{100}$ or 0.01 **7.** Answers will vary; about 8 to 10 cm. **9.** Answers will
vary; about 20 to 25 mm. **11.** cm **13.** m **15.** km **17.** mm **19.** cm
21. m **23.** Some possible answers are: track and field events, metric
auto parts, and lead refills for mechanical pencils. **25.** 700 cm
27. 0.040 m or 0.04 m **29.** 9400 m **31.** 5.09 m **33.** 40 cm
35. 910 mm **37.** less; 18 cm or 0.18 m

39. ⌒ 5 mm = 0.5 cm
▭ ← 1 mm = 0.1 cm

41. 0.018 km **43.** 164 cm; 1640 mm **45.** 0.0000056 km

Section 7.3 (pages 509–512)

1. mL **3.** L **5.** kg **7.** g **9.** mL **11.** mg **13.** L **15.** kg
17. unreasonable; too much **19.** unreasonable; too much **21.** reasonable
23. reasonable **25.** Some capacity examples are 2 L bottles of soda and
shampoo bottles marked in mL; weight examples are grams of fat listed
on food packages and vitamin doses in milligrams. **27.** Unit for your
answer (g) is in numerator; unit being changed (kg) is in denominator so
it will divide out. The unit fraction is $\dfrac{1000 \text{ g}}{1 \text{ kg}}$. **29.** 15,000 mL **31.** 3 L
33. 0.925 L **35.** 0.008 L **37.** 4150 mL **39.** 8 kg **41.** 5200 g
43. 850 mg **45.** 30 g **47.** 0.598 g **49.** 0.06 L **51.** 0.003 kg
53. 990 mL **55.** mm **57.** mL **59.** cm **61.** mg **63.** 0.3 L
65. 1340 g **67.** 0.9 L **69.** 3 kg to 4 kg **71.** greater; 5 mg or 0.005 g
73. 200 nickels **75.** **(a)** 1,000,000

(b) $\dfrac{3.5 \text{ Mm}}{1} \cdot \dfrac{1,000,000 \text{ m}}{1 \text{ Mm}} = 3,500,000 \text{ m}$

76. **(a)** 1,000,000,000

(b) $\dfrac{2500 \text{ m}}{1} \cdot \dfrac{1 \text{ Gm}}{1,000,000,000 \text{ m}} = 0.0000025 \text{ Gm}$

77. **(a)** 1,000,000,000,000 **(b)** 1000; 1,000,000
78. 1,000,000; 1,000,000,000; $2^{20} = 1,048,576$; $2^{30} = 1,073,741,824$

Summary Exercises on U.S. Customary and Metric Units (pages 513–514)

1. Length: inch, foot, yard, mile; Weight: ounce, pound, ton; Capacity:
fluid ounce, cup, pint, quart, gallon. **2.** Length: millimeter, centimeter,
meter, kilometer; Weight: milligram, gram, kilogram; Capacity: milliliter,
liter **3.** **(a)** ft **(b)** yd **(c)** 5280 **4.** **(a)** min **(b)** 60 **(c)** 24
5. **(a)** 8 **(b)** gal **(c)** 2 **6.** **(a)** lb **(b)** 2000 **(c)** 16 **7.** mL **8.** m
9. kg **10.** mg **11.** cm **12.** mm **13.** L **14.** g **15.** 0.45 m

16. 45 sec **17.** 600 mL **18.** 8000 mg **19.** 30 cm **20.** $3\frac{3}{4}$ or 3.75 ft

21. 0.050 or 0.05 L **22.** $4\frac{1}{2}$ or 4.5 gal **23.** 7280 g **24.** 36 oz

25. 0.009 kg **26.** 180 in. **27.** 272 cm; 2720 mm **28.** 2.64 m;
0.00264 km **29.** 0.04 m; 4 cm; 40 mm **30.** 14 cm; 0.14 m; 140 mm
31. $4.65 (rounded) **32.** 10.3 ft (rounded)

Section 7.4 (pages 517–518)

1. $0.83 (rounded)   **3.** 89.5 kg   **5.** 71 beats (rounded)   **7.** 180 cm;
$0.02/cm (rounded)   **9.** 5.03 m   **11.** 4 bottles   **13.** 1.89 g
15. (a) 10 km   **(b)** 600 km   **(c)** 36,000 km   **17.** 215 g; 4.3 g; 4300 mg
18. 330 g; 0.33 g; 330 mg   **19.** 1550 g; 1500 g; 3 g   **20.** 55 g; 50 g; 0.5 g

Section 7.5 (pages 523–526)

1. 21.8 yd   **3.** 262.4 ft   **5.** 4.8 m   **7.** 5.3 oz   **9.** 111.6 kg   **11.** 30.3 qt
13. (a) about 0.2 oz   **(b)** probably not   **15.** about 31.8 L   **17.** about
1.3 cm   **19.** $-8\,°C$ **21.** 40 °C   **23.** 150 °C   **25.** More. There are
180 degrees between freezing and boiling on the Fahrenheit scale, but
only 100 degrees on the Celsius scale, so each Celsius degree is a greater
change in temperature.   **27.** 16° C (rounded)   **29.** 40° C   **31.** 46° C
(rounded)   **33.** 95° F   **35.** 58° C (rounded)   **37. (a)** pleasant weather,
above freezing but not hot   **(b)** 75° F to 39° F (rounded)   **(c)** Answers
will vary. In Minnesota, it's 0 °C to -40 °C; in California, 24 °C to 0 °C.
39. about 1.8 m   **40.** about 5.0 kg   **41.** about 3040.2 km
42. 38 hr 23 min   **43.** about 300 m   **44.** less than 3790 mL
45. about 59.4 mL   **46.** about 1.6%

Chapter 7 Review Exercises (pages 533–536)

1. 16   **2.** 3   **3.** 2000   **4.** 4   **5.** 60   **6.** 8   **7.** 60   **8.** 5280
9. 12   **10.** 48   **11.** 3   **12.** 4   **13.** $\frac{3}{4}$ or 0.75   **14.** $2\frac{1}{2}$ or 2.5
15. 28   **16.** 78   **17.** 112   **18.** 345,600   **19. (a)** $4153\frac{1}{3}$ yd
(exact) or 4153.3 yd (rounded)   **(b)** 2.4 mi (rounded)
20. $2465.20   **21.** mm   **22.** cm   **23.** km   **24.** m   **25.** cm
26. mm   **27.** 500 cm   **28.** 8500 m   **29.** 8.5 cm   **30.** 3.7 m
31. 0.07 km   **32.** 930 mm   **33.** mL   **34.** L   **35.** g   **36.** kg
37. L   **38.** mL   **39.** mg   **40.** g   **41.** 5 L   **42.** 8000 mL
43. 4580 mg   **44.** 700 g   **45.** 0.006 g   **46.** 0.035 L   **47.** 31.5 L
48. 357 g (rounded)   **49.** 87.25 kg   **50.** $1.42 (rounded)
51. 6.5 yd (rounded)   **52.** 11.7 in. (rounded)
53. 67.0 mi (rounded)   **54.** 1288 km   **55.** 21.9 L (rounded)
56. 44.0 qt (rounded)   **57.** 0 °C   **58.** 100 °C   **59.** 37 °C
60. 20 °C   **61.** 25 °C   **62.** 33 °C (rounded)   **63.** 43 °F (rounded)
64. 120 °F (rounded)   **65.** L   **66.** kg   **67.** cm   **68.** mL
69. mm   **70.** km   **71.** g   **72.** m   **73.** mL   **74.** g   **75.** mg
76. L   **77.** 105 mm   **78.** $\frac{3}{4}$ hr or 0.75 hr   **79.** $7\frac{1}{2}$ ft or 7.5 ft
80. 130 cm   **81.** 77 °F
82. 14 qt   **83.** 0.7 g   **84.** 810 mL
85. 80 oz   **86.** 60,000 g   **87.** 1800 mL   **88.** 30 °C   **89.** 36 cm
90. 0.055 L   **91.** 1.26 m   **92.** 18 tons   **93.** 113 g; 177 °C
(both rounded)   **94.** 178.0 lb; 6.0 ft (both rounded)   **95.** 13.2 m
96. 22.0 ft   **97.** 3.6 m   **98.** 39.8 in.   **99.** 484,000 lb
100. 19.0 to 22.7 L   **101. (a)** 242 tons   **(b)** 144 in.
102. (a) 1.02 m   **(b)** 670 cm

Chapter 7 Test (pages 537–538)

1. 36 qt   **2.** 15 yd   **3.** 2.25 hr or $2\frac{1}{4}$ hr   **4.** 0.75 ft or $\frac{3}{4}$ ft
5. 56 oz   **6.** 7200 min   **7.** kg
8. km   **9.** mL   **10.** g   **11.** cm
12. mm   **13.** L   **14.** cm   **15.** 2.5 m   **16.** 4600 m   **17.** 0.5 cm
18. 0.325 g   **19.** 16,000 mL   **20.** 400 g   **21.** 1055 cm
22. 0.095 L   **23.** 3.2 ft (rounded)   **24. (a)** 2310 mg
(b) 500 mg or 0.5 g more   **25.** 95 °C   **26.** 0 °C
27. 1.8 m   **28.** 56.3 kg (rounded)   **29.** 13 gal
30. 5.0 mi (rounded)   **31.** 23 °C (rounded)   **32.** 36 °F (rounded)
33. 6 m $\approx$ 6.54 yd; $26.03 (rounded)   **34.** Possible answers: Use
same system as rest of the world; easier system for children to
learn; less use of fractional numbers; compete internationally.

Cumulative Review Exercises: Chapters 1–7 (pages 539–540)

1. *Estimate:* $60 \times 5 = 300$; *Exact:* 265.644
2. *Estimate:* $20,000 \div 40 = 500$; *Exact:* 530
3. *Estimate:* $2 + 4 = 6$; *Exact:* $5\frac{1}{2}$
4. $\frac{11}{16}$ **5.** $26\frac{2}{3}$ **6.** 13.0 (rounded)   **7.** $1\frac{5}{24}$ **8.** 7.0793   **9.** $\frac{5}{8}$
10. 0.00762   **11.** 8   **12.** 265   **13.** 0.067, 0.6, 0.6007, 0.67
14. 16 diapers for $3.87, about $0.242 per diaper   **15.** $x = 12$
16. $x \approx 1.67$ (rounded)   **17.** 5%   **18.** 30 hours   **19.** 30 in.
20. $1\frac{3}{4}$ or 1.75 min   **21.** 280 cm   **22.** 122.76 mi   **23.** g   **24.** mL
25. m   **26.** kg   **27.** $18.99 (rounded); $170.95   **28.** $6.15 total;
$0.14 per print (both rounded)   **29.** $12\frac{3}{4}$ or 12.75 lb   **30. (a)** 93.6 km
(b) 58.0 mi (rounded)   **31.** $2756.25; $11,506.25   **32.** 88.6% (rounded)
33. $9.74 (rounded)   **34.** $\frac{1}{6}$ cup more than the amount needed
35. 120 rows   **36.** 2100 students   **37.** $0.15   **38.** 37.5% decrease

Chapter 8

Section 8.1 (pages 547–548)

1. line named $\overleftrightarrow{CD}$ or $\overleftrightarrow{DC}$ **3.** line segment named $\overline{GF}$ or $\overline{FG}$
5. ray named $\overrightarrow{PQ}$ **7.** perpendicular   **9.** parallel   **11.** intersecting
13. $\angle AOS$ or $\angle SOA$ **15.** $\angle CRT$ or $\angle TRC$ **17.** $\angle AQC$ or $\angle CQA$
19. right (90°)   **21.** acute   **23.** straight (180°)   **25. (a)** The car
turned around in a complete circle.   **(b)** The governor took the opposite
view, for example, having once opposed taxes but now supporting them.
27. True, because $\overrightarrow{UQ}$ is perpendicular to $\overleftrightarrow{ST}$.   **28.** True, because they
form a 90° angle, as indicated by the small red square.   **29.** False; the
angles have the same measure (both are 180°).   **30.** False; $\overleftrightarrow{ST}$ and $\overleftrightarrow{PR}$
are parallel.   **31.** False; $\overleftrightarrow{QU}$ and $\overleftrightarrow{TS}$ are perpendicular.   **32.** True; both
angles are formed by perpendicular lines, so they both measure 90°.

Section 8.2 (pages 553–554)

1. $\angle EOD$ and $\angle COD$; $\angle AOB$ and $\angle BOC$ **3.** $\angle HNE$ and $\angle ENF$;
$\angle ACB$ and $\angle KOL$ **5.** 50°   **7.** 4°   **9.** 50°   **11.** 90°
13. $\angle SON \cong \angle TOM$; $\angle TOS \cong \angle MON$ **15.** $\angle GOH$ measures 63°;
$\angle EOF$ measures 37°; $\angle AOC$ and $\angle GOF$ both measure 80°.
17. Two angles are complementary if the sum of their measures is 90°.
Two angles are supplementary if the sum of their measures is 180°.
Drawings will vary; examples are:

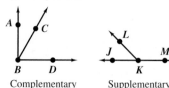

Complementary   Supplementary

19. $\angle ABF \cong \angle ECD$; Both are 138°. $\angle ABC \cong \angle BCD$; Both are 42°.
21. No; obtuse angles are $>90°$, so their sum would be $>180°$.

Section 8.3 (pages 561–564)

1. $P = 28$ yd; $A = 48$ yd^2 **3.** $P = 3.6$ km; $A = 0.81$ km^2
5. $P = 40$ ft; $A = 100$ ft^2 **7.** $P = 29$ m; $A = 7$ m^2 **9.** $P = 196.2$ ft;
$A = 1674.2$ ft^2 **11.** $P = 12$ mi; $A = 9$ mi^2 **13.** $P = 38$ m; $A = 39$ m^2
15. $P = 98$ m; $A = 492$ m^2 **17.** unlabeled side $= 12$ in.; $P = 78$ in.;
$A = 234$ in.^{2} **19.** $P = 48$ m; $A = 144$ m^2 **21.** $460   **23.** $94.81

25. Area of largest field is 13,000 yd²; Area of smallest field is 5000 yd²; difference is 13,000 yd² − 5,000 yd² = 8000 yd². **27.** 53 yd
29. A = 6528 ft²

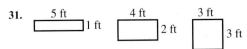

31.

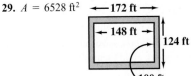

32. (a) 5 ft by 1 ft has area of 5 ft²; 4 ft by 2 ft has area of 8 ft²; 3 ft by 3 ft has area of 9 ft² (b) The square plot 3 ft by 3 ft has greatest area.
33.

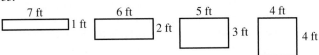

34. (a) 7 ft²; 12 ft²; 15 ft²; 16 ft² (b) Square plots have the greatest area.
35. (a) P = 10 ft; A = 6 ft²

(b) P = 20 ft; A = 24 ft²

(c) Perimeter is twice the original; area is four times the original.
36. (a) P = 30 ft; A = 54 ft²

(b) Perimeter is three times the original; area is nine times the original.
(c) Perimeter will be four times the original; area will be 16 times the original.

Section 8.4 (pages 569–570)

1. P = 208 m **3.** P = 207.2 m **5.** P = 6.06 km **7.** A = 775 mm²
9. A = 19.25 ft² **11.** A = 3099.6 cm²
13. $1410.75

15. A = 437.5 in.²

17. Height is not part of perimeter; square units are used for area, not perimeter. P = 2.5 cm + 2.5 cm + 2.5 cm + 2.5 cm = 10 cm
19. A = 3.02 m² **21.** A = 25,344 ft²

Section 8.5 (pages 575–577)

1. P = 202 m; A = 1914 m² **3.** P = 58.9 cm; A = 139.15 cm²

5. P = $26\frac{1}{4}$ yd or 26.25 yd; A = $30\frac{3}{4}$ yd² or 30.75 yd²

7. P = 85.2 cm; A = 302.46 cm² **9.** A = 198 m² **11.** A = 1664 m²
13. 32° **15.** 48° **17.** No. Right angles are 90°, so two right angles are 180°, and the sum of all *three* angles in a triangle equals 180°.

19. $7\frac{7}{8}$ ft² or 7.875 ft² **21.** (a) 126.8 m of curb (b) 672 m² of sod
23. (a) A = 32 m² (b) A = 13.41 m² **25.** (a) 175 yd of frontage
(b) A = 8750 yd²

Section 8.6 (pages 585–588)

1. d = 18 mm **3.** r = 0.35 km **5.** C ≈ 69.1 ft; A ≈ 379.9 ft²
7. C ≈ 8.2 m; A ≈ 5.3 m² **9.** C ≈ 47.1 cm; A ≈ 176.6 cm²
11. C ≈ 23.6 ft; A ≈ 44.2 ft² **13.** C ≈ 27.2 km; A ≈ 58.7 km²
15. A ≈ 76.9 in.² **17.** A ≈ 57 cm²
19. π is the ratio of circumference of a circle to its diameter. If you divide the circumference of any circle by its diameter, the answer is always a little more than 3. The approximate value is 3.14, which we call π (pi). Your test question could involve finding the circumference or the area of a circle. **21.** A ≈ 7850 yd² **23.** C ≈ 91.4 in.;
Bonus: 693 revolutions/mile (rounded)
25. A ≈ 70,650 mi²

27. Watch C ≈ 3.1 in. Wall Clock C ≈ 18.8 in.
A ≈ 0.8 in.² A ≈ 28.3 in.²

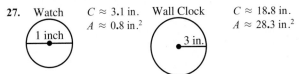

29. A ≈ 78.5 mi² − 12.6 mi²
Difference ≈ 65.9 mi²

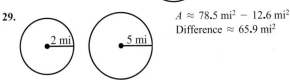

31. (a) d ≈ 45.9 cm (b) Divide the circumference by π (3.14).
33. $325.83 (rounded). **35.** The prefix *rad* tells you that radius is a ray from the center of the circle. The prefix *dia* means the diameter goes through the circle, and the prefix *circum* means the circumference is the distance around. **37.** A ≈ 44.2 in.² **38.** A ≈ 132.7 in.²
39. A ≈ 201.0 in.² **40.** small: $0.063 (rounded); medium: $0.049 (rounded); large: $0.046 (rounded) Best Buy **41.** small: $0.084 (rounded); medium: $0.067 (rounded) Best Buy; large: $0.071 (rounded)
42. small: $0.077 (rounded) Best Buy; medium: $0.083 (rounded); large: $0.078 (rounded)

Summary Exercises on Perimeter, Circumference, and Area (pages 589–590)

1. (a) All sides have the same length. (b) Opposite sides have the same length.
(c) Opposite sides have the same length.

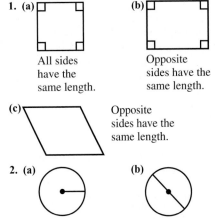

2. (a) (b)

(c) The diameter is twice the radius, or the radius is half the diameter.
3. Add up the lengths of all the sides to find the perimeter. **4.** Perimeter is the total distance around the outside edges of a shape. Area is the number of square units needed to cover the space inside the shape. **5.** f, e, d, b, c, a **6.** (a) C = 2 · π · r (b) C = π · d (c) Divide the diameter by 2 to find the radius. **7.** P = 63 ft; A = 180 ft² P = 32.4 cm;
A ≈ 45.1 cm² (rounded) **8.** (a) A = 12 cm² (b) P = $6\frac{1}{2}$ ft

(c) C ≈ 28.5 m (d) A = 307 in.²

9. Rectangle; $P = 27$ in.; $A = 31\frac{1}{2}$ in.² or 31.5 in.² **10.** Square;
$P = 48$ yd; $A = 144$ yd² **11.** Parallelogram; $P = 6.4$ m;
$A \approx 2.2$ m² (rounded) **12.** Trapezoid; $P = 33.5$ mm; $A = 64$ mm²
13. $d = 12$ cm; $C \approx 37.7$ cm (rounded); $A \approx 113.0$ cm² (rounded)
14. $r = 15$ mi; $C \approx 94.2$ mi; $A \approx 706.5$ mi²
15. $r = 4.5$ ft; $C \approx 28.3$ ft; (rounded) $A \approx 63.6$ ft² (rounded)
16. $A \approx 4.5$ m² (rounded) **17.** $A = 217.5$ yd² **18.** $A \approx 86$ in.²
19. $C \approx 14.6$ ft; Bonus: About 361.6 revolutions; the Mormons counted
360 revolutions, which is the answer you get using $2\frac{1}{3}$ ft instead of 2.33 ft
as the radius. **20.** \$4.47 (rounded)

Section 8.7 (pages 597–598)

1. Rectangular solid; $V = 550$ cm³ **3.** Sphere; $V \approx 44,579.6$ m³
5. Hemisphere; $V \approx 3617.3$ in.³ **7.** Cylinder; $V \approx 471$ ft³ **9.** Cone;
$V \approx 418.7$ m³ **11.** Pyramid; $V = 800$ cm³ **13.** $V = 18$ in.³
15. $V \approx 2481.5$ cm³ **17.** $V \approx 651,775$ m³ **19.** $V \approx 3925$ ft³
21. Student used diameter of 7 cm; should use radius of 3.5 cm in formula.
Units for volume are cm³, not cm² Correct answer is $V \approx 192.3$ cm³.
23. $V \approx 513$ cm³

Section 8.8 (pages 603–606)

1. 4 **3.** 8 **5.** 3.317 **7.** 2.236 **9.** 8.544 **11.** 10.050 **13.** 13.784
15. 31.623 **17.** 30 is about halfway between 25 and 36, so $\sqrt{30}$ should
be about halfway between 5 and 6, or about 5.5. Using a calculator,
$\sqrt{30} \approx 5.477$. Similarly, $\sqrt{26}$ should be a little more than $\sqrt{25}$;
by calculator $\sqrt{26} \approx 5.099$. And $\sqrt{35}$ should be a little less than $\sqrt{36}$;
by calculator $\sqrt{35} \approx 5.916$. **19.** $\sqrt{1521} = 39$ ft **21.** $\sqrt{289} = 17$ in.
23. $\sqrt{144} = 12$ mm **25.** $\sqrt{73} \approx 8.5$ in. **27.** $\sqrt{65} \approx 8.1$ yd
29. $\sqrt{195} \approx 14.0$ cm **31.** $\sqrt{7.94} \approx 2.8$ m **33.** $\sqrt{65.01} \approx 8.1$ cm
35. $\sqrt{292.32} \approx 17.1$ km **37.** hypotenuse = $\sqrt{65} \approx 8.1$ ft
39. leg = $\sqrt{360,000} = 600$ m **41.** hypotenuse = $\sqrt{32.5} \approx 5.7$ ft
43. leg = $\sqrt{135} \approx 11.6$ ft

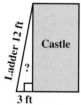

45. The student did not square the numbers correctly: 9^2 is 81 and 7^2 is 49.
Also, the final answer is rounded to thousandths instead of tenths.
Correct answer is $\sqrt{130} \approx 11.4$ in. **47.** $\sqrt{16200} \approx 127.3$ ft
48. (a)

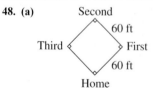

(b) $\sqrt{7200} \approx 84.9$ ft

49. The distance from third to first is the same as the distance from home
to second because the baseball diamond is a square.
50. One possibility is:

$$\text{major league} \left\{ \frac{90 \text{ ft}}{127.3 \text{ ft}} = \frac{60 \text{ ft}}{x} \right\} \text{softball}$$

$x \approx 84.9$ ft

Section 8.9 (pages 611–614)

1. similar **3.** not similar **5.** similar **7.** $\angle 1$ and $\angle 4$; $\angle 2$ and $\angle 5$;
$\angle 3$ and $\angle 6$; $\overline{AB}$ and $\overline{PQ}$; $\overline{BC}$ and $\overline{QR}$; $\overline{AC}$ and $\overline{PR}$ **9.** $\angle 1$ and $\angle 6$;
$\angle 2$ and $\angle 5$; $\angle 3$ and $\angle 4$; $\overline{MP}$ and $\overline{QS}$; $\overline{MN}$ and $\overline{QR}$; $\overline{NP}$ and $\overline{RS}$
11. $\frac{3}{2}; \frac{3}{2}; \frac{3}{2}$ **13.** $a = 6$ cm; $b = 15$ cm **15.** $a = 5$ mm; $b = 3$ mm

17. $x = 24.8$ m; $P = 72.8$ m; $y = 15$ m; $P = 54.6$ m
19. $P = 8$ cm + 8 cm + 8 cm = 24 cm;
$A \approx 0.5 \cdot 8$ cm $\cdot 6.9$ cm ≈ 27.6 cm² **21.** $h = 24$ ft **23.** $\frac{3}{4}$ ft or
0.75 ft **25.** One dictionary definition is "resembling, but not identical."
Examples of similar objects are sets of different-size pots or measuring
cups; small- and large-size cans of beans; child's tennis shoe and adult's
tennis shoe. **27.** $x = 50$ m **29.** $n = 110$ m

Chapter 8 Review Exercises (pages 623–630)

1. line segment named $\overline{AB}$ or $\overline{BA}$ **2.** line named $\overleftrightarrow{CD}$ or $\overleftrightarrow{DC}$
3. ray named $\overrightarrow{OP}$ **4.** parallel **5.** perpendicular **6.** intersecting
7. acute **8.** obtuse **9.** straight; 180° **10.** right; 90°
11. $\angle 1$ and $\angle 3$ measure 30°; $\angle 2$ measures 90°; $\angle 4$ measures 60°
12. $\angle 1$ measures 100°; $\angle 2$ and $\angle 4$ measure 45°; $\angle 3$ measures 35°
13. $\angle AOB$ and $\angle BOC$; $\angle BOC$ and $\angle COD$; $\angle COD$ and $\angle DOA$;
$\angle DOA$ and $\angle AOB$ **14.** $\angle ERH$ and $\angle HRG$; $\angle HRG$ and $\angle GRF$;
$\angle FRG$ and $\angle FRE$; $\angle FRE$ and $\angle ERH$ **15. (a)** 10° **(b)** 45°
(c) 83° **16. (a)** 25° **(b)** 90° **(c)** 147° **17.** $P = 4.84$ m
18. $P = 128$ in. **19.** 152 cm **20.** 41 ft **21.** $A = 486$ mm²
22. $A = 16.5$ ft² or $16\frac{1}{2}$ ft² **23.** $A \approx 39.7$ m² **24.** $P = 50$ cm;
$A = 140$ cm² **25.** $P = 102.1$ ft; $A = 567$ ft² **26.** $P = 200.2$ m;
$A \approx 2206.1$ m² **27.** $P = 518$ cm; $A = 11,660$ cm²
28. $P = 27.1$ m; $A = 23.52$ m² **29.** $P = 20\frac{1}{4}$ ft or 20.25 ft;
$A = 14$ ft² **30.** 70° **31.** 24° **32.** $d = 137.8$ m **33.** $r = 1\frac{1}{2}$ in.
or 1.5 in. **34.** $C \approx 6.3$ cm; $A \approx 3.1$ cm² **35.** $C \approx 109.3$ m;
$A \approx 950.7$ m² **36.** $C \approx 37.7$ in.; $A \approx 113.0$ in.²
37. $A \approx 20.3$ m² **38.** $A = 64$ in.² **39.** $A = 673$ km²
40. $A = 1020$ m² **41.** $A = 229$ ft² **42.** $A = 132$ ft²
43. $A = 5376$ cm² **44.** $A \approx 498.9$ ft² **45.** $A \approx 447.9$ yd²
46. Rectangular solid; $V = 30$ in.³ **47.** Rectangular solid; $V = 96$ cm³
48. Rectangular solid; $V = 45,000$ mm³
49. Sphere; $V \approx 267.9$ m³ **50.** Hemisphere; $V \approx 452.2$ ft³
51. Cylinder; $V \approx 549.5$ cm³ **52.** Cylinder; $V \approx 1808.6$ m³
53. Cone; $V \approx 512.9$ m³ **54.** Pyramid; $V = 16$ yd³ **55.** 7
56. 2.828 (rounded) **57.** 54.772 (rounded) **58.** 12
59. 7.616 (rounded) **60.** 25 **61.** 10.247 (rounded)
62. 8.944 (rounded) **63.** $\sqrt{289} = 17$ in. **64.** $\sqrt{49} = 7$ cm
65. $\sqrt{104} \approx 10.2$ cm **66.** $\sqrt{52} \approx 7.2$ in. **67.** $\sqrt{6.53} \approx 2.6$ m
68. $\sqrt{71.75} \approx 8.5$ km **69.** $y = 30$ ft; $x = 34$ ft; $P = 104$ ft
70. $y = 7.5$ m; $x = 9$ m; $P = 22.5$ m **71.** $x = 12$ mm;
$y = 7.5$ mm; $P = 38$ mm **72.** Square; $P = 18$ in.; $A \approx 20.3$ in.²
or $A = 20\frac{1}{4}$ in.² **73.** Trapezoid; $P = 10.3$ cm; $A \approx 6.2$ cm²
74. Circle; $C \approx 40.8$ m; $A \approx 132.7$ m² **75.** Parallelogram;
$P = 54$ ft; $A = 140$ ft² **76.** Triangle; $P = 20$ yd; $A = 18\frac{3}{4}$ yd² or
18.8 yd² (rounded) **77.** Rectangle; $P = 7$ km; $A \approx 2.0$ km²
78. Circle; $C \approx 53.4$ m; $A \approx 226.9$ m² **79.** Parallelogram;
$P = 78$ mm; $A = 288$ mm² **80.** Triangle; $P = 37.8$ mi;
$A = 58.5$ mi² **81.** parallel lines **82.** line segment
83. acute angle **84.** intersecting lines **85.** right angle; 90°
86. ray **87.** straight angle; 180° **88.** obtuse angle
89. perpendicular lines **90.** 81° **91.** 138° **92.** $P = 90$ m;
$A = 92$ m² **93.** $P = 282$ cm; $A = 4190$ cm² **94.** $V \approx 100.5$ ft³
95. $V \approx 3.4$ in.³ or $V = 3\frac{3}{8}$ in.³ **96.** $V \approx 7.4$ m³
97. $V = 561$ cm³ **98.** $V \approx 1271.7$ cm³ **99.** $V \approx 1436.0$ m³
100. $x \approx 12.6$ km **101.** $\angle D = 72°$ **102.** $x = 12$ mm;
$y = 14$ mm **103.** The prefix *dec* in *dec*ade means 10 and the
prefix *cent* in *cent*ury means 100, so divide 200 (two centuries)
by 10. The answer is 20 decades.

Chapter 8 Test (pages 631–632)

1. (e) **2.** (a); 90° **3.** (d) **4.** (g); 180° **5.** Parallel lines are lines in the same plane that never intersect. Perpendicular lines intersect to form a right angle.

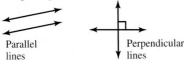

Parallel lines

Perpendicular lines

6. 9° **7.** 160° **8.** $\angle 1$ measures 50°; $\angle 3$ measures 90°; $\angle 2$ and $\angle 4$ measure 40° **9.** Rectangle; $P = 23$ ft; $A = 30$ ft^2
10. Square; $P = 72$ mm; $A = 324$ mm^2 **11.** Parallelogram;
$P = 26.2$ m; $A = 33.12$ m^2 **12.** Trapezoid; $P = 169.4$ cm;
$A = 1591$ cm^2 **13.** $P = 32.45$ m; $A = 48$ m^2 **14.** $P = 37.8$ yd or
$37\frac{4}{5}$ yd; $A = 58.5$ yd^2 **15.** 55° **16.** $r = 12.5$ in. or $12\frac{1}{2}$ in.
17. $C \approx 5.7$ km (rounded) **18.** $A \approx 206.0$ cm^2 (rounded)
19. $A \approx 39.3$ m^2 (rounded) **20.** Rectangular solid; $V = 6480$ m^3
21. Sphere; $V \approx 33.5$ ft^3 (rounded) **22.** Cylinder; $V \approx 5086.8$ ft^3
23. $\sqrt{85} \approx 9.2$ cm (rounded) **24.** $y = 12$ cm; $z = 6$ cm
25. Linear units like cm are used to measure perimeter, radius, diameter, and circumference. Area is measured in square units like cm^2 (squares that measure 1 cm on each side). Volume is measured in cubic units like cm^3.

Cumulative Review Exercises: Chapters 1–8 (pages 633–634)

1. $\frac{9}{20}$ **2.** 0.9132 **3.** 0.000078 **4.** 19.44 (rounded) **5.** $4\frac{5}{12}$ **6.** 9
7. two hundred eight ten-thousandths **8.** 2.055; 2.5005; 2.505; 2.55
9. $x = 35$ **10.** $x \approx 2.01$ (rounded) **11.** 160% **12.** $600 **13.** 135 min
14. $2\frac{1}{2}$ or 2.5 lb **15.** 0.08 m **16.** 1800 mL **17.** mm **18.** L
19. kg **20.** cm **21.** Rectangle; $P = 11$ in.; $A = 7$ in.^{2} **22.** Triangle;
$P = 5.8$ m; $A = 1.575$ m^2 **23.** Circle; $C \approx 31.4$ ft; $A \approx 78.5$ ft^2
24. Parallelogram; $P = 76$ cm; $A = 264$ cm^2 **25.** $y \approx 24.2$ mm
26. 59% (rounded) **27.** $V \approx 2255.3$ cm^3 **28.** Brand T at 15.5 oz for
$2.99 − $0.30 coupon, about $0.174/oz. **29.** $1\frac{1}{12}$ yd **30.** 35.7 lb
meat (rounded); 50 lb potatoes; 21.4 lb carrots
(rounded) **31.** 3.4 m **32.** $22.40

Chapter 9

Section 9.1 (pages 641–642)

1. +32 or 32 degrees **3.** −$12 **5.** $-6\frac{1}{2}$ lb **7.** +20,320 ft or 20,320 ft
9.
$$-5\ -4\ -3\ -2\ -1\ 0\ 1\ 2\ 3\ 4\ 5$$
11.
$$-5\ -4\ -3\ -2\ -1\ 0\ 1\ 2\ 3\ 4\ 5$$
13.
$$-5\ -4\ -3\ -2\ -1\ 0\ 1\ 2\ 3\ 4\ 5$$
15. < **17.** > **19.** < **21.** > **23.** < **25.** >
27.   A   C D B
$$-2\quad -1\quad 0\quad 1$$
28. $-1.2, -0.5, 0, 0.6$ **29.** A: may be at risk; B: above normal;
C: normal; D: normal **30. (a)** The patient would think the
interpretation was "above normal" and wouldn't get treatment.
(b) Patient D's score of 0; because 0 is neither positive nor negative.
31. 3 **33.** 10 **35.** 0 **37.** −18 **39.** −32 **41.** −7 **43.** 14

45. $-\frac{2}{3}$ **47.** 8.3 **49.** $\frac{1}{6}$ **51.** true **53.** true **55.** false

Section 9.2 (pages 651–654)

1. 3
3. −7
5. −1

7. −3 **9.** 7 **11.** −7 **13.** 1 **15.** −8 **17.** −20
19. $13 + (-17) = -4$ yd **21.** $-$52.50 + $50 = -$2.50
23. Jeff: $-20 + 75 + (-55) = 0$ pts; Terry: $42 + (-15) + 20 = 47$ pts
25. −6.8 **27.** $\frac{1}{4}$ **29.** $-\frac{3}{10}$ **31.** $-\frac{26}{9}$ **33.** −3 **35.** 9 **37.** $-\frac{1}{2}$
39. 6.2 **41.** 14 **43.** −2 **45.** −12 **47.** −25 **49.** −23 **51.** 5
53. 20 **55.** 11 **57.** −60 **59.** 0 **61.** $-\frac{3}{2}$ **63.** $-\frac{2}{5}$ **65.** 0.7
67. (a) Change 6 to its opposite, −6. Correct answer is $-6 + (-6) = -12$.
(b) Do *not* change 9 to its opposite. Correct answer is $-9 + (-5) = -14$.
69. (a) 21 °F; $30 - 21 = 9$ degrees difference **(b)** 0 °F; $15 - 0 = 15$
degrees difference **71. (a)** −17 °F; $5 - (-17) = 22$ degrees
difference **(b)** −41 °F; $-10 - (-41) = 31$ degrees difference
73. −10 **75.** 12 **77.** −7 **79.** 5 **81.** −1 **83.** −4.4 **85.** −11
87. −10 **89.** $196 **91.** $0 **93. (a)** −2; −2 **(b)** −8; −8
(c) 10; 10 Addition is commutative, so changing the order of the addends
does *not* change the sum. **94. (a)** −8; 8 **(b)** 2; −2 **(c)** −7; 7
Changing the order of the numbers in subtraction *does change* the result.
95. (a) The answers have the same absolute value but opposite signs.
(b) When the order of the numbers in a subtraction problem is switched,
change the sign on the answer to its opposite.
96. (a) −18; 20; −5; 4. Adding 0 to any number leaves the number
unchanged. **(b)** 2; −10; −3; 7. Subtracting 0 from a number leaves
the number unchanged, but subtracting a *number from 0* changes the
sign of the number.

Section 9.3 (pages 657–660)

1. −35 **3.** −45 **5.** −18 **7.** −50 **9.** −40 **11.** 32 **13.** 77
15. 133 **17.** 13 **19.** 0 **21.** 4 **23.** −4 **25.** $-\frac{1}{10}$ **27.** $\frac{14}{3}$
29. $-\frac{5}{6}$ **31.** $\frac{7}{4}$ **33.** −42.3 **35.** 6 **37.** −31.62 **39.** 4.5 **41.** 8.23
43. 0 **45.** −2 **47.** −5 **49.** undefined **51.** −14 **53.** 10 **55.** 4
57. −1 **59.** 191 **61.** 0 **63.** 1 **65.** $\frac{2}{3}$ **67.** $\frac{1}{3}$ **69.** −8 **71.** $-\frac{14}{3}$
73. 4.73 **75.** −4.7 **77.** −5.3 **79.** 12 **81.** −0.36 **83.** $-\frac{7}{40}$
85. −2 **87.** −10 **89.** −$119,400 **91.** −2011 ft **93.** $2366
95. −22 degrees **97. (a)** Examples will vary. Some possibilities:
$6(-1) = -6; 2(-1) = -2; 15(-1) = -15$ **(b)** Examples will vary.
Some possibilities: $-6(-1) = 6; -2(-1) = 2; -15(-1) = 15$.
The result of multiplying any nonzero number times −1 is the number
with the opposite sign. **98. (a)** Examples will vary.
Some possibilities: $\frac{-6}{-1} = 6; \frac{-2}{-1} = 2; \frac{-15}{-1} = 15$

(b) Some possibilities: $\dfrac{6}{-1} = -6; \dfrac{2}{-1} = -2; \dfrac{15}{-1} = -15$

(c) Some possibilities: $\dfrac{-6}{-6} = 1; \dfrac{-2}{-2} = 1; \dfrac{-15}{-15} = 1$

When dividing by -1, the sign of the number changes to its opposite. A negative number divided by itself gives a result of 1.

Section 9.4 (pages 665–668)

1. -6 **3.** 0 **5.** 52 **7.** -39 **9.** -30 **11.** 17 **13.** 16 **15.** 30 **17.** 23 **19.** -43 **21.** 19 **23.** -3 **25.** 0 **27.** 47 **29.** 41 **31.** -2 **33.** 13 **35.** 126 **37.** $\dfrac{27}{-3} = -9$ **39.** $\dfrac{-48}{-4} = 12$ **41.** $\dfrac{-60}{-1} = 60$ **43.** 8 **45.** 1.24 **47.** -1.47 **49.** -2.31 **51.** $-\dfrac{13}{10}$ **53.** $\dfrac{5}{4}$ **55.** $\dfrac{4}{5}$ **57.** -1200 **59.** 1 **61.** $4, -8, 16, -32, 64, -128, 256, -512$.

(a) When a negative number is raised to an even power, the answer is positive; when raised to an odd power, the answer is negative.

(b) positive; negative; positive **63.** $\dfrac{27}{-27} = -1$ **65.** 7 **67.** 6

Summary Exercises on Operations with Signed Numbers (pages 669–670)

1. -6 **2.** 0 **3.** -7 **4.** -7 **5.** 63 **6.** -1 **7.** -56 **8.** -22 **9.** 12 **10.** 6 **11.** -13 **12.** 0 **13.** -80 **14.** -17 **15.** -48 **16.** -1 **17.** $-\dfrac{1}{6}$ **18.** undefined **19.** -14.6 **20.** -0.6 **21.** 48 **22.** -19 **23.** 2 **24.** -20 **25.** 0 **26.** 16 **27.** -3 **28.** -36 **29.** 6 **30.** -2 **31.** -32 **32.** -5 **33.** -4 **34.** -9732 **35.** 100 **36.** 4 **37.** -343 **38.** 5 **39.** -6 **40.** -10 **41.** -5 **42.** 8 **43.** 1 **44.** $\dfrac{27}{0}$ is undefined. **45.** $\dfrac{-8}{8} = -1$ **46.** $+\$212$ or $\$212$ **47.** $-\$24$ **48.** -13 degrees **49.** $+27{,}010$ ft or $27{,}010$ ft **50.** -43 yd **51.** -90 ft **52.** January: $-\$700$; February: $-\$100$; March: $\$700$; April: $\$900$; May: $\$0$; June: $\$700$ **53.** January; April **54.** $\$2100$ **55.** $-\$1850$ **56.** Subtract the lesser absolute value from the greater absolute value. The answer has the same sign as the addend with the greater absolute value. Examples: $-6 + 2 = -4$ and $6 + (-2) = 4$ **57.** Similar: If the signs match, the result is positive. If the signs are different, the result is negative. Different: Multiplication is commutative, division is not. You can multiply by 0, but dividing by 0 cannot be done.

Section 9.5 (pages 673–676)

1. 28 **3.** -10 **5.** 8 **7.** -30 **9.** -8 **11.** 0 **13.** 2 **15.** -22 **17.** $-\dfrac{13}{8}$ **19.** 28 **21.** -5 **23.** 3 **25.** $\dfrac{12}{-6} = -2$ **27.** $P = 30$ **29.** $P = 28$ **31.** $A \approx 78.5$ **33.** $A = \dfrac{45}{2}$ or $22\dfrac{1}{2}$ **35.** $V = 600$ **37.** $d = 318$ **39.** $C \approx 25.12$ **41. (a)** $P = 33$ in. **(b)** $P = 9$ ft

43. (a) average score $= 83$ **(b)** average score $= 91$

45. The error was made when replacing x with -3; should be $-(-3)$, not -3.

$$-x - 4y \qquad \text{Replace } x \text{ with } (-3) \text{ and } y \text{ with } (-1).$$
$$-(-3) - 4(-1) \qquad \text{Opposite of } (-3) \text{ is } +3.$$
$$(+3) - (-4) \qquad \text{Change subtraction to addition.}$$
$$3 + (+4) \qquad \text{Add.}$$
$$7$$

47. $F = -40$; convert a Celsius temperature to Fahrenheit **48.** $C = -20$; convert a Fahrenheit temperature to Celsius **49.** $V \approx 113.04$; find the volume of a sphere

50. $c = 5$; Pythagorean Theorem for finding the hypotenuse or a leg in a right triangle **51.** $A = 56$; find the area of a trapezoid **52.** $V \approx 376.8$; find the volume of a cone

Section 9.6 (pages 683–686)

1. yes **3.** yes **5.** no **7.** $p = 4$ **9.** $k = -15$ **11.** $z = 8$ **13.** $r = 10$ **15.** $n = -8$ **17.** $r = -6$ **19.** $k = 18$ **21.** $x = 11$ **23.** $r = -3$ **25.** $d = \dfrac{7}{3}$ or $2\dfrac{1}{3}$ **27.** $z = \dfrac{87}{8}$ or $10\dfrac{7}{8}$ **29.** $k = \dfrac{5}{2}$ or $2\dfrac{1}{2}$ **31.** $m = \dfrac{83}{20}$ or $4\dfrac{3}{20}$ **33.** $x = 5.87$ **35.** $r = -1.51$ **37.** $z = 2$ **39.** $r = 4$ **41.** $y = 0$ **43.** $k = -6$ **45.** $p = 9$ **47.** $m = -7$ **49.** $p = 1.1$ **51.** $k = 34$ **53.** $a = 66$ **55.** $r = -36$ **57.** $p = -20$ **59.** $m = 4$ **61.** $x = 16$ **63.** $y = 1.3$ **65.** $z = -4.94$ **67.** You may add or subtract the same number on both sides of an equation. Many different equations could have -3 as the solution. One possibility is:

$$x + 5 = 2$$
$$x + 5 - 5 = 2 - 5$$
$$x = -3$$

69. $x = 19$ **71.** $x = 9$ **73.** $x = \dfrac{8}{21}$ **75.** $a = -\dfrac{5}{4}$ **77.** $m = -4$

Section 9.7 (pages 691–694)

1. $p = 1$ **3.** $y = 1$ **5.** $m = 0$ **7.** $a = -2$ **9.** $x = -4$ **11.** $z = 6$ **13.** $b = 0.9$ **15.** $6x + 24$ **17.** $7p - 56$ **19.** $-3m - 18$ **21.** $-2y + 6$ **23.** $-8c - 64$ **25.** $-10w + 90$ **27.** $17r$ **29.** $1z$ or z **31.** $-2y$ **33.** $-7t$ **35.** $0p = 0$ **37.** $-3b$ **39.** $k = 5$ **41.** $m = 6$ **43.** $b = -6$ **45.** $y = 1$ **47.** $p = 4$ **49.** $z = -2$ **51.** $y = 2$ **53.** $b = -1.05$ **55.** $r = -4$ **57.** $y = -9$ **59.** $t = -5$ **61.** $x = 0$

63.

$$-2t - 10 = 3t + 5$$
$$-2t - 3t - 10 = 3t - 3t + 5 \qquad \text{Subtract } 3t \text{ from both sides (addition property).}$$
$$-5t - 10 = 5$$
$$-5t - 10 + 10 = 5 + 10 \qquad \text{Add 10 to both sides (addition property).}$$
$$-5t = 15$$
$$\dfrac{-\cancel{5} \cdot t}{-\cancel{5}} = \dfrac{15}{-5} \qquad \text{Divide both sides by } -5 \text{ (multiplication property).}$$
$$t = -3$$

65. $x = -10$ **67.** $y = 2$ **69.** $y = 20$ **71.** $w = -1$ **73.** $a = -6.75$

Section 9.8 (pages 701–704)

1. $14 + x$ or $x + 14$ **3.** $-5 + x$ or $x + (-5)$ **5.** $20 - x$ **7.** $x - 9$ **9.** $x - 4$ **11.** $6x$ **13.** $2x$ **15.** $\dfrac{x}{2}$ **17.** $8 + 2x$ or $2x + 8$ **19.** $7x - 10$ **21.** $2x + x$ or $x + 2x$ **23. (a)** A variable is a letter that represents an unknown quantity. Examples: x, w, p **(b)** An expression is a combination of operations on variables and numbers. Examples:

$6x, w - 5, 2p + 3x$ **(c)** An equation has an $=$ sign and shows that two expressions are equal. Examples: $2y = 14$; $x + 5 = 2x$; $8p - 10 = 54$ **25.** $4n - 2 = 26$; $n = 7$ **27.** $2n + n = -15$; $n = -5$ **29.** $5n + 12 = 7n$; $n = 6$ **31.** $30 - 3n = 2 + n$; $n = 7$ **33.** $\dfrac{1}{2}n + 2n = 50$; $n = 20$ **35.** Let x be Ricardo's original weight.

$x + 15 - 28 + 5 = 177$. He weighed 185 pounds originally.

37. Let x be the amount Brenda spent. $81 = 2x - 3$. Brenda spent $\$42$. **39.** Let x be my age; $x - 9$ is my sister's age. $x + x - 9 = 51$. I am 30; my sister is 21. **41.** Let x be husband's earnings; $x + 1500$ is Lien's earnings. $x + x + 1500 = 37{,}500$. Husband earned $\$18{,}000$; Lien earned $\$19{,}500$. **43.** Let x be printer's cost; $5x$ is computer's cost. $x + 5x = 1320$. The printer cost $\$220$; the computer cost $\$1100$. **45.** Let x be length of shorter piece; $x + 10$ is length of longer piece. $x + x + 10 = 78$. Shorter piece is 34 cm; longer piece is 44 cm. **47.** Let x be one piece; $x - 7$ is the other piece. $x + x - 7 = 31$. One piece is 19 ft; the other piece is 12 ft.

49. $48 = 2(5) + 2x$; length is 19 yd.

51. $36 = 2(2w) + 2w$; length is 12 ft; width is 6 ft.

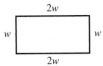

53. $36 = 2(2x + 3) + 2x$; length is 13 in.; width is 5 in.

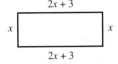

Chapter 9 Review Exercises (pages 709–714)

1.

2.

3.

4.

5. $>$ **6.** $<$ **7.** $>$ **8.** $<$ **9.** 8 **10.** 19 **11.** -7 **12.** -15 **13.** 2
14. -7 **15.** -19 **16.** -33 **17.** 1 **18.** -19 **19.** $\frac{3}{10}$ **20.** $-\frac{3}{8}$
21. -5.2 **22.** -1.5 **23.** -6 **24.** 14 **25.** $\frac{5}{8}$ **26.** -3.75 **27.** -6
28. -8 **29.** -7 **30.** -17 **31.** 11 **32.** 11 **33.** 13 **34.** -6
35. -80 **36.** 0 **37.** $-\frac{1}{2}$ **38.** 9 **39.** -24 **40.** -20 **41.** 15
42. 64 **43.** -8 **44.** -3 **45.** 5 **46.** 20 **47.** 0 **48.** -81 **49.** 0
50. undefined **51.** $-\frac{4}{7}$ **52.** 6 **53.** 1.4 **54.** -6.6 **55.** 57 **56.** 23
57. -1 **58.** -2 **59.** -34 **60.** -32 **61.** -100 **62.** 117 **63.** -1
64. 1.328 **65.** 0 **66.** 2 **67.** 27 **68.** -8 **69.** 0 **70.** 19 **71.** 23
72. -1 **73.** $P = 35$ **74.** $A = 27$ **75.** $y = -3$ **76.** $a = 16$
77. $z = 1$ **78.** $r = 1$ **79.** $x = -\frac{5}{4}$ or $-1\frac{1}{4}$ **80.** $k = -8.05$
81. $r = -7$ **82.** $p = 8$ **83.** $z = 20$ **84.** $a = -55$ **85.** $y = 9$
86. $b = -4$ **87.** $6r - 30$ **88.** $11p + 77$ **89.** $-9z + 27$
90. $-8x - 32$ **91.** $11r$ **92.** $-5z$ **93.** $-8p$ **94.** $2x$ **95.** $z = -9$
96. $k = -5$ **97.** $y = -5$ **98.** $b = 7$ **99.** $a = -4$ **100.** $t = 0$
101. $18 + x$ or $x + 18$ **102.** $\frac{1}{2}x$ or $\frac{x}{2}$ **103.** $-5x$ **104.** $20 - x$
105. $4x + 6 = -14$; $x = -5$ **106.** $5x - x = 100$; $x = 25$
107. Let x be one student's scholarship; $3x$ is the other student's scholarship. $x + 3x = 30,000$. One student receives $7500; the other student receives $22,500.
108. $48 = 2(w + 4) + 2w$; width is 10 in.; length is 14 in.

109. 3 **110.** 40 **111.** -1 **112.** -7 **113.** -16 **114.** -9 **115.** 6
116. 5 **117.** -20 **118.** undefined **119.** $-\frac{5}{9}$ **120.** -0.35 **121.** 7

122. -19 **123.** $y = 9$ **124.** $b = -4$ **125.** $z = 4$ **126.** $r = -10$
127. $x = -11$ **128.** $z = 3$ **129.** $k = 4$ **130.** $t = 0$
131. $a = -40$ **132.** $p = 3$ **133.** Let x be the daily amount for an infant. $2x - 4 = 18$. An infant should receive 11 mg of iron.
134. Let x be the number of Senators; $5x - 65$ is the number of Representatives. $x + 5x - 65 = 535$. 100 Senators; 435 Representatives
135. Let x be the zebra's speed; $x + 25$ is the cheetah's speed. $x + x + 25 = 111$. Zebra's speed is 43 mph; Cheetah's speed is 68 mph.
136. Let w be the width, $5w$ is the length. $1320 = 2(5w) + 2(w)$. Width is 110 yd; length is 550 yd.

Chapter 9 Test (pages 715–716)

1.

2. $<, >$ **3.** 7, 15 **4.** -1 **5.** -13 **6.** 5.3 **7.** -7 **8.** 16 **9.** $\frac{1}{4}$
10. -32 **11.** 84 **12.** 0 **13.** -25 **14.** 8 **15.** $-\frac{3}{5}$ **16.** 1 **17.** -21
18. -38 **19.** 27 **20.** When evaluating, you are given specific values to replace each variable. When solving an equation, you are not given the value of the variable. You must find a value that "works"; that is, when your solution is substituted for the variable, the two sides of the equation are equal. **21.** 110 **22.** $x = 5$ **23.** $r = 31$ **24.** $t = 5$
25. $p = -15$ **26.** $a = -3$ **27.** $m = 2$ **28.** Let x be the shorter piece; $x + 4$ is longer piece. $x + x + 4 = 118$. Shorter piece is 57 cm; longer piece is 61 cm.
29. $420 = 2(4w) + 2w$; width is 42 ft; length is 168 ft.

30. Let x be Marcella's time; $x - 3$ is Tim's time. $x + x - 3 = 19$. Marcella spent 11 hr; Tim spent 8 hr.

Cumulative Review Exercises: Chapters 1–9 (pages 717–718)

1. *Estimate:* $40,000 \div 80 = 500$; *Exact:* 503
2. *Exact:* $5\frac{1}{4}$; *Estimate:* $6 \cdot 1 = 6$
3. *Exact:* $2\frac{1}{2}$; *Estimate:* $4 \div 2 = 2$
4. 8.906 **5.** 56.4 (rounded) **6.** -5 **7.** 4 **8.** -1.3 **9.** $-\frac{1}{2}$ **10.** 5
11. 4 **12.** $x = 0.4$ **13.** 50 cars **14.** 4% **15.** 14 qt **16.** 3 days
17. 3750 g **18.** 0.4 m **19.** Square; $P = 9$ ft; $A = 5\frac{1}{16}$ or 5.0625 ft²
20. Circle; $C \approx 28.26$ mm; $A \approx 63.585$ mm² **21.** Right triangle; $P = 56$ mi; $A = 84$ mi² **22.** Parallelogram; $P = 36$ ft; $A = 70$ ft²
23. $t = -5$ **24.** $x = 8$ **25.** $p = -4$ **26.** $4x + 40 = 0$; $x = -10$
27. Let m be Reggie's money; $m + 300$ is Donald's money. $m + m + 300 = 1000$; $350 for Reggie; $650 for Donald
28. $82 = 2(w + 5) + 2w$; length is 23 cm; width is 18 cm

29. $y \approx 13.2$ yd (rounded) **30.** $31.91 (rounded)
31. 54 min (rounded) **32.** 0.544 m³ less
33. Naomi's car; 25.2 miles per gallon (rounded)
34. 124% **35.** $8\frac{1}{3}$ cups

Chapter 10 Statistics

Section 10.1 (pages 725–730)

1. 360 million or 360,000,000 pets **3.** $\frac{90}{360} = \frac{1}{4}$ **5.** $\frac{90}{75} = \frac{6}{5}$ **7.** Don't know; 180 people **9.** $\frac{720}{6000} = \frac{3}{25}$ **11.** $\frac{1020}{1200} = \frac{17}{20}$ **13.** $\frac{1740}{1020} = \frac{29}{17}$ **15.** 160 people **17.** 96 people **19.** 960 people **21.** 1219 people (rounded) **23.** 1828 people (rounded) **25.** 277 people **27.** First find the percent of the total that is to be represented by each item. Next, multiply the percent by 360° to find the size of each sector. Finally, use a protractor to draw each sector. **29.** 90° **31.** 10%; 36° **33.** 54° **35.** 15% **37. (a)** $200,000 **(b)** 22.5°; 72°; 108°; 90°; 67.5° **(c)**

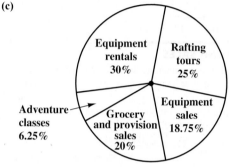

39. (a) 2464; 55% (rounded); 198°; 220; 5% (rounded); 18°; 536; 12% (rounded); 43° (rounded); 520; 12% (rounded); 43° (rounded); 748; 17% (rounded); 61° (rounded)

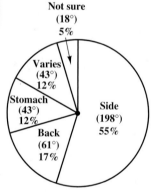

(b) No. The total is 101% due to rounding.
(c) No. The total is 363° due to rounding.

Section 10.2 (pages 735–740)

1. India; 51.4% **3.** USA, Britain, and Australia **5.** 121 days (rounded) **7.** May; 10,000 plants **9.** 1500 plants **11.** 4000 plants **13.** 150,000 gal **15.** 2006; 250,000 gal **17.** 550,000 gal **19.** 24.1 million or 24,100,000 PCs **21.** 144.5 million or 144,500,000 PCs **23.** Answers will vary. Some possibilities are: greater demand as a result of lower price; more uses and applications for the owner; improved technology; multiple computers in each location; more laptop computers sold. **25.** 300,000 iPods **27.** 150,000 iPods **29.** 350,000 iPods **31.** Probably Store B with greater sales. Predicted sales might be 450,000 iPods to 500,000 iPods in 2011. **33.** A single bar or a single line must be used for each set of data. To show multiple sets of data, multiple sets of bars or lines must be used. **35.** $40,000 **37.** $25,000 **39.** $5000 **41.** Answers will vary. Some possibilities are that the decrease in sales may have resulted from poor service or greater competition. The increase in sales may have been a result of more advertising or better service. **43.** 23 years **44.** 20 additional

flavors **45.** 214,286 rolls each day (rounded) **46.** 1.17 lb (rounded) **47.** The weight of one roll of LifeSavers is much less than 1.17 lb. The answer is "correct" using the information given, but some of the data given must not be accurate. Perhaps 3 million **rolls** of LifeSavers are produced daily. **48.** Answers will vary. Possible answers are: Misprints or typographical errors; careless reporting of data; math errors.

Summary Exercises on Graphs (pages 741–742)

1. 2000 people **2.** vanilla; 660 people **3.** $\frac{380}{2000} = \frac{19}{100}$ **4.** $\frac{100}{2000} = \frac{1}{20}$ **5.** $\frac{640}{80} = \frac{8}{1}$ **6.** $\frac{660}{140} = \frac{33}{7}$ **7.** 2100; 9.1 billion or 9,100,000,000 people **8.** 2.9 billion or 2,900,000,000 people **9.** 0.6 billion or 600,000,000 people **10.** 45.9% (rounded) **11.** 2.112 billion or 2,112,000,000 in China; 0.396 billion or 396,000,000 in the United States **12.** 1960, about 68% **13.** about $2\frac{1}{2}$%; yes **14.** widowed **15.** about 4% **16.** Answers will vary. Some possibilities are: no one lived together in 1970; this category was very small; no one kept track of this group; no one would admit it.

Section 10.3 (pages 745–748)

1. 61–65 years; 16,000 members **3.** 13,000 members **5.** 50,000 members **7.** $4100 to $5000; 16 employees **9.** 11 employees **11.** 49 employees **13.** Class intervals are the result of combining data into groupings. Class frequency is the number of data items that fit in each class interval. These are used to group data and to have multiple responses (frequency) in a class interval—this makes the data easier to interpret. **15.** 𝅗𝅥; 6 **17.** |; 1 **19.** ||; 2 **21.** ||||; 4 **23.** 𝅗𝅥; 5 **25.** |||; 3 **27.** 𝅗𝅥; 7 **29.** 𝅗𝅥𝅗𝅥||||; 14 **31.** 𝅗𝅥𝅗𝅥𝅗𝅥; 16 **33.** 𝅗𝅥𝅗𝅥; 11 **35.** 𝅗𝅥; 5

Section 10.4 (pages 753–756)

1. 11 years **3.** 2.5 in. of rain (rounded) **5.** $37,127 **7.** $72.53 **9.** 12.5 customers (rounded) **11.** 17.2 fish (rounded) **13.** $35,500 **15.** 125 books **17.** 508 calories **19.** 21% **21.** 68 and 74 yr (bimodal) **23.** The median is a better measure of central tendency when the list contains one or more extreme values. Find the mean and the median of the following home values. $182,000; $164,000; $191,000; $115,000; $982,000; mean home value = $326,800; median home value = $182,000 **25.** 2.42 (rounded) **27.** 2.60 **29.** 3.14 (rounded) **31.** 176 sales calls **32.** mean for Samuels: 22; mean for Stricker: 22 **33.** median for Samuels: 19.5; median for Stricker: 22 **34.** mode for Samuels: 22; mode for Stricker: 22 **35.** The mean and mode are identical for both sales representatives and the medians are close. **36.** The number of weekly sales calls made by Samuels varies greatly from week to week while the number of weekly sales calls made by Stricker remains fairly constant. **37.** range = 40 − 8 = 32 **38.** range = 25 − 19 = 6 **39.** No, not with any certainty. There probably are additional questions that need to be answered before a determination could be made, such as the dollar amount of sales, number of repeat customers, and so on. **40.** Answers will vary. Some possible answers are: He works hard one week, then takes it easy the next week; the characteristics of the sales territories vary greatly; illness or personal problems may be affecting performance.

Chapter 10 Review Exercises (pages 763–768)

1. Basketball; 18,000 teams **2.** $\frac{16,000}{76,000} = \frac{4}{19}$ **3.** $\frac{15,000}{76,000} = \frac{15}{76}$ **4.** $\frac{14,000}{76,000} = \frac{7}{38}$ **5.** $\frac{18,000}{16,000} = \frac{9}{8}$ **6.** $\frac{16,000}{14,000} = \frac{8}{7}$ **7.** 3936 companies

8. 1728 companies   **9.** 720 companies   **10.** 2904 companies
11. On-site child care and Fitness centers   Answers will vary. Perhaps employers feel that they are not needed or would not be used. Or, it may be that they would be too expensive for the benefit derived.   **12.** Flexible hours and casual dress   Answers will vary. Perhaps employees request them and appreciate them. Or, it may be that neither of them cost the employer anything to offer.   **13.** March; 8,000,000 acre-feet   **14.** June; 2,000,000 acre-feet   **15.** 5,000,000 acre-feet   **16.** 4,000,000 acre-feet
17. 5,000,000 acre-feet   **18.** 2,000,000 acre-feet   **19.** $50,000,000
20. $20,000,000   **21.** $20,000,000   **22.** $40,000,000   **23.** The floor-covering sales decreased for 2 years and then moved up slightly. Answers will vary. Perhaps there is less new construction, remodeling, and home improvement in the area near Center A, or a better product selection and service have reversed the decline in sales.   **24.** The floor-covering sales are increasing. Answers will vary. Perhaps new construction and home re-modeling have increased in the area near Center B, or greater advertising has attracted more customers.   **25.** 28.5 digital cameras   **26.** 35.2 complaints   **27.** $51.05   **28.** 257.3 points (rounded)   **29.** 39 forms
30. $562   **31.** $79   **32.** 18 and 32 launchings (bimodal)   **33.** 36°
34. 35%   **35.** 20%; 72°   **36.** 25%; 90°   **37.** 10%; 36°
38.

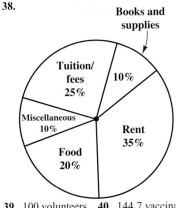

39. 100 volunteers   **40.** 144.7 vaccinations (rounded)
41. 48 applicants   **42.** 31 and 43 two-bedroom apartments (bimodal)   **43.** 5.0 hr   **44.** 21 e-mails   **45.** ||||; 4   **46.** |; 1
47. 𝍸𝍸|; 6   **48.** ||; 2   **49.** 𝍸𝍸𝍸𝍸|||; 13   **50.** 𝍸𝍸; 7   **51.** 𝍸𝍸||; 7
52.

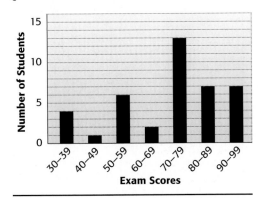

53. 64.5 (rounded)   **54.** 118.8 units (rounded)

Chapter 10 Test (pages 769–772)

1. $59.0 billion (rounded)   **2.** $21.2 billion (rounded)   **3.** $5.4 billion (rounded)   **4.** $57.2 billion (rounded)   **5.** $146.0 billion (rounded)
6. $9.2 billion (rounded)   **7.** 108°   **8.** 36°   **9.** 72°   **10.** 126°   **11.** 5%

12.

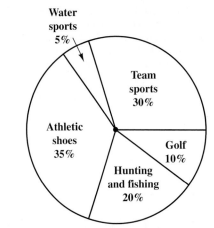

13. 3   **14.** 2   **15.** 4   **16.** 3   **17.** 3   **18.** 5
19.

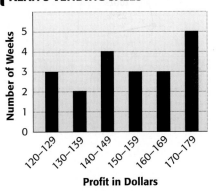

20. 75 mi   **21.** 15.3 lb (rounded)   **22.** 478.9 mph
23. The weighted mean must be used because different classes are worth different numbers of credits. GPA problems will vary; one possibility is shown.

Credits	Grade	
3	A	$3 \times 4 = 12$
2	C	$2 \times 2 = 4$
$\dfrac{4}{9}$	B	$\dfrac{4 \times 3 = 12}{28}$

$28 \div 9 \approx 3.11$

24. Arrange the values in order, from smallest to largest. When there is an odd number of values in a list, the median is the middle value.
8, 17, 23, 32, 64
        Median
25. $26 (rounded)   **26.** 174 points (rounded)   **27.** 31.5 degrees
28. 10.0 meters   **29.** 52 milliliters   **30.** 103 and 104 degrees (bimodal)

Cumulative Review Exercises: Chapters 1–10 (pages 773–774)

1. *Estimate:* $900 - 60 = 840$; *Exact:* 811.863
2. *Estimate:* $7000 \times 600 = 4,200,000$; *Exact:* 4,485,640
3. *Estimate:* $60 \div 4 = 15$; *Exact:* 14.72
4. $10\dfrac{4}{15}$ **5.** $1\dfrac{11}{12}$ **6.** $44\dfrac{2}{5}$ **7.** $8\dfrac{4}{5}$ **8.** $\dfrac{8}{21}$ **9.** $\dfrac{1}{12}$ **10.** 23
11. 10   **12.** 108   **13.** 0.199, 0.207, 0.218, 0.22, 0.2215
14. $0.58, 0.608, \dfrac{5}{8}, 0.6319, \dfrac{13}{20}$ **15.** $\dfrac{1}{8}$ **16.** $\dfrac{4}{1}$ **17.** $x = 6$ **18.** $x = 6$
19. $x = 16.2$ **20.** $x = 52$ **21.** 324 homes   **22.** 2800 people
23. 260%   **24.** 7   **25.** 40 cm   **26.** 0.23 kg   **27.** 12,000 lb
28. Rectangle: A ≈ 47.3 m² (rounded)   **29.** Triangle: A ≈ 9.6 ft² (rounded)   **30.** Circle: A ≈ 132.7 cm² (rounded)   **31.** Cylinder: V ≈ 549.8 cm³ (rounded)   **32.** Rectangular solid: V = 199.5 m³

33. 34 cm **34.** −22 **35.** 6.9 **36.** 83.22 **37.** −2.3 **38.** $x = 7$
39. $x = -6$ **40.** $x = -7$ **41.** mean = 27 hookups; median = 26
hookups; mode = 19 hookups **42.** mean = 4.4 acres; median = 4.3
acres; mode = 2.85 acres **43.** 42.0% **44.** **(a)** $889.80 **(b)** $977
45. 38 nasal strips for $9.95 **46.** $59,225.25 **47.** 17.5 L
48. $11,761

Chapter 1 Whole Numbers

Section 1.1 (pages 7–8)

9. 60,000,502,109; billions: 60; millions: 000; thousands: 502; ones: 109

15. 346,009 is three hundred forty-six thousand, nine.

31. The least-used method of transportation is public transportation. 6,069,589 in words is six million, sixty-nine thousand, five hundred eighty-nine.

Section 1.2 (pages 17–20)

11. Line up the numbers in columns. Then start at the right and add the ones digits. Add the ten digits next, and, finally, the hundreds digits.

$$\begin{array}{r} 932 \\ 44 \\ +\ 613 \\ \hline 1589 \end{array}$$

49. *Step 1*
Add the digits in the ones column.

Step 2
Add the digits in the tens column, including the regrouped 2.

Step 3
Add the hundreds column, including the regrouped 1.

Step 4
Add the thousands column, including the regrouped 1.

$$\begin{array}{r} {}^{112} \\ 18 \\ 708 \\ 9\,286 \\ +\ \ 636 \\ \hline 10{,}648 \end{array}$$

63. Add up to check addition.

$$\begin{array}{r} 769 \\ \hline 179 \\ 214 \\ +\ 376 \\ \hline 759 \end{array}$$
759 incorrect; should be 769

77. To find the total amount raised, add the amounts raised at each event.

$$\begin{array}{rl} \$3482 & \text{flea market} \\ +\ 12{,}860 & \text{annual auction} \\ \hline \$16{,}342 & \text{total amount raised} \end{array}$$

Section 1.3 (pages 27–30)

31. In the ones column, 5 is less than 7, so in order to subtract, regroup 1 ten as 10 ones. Then subtract 7 ones from 15 ones in the ones column. Finally, subtract 3 tens from 6 tens in the tens column.

$$\begin{array}{r} {}^{6}\!\!\!\not{7}{}^{15}\!\!\not{5} \\ -3\,7 \\ \hline 3\,8 \end{array}$$

67.
$$\begin{array}{r} {}^{5}\!\!\not{6}\,{}^{9}\!\!\not{0}{}^{9}\!\!\not{0}\,{}^{10}\!\!\not{0}\,{}^{10}\!\!\not{0} \\ -34,\,4\,4\,4 \\ \hline 31,\,5\,5\,6 \end{array}$$

81. To find how many fewer calories a woman burns, subtract the number of calories a woman burns from the number of calories a man burns.

$$\begin{array}{rl} 187 & \text{calories man burns} \\ -140 & \text{calories woman burns} \\ \hline 47 & \text{fewer calories for a woman} \end{array}$$

The woman burned 47 fewer calories.

Section 1.4 (pages 37–40)

9. (4)(5)(2) = [(4)(5)](2) = 20(2) = 40 or (4)[(5)(2)] = (4)(10) = 40.

37.
$$\begin{array}{ccc} 600 & 6 & 600 \\ \times\ 6 & \times\ 6 & \times\ 6 \\ \hline & 36 & 3600 \end{array}$$ Attach 00.

73. First multiply 9352 by 4. Then multiply 9352 by 6, making sure to line up the tens. Then multiply 9352 by 2, making sure to line up the hundreds. Then add the partial products.

$$\begin{array}{r} 9\,3\,5\,2 \\ \times\ \ 2\,6\,4 \\ \hline 3\,7\,4\,0\,8 \quad \leftarrow\ 4 \times 9352 \\ 5\,6\,1\,1\,2 \quad\ \leftarrow\ 6 \times 9352 \\ 1\,8\,7\,0\,4 \quad\quad \leftarrow\ 2 \times 9352 \\ \hline 2{,}4\,6\,8{,}9\,2\,8 \end{array}$$

85. *Approach* To find the number of balls purchased, multiply the number of cartons (300) by the number of balls per carton (10).

Solution

$$\begin{array}{rl} 300 & \text{cartons} \\ \times\ 10 & \text{balls per carton} \\ \hline 3000 & \text{balls} \end{array}$$
3000 balls were purchased.

103. Multiply the number of laptop computers purchased (6) by the price per laptop ($880). Then multiply the number of printers purchased (6) by the price per printer ($235). Then multiply the number of fax machines purchased (5) by the price per fax machine ($140).

$$\begin{array}{ccc} 880 & 235 & 140 \\ \times\ 6 & \times\ 6 & \times\ 5 \\ \hline 5280 & 1410 & 700 \end{array}$$

The total cost is the sum of these values:
$5280 + $1410 + $700 = $7390

Section 1.5 (pages 51–54)

1. $24 \div 4 = 6$: $4\overline{\smash{\big)}24}$ with 6 above, or $\dfrac{24}{4} = 6$

31.
$$6\,\overline{\smash{\big)}9\,{}^{3}3\,1\,{}^{1}1\,{}^{3}3\,{}^{1}1\,7}\quad 1\,5\,2\,2\,\textbf{R5}$$

Check $(6 \times 1522) + 5 = 9132 + 5 = 9137$

51.
$$7\,\overline{\smash{\big)}7\,1,\,{}^{1}1\,7\,\,{}^{3}3\,7\,\,{}^{2}2\,6}\quad 1\,0,\,2\,5\,3\,\textbf{R5}$$

Check $(7 \times 10{,}253) + 5 = 71{,}771 + 5 = 71{,}776$

67.
$$9\,\overline{\smash{\big)}8\,6,\,6\,5\,5}\quad 9\,6\,2\,8\,\textbf{R7}$$

Check $(9 \times 9628) + 7 = 86{,}652 + 7 = 86{,}659$

Does not match dividend, incorrect

Rework:
$$9\,\overline{\smash{\big)}8\,6,\,{}^{5}5\,6\,\,{}^{2}2\,5\,\,{}^{7}7\,5}\quad 9\,6\,2\,8\,\textbf{R3}$$

Check $(9 \times 9628) + 3 = 86{,}652 + 3 = 86{,}655$

Matches dividend, correct

73. To find the number of tables that can be set, divide the number of napkins (2624) by the number of napkins it takes to set each table (8).

$$8\,\overline{\smash{\big)}2\,6\,\,{}^{2}2\,\,{}^{6}6\,4}\quad 3\,2\,8$$

328 tables can be set.

Section 1.6 (pages 61–63)

13. Use 2 as a trial divisor since 18 is closer to 20 than to 10.

$$\dfrac{131}{2} = 6 \text{ with } 1 \text{ left over}$$

Because $6 \times 18 = 108$ and $131 - 108 = 23$, which is greater than the divisor, use 7 instead. To find the next digit in the quotient, use 2 as a trial divisor again.

$$\dfrac{5}{2} = 2 \text{ with } 1 \text{ left over}$$

Because $2 \times 18 = 36$ and $59 - 36 = 23$, which is greater than the divisor, use 3 instead.

$$\begin{array}{r} 7\,3\,\textbf{R5} \\ 18\,\overline{\smash{\big)}1\,3\,1\,9} \\ 1\,2\,6 \\ \hline 5\,9 \\ 5\,4 \\ \hline 5 \end{array} \quad \textbf{Check} \quad \begin{array}{r} 7\,3 \\ \times\ \ 1\,8 \\ \hline 5\,8\,4 \\ 7\,3 \\ \hline 1\,3\,1\,4 \\ +\ \ 5 \\ \hline 1\,3\,1\,9 \end{array}$$

25.

$$35\overline{\smash{\big)}3549} \quad \underset{\text{R4}}{101}$$

Check

$$\begin{array}{r} 101 \\ \times\ 35 \\ \hline 505 \\ 303 \\ \hline 3535 \\ +\quad 4 \\ \hline 3539 \end{array} \text{ incorrect}$$

Rework:

$$\begin{array}{r} 101\ \text{R}14 \\ 35\overline{\smash{\big)}3549} \\ 35 \\ \hline 049 \\ 35 \\ \hline 14 \end{array}$$

Check

$$\begin{array}{r} 101 \\ \times\ 35 \\ \hline 505 \\ 303 \\ \hline 3535 \\ +\quad 14 \\ \hline 3549 \end{array} \text{ Matches}$$

The correct answer is 101 **R**14.

41. Divide the yearly amount spent for eating away from home by 52 (the number of weeks in one year).

$$\begin{array}{r} 39 \\ 52\overline{\smash{\big)}2028} \\ 156 \\ \hline 468 \\ 468 \\ \hline 0 \end{array}$$

The average weekly amount spent for eating away from home is $39 per household.

Section 1.7 (pages 73–76)

1. 624 rounded to the nearest ten: 620

6$\underline{2}$4 Underline the 2 in the tens place. Next digit is 4 or less, so tens place does not change. All digits to the right of the underlined place change to zero.

25. To the nearest ten: 447$\underline{6}$

Underline the 7 in the tens place. Next digit is 5 or more, so add 1 to the 7. All digits to the right of the underlined place are changed to zero.

4480

To the nearest hundred: 4$\underline{4}$76

Underline the 4 in the hundreds place. Next digit is 5 or more, so add 1 to the 4. All digits to the right of the underlined place are changed to zero.

4500

To the nearest thousand: $\underline{4}$476

Underline the 4 in the thousands place. Next digit is 4 or less, so leave 4 as 4. All digits to the right of the underlined place are changed to zero.

4000

41.

Estimate:	Exact:
30	25
60	63
50	47
+ 80	+ 84
220	219

47.

Estimate:	Exact:
900	863
700	735
400	438
+ 800	+ 792
2800	2828

53.

Estimate:	Exact:
8000	8 215
60	56
700	729
+ 4000	+ 3 605
12,760	12,605

61. 7$\underline{6}$,000,000 Next digit is 5 or more, so add 1 to 7. Change digit to the right of the underlined place to zero. Add 1 to 7.

80 million people

3$\underline{0}$3,000,000 Next digit is 4 or less, so leave 0 as 0. Change digit to the right of the underlined place to zero.

300 million people

Section 1.8 (pages 79–82)

1. 3^2: exponent is 2, base is 3.

$3^2 = 3 \cdot 3 = 9$

21. $35^2 = \underline{1225}$, so $\sqrt{1225} = 35$.

37.

$5 \cdot 3^2 + \dfrac{0}{8}$	Exponent
$5 \cdot 9 + \dfrac{0}{8}$	Multiply
$45 + \dfrac{0}{8}$	Divide
$45 + 0 = 45$	Add

67.

$8 \cdot \sqrt{49} - 6(9 - 4)$	Parentheses
$8 \cdot \sqrt{49} - 6(5)$	Square root
$8 \cdot 7 - 6(5)$	Multiply
$56 - 6(5)$	Multiply.
$56 - 30 = 26$	Subtract

85. $8 \cdot 9 \div \sqrt{36} - 4 \div 2 + (14 - 8)$

	Parentheses
$8 \cdot 9 \div \sqrt{36} - 4 \div 2 + 6$	Square root
$8 \cdot 9 \div 6 - 4 \div 2 + 6$	Multiply
$72 \div 6 - 4 \div 2 + 6$	Divide
$12 - 4 \div 2 + 6$	Divide
$12 - 2 + 6$	Subtract
$10 + 6 = 16$	Add

Section 1.9 (pages 87–90)

5. According the pictograph, Family Dollar has 4500 stores. Subtract 4500 from 5750 to find the difference.

$5750 - 4500 = 1250$ fewer stores

11. According to the bar graph, $18 - 9 = 9$ more people out of 100 found their career as a result of "Studied in school" than "Luck or chance."

15. According to the line graph the increase in the number of trees planted from 2010 to 2011 is $7000 - 2500 = 4500$.

19. We want to insert parentheses in $7 - 2 \cdot 3 - 6$ so that it simplifies to 9.

There isn't a method to use, so just use trial and error.

$(7 - 2) \cdot 3 - 6 = 5 \cdot 3 - 6$
$ = 15 - 6 = 9$

Section 1.10 (pages 95–98)

7. Step 1

Find the number of toys each child will receive.

Step 2

"Same number of toys to each" indicates division should be used.

Step 3

3000 divided by 700 gives an estimate of about 4 toys.

Step 4

$$\begin{array}{r} 4 \\ 657\overline{\smash{\big)}2628} \\ 2628 \\ \hline 0 \end{array}$$

Step 5

Each child will receive 4 toys.

Step 6

Exact answer is close to estimate. Also, $4 \times 657 = 2628$ which matches the number given in the problem.

17. Step 1

Find her monthly savings.

Step 2

Her monthly take home pay and expenses are given. "Remainder" indicates subtraction may be used.

Step 3

Estimate:

$2000 - $700 - $300 - $400 - $200 - $200 = $200

Step 4

$695	$2240
340	−1890
435	$350
240	
+ 180	
1890	

Step 5

Her monthly savings are $350.

Step 6

The exact answer seems reasonable compared to the estimate.

Check: $350 + $1890 = $2240. Re-add the expenses to check the sum of $1890.

25. Step 1

The number and cost of wheelchairs and recorder-players are given, and the total cost of all items must be found.

Step 2

Find the cost of all wheel chairs and the cost of all recorder-players. Then add these costs to get the total cost.

Step 3

The cost of the wheelchairs is about $1000 \times 6 = \$6000$.

The cost of the recorder-players is about $\$900 \times 20 = \$18,000$.

Estimate: $\$6000 + \$18,000 = \$24,000$

Step 4

cost of wheelchairs: $\$1256$

$$\begin{array}{r} \$1256 \\ \times\ 6 \\ \hline \$7536 \end{array}$$

cost of recorder-players:

$$\begin{array}{r} \$8\,9\,5 \\ \times\ 1\,5 \\ \hline 4\,4\,7\,5 \\ 8\,9\,5 \\ \hline \$1\,3,4\,2\,5 \end{array}$$

Step 5

The total cost is $\$13,425 + \$7536 = \$20,961$.

Step 6

The exact answer is reasonably close to the estimate. Check by repeating Step 4.

39. *Step 1*

The total seating is given along with the information to find the number of seats on the main floor. The number of rows of seats in the balcony is given, and the number of seats in each row of the balcony must be found.

Step 2

First, the number of seats on the main floor must be found. Next, the number of seats on the main floor must be subtracted from the total number of seats to find the number of seats in the balcony. Finally, the number of seats in the balcony must be divided by the number of rows of seats in the balcony.

Step 3

Seats on main floor: $30 \times 25 = 750$

Seats in balcony: $1250 - 750 = 500$

Seats in each row in balcony:
$500 \div 25 = 20$

Since we didn't round, our estimate matches our exact answer.

Step 4

See the calculations in Step 3.

Step 5

The number of seats in each row of the balcony is 20.

Step 6

The answer matches the estimate, as expected.

Check

Seats in balcony: $20 \times 25 = 500$
Seats on main floor: $30 \times 25 = 750$
Total seats: $500 + 750 = 1250$, which matches the number given in the problem.

Chapter 2 Multiplying and dividing fractions

Section 2.1 (pages 117–118)

5. Each of the two figures is divided into 5 parts and 7 are shaded: $\dfrac{7}{5}$

Three are unshaded: $\dfrac{3}{5}$

17. Proper fractions: numerator *smaller* than denominator.

$$\frac{1}{3}, \frac{5}{8}, \frac{7}{16}$$

Improper fractions: numerator *greater than or equal to* denominator.

$$\frac{8}{5}, \frac{6}{6}, \frac{12}{2}$$

Section 2.2 (pages 123–126)

16. Write $5\dfrac{4}{5}$ as an improper fraction.

$5 \cdot 5 = 25$ Multiply 5 and 5.

$25 + 4 = 29$ Add 4. The numerator is 29.

$5\dfrac{4}{5} = \dfrac{29}{5}$ Use the same denominator.

43. $\dfrac{63}{4}$

Divide 63 by 4.

$$\begin{array}{r} 1\,5 \leftarrow \text{Whole number part} \\ 4\,\overline{)6\,3} \\ 4 \\ \hline 2\,3 \\ 2\,0 \\ \hline 3 \leftarrow \text{Remainder} \end{array}$$

The quotient 15 is the whole number part of the mixed number. The remainder 3 is the numerator of the fraction, and the denominator remains as 4.

$$\frac{63}{4} = 15\frac{3}{4}$$

61. Write $522\dfrac{3}{8}$ as an improper fraction.

$522 \cdot 8 = 4176$

$4176 + 3 = 4179$ Add 3. The numerator is 4179.

$522\dfrac{3}{8} = \dfrac{4179}{8}$ Use the same denominator.

Section 2.3 (pages 131–132)

5. Factorizations of 48:

$1 \cdot 48 = 48 \quad 2 \cdot 24 = 48 \quad 3 \cdot 16 = 48$
$4 \cdot 12 = 48 \quad 6 \cdot 8 = 48$

The factors of 48 are 1, 2, 3, 4, 6, 8, 12, 16, 24, and 48.

41. 44

$$\begin{array}{r} 22 \\ 2\,\overline{)44} \end{array} \quad \text{Divide 44 by 2.}$$

$$\begin{array}{r} 11 \\ 2\,\overline{)22} \end{array} \quad \text{Divide 22 by 2.}$$

$$\begin{array}{r} 1 \\ 11\,\overline{)11} \end{array} \quad \text{Divide 11 by 11.}$$

Quotient is 1.

Because all factors (divisors) are prime, the prime factorization of 44 is

$2 \cdot 2 \cdot 11 = 2^2 \cdot 11$

Section 2.4 (pages 137–138)

17. Because the greatest common factor of 56 and 64 is 8, divide both numerator and denominator by 8.

$$\frac{56}{64} = \frac{56 \div 8}{64 \div 8} = \frac{7}{8}$$

29. Write the prime factorization of both numerator and denominator. Then divide both numerator and denominator by any common factors, and write a 1 by each factor that has been divided. Finally, multiply the remaining factors in both numerator and denominator.

$$\frac{18}{24} = \frac{\overset{1}{\cancel{2}} \cdot \overset{1}{\cancel{3}} \cdot 3}{\underset{1}{\cancel{2}} \cdot 2 \cdot 2 \cdot \underset{1}{\cancel{3}}} = \frac{1 \cdot 1 \cdot 3}{1 \cdot 2 \cdot 2 \cdot 1} = \frac{3}{4}$$

39. $\dfrac{3}{6}$ and $\dfrac{18}{36}$

Write each fraction in lowest terms.

$$\frac{3}{6} = \frac{1 \cdot \overset{1}{\cancel{3}}}{2 \cdot \underset{1}{\cancel{3}}} = \frac{1}{2}$$

$$\frac{18}{36} = \frac{\overset{1}{\cancel{2}} \cdot \overset{1}{\cancel{3}} \cdot \overset{1}{\cancel{3}}}{\underset{1}{\cancel{2}} \cdot 2 \cdot \underset{1}{\cancel{3}} \cdot \underset{1}{\cancel{3}}} = \frac{1}{2}$$

The fractions are *equivalent* $\left(\dfrac{1}{2} = \dfrac{1}{2}\right)$.

53. Write the prime factorization of both numerator and denominator. Then divide both numerator and denominator by any common factors, and write a 1 by each factor that has been divided. Finally, multiply the remaining factors in both numerator and denominator.

$$\frac{160}{256} = \frac{\overset{1}{\cancel{2}} \cdot \overset{1}{\cancel{2}} \cdot \overset{1}{\cancel{2}} \cdot \overset{1}{\cancel{2}} \cdot \overset{1}{\cancel{2}} \cdot 5}{\underset{1}{\cancel{2}} \cdot \underset{1}{\cancel{2}} \cdot \underset{1}{\cancel{2}} \cdot \underset{1}{\cancel{2}} \cdot \underset{1}{\cancel{2}} \cdot 2 \cdot 2 \cdot 2} = \frac{5}{8}$$

Section 2.5 (pages 149–152)

7. Divide a numerator and denominator by the same number to use the multiplication shortcut.

$$\frac{2}{3} \cdot \frac{7}{12} \cdot \frac{9}{14} = \frac{\overset{1}{\cancel{2}}}{\underset{1}{\cancel{3}}} \cdot \frac{7}{\underset{2}{\cancel{12}}} \cdot \frac{\overset{\overset{1}{\cancel{3}}}{\cancel{9}}}{\underset{2}{\cancel{14}}}$$

$$= \frac{1 \cdot 1 \cdot 1}{1 \cdot 2 \cdot 2} = \frac{1}{4}$$

25. Write 36 as $\dfrac{36}{1}$ and multiply.

$$36 \cdot \frac{5}{8} \cdot \frac{9}{15} = \frac{\overset{9}{\cancel{36}}}{1} \cdot \frac{\overset{1}{\cancel{5}}}{\underset{2}{\cancel{8}}} \cdot \frac{\overset{3}{\cancel{9}}}{\underset{\cancel{3}}{\cancel{15}}} = \frac{27}{2} = 13\frac{1}{2}$$

41. Area = length • width

$$\text{Area} = \frac{7}{5} \cdot \frac{3}{14}$$

$$= \frac{\overset{1}{\cancel{7}}}{5} \cdot \frac{3}{\underset{2}{\cancel{14}}} \qquad \text{Divide numerator and denominator by 7.}$$

$$= \frac{3}{10} \text{ in.}^2$$

45. Area = length • width

$$\text{Area} = 2 \cdot \frac{3}{4}$$

$$= \frac{\overset{1}{\cancel{2}}}{1} \cdot \frac{3}{\underset{2}{\cancel{4}}} \qquad \text{Divide numerator and denominator by 2.}$$

$$= \frac{3}{2} \qquad \text{Write as a mixed number.}$$

$$= 1\frac{1}{2} \text{ yd}^2$$

Section 2.6 (pages 157–160)

1. *Step 1*
The problem asks for the area of the top of a file cabinet.

Step 2
To find area, multiply the length (depth) $\frac{3}{4}$ yd by the width $\frac{2}{3}$ yd.

Step 3
An estimate is $\frac{6}{12}$ yd^2.

$$\frac{3}{4} \cdot \frac{2}{3} = \frac{\overset{1}{\cancel{3}} \cdot \overset{1}{\cancel{2}}}{\underset{2}{\cancel{4}} \cdot \underset{1}{\cancel{3}}} = \frac{1}{2}$$

Step 4
The exact value is $\frac{1}{2}$, which is the same as the estimate since we didn't round.

Step 5
The area of the file cabinet top is $\frac{1}{2}$ yd^2.

Step 6
The answer, $\frac{1}{2}$ yd^2, matches our estimate.

11. *Step 1*
The problem asks for the number of runners who are women.

Step 2
$\frac{7}{12}$ of the 1560 runners are women, so

multiply $\frac{7}{12}$ by 1560 to find the number of runners who are women.

Step 3
Round the number of runners to 1600 then, $\frac{1}{2}$ of 1600 is 800. Since $\frac{7}{12}$ is more than $\frac{1}{2}$, our estimate is that "more than 800 runners" are women.

Step 4

$$\frac{7}{12} \cdot 1560 = \frac{7 \cdot \overset{130}{\cancel{1560}}}{\underset{1}{\cancel{12}} \cdot 1} = 910$$

Step 5 910 runners are women.

Step 6
The exact answer, 910, fits our estimate of "more than 800."

21. From Exercise 19 the total income is $58,000. The circle graph shows that $\frac{1}{5}$ of the income is for rent.

$$\frac{1}{5} \cdot 58{,}000 = \frac{1}{\underset{1}{\cancel{5}}} \cdot \frac{\overset{11{,}600}{\cancel{58{,}000}}}{1} = 11{,}600$$

The amount of their rent is $11,600.

33. To find the remaining amount of the estate subtract $\frac{7}{8}$ from 1.

$$1 - \frac{7}{8} = \frac{8}{8} - \frac{7}{8} = \frac{1}{8}$$

Multiply the remaining $\frac{1}{8}$ of the estate by the fraction going to the American Cancer Society.

$$\frac{1}{4} \cdot \frac{1}{8} = \frac{1}{32}$$

$\frac{1}{32}$ of the estate goes to the American Cancer Society.

Section 2.7 (pages 167–170)

1. The reciprocal of $\frac{3}{8}$ is $\frac{8}{3}$ because

$$\frac{3}{8} \cdot \frac{8}{3} = \frac{24}{24} = 1.$$

25. $\dfrac{\frac{18}{3}}{\frac{3}{4}} = 18 \div \frac{3}{4}$ Rewrite by using the $\div$ symbol for division.

$$= \frac{18}{1} \div \frac{3}{4} \qquad \text{Write 18 as } \frac{18}{1}.$$

$$= \frac{\overset{6}{\cancel{18}}}{1} \cdot \frac{4}{\underset{1}{\cancel{3}}} \qquad \begin{array}{l}\text{The reciprocal of } \frac{3}{4} \text{ is } \frac{4}{3}.\\ \text{Change "}\div\text{" to " } \cdot \text{ ".}\end{array}$$

$$\qquad\qquad \begin{array}{l}\text{Divide the numerator and}\\ \text{denominator by 3.}\end{array}$$

$$= \frac{6 \cdot 4}{1 \cdot 1} \qquad \text{Multiply.}$$

$$= 24$$

31. *Step 1*
The problem asks for the number of times a measuring cup needs to be filled.

Step 2
Solve the problem by dividing the total number of cups (5) by the size of the measuring cup $\left(\frac{1}{3}\right)$.

Step 3
Because there are three $\frac{1}{3}$-cups in one cup, multiply 3 by 5 to get 15. So, our estimate is 15.

Step 4

$$5 \div \frac{1}{3} = \frac{5}{1} \div \frac{1}{3} = \frac{5}{1} \cdot \frac{3}{1} = 15$$

Step 5
They need to fill the measuring cup 15 times.

Step 6
The exact answer, 15, is the same as our estimate.

45. *Step 1*
The problem asks for the number of towels that can be made.

Step 2
Solve the problem by dividing the 912 yards of fabric by the fraction of a yard $\left(\frac{3}{8}\right)$ needed for each dish towel.

Step 3
Round 912 yards to 900 yards. Round $\frac{3}{8}$ to $\frac{1}{2}$.

Multiply 900 by 2, the reciprocal of $\frac{1}{2}$, to get the estimate of 1800 towels.

Step 4

$$912 \div \frac{3}{8} = \frac{912}{1} \cdot \frac{8}{3} = \frac{\overset{304}{\cancel{912}}}{1} \cdot \frac{8}{\underset{1}{\cancel{3}}} = 2432$$

Step 5 2432 towels can be made.

Step 6
The exact answer, 2432, is close to our estimate, 1800.

Section 2.8 Exercises (pages 177–180)

1. $4\frac{1}{2} \cdot 1\frac{3}{4}$

Estimate: $5 \cdot 2 = 10$

To find the exact answer, change each mixed number to an improper fraction and then multiply.

Exact: $4\frac{1}{2} \cdot 1\frac{3}{4} = \frac{9}{2} \cdot \frac{7}{4} = \frac{63}{8} = 7\frac{7}{8}$

13. $1\frac{1}{4} \div 3\frac{3}{4}$

Estimate: $1 \div 4 = \frac{1}{4}$

To find the exact answer, first change each mixed number to an improper fraction. Then, use the reciprocal of the second fraction and multiply.

Exact:

$$1\frac{1}{4} \div 3\frac{3}{4} = \frac{5}{4} \div \frac{15}{4} = \frac{\overset{1}{\cancel{5}}}{\underset{1}{\cancel{4}}} \cdot \frac{\overset{1}{\cancel{4}}}{\underset{3}{\cancel{15}}} = \frac{1}{3}$$

Chapter 3 Adding and Subtracting Fractions

Section 3.1 (pages 203–204)

13. Add the numerators. Keep the denominator.

$$\frac{4}{15} + \frac{2}{15} + \frac{5}{15} = \frac{4+2+5}{15} = \frac{11}{15}$$

29. First, subtract the numerators and keep the denominator. Then write the fraction in lowest terms.

$$\frac{47}{36} - \frac{5}{36} = \frac{47-5}{36} = \frac{42}{36}$$

$$= \frac{42 \div 6}{36 \div 6} = \frac{7}{6} = 1\frac{1}{6}$$

39. First, add the two fractions of land that were purchased.

$$\frac{9}{10} + \frac{3}{10} = \frac{9+3}{10} = \frac{12}{10}$$

Next, subtract the fraction of land that was planted in carrots.

$$\frac{12}{10} - \frac{7}{10} = \frac{12-7}{10} = \frac{5}{10} = \frac{1}{2}$$

$\frac{1}{2}$ acre is planted in squash.

Section 3.2 (pages 213–216)

11. 20 and 50

Multiples of 50:

50, 100, 150, 200, 250, . . .

100 is the first multiple of 50 that is divisible by 20.   ($100 \div 20 = 5$)

The least common multiple of 20 and 50 is 100.

19. Find the prime factorization for each number.

4, 6, 8, 10

$4 = 2 \cdot 2$
$6 = 2 \cdot 3$
$8 = 2 \cdot 2 \cdot 2$
$10 = 2 \cdot 5$

LCM $= 2 \cdot 2 \cdot 2 \cdot 3 \cdot 5 = 120$

The LCM of 4, 6, 8, and 10 is 120.

37. $\frac{3}{16} = \frac{?}{64}$ Divide 64 by 16, getting 4. Now multiply both numerator and denominator of $\frac{3}{16}$ by 4.

$$\frac{3}{16} = \frac{3 \cdot 4}{16 \cdot 4} = \frac{12}{64}$$

53. $\frac{109}{1512}, \frac{23}{392}$

```
2 | 1512   392
2 |  756   196
2 |  378    98
3 |  189    49
3 |   63    49
3 |   21    49
7 |    7    49
7 |    1     7
        1     1
```

The LCM of the denominators is
$2 \cdot 2 \cdot 2 \cdot 3 \cdot 3 \cdot 3 \cdot 7 \cdot 7 = 10,584$.

59. Find the prime factorization for each number.

5, 7, 14, 10

$5 = 1 \cdot 5$
$7 = 1 \cdot 7$
$14 = 2 \cdot 7$
$10 = 2 \cdot 5$

LCM $= 2 \cdot 5 \cdot 7 = 70$

The LCM of 5, 7, 14, and 10 is 70.

Section 3.3 (pages 221–224)

1. The least common multiple of 4 and 8 is 8. Rewrite both fractions as fractions with a least common denominator of 8. Then add the numerators.

$$\frac{3}{4} + \frac{1}{8} = \frac{6}{8} + \frac{1}{8}$$

$$= \frac{6+1}{8}$$

$$= \frac{7}{8}$$

29. The least common multiple of 12 and 4 is 12. Rewrite both fractions as fractions with a least common denominator of 12. Then subtract the numerators and write the resulting fraction in lowest terms.

$$\frac{5}{12} - \frac{1}{4} = \frac{5}{12} - \frac{3}{12}$$

$$= \frac{5-3}{12}$$

$$= \frac{2}{12} = \frac{1}{6}$$

41. Add the fractions to find the total length of the screw. The least common denominator of the fractions is 40. Rewrite all three fractions with a denominator of 40. Then add the numerators.

$$\frac{1}{8} + \frac{1}{4} + \frac{2}{5} = \frac{5}{40} + \frac{10}{40} + \frac{16}{40}$$

$$= \frac{5+10+16}{40}$$

$$= \frac{31}{40}$$

The total length of the screw is $\frac{31}{40}$ inch.

49. One way to compare fractions accurately is to rewrite each fraction with a common denominator.

$$\frac{1}{3} = \frac{8}{24}, \frac{1}{6} = \frac{4}{24}, \frac{1}{8} = \frac{3}{24}, \frac{1}{12} = \frac{2}{24}, \frac{7}{24}$$

The fraction with the largest numerator is the largest fraction. This is $\frac{8}{24}$ or $\frac{1}{3}$, which is "Work and Travel."

To find the number of hours, multiply.

$$\frac{1}{3} \cdot 24 = \frac{1}{\overset{}{\underset{1}{3}}} \cdot \frac{\overset{8}{24}}{1} = \frac{8}{1} = 8$$

8 hours were spent on work and travel.

Note that this is just the numerator in the equivalent fraction $\frac{8}{24}$.

To find what fraction of the day was spent on "Work and Travel" and "Class," add the fractions $\frac{1}{3}$ and $\frac{1}{6}$. The least common denominator is 6. Rewrite $\frac{1}{3}$ with a denominator of 6. Then add the numerators and write in lowest terms.

$$\frac{1}{3} + \frac{1}{6} = \frac{2}{6} + \frac{1}{6} = \frac{2+1}{6} = \frac{3}{6} = \frac{1}{2}$$

Section 3.4 (pages 231–238)

11. *Estimate:*   *Exact:*

$$23 \xleftarrow{\text{Rounds to}} \begin{cases} 22\frac{3}{4} = 22\frac{21}{28} \end{cases}$$

$$+15 \xleftarrow{\text{Rounds to}} \begin{cases} +15\frac{3}{7} = 15\frac{12}{28} \end{cases}$$

$$38 \qquad\qquad 37\frac{33}{28}$$

$$37\frac{33}{28} = 37 + 1\frac{5}{28} = 38\frac{5}{28}$$

21. *Estimate:*   *Exact:*

$$17 \xleftarrow{\text{Rounds to}} \begin{cases} 17 \end{cases}$$

$$-7 \xleftarrow{\text{Rounds to}} \begin{cases} -6\frac{5}{8} \end{cases}$$

$$10$$

To subtract $\frac{5}{8}$, first regroup the whole number 17 into $16 + 1$.

$$17 = 16 + 1 = 16 + \frac{8}{8}$$

Now you can subtract.

$$16\frac{8}{8}$$
$$-6\frac{5}{8}$$
$$\overline{10\frac{3}{8}}$$

39.

$$8\frac{3}{4} = \frac{35}{4} = \frac{70}{8}$$
$$-5\frac{7}{8} = \frac{47}{8} = \frac{47}{8}$$
$$\overline{\frac{23}{8} = 2\frac{7}{8}}$$

55. Subtract the size of the smallest hose clamp from the size of the largest hose clamp.

Estimate: $3 - 1 = 2$ in.

Exact: $2\frac{3}{4} = \frac{11}{4} = \frac{44}{16}$

$$-\frac{9}{16} = \frac{9}{16} = \frac{9}{16}$$

$$\overline{\frac{35}{16} = 2\frac{3}{16}}$$

The difference in size is $2\frac{3}{16}$ inches.

SOLUTIONS

65. Add the weights.

Estimate:

$$59 + 24 + 17 + 29 + 58 = 187 \text{ tons}$$

Exact:

$$58\frac{1}{2} = 58\frac{12}{24}$$

$$23\frac{5}{8} = 23\frac{15}{24}$$

$$16\frac{5}{6} = 16\frac{20}{24}$$

$$29\frac{1}{4} = 29\frac{6}{24}$$

$$+\ 58\frac{1}{3} = 58\frac{8}{24}$$

$$184\frac{61}{24} = 184 + 2\frac{13}{24} = 186\frac{13}{24}$$

The total weight is $186\frac{13}{24}$ tons.

Section 3.5 (pages 243–246)

7.

33. $\left(\dfrac{3}{2}\right)^4 = \dfrac{3}{2} \cdot \dfrac{3}{2} \cdot \dfrac{3}{2} \cdot \dfrac{3}{2} = \dfrac{81}{16} = 5\dfrac{1}{16}$

49. $6\left(\dfrac{2}{3}\right)^2\left(\dfrac{1}{2}\right)^3 = 6\left(\dfrac{2}{3} \cdot \dfrac{2}{3}\right)\left(\dfrac{1}{2} \cdot \dfrac{1}{2} \cdot \dfrac{1}{2}\right)$

Simplify the expressions with exponents.

$$= \dfrac{\overset{1}{\cancel{\overset{2}{\cancel{6}}}} \cdot \overset{1}{\cancel{2}} \cdot \overset{1}{\cancel{2}} \cdot 1 \cdot 1 \cdot 1}{\underset{1}{\cancel{3}} \cdot 3 \cdot \underset{1}{\cancel{2}} \cdot \underset{1}{\cancel{2}} \cdot \underset{1}{\cancel{2}}}$$

$$= \dfrac{1}{3}$$

63. $\left(\dfrac{3}{4}\right)^2 - \left(\dfrac{1}{2} - \dfrac{1}{6}\right) \div \dfrac{4}{3}$

$$= \left(\dfrac{3}{4}\right)^2 - \left(\dfrac{3}{6} - \dfrac{1}{6}\right) \div \dfrac{4}{3}$$

Work inside parentheses first.

$$= \left(\dfrac{3}{4}\right)^2 - \left(\dfrac{2}{6}\right) \cdot \dfrac{3}{4}$$

Simplify the expression with the exponent.

$$= \dfrac{3}{4} \cdot \dfrac{3}{4} - \left(\dfrac{2}{6}\right) \cdot \dfrac{3}{4} \quad \text{Multiply.}$$

$$= \dfrac{9}{16} - \dfrac{\overset{1}{\cancel{2}}}{\underset{2}{\cancel{6}}} \cdot \dfrac{\overset{1}{\cancel{3}}}{\underset{2}{\cancel{4}}}$$

$$= \dfrac{9}{16} - \dfrac{1}{4} \quad \begin{array}{l}\text{The LCD is 16. Rewrite } \tfrac{1}{4} \\ \text{as a fraction with a} \\ \text{denominator of 16.}\end{array}$$

$$= \dfrac{9}{16} - \dfrac{4}{16} \quad \text{Subtract.}$$

$$= \dfrac{5}{16}$$

Chapter 4 Decimals

Section 4.1 (pages 271–274)

1.

tens	ones		tenths		
7	0	.	4	8	9

---

17. Reorder as: 5 thousands, 4 hundreds, 0 tens, 0 tenths, 4 hundredths, 5 thousandths: 5406.045

31. $0.205 = \dfrac{205}{1000} = \dfrac{205 \div 5}{1000 \div 5} = \dfrac{41}{200}$

45. Three decimal places is *thousandths,* so 1.075 is one and seventy-five *thousandths.*

53. seven hundred three ten-thousandths:

$$\dfrac{703}{10,000} = 0.0703$$

59. 8-pound test line has a diameter of 0.010 inch. 0.010 inch is read ten thousandths of an inch.

$$0.010 = \dfrac{10}{1000} = \dfrac{10 \div 10}{1000 \div 10} = \dfrac{1}{100} \text{ inch}$$

Section 4.2 (pages 281–282)

9. Draw a cut-off line after the tenths place: 793.9|88

The first digit cut is 8, which is 5 or more, so round up the tenths place.

$$\begin{array}{r} 793.9 \\ +\ \ 0.1 \\ \hline 794.0 \end{array}$$

Answer: $793.988 \approx 794.0$

27. Round $310.08 to the nearest dollar. Draw a cut-off line: $310.|08

The first digit cut is 0, which is 4 or less. The part you keep stays the same. Union dues: $310.08 \approx 310

35. Round $999.73 to the nearest dollar. Draw a cut-off line: $999.|73

The first digit cut is 7, which is 5 or more, so round up.

$$\begin{array}{r} \$999 \\ +\ \ \ 1 \\ \hline \$1000 \end{array}$$

Answer: $999.73 \approx 1000

Section 4.3 (pages 287–290)

11. Subtract 0.291 from 0.4

$$\begin{array}{r} \text{Line up decimal points.} \\ 0.400 \quad \text{Write two zeros.} \\ -\ 0.291 \\ \hline 0.109 \end{array}$$

23.

Estimate:	*Exact:*
400	392.700
1	0.865
+ 20	+ 21.080
421	414.645

41. First add the three players' heights.

Estimate:	*Exact:*
2	1.83
2	2.16
+ 2	+ 2.11
6 meters	6.10 meters

Now subtract the players' combined height of 6.1 meters from the rhino's height of 6.4 meters.

$$6.4 - 6.1 = 0.3$$

So, the NBA stars' combined height is 0.3 meter less than the rhino's height.

47. Subtract the price of the regular fishing line, $4.84, from the price of the fluorescent fishing line, $5.14.

Estimate:	*Exact:*
$5	$5.14
− 5	− 4.84
$0	$0.30

The fluorescent fishing line costs $0.30 more than the regular fishing line.

Section 4.4 (pages 293–296)

11. (7.2) (0.06)

72	7.2 ← 1 decimal place
× 6	× 0.06 ← 2 decimal places
432	0.432 ← 3 decimal places

17. (0.006) (0.0052)

$$\begin{array}{r} 0.0052 \quad \leftarrow 4 \text{ decimal places} \\ \times\ 0.006 \quad \leftarrow 3 \text{ decimal places} \\ \hline 0.0000312 \quad \leftarrow 7 \text{ decimal places} \end{array}$$

43. Multiply the number of gallons that she pumped into her pickup truck by the price per gallon. Use a calculator.

$$20.510 \times 4.389 = 90.01839$$

Round 90.01839 to the nearest cent. Michelle paid $90.02 for the gas.

59. (a) Find the cost for three short-sleeved, solid-color shirts

$$\begin{array}{r} \$14.75 \\ \times\ \ \ 3 \\ \hline \$44.25 \end{array}$$

Based on this subtotal, shipping is $5.95. Next, find the cost of the three mono-grams.

$$\begin{array}{r} \$4.95 \\ \times\ \ \ 3 \\ \hline \$14.85 \end{array}$$

Add these amounts, plus $5.00 for a gift box.

$$\begin{array}{rl} \$44.25 & \leftarrow \text{ Shirts} \\ 5.95 & \leftarrow \text{ Shipping} \\ 14.85 & \leftarrow \text{ Monograms} \\ +\ \ 5.00 & \leftarrow \text{ Gift box} \\ \hline \$70.05 \end{array}$$

The total cost is $70.05.

(b) Subtract the cost of the shirts to find the difference.

$$\begin{array}{r} \$70.05 \\ -\ 44.25 \\ \hline \$25.80 \end{array}$$

The monograms, gift box, and shipping added $25.80 to the cost of the gift.

Section 4.5 (pages 305–308)

7.

Move decimal point in divisor and dividend 1 place; write 0 in the dividend.

9. Given: $108 \div 18 = 6$

Find: $0.108 \div 1.8$ by moving the decimal points.

$$1.8 \overline{)0.1\,08} \quad \overset{.06}{}$$

So, $0.108 \div 1.8 = 0.06$

15. $\dfrac{3.1}{0.006}$

$$0.006 \overline{)3.1\,0\,0\,0\,0\,0} \quad \overset{5\,1\,6.\,6\,6\,6}{}$$

Line up decimal points.

Move decimal point in divisor and dividend three places. Write 00 in dividend.

$$\begin{array}{r} 3\ 0 \\ \hline 1\ 0 \\ 6 \\ \hline 4\ 0 \\ 3\ 6 \\ \hline 4\ 0 \\ 3\ 6 \\ \hline 4\ 0 \\ 3\ 6 \\ \hline 4\ 0 \\ 3\ 6 \\ \hline 4 \end{array}$$

Write 0 in dividend.

Write 0 in dividend.

Write 0 in dividend.

Stop and round answer to the nearest hundredth.

The quotient is 516.67 (rounded).

27. $307.02 \div 5.1 = 6.2$

Estimate: $300 \div 5 = 60$

The answer 6.2 is *unreasonable* because it is so much less than 60.

$$5.1 \overline{)3\,0\,7.0\,2} \quad \overset{6\,0.\,2}{}$$

$$\begin{array}{r} 3\ 0\ 6 \\ \hline 1\ 0 \\ 0 \\ \hline 1\ 0\ 2 \\ 1\ 0\ 2 \\ \hline 0 \end{array}$$

The correct answer is 60.2, which is close to the estimate of 60.

53. $33 - 3.2\underbrace{(0.68 + 9)} - 1.3^2$ Parentheses; Exponent

$33 - 3.2(9.68) - 1.69$ Multiply

$33 - 30.976 - 1.69$ Subtract

$2.024 - 1.69 = 0.334$ Subtract

61. Divide 100,000 box tops (the answer from Exercise 59) by 38 weeks.

$$38 \overline{)1\,0\,0,\,0\,0\,0.\,0} \quad \overset{2\,6\,3\,1.\,5}{}$$

$$\begin{array}{r} 7\ 6 \\ \hline 2\ 4\ 0 \\ 2\ 2\ 8 \\ \hline 1\ 2\ 0 \\ 1\ 1\ 4 \\ \hline 6\ 0 \\ 3\ 8 \\ \hline 2\ 2\ 0 \\ 1\ 9\ 0 \\ \hline 3\ 0 \end{array}$$

2631.5 rounds to 2632.

The school needs to collect 2632 box tops (rounded) during each of the 38 weeks.

Section 4.6 (pages 313–316)

17. $3\dfrac{5}{8} = \dfrac{29}{8} = 3.625$

$$8 \overline{)2\,9.\,0\,0\,0} \quad \overset{3.\,6\,2\,5}{}$$

$$\begin{array}{r} 2\ 4 \\ \hline 5\ 0 \\ 4\ 8 \\ \hline 2\ 0 \\ 1\ 6 \\ \hline 4\ 0 \\ 4\ 0 \\ \hline 0 \end{array}$$

33. $0.35 = \dfrac{35}{100} = \dfrac{35 \div 5}{100 \div 5} = \dfrac{7}{20}$

49. Write zeros so that all the numbers have four decimal places. The acceptable lengths must be greater than 0.9980 cm and less than 1.0020 cm.

$1.0100 > 1.0020$ *unacceptable*

$0.9991 > 0.9980$ and

$0.9991 < 1.0020$ *acceptable*

$1.0007 > 0.9980$ and

$1.0007 < 1.0020$ *acceptable*

$0.9900 < 0.9980$ *unacceptable*

The lengths of 0.9991 cm and 1.0007 cm are acceptable.

63. $\dfrac{3}{8}, \dfrac{2}{5}, 0.37, 0.4001$

$\dfrac{3}{8} = 0.3750$

$\dfrac{2}{5} = 0.4000$

$0.37 = 0.3700 \leftarrow$ least

$0.4001 = 0.4001 \leftarrow$ greatest

From least to greatest: $0.37, \dfrac{3}{8}, \dfrac{2}{5}, 0.4001$

Chapter 5 Ratio and Proportion

Section 5.1 (pages 337–340)

3. $100 to $50

$$\dfrac{\$100}{\$50} = \dfrac{100}{50} = \dfrac{100 \div 50}{50 \div 50} = \dfrac{2}{1}$$

15. $1\dfrac{1}{4}$ to $1\dfrac{1}{2}$

$$\dfrac{1\frac{1}{4}}{1\frac{1}{2}} = \dfrac{\frac{5}{4}}{\frac{3}{2}} = \dfrac{5}{4} \div \dfrac{3}{2} = \dfrac{5}{\overset{}{4}} \cdot \dfrac{\overset{1}{2}}{3} = \dfrac{5}{6}$$

35. **(a)** rent to utilities

$$\dfrac{\$750}{\$125} = \dfrac{750}{125} = \dfrac{750 \div 125}{125 \div 125} = \dfrac{6}{1}$$

(b) rent to food

$$\dfrac{\$750}{\$300} = \dfrac{750}{300} = \dfrac{750 \div 150}{300 \div 150} = \dfrac{5}{2}$$

(c) rent and utilities to total budget

rent and utilities: $\$750 + \$125 = \$875$

total budget

$\$225 + \$125 + \$200 + \$400$

$+ \$300 + \$750 = \$2000$

$$\dfrac{\$875}{\$2000} = \dfrac{875}{2000} = \dfrac{875 \div 125}{2000 \div 125} = \dfrac{7}{16}$$

43. The increase in price is

$\$12.50 - \$10 = \$2.50$

$$\dfrac{\$2.50}{\$10} = \dfrac{2.50}{10} = \dfrac{2.50 \cdot 10}{10 \cdot 10} = \dfrac{25}{100}$$

$$= \dfrac{25 \div 25}{100 \div 25} = \dfrac{1}{4}$$

The ratio of the increase in price to the original price is $\dfrac{1}{4}$.

Section 5.2 (pages 345–348)

15. Miles traveled:

$28,396.7 - 28,058.1 = 338.6$

Miles per gallon:

$$\dfrac{338.6 \text{ miles} \div 16.2}{16.2 \text{ gallons} \div 16.2} \approx 20.90 \approx 20.9$$

19.

Size	Cost per Unit
12 ounces	$\dfrac{\$2.49}{12 \text{ ounces}} \approx \$0.208$
14 ounces	$\dfrac{\$2.89}{14 \text{ ounces}} \approx \$0.206$
18 ounces	$\dfrac{\$3.96}{18 \text{ ounces}} = \$0.22$

Because $0.206 is the lowest cost per ounce, the best buy is 14 ounces for $2.89.

33. One battery for $1.79 is like getting 3 batteries for $1.79 \div 3 \approx \$0.597$ per battery.

An eight-pack of AAA batteries for $4.99 is $4.99 \div 8 \approx \$0.624$ per battery.

Because $0.597 is the lower cost per battery, the better buy is the one battery package.

Section 5.3 (pages 353–354)

11. $\dfrac{150}{200} = \dfrac{200}{300}$

$$\dfrac{150 \div 50}{200 \div 50} = \dfrac{3}{4} \quad \text{and} \quad \dfrac{200 \div 100}{300 \div 100} = \dfrac{2}{3}$$

Because $\dfrac{3}{4}$ is *not* equivalent to $\dfrac{2}{3}$, the proportion is *false*.

31. $\dfrac{2\frac{5}{8}}{3\frac{1}{4}} = \dfrac{21}{26}$

Cross products:

$$2\dfrac{5}{8} \cdot 26 = \dfrac{21}{\overset{}{8}} \cdot \dfrac{\overset{13}{26}}{1} = \dfrac{273}{4} = 68\dfrac{1}{4}$$

$$3\dfrac{1}{4} \cdot 21 = \dfrac{13}{4} \cdot \dfrac{21}{1} = \dfrac{273}{4} = 68\dfrac{1}{4}$$

The cross products are *equal*, so the proportion is *true*.

35. $\dfrac{2\frac{3}{10}}{8.05} = \dfrac{1\frac{1}{4}}{0.9}$

SOLUTIONS

(continued)

Cross products:

$$2\frac{3}{10}(0.9) = \frac{23}{10} \cdot \frac{9}{10} = \frac{207}{100} = 2\frac{7}{100}$$

or $(2.3)(0.9) = 2.07$

$$(8.05)\frac{1}{4} = (8.05)(0.25) = 2.0125$$

or $8\frac{5}{100} \cdot \frac{1}{4} = \frac{805}{100} \cdot \frac{1}{4} = \frac{805}{400} = 2\frac{1}{80}$

The cross products are *unequal*, so the proportion is *false*.

$$\left(2\frac{7}{100} \neq 2\frac{1}{80} \quad \text{or} \quad 2.07 \neq 2.0125\right)$$

Section 5.4 (pages 359–360)

7. $\dfrac{42}{x} = \dfrac{18}{39}$

$x \cdot 18 = 42 \cdot 39$ — Cross products are equivalent

$\dfrac{x \cdot \cancel{18}^{1}}{\cancel{18}_{1}} = \dfrac{1638}{18}$ — Divide both sides by 18.

$x = 91$

Check

$42 \cdot 39 = 1638$

$91 \cdot 18 = 1638$

13. $\dfrac{99}{55} = \dfrac{44}{x}$ **OR** $\dfrac{9}{5} = \dfrac{44}{x}$

$9 \cdot x = 5 \cdot 44$ — Cross products are equivalent

$\dfrac{\cancel{9}^{1} \cdot x}{\cancel{9}_{1}} = \dfrac{220}{9}$ — Divide both sides by 9.

$x = \dfrac{220}{9}$

$x \approx 24.44$

Check

$55 \cdot 44 = 2420$

$99 \cdot 24.44 = 2419.56$

Slightly different because of rounding.

21. $\dfrac{2\frac{1}{3}}{1\frac{1}{2}} = \dfrac{x}{2\frac{1}{4}}$

$1\frac{1}{2} \cdot x = 2\frac{1}{3} \cdot 2\frac{1}{4}$

$\dfrac{3}{2} \cdot x = \dfrac{7}{\cancel{3}} \cdot \dfrac{\cancel{9}^{3}}{4}$

$\dfrac{3}{2} \cdot x = \dfrac{21}{4}$

$\dfrac{\frac{3}{2} \cdot x}{\frac{3}{2}} = \dfrac{\frac{21}{4}}{\frac{3}{2}}$

$x = \dfrac{21}{4} \div \dfrac{3}{2} = \dfrac{\cancel{21}^{7}}{\cancel{4}_{2}} \cdot \dfrac{\cancel{2}^{1}}{\cancel{3}_{1}} = \dfrac{7}{2} = 3\frac{1}{2}$

25. $\dfrac{x}{\frac{3}{50}} = \dfrac{0.15}{1\frac{4}{5}}$

Change to decimals:

$$\frac{3}{50} = 3 \div 50 = 0.06$$

$$1\frac{4}{5} = \frac{9}{5} \text{ and } 9 \div 5 = 1.8$$

$$\frac{x}{0.06} = \frac{0.15}{1.8}$$

$x \cdot 1.8 = (0.06)(0.15)$

$\dfrac{x \cdot \cancel{1.8}}{\cancel{1.8}} = \dfrac{0.009}{1.8}$

$x = 0.005$

Change to fractions:

$$0.15 = \frac{15 \div 5}{100 \div 5} = \frac{3}{20}$$

$$\frac{x}{\frac{3}{50}} = \frac{\frac{3}{20}}{1\frac{4}{5}}$$

$1\frac{4}{5} \cdot x = \dfrac{3}{50} \cdot \dfrac{3}{20}$

$\dfrac{1\frac{4}{5} \cdot x}{1\frac{4}{5}} = \dfrac{\frac{9}{1000}}{1\frac{4}{5}}$

$x = \dfrac{9}{1000} \div 1\frac{4}{5} = \dfrac{\cancel{9}^{1}}{\cancel{1000}_{200}} \cdot \dfrac{\cancel{5}^{1}}{\cancel{9}_{1}}$

$= \dfrac{1}{200}$

Compare answers.

$$0.005 = \frac{5 \div 5}{1000 \div 5} = \frac{1}{200} \leftarrow \text{Matches}$$

Section 5.5 (pages 367–370)

9. $\dfrac{6 \text{ ounces}}{7 \text{ servings}} = \dfrac{x \text{ ounces}}{12 \text{ servings}}$

$7 \cdot x = 6 \cdot 12$

$\dfrac{\cancel{7} \cdot x}{\cancel{7}} = \dfrac{72}{7}$

$x \approx 10.3 \approx 10$

You need about 10 ounces for 12 servings.

15. The length of the dining area is the same as the length of the kitchen, which is 14 feet (from Exercise 13).

Find the width of the dining area.

$4.5 \text{ inches} - 2.5 \text{ inches} = 2 \text{ inches}$ on the floor plan

$$\frac{1 \text{ inch}}{4 \text{ feet}} = \frac{2 \text{ inches}}{x \text{ feet}}$$

$1 \cdot x = 4 \cdot 2$

$x = 8$

The dining area is 8 feet wide.

19. $\dfrac{7 \text{ refresher}}{10 \text{ entering}} = \dfrac{x \text{ refresher}}{2950 \text{ entering}}$

$10 \cdot x = 7 \cdot 2950$

$\dfrac{\cancel{10} \cdot x}{\cancel{10}} = \dfrac{20{,}650}{10}$

$x = 2065$

2065 students will probably need a refresher course. This is a reasonable answer because it's more than half the students, but not all the students.

Incorrect setup

$$\frac{10 \text{ entering}}{7 \text{ refresher}} = \frac{x \text{ refresher}}{2950 \text{ entering}}$$

$7 \cdot x = 10 \cdot 2950$

$\dfrac{\cancel{7} \cdot x}{\cancel{7}} = \dfrac{29{,}500}{7}$

$x \approx 4214$

The *incorrect* setup gives an *unreasonable* estimate of 4214 entering students since there are only 2950 entering students.

31. Coretta $\left\{\dfrac{1.05 \text{ meters}}{1.68 \text{ meters}} = \dfrac{6.58 \text{ meters}}{x \text{ meters}}\right\}$ tree

$1.05 \cdot x = 1.68(6.58)$

$\dfrac{\cancel{1.05} \cdot x}{\cancel{1.05}} = \dfrac{11.0544}{1.05}$

$x = 10.528 \approx 10.53$

The height of the tree is about 10.53 meters.

37. First find the number of calories in a $\frac{1}{2}$-cup serving of bran cereal.

$$\frac{\frac{1}{3} \text{ cup}}{80 \text{ calories}} = \frac{\frac{1}{2} \text{ cup}}{x \text{ calories}}$$

$\dfrac{1}{3} \cdot x = 80 \cdot \dfrac{1}{2}$

$\dfrac{\frac{1}{3} \cdot x}{\frac{1}{3}} = \dfrac{40}{\frac{1}{3}}$

$x = \dfrac{40}{1} \cdot \dfrac{3}{1}$

$x = 120 \text{ calories}$

Then find the number of grams of fiber in a $\frac{1}{2}$-cup serving of bran cereal.

$$\frac{\frac{1}{3} \text{ cup}}{8 \text{ grams}} = \frac{\frac{1}{2} \text{ cup}}{x \text{ grams}}$$

$\dfrac{1}{3} \cdot x = 8 \cdot \dfrac{1}{2}$

$\dfrac{\frac{1}{3} \cdot x}{\frac{1}{3}} = \dfrac{4}{\frac{1}{3}}$

$x = \dfrac{4}{1} \cdot \dfrac{3}{1}$

$x = 12 \text{ grams}$

A $\frac{1}{2}$-cup serving of bran cereal provides 120 calories and 12 grams of fiber.

Chapter 6 Percent

Section 6.1 (pages 391–396)

13. $0.5\% = 0.005$

Drop the percent sign. Attach two zeros so the decimal point can be moved two places to the left.

27. $2 = 200\%$

Two zeros are attached so the decimal point can be moved two places to the right. Attach a percent sign.

51. 100% is all of the children.

So, 100% of 20 children is 20 children.

20 children are served both meals.

77. 16.8% of the population of Spain is 65 or older. To write 16.8% as a decimal, drop the percent sign and move the decimal point two places to the left, resulting in 0.168. So, 0.168 of the population of Spain is 65 or older.

Section 6.2 (pages 403–408)

9. First write 6.25 over 100. Then to get a whole number in the numerator, multiply the numerator and denominator by 100. Finally, write the fraction in lowest terms.

$$6.25\% = \frac{6.25}{100} = \frac{6.25(100)}{100(100)}$$

$$= \frac{625 \div 625}{10,000 \div 625} = \frac{1}{16}$$

41. $\dfrac{5}{9} = \dfrac{p}{100}$

$9 \cdot p = 5 \cdot 100$ Find cross products.

$\dfrac{\cancel{9} \cdot p}{\cancel{9}} = \dfrac{500}{9}$ Divide both sides by 9.

$p = \dfrac{500}{9}$

$p \approx 55.5\overline{5}$

Thus, $\dfrac{5}{9} = 55.6\%$ (rounded).

47. $87.5\% = 0.875$ decimal

$= \dfrac{875 \div 125}{1000 \div 125} = \dfrac{7}{8}$ fraction

63. $\dfrac{1}{200} = \dfrac{1 \cdot 5}{200 \cdot 5} = \dfrac{5}{1000} = 0.005$ decimal

$= 0.5\%$ percent

67. $3\dfrac{1}{4} = 3\dfrac{1 \cdot 25}{4 \cdot 25} = 3\dfrac{25}{100} = 3.25$ decimal

$= 325\%$ percent

77. 64 out of 80 employees have cell phones. Therefore, $80 - 64 = 16$ employees do *not* have cell phones.

$\dfrac{16}{80} = \dfrac{16 \div 16}{80 \div 16} = \dfrac{1}{5}$ fraction

$\dfrac{1}{5} = \dfrac{1 \cdot 2}{5 \cdot 2} = \dfrac{2}{10} = 0.2$ decimal

$0.2 = 0.20 = 20\%$ percent

Section 6.3 (pages 413–416)

7. part = 15, whole = 60

$\dfrac{15}{60} = \dfrac{x}{100}$

$\dfrac{1}{4} = \dfrac{x}{100}$ $\dfrac{15}{60}$ is $\dfrac{1}{4}$ in lowest terms.

$4 \cdot x = 1 \cdot 100$ Find cross products.

$\dfrac{\cancel{4} \cdot x}{\cancel{4}} = \dfrac{100}{4}$ Divide both sides by 4.

$x = 25$

The percent is 25, written as 25%.

21. whole = 5000, part = 20

$\dfrac{20}{5000} = \dfrac{x}{100}$ Percent proportion.

$5000 \cdot x = 20 \cdot 100$ Find cross products.

$\dfrac{\cancel{5000} \cdot x}{\cancel{5000}} = \dfrac{2000}{5000}$ Divide both sides by 5000.

$x = 0.4$

The percent is 0.4, written as 0.4%.

33. $\underbrace{\text{12 injections}}_{\text{part}}$ is $\underbrace{20\%}_{\text{percent}}$ of what number of injections?
$\underbrace{}_{\substack{\text{whole} \\ \text{(unknown)}}}$

$\dfrac{12}{\text{unknown}} = \dfrac{20}{100}$

47. $\underbrace{86}_{\text{part}}$ of $\underbrace{\text{142 people}}_{\text{whole}}$ is $\underbrace{\text{what percent?}}_{\substack{\text{percent} \\ \text{(unknown)}}}$

$\dfrac{86}{142} = \dfrac{\text{unknown}}{100}$

Section 6.4 (pages 495–430)

13. Write the percent 225% as a decimal, 2.25.

225% of 680 tables

$(2.25)(680) = 1530$

part = 1530 tables

27. part is 350; percent is 12.5

$\dfrac{\text{part}}{\text{whole}} = \dfrac{\text{percent}}{100}$

so $\dfrac{350}{x} = \dfrac{12.5}{100}$

$12.5 \cdot x = 35{,}000$ Cross products

$\dfrac{\cancel{12.5} \cdot x}{\cancel{12.5}} = \dfrac{35{,}000}{12.5}$ Divide both sides by 12.5.

$x = 2800$

$12\dfrac{1}{2}\%$ of 2800 is 350.

37. part is 64; whole is 344

$\dfrac{\text{part}}{\text{whole}} = \dfrac{\text{percent}}{100}$

$\dfrac{64}{344} = \dfrac{x}{100}$ **OR** $\dfrac{8}{43} = \dfrac{x}{100}$

Cross products $43 \cdot x = 8 \cdot 100$

Divide both sides by 43. $\dfrac{\cancel{43} \cdot x}{\cancel{43}} = \dfrac{800}{43}$

$x \approx 18.6$

$64 is 18.6% (rounded) of $344.

Section 6.3 continued — right column

43. part is unknown; whole is 240; percent is 22

$\dfrac{\text{part}}{\text{whole}} = \dfrac{\text{percent}}{100}$

$\dfrac{x}{240} = \dfrac{22}{100}$ **OR** $\dfrac{x}{240} = \dfrac{11}{50}$

Cross products $x \cdot 50 = 240 \cdot 11$

Divide both sides by 50. $\dfrac{x \cdot \cancel{50}}{\cancel{50}} = \dfrac{2640}{50}$

$x = 52.8$

The amount withheld is $52.80.

55. part is 960; whole is 48,000; percent is unknown

$\dfrac{\text{part}}{\text{whole}} = \dfrac{\text{percent}}{100}$

$\dfrac{960}{48{,}000} = \dfrac{x}{100}$ **OR** $\dfrac{1}{50} = \dfrac{x}{100}$

Cross products $50 \cdot x = 1 \cdot 100$

Divide both sides by 50. $\dfrac{50 \cdot x}{50} = \dfrac{100}{50}$

$x = 2$

There are 2% of these jobs filled by women.

67. percent is 20; whole is 3100; part is unknown

$\dfrac{\text{part}}{\text{whole}} = \dfrac{\text{percent}}{100}$

$\dfrac{x}{3100} = \dfrac{20}{100}$ **OR** $\dfrac{x}{3100} = \dfrac{1}{5}$

Find cross products. $\dfrac{\cancel{5} \cdot x}{\cancel{5}} = \dfrac{3100}{5}$

Divide both sides by 5.

$x = 620$

Add the application fee.

$620 + $25 = 645

The total charge is $645.

Section 6.5 (pages 435–438)

1. Write 25% as the decimal 0.25. The whole is 1080. Let x represent the unknown part.

part = percent • whole

$x = (0.25)(1080)$

$x = 270$

25% of 1080 blood donors is 270 blood donors.

21. The part is 3.75 and the percent is

$1\dfrac{1}{4}\% = 1.25\%$ or 0.0125 as a decimal. The whole is unknown.

part = percent • whole

$3.75 = (0.0125)(x)$

$\dfrac{3.75}{0.0125} = \dfrac{(0.0125)(x)}{0.0125}$

$300 = x$

$1\dfrac{1}{4}\%$ of 300 gallons in 3.75 gallons.

29. Because 160 follows *of*, the whole is 160. The part is 2.4, and the percent is unknown.

part = percent • whole

$$2.4 = x \cdot 160$$

$$\frac{2.4}{160} = \frac{x \cdot 160}{160}$$

$$0.015 = x$$

0.015 is 1.5%

1.5% of 160 liters is 2.4 liters.

43. Because 1250 follows *of*, the whole is 1250. The part is 461, and the percent is unknown.

part = percent • whole

$$461 = x \cdot 1250$$

$$\frac{461}{1250} = \frac{x \cdot 1250}{1250}$$

$$0.369 \approx x$$

0.369 is 36.9%

36.9% (rounded) of these Americans rate their health as excellent.

55. Find the sales tax on the new Polaris unit. The whole is 524. The percent is $7\frac{3}{4}\% = 7.75\%$, which is 0.0775 as a decimal

part = percent • whole

$$x = (0.0775)(524)$$

$$x = 40.61$$

Add the sales tax to the purchase price.

$$\$524 + \$40.61 = \$564.61$$

Subtract the trade-in.

$$\$564.61 - \$125 = \$439.61$$

The total cost to the customer is $439.61.

Section 6.6 (pages 447–450)

3. Find the sales tax rate using the sales tax formula. The sales tax is $12.75, and the cost of the item is $425.

sales tax = rate of tax • cost of item

$$\$12.75 = r \cdot \$425$$

$$\frac{12.75}{425} = \frac{r \cdot 425}{425} \quad \text{Divide both sides by 425.}$$

$$0.03 = r$$

0.03 is 3%   Write the decimal as a percent.

The tax rate is 3% and the total cost is $425 + \$12.75 = \$437.75.$

9. The problem asks for the amount of commission. Use the commission formula. The rate of commission is 8%, and the sales are $280.

commission = rate of commission • sales

$$= (8\%)(\$280)$$

$$= (0.08)(\$280)$$

$$= \$22.40$$

The amount of commission is $22.40.

23. The problem asks for the amount of the discount and the sale price. First, find the amount of discount using the discount formula. The rate of discount is 15%, and the original price is $58.40.

amount of discount = rate of discount • original price

$$= (15\%)(\$58.40)$$

$$= (0.15)(\$58.40) \quad \text{Write 15\% as a decimal.}$$

$$= \$8.76 \quad \text{Amount of discount}$$

Now find the sale price by subtracting the amount of the discount ($8.76) from the original price.

The amount of discount is $8.76 and the sale price is $49.64.

31. The problem asks for the rate of sales tax. Use the sales tax formula. The sales tax is $99. The cost of the door is $1980. Use *r* to represent the unknown rate of tax.

sales tax = rate of tax • cost of item

$$\$99 = r \cdot \$1980$$

$$\frac{99}{1980} = \frac{r \cdot 1980}{1980} \quad \text{Divide both sides by 1980.}$$

$$0.05 = r$$

0.05 is 5%   Write the decimal as a percent.

The rate of sales tax is 5%.

47. The problem asks for the cost of the dictionary. First find the amount of discount using the discount formula. The rate of discount is 6%, and the original price is $18.50.

discount = rate of discount • original price

$$= (6\%)(\$18.50)$$

$$= (0.06)(\$18.50) \quad \text{Write 6\% as a decimal.}$$

$$= \$1.11 \quad \text{Amount of discount}$$

To find the sale price, subtract the amount of discount ($1.11) from the original price.

Sale price

$$= \text{original price} - \text{amount of discount}$$

$$= \$18.50 - \$1.11 = \$17.39 \quad \text{sale price}$$

To find the sales tax, use the sales tax formula. The rate of tax is 6% and the cost of the dictionary is $17.39.

sales tax = rate of tax • cost of item

$$= (6\%)(\$17.39)$$

$$= (0.06)(\$17.39) \quad \text{Write 6\% as a decimal.}$$

$$\approx \$1.04 \quad \text{Sales tax}$$

The total cost of the dictionary is $17.39 + \$1.04 = \$18.43.$

57. To find the sales tax, use the sales tax formula.

sales tax = rate of tax • cost of item

$$= \left(7\frac{3}{4}\%\right)(\$1248)$$

$$= (0.0775)(\$1248) \quad \text{Write } 7\frac{3}{4}\% \text{ as a decimal.}$$

$$= \$96.72 \quad \text{Sales tax}$$

The total cost is the sum of the ticket, the sales tax, and the excise tax.

$$\$1248 + \$96.72 + \$13.40 = \$1358.12$$

Section 6.7 (pages 453–456)

7. $2300 at $8\frac{1}{2}\%$ for $2\frac{1}{2}$ years

The principal (*p*) is $2300. The rate (*r*) is $8\frac{1}{2}\%$, or 0.085 as a decimal, and the time (*t*) is $2\frac{1}{2}$ or 2.5 years.

$$I = p \cdot r \cdot t$$

$$= (2300)(0.085)(2.5)$$

$$= 488.75$$

The interest is $488.75.

15. $940 at 3% for 18 months

The principal is $940. The rate is 3% or 0.03, and the time is $\frac{18}{12}$ of a year.

$$I = p \cdot r \cdot t$$

$$= \underbrace{(940)(0.03)}\left(\frac{18}{12}\right)$$

18 months = $\frac{18}{12}$ of a year.

$$= (28.2)(1.5) \quad \frac{18}{12} = (1.5).$$

$$= 42.3$$

The interest is $42.30.

29. $16,850 at $7\frac{1}{2}\%$ for 9 months

First find the interest. The principal is $16,850. The rate is $7\frac{1}{2}\%$ or 0.075, and the time is $\frac{9}{12}$ of a year.

$$I = p \cdot r \cdot t$$

$$= (16,850)(0.075)\left(\frac{9}{12}\right)$$

$$= (1263.75)\left(\frac{3}{4}\right) \quad \text{Write } \frac{9}{12} \text{ in lowest terms as } \frac{3}{4}.$$

$$= 947.81 \text{ (rounded)}$$

The interest is $947.81.

To find the total amount due, add the principal and the interest.

amount due = principal + interest

$$= \$16,850 + \$947.81$$

$$= \$17,797.81$$

The total amount due is $17,797.81.

39. $14,800 at $2\frac{1}{4}\%$ for 10 months

The principal is $14,800. The rate is $2\frac{1}{4}\%$ or 0.0225, and the time is $\frac{10}{12}$ of a year.

$$I = p \cdot r \cdot t$$

$$= \underbrace{(14,800)(0.0225)}\left(\frac{10}{12}\right)$$

$$= (333)\left(\frac{5}{6}\right) \quad \text{Write } \frac{10}{12} \text{ in}$$

$$\text{lowest terms as } \frac{5}{6}.$$

$$= 277.50$$

She will earn $277.50 in interest.

Section 6.8 (pages 461–464)

1. $500 at 4% for 2 years

Year	Interest	Compound Amount
1	($500)(0.04)(1) = $20	
		$500 + $20 = $520
2	($520)(0.04)(1) = $20.80	
		$520 + $20.80 = $540.80

The compound amount is $540.80.

9. $1400 at 6% for 5 years

Year 1 Year 2 Year 3 Year 4 Year 5

($1400) $\underbrace{(1.06)(1.06)(1.06)(1.06)(1.06)} \approx \$1873.52$

↑   ↑

Original   100% + 6% = 106% = 1.06   Compound

deposit   amount

The compound amount is $1873.52.

15. $1000 at 4% for 5 years

Look down the column headed 4%, and across to row 5 (because 5 years = 5 time periods). At the intersection of the column and the row, read the compound amount, 1.2167.

$$compound\ amount = (\$1000)(1.2167)$$

$$= \$1216.70$$

Find the interest by subtracting the principal ($1000) from the compound amount.

$$interest = \$1216.70 - \$1000$$

$$= \$216.70$$

25. (a) 6% column, row 9 from the table is 1.6895.

$$compound\ amount = (\$76,000)(1.6895)$$

$$= \$128,402$$

The total amount that should be repaid is $128,402.

(b) To find the amount of interest earned, subtract the principal ($76,000) from the compound amount.

The amount of interest earned is

$$\$128,402 - \$76,000 = \$52,402$$

Chapter 7 Measurement

Section 7.1 (pages 491–494)

25. 3 in. to feet

$$\frac{\overset{1}{\cancel{3}}\ \cancel{in.}}{1} \cdot \frac{1\ ft}{\underset{4}{\cancel{12}\ in.}} = \frac{1}{4}\ ft \text{ or } 1 \div 4 = 0.25\ ft$$

35. $4\frac{1}{4}$ gal to quarts

$$\frac{4\frac{1}{4}\ \cancel{gal}}{1} \cdot \frac{4\ qt}{1\ \cancel{gal}} = \frac{17}{\underset{1}{\cancel{4}}} \cdot \frac{\overset{1}{\cancel{4}}}{1}\ qt = 17\ qt$$

43. 6 days to seconds

$$\frac{6\ \cancel{days}}{1} \cdot \frac{24\ \cancel{hr}}{1\ \cancel{day}} \cdot \frac{60\ \cancel{min}}{1\ \cancel{hr}} \cdot \frac{60\ sec}{1\ \cancel{min}}$$

$$= 6 \cdot 24 \cdot 60 \cdot 60\ sec$$

$$= 518,400\ sec$$

55. *Step 1*

The problem asks for the price per pound of strawberries.

Step 2

Convert ounces to pounds. Then divide the cost by the number of pounds.

Step 3

To estimate, round $2.29 to $2. Then, there are 16 oz in a pound, so 20 oz is a little more than 1 lb. Thus, $2 ÷ 1 = $2 per pound as our estimate.

Step 4

Use a unit fraction to convert 20 oz to pounds.

$$\frac{\overset{5}{\cancel{20}}\ \cancel{oz}}{1} \cdot \frac{1\ lb}{\underset{4}{\cancel{16}\ \cancel{oz}}} = \frac{5}{4}\ lb = 1.25\ lb$$

$$\begin{array}{c}\text{Cost} \longrightarrow \\ \text{Per} \longrightarrow \\ \text{Pound} \longrightarrow\end{array}\ \frac{\$2.29}{1.25\ lb} = 1.832 \approx 1.83$$

Step 5

The strawberries are $1.83 per pound (to the nearest cent).

Step 6

The exact answer, $1.83, is close to our estimate of $2.

61. (a) Find the total number of cups per week. Then convert cups to quarts.

$$\frac{2}{3} \cdot 15 \cdot 5 = \frac{2}{\underset{1}{\cancel{3}}} \cdot \frac{\overset{5}{\cancel{15}}}{1} \cdot \frac{5}{1} = 50\ cups$$

$$\frac{\overset{25}{\cancel{50}}\ \cancel{cups}}{1} \cdot \frac{1\ pt}{\underset{1}{\cancel{2}\ \cancel{cups}}} \cdot \frac{1\ qt}{2\ \cancel{pt}}$$

$$= \frac{25}{2}\ qt = 12\frac{1}{2}\ qt$$

The center needs $12\frac{1}{2}$ qt of milk per week.

(b) Convert quarts to gallons.

$$\frac{12\frac{1}{2}\ \cancel{qt}}{1} \cdot \frac{1\ gal}{4\ \cancel{qt}}$$

$$= \frac{25}{2} \cdot \frac{1}{4}\ gal = 3.125\ gal$$

The center should order 4 gallon containers, because you can't buy part of a container.

Section 7.2 (pages 501–502)

19. A paper clip is about 3 <u>cm</u> long. Both meters and kilometers would be much too long, and 3 mm is a very tiny length (see ruler in Section 7.2).

---

25. 7 m to cm

$$\frac{7\ \cancel{m}}{1} \cdot \frac{100\ cm}{1\ \cancel{m}} = 7 \cdot 100\ cm = 700\ cm$$

33. 400 mm to cm

From mm to cm is *one* place to the *left* on the conversion line, so move the decimal point *one* place to the *left* also.

40 0. mm = 40.0 cm   or   40 cm

37. 82 cm to m

From cm to m is *two* places to the *left* on the conversion line, so move the decimal point *two* places to the *left* also.

82. cm = 0.82 m

0.82 m is less than 1 m, so 82 cm is **less than** 1 m. The difference in length is

1 m − 0.82 m = 0.18 m or

100 cm − 82 cm = 18 cm.

45. 5.6 mm to km

$$\frac{5.6\ \cancel{mm}}{1} \cdot \frac{1\ \cancel{m}}{1000\ \cancel{mm}} \cdot \frac{1\ km}{1000\ \cancel{m}}$$

$$= \frac{5.6}{1,000,000}\ km$$

$$= 0.0000056\ km$$

Section 7.3 (pages 509–512)

7. Lori caught a small sunfish weighing 150 g. (Weight is measured in mg, g, or kg.)

35. 8 mL to L

$$\frac{8\ \cancel{mL}}{1} \cdot \frac{1\ L}{1000\ \cancel{mL}} = \frac{8}{1000}\ L = 0.008\ L$$

41. 5.2 kg to g

From kg to g is *three* places to the *right* on the conversion line.

5.200 kg = 5200. g   or   5200 g

67. Convert 900 mL to L.

$$\frac{900\ \cancel{mL}}{1} \cdot \frac{1\ L}{1000\ \cancel{mL}} = \frac{900}{1000}\ L = 0.9\ L$$

On average, we breathe in and out roughly 0.9 L of air every 10 seconds.

73. Convert 1 kg to g.

$$\frac{1\ \cancel{kg}}{1} \cdot \frac{1000\ g}{1\ \cancel{kg}} = 1000\ g$$

Divide the 1000 g by the weight of 1 nickel.

$$\frac{1000\ \cancel{g}}{5\ \cancel{g}} = 200$$

There are 200 nickels in 1 kg of nickels.

Section 7.4 (pages 517–518)

7. Multiply to find the total length of pencil lead in mm, then convert mm to cm.

60 mm • 30 = 1800 mm of pencil lead

$$\frac{\overset{180}{\cancel{1800}}\ \cancel{mm}}{1} \cdot \frac{1\ cm}{\underset{1}{\cancel{10}\ \cancel{mm}}} = 180\ cm$$

The total length of pencil lead is 180 cm. Divide to find the cost per cm.

$$\frac{\$3.29}{180\ cm} \approx \$0.0183/cm \approx \$0.02/cm$$

The cost is $0.02/cm (rounded).

13. Three cups a day for one week is
$3 \cdot 7 = 21$ cups in one week.

$$\frac{21 \text{ cups}}{1} \cdot \frac{90 \text{ mg}}{1 \text{ cup}} = 1890 \text{ mg}$$

Convert mg to g.

$$\frac{1890 \text{ mg}}{1} \cdot \frac{1 \text{ g}}{1000 \text{ mg}} = 1.89 \text{ g}$$

Agnete consumes 1.89 g of caffeine in one week.

Section 7.5 (pages 523–526)

3. 80 m to feet

$$\frac{80 \text{ m}}{1} \cdot \frac{3.28 \text{ ft}}{1 \text{ m}} \approx 262.4 \text{ ft}$$

17. Convert 0.5 in. to centimeters.

$$\frac{0.5 \text{ in.}}{1} \cdot \frac{2.54 \text{ cm}}{1 \text{ in.}} = 1.27 \text{ cm} \approx 1.3 \text{ cm}$$

The dwarf gobie is about 1.3 cm long.

27. 60 °F

$$C = \frac{5(60 - 32)}{9} = \frac{5 \cdot 28}{9}$$

$$= \frac{140}{9} \approx 15.6 \approx 16$$

60 °F $\approx$ 16 °C

37. (a) Since the comfort range of the boots is from 24 °C to 4 °C, you would wear these boots in pleasant weather—above freezing, but not hot.

(b) Change 24 °C to Fahrenheit.

$$F = \frac{9 \cdot C}{5} + 32 = \frac{9 \cdot 24}{5} + 32$$

$$= \frac{216}{5} + 32$$

$$= 43.2 + 32 = 75.2$$

Thus, 24 °C $\approx$ 75 °F.

Change 4 °C to Fahrenheit.

$$F = \frac{9 \cdot C}{5} + 32 = \frac{9 \cdot 4}{5} + 32$$

$$= \frac{36}{5} + 32$$

$$= 7.2 + 32 = 39.2$$

Thus, 4 °C $\approx$ 39 °F.

The boots are designed for Fahrenheit temperatures of about 75 °F to about 39 °F.

(c) The range of metric temperatures in January would depend on where you live. In Minnesota, it's 0 °C to −40 °C, and in California it's 24 °C to 0 °C.

45. Fuel left after landing

$$\frac{2 \text{ fl oz}}{1} \cdot \frac{1 \text{ qt}}{32 \text{ fl oz}} \cdot \frac{0.95 \text{ L}}{1 \text{ qt}} \cdot \frac{1000 \text{ mL}}{1 \text{ L}}$$

$$= \frac{950}{16} \text{ mL} = 59.375 \text{ mL} \approx 59.4 \text{ mL}$$

Chapter 8 Geometry

Section 8.1 (pages 547–548)

5. This is a *ray* named $\overrightarrow{PQ}$. A ray is a part of a line that has only one endpoint and goes on forever in one direction.

7. The lines are *perpendicular* because they intersect at right angles; the small red square indicates a right angle.

23. Two rays in a straight line pointing in opposite directions measure 180°. An angle that measures 180° is called a *straight angle*.

Section 8.2 (pages 553–554)

7. Find the complement of 86° by subtracting. $90° − 86° = 4°$, so the complement is 4°.

11. Find the supplement of 90° by subtracting. $180° − 90° = 90°$, so the supplement is 90°.

13. $\angle SON \cong \angle TOM$ because the angles are vertical angles.

$\angle TOS \cong \angle MON$ because the angles are vertical angles.

15. Because $\angle COE$ and $\angle GOH$ are vertical angles, they are also congruent. This means they have the same measure. $\angle COE$ measures 63°, so $\angle GOH$ measures 63°.

The sum of the measures of $\angle COE$, $\angle AOC$, and $\angle AOH$ equals 180° because $\angle EOH$ is a straight angle. Therefore, $\angle AOC$ measures $180° − (63° + 37°) = 180° − 100° = 80°$.

Since $\angle AOC$ and $\angle GOF$ are vertical, they are congruent, so $\angle GOF$ measures 80°.

Since $\angle AOH$ and $\angle EOF$ are vertical, they are congruent, so $\angle EOF$ measures 37°.

Section 8.3 (pages 561–564)

3. $P = 4 \cdot s$
$= 4 \cdot 0.9 \text{ km}$
$= 3.6 \text{ km}$

$A = s \cdot s$
$= 0.9 \text{ km} \cdot 0.9 \text{ km}$
$= 0.81 \text{ km}^2$ ← Write km² for area
(**not** km).

9. A storage building that is 76.1 ft by 22 ft

76.1 ft

22 ft

The length of this rectangle is 76.1 ft and the width is 22 ft. Use the formula $P = 2 \cdot l + 2 \cdot w$ to find the perimeter.
$P = 2 \cdot 76.1 \text{ ft} + 2 \cdot 22 \text{ ft}$
$= 152.2 \text{ ft} + 44 \text{ ft}$
$= 196.2 \text{ ft}$

Use the formula $A = l \cdot w$ to find the area.
$A = 76.1 \text{ ft} \cdot 22 \text{ ft}$
$= 1674.2 \text{ ft}^2$

17. The unlabeled side is 6 in. + 6 in. or 12 in. Add the lengths of the sides to find the perimeter.
$P = 15 \text{ in.} + 6 \text{ in.} + 6 \text{ in.} + 6 \text{ in.}$
$+ 6 \text{ in.} + 9 \text{ in.} + 12 \text{ in.} + 6 \text{ in.} + 6 \text{ in.}$
$= 78 \text{ in.}$

Draw two horizontal lines to break up the figure into two smaller squares and one larger rectangle. Add the areas of these squares and rectangle to find the total area.
$A = (6 \text{ in.} \cdot 6 \text{ in.}) + (18 \text{ in.} \cdot 9 \text{ in.})$
$+ (6 \text{ in.} \cdot 6 \text{ in.})$
$= 36 \text{ in.}^2 + 162 \text{ in.}^2 + 36 \text{ in.}^2$
$= 234 \text{ in.}^2$

You could also draw two vertical lines to break up the figure into three rectangles.

$A = (15 \text{ in.} \cdot 6 \text{ in.}) + (15 \text{ in.} \cdot 6 \text{ in.})$
$+ (9 \text{ in.} \cdot 6 \text{ in.})$
$= 90 \text{ in.}^2 + 90 \text{ in.}^2 + 54 \text{ in.}^2$
$= 234 \text{ in.}^2$ ← Same result

23. Find the perimeter of the room.
$P = 2 \cdot 4.4 \text{ m} + 2 \cdot 5.1 \text{ m}$
$= 8.8 \text{ m} + 10.2 \text{ m} = 19 \text{ m}$

She will need 19 m of the strip. Multiply by the cost per meter to find the total cost.

$$\text{Cost} = \frac{19 \text{ m}}{1} \cdot \frac{\$4.99}{1 \text{ m}} = \$94.81$$

Tyra will have to spend \$94.81 for the strip.

27. $A = \text{length} \cdot \text{width}$
$5300 \text{ yd}^2 = 100 \text{ yd} \cdot \text{width}$

$$\frac{5300 \text{ yd} \cdot \text{yd}}{100 \text{ yd}} = \frac{100 \text{ yd} \cdot \text{width}}{100 \text{ yd}}$$

$53 \text{ yd} = \text{width}$

The width is 53 yd.

Section 8.4 (pages 569–570)

11. The figure is a trapezoid. The height (h) is 42 cm, the short base (b) is 61.4 cm, and the long base (B) is 86.2 cm.

$A = \frac{1}{2} \cdot h \cdot (b + B)$
$= 0.5 \cdot 42 \text{ cm} \cdot (61.4 \text{ cm} + 86.2 \text{ cm})$
$= 0.5 \cdot 42 \text{ cm} \cdot (147.6 \text{ cm})$
$= 3099.6 \text{ cm}^2$

15. Area of one parallelogram:
$A = b \cdot h$
$= 5 \text{ in.} \cdot 3.5 \text{ in.}$
$= 17.5 \text{ in.}^2$

To find the total area of the 25 parallelogram pieces, multiply the area of one parallelogram (17.5 in.^2) by the number of pieces (25).

Total area $= 25 \cdot 17.5 \text{ in.}^2 = 437.5 \text{ in.}^2$

19. Break the figure into two pieces, a rectangle on the left side and a parallelogram. Find the area of each piece, and then add the areas.

Area of rectangle:
$A = \text{length} \cdot \text{width}$
$= 1.3 \text{ m} \cdot 0.8 \text{ m}$
$= 1.04 \text{ m}^2$

Area of parallelogram:
$A = \text{base} \cdot \text{height}$
$= 1.8 \text{ m} \cdot 1.1 \text{ m}$
$= 1.98 \text{ m}^2$

Total area $= 1.04 \text{ m}^2 + 1.98 \text{ m}^2$
$= 3.02 \text{ m}^2$

Section 8.5 (pages 575–577)

7. To find the perimeter, add the lengths of the three sides.

$P = 35.5 \text{ cm} + 21.3 \text{ cm} + 28.4 \text{ cm}$
$= 85.2 \text{ cm}$

When finding the area of this right triangle, the base and height are the perpendicular sides. So the base is 28.4 cm and the height is 21.3 cm.

$A = \frac{1}{2} \cdot b \cdot h$

$= \frac{1}{2} \cdot 28.4 \text{ cm} \cdot 21.3 \text{ cm}$

$= 302.46 \text{ cm}^2$

11. First, find the area of the entire figure, which is a rectangle.
$$A = l \cdot w$$
$$= 52 \text{ m} \cdot 37 \text{ m}$$
$$= 1924 \text{ m}^2$$

Find the area of the unshaded triangle.
$$A = \frac{1}{2} \cdot b \cdot h$$
$$= \frac{1}{2} \cdot 52 \text{ m} \cdot 10 \text{ m}$$
$$= 260 \text{ m}^2$$

Subtract the unshaded area from the area of the entire figure to find the shaded area.
$$\text{Shaded area} = 1924 \text{ m}^2 - 260 \text{ m}^2$$
$$= 1664 \text{ m}^2$$

13. *Step 1* Add the two angles given.
$$90° + 58° = 148°$$

Step 2 Subtract the sum from 180°.
$$180° - 148° = 32°$$

The third angle measures 32°.

21. **(a)** To find the amount of curbing needed to go around the triangular space, find the perimeter.
$$P = 42 \text{ m} + 32 \text{ m} + 52.8 \text{ m}$$
$$= 126.8 \text{ m}$$
126.8 m of curbing will be needed.

(b) To find the amount of sod needed to cover the space, find the area. It is a right triangle, so the perpendicular sides are the base and height.
$$A = \frac{1}{2} \cdot 32 \text{ m} \cdot 42 \text{ m} = 672 \text{ m}^2$$
672 m² of sod will be needed.

Section 8.6 (pages 585–588)

11. $d = 7\frac{1}{2}$ ft or 7.5 ft
$$C = \pi \cdot d$$
$$\approx 3.14 \cdot 7.5 \text{ ft}$$
$$\approx 23.6 \text{ ft}$$

To find the area, first find the radius.
$$r = \frac{7.5 \text{ ft}}{2} = 3.75 \text{ ft}$$
$$A = \pi \cdot r \cdot r$$
$$\approx 3.14 \cdot 3.75 \text{ ft} \cdot 3.75 \text{ ft}$$
$$\approx 44.2 \text{ ft}^2$$

17. Find the area of a whole circle with a radius of 10 cm:
$$A = \pi \cdot r \cdot r$$
$$\approx 3.14 \cdot 10 \text{ cm} \cdot 10 \text{ cm}$$
$$= 314 \text{ cm}^2$$

Divide the area of the whole circle by 2 to find the area of the semicircle:
$$\frac{314 \text{ cm}^2}{2} = 157 \text{ cm}^2$$

Area of the triangle:
$$A = \frac{1}{2} \cdot b \cdot h$$
$$= \frac{1}{2} \cdot 20 \text{ cm} \cdot 10 \text{ cm}$$
$$= 100 \text{ cm}^2$$

Subtract to find the shaded area.
$$157 \text{ cm}^2 - 100 \text{ cm}^2 = 57 \text{ cm}^2$$

23. A point on the tire tread moves the length of the circumference in one complete turn.
$$C = \pi \cdot d$$
$$\approx 3.14 \cdot 29.10 \text{ in.}$$
$$\approx 91.4 \text{ in.}$$

Bonus question:
$$\frac{1 \text{ revolution}}{91.4 \text{ inches}} \cdot \frac{12 \text{ inches}}{1 \text{ foot}} \cdot \frac{5280 \text{ feet}}{1 \text{ mile}}$$
$$\approx 693 \text{ revolutions/mile}$$

25.

Station

Find the area of a circle with a radius of 150 miles.
$$A = \pi \cdot r \cdot r$$
$$\approx 3.14 \cdot 150 \text{ mi} \cdot 150 \text{ mi}$$
$$= 70,650 \text{ mi}^2$$

There are about 70,650 mi² in the broadcast area.

31. **(a)** $C = 144$ cm
$$C = \pi \cdot d$$
$$144 \text{ cm} = \pi \cdot d$$
$$144 \text{ cm} \approx 3.14 \cdot d$$
$$\frac{144 \text{ cm}}{3.14} \approx \frac{3.14 \cdot d}{3.14}$$
$$45.9 \text{ cm} \approx d$$

The diameter is about 45.9 cm.

(b) Divide the circumference by π (3.14).

Section 8.7 (pages 597–598)

5. The figure is a hemisphere.
$$V = \frac{2}{3} \cdot \pi \cdot r^3 \quad \text{or} \quad \frac{2 \cdot \pi \cdot r^3}{3}$$
$$\approx \frac{2 \cdot 3.14 \cdot 12 \text{ in.} \cdot 12 \text{ in.} \cdot 12 \text{ in.}}{3}$$
$$\approx 3617.3 \text{ in.}^3$$

9. The figure is a cone.
First find B, the area of the circular base.
$$B = \pi \cdot r \cdot r$$
$$\approx 3.14 \cdot 5 \text{ m} \cdot 5 \text{ m}$$
$$\approx 78.5 \text{ m}^2$$
Now find the volume of the cone.
$$V = \frac{B \cdot h}{3}$$
$$\approx \frac{78.5 \text{ m}^2 \cdot 16 \text{ m}}{3}$$
$$\approx 418.7 \text{ m}^3$$

17. Use the formula for the volume of a pyramid. First find B, the area of the square base.
$$B = s \cdot s$$
$$= 145 \text{ m} \cdot 145 \text{ m}$$
$$= 21,025 \text{ m}^2$$
Now find the volume of the pyramid.
$$V = \frac{B \cdot h}{3}$$
$$= \frac{21,025 \text{ m}^2 \cdot 93 \text{ m}}{3}$$
$$= 651,775 \text{ m}^3$$
The volume of the ancient stone pyramid is 651,775 m³.

19. Use the formula for the volume of a cylinder.
First find the radius of the pipe.
$$\text{radius} = \frac{5 \text{ ft}}{2} = 2.5 \text{ ft}$$
Now find the volume.
$$V = \pi \cdot r^2 \cdot h$$
$$\approx 3.14 \cdot 2.5 \text{ ft} \cdot 2.5 \text{ ft} \cdot 200 \text{ ft}$$
$$\approx 3925 \text{ ft}^3$$
The volume of the city sewer pipe is about 3925 ft³.

Section 8.8 (pages 603–606)

25. The unknown length is the side opposite the right angle, which is the hypotenuse. The lengths of the legs are 8 in. and 3 in.
$$\text{hypotenuse} = \sqrt{(\text{leg})^2 + (\text{leg})^2}$$
$$= \sqrt{(8)^2 + (3)^2}$$
$$= \sqrt{64 + 9}$$
$$= \sqrt{73}$$
$$\approx 8.5 \text{ in.}$$

35. The length of the hypotenuse is 21.6 km. The length of one of the legs is 13.2 km.
$$\text{leg} = \sqrt{(\text{hypotenuse})^2 - (\text{leg})^2}$$
$$= \sqrt{(21.6)^2 - (13.2)^2}$$
$$= \sqrt{466.56 - 174.24}$$
$$= \sqrt{292.32}$$
$$\approx 17.1 \text{ km}$$

41. The diagonal brace is the hypotenuse. The lengths of the legs are 4.5 ft and 3.5 ft.
$$\text{hypotenuse} = \sqrt{(\text{leg})^2 + (\text{leg})^2}$$
$$= \sqrt{(4.5)^2 + (3.5)^2}$$
$$= \sqrt{20.25 + 12.25}$$
$$= \sqrt{32.5}$$
$$\approx 5.7 \text{ ft}$$
The diagonal brace is about 5.7 ft long.

43.

Ladder 12 ft | **Castle**
?
3 ft

The ladder is opposite the right angle so it is the hypotenuse. Thus, the length of the hypotenuse is 12 ft and the length of one of the legs is 3 ft.
$$\text{leg} = \sqrt{(\text{hypotenuse})^2 - (\text{leg})^2}$$
$$= \sqrt{(12)^2 - (3)^2}$$
$$= \sqrt{144 - 9}$$
$$= \sqrt{135}$$
$$\approx 11.6 \text{ ft}$$
The ladder will reach about 11.6 ft high on the building.

Section 8.9 (pages 611–614)

15. Set up a ratio of corresponding sides, using the longest side in each triangle.
$$\frac{6 \text{ mm}}{12 \text{ mm}} = \frac{6}{12} = \frac{1}{2}$$

(continued)

SOLUTIONS

Write a proportion to find a.

$$\frac{a}{10} = \frac{1}{2}$$

$$a \cdot 2 = 10 \cdot 1$$

$$\frac{a \cdot \cancel{2}}{\cancel{2}} = \frac{10}{2}$$

$$a = 5 \text{ mm}$$

Write a proportion to find b.

$$\frac{b}{6} = \frac{1}{2}$$

$$b \cdot 2 = 6 \cdot 1$$

$$\frac{b \cdot \cancel{2}}{\cancel{2}} = \frac{6}{2}$$

$$b = 3 \text{ mm}$$

19. First, find the lengths of $\overline{FH}$ and $\overline{FG}$ in triangle FGH. Set up a ratio of corresponding sides where x is the length of side FH.

$$\frac{DE}{GH} = \frac{CE}{FH}$$

$$\frac{12}{8} = \frac{12}{x}$$

$$12 \cdot x = 8 \cdot 12$$

$$\frac{\cancel{12} \cdot x}{\cancel{12}} = \frac{96}{12}$$

$$x = 8 \text{ cm}$$

Note that when you set up a proportion to find the length of $\overline{FG}$ it is the same as that used to find the length of $\overline{FH}$. Therefore the length of $\overline{FG}$ is also 8 cm.

Perimeter of triangle FGH.

$$= 8 \text{ cm} + 8 \text{ cm} + 8 \text{ cm}$$

$$= 24 \text{ cm}$$

Set up a ratio of corresponding parts to find the height h of triangle FGH.

$$\frac{10.4}{12} = \frac{h}{8}$$

$$12 \cdot h = 8 \cdot 10.4$$

$$\frac{\cancel{12} \cdot h}{\cancel{12}} = \frac{83.2}{12}$$

$$h \approx 6.9 \text{ cm}$$

Area of triangle FGH

$$= 0.5 \cdot b \cdot h$$

$$\approx 0.5 \cdot 8 \text{ cm} \cdot 6.9 \text{ cm}$$

$$= 27.6 \text{ cm}^2$$

23. Let s be the length of the stick's shadow, and h is the height of the house from Exercise 21. Write a proportion to find s.

$$\frac{h}{3} = \frac{6}{s}$$

$$\frac{24}{3} = \frac{6}{s}$$

$$\frac{8}{1} = \frac{6}{s} \qquad \text{Write } \frac{24}{3} \text{ in lowest terms as } \frac{8}{1}$$

$$8 \cdot s = 1 \cdot 6$$

$$\frac{\cancel{8} \cdot s}{\cancel{8}} = \frac{6}{8}$$

$$s = \frac{3}{4} \text{ ft or } 0.75 \text{ ft}$$

29. Write a proportion to find n.

$$\frac{50}{n} = \frac{100}{100 + 120} \quad \textbf{or} \quad \frac{50}{n} = \frac{100}{220}$$

$$100 \cdot n = 50 \cdot 220$$

$$\frac{\cancel{100} \cdot n}{\cancel{100}} = \frac{11,000}{100}$$

$$n = 110 \text{ m}$$

The length of the lake is 110 m.

Chapter 9 Basic Algebra

Section 9.1 (pages 641–642)

21. Compare 1 and -1. Because 1 is to the *right* of -1 on a number line, $1 > -1$.

23. Compare -11 and -2. Because -11 is to the *left* of -2 on a number line, $-11 < -2$.

37. $-|-18|$

First, $|-18| = 18$. But there is also a negative sign *outside* the absolute value bars. So, -18 is the simplified expression.

47. The opposite of -8.3 is $-(-8.3) = 8.3$ because the opposite of any negative number is positive.

55. Because $|-4| = 4$, then $-|-4| = -4$. Because $|-7| = 7$, then $-|-7| = -7$. Since -4 is to the right of -7 on the number line, $-4 > -7$, therefore $-|-4| < -|-7|$ is false.

Section 9.2 (pages 651–654)

25. $7.8 + (-14.6)$

The signs are different, so subtract the absolute values.

First, find the absolute values.

$$|7.8| = 7.8 \text{ and } |-14.6| = 14.6$$

Subtract the lesser absolute value from the greater absolute value.

$$14.6 - 7.8 = 6.8$$

The *negative* number, -14.6, has the greater absolute value, so the answer is *negative*.

$$7.8 + (-14.6) = -6.8$$

29. $-\dfrac{7}{10} + \dfrac{2}{5}$

The signs are different, so subtract the absolute values.

First, find the absolute values.

$$\left|-\frac{7}{10}\right| = \frac{7}{10} \text{ and } \left|\frac{2}{5}\right| = \frac{2}{5}$$

Subtract the lesser absolute value from the greater absolute value.

$$\frac{7}{10} - \frac{2}{5} = \frac{7}{10} - \frac{4}{10} = \frac{3}{10}$$

The *negative* number, $-\dfrac{7}{10}$, has the greater absolute value, so the answer is *negative*.

$$-\frac{7}{10} + \frac{2}{5} = -\frac{3}{10}$$

45. The first number, 7, stays the same. Change the subtraction sign to addition. Change the sign of the second number to its opposite.

$$7 - 19 = 7 + (-19)$$

Now add.

$$7 + (-19) = -12$$

So $7 - 19 = -12$ also.

51. The first number, -3, stays the same. Change the subtraction sign to addition. Change the sign of the second number to its opposite.

$$-3 - (-8) = -3 + (+8)$$

Now add.

$$-3 + 8 = 5$$

So $-3 - (-8) = 5$ also.

57. The first number, -30, stays the same. Change the subtraction sign to addition. Change the sign of the second number to its opposite.

$$-30 - 30 = -30 + (-30)$$

Now add.

$$-30 + (-30) = -60$$

So $-30 - 30 = -60$ also.

63. The common denominator is 10.

Rewrite $\dfrac{1}{2}$ as $\dfrac{5}{10}$. Change the subtraction sign to addition. Positive $\dfrac{9}{10}$ is changed to its opposite, $\left(-\dfrac{9}{10}\right)$.

$$\frac{1}{2} - \frac{9}{10} = \frac{5}{10} + \left(-\frac{9}{10}\right)$$

$$= -\frac{4}{10} = -\frac{2}{5}$$

Once the subtraction is changed to addition, notice that the numbers have different signs. So, subtract the lesser absolute value from the greater absolute value.

$$\left|\frac{5}{10}\right| \text{ is } \frac{5}{10} \text{ and } \left|-\frac{9}{10}\right| = \frac{9}{10}.$$

Then $\dfrac{9}{10} - \dfrac{5}{10} = \dfrac{4}{10} = \dfrac{2}{5}$.

The *negative* number $-\dfrac{9}{10}$ has the greater absolute value so the answer is *negative*.

75. Change subtraction to addition. Change -13 to its opposite, $(+13)$.

$$4 - (-13) + (-5) = 4 + (+13) + (-5)$$

$$= 17 + (-5)$$

$$= 12$$

81. Change subtraction to addition. Change $\dfrac{2}{3}$ to its opposite, $\left(-\dfrac{2}{3}\right)$.

$$\frac{1}{2} - \frac{2}{3} + \left(-\frac{5}{6}\right) = \frac{1}{2} + \left(-\frac{2}{3}\right) + \left(-\frac{5}{6}\right)$$

Common denominator is 6.

$$= \frac{3}{6} + \left(-\frac{4}{6}\right) + \left(-\frac{5}{6}\right)$$

$$= -\frac{1}{6} + \left(-\frac{5}{6}\right)$$

$$= -\frac{6}{6} = -1$$

87. $-3 - (-2 + 4) + (-5)$

$$= -3 - (+2) + (-5)$$

Change subtraction to addition. Change 2 to its opposite, (-2).

$$-3 - (+2) + (-5)$$
$$= -3 + (-2) + (-5)$$
$$= -5 + (-5)$$
$$= -10$$

Section 9.3 (pages 657–660)

17. $-13(-1) = 13$

The numbers have the *same* sign, so the product is *positive*.

21. $-\dfrac{1}{2}(-8) = -\dfrac{1}{\cancel{2}_{1}}\left(-\dfrac{\cancel{8}^{4}}{1}\right) = \dfrac{4}{1} = 4$

The numbers have the *same* sign, so the product is *positive*.

29. $-\dfrac{7}{15} \cdot \dfrac{25}{14} = -\dfrac{\cancel{7}^{1}}{\cancel{15}_{3}} \cdot \dfrac{\cancel{25}^{5}}{\cancel{14}_{2}} = -\dfrac{5}{6}$

The numbers have *different* signs, so the product is *negative*.

39. $-1.25(-3.6) = 4.5$

The numbers have the *same* sign, so the product is *positive*.

65. Rewrite the division as multiplication that uses the reciprocal of $-\dfrac{15}{14}$, which is $-\dfrac{14}{15}$. Divide out all the common factors. Then multiply.

$$\dfrac{-\dfrac{5}{7}}{-\dfrac{15}{14}} = -\dfrac{\cancel{5}^{1}}{\cancel{7}_{1}} \cdot \left(-\dfrac{\cancel{14}^{2}}{\cancel{15}_{3}}\right) = \dfrac{2}{3}$$

The numbers have the *same* sign, so the quotient is *positive*.

67. Rewrite division as multiplication using the reciprocal of -2, which is $-\dfrac{1}{2}$.

Divide out the common factor. Then multiply.

$$-\dfrac{2}{3} \div (-2) = -\dfrac{\cancel{2}^{1}}{3} \cdot \left(-\dfrac{1}{\cancel{2}_{1}}\right) = \dfrac{1}{3}$$

The numbers have the *same* sign, so the quotient is *positive*.

77. $\dfrac{45.58}{-8.6} = -5.3$

The numbers have *different* signs, so the quotient is *negative*.

$$\begin{array}{r} 5.3 \\ 8.6\,\overline{)45.5\,8} \\ 43\,0 \\ \hline 2\,5\,8 \\ 2\,5\,8 \\ \hline 0 \end{array}$$

81. $(-0.6)(-0.2)(-3) = (0.12)(-3)$
$$= -0.36$$

87. $|-8| \div (-4) \cdot |-5|$
$$= 8 \div (-4) \cdot (5)$$
$$= -2 \cdot 5$$
$$= -10$$

95. First, divide 24,000 by 1000.
$$\dfrac{24{,}000}{1000} = 24$$

Then, multiply 24 by the number of degrees the temperature changes per 1000 feet.

The temperature will change $24(-3) = -72$ degrees in 24,000 ft.

Finally, add the ground temperature and the change in temperature.

The temperature will be $50 + (-72) = -22$ degrees.

Section 9.4 (pages 665–668)

19. $-7 + 6(8 - 14)$ Work inside parentheses
$$= -7 + 6(-6) \quad \text{Multiply } 6(-6)$$
$$= -7 + (-36) \quad \text{Add}$$
$$= -43$$

31. $30 \div (-5) - 36 \div (-9)$ Divide 30 by -5.
$$= -6 - 36 \div (-9) \quad \text{Divide 36 by } -9.$$
$$= -6 - (-4)$$

Change subtraction to adding the opposite.
$$= -6 + 4$$
$$= -2$$

35. $4(3^2) + 7(3 + 9) - (-6)$ Work inside parentheses.
$$= 4 \cdot 3^2 + 7(12) - (-6) \quad \text{Apply exponent: } 3^2 \text{ is 9.}$$
$$= 4 \cdot 9 + 7 \cdot 12 - (-6) \quad \text{Multiply } 4 \cdot 9.$$
$$= 36 + 7 \cdot 12 - (-6) \quad \text{Multiply } 7 \cdot 12.$$
$$= 36 + 84 - (-6)$$

Add $36 + 84$. Change subtraction to adding the opposite.
$$= 120 + (+6)$$
$$= 126$$

41. $\dfrac{2^3(-2 - 5) + 4(-1)}{4 + 5(-6 \cdot 2) + (5 \cdot 11)}$

First do the work in the numerator.
$$2^3(-2 - 5) + 4(-1) \quad \text{Parentheses}$$
$$= 2^3(-7) + 4(-1) \quad \text{Exponent: } 2^3 \text{ is 8.}$$
$$= 8(-7) + 4(-1) \quad \text{Multiply } 8(-7).$$
$$= -56 + 4(-1) \quad \text{Multiply } 4(-1).$$
$$= -56 + (-4) \quad \text{Add.}$$
$$= -60$$

Then do the work in the denominator.
$$4 + 5(-6 \cdot 2) + (5 \cdot 11) \quad \text{First set of parentheses}$$
$$= 4 + 5(-12) + (5 \cdot 11) \quad \text{Second set of parentheses}$$
$$= 4 + 5(-12) + 55 \quad \text{Multiply } 5(-12).$$
$$= 4 + (-60) + 55 \quad \text{Add } 4 + (-60).$$
$$= -56 + 55 \quad \text{Add.}$$
$$= -1$$

The last step is the division.
$$\dfrac{-60}{-1} = 60$$

55. $\dfrac{3}{5}\left(-\dfrac{7}{6}\right) - \left(\dfrac{1}{6} - \dfrac{5}{3}\right)$ Rewrite $\dfrac{5}{3}$ as $\dfrac{10}{6}$

$$= \dfrac{3}{5}\left(-\dfrac{7}{6}\right) - \left(\dfrac{1}{6} - \dfrac{10}{6}\right) \quad \text{Parentheses}$$
$$= \dfrac{3}{5}\left(-\dfrac{7}{6}\right) - \left(-\dfrac{9}{6}\right) \quad \text{Divide out the common factor.}$$
$$= \dfrac{\cancel{3}^{1}}{5} \cdot \left(-\dfrac{7}{\cancel{6}_{2}}\right) - \left(-\dfrac{9}{6}\right) \quad \text{Multiply.}$$
$$= -\dfrac{7}{10} - \left(-\dfrac{9}{6}\right) \quad \begin{array}{l}\text{Change sub-}\\\text{traction to}\\\text{addition and}\\-\dfrac{9}{6} \text{ to } \dfrac{3}{2}\end{array}$$
$$= -\dfrac{7}{10} + \left(+\dfrac{3}{2}\right) \quad \begin{array}{l}\text{Rewrite}\\\dfrac{3}{2} \text{ as } \dfrac{15}{10}\end{array}$$
$$= -\dfrac{7}{10} + \dfrac{15}{10} \quad \text{Add.}$$
$$= \dfrac{8}{10} \quad \text{Simplify.}$$
$$= \dfrac{4}{5}$$

65. $-7\left(6 - \dfrac{5}{8} \cdot 24 + 3 \cdot \dfrac{8}{3}\right)$ Inside parentheses: multiply $\dfrac{5}{8}$ and 24 to get 15.

$$= -7\left(6 - 15 + 3 \cdot \dfrac{8}{3}\right) \quad \begin{array}{l}\text{Multiply}\\ 3 \text{ and } \dfrac{8}{3} \text{ to}\\ \text{get 8.}\end{array}$$
$$= -7(6 - 15 + 8) \quad \begin{array}{l}\text{Subtract}\\ 6 - 15.\end{array}$$
$$= -7(-9 + 8) \quad \begin{array}{l}\text{Add}\\ -9 + 8.\end{array}$$
$$= -7(-1) \quad \text{Multiply.}$$
$$= 7$$

Section 9.5 (pages 673–676)

17. $-m - 3n;\ m = \dfrac{1}{2},\ n = \dfrac{3}{8}$

Replace m with $\dfrac{1}{2}$. Replace n with $\dfrac{3}{8}$.

$$-m - 3n$$
$$= -\dfrac{1}{2} - 3\left(\dfrac{3}{8}\right) \quad \begin{array}{l}\text{Multiply } 3\left(\dfrac{3}{8}\right) \text{ to}\\ \text{get } \dfrac{9}{8}\end{array}$$
$$= -\dfrac{1}{2} - \dfrac{9}{8} \quad \begin{array}{l}\text{Change subtraction to}\\ \text{addition and } \dfrac{9}{8} \text{ to } -\dfrac{9}{8}\end{array}$$
$$= -\dfrac{1}{2} + \left(-\dfrac{9}{8}\right) \quad \text{Rewrite } -\dfrac{1}{2} \text{ as } -\dfrac{4}{8}$$

(continued)

SOLUTIONS

$$= -\frac{4}{8} + \left(-\frac{9}{8}\right) \quad \text{Add.}$$

$$= -\frac{13}{8}$$

19. $-c - 5b;\ c = -8,\ b = -4$

Replace c with -8. Replace b with -4.

$$-c - 5b = -(-8) - 5(-4)$$

$-(-8)$ is the opposite of -8 or 8. Multiply $5(-4)$ to get -20.

$$= 8 - (-20)$$

Change subtraction to addition and -20 to 20.

$$= 8 + (+20)$$
$$= 28$$

25. $\dfrac{-3s - t - 4}{-s + 6 + t};\ s = -1,\ t = -13$

Replace s with -1. Replace t with -13.

$$= \frac{-3(-1) - (-13) - 4}{-(-1) + 6 + (-13)}$$

Multiply $-3(-1)$ in the numerator; $-(-1)$ is 1 in the denominator.

$$= \frac{3 + (+13) - 4}{1 + 6 + (-13)} \quad \begin{array}{l}\text{Add 3 and 13 in the}\\ \text{numerator.}\end{array}$$

$$= \frac{16 - 4}{1 + 6 + (-13)} \quad \begin{array}{l}\text{Subtract in the}\\ \text{numerator.}\end{array}$$

$$= \frac{12}{1 + 6 + (-13)} \quad \begin{array}{l}\text{Add } 1 + 6 \text{ in the}\\ \text{denominator.}\end{array}$$

$$= \frac{12}{7 + (-13)} \quad \begin{array}{l}\text{Add in the}\\ \text{denominator.}\end{array}$$

$$= \frac{12}{-6} \quad \text{Divide 12 by } -6.$$

$$= -2$$

31. $A = \pi r^2;\ \pi \approx 3.14,\ r = 5$

Replace π with 3.14. Replace r with 5.

$$A \approx 3.14 \cdot 5^2$$
$$= 3.14 \cdot 5 \cdot 5$$
$$= 78.5$$

Section 9.6 (pages 683–686)

9. $k + 15 = 0$

Add -15 to both sides.

$$k + 15 + (-15) = 0 + (-15)$$
$$k + 0 = -15$$
$$k = -15$$

Check the solution by replacing k with -15 in the original equation.

$$k + 15 = 0$$
$$-15 + 15 = 0$$
$$0 = 0 \quad \text{True}$$

The solution is -15.

23. $-5 = -2 + r$

Add 2 to both sides.

$$-5 + 2 = -2 + 2 + r$$
$$-3 = 0 + r$$
$$-3 = r$$

Check the solution by replacing r with -3 in the original equation.

$$-5 = -2 + r$$
$$-5 = -2 + (-3)$$
$$-5 = -5 \quad \text{True}$$

The solution is -3.

29. $\dfrac{1}{2} = k - 2$

Add 2 to both sides.

$$\frac{1}{2} + 2 = k - 2 + 2$$
$$\frac{1}{2} + \frac{4}{2} = k + 0$$
$$\frac{5}{2} = k$$

Check the solution by replacing k with $\dfrac{5}{2}$ in the original equation.

$$\frac{1}{2} = k - 2$$
$$\frac{1}{2} = \frac{5}{2} - 2$$
$$\frac{1}{2} = \frac{1}{2} \quad \text{True}$$

The solution is $\dfrac{5}{2}$ or $2\dfrac{1}{2}$.

41. $3y = 0$

Divide both sides by 3.

$$\frac{\cancel{3} \cdot y}{\cancel{3}} = \frac{0}{3}$$
$$y = 0$$

Check the solution by replacing y with 0 in the original equation.

$$3y = 0$$
$$3(0) = 0$$
$$0 = 0 \quad \text{True}$$

The solution is 0.

45. $-36 = -4p$

Divide both sides by -4.

$$\frac{-36}{-4} = \frac{\cancel{-4} \cdot p}{\cancel{-4}}$$
$$9 = p$$

Check the solution by replacing p with 9 in the original equation.

$$-36 = -4p$$
$$-36 = -4(9)$$
$$-36 = -36 \quad \text{True}$$

The solution is 9.

55. $\dfrac{r}{3} = -12$ can be written as $\dfrac{1}{3}r = -12$

Multiply both sides by $\dfrac{3}{1}$, the reciprocal of $\dfrac{1}{3}$.

$$\frac{\cancel{3}}{1} \cdot \frac{1}{\cancel{3}}r = -12 \cdot 3$$
$$1r = -36$$
$$r = -36$$

Check the solution by replacing r with -36 in the original equation.

$$\frac{r}{3} = -12$$
$$\frac{-36}{3} = -12$$
$$-12 = -12 \quad \text{True}$$

The solution is -36.

59. $-\dfrac{3}{4}m = -3$

Multiply both sides by $-\dfrac{4}{3}$, the reciprocal of $-\dfrac{3}{4}$.

$$-\frac{\cancel{4}}{\cancel{3}} \cdot \left(-\frac{\cancel{3}}{\cancel{4}}m\right) = -\cancel{3} \cdot \left(-\frac{4}{\cancel{3}}\right)$$
$$1m = 4$$
$$m = 4$$

Check the solution by replacing m with 4 in the original equation.

$$-\frac{3}{4}m = -3$$
$$-\frac{3}{4}(4) = -3$$
$$-3 = -3 \quad \text{True}$$

The solution is 4.

75. $\dfrac{1}{2} - \dfrac{3}{4} = \dfrac{a}{5} \quad$ Rewrite $\dfrac{1}{2}$ as $\dfrac{2}{4}$

$$\frac{2}{4} - \frac{3}{4} = \frac{a}{5} \quad \text{Subtract } \frac{2}{4} - \frac{3}{4}$$

$$-\frac{1}{4} = \frac{a}{5} \quad \begin{array}{l}\text{Multiply both sides by}\\ \frac{5}{1}, \text{ the reciprocal of } \frac{1}{5}\end{array}$$

$$\frac{5}{1} \cdot \left(-\frac{1}{4}\right) = \frac{a}{\cancel{5}} \cdot \frac{\cancel{5}}{1}$$

$$-\frac{5}{4} = a$$

The solution is $-\dfrac{5}{4}$.

77.
$$m - 2 + 18 = |-3 - 4| + 5$$
$$m + 16 = |-3 - 4| + 5$$
$$m + 16 = |-7| + 5$$
$$m + 16 = 7 + 5$$
$$m + 16 = 12$$
$$m + 16 + (-16) = 12 + (-16)$$
$$m = -4$$

The solution is -4.

Section 9.7 (pages 691–694)

3. $2 = 8y - 6$

Add 6 to both sides.

$$2 + 6 = 8y - 6 + 6$$
$$8 = 8y$$

Divide both sides by 8.

$$\frac{8}{8} = \frac{\cancel{8}y}{\cancel{8}}$$
$$1 = y$$

Check Replace y with 1 in the original equation.

$$2 = 8y - 6$$
$$2 = 8(1) - 6$$
$$2 = 2 \quad \text{True}$$

The solution is 1.

11. $-\dfrac{1}{2}z + 2 = -1$

Subtract 2 from both sides.

$-\dfrac{1}{2}z + 2 - 2 = -1 - 2$

$-\dfrac{1}{2}z = -3$

Multiply both sides by $-\dfrac{2}{1}$, the reciprocal of $-\dfrac{1}{2}$.

$-\dfrac{\overset{1}{\cancel{2}}}{1} \cdot \left(-\dfrac{1}{\underset{1}{\cancel{2}}}z\right) = -3 \cdot \left(-\dfrac{2}{1}\right)$

$z = 6$

Check   Replace z with 6 in the original equation.

$-\dfrac{1}{2}z + 2 = -1$

$-\dfrac{1}{2}(6) + 2 = -1$

$-3 + 2 = -1$

$-1 = -1$ True

The solution is 6.

19. $-3(m + 6) = -3 \cdot m + (-3) \cdot 6$

$= -3m - 18$

25. $-10(w - 9) = -10 \cdot w - (-10) \cdot 9$

$= -10w - (-90)$

$= -10w + 90$

37. $\dfrac{5}{2}b - \dfrac{11}{2}b = \left(\dfrac{5}{2} - \dfrac{11}{2}\right)b$

$= -\dfrac{6}{2}b = -3b$

45. $-12 = 6y - 18y$

Use the distributive property to combine $6y$ and $-18y$.

$-12 = (6 - 18)y$

$-12 = -12y$

Divide both sides by -12.

$\dfrac{-12}{-12} = \dfrac{\cancel{-12}y}{\cancel{-12}}$

$1 = y$

Check   Replace y with 1 in the original equation.

$-12 = 6y - 18y$

$-12 = 6(1) - 18(1)$

$-12 = 6 - 18$

$-12 = -12$ True

The solution is 1.

51. $-2y + 6 = 6y - 10$

Add 10 to both sides.

$-2y + 6 + 10 = 6y - 10 + 10$

$-2y + 16 = 6y$

Add $2y$ to both sides.

$-2y + 2y + 16 = 6y + 2y$

Combine $6y$ and $2y$ because they are like terms.

$16 = 8y$

Divide both sides by 8.

$\dfrac{16}{8} = \dfrac{\cancel{8}y}{\cancel{8}}$

$2 = y$

Check   Replace y with 2 in the original equation.

$-2y + 6 = 6y - 10$

$-2(2) + 6 = 6(2) - 10$

$(-4) + 6 = 12 - 10$

$2 = 2$ True

The solution is 2.

59. $-4(t + 2) = 12$

Use the distributive property on the left side of the equation.

$-4t + (-4) \cdot 2 = 12$

$-4t + (-8) = 12$

Add 8 to both sides.

$-4t - 8 + 8 = 12 + 8$

$-4t = 20$

Divide both sides by -4.

$\dfrac{\cancel{-4}t}{\cancel{-4}} = \dfrac{20}{-4}$

$t = -5$

Check   Replace t with -5 in the original equation.

$-4(t + 2) = 12$

$-4(-5 + 2) = 12$

$-4(-3) = 12$

$12 = 12$ True

The solution is -5.

67. $0 = -2(y - 2)$

Use the distributive property on the right side of the equation.

$0 = -2y - (-2) \cdot 2$

$0 = -2y + 4$

Add $2y$ to both sides.

$0 + 2y = -2y + 2y + 4$

$2y = 4$

Divide both sides by 2.

$\dfrac{2y}{2} = \dfrac{4}{2}$

$y = 2$

The solution is 2.

69. $\dfrac{y}{2} - 2 = \dfrac{y}{4} + 3$

Add 2 to both sides.

$\dfrac{y}{2} - 2 + 2 = \dfrac{y}{4} + 3 + 2$

$\dfrac{y}{2} = \dfrac{y}{4} + 5$

$\dfrac{1}{2}y = \dfrac{1}{4}y + 5$

Subtract $\dfrac{1}{4}y$ from both sides.

$\dfrac{1}{2}y - \dfrac{1}{4}y = \dfrac{1}{4}y - \dfrac{1}{4}y + 5$

$\dfrac{1}{4}y = 5$

Multiply both sides by $\dfrac{4}{1}$, the reciprocal

of $\dfrac{1}{4}$.

$\dfrac{\overset{1}{\cancel{4}}}{1} \cdot \dfrac{1}{\underset{1}{\cancel{4}}}y = \dfrac{4}{1} \cdot 5$

$y = 20$

The solution is 20.

73. $2(a + 0.3) = 1.2(a - 4)$

Use the distributive property on both sides.

$2 \cdot a + 2(0.3) = 1.2 \cdot a - 1.2 \cdot 4$

$2a + 0.6 = 1.2a - 4.8$

Subtract 0.6 from both sides.

$2a + 0.6 - 0.6 = 1.2a - 4.8 - 0.6$

$2a = 1.2a - 5.4$

Subtract $1.2a$ from both sides.

$2a - 1.2a = 1.2a - 5.4 - 1.2a$

Combine like terms.

$0.8a = -5.4$

Divide both sides by 0.8.

$\dfrac{\cancel{0.8}a}{\cancel{0.8}} = \dfrac{-5.4}{0.8}$

$a = -6.75$

The solution is -6.75.

Section 9.8 (pages 701–704)

31. Let n represent the unknown number.

30	subtract	3 times a number	is	2	plus	the number
↓	↓	↓	↓	↓	↓	↓
30	−	$3n$	=	2	+	n

$30 - 3n = 2 + n$

Add $3n$ to both sides.

$30 - 3n + 3n = 2 + n + 3n$

Combine like terms on each side.

$30 = 2 + 4n$

Subtract 2 from both sides.

$30 - 2 = 2 + 4n - 2$

$28 = 4n$

Divide both sides by 4.

$\dfrac{28}{4} = \dfrac{\cancel{4}n}{\cancel{4}}$

$7 = n$

The number is 7.

Check   Three times 7 subtracted from 30 is $[30 - (3 \cdot 7)]$ is 9, which does equal 2 plus 7.   True

37. *Step 2* Let x be the amount Brenda spent.

Step 3

Twice what Brenda spent	less	$3	is	the amount Consuelo spent
↓	↓	↓	↓	↓
$2x$	−	3	=	81

Step 4

$2x - 3 = 81$

Add 3 to both sides.

$2x - 3 + 3 = 81 + 3$

$2x = 84$

Divide both sides by 2.

$\dfrac{2x}{2} = \dfrac{84}{2}$

$x = 42$

(continued)

Step 5 Brenda spent $42.

Step 6

Check Consuelo spent $3 less than twice $42 or $[(2 \cdot \$42) - \$3]$, which is $81. True

39. *Step 2*

You know the least about my age, so let x be my age. Then $x - 9$ is my sister's age.

Step 3

$$
\begin{array}{ccc}
\text{The sum of the ages} & \text{is} & 51 \\
\downarrow & \downarrow & \downarrow \\
x + (x - 9) & = & 51
\end{array}
$$

Step 4

$x + x - 9 = 51$
$2x - 9 = 51$

Add 9 to both sides.

$2x - 9 + 9 = 51 + 9$
$2x = 60$

Divide both sides by 2.

$$\frac{\cancel{2}x}{\cancel{2}} = \frac{60}{2}$$
$$x = 30$$

Step 5

x is my age so I am 30; my sister is $x - 9$ or $30 - 9$, so she is 21.

Step 6

Check 21 is 9 less than 30 and the sum of 30 and 21 is 51. True

43. *Step 2*

You know the least about the cost of the printer, so let x be the cost of the printer. Then $5 \cdot x$ or $5x$ is the cost of the computer.

Step 3

$$
\begin{array}{ccc}
\text{The total cost (add)} & \text{is} & \$1320. \\
\downarrow & \downarrow & \downarrow \\
x + 5x & = & 1320
\end{array}
$$

Step 4

$x + 5x = 1320$
$6x = 1320$

Divide both sides by 6.

$$\frac{\cancel{6}x}{\cancel{6}} = \frac{1320}{6}$$
$$x = 220$$

Step 5

The cost of the printer is x, so the printer cost $220. The computer cost 5 times as much, so $5(\$220) = \$1100$ for the computer.

Step 6

Check Five times $220 is $1100 and the sum of $220 and $1100 is $1320. True

45. *Step 2*

You know the least about the shorter piece, so let x be the length of the shorter piece. Then $x + 10$ is the length of the longer piece.

Step 3

$$
\begin{array}{ccccc}
\text{The length} & & \text{the length} & & \\
\text{of the} & & \text{of the} & & \\
\text{shorter} & & \text{longer} & & \\
\text{piece} & \text{and} & \text{piece} & \text{are} & 78 \text{ cm.} \\
\downarrow & \downarrow & \downarrow & \downarrow & \downarrow \\
x & + & x + 10 & = & 78
\end{array}
$$

Step 4

$x + x + 10 = 78$

Combine like terms.

$2x + 10 = 78$

Subtract 10 from both sides.

$2x + 10 - 10 = 78 - 10$
$2x = 68$

Divide both sides by 2.

$$\frac{\cancel{2}x}{\cancel{2}} = \frac{68}{2}$$
$$x = 34$$

Step 5

The length of the shorter piece is x, so the shorter piece is 34 cm. The length of the longer piece is $x + 10$, so $34 + 10 = 44$ cm.

Step 6

Check 44 cm is 10 cm longer than 34 cm and the sum of 34 cm and 44 cm is 78 cm. True

53. *Step 2*

You know the least about the width, so let x be the width. Then $2x + 3$ is the length.

$$
\begin{array}{c}
2x + 3 \\
\boxed{} \\
x \qquad\qquad x \\
2x + 3
\end{array}
$$

Step 3

$P = 2 \cdot l + 2 \cdot w$
$36 = 2 \cdot (2x + 3) + 2 \cdot x$

Step 4

$36 = 2 \cdot (2x + 3) + 2 \cdot x$
$36 = 4x + 6 + 2x$

Combine like terms.

$36 = 6x + 6$

Subtract 6 from both sides.

$36 - 6 = 6x + 6 - 6$
$30 = 6x$

Divide both sides by 6.

$$\frac{30}{6} = \frac{\cancel{6}x}{\cancel{6}}$$
$$5 = x$$

Step 5

The width is x, so the width is 5 in. The length is $2x + 5$, so $2(5) + 3$ or 13 in. is the length.

Step 6

Check $P = 2(13) + 2(5)$
$P = 26 + 10$
$P = 36$ ← Matches perimeter given in problem.

Chapter 10 Statistics

Section 10.1 (pages 725–730)

1. The number of pets owned in the United States is 11 million + 150 million + 90 million + 75 million + 18 million + 16 million = 360 million.

13. "Wanted food they couldn't cook at home" to "Less work/No clean up":

$$\frac{1740}{1020} = \frac{1740 \div 60}{1020 \div 60} = \frac{29}{17}$$

25. Don't know:

$x = 5\%$ of 5540
$= (0.05)(5540)$
$= 277$ people

31. Percent for day camp $= \dfrac{\$546}{\$5460}$

$= 0.10 = 10\%$

Degrees of a circle $= 10\%$ of $360°$
$= (0.10)(360°)$
$= 36°$

37. (a) Total sales $= \$12,500 + \$40,000 + \$60,000 + \$50,000 + \$37,500$
$= \$200,000$

(b) Adventure classes $= \$12,500$

percent of total $= \dfrac{12,500}{200,000} = 0.0625$
$= 6.25\%$

number of degrees $= (0.0625)(360°)$
$= 22.5°$

Grocery and provision sales $= \$40,000$

percent of total $= \dfrac{40,000}{200,000} = 0.2 = 20\%$

number of degrees $= (0.2)(360°) = 72°$

Equipment rentals $= \$60,000$

percent of total $= \dfrac{60,000}{200,000} = 0.3 = 30\%$

number of degrees $= (0.3)(360°) = 108°$

Rafting tours $= \$50,000$

percent of total $= \dfrac{50,000}{200,000} = 0.25 = 25\%$

number of degrees $= (0.25)(360°) = 90°$

Equipment sales $= \$37,500$

percent of total $= \dfrac{37,500}{200,000} = 0.1875$
$= 18.75\%$

number of degrees $= (0.1875)(360°)$
$= 67.5°$

(c)

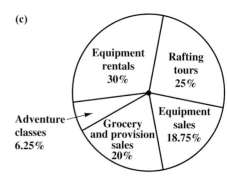

Section 10.2 (pages 735–740)

3. The countries in which less than 15% of household income is spent, on average, for food are USA, Britain, and Australia.

9. There are two bars for February. One is for 2009 and one is for 2010. The bar for 2010 rises to 7. Multiply 7 by 1000 because the label on the left side of the graph says *in thousands*. The bar for 2009 rises to halfway between 5 and 6. So multiply 5.5 by 1000.

Plants shipped in February of 2010 = 7000

Plants shipped in February of 2009 = 5500

7000 − 5500 = 1500

There were 1500 more plants shipped in February of 2010.

19. Above the dot for 1990 is 24.1. Looking at the left edge of the graph, notice that this number is in millions.

The number of PCs shipped in 1990 was 24.1 million or 24,100,000.

29. Find the dot above 2009 on the line for Store B. Use a ruler or straightedge to line up the dot with the numbers along the left edge. The dot is aligned with 350, so chain Store B sold 350 • 1000 = 350,000 iPods in 2009.

35. Find the dot above 2010 on the line for total sales. Use a ruler or straightedge to line up the dot with the numbers along the left edge. The dot is aligned with 40. Multiply 40 by 1000 because the label on the left side of the graph says *in thousands*. The total sales in 2010 were

40 • $1000 = $40,000.

Section 10.3 (pages 745–748)

1. Because the bar associated with 61–65 is the highest, the age group of 61–65 years has the greatest number of members, which is 16,000.

9. The bar for 31–40 rises to 11, so there are 11 employees who earn $3100 to $4000.

21. 0–5; 4 tally marks and a class frequency of 4

Section 10.4 (pages 753–756)

1. Mean = $\dfrac{\text{sum of all values}}{\text{number of values}}$

$= \dfrac{6 + 22 + 15 + 2 + 8 + 13}{6}$

$= \dfrac{66}{6}$

$= 11$ years

The mean (average) age of the shopping centers was 11 years.

9.

Customers Each Hour	Frequency	Product
8	2	(8 • 2) = 16
11	12	(11 • 12) = 132
15	5	(15 • 5) = 75
26	1	(26 • 1) = 26
Totals	20	249

weighted mean

$= \dfrac{\text{sum of products}}{\text{total number of customers}}$

$= \dfrac{249}{20} \approx 12.5$ customers (rounded)

17. Arrange the numbers in numerical order from least to greatest.

298, 346, 412, 501, 515, 521, 528, 621

The list has 8 numbers. The middle numbers are the 4th and 5th numbers, so the median is

$\dfrac{501 + 515}{2} = \dfrac{1016}{2} = 508$ calories.

21. 74, 68, 68, 68, 75, 75, 74, 74, 70

Because both 68 and 74 occur three times, each is a mode. This list is *bimodal*.

29.

Credits	Grade	Credits • Grade
2	A (= 4)	2 • 4 = 8
3	C (= 2)	3 • 2 = 6
4	A (= 4)	4 • 4 = 16
1	C (= 2)	1 • 2 = 2
4	B (= 3)	4 • 3 = 12
14		**44**

GPA $= \dfrac{\text{sum of (Credits • Grade)}}{\text{total number of credits}}$

$= \dfrac{44}{14} \approx 3.14$ (rounded)

SOLUTIONS

AP Wideworld, pp. 118 right, 298 top, 514 top, 563; **Beth Anderson,** pp. 2, 8 top right, 53 right, 98 left, 236 left, 281, 322 right, 324, 406 right, 415 top right and bottom left and right; 437 left; 440 bottom left, 509 top right, middle, and bottom right, 510, 511 top left, 512 bottom left, 518, 534 middle right, 586 left, 714 right, 722, 740, 765; **Blend Images/Getty RF,** pp. 298 right, 331, 439; **Bohemian Nomad Picturemakers/Corbis,** p. 562 right; **Brand X Pictures,** pp. 492 bottom left, 525 right; **Bruce Anderson Photos,** p. 416; **Courtesy of Caterpillar, Inc.,** p. 536; **Comstock,** p. 308 left; **Corbis,** pp. 14, 60, 282, 315 bottom left, 509 top left, 512 bottom right, 517 middle, 540, 562 left, 588, 634, 714, 745, 768, 774; **Custom Medical Stock Photo,** p. 642; **Digital Vision,** pp. 151 right, 234 right, 314 right, 315 top left and bottom right, 330, 392 right, 393 left, 402 right, 428 right, 494 top right, 514 bottom, 726; **Earl Orf/Heart and Mind, Inc.,** p. 366; **Getty,** pp. 253 left, 254 right, 256 left, 449 bottom left, 476 left, 729 right, 738; **Getty Editorial,** 160 right, 176, 472, 578, 640; **Getty Sports,** p. 354; **Golden Horseshoe Mustang Association,** p. 438; **Image Source/Getty RF,** p. 320; **iStockphoto,** pp. 429, 449 top left, 511 bottom right; **Courtesy of Jelly Belly Candy Company,** pp. 113, 177; **Johner Images/Getty RF,** p. 8 bottom left; **Kara Salzman,** p. 437; **kcarlson@cardstacker.com,** pp. 199, 234 left, 245 right; **NASA,** pp. 315 top right, 586 right; **NationBill/Corbis Sygma,** p. 308 right; **PhotoDisc,** pp. 8 top left, 39 bottom, 110, 256 right, 368, 370, 379 left, 380 bottom, 416 right, 427 right, 438 right, 456 left, 476 right, 509 bottom left, 511 top right, 512 top left and right, 517 left, 534 bottom, 539, 598, 641, 670, 729 left; **PhotoDisc Blue,** pp. 314 left, 322 left, 406 left, 440 bottom right, 472 right, 483 bottom, 494 top left, 502 left, 587 right; **PhotoDisc Red,** pp. 60 right, 379 right, 393 right, 525 left, 534 middle left; **PhotoDisc/Getty RF,** p. 376; **Photographer's Choice/Getty RF,** p. 552; **Photo Researchers,** pp. 483, 494 bottom; **Photonica/Getty Images,** p. 456; **Picture Quest/Brand X Pictures,** p. 338; **Picture Quest/PhotoDisc,** p. 340; **Reuters/Corbis,** pp. 307, 361; **Richard Carson/Reuters/Corbis,** pp. 493 bottom right; **Shutterstock,** pp. 1, 6, 8 bottom right, 19, 29 left, 39 top, 50, 53 left, 62 left, 63, 92, 98 right, 108, 109, 117, 118 left, 151 left, 157, 158, 166, 168, 169, 223, 224, 236 right, 237 left, 245, 253 right, 254 left, 258, 264, 265, 273, 280, 347, 383, 285, 394, 407 left, 415 top left, 419, 424, 440 top left and right, 449 top right, 463, 490, 491, 492 top, 500, 502 right, 511 bottom left, 517 right, 534 top left and right, 635, 653, 682, 719, 724, 734, 747, 748, 755, 756, 759, 763, 767, 769; **Stan Salzman,** pp. 29 right, 93, 159, 204, 392 left, 428 left, 750; **Stockbyte/Getty RF,** p. 156; **Thierry Orban/Corbis Sygma,** p. 523; **ThinkStock,** pp. 274, 380; **Twentieth Century Fox/Photofest,** p. 352; **U.S. Department of Agriculture,** pp. 541, 587 left; **U.S. Navy,** pp. 160 left, 427; **U.S. Postal Service,** pp. 64, 85, 170, 190